MICROPROGRAMMING
Principles and Practices

SAMIR S. HUSSON

International Business Machines Corporation
Systems Research Institute
New York, N.Y.

PRENTICE-HALL, INC., *Englewood Cliffs, N.J.*

PRENTICE-HALL INTERNATIONAL, INC., *London*
PRENTICE-HALL OF AUSTRALIA, PTY. LTD., *Sydney*
PRENTICE-HALL OF CANADA, LTD., *Toronto*
PRENTICE-HALL OF INDIA PRIVATE LTD., *New Delhi*
PRENTICE-HALL OF JAPAN, INC., *Tokyo*

Current printing (last digit):
10 9 8 7 6 5 4 3 2 1

13-581454-5

Library of Congress Catalog Card Number 72-122612

PRINTED IN THE UNITED STATES OF AMERICA

FOREWORD

A design technique may be said to be born when it is first set out in detail on paper. It is out of the nursery when it begins to be applied in practice. Not, however, until it is made the subject of a full-length book can it be said to be grown up. Thanks to Mr. Husson's work, microprogramming has now reached the latter stage.

It is hard for the current generation of computer engineers to realize what things were like in 1951, when I published my first note on microprogramming. Vacuum tubes were all we had, and there were only a handful of digital computers in existence. Even germanium diodes had not yet become available, and no one would have believed that in a few short years semiconductor devices would dominate the scene.

Microprogramming was originally put forward as a means whereby the design of a processor might be rendered more systematic. Mr. Husson brings out very clearly the relationship behind the design of the microprogram sequencing circuits of a processor and the specification of the data paths. A particularly valuable feature of his book is the wealth of material it contains on current practice in both these respects. The detailed information given about processors from a number of different manufacturers will provide a much needed basis for a comparative study of processor morphology.

While microprogramming remains of prime importance to the hardware designer, the coming of efficient, easily changed, read-only control stores, and above all of writable control stores, has made it of interest to a much wider circle. This includes designers of operating systems, special application packages, and general software, together with students taking courses in computer science.

It is a pleasure to commend Mr. Husson's book to the reader.

Cambridge, England M. V. WILKES

PREFACE

A significant trend in computer design in recent years has been the replacement of the conventional transistor logic circuit control (hardware control) section by a microprogrammed control. The microroutines which specify the actions and the flow of data over the data paths in the digital computer are stored in a permanent, semipermanent, or changeable control storage.

Since first proposed by Dr. M. V. Wilkes in the early 1950s, microprogramming has appealed to many special interest groups in the data processing and computer manufacturing fields. At first, it appealed primarily to the theoretician who was intrigued by its implications and its systematic structure. Although a number of microprogrammed computers were built in the late fifties and early sixties, the concept did not excite the computer manufacturer, possibly because of the absence of a reliable, inexpensive, and fast nondestructive read storage technology. Today microprogrammed computers are in widespread use, and the interest in microprogramming cuts across the specialties of logic design, operating system design, finite state machine theory, application system design, and system architecture. It has also proved to be a successful pedagogical tool for teaching a number of computer architecture and design implementation courses.

The biggest factors which lead to the present level of interest in microprogramming is the inherent flexibility of a microprogrammed computer and also its economic superiority over the conventional hardware control implementation. This flexibility may be found in the economic implementation of a whole line of processors that are architecturally compatible, or in its use in specializing the computer and increasing its efficiency in specific applications.

This book introduces and discusses the general concepts and technology related to microprogramming. It compares the conventional sequential hardware control to the microprogrammed control. By presenting the simplest of machine organizations with minimum functions and minimum data flow and control gates, then graduating to the more complex and more sophisticated modern computer systems, it introduces the fundamental microprogramming concepts in a logical and systematic way. Special emphasis is given to the control memory and the control word structures, their design parameters and tradeoffs. The best known microprogrammable systems are discussed in order to evaluate respective microprogrammed design approaches and the impact of these approaches on the performance and flexibility of these systems. The systems that are discussed are the IBM System/360, Models 40 and 50 (to explore and analyze the notion of architectural compatibility); the RCA Spectra 70/Model 45; and the Honeywell H4200. The last two are contrasted to the IBM System/360. The RCA Spectra 70/Model 45 was chosen because of its architectural compatibility with System/360; yet it is strikingly different in microprogramming design implementation from the System/360. The Honeywell H4200 applies still another, totally different approach to solving the microprogramming design, and is an excellent means for studying the complexity and the microprogramming difficulties to be anticipated in a variable field length computer.

The text is divided into three basic parts: concepts (Chapters 1–4), technology and read-only storage media (Chapter 5), and a detailed analysis of the above-mentioned system implementations (Chapters 6–10). A novel feature of the book is its extensive bibliography, organized alphabetically by the authors' names and then chronologically. The latter organization is designed to show the trends and shifts in emphasis since the birth of this field.

Much of the information for this book was assembled from the huge amount of literature buried in the archives of many a manufacturer.

I wish to express my sincere thanks to the IBM Corporation; to the staff of the IBM Systems Research Institute under the direction of Dr. E. S. Kopley for their encouragement; to the efficient secretarial staff of the SRI under the leadership of Miss Cora Tangney for active support in the preparation of the manuscript; to the Honeywell Corporation, particularly to Mr. T. H. Bonn and Mr. George S. Hoff for their valuable assistance, cooperation, and counsel during the preparation of the chapter on the Honeywell H4200; to the Radio Corporation of America and Mr. R. J. Quigley, Mr. Anthony G. McDonnell, and Mr. J. B. Douglas for their help in the preparation of the chapter on the RCA Spectra 70/Model 45.

Many other people assisted in this work. I am grateful to my friend, Professor Maurice V. Wilkes, for his encouragement and valuable suggestions. Finally, thanks also go to the many IBM Systems Research Institute students who proofread the many versions of the manuscript.

SAMIR S. HUSSON

CONTENTS

PRINCIPLES OF MICROPROGRAMMING

1

1.1 INTRODUCTION

A significant trend in computer design in recent years has been the replacement of the conventional transistor logic control section of a digital computer with "stored logic" or microprogrammed control, stored in a high-speed, nondestructive read-only storage (ROS). This means of controlling and sequencing of individual operations to accomplish the execution of a program instruction has received a heavy emphasis in several models of the IBM System/360, RCA Spectra/70, and the English Electric System 4.

The main reason for this shift in computer control implementation is the economic superiority of microprogramming over conventional logic control. Microprogramming has made it economically feasible, for example, to have the same comprehensive instruction set (operation codes) built into a whole line of new computers, even the smallest ones. Thus, we now have computer lines that are architecturally* compatible; yet their internal hardware, organization, and structure are drastically different. Microprogrammed control offers many

*The term "architecture" is used to describe the attributes of the system as seen by the programmer. It represents the conceptual structure of the functional behavior of the system as distinct from the organization of the data flow, controls, logical design, and physical implementation.

other advantages besides the much publicized upwards and downwards compatibility in the IBM System/360 and RCA Spectra/70. Among these is the inherent flexibility of a microprogrammed system to permit architectural extensions and modifications that will make the system perform a specific data processing function with maximum efficiency. This capability has not been widely exploited.

All present indicators point to an almost explosive increase in microprogramming activity by system engineers and users. This projection is based on the growing number of microprogrammed systems announced by the major computer designers in a few short years.*

The biggest factor that will lead to the projected increased exploitation of microprogrammed control is the inclusion of a *writable* control storage. This would remove the restriction of permanent fixed operation codes (instruction set) locked into the system architecture, and it would lead to a number of fascinating possibilities. The programmer himself would be able to choose his own operation codes to suit his requirement. The IBM System/360, Model 25, the Standard Computer Corporation IC6000E, the Honeywell H4200/H8200, and the Interdata 2, 3, and 4, are but a few of the announced systems having this flexibility. These systems are not the first machines with a writable control storage, however. At least four other machines, the Collins Radio C8401[1], the TRW-133 Computer[2], the PB440[3], and the stored logic computer of Ramo-Wooldridge[4], were also designed with a writable control storage.

Microprogramming can enable the designer to implement a system that can operate on a wide variety of problems with varied degrees of efficiency. Most computers designed recently are of the general purpose data processing type intended for commercial and scientific applications.

To be sure, microprogrammed control has some disadvantages also. The advantages and disadvantages will be discussed later in this chapter; but first, *what is microprogramming*? A more basic question is: What is the control section, and where does it fit in the computer system? It is imperative that we answer the second part of the question first.

A modern digital computer (in general, any data processing system) can be partitioned into five distinct functional sections: input, storage, arithmetic and logic unit (ALU), output, and control. These five sections communicate with each other through electronic signals that represent data, instructions, and control signals as shown in Fig. 1.1. The first two, namely, instructions and data, are well known to almost every person who has casual familiarity with a digital computer. However, the control signals are generally known only to the system designer.

The order, timing, and direction in which this information flows within and between the five principal functional sections in a data processing system, are

*Other microprogrammed computers are: Burroughs 2500 and 3500; Interdata 2, 3, and 4; Standard Computer 670, 2700, IC7000, and IC9000; TRW-130 and 530; IBM 4-PI series, and the IBM 2841.

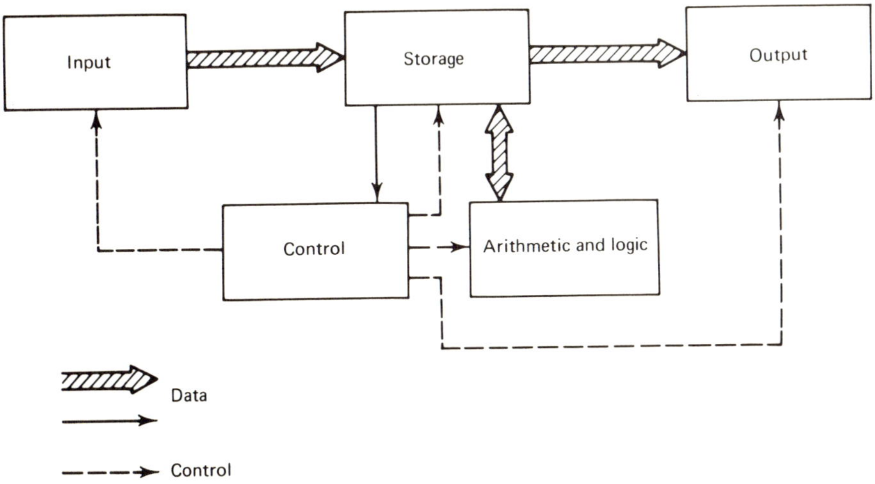

Figure 1.1 Functional sections of a data processing system.

effected by the control section. It, in turn, is directed by the sequence of machine instructions or by the people who wire the control panel (back panel) and punch the control tapes and/or the control information in the input cards.

The control section of the computer directs the operations of the entire computer. It receives units of information from the storage section which tell it what operations are to be performed and where the data to be operated upon are located in the storage section. After the control section (a) determines the exact instruction to be executed, (b) it then issues control signals to open and close the specific gates throughout the system, thus permitting the necessary data in the form of electrical signals to flow from one functional unit to another in the execution of that operation. Once the arithmetic and logic section is finished with its part in this operation, (c) the control section may issue the necessary control to permit the results to be transmitted back into the storage section or be channeled to some output device to be stored on another storage media such as a disk or drum. At the end of the execution of an instruction, (d) the control section causes the computer to proceed to fetch and decode the next instruction. Hence, every computer must be directed by a series of machine instructions programmed to solve a particular logic or arithmetic problem which is of interest to the user. Each of these instructions (operations codes), is simply a request made to the control section to activate the necessary hardware functions built into the system, such as the adder, shifter, and counters, and to open and close certain gates in the data path that will allow the operands to flow into these more basic arithmetic and logic static functions in a predetermined order and a well-defined sequence of machine cycles.

The general nature of the control function, then, is fourfold: the first two consist of fetching and decoding of the sequence of machine instruction; the third function is the gating of data paths to perform operations on data fields;

and the fourth is changing of the state of the computer so as to allow the next required operation to be performed.

1. Fetch
2. Decode I
3. $C_1 \rightarrow C_2$ Change Controls
4. $S_1 \rightarrow S_2$ Change State of System

Since the data being processed may often by its attributes such as its size or magnitude affect the next operation to be performed, it is necessary to provide for conditional testing and branching. By branching we mean the alteration of the control sequence from the immediate sequence to an alternate or a different sequence. This branching operation may be conditional or unconditional; if it is conditional, testing is involved. In other words, an instruction may result in more than one possible control sequence, i.e., the sequential control logic may be a function of data testing as well as operation code.

A digital computer may simply be described as an elaborate array of logic network called *data paths*. These data paths consist of the static microspec* functions (functional stations) such as adders, shifters, indicators, registers parity circuits, priority circuits, mask circuits, and other Boolean functions. Each of these is designed to perform a specific (sub or elementary) micro-operation. These static microspec functions are interconnected by data buses which permit the information to flow from one functional station (microspec function) to another and to be stored or to be operated upon. The interconnections are sprinkled liberally with control gates that permit the information to flow between these functional stations. All these data paths are static in nature and can be activated by the enabling and disabling signals which emanate from the control section. The control signals are made up of clock pulses for timing, decoding, sequencing, and decision logic, and they direct and control the operation of the total system over any number of consecutive cycles. This is much the same as an orchestra and its conductor, where the orchestra consists of a group of individuals, each capable of producing a series of notes or playing a particular instrument when asked to do so; the conductor (the control section) directs them and leads them to harmonize and synchronize their actions into a musical composition or structure.

The control function can be described as combinational (static) and sequential. The first type performs the operations on the data fields. The sequential control places the computer in a state to perform the next required operation in order to execute the instruction(s) given to the computer. In a data processing system, the sequential control logic fetches the machine instruction from main storage and *decodes* it to determine what *sequencing* is to be

*A microspec function is defined as a logical entity in the data flow of a system that performs a specific operation in one or more machine cycles. These functional stations are sequential or combinational circuits designed to store or to perform a basic boolean operation on a specified input data.

done. It is this sequencing, synchronized by the timing pulses and modified by the branching control, which directly controls the computer. The sequencing determines which control gates in the data paths are to be activated, what decision tests should be performed, and what successor sequencing is to be done. The timing determines when the selected gates in the data paths are to be opened and for how long they remain open. The testing and associated decision logic determine the state of the machine and set up the conditional controls which are used to select one of two or more alternate sequencing controls.

To illustrate this control function and how it interacts with the total system, consider the following hypothetical system. Assume that it is of fixed word length and fixed instruction length. The instructions consist of the operation code field, followed by the two address fields indicating the addresses of the two operand fields in main storage. Assume also that the width of the hardware storage registers and all other microspec functions, data paths, and associated logic are equal to the width of operand fields. Finally, assume that the arithmetic and logic unit cycle time is one-fourth the main storage memory cycle time (i.e., the duration of the main storage cycle time is four times that of the CPU cycle time). Figure 1.2 shows the data flow of this hypothetical system. It contains the following microspec functions:

1. Instruction-address register, IAR.
2. Storage address register, SAR.
3. A one-bit register (incrementor) containing a 1 to be used for incrementing IAR.

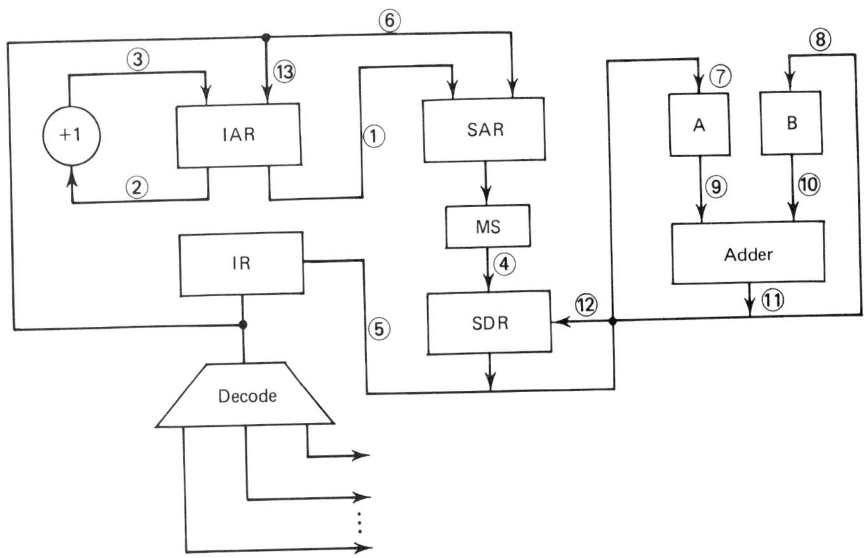

Figure 1.2

4. A parallel adder.
5. Instruction register, IR.
6. Instruction decoder.
7. Two hardware registers, A and B.
8. Storage data register, SDR.
9. Main storage, MS.

These functions, together with the data buses which interconnect them, make up the computer data paths. These interconnections are:

1. IAR to SAR.
2. IAR to incrementor.
3. Incrementor to IAR.
4. MS to SDR.
5. SDR to IR.
6. Address-one and address-two fields of IR to SAR.
7. SDR to A register.
8. SDR to B register.
9. A register to adder.
10. B register to adder.
11. Adder to A register and adder to B register.
12. Adder to SDR.
13. IR to IAR.

We will further assume that, once an address is transmitted to SAR, the main storage clock begins cycling through the necessary time to read the contents of the selected instruction or operand into the proper hardware storage register in the ALU; it then continues in a regeneration cycle which will ensure that the selected location in main storage is left with the desired stored information.

In this hypothetical machine, the fixed-point *add* instruction format is shown in Fig. 1.3 and the instruction is executed as follows:

Instruction	Op Code	Address of Operand one	Address of Operand two

Figure 1.3 Instruction formats.

1. A control signal is issued which causes the contents of the instruction address register (IAR) to be transferred to the storage address register (SAR). This starts the main storage clock running. The contents of IAR are incremented by one, and the results are transferred back to IAR.

$$\begin{bmatrix} \text{IAR} \rightarrow \text{SAR (get instruction)} \\ \text{IAR} + 1 \rightarrow \text{IAR} \end{bmatrix}$$

2. After a fixed increment of time (two ALU cycles), the contents of the

selected location in main storage (MS) are transferred to the instruction register (IR). (The SAR will not receive a new address for two additional machine cycles to permit the MS regeneration phase.)

$$[\text{MS(SAR)} \rightarrow \text{IR (instruction out)}]$$

3. The Op Code portion of the IR is decoded. Since it was found to be a fixed-point, add instruction, the second address field of IR is transferred to SAR to fetch the second operand in main storage.

$$\begin{bmatrix} \text{Decode Op Code} \\ \text{IR [Op (2)]} \rightarrow \text{SAR} \end{bmatrix}$$

4. After two ALU cycles, the second operand is transferred to a preselected hardware storage register B via the SDR register.

$$\begin{bmatrix} \text{MS[IR[Op(2)]]} \rightarrow \text{SDR} \\ \text{SDR} \rightarrow \text{B} \end{bmatrix}$$

5. At the end of this MS cycle, the first IR operand is transferred to a second hardware register in the ALU, register A.

$$\begin{bmatrix} \text{IR [Op (1)]} \rightarrow \text{SAR} \\ \text{MS [IR [Op (1)]]} \rightarrow \text{SDR} \\ \text{SDR} \rightarrow \text{A} \end{bmatrix}$$

6. The contents of both registers A and B are simultaneously gated into the adder to be operated upon.

$$\begin{bmatrix} \text{A, B} \rightarrow \text{Adder} \\ \text{A + B} \rightarrow \text{A} \end{bmatrix}$$

7. The resultant sum is transferred into register A, and the first operand address is once again gated to the SAR, with a command to transfer the contents of register A into the selected location in main storage. At the end of this main storage cycle, the sum will have replaced the first operand in main storage.

$$\begin{bmatrix} \text{IR [Op (1)]} \rightarrow \text{SAR} \\ \text{A} \rightarrow \text{SDR} \\ \text{SDR} \rightarrow \text{MS [IR [Op (1)]]} \end{bmatrix}$$

This sequence of steps suggests that a minimum of hardware facilities, a number of interconnections between them, and a well-defined sequence of operations must be designed into this system in order for it to sequence through these steps and through the many consecutive ALU cycles it takes to decode and execute the add instruction.

The hypothetical computer shown in Fig. 1.2 has all the necessary elements and interconnections to perform this add operation. What remains to be defined is the control section.

The static elements and data paths are dynamically controlled by the control section. Any transfer of information between these elements over the thirteen data paths which interconnect them must be activated at the proper time by the control section in order for the information to flow according to the well-defined sequence for this hypothetical machine during execution of the add instruction.

The sequencing of the micro-operations stated before can be defined in terms of computer states, that is, the suboperations that are to be carried out during every machine (ALU) cycle in the process of the instruction execution. This sequence of computer states is shown in Fig. 1.4.

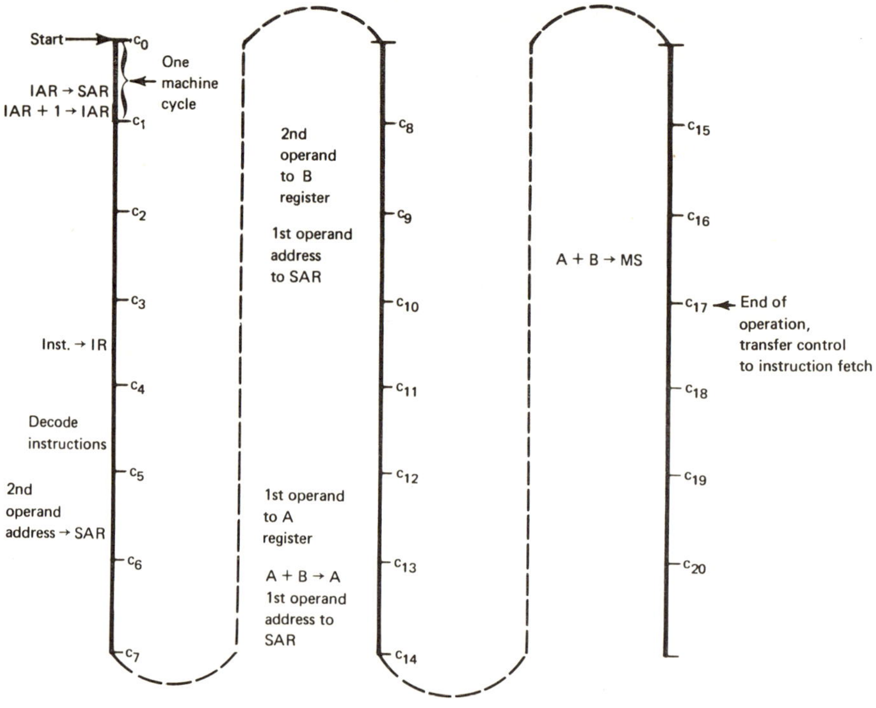

Figure 1.4 Timing chart.

This process of defining the sequence of machine states, the static hardware functions and the interconnections between them, is repeated for every machine instruction code in order to define the characteristics of every functional unit and their interfaces. The control function is best described and analyzed by the timing charts which show the sequencing of machine states from the first cycle in the instruction fetch through the decoding and execution of the Op Code.

Historically, the control section of a digital computer has been constructed from conventional switching devices such as relays, vacuum tubes, diodes, and transistors. These, in turn, are designed into the basic logic circuits such as AND, OR, and INVERT, or into more complex logical building blocks such as the NAND, NOR, AND-OR-INVERT, and OR-AND-INVERT blocks. In the control section, these binary logic circuits are grouped together to form the conventional storage elements ("flip-flops," latches, triggers, etc.) which are set or reset during each machine interval "cycle" to set the machine in a specific state. The outputs of these storage elements or status switches control the execution of arithmetic or logic operations by issuing control pulses over the control lines to the specific gates in the data flow. The opening of these selected gates allows the information to flow between functional units, for example, from register to register, register to adder, adder to shifter, etc. Thus, in relatively few consecutive machine cycles, information is guided to flow over many data paths and through many functional units in a given or specific order required for the execution of the machine instruction.

Before illustrating a hardware control section, it is important that we establish a clear understanding of the *control gate*.

A gate is a logic circuit, containing two or more input lines, which is designed to perform a specific logic function on these inputs. A control gate may be an AND gate as shown in Fig. 1.5, which produces an output only when there is an input signal on *both* input lines.

In a given data path, the control gate is merely a switch which is normally closed; thus, any change in the level of the binary information signal at its

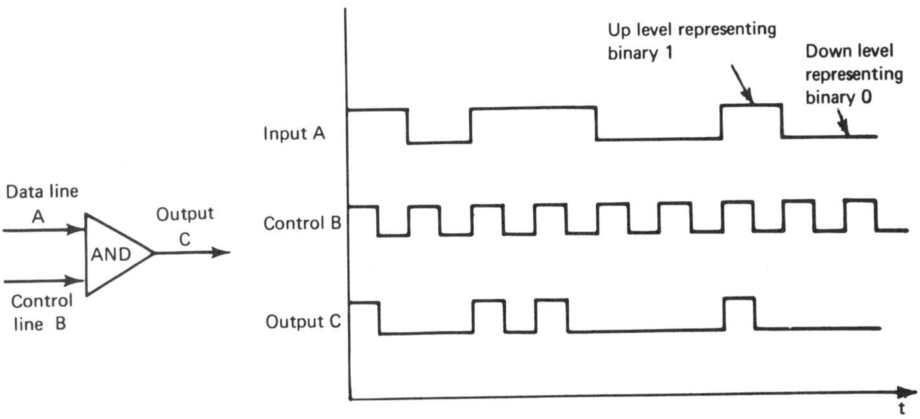

Figure 1.5 Control gate.

input is not reflected at the output, i.e., is not permitted to flow through the gate, unless a second signal line, called the *control line*, is directly or indirectly activated by the sequencing logic in the control section. Normally, in a synchronous* computer, this control line is activated for a short increment of time during the machine cycle when it is required. Thus, a control gate can be thought of as an AND gate with the following types of inputs: one or more inputs represent information to be transmitted over a common bus to another register or hardware function; a controlled input that restricts the flow of this information between the functional units to a predetermined timing point in the sequence and for the duration of time during which the given control line is active.

In Fig. 1.5, input A represents an information bit, and the train of pulses at that input A may represent the binary contents of a given trigger or latch during a few consecutive machine cycles. The train of pulses at the control input line B represents the time in this specific sequence of machine cycles when the control line is active. This determines the time when the information signal at input A will be permitted to flow through that control gate. If we monitor the output of this control gate at the output C, we will observe for these consecutive machine cycles a train of pulses as shown in Fig. 1.5.

As shown in Fig. 1.4, the add instruction execution can be described by the sequence of suboperations which takes place if the specific subsets of the control gates are activated in the proper machine cycles. The sequences of suboperations are called the machine states. These states can be completely described and specified by what subset of the control gates, therefore control lines (emanating from the control section), is activated each machine cycle. Traditionally, the hardware control has consisted of a set of storage elements connected (Fig. 1.6) in such a way that, either individually or in combination, they can activate the subsets of control lines which represent the different sequences of machine states.

To illustrate the hardware control, let us assume that the six gates, 7, 8, 9, 10, 11, and 12 control the flow of information in the ALU and between the ALU and the SDR of the MS section of the hypothetical computer shown in Fig. 1.2. Assume also that a process involves the following steps: Transfer the contents of SDR to register A; add the new contents of A to an accumulator register B; then store the results in both B and SDR.

This process will require three separate machine cycles. The states of the control lines during each of these three consecutive machine cycles (N, N + 1, and N + 2) are shown in the table at the top of page 12.

Figure 1.6 shows the timing ring* designed to activate the control lines in the specified manner. Note that one and only one of the storage elements, or flip-flops, is set to 1 at any machine cycle. That is, if at the (N − 1) cycle, the

*A synchronous computer is made up of logical elements (microspec functions) in which the propagation delay for such elements is fixed and is independent of function variations; the timing relationship between the elements is controlled by well-defined periodic pulses emanating from a central clock.

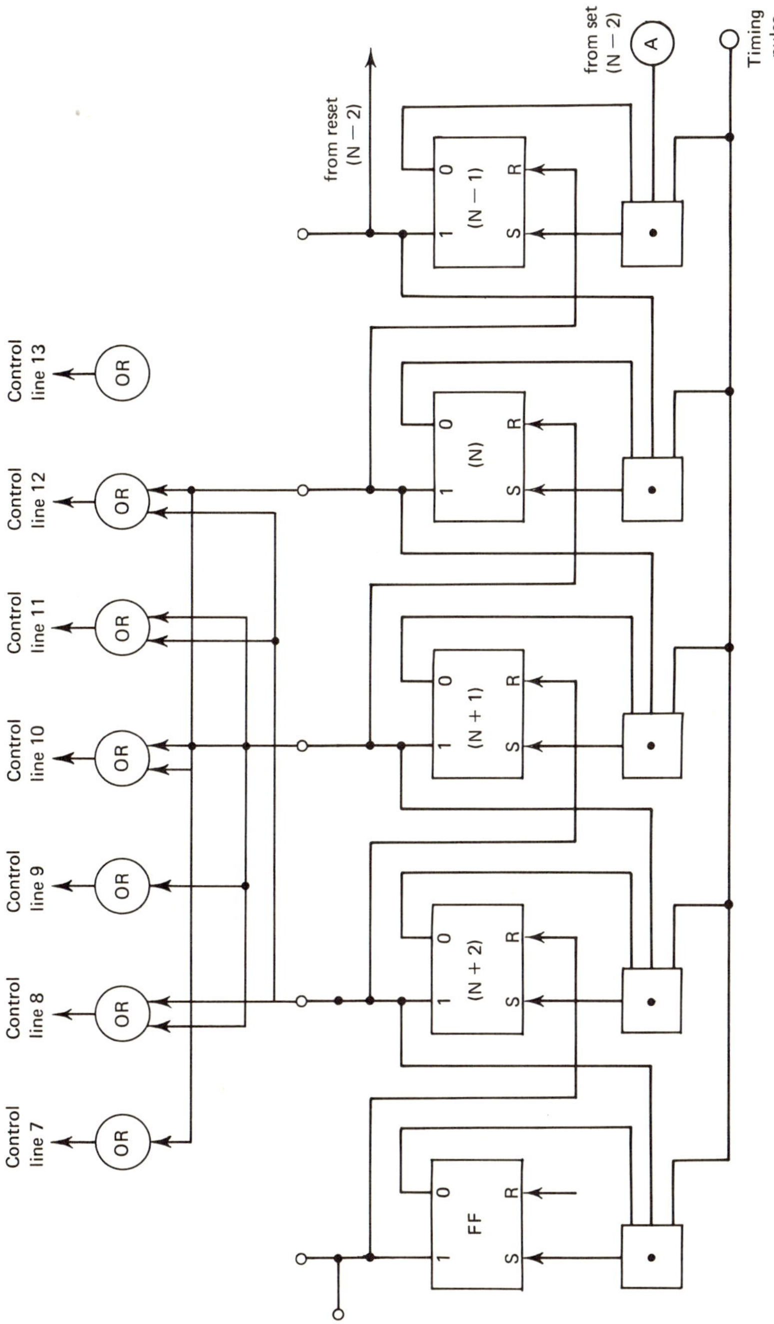

Note: S = set; R = reset.

Figure 1.6 Sequencing through machine states.

	Time (N)	*Time (N + 1)*	*Time (N + 2)*
Control line 7	On	Off	Off
Control line 8	Off	On	On
Control line 9	Off	On	Off
Control line 10	Off	On	On
Control line 11	Off	On	On
Control line 12	On	Off	On

(N − 1) flip-flop is set to 1 and the other five are 0, then at the beginning of the next machine cycle, (N − 1) is turned off as a result of turning on the *N*th stage. The remaining flip-flops are not affected in this cycle and will remain off.

Note also how the true outputs of the storage elements, which are activated one at a time and in sequence from right to left, are ORed into the required control lines. This implementation (Fig. 1.6) has been simplified so as to convey the concept without getting too deep in the circuitry and signal timing.

1.2 WHAT IS MICROPROGRAMMING?

We have seen in the previous discussion that the control section produces essentially a sequential group of pulses that activate the otherwise static data paths and microspec functions in the data processing system. *Microprogramming*, then, can be viewed as was originally conceived by Wilkes[6] as *a systematic approach and an orderly approach to designing the control section of any computing system.*

The main function of the control section is to specify the conditions under which sets of gates have to be opened. Traditionally, as mentioned earlier, this information has been permanently built into the system by connecting a set of decoders and flip-flops in an ad hoc manner—control section designers draw block diagrams until they see a satisfactory compromise of a whole array of logic circuits tied together that would satisfy the machine state and sequencing requirements. Generally, this type of control section offers the greatest degree of complexity, especially if the particular computer has a comprehensive instruction set. This complexity in designing the control section, and the ad hoc design procedures that are applied to it, are due to the lack of modularity (in contrast to the arithmetic and logic section in the same data processing system).

The conventional hardware control logic, when implemented, can therefore be viewed as a tree of storage elements and signal wires tied together in an untidy, unsystematic way. A decision to make the slightest modification to the instruction set could entail a major modification to the whole structure of the tree. This lack of systematic structure or modularity in the design must be borne in mind when we look into the difficulties in testing and maintaining that section. Bear in mind, also, that it is this complexity of the control section design that has prevented the computer designer in the past from offering a rich and efficient instruction set that would characterize it as a general purpose computer.

*The timing ring considered here is a device which permits a series of N unique pulses during the execution of each unique state of the machine function.

Microprogramming was originally conceived and proposed as an orderly, alternative design procedure to the ad hoc procedure applied to conventional hardware. It eventually became an alternate design and implementation tool for the control section, where the hardware control is replaced by a "stored logic" section, or "microprogram control" section, stored in a high-speed, nondestructive read-only storage (ROS). The information stored therein (zeros and ones) is designed to control each microspec function (adder, mover, shifter, and other hardware facilities) for each consecutive machine cycle.

The above discussion of the sequential logic design may be undeservedly negative. There are a number of cases when a conventional hardware control is preferred over a microprogrammed control. Some of these cases will be reviewed in a later section that deals with the advantages and disadvantages of the microprogrammed control implementation. It should also be pointed out that the difficulty in designing conventional hardware is not entirely due to the lack of disciplined techniques. There are a number of other constraints such as cost minimization, path length, circuit loading, and other technological restrictions which complicate the process and the final design.

Microprogramming—interpreted as implementing control logic, primarily by read-only storage—cuts across the specialties of electronic module design, mechanical languages, programming, and systems architecture. It is therefore a promising means for designing integrated hardware-software systems. In the remainder of this chapter we will consider microprogramming and its applications to a number of different special interest groups and disciplines which make up the data processing community. These will include the programmer, the system architect and system designer, the system engineer,* and the user.

This method of control, from the *programmer's* point of view, is similar to writing a program in which a given arithmetic or logic operation is executed by giving the CPU or the system a step-by-step description of the job to be done. This program, then, is a series of subcommands for the microspec functions built into the system, and thus, the word *microprogramming* was coined.

Microprogramming is analogous to conventional programming. The user programmer tells the system or device *what to do* by placing instructions in the high-speed main storage. The microprogrammer tells the system or device *how to do it* by controlling which storage and logic elements are used and how they are used for each operation. Thus, the machine instruction which the programmer considered to be the lowest level of communication with the system can now be viewed as a closed subroutine broken down into sequence of more elementary functions called *microinstructions*. Each microinstruction is designed to specify the control gates that are opened at a particular point during the machine cycle.

The concepts of *loops, subroutines, conditional* and *nonconditional branching, multiprogramming,* and *multiprocessing* have the same connotations in both fields. However, in microprogramming, one is faced with a few additional com-

*A systems engineer is the manufacturer's technical consultant to the customer.

plications in preparing a sound and efficient working microprogram. A microprogrammer must have a thorough knowledge and understanding of every microspec function, every data path, every gate, and the amount of parallelism that exists in the system. He must be familiar with every in-gate and out-gate position associated with every piece of hardware, the timing pulses and the gates affected by each pulse, and the relationship between CPU clocks and the main storage, the local storage, and all other units of hardware in the system. It is important to realize that the microprogram controls the operations of the machine at the circuit level; therefore, the microroutines established to execute any user program instruction can be very complex. The microprogrammer must have sufficient machine-programming experience as well as sufficient knowledge of the internal logic design philosophy of the system in order to properly and effectively produce an efficient program. Thus, microprogramming from the programmer's point of view is the art of coding the controls for all the functional units in the processor that will implement each machine operation in a minimum number of machine cycles. The microprogram is composed of microinstructions that are read out (in this case from a control store) one at a time, decoded, and executed.

From a *systems designer's* point of view, microprogramming is a method of machine control which entails discretely storing the states of the computer control lines in storage media organized on a word basis. Each word is designed to specify the operation of every data path and every functional unit in the system for one or more machine cycles. Hence, by fetching these words in a predetermined sequence and then decoding them, we would activate a subset of the control lines in the system that would open a subset of control gates throughout the system. The sequencing between control words here is done by allowing the control word being executed in this machine cycle to specify the address of the ROS word that would control the ensuing machine cycle.

To summarize: The systems designer describes the control section of a data processing system in terms of sequencing, timing, decoding, and decision logic[8]. These form the four major requirements of any control section regardless of its implementation. The *decoders* determine which set of control lines is to be activated; the *sequencing* (synchronized by well-defined *timing pulses*) is designed to close or open a set of gates throughout the system to allow the data to flow out of one functional unit and into another. The *decision logic* is designed to perform certain tests on the data stored in the various machine functions, and the results of these tests can be used to alter the selection of the next ROS word which will control the next machine cycle.

Before we illustrate how the microprogram and data paths interact in a system, we shall view microprogramming from yet another field, that of the *systems engineer.* Keep in mind that a systems engineer's primary concern is the efficiency of the system in performing the mission for which it is acquired; that is, the systems engineer is more application-oriented than the other two disciplines we have just reviewed.

From a systems engineer's point of view, then, microprogramming may become a significant option that is well-suited to applications where very specialized tasks require major portions of the total computer time. This is also the *user's* point of view. The feasibility of the user doing microprogramming is primarily related to the characteristic ease with which specialization and simulation can be achieved and to overall cost advantage. Specialization is the incorporation of special instructions, micro-operations, and hardware functions to facilitate a particular application or class of applications.[9]

Specialization and simulation from a systems engineer's point of view are not incompatible and, in some applications, they are complementary. The systems engineer may view microprogramming as a tool that would help decrease the execution time of a given subroutine by replacing it with a single macroinstruction, executed directly from ROS storage. This decreased execution time, when compared to the subroutine written in machine code, can be the direct result of one or more of the following factors:

1. Elimination of all the fetch and decode cycles for every machine operation code in the subroutine.
2. Elimination of all intermediate main-storage access times or access cycles for storing intermediate results which are normally of little value to the user.
3. Elimination of costly and yet necessary subroutine linkages. (This is because the microprogrammer will use the available hardware registers and oftentimes the high-speed local storage or scratch pad present in most modern, high-speed, digital computers, rather than the often limited, high-speed, general purpose registers available to the programmer.)
4. Microprogramming has an inherent capability to facilitate efficient emulation. Here, emulation is defined as the ability of one system to execute machine language programs written for a different system. This inherent compatibility permits a significant reduction in cost, manpower, and time otherwise necessary for the customer to update his equipment and his data processing capability. For example, the System/360, Model 40, was microprogrammed to execute 1401 programs efficiently. This was done despite the tremendous architectural difference between the two systems. (The term, architecture, is used to describe the attributes of a system as seen by the programmer, and it represents the conceptual structure or the functional behavior of the system as distinct from the organization of the data flow, controls, logical design, and physical implementation.) System/360, which is functionally structured to use eight-bit bytes, was made to recognize the 1401 architecture which uses six-bit binary-coded decimal characters. This inherent capability in a microprogrammed system to emulate other system architectures is the result of the direct access to selected portions of the information stored in the internal registers.

Thus, the systems engineer and the user view a microprogrammable system as a flexible tool whose architecture can be easily modified in terms of a specified application at an incremental cost much less than that required to implement the same modification on a conventional hardware-controlled system. Microprogramming offers the ability to customize systems features by modifying the architectural characteristics of the system with minimum impact on its logic design and data paths. This flexibility may mean additional macroinstructions such as trigonometric functions, extended precision arithmetic, matrix operations and manipulations, data correlation, and many other specialized macros.

Similarly, the *systems programmer* will be able to increase the efficiency of his operating system by speeding up the execution of the different control programs in the total operating system. Such functions as scheduling, dispatching, allocating the necessary storage, and other functions that enjoy the highest utilization can be microprogrammed and executed from ROS directly. Because of its nondestructive read-only characteristic, the ROS offers a convenient solution to the storage protection problem for some permanent and semipermanent tables of information, dictionaries, I/O device information, etc.

We have seen that microprogramming offers many advantages over a conventional hardware control in factors such as cost, performance, flexibility and tailorability, ease of maintenance, and many others which will be reviewed in more detail in a later chapter. Yet in reality, except for few isolated cases, microprogramming remains in the domain of the design engineer. Why? What is holding the different interested disciplines from taking advantage of the flexibility and efficiency microprogramming can offer? The following is a partial list of observations on this question.

1. Microprogramming was not intended for the novice programmer. To design efficient microroutines, a microprogrammer must be thoroughly familiar with the design philosophy, the architectural characteristics, and the technological capability that govern the design of such a computing system. He has to know the data flow, every in-gate and out-gate position, every control gate, and the timing points that control the transmission of data between the static units in the system. This information oftentimes is not disclosed by the designer.

2. Except for few special system designs, the control programs are stored in read-only storage devices that are difficult and expensive to modify. The production of the bit planes for existing ROS memories requires specialized equipment, and specialized talents are needed for their manufacture and installation in the system. The user need for such capability is too limited to justify making such a large investment, and conversely, the designers are not organized or prepared to fill the different customer's orders for these changes.

With the advent of writable control memories, the user will have this option of transferring the microcontrol routines from any other storage medium into this control storage, without any concern about physical or electrical problems.

3. The lack of standard assembly language and standard micro-orders and microinstructions discourages the users from attempting to apply microprogramming.

Today, every system design group, even in the same manufacturer's house, is responsible for specifying its microprogramming requirements, micro-orders, assemblers, ROS word organization, ROS sequencing schemes, and the training of its own personnel to carry out this function. Hence, within the manufacturer's own house, a microprogrammer assigned for one machine can be totally unaware of the language and microprogramming requirements and approaches designed for an adjacent machine within a compatible line of computer systems.

The assemblers designed by separate groups are often not suitable for general release since the documentation on the language and assembler are often not complete or up-to-date. Consequently, a higher-level language and the associated compilers are needed if this microprogramming application field is to flourish. Such a language must be machine-independent, procedural as well as descriptive. It must have the facilities to transform one machine architecture and map it into any other machine. It must also have the facilities to permit data manipulation, even at the register-to-register and bit levels. This high-level language can be compiled to an intermediate assembly or macro level, which can then be translated by separate machine-oriented assembly programs into the ROS bit patterns for that particular machine.

Many such attempts to define these higher-level microprogramming languages have been made. All of these projects have enjoyed a limited degree of success since they were able to shed more light on the problem. Yet, the biggest drawback from using the existing languages, and the most difficult and most challenging problem yet to be solved, is the optimization problem.

4. A fourth major problem is the lack of sufficient educational effort in preparing the potential user to cope with the problems he is to be confronted with, in instructing him of the available means for solving them, and in acquainting him with the advantages and disadvantages of this additional option for any given class of problems. Basically, microprogramming has been treated as an adjunct to machine design; no particular effort has been made to separate the related information or to make microprogramming itself convenient. Clearly, such a responsibility does not all fall on the designer. It must also fall on the leadership of the technical organizations, whose function is to keep its membership abreast of the latest technological changes. It is encouraging to note that more and more articles on this and related subjects are beginning to appear in the technical publications and trade journals.

5. *Cost Factor.* The biggest factor that will spur the potential users, which will lead to a tremendous growth in microprogramming, is the inclusion of a writable control storage. As long as the user is dependent on the manufacturer to implement any such changes into a permanent storage, microprogramming will still be an option exercised in isolated cases.

For small-scale, slow-speed computers, core storage can be shared by the program and control storage. Both sections can be accessed by a common set of storage address registers, with certain provisions to prevent the user from making inadvertent access to control areas. The microprogrammed control storage section can be loaded from the program storage by a control storage load procedure. A small portion of control storage is reserved for a routine to read control program and auxiliary storage information from an input device. If this special program is disturbed, a limited amount of data must be manually entered.

Since cost is of major importance in determining the marketing potential of any commercial product, and since the cost of a 500-ns memory is reasonably low (estimated at three cents per bit for a complete memory), it is feasible to have the control program stored in such a destructive read medium without degrading the performance of the system. By providing a wider memory data path, it is possible to organize the control word to direct the operation of the system for two or more consecutive machine cycles, thus avoiding some performance degradation that might have otherwise resulted from the write cycle imposed by the destructive read characteristic of a core memory.

The deposited planar-magnetic films and the woven beryllium–copper-plated wire memories may offer a brighter cost/speed trade-off for such a writable control storage. The cost is expected to level off at about two cents per bit. Yet the information stored therein can be read out without altering its contents. The speed of these thin-film memories is projected to be three times faster than that of magnetic cores for the same cost per bit.

Much has been written recently about the large-scale integration phenomenon. There are two approaches proposed: the bipolar integrated-circuit approach, and the metal-oxide, semiconductor approach (MOS). The MOS technology is slower due to its high impedance, and it is relatively less stable than the bipolar device. MOS has more potential for memory applications because it has a greater bit density than can be achieved in bipolar. This is due to the isolation problem inherent in the bipolar technology.[10]

The cost of such monolithic memories is still prohibitive, except for large-scale computers. However, cost projections show a substantial drop in cost, much below that of any magnetic core or thin film. Recent projections show a monolithic storage bit price drop by the mid 1970's to below the one cent per bit, a price that neither the core nor thin-film technology can match. This optimistic projection should spur more computer designers and architects to provide a writable control storage into the system, thus opening the field for specialized applications.

6. A sixth reason for the lack of widespread usage of the microprogramming option is the manufacturer's concern for the preservation of the architectural identity of the system and with preserving its effectiveness and its compatibility with other models in the product line. Users today are applying the computer in more sophisticated ways. They are placing it in the midst of their minute-by-minute, decision-making process; hence, the reliability of the system and its availability characteristics are of utmost concern to both the user and the designer. Consequently, a number of diagnostic routines and serviceability aids are designed for a fixed architecture and fixed hardware implementation. Once this is changed, the tools just mentioned become

obsolete, and maintaining the system components, both software and hardware, becomes an impossible task because the cost, manpower, and the experience level required will be beyond the capability of the user or the manufacturer.

This problem becomes a simple one if the system's original identity or affiliation with any product line or any operating system is not needed, that is, if the system is to become completely and permanently a slave to one fixed mission or task.

It has been demonstrated on many occasions that an architectural extension of the existing system can be achieved without affecting the original system identity. By providing another ROS section to contain new architectural features and the proper linkage between the two ROS devices, one can have a system that can operate in one or two modes of operations. This was accomplished in many emulators in which the architecture of the target machine to be emulated is mapped on the host or emulating machine. It was also successfully demonstrated in the design of the conversational programming system (CPS) by the joint efforts of IBM Corporation and the Allen-Babock Computing Corporation. This system and the emulator applications will be discussed in a later chapter.

7. Another reason closely associated with the cost factor is finding a satisfactory answer to the question of how much additional ROS the manufacturer should provide for the user's microprogramming needs. This question can only be answered for a specific customer with a limited and well-defined, well-documented system architectural extension.

Clearly, the answer to this question is to provide a writable control storage which can be used by the user to satisfy any number of his needs while preserving the original identity of the system in a read-only storage device.

1.3 DEFINITION OF MICROPROGRAMMING

In Sections 1.1 and 1.2 we discussed the general characteristics of microprogramming and its interpretation by the different special interest groups in the computer field. Any definition of microprogramming, to be universally acceptable, must satisfy each of these special interest groups. Considering this broad range of interests and backgrounds, the definition of microprogramming must cut across the specialties of electronic circuit design, logic design, mechanical languages, programming system architecture, and systems engineering. The danger here is that to satisfy all these disciplines and special interests with one single and concise definition would mean compromising the definition to the point where it has little or no meaning.

There are other difficulties in proposing a universally acceptable definition. Among these are:

1. Microprogramming was proposed as an orderly technique for designing and organizing the control signals which represent the states of the computer

and the orderly transition between these finite states. This was a direct replacement of the classical sequential logic circuits which were designed to interpret fixed data processing functions and execute them in a fixed hardware environment. This view places microprogramming within the domain of the system designer. The user, programmer, and other interested persons view this definition of microprogramming as too restrictive. They see microprogramming as an analytical procedure, a control automaton, designed to interpret fixed or variable data processing requirements into a language commensurate with the executing functions. That is, their definition tends to emphasize the concept of interpretation at a higher level without specifying the role and the organization of the hardware functions and the basic elementary operations that manipulate these hardware functions.

2. Another difficulty stems from the desire to exploit the concept of dynamically microprogrammable systems. The definition, therefore, must encompass this new environment. That is, it must be flexible enough to permit a wide interpretation of elementary operation, word size, and variable definitions of the basic number and arithmetic conventions without any adverse effects on the system performance.

Keeping these difficulties in mind, I propose the following definition:

> *Microprogramming* is a technique for designing and implementing the control function of a data processing system as a sequence of control signals, to interpret fixed or dynamically changeable data processing functions. These control signals, organized on a word basis and stored in a fixed or dynamically changeable control memory, represent the states of the signals which control the flow of information between the executing functions and the orderly transition between these signal states.

REVIEW QUESTIONS

1. What are the basic functions of the control section of a digital computer system?
2. How do the five basic functional units in a data processing system communicate with each other?
3. In what respect does the conventional computer hardware differ from a microprogrammed machine?
4. Define the relationship between the machine instruction and microinstruction.
5. What are some of the reasons that kept computer scientists from applying microprogramming sooner than they did?
6. List the characteristics of a data process in which some performance gains can be expected if such a process is microprogrammed rather than machine programmed.

7. Define the following terms:
   (a) Microprogramming
   (b) System architecture
   (c) Control gate
   (d) Microroutine
   (e) Writable control storage

8. What are the differences between synchronous and asynchronous computers?

9. What is the difference between a Boolean function and a static microspec function?

10. Can you draw any analogy between the relationship of the control function to the data processing system and the human brain to the human system?

REFERENCES

1. Beck, L., and Keeler, F., "The C8401 Data Processor," *Datamation*, 5, No. 2 (February, 1964), pp. 33–35.

2. McGee, W. C., "The TRW-133 Computer," *Datamation*, 5, No. 2 (February, 1964), pp. 27–29.

3. Boutwell, E. O., Jr., "The PB440," *Datamation*, 5, No. 2 (February, 1964), pp. 30–32.

4. Semarne, H. M., and Porter, R. E., "A Stored Logic Computer," *Datamation*, 2, No. 5 (May, 1961), pp. 33–36.

5. Hill, Richard H., "Stored Logic Programming and Applications," *Datamation*, 5, No. 2 (February, 1964), pp. 36–39.

6. Wilkes, M. V., "The Best Way to Design an Automatic Calculating Machine." Report of Manchester University Computer Inaugural Conference, (July, 1951), pp. 16–18.

7. Hellerman, L., "An Introduction to IBM System/360, Model 40 Microprogramming." Class Notes.

8. Pitkowsky, S. H., "Control Storage System." IBM Internal document.

9. Patzer, W. J., and Vanderling, G. C., "Systems Engineering Implications of Microprogramming," *IBM Technical Report No. 66-825-1990.*

10. Mayne, D., "What's Next in Memories?" *Datamation*, 5, (February, 1968), pp. 30–32.

11. Wilkes, M. V., and Stringer, J. B., "Microprogramming and the Design of the Control Circuits in Electronic Digital Computers," *Proc. Camb. Phil. Soc.*, 49, Part 2 (1953) pp. 230–238.

12. Glantz, H. T., "A Note on Microprogramming," *Journal ACM*, 3 (1956), pp. 77–84.

MICROPROGRAM CONTROL

2

2.1 HISTORICAL REVIEW

At this point, a historical review of microprogramming development over the past two decades may be in order. The original concept of a programmed control stored in a nondestructive read-only storage was proposed by M. V. Wilkes of the Cambridge University Mathematical Laboratory in a speech prepared for the Manchester University computer conference[1] and entitled "The Best Way to Design an Automatic Calculating Machine." Wilkes and his colleagues were clearly concerned with the ad hoc and nonsystematic approach that was applied to design the computer control. They sought a means for rearranging the computer components into a systematic order, easy to implement and maintain. They were more interested in simplifying the design task than in any hardware savings that can be realized.

Wilkes noted the extent to which the cross-connections between the various static units, especially in computers with more complex instruction sets, tends to obscure their logical interconnection. This is especially true in the general purpose machine that is intended to provide the ideal instruction set for every user and every field of application. While Wilkes did not embrace the concept of a general purpose machine on the grounds that there is no need for it, he proposed a scheme that made the concept feasible. Wilkes' team noted that each machine instruction can be broken down into a sequence of more elementary operations, which specify the transfer of information between the different microspec functions in the arithmetic and logic unit. Such micro-operations would, for example, specify a register-to-adder transfer, register-to-register transfer, etc. By providing and activating the necessary control gates between these microspec functions, the machine designer would be able to control the data flow through them. That is, each gate would be controlled by a control line whose state can be represented by a (0 or 1) stored in a non-destructive read-only matrix. A ferrite core matrix with a linear selection scheme was employed in the first Wilkes model[2]. This ferrite core matrix was permanently wired as a set of OR gates that would activate a subset of the control lines which form the output of the matrix. The original scheme proposed by Wilkes is shown in Fig. 2.1. It consists of the control matrix, C, the sequencing matrix, S, a well-defined timing pulse along with a decoding tree, and the decision logic.

The binary decoding tree would accept the output of register R as an input, decode the n-bits (width of R register), and select one of 2^n unique output lines, i.e., the decoding matrix would energize a single output line.

Matrix C consists of a set of horizontal wires (the 2^n outputs of the decoding tree) and another set of vertical wires orthogonal to the first set. The black dots shown in matrix C represent impedance coupling between the two sets, that is, any electrical signal propagating along any horizontal wire will be coupled through to propagate along the vertical wires where an impedance coupling (black dots) is indicated. If every vertical line in matrix C is connected to a unique control gate in the system, these control gates would be opened by simply energizing the corresponding control lines in matrix C.

A horizontal line can now be thought of as a microinstruction. When it is activated, it will, in turn, activate the predetermined set of control lines.

The pulse from the decoding tree passes also through the sequencing matrix S. The output lines are made to set a corresponding set of storage elements which make up the address register R. The delay circuit is provided to control the timing for changing the contents of the address register R.

Thus, a microinstruction is selected according to the contents of the R register. The microinstruction will issue the gating pulses according to the control code stored in that word, and it will also select the address of the successor microinstruction, which will control the function of the system during the ensuing fixed cycle of operation.

The fourth requirement of the control function, that of decision and branching logic, is provided by a conditional flip-flop* whose output is made dependent on satisfying a given predetermined test in the system. Depending on whether or not a predetermined test is successful, the conditional flip-flop would select one of two alternate paths, shown in Fig. 2.1(a), between the two matrices, C and S. Figure 2.1(b) shows the logic net used for executing the

*A flip-flop is a temporary storage element which is automatically reset (to zero) at a fixed point in time in every machine cycle.

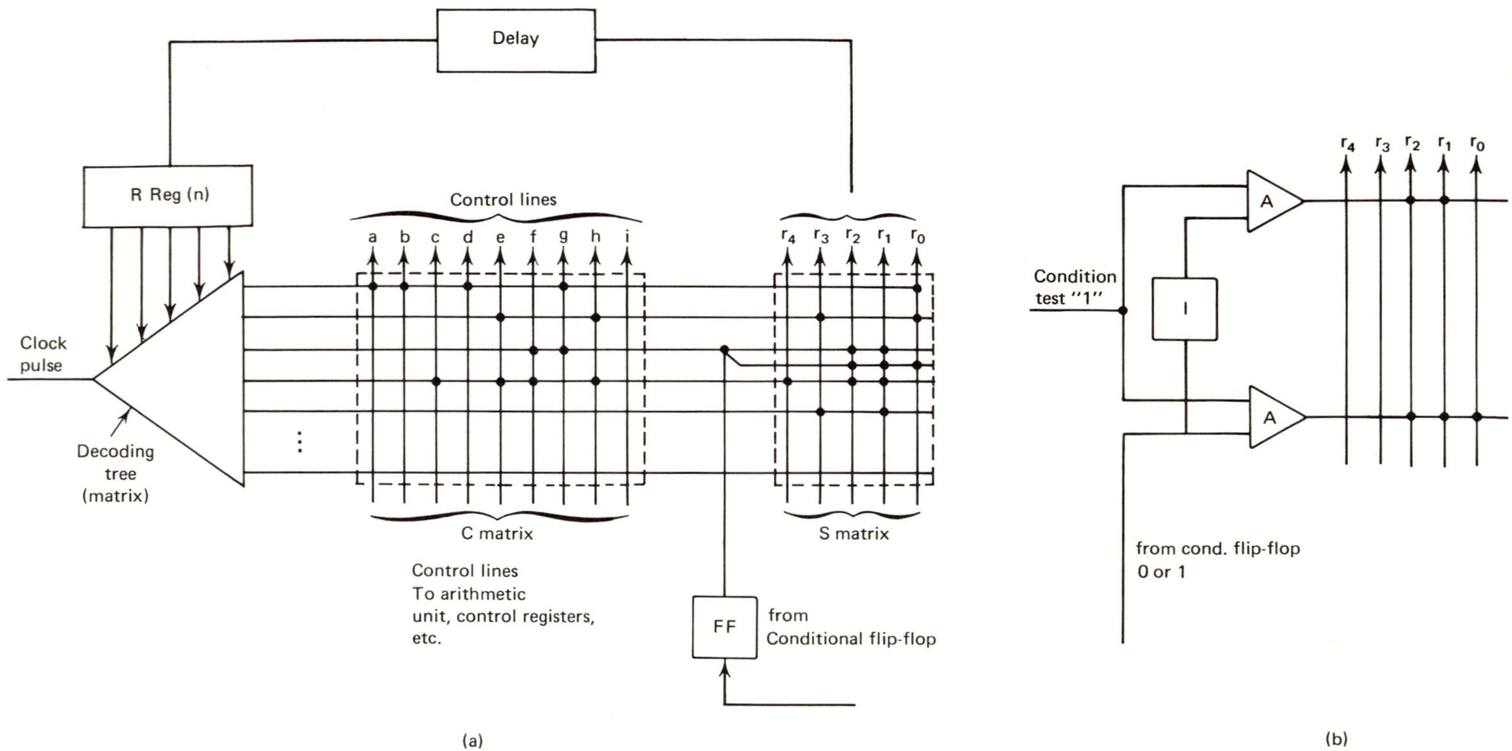

Figure 2.1 Wilkes' original microprogram control.

final sequencing decision and selecting between two potential successor microinstructions. A complete discussion on the sequencing of ROS storage will appear in a later section. Note the following points:

1. The number of horizontal lines emanating from the decode matrix represents the number of microinstructions that can be stored in this matrix.
2. The number of vertical lines emanating from the C matrix, plus the number of lines emanating from the S matrix, form the fixed width in the ROS word organization.
3. The static control lines emanating from matrix C do not need to be decoded but are directly mapped into the control gates in the system on a one-to-one basis.
4. In a similar way, by inserting the output of the decision element (flip-flop) between the decoding tree and the control matrix, the micro-operation itself is made dependent on the setting of the flip flops in question. This possibility may provide the solution to some critical timing problems the system designer may encounter.

The ideas outlines by Wilkes were elaborated in many other papers by himself and his colleagues at the University of Cambridge[3,4,5]. Many modifications and alternate schemes were proposed in these papers.

In 1955, Billing and Hopmann published a paper[6] in which they outlined the general concept and listed some practical applications. In 1956 and 1957, two papers by Glantz[7] and Mercer[8] were published in which the stored logic concept was presented as a means for altering the microprogram, and thus the instruction set recognized by the processor. The following year, Blankenbaker[9,10] published two articles in which he proposed a very simple, logically microprogrammed computer. In 1957 Dinneen[11], and two years later Kamp[12], described simple computers based on the Wilkes model.

One concludes that, during the 1950s, the microprogramming concept was not widely accepted; rather, it generated some modest theoretical interest. Many reasons can be offered, but it is sufficient to cite one: the cost of implementing this concept was prohibitive.

It is believed that Wilkes introduced the term "microprogramming" with its present meaning. However, the Lincoln Laboratory used it prior to 1950 to describe a system in which the individual bits in an instruction directly control certain gates in the processor[25]. Van der Poel[26,27] published two papers related to this subject; however, his emphasis was different than Wilkes'. Van der Poel sought to give the programmer a larger instruction set and argued for the programmer's right in voicing his opinion in the selection of the instruction repertoire. Wilkes was more hardware design oriented; he sought every opportunity to clarify the difference between these two types of microprogramming. The Wilkes scheme was first implemented in the control system built under his direction at the University of Cambridge and known as EDSAC 2.

Early in the 1960s, more interest was shown not only in the United States and England, but also in many other countries. There were two Italian papers:

Gerace[13] described a CEP microprogrammed processor built in Pisa, and Grasselli[14] explored certain aspects of microprogramming changeability.

A Japanese paper by Hagiwace et al.[15] was published in 1962, in which a diode matrix was used with a diode at each possible bit position in the control matrix. Each diode was connected in series with a phototransistor. An information document in the form of a punched card was used to mask out and illuminate certain subsets of the phototransistors, thus switching the corresponding diodes in the control and sequencing matrices.

Other Russian, Australian, and French papers[16,17,18] began to appear on the general concept of microprogramming. The February 1964 issue of *Datamation* contained a number of articles describing specific "stored logic" systems designed to extend the computing capacity of small machines.

The first microprogramming application by any major computer manufacturer was made by IBM. The microprogramming of the IBM 7950 was presented in several papers at the 1961 ACM National Conference[19]. However, the major burst of interest came with the unveiling of the IBM System/360, where all but the largest system included in their original announcement were microprogrammed. In 1966, the RCA Spectra/70 was described in a number of articles, including one by Campbell[24] in which the microprogramming of the Model 35 was disclosed.

It is interesting to note that, since the original disclosure of this concept by Wilkes, little in the way of additions, or modifications to his proposal have been made; in fact, most of the modifications came from Wilkes' group itself.

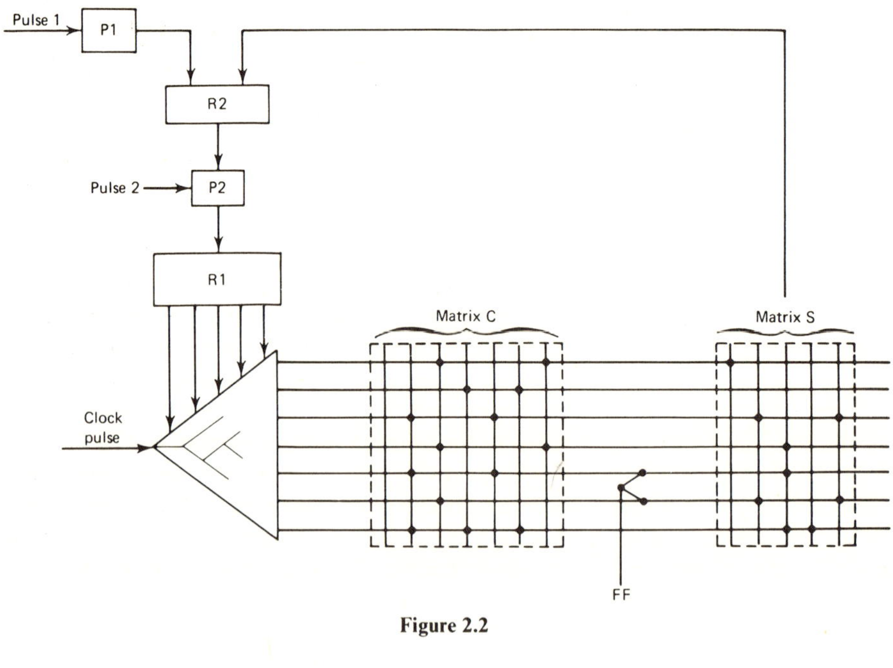

Figure 2.2

We will examine a few of these modifications in the following section. The first modification is shown in Fig. 2.2. The main extension here is the inclusion of another register, R2, to hold the address of the next microinstruction to be executed. R1 holds the address of the current microinstruction being executed. At the start of the next cycle, a P2 pulse would gate the contents of R2 into R1 to be decoded and then select the next microinstruction. With this arrangement, the sequencing of microinstructions could come from either the sequencing matrix or by the insertion of a starting address at a fixed time during the machine cycle designated by clock pulse P1. This starting address may be the operation code or subset of the binary bits in the operation code field which are uniquely assigned to distinguish it from all other operation codes.

At this point, it is advisable to classify these modifications and extensions into three categories:

1. The microinstruction address selection (Sequencing and branching).
2. Elementary operation definition.
3. The organization of the control word (ROS word).

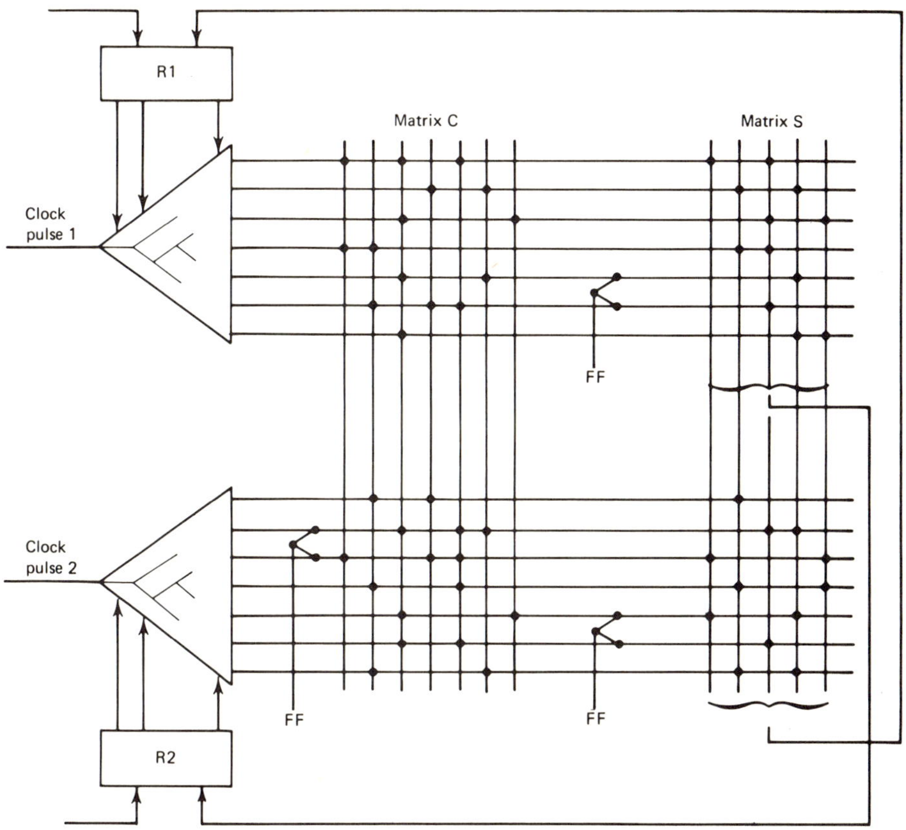

Figure 2.3

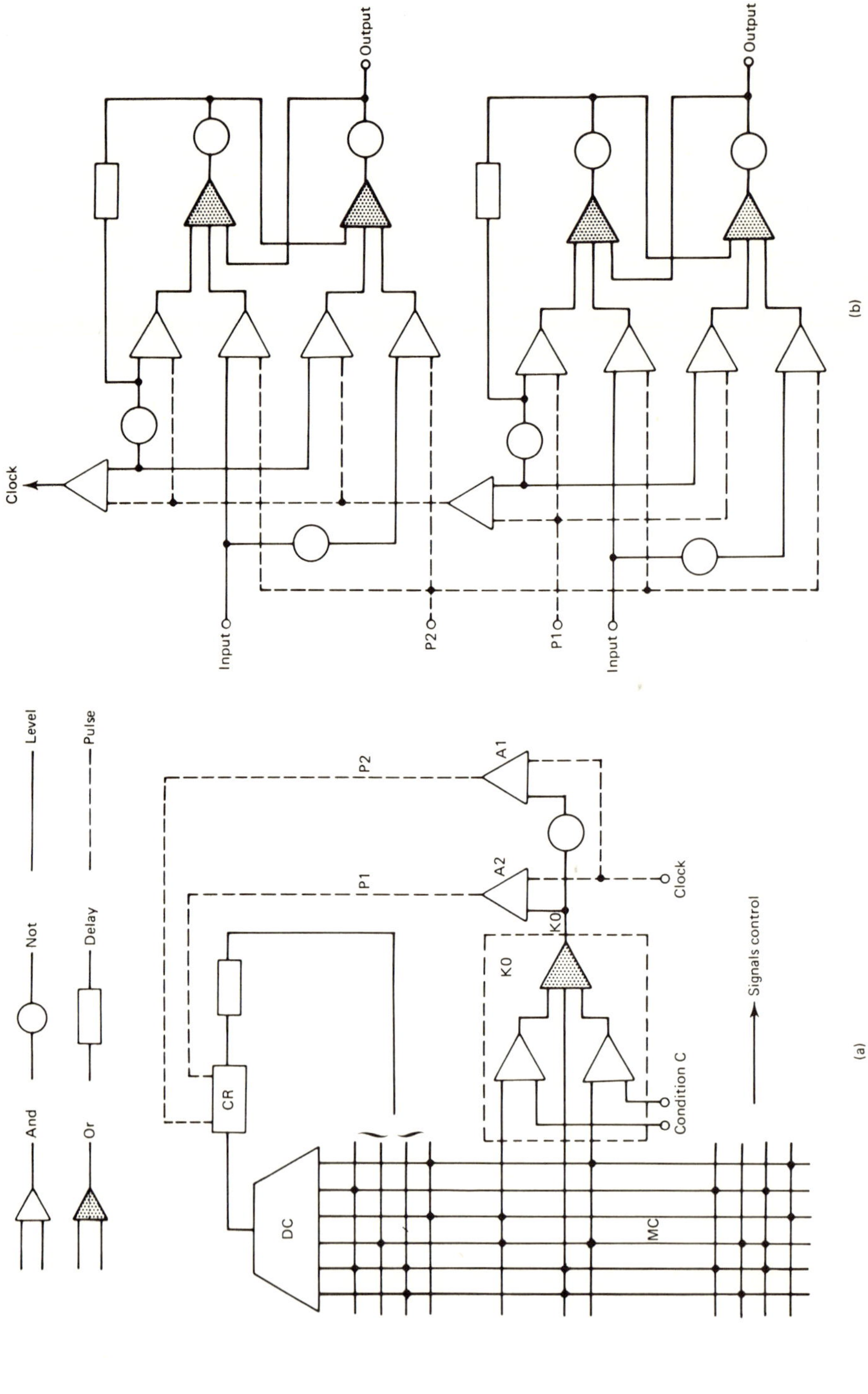
And
Or
Not
Delay
Level
Pulse
CR
DC
MC
K0
K0
P1
P2
A1
A2
Clock
Condition C
Signals control
(a)
Clock
Output
Output
Input
P2
P1
Input
(b)

Figure 2.4

In the *first class*, a wide variety of modifications were put forth. The first modification shown in Fig. 2.2 belongs to this class. Figure 2.3 gives another computer control scheme[4]. The new concept here is the organization of the control program into two distinct read-only storage matrices (stacks) and the way the buffering of addresses is achieved. Actually, the new scheme avoids the necessity of buffering by providing two decoding trees, each containing half the original memory capacity. The main advantage of this scheme over the one shown in Fig. 2.2 is faster operation at lower cost. These advantages must be carefully weighed against the disadvantage of the programming difficulty. This organization was implemented in the RCA Spectra/70, Model 45, which will be discussed in Chapter 9.

A fourth scheme, proposed by Gerace[13] (Fig. 2.4), is distinguished by the inclusion of a logical device, CR, which operates as a register or a counter, depending on which sequencing pulse is applied. When the address of the next microinstruction immediately follows the address of the current one, the CR functions as a counter, gate A2 is selected by the k_0 output from K_0, and the counter is operated by pulse P1. When the current micro-order specifies a conditional or an unconditional branch, gate A1 is selected, and the register is operated by pulse P2. This conditional branch could have as many successor microinstructions as the number of conditions to be tested. This organization would, in fact, increase the flexibility of the internal organization of the computer and the processing capability of the microprogram. It also reduces the number of repeated microwords in ROS.

In his paper, Gerace shows other extensions of this basic idea, gradually working up to a more general scheme, where one of four addresses may occur at the end of each microinstruction, each of which can select any microinstruction.

A fifth and final sequencing approach to be reviewed is the scheme suggested by H. P. Schlaeppi of IBM (Fig. 2.5). The scheme eliminates the need for an explicit address for every successor microinstruction. This represents an appreciable net saving in the size and organization of the read-only storage matrix. The scheme makes use of the fact that the microinstructions in each microroutine are stored in consecutive addresses in the ROS matrix, in the order in which they are executed, and in much the same fashion as the user's program written in machine language is stored in main storage. Except for some conditional or nonconditional branches, the address of the next microinstruction is obtained by an instruction counter incremented by 1, or by providing the necessary circuitry to add 1 to the contents of the ROS address register. A special return register, RR, was introduced to permit the design of a closed microroutine loop. The sequencing matrix of the ROS control matrix (S) is replaced by two output lines, S and R, which are designed to control different sequencing circuit paths in this scheme.

The microprogram is initiated by transferring the operation code of a machine instruction into register AR. Depending on the settings of the two variables, R and S, one of four distinct operational modes can be executed.

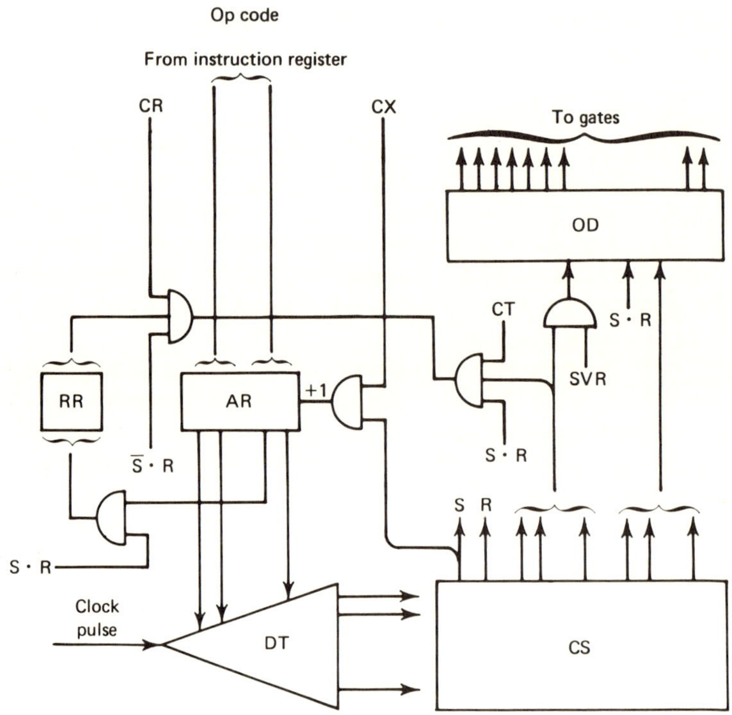

Figure 2.5

In the *normal* mode, where the microinstructions are stored in the consecutive ROS word addresses, the contents of AR would be incremented by 1. This mode is designated by S = 1, CX = 1 independent of signal line R. If CX = 0, however, the current microinstruction is repeated. If such an instruction does not activate any control lines, this would be equivalent to putting the computer in a *wait* state until a given condition is satisfied.

If both S and R lines are up (S = 1, R = 1), then the current microinstruction address is transmitted to the RR register (effected by the variable R = 1), and AR is incremented by 1, or by 0, depending on the status of the CX line, as was discussed in the normal mode. Here we effectively stored a return address in the special return register RR. Hence, the name given to this mode of operation is the *prepare-return* mode.

A third mode called the *execute-return* mode is implemented with R = 1 and S = 0. The contents of the RR register are transmitted to the AR register. This, in effect, is how a loop in the microroutine is formed.

The final mode of operation in this scheme is called the *jump* mode. This implies that the succeeding microinstruction is not stored in the next consecutive address. Hence, a new microinstruction address must be gated into AR. Schlaeppi proposed that a subset of the function field of the microinstruction currently emanating from CS be interpreted as a full-length (or truncated) microaddress. The values of the R and S signals, which would imply this mode of operation, are 0 and 0. The main advantage of this scheme is that it conserves the storage capacity of the ROS storage matrix.

The five sequencing and branching schemes were presented only to give an appreciation of some of the timing, cost, and microprogramming efficiency trade-offs that have to be made. There are many other interesting schemes that were designed as the best schemes for specific system organization and data flow performing specific functions. Among these are the addressing and branching logic implemented for the IBM System/360, the RCA Spectra/70, Model 45, and, finally, the Honeywell H4200. The main features of these three totally different schemes will be described next. Detailed discussion for the IBM System/360 will be given in Chapters 6, 7, and 8; the RCA Spectra/70, Model 45, is discussed in Chapter 9; and the Honeywell H4200, in Chapter 10.

The System/360 follows the original Wilkes sequencing scheme with a slight modification. In the System/360, each ROS word contains the address of its successor microinstruction. This is done for all sequential microinstruction fetches, except for conditional or unconditional testing and branching. In the case of conditional testing, the ROS word contains one or two test specification fields. These fields were organized to specify *one* of a *set* of mutually exclusive tests that can be performed in the system. Each test field is associated with one of the low order bits in the ROS address field. IBM designers introduced the concept of a branch set, which is the set of addresses that a branch node could select from. It was stated that these branch sets are to be stored in consecutive locations in ROS, and hence, the low-order two-bits can be logically set from the results of the specified tests. If, for example, a test is specified in the test field associated with the low-order bit, then the high-order ($n - 1$) bits are completely specified by the ROS word in the ROS address field. A zero is forced into the low-order bit position of the ROSAR; however, its final setting depends on the result of the branch test specified. If the test is satisfied, a binary 1 is ORed into that bit position; otherwise, it remains 0. Figure 2.6 illustrates this general concept of conditional branching based on two different sets of mutually exclusive tests.

The two test fields, when used in the same microinstruction, can specify one of four successor microinstructors whose addresses vary only in the binary bit designation of the two low-order bit positions.

In the case of nonconditional branch, the System/360 designers reserved the next four low-order bits to which four bits can be gated from a number of sources. This is called a *functional branch*. In this case, only the high-order base and the low-order two bits (if no test is specified) are set from the ROS field, and the functional branch bits are transmitted from a source specified by the microprogram. Note that the functional branch can be coupled with the bit test fields to give a one of 64-way branch.

Another feature of the sequencing and branching is that of timing. This problem will be discussed in general later in this chapter, and in more detail with each computer system that will be presented. Therefore, it is sufficient

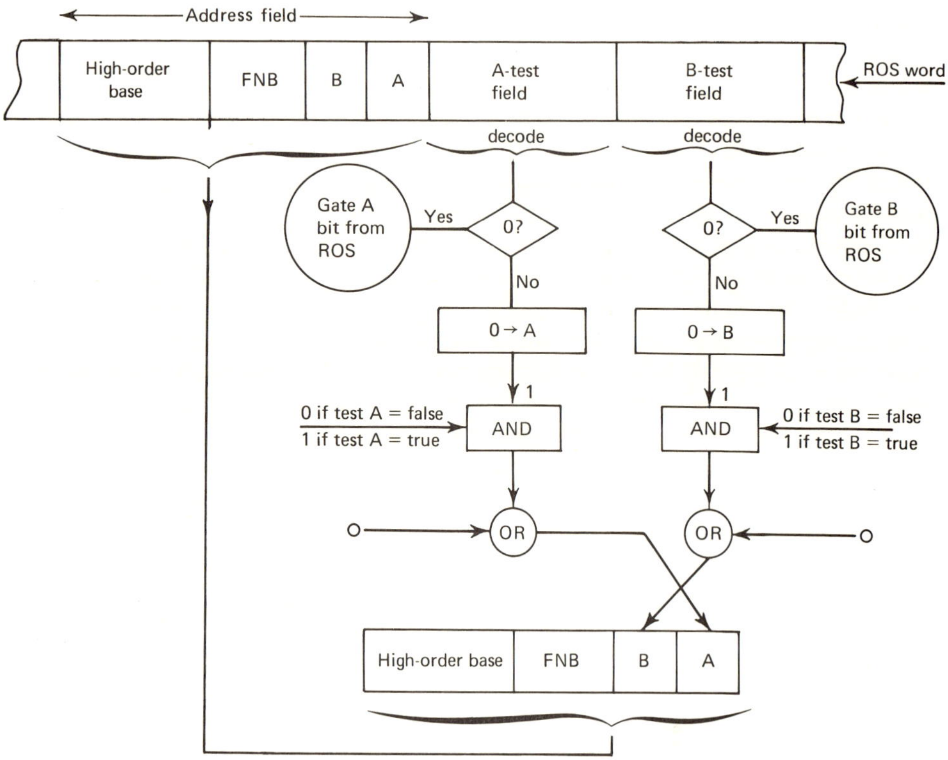

Figure 2.6

to state that the branching specified in one machine cycle is generally based on the contents of indicators previously set to reflect the state of the machine. That is, the result of a specified test, in order to be effective in the selection of the next ROS address, cannot be made to depend on the results of any arithmetic and logic operations specified in the same microinstruction.

The RCA Spectra/70, Model 45, has a different arrangement altogether. The ROS word has a six-bit field, T, to designate the test condition. It also has two six-bit A and N fields which will contain the low-order six-bits of the addresses of the two successor microinstructions. If T is true, then the contents of the A field are transmitted to the six low-order bit positions of ROSAR; the contents of the N field are used if T is false. There are a number of overriding considerations such that, in an unconditional test, the microprogrammer has to specify the two halves of the next ROS word address in the A and N fields. A third field (S field) can force the address to be taken from the C and V fields.

The Honeywell H4200 branching and sequencing method was completely based on the specific processing requirements. Since most algorithms to be executed in the H4200 include several sequential steps, an incrementor box (similar to the CR box in Geraces' scheme) was provided in the ROS addressing loop. Another requirement was imposed on the system by the critical algorithms in the system. It became evident that there are several critical routines where six-way branch capability is necessary to ensure the system efficiency. The branching scheme which evolved from these requirements consists of a set of five full-width ROM (ROS) address registers (A to E registers) which could be loaded by the microprogram from a specified field in the ROS word. The test micro-orders that sample over 100 indicators in the system would result in setting one of five indicators (A–E). A given branch command would then sample these five indicators with a fixed priority, A highest and E lowest. When the highest-priority indicator is found, the address in its corresponding ROS address register is transmitted to the ROSAR to be decoded. The timing for testing and setting the indicators and their associated address registers is too complicated to describe yet. It will be thoroughly examined and analyzed in Chapter 10.

Each of the sequencing schemes discussed above sought to improve the flexibility in microinstruction sequencing, the width of the ROS word, and the timing to access the next microinstruction. Most of these schemes sought solutions to the problems concerning the conditions that intervene at the end of the microinstructions; they seek to reduce the timing for sequencing and execution of these microinstructions; and finally, these schemes are concerned with time-sharing and multiprogramming the facilities in the processor.

The basic Wilkes approach (Figs. 2.1 and 2.2) required the full address to be transmitted to the ROS address register. This requirement eliminated any possibility of overlapping of ROS word access and execution. Wilkes' second modification (Fig. 2.3) provided some overlap when the system executes sequential microinstructions. This scheme had the additional advantage of reducing the size of the operation decoder, thus speeding the access time.

The main disadvantages to Wilkes' approach are that of the complexity in assigning the addresses and that, when a conditional branching is indicated, the system reverts back to the serial access and execution process. The last two schemes sought to provide a counter register with the logical input to step up the ROS address therein as long as the microinstructions were executed sequentially. This approach would reduce the width of the S matrix in the ROS word essentially to few bits that provide an indication of the exceptions to the implied normal sequential process, thus reducing the cost of the control system.

2.2 ROS WORD ORGANIZATION

Direct Control ROS Word Organization

A second instance where the process of modifications and extensions occurred was in the area of what the elementary micro-operations are or should be. Since the third type of change, namely, the word organization, is definitely related to the second type, we shall introduce them together in a manner that would permit us to study their tradeoffs. What are the true basic micro-operations? Many different answers to this basic question were formulated by different practitioners in the field. Microprogramming was defined earlier as the technique of designing the control circuits of an electronic digital computer to formally interpret and execute a given set of machine operations, M_i, as an equivalent set of sequences of *micro-operations that can be executed in one pulse time*[4]. Examples are: transfer data from the storage to arithmetic registers, or vice versa; and a transfer from one register to another. Blankenbaker's[9] concept of this, while close to Wilkes', seems to imply a yet more basic operation, that is, closer to the bit-by-binary-bit operation. Tompkins[20], on the other hand, had a less restrictive definition of micro-operations. He considered any operation "whose total course of action is dictated by pulses occurring simultaneously without requirement for later generation of stimulus pulses" to be a micro-operation.

What is clearly common to all these definitions is that a micro-operation is basically an elementary data-processing operation that permits the many similar operations which are mutually exclusive to occur simultaneously and during a well-defined timing point during a given machine cycle. This may involve the whole register or a selected group of bits in a register. It also implies that all other microspec functions (adder, shifter, etc.) are considered as registers. In 1958, Dineen, et al., of MIT[11] described the design of the CG24 general purpose computer in which this concept of register-to-register transfer was affirmed; yet the CG24 sequencing of the ROS storage was not done every cycle. The diode matrix was read out once every four machine cycles. This implies that an ROS word was designed to control and sequence the machine for four machine cycles.

One may think of the ROS word here as a microroutine designed to sequence the machine through more complex micro-operations for a number of machine cycles. In the TRW-133 computer, the control section was designed so that each control fetch would sequence the machine for a number of machine cycles. These more complex microsteps were defined as *logands*. Each logand may control any number of cycles[21].

All previously stated references did not deviate much from Wilkes' original approach. They all expressed the same basic concept of direct control. For each control gate in the system there exists a unique signal line (control line) emanating from the control matrix. We will call this scheme *direct control.*

In a complex, modern, digital computer, where the number of control gates in the system and, hence, the number of control lines emanating from the control section is very large (over four hundred), such a scheme would have an adverse implication on the speed, cost, and efficiency of the system. This is not to suggest that the direct control organization is always undesirable. In a small scale processor where the number of control gates is small and the microspec functions are organized into one data path, the direct control scheme will be superior in terms of cost-speed trade-off than the encoded control scheme which will be discussed next.

2.3 ENCODED CONTROL

The clue to this solution lies in the fact that modern digital computer systems are built of functions and data paths that are independent and can be operated in parallel. The inputs to all logical and arithmetic functions are furthermore mutually exclusive since, regardless of the number of sources that can provide the input to a specific arithmetic or logic function, one and only one source can be selected. That is, if the number of sources to such a function is N, then one unique control line out of N possible ones will be energized in a given machine cycle. For example, in a binary adder, the addend or augend values may come from M sources; but in a given add cycle (where the add function is specified) one and only one register will be designated to emit either of the two operands to the adder. This suggests the possibility that the states of these mutually exclusive gates, say M gates (which, in the scheme originally suggested by Wilkes, would have a unique control line per gate emanating from the control matrix), can now be encoded in a control field in the ROS word designated for this function. In the adder function example, rather than reserving M control bits in the control matrix, we would only need x bits, where $M = 2^x$. Upon access, the x bit in the control field will be decoded to activate the desired one of the possible 2^x control lines. This method requires a unique decoder for each field. Each decoder is designated to control a given static function or data path. In spite of the additional cost for the encoded field decoders, such a control system organization may prove to be optimum from the standpoint of cost and ease of programming. This is especially true in designing the control section of a large data processing system. These assertions are difficult to prove since they are dependent on the technology and the internal data flow structure of the system. Section 2.4 will illustrate these two ROS word organization schemes, their advantages, and the trade-offs implied in each without attempting to prove these assertions.

To design an efficient ROS word based on this concept of mutually exclusive fields of control signals, a few basic steps must be observed.

1. The microprogrammer must have a complete understanding of the architectural specifications of the data processing system for which the control is being designed. He must also possess an equal understanding of the hardware design philosophy of the system proposed to implement such an architecture.

On the hardware level, the microprogrammer must be aware of every gate, every data path, all the interfaces between the data paths within a given box, and the interfaces between boxes.

2. Most modern digital computers are characterized as synchronous digital computers. Yet, upon careful examination, one would find that the internal data paths, except for their interfacings, do operate asynchronously. For example, the I/O channels and CPU data paths do operate asynchronously; in some computer systems, main storage and the CPU also operate asynchronously. It is imperative for the microprogrammer to realize that, in such

a system, there exists in fact a combination of hardware and microprogrammed control operating together. This mixture of control may demand a rigid specification of priorities and communication between the hardware control and the microprogrammed control.

3. A given box, for economic reasons, may serve more than one function. For example, the adder data path in the IBM System/360, Model 50, in the normal mode, is used for performing arithmetic and logic operations. However, when an I/O command is given to transfer information from an I/O device to main storage, certain CPU facilities become part of the common-channel facilities. Thus the definition of the control fields, their width and function, may be different in this I/O mode of operation. This dual role suggests additional hardware and different micro-order specifications for the different modes of operations. It also suggests that the function operating in a certain mode, when requested to operate in another mode, must have a well-defined procedure to follow in preserving its state so that it can shift back and resume operating where it left off without any undue loss in performance.

4. The microprogrammer must prepare a list of all the control gates that are to be controlled by the ROS storage contents. This list will contain all the register in-gates and out-gates, adder input and output lines, shifter control lines, status registers, sign control, sequencing control, etc.

5. This list is now divided into three major categories:

(a) *Static control lines* which control the gates that facilitate the flow of information between the static (microspec) functions in the data processing system.

(b) *Sequencing controls* which are used to specify the address of the next ROS word, the conditional branch tests available, and the means for modifying the sequencing scheme based on the results of the branch tests specified. The options available here are many, and they have some far-reaching implications on the performance and cost of the system. Hence, this sequencing will be dealt with in a later section.

(c) *Emit field.* In 1960, Kampe[12], who was perhaps the first to propose the explicit grouping of control gates which are assigned to control a particular machine function, also proposed the concept of emit field. This concept implies reserving a field of fixed width in each ROS word, into which a string of binary bits (constant) specified by the microprogrammer is stored. These bits are designed to assist in the execution of a function. The width of this field is a fundamental design parameter with cost/performance overtones that are not easily identified. Such a field can be used to sequence the ROS storage, set a group of states, modify contents of data registers, etc.

6. The static control set of gates can be now divided into subsets, each of which is designed to control a certain identifiable function; and each subset is mutually exclusive, i.e., one and only one gate in each subset is required to operate during a given cycle. Each subset is encoded into a field.

7. Unique micro-orders, i.e., mnemonics that are descriptive of the function each micro-order performs, are assigned to each encoded value. The procedure for assigning these mnemonics is to some extent dictated by the assemblers and the processing systems.

Once these steps or similar ones are carried out, we are then in a good position to specify the characteristics of the ROS memory, the speed and the technology from which it is to be implemented.

The width of the control word is then equal to the sum of the widths of all the static fields, plus the width of the emit field, plus the sequencing and branch functions.

It is important to point out that certain data paths may be idle during any given cycle; hence, in the process of assigning bit strings for any field, a no operation (NOP) option must be added. For example, if the adder data path is to be idle in a given cycle, control fields in the ROS word that specify the left and right inputs (the addend and augend) to the adder must contain a unique bit string which, when decoded, will indicate this no operation option.

The encoded control organization can reduce the ROS bit requirement by one-third to one-fourth of that required for direct control. Care must be exercised early in the design cycle so that the ability to introduce new micro-orders without rewiring the decoders is not jeopardized. In its ideal implementation, the encoded control can produce a most general microprogrammable system, totally unspecified, and one which permits any modification and architectural extensions during any time in the useful lifetime of the system.

Between these two approaches to the ROS word organizations, there exist many other compromises. These compromises were imposed by the system's hardware organization and the process it is required to perform.

The RCA Spectra/70, Model 45, and the Honeywell H4200 do provide an excellent contrast to the IBM System/360 in their definition of the elementary operations and ROS word organization. The System/360 is a totally unspecified system. Its ROS word organization is that of the encoded control. The System/360 micro-orders are designed to specify the most basic data transfers. Its micro-order set is rich and does permit a wide class of architectural modifications and extensions without a need for any modification in the ROS word organization or field decoder rewiring.

The System/360 is a micro-order-oriented system. The Honeywell H4200 and the RCA Spectra/70, Model 45, while they differ in their basic ROS word organization, are basically microinstruction machines. Both systems resulted in a ROS word organization which is a mixture of the direct and encoded control schemes. The H4200 designers introduced the concept of ROS-bit steering, which behaves as follows: A few bits are set aside to specify how the bits in other fields in the word are to be interpreted. This method of bit steering can result in a five-to-one saving of ROS bits over the direct control organization, and it generates the largest number of simultaneous subcommands from a relatively small number of bits. Each configuration of these specification bits selects a set of micro-orders that are represented in the general field in a direct bit control basis without any decoding. The Spectra/70 and the Honeywell H4200 ROS word organization can be described as process-oriented. The microroutines were flow-charted first to determine the best ROS control organization for that process. This method of design tends to reduce the universality of the micro-orders. It also produces a system which is highly specified; one that is difficult to modify without reexamining and redesigning the control system to suit the new requirements.

Before discussing the control memory organizations and specifications, we shall demonstrate the control memory function, using both the direct control and the encoded control concepts.

Conventionally, the static function as well as the sequencing function in the control section of the digital computer are both functions of the operation field in the instruction format. Before the fetched instruction is executed, it must first be decoded. This step would affect both the static and sequential control functions. One may think of the operation code as an A input field which is decoded into one of 2^A decoder outputs, and each output would then be encoded into a unique machine state which, as indicated earlier, would continue the sequencing process until the whole function is performed.

2.4 EXAMPLE 1. MICROPROGRAMMING A HYPOTHETICAL COMPUTER

To illustrate the control function as it relates to these two processes, namely, instruction fetch and instruction execution, we will propose a hypothetical computer data flow, shown in Fig. 2.7. It consists of three sections: the arithmetic and logic unit (ALU), main storage (MS), and the control sections (CS). The main storage is of the coincident-current type, randomly accessed, and it contains both data and instructions. We will assume that the operation of storage or retrieval of information from main storage is asynchronous from the ALU unit. Assume also that the main storage is activated by providing it with an address to the storage address register (SAR) and a start signal. The address is decoded within the storage unit to select X and Y drive lines for the addressed location. The start signal initiates a storage cycle to read out or store the data.

In the read operation, the contents of the selected location are placed in the storage data register (SDR). Since this is a destructive read out, if the original information is to remain at the same storage location, the contents of the SAR must not be modified until a memory regenerate (write) cycle is executed. If, between the read cycle and the regenerate cycle, the contents of the SDR are modified (by transferring information into the SDR at this point in time), then the result of the regenerate cycle would be to store the new information (write) into the selected core locations. Hence, the microprogrammer can specify to

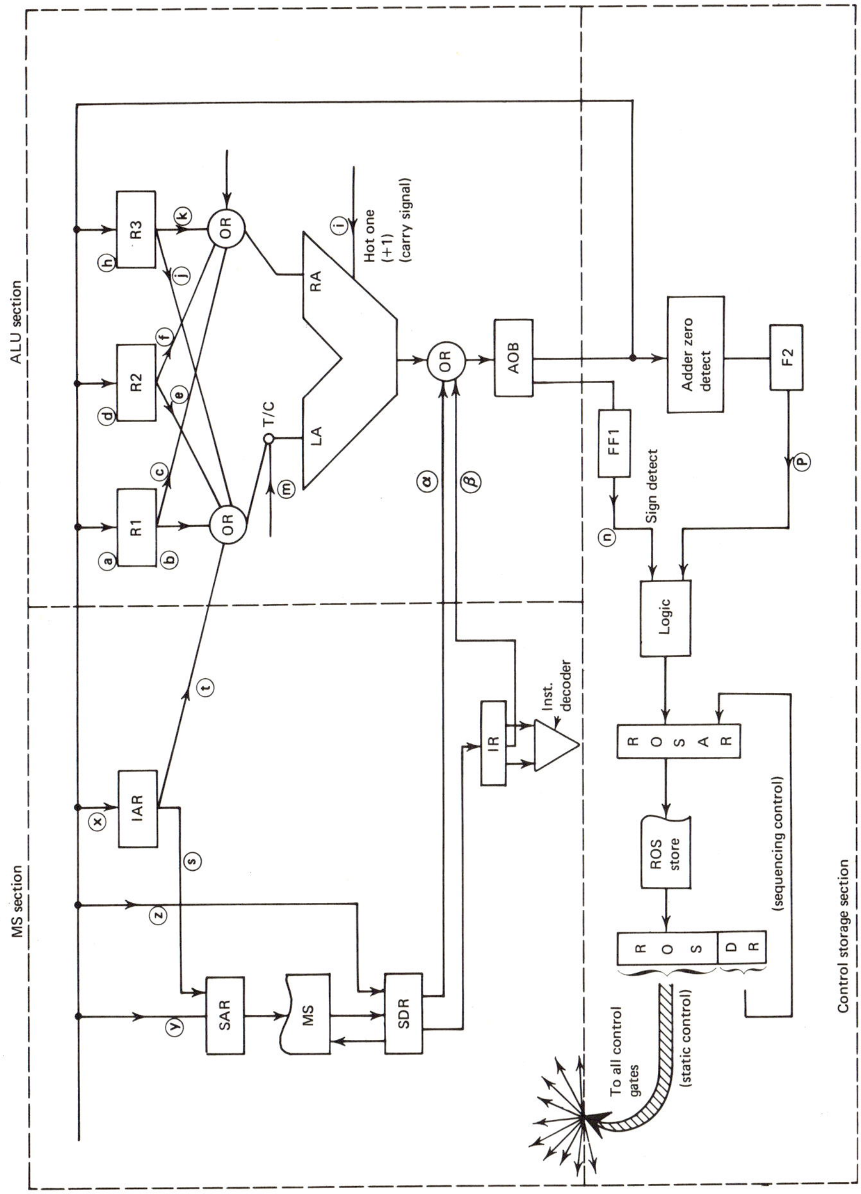

Figure 2.7 Hypothetical data processing system.

the main storage a write or read operation by transferring information into or transferring out of the SDR at the appropriate time. The contents of the SDR can be transferred into the adder output latches* (AOB) to be distributed at the appropriate point in the timing cycle to any destination registers specified by the microprogrammer. In case of an instruction being fetched, the operation-code portion of the instruction is gated into the ROS storage address register. The ROSAR is in the box labeled Control Storage Section.

The ALU section is a very limited one. It consists of three hardware registers, and a parallel-input adder with a true/complement gate (1's complement) at the left adder inputs (LA). A hot 1 (+1) is provided to the adder low-order bit input and is controlled by gate *i*; the output of the adder feeds into the AOB (adder output) latches, from where the result of an operation is distributed to the storage registers in the processing system. Note the two tests which are provided at the outputs of the AOB. The sign check is provided to test the sign bit of the AOB. Also, an adder zero-detect test that emits a positive signal (1) when every bit in the AOB latches is zero. The outputs of the FF1 and FF2 flip-flops will combine logically with the sequencing control logic in the control section in order to determine the state of the machine in the following machine cycle. Note also that the Instruction Address Register (IAR) is incremented by the adder in the ALU. This method of sequencing the main storage is arbitrary; we would have a better performance machine if an instruction address counter is implemented into the storage address register (SAR) to IAR data loop.

Figure 2.7 shows the basic functional blocks which make up the control storage section. These are the ROS memory, ROSAR, ROSDR, and the sequencing logic. The details will be presented later. Note that in this data processing system there exists a total of seventeen gates, designated by an arrow (→) and identified alphabetically. Regardless of the ROS storage implementation, it should be clear by now that the outputs of the control box would contain a unique control line for each of the seventeen control gates.

To simplify the understanding of the data flow shown in Fig. 2.7 it would be helpful to make few general observations and definitions.

1. Data gates into the top of all the facilities, or functions, and out of the bottom of these facilities.

Definitions

> *In-gates* are the control gates which control the data flow *into* the facilities. For example, gate *a* is an in-gate to register R1; similarly, *d* and *h* are in-gates to registers R2 and R3, respectively.

*A *latch* is a temporary storage element in the system. It differs from the register in that it is automatically reset at the same timing point in every machine cycle, whereas a register retains its contents indefinitely. The microprogrammer has complete control of the register gating and no control on the latch.

Out-gates are the control gates which permit information to flow *out of* the facilities. For example, gates (b, c), (e, f), and (j, k) are sets of out-gates from registers R1, R2, and R3, respectively.

2. Data gates into all the registers in the ALU and MS at the same timing point. The AOB latches are reset to zero just prior to the set latch timing point (Fig. 2.8). A register is changed only by a direct command from the control program to modify its contents, whereas the latch is reset automatically and at the same timing point in every cycle.

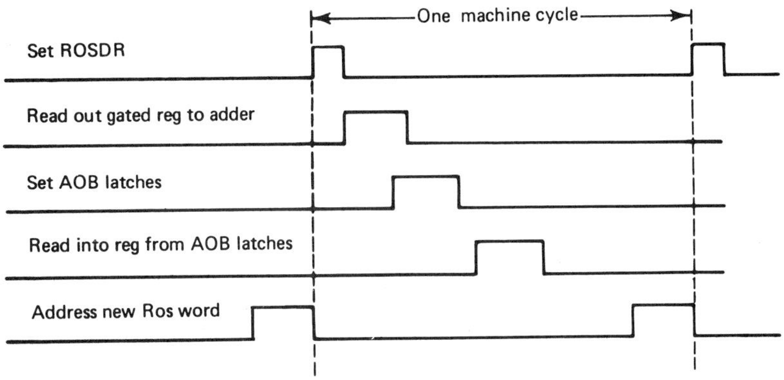

Figure 2.8

3. Based on the preceding observations, if no data are gated into AOB, then the contents of these latches at the end of that cycle will be zero. Thus, in the next machine cycle, if a register in-gate is activated, that particular register will be set to zero.

4. Fixed instruction and data formats are assumed. The high-order bits of the data format are reserved for the sign bit. If S = 1, the field is negative; otherwise, it is positive.

5. When a machine instruction or macro-operation is fetched from main storage, the unique operation code assigned to it is gated into the ROSAR to select the microprogram routine to handle its execution.

6. The widths of the R1, R2, R3, adder facility, AOB, SDR, and the data paths which connect them are equal.

Assume that a fixed-point multiply instruction has been decoded, and the address of the first microinstruction in the multiply instruction execution routine is gated into the ROSAR. Assume also that the two operands (multiplier and multiplicand) have already been brought to the R1 and R2 registers. To further simplify our problem, assume that the two operands are positive, fixed-point integers, and the sum of the significant bits in the multiplier and the multiplicand is less than or equal to the width of the register R3; that is, the overflow problem is eliminated.

The problem is to multiply the contents of R1 (the multiplier) by the contents of R2 (the multiplicand) and store the product in R3. To illustrate the two main ROS storage organizations discussed earlier, the solution will be implemented twice.

The common areas shared by both implementations are the sequencing and timing. In a *synchronous* computer as previously defined, each function is given a well-defined segment of the clock cycle in which to fulfill its mission. This, in fact, distinguishes it from an asynchronous process where the start of the next function in the process is signaled by the current function once this current function is completed. Since the solution to the proposed multiplication problem deals with the hardware in the arithmetic and logic unit (ALU) and in the ROS memory, only the timing points related to these two sections of the system will be defined. Figure 2.8 gives the following main timing points.

1. *Set ROS Data Register (ROSDR) Pulse.* Since the output of ROS is usually only pulses, a register (ROSDR) is provided to hold these pulses for the duration of the machine cycle. In the first implementation, that of direct gating, the control lines are immediately available. In the second implementation, that of the encoded ROS fields, a fixed delay time is required for decoding before the control lines are activated. Assume that the set ROSDR pulse is sufficiently long for either implementation, and that, by the end of this pulse, the control lines are activated.

2. *Register Out-gate Pulse.* The selected register out-gate positions are opened to permit information to flow into the adder to be operated upon.

3. *Set AOB Latches.* Data from the adder output or main storage data path are gated into the AOB latches during this timing period. Note that the AOB latches are first reset just prior to this timing point. Hence, if no information is gated into the AOB latches, their contents will be zeros.

4. *Register Set Pulse.* The information which now resides in the AOB latches is transmitted to the set of registers specified by the ROS words.

5. *Address New ROS Pulse.* Parts of the read-only storage address register that select the next ROS word may be gated prior to this time in the cycle. However, since this address may be dependent on the status of the computer, the gating of the branch condition states which reflect the result of the arithmetic and logic operations in this cycle is withheld until the last possible moment in the machine cycle. Hence, the conditional branch specified in one machine cycle would affect the fetching of the next microinstructions.

The limited functions in the arithmetic and logic unit dictate the algorithm in which the multiply instruction is to be executed. This algorithm can be characterized as multiplication through successive additions, i.e., the multiplicand is added to an accumulator register as many times as indicated by multiplier. A flow chart for the multiply instruction execution on this hypothetical machine is shown in Fig. 2.9. The first microinstruction ensures that the R3 register (the

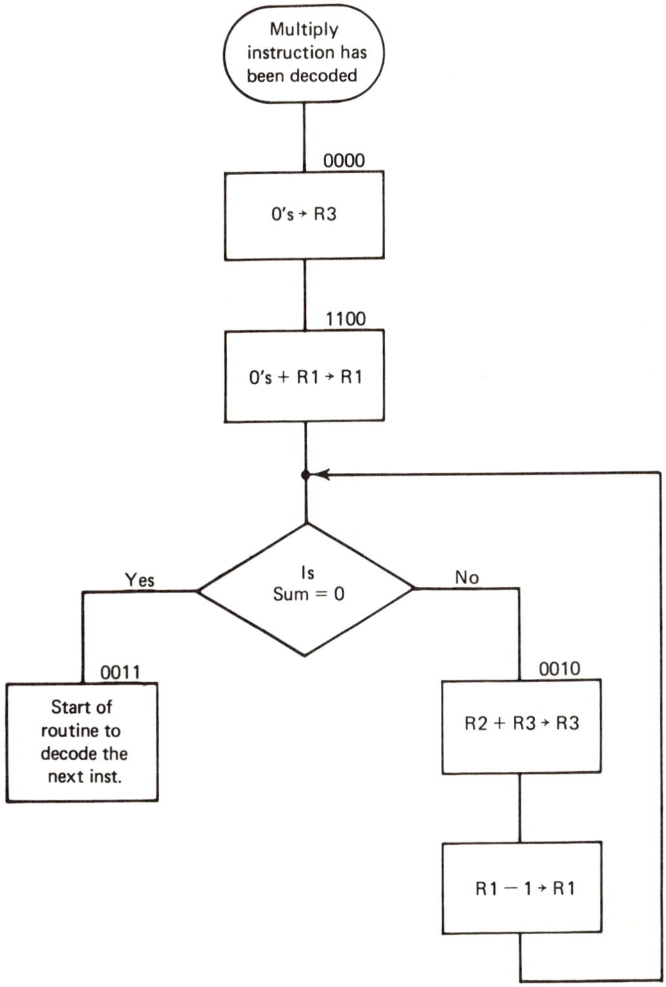

Figure 2.9

accumulator register) contains zero. Many options are available for the microprogrammer. One way is not to gate any inputs to the adder; hence, by the end of the set-AOB-latches pulse, the adder output bus would have zeros. Therefore, by activating gate *h* in the same microinstruction, register R3 would be initialized to zero.

Another method which could be implemented in one microinstruction is to subtract the contents of R3 from itself, by gating R3 to the right input (open gate *k*) and simultaneously gating R3 to the left input to the adder (open gate *j*), with the 1's complement gate *m* activated and the hot one (gate *i*) gated to the rightmost bit in the adder. This is equivalent to subtracting R3 from itself by adding the 2's complement of R3 to the true representation of R3. By activating gate *h* in the same microinstruction, the AOB contents (zeros) are set into the accumulator register R3. The first method of setting R3 to zero is a simpler one and would only require activating gate *h*.

The next logical step in designing this microprogram is to test the contents of the multiplier (R1) and the multiplicand (R2). If either of them is zero, then the result (zero) is already in the accumulator. Testing of the contents of a register for zero is accomplished by gating the contents of that register to one input of the adder, and zeros (or no input) to the other input of the adder. This results in setting the contents of the specified register in the AOB latches. The adder zero-detect would gate a 1 to the conditional flip-flop, FF2, if the contents of the AOB latches are zero. Otherwise, FF2 would remain zero. As shown in Fig. 2.8, the bit in FF2 will combine logically with other inputs to resolve the address of the next ROS word. Assuming that neither of the two operands is zero, the microprogrammer would then define the loop for adding R2 to R3, using R1 as a counter that is decremented by one at the end of each addition cycle. Figure 2.10 shows the ROS memory implementation for this

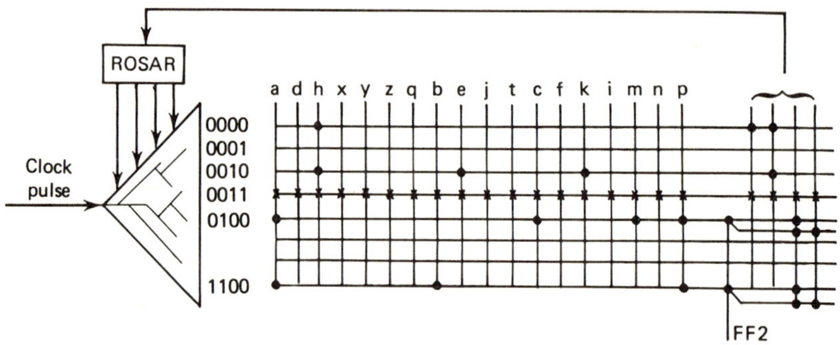

Figure 2.10 Multiplication (Fig. 2.9) microprogram.

microroutine using the original Wilkes scheme with direct control. Notice that the addressing here is completely random.

Contrast this with the following solution based on the encoded ROS word organization. The seventeen control gates can be divided into the following mutually exclusive fields.

1. *AOB Destination Field, DF.* It contains *a*, *d*, *h*, *x*, *y* and *z* gates. Therefore, we have a minimum of six unique micro-orders, plus a seventh which can specify the null option (i.e., no adder output destination register is specified this cycle). These seven unique micro-orders can be encoded into the three-bit field, DF, and some descriptive mnemonics are assigned to each, as shown in Fig. 2.11.

2. *Right Input to the Adder, AR Field.* Gates *c*, *f*, and *k*, plus the null option, are encoded into a two-bit-wide field AR as shown in Fig. 2.12.

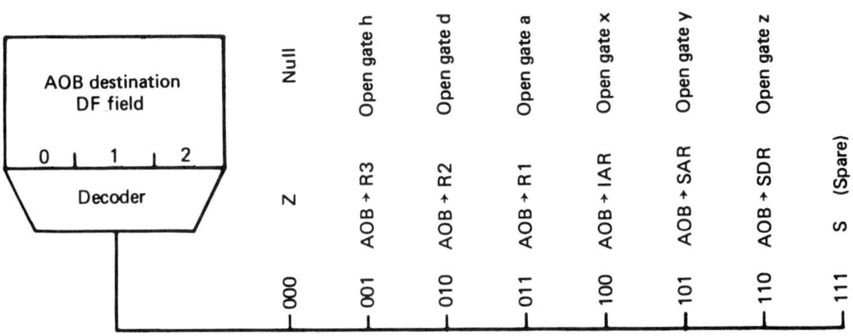

Figure 2.11

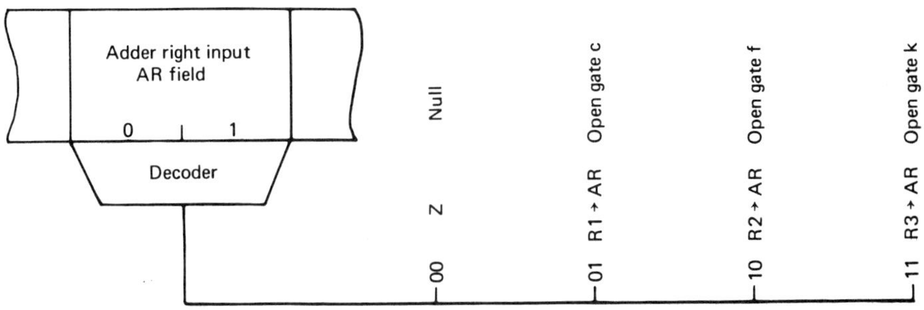

Figure 2.12

3. *Adder Left-Input Designation Field, AL.* This controls gates *b*, *e*, *j*, and *t*. These four, plus the null option, can also be encoded into a three-bit-wide field AL. The micro-orders and their descriptions are shown in Fig. 2.13. Note that

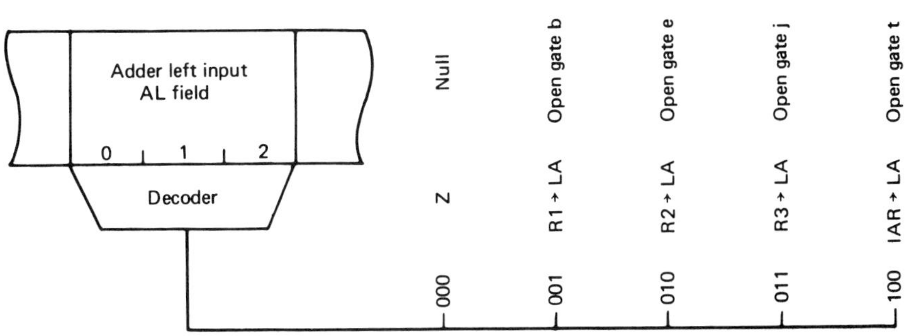

Figure 2.13

the left input to the adder is controlled by this field and by the field which designates the status of the true/complement functions which, when activated, will invert every bit from the designated input register prior to gating it to the left input to the adder.

4. *Carry/No-Carry Field, C.* This field specifies the state of gate *i*. Hence, a one-bit field is sufficient to encode the two possibilities, as shown in the control field decoding chart in Fig. 2.14.

5. *The True or Complement Gate*, T/C. Again, a one-bit field is sufficient to encode these two possibilities. The specific micro-orders are shown in Fig. 2.14.

6. *Field BZ.* This is also a one-bit wide field and is designed to gate the result of the adder zero-detect function out of flip flop FF2 for a possible conditional branch. If gate *p* is activated, the content of FF2 (0 if AOB $\neq$ 0, 1 if AOB = 0) is Or'ed into the low-order bit position of ROAR. (Fig. 2.14.)

Field	*Binary value*	*Micro-order*	*Function*	*Gate*
C	0	NC	Insert no carry	*i* inactive
C	1	C	Insert carry (Hot 1)	*i* active
T/C	0	TA	Transmit true input	*m* inactive
T/C	1	CA	Perform 1's complement on input	*m* active
BZ	0	NBZ	No branch (address from ROSDR)	*p* inactive
BZ	1	BZ	Branch test Or 1 into ROAR; low-order bit if AOB = 0	*p* active
BS	0	PS	No branch (address from ROSDR)	*n* inactive
BS	1	NS	Branch test Or 1 into ROAR; low-order bit if AOB negative	*n* active

Figure 2.14 Partial list of control field decode chart.

7. *BS Field.* This field is another branch test condition that will effect the low-order bit position of the ROAR. It is one bit wide. It tests the sign bit of the data field and is performed automatically every time new data is gated into the AOB. If the sign is positive, the FF1 is set to 1; otherwise, it is set to zero. If gate *n* is activated, its content is Or'ed into the low-order bit position of ROAR.

Note that by assigning both tests to set the same bit position in ROAR, we made them mutually exclusive. One and only one test is performed in any given machine cycle. Therefore, these two fields can be grouped together. In this example, however, no saving is realized by such a grouping. In either case, we have limited the branching capability to a two-way branch. However, if the BS branch test field is made to set ROAR bit ($n - 1$) the second least significant bit in the address field, it will provide two conditional tests which can be performed and made to set two different ROAR bits. This will provide the system with a four-way branch capability.

Figure 2.15 shows the complete ROS word organization and shows the same microprogram that was earlier implemented on the direct control system. The microprogram here is expressed symbolically as well as in the bit representation which would permanently reside in the ROS storage device.

	Microprogram word								
ROS Word address	*AL field*	*AR field*	*DF field*	*C field*	*T/C field*	*BZ field*	*BS field*	*Next ROS Address*	*Resulting micro-orders*
0 0 0 0	0 0 0	0 0	0 0 1	0	0	0	0	1 1 0 0	Z, Z, AOB → R3, NC, TA, NBZ, PS
1 1 0 0	0 0 1	0 0	0 1 1	0	0	1	0	0 0 1 0	R1 → LA, Z, AOB → R1, NC, TA, BZ, PS
0 0 1 0	0 1 0	1 1	0 0 1	0	0	0	0	0 1 0 0	R2 → LA, R3 → AR, AOB → R3, NC, TA, NBZ, PS
0 1 0 0	0 0 0	1 0	0 1 1	0	1	1	0	0 0 1 0	Z, R2 → AR, AOB → R1, NC, CA, BZ, PS
0 0 1 1	X X X	X X	X X X	X	X	X	X	X X X X	Next instruction decode routine

Figure 2.15 Multiply operation microroutine using encoded field control.

The superiority of the encoded field concept in the ROS word organization over the direct control line scheme can be attributed to the following factors:

1. Economics: The resulting ROS word in the encoded control scheme is smaller than the direct control for the same function. This saving in the cost of the ROS memory arrays may be offset by the investment in the cost of the binary decoders needed for each field. Intuitively, the cost advantage is enhanced as the number of control gates in the system is increased. Moreover, this cost advantage begins to improve as the width of the individual field, i.e., the number of control gates that can be grouped under the mutually exclusive concept, increases. This suggests that direct control is better suited for smaller machines, while encoded control is better suited for the more complex general purpose processors. The choice, however, is a complex and a fundamental one. It is dependent on the technology, the internal system data flow structure and the desired speed.

2. The encoded ROS word scheme offers one other advantage from the microprogrammer's point of view. This is the ability to define the actions of many gates or data paths in one single micro-order. Whereas in the direct control scheme, the concept of micro-order may be very rigid and can be related to one single control line, the second ROS word organization scheme permits the microprogrammer to define more powerful micro-orders.

3. Ease of tracing and maintaining: In the latter scheme of ROS word organization, the micro-orders and their symbolic representations are related primarily to the individual data path and then to an obscure control gate that exists somewhere in the system. Hence, a microprogram trace from the symbolic (assembly) language level would be much easier to perform.

To be sure, there are some disadvantages as well. One is the investment that has to be made for an assembler program to translate from the symbolic level to the ROS bit level. Another disadvantage is that, traditionally, the number of micro-orders resulting from the second implementation tends to get much larger than the number of gates in the system; consequently, an added difficulty exists in teaching and microprogramming a data processing system. These advantages and disadvantages are not fixed. They are functions of the system organization and of the algorithms the system is required to perform.

2.5 EXAMPLE 2. MULTIPLICATION MICROROUTINE WITH DIRECT CONTROL IMPLEMENTATION

In the previous example, many assumptions were made which forced the solution in the direction we discussed. These assumptions were made because of the limitation of the hardware functions present, and also to make the comparison between the two types of ROS word organization, i.e., the direct control and the encoded control fields, easy to comprehend. It would be advantageous if some of these constraints were removed and if the arithmetic and logic functions were augmented in such a way that a solution to the more general binary multiplication problem can be proposed.

It was assumed that the multiplier and the multiplicand were positive integers and that the sum of the significant bits in both fields was less than or equal to the width of the R3 register which contains the final product. Since these assumptions were too restrictive, we will replace them with the following:

1. The multiplier and multiplicand may have any algebraic sign.
2. Negative numbers are stored in their 2's complement notation.
3. The number of significant bits in the multiplier or the multiplicand field may be equal to the width of the hardware register.

This last assumption implies that the width of the resulting product field may exceed the width of a single hardware register. Therefore, a fourth hardware register, R4, must be added in the ALU section to store the least significant part of the product.

The first two assumptions suggest that a multiplication scheme be selected which is completely memoryless with regard to the initial sign of either the multiplier or multiplicand. That is, the scheme must dynamically compensate for the product sign. Furthermore, the multiplication scheme must preserve the contents of the multiplier and the multiplicand registers.

A number of binary multiplication algorithms are presented in Richard's book, *Arithmetic Operations in Digital Computers* (Van Nostrand, 1955). Probably the most frequently used scheme is the multiplication by successive shifting and testing of the multiplier bits and accumulation of the product.

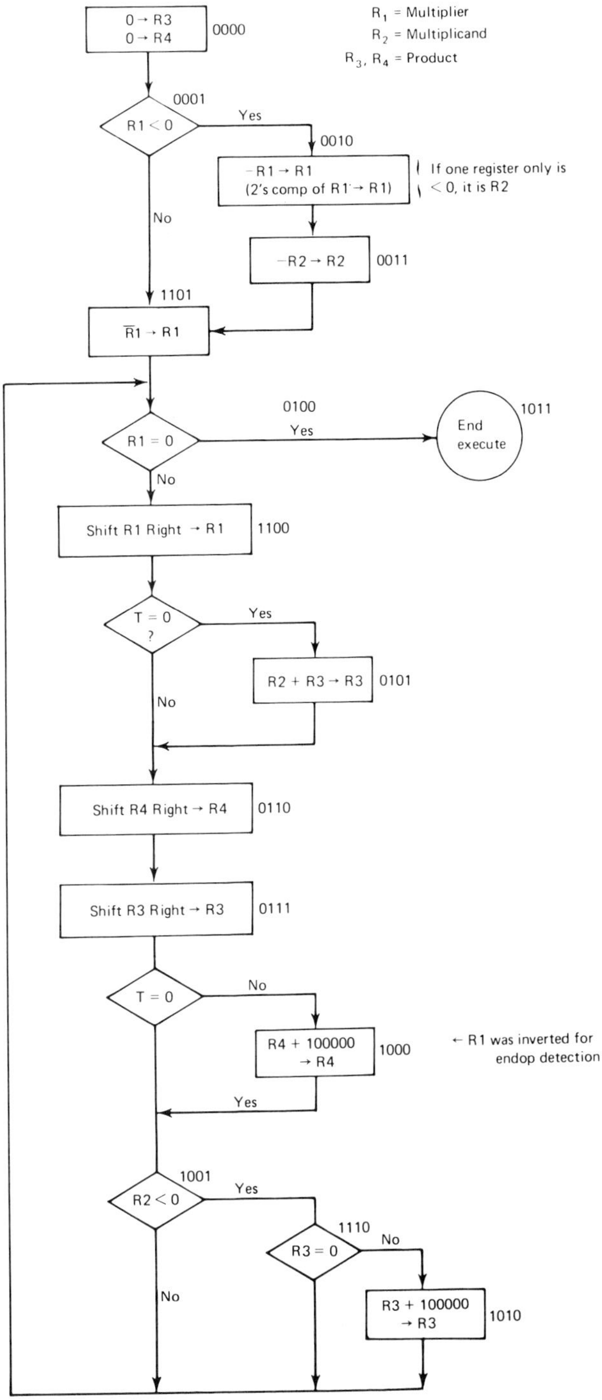
0 → R3
0 → R4
0000
R1 = Multiplier
R2 = Multiplicand
R3, R4 = Product
0001
R1 < 0
Yes
0010
−R1 → R1
(2's comp of R1 → R1)
If one register only is < 0, it is R2
−R2 → R2
0011
No
1101
R̄1 → R1
0100
Yes
R1 = 0
End
execute
1011
No
Shift R1 Right → R1
1100
T = 0
?
Yes
R2 + R3 → R3
0101
No
Shift R4 Right → R4
0110
Shift R3 Right → R3
0111
T = 0
No
R4 + 100000
→ R4
1000
← R1 was inverted for endop detection
Yes
1001
R2 < 0
Yes
1110
R3 = 0
No
R3 + 100000
→ R3
1010
No

Figure 2.16

Assume that the multiplier is stored in register R1 and the multiplicand in R2. Assume that the R3 and R4 registers are used to store the most and least significant bits in the product. This method (Fig. 2.16) begins by initializing the R3 and R4 registers to zero. Next, the multiplier is tested. If the multiplier is negative, its 2's complement is stored back in R1, and the multiplicand's 2's complement is stored in R2. That is, if only one of the two registers, R1 or

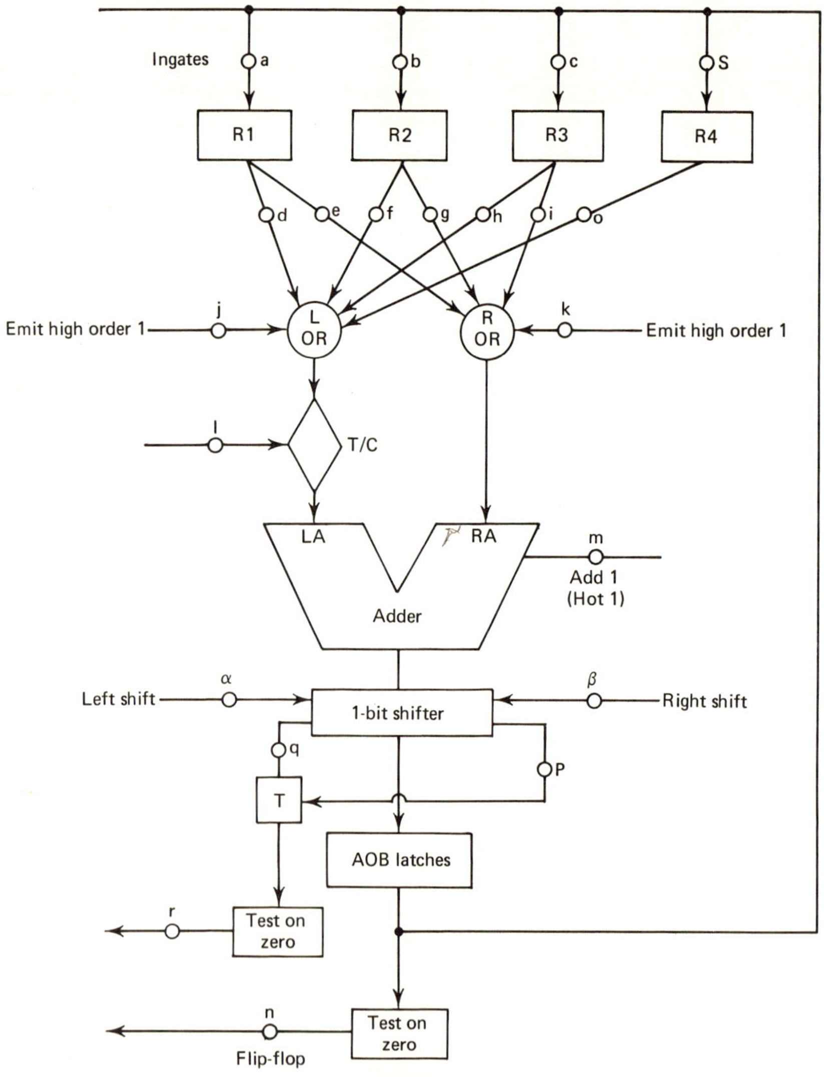

Figure 2.17

R2, is negative, we would do what is necessary to ensure that R1 is the positive one.

Let the $\overline{R}i$ denote the 1's complement of the contents of Ri register. At this point the 1's complement of R1 is stored in R1. It is true that, if either the multiplier or the multiplicand field is zero, then the product in the R3 and R4 registers must be zero. However, it is sufficient to test one or the other for zero. In the flow chart shown in Fig. 2.16, this zero test will be performed on the multiplicand. If the multiplier in R1 is zero, then the contents of R3 and R4, which have been set to zero, reflect the true result of the multiplication and, hence, terminate the execution of the multiplication operation. If the multiplier is not equal to zero, the operation will continue as follows.

Gate the contents of the R1 register to the one-bit shift register. If the overflow bit resulting from shifting the R1 field one bit to the right is zero, then *add* the contents of the multiplicand, which is stored in R2, to R3 (the most significant bits of the product field). The next step is to shift R4 and R3 right one bit each. Again, if the overflow bit resulting from shifting the R3 register is zero, then add one to the most significant bit position in R4, and add zeros to all other bit positions of R4. Similarly, if the multiplicand field is negative, add one to the most significant bit in the R3 register and zeros to every other bit position in R3. R2 is added to R3 only if the bit shifted out of R1 is zero. Also the addition of a one to the most significant bit in R3 takes place only if the multiplicand is negative and R3 already contains this multiplicand. At the

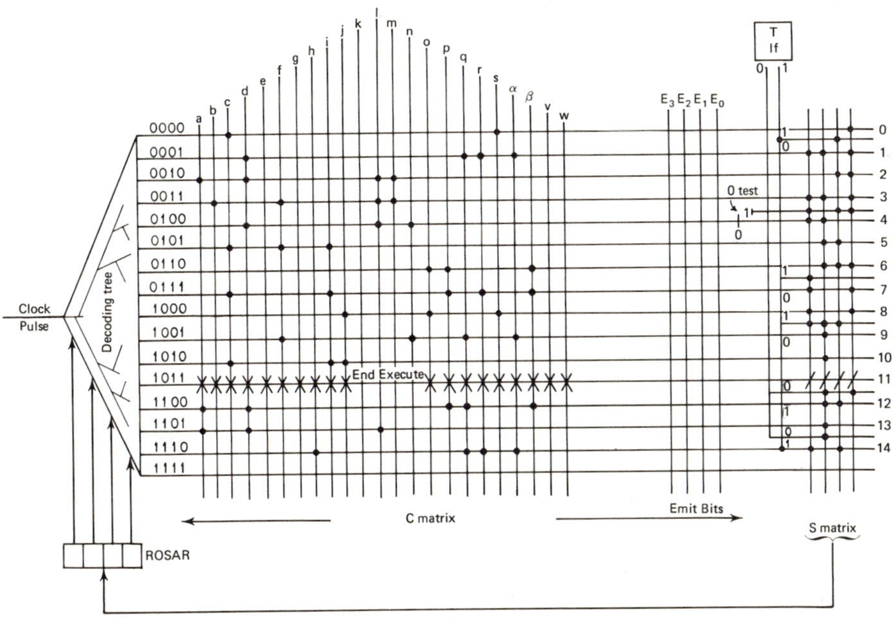

Figure 2.18

Step	Action	*Register Contents* R1	R2	R3	R4
0 0 0 0	0 → R3, R4	1 1 1 0 1 1	0 0 1 1 1 1	0 0 0 0 0 0	0 0 0 0 0 0
0 0 1 0	−R1 → R1	0 0 0 1 0 1			
0 0 1 1	−R2 → R2		1 1 0 0 0 1		
1 1 0 1	$\overline{\text{R1}}$ → R1	1 1 1 0 1 0			
1 1 0 0	Shift R1 Right → R1	0 1 1 1 0 1			
0 1 0 1	R2 + R3 → R3			1 1 0 0 0 1	
0 1 1 0	Shift R4 Right → R4				0 0 0 0 0 0
0 1 1 1	Shift R3 Right → R3			0 1 1 0 0 0	
1 0 0 0	R4 + 100000 → R4				1 0 0 0 0 0
1 0 1 0	R3 + 100000 → R3			1 1 1 0 0 0	
1 1 0 0	Shift R1 Right → R1	0 0 1 1 1 0			
0 1 1 0	Shift R4 Right → R4				0 1 0 0 0 0
0 1 1 1	Shift R3 Right → R3			0 1 1 1 0 0	
1 0 1 0	R3 + 100000 → R3			1 1 1 1 0 0	
1 1 0 0	Shift R1 Right → R1	0 0 0 1 1 1			
0 1 0 1	R2 + R3 → R3			1 0 1 1 0 1	
0 1 1 0	Shift R4 Right → R4				0 0 1 0 0 0
0 1 1 1	Shift R3 Right → R3			0 1 0 1 1 0	
1 0 0 0	R4 + 100000 → R4				1 0 1 0 0 0
1 0 1 0	R3 + 10000 → R3			1 1 0 1 1 0	
1 1 0 0	Shift R1 Right → R1	0 0 0 0 1 1			
0 1 1 0	Shift R4 Right → R4				0 1 0 1 0 0
0 1 1 1	Shift R3 Right → R3			0 1 1 0 1 1	
1 0 1 0	R3 + 100000 → R3			1 1 1 0 1 1	
1 1 0 0	Shift R1 Right → R1	0 0 0 0 0 1			
0 1 1 0	Shift R4 Right → R4				0 0 1 0 1 0
0 1 1 1	Shift R3 Right → R3			0 1 1 1 0 1	
1 0 0 0	R4 + 100000 → R4				1 0 1 0 1 0
1 0 1 0	R3 + 10000 → R3			1 1 1 1 0 1	
1 1 0 0	Shift R1 Right → R1	0 0 0 0 0 0			
0 1 1 0	Shift R4 Right → R4				0 1 0 1 0 1
0 1 1 1	Shift R3 Right → R3			0 1 1 1 1 0	
1 0 0 0	R4 + 100000 → R4				1 1 0 1 0 1
1 0 1 0	R3 + 100000 → R3			1 1 1 1 1 0	
0 1 0 0	Go To End Execute				
Final Register Contents		0 0 0 0 0 0	1 1 0 0 0 1	1 1 1 1 1 0	1 1 0 1 0 1

Result is −75

Figure 2.19 Solution to −5 * 15.

end of this iteration, we test once again to see whether the remaining bits in the R1 field (multiplier) are zero to signal the end of execution. Otherwise, we continue this iteration process.

The 1's complement of R1 gives it dual functions. The first function is associated with its original designation as the multiplier register. Note that, by taking its 1's complement, the decision to add and shift the accumulator register, or simply to shift without adding the contents of the multiplicand, would have to be reversed. The flow chart is altered to add and shift when R1 right shifted-out bit is zero, and to just shift if the same bit is 1. The second

Step	Action	Register Contents R1	R2	R3	R4
0 0 0 0	0 → R3, R4	1 0 0 1 0 1	1 1 0 1 0 0	0 0 0 0 0 0	0 0 0 0 0 0
0 0 1 0	−R1 → R1	0 1 1 0 1 1			
0 0 1 1	−R2 → R2		0 0 1 1 0 0		
1 1 0 1	$\overline{R}$1 → R1	1 0 0 1 0 0			
1 1 0 0	Shift R1 Right → R1	0 1 0 0 1 0			
0 1 0 1	R2 + R3 → R3			0 0 1 1 0 0	
0 1 1 0	Shift R4 Right → R4				0 0 0 0 0 0
0 1 1 1	Shift R3 Right → R3			0 0 0 1 1 0	
1 1 0 0	Shift R1 Right → R1	0 0 1 0 0 1			
0 1 0 1	R2 + R3 → R3			0 1 0 0 1 0	
0 1 1 0	Shift R4 Right → R4				0 0 0 0 0 0
0 1 1 1	Shift R3 Right → R3			0 0 1 0 0 1	
0 1 0 0	Shift R1 Right → R1	0 0 0 1 0 0			
0 1 1 0	Shift R4 Right → R4				0 0 0 0 0 0
0 1 1 1	Shift R3 Right → R3			0 0 0 1 0 0	
1 0 0 0	R4 + 10000 → R4				1 0 0 0 0 0
1 1 0 0	Shift R1 Right → R1	0 0 0 0 1 0			
0 1 0 1	R2 + R3 → R3			0 1 0 0 0 0	
0 1 1 0	Shift R4 Right → R4				0 1 0 0 0 0
0 1 1 1	Shift R3 Right → R3			0 0 1 0 0 0	
1 1 0 0	Shift R1 Right → R1	0 0 0 0 0 1			
0 1 0 1	R2 + R3 → R3			0 1 0 1 0 0	
0 1 1 0	Shift R4 Right → R4				0 0 1 0 0 0
0 1 1 1	Shift R3 Right → R3			0 0 1 0 1 0	
1 1 0 0	Shift R1 Right → R1	0 0 0 0 0 0			
0 1 1 0	Shift R4 Right → R4				0 0 0 1 0 0
0 1 1 1	Shift R3 Right → R3			0 0 0 1 0 1	
0 1 0 0	Go To End Execute				
Final Register Contents		0 0 0 0 0 0	0 0 1 1 0 0	0 0 0 1 0 1	0 0 0 1 0 0

Result is + 324

Figure 2.20 Solution to −27 * −12.

function for $\overline{R}$1 is as a counter to signal the end of the operation. Remember that in the initialization steps, we made sure that R1 was positive; hence, its sign bit is zero. By taking the 1's complement of R1, we assured that at least the leftmost bit in the R1 register is set to zero. As we shift R1 right one bit, the vacated bit is set to zero. Hence, for the whole register (R1) to be zero, we would have to repeat this shifting R1 and testing the overflow bits a number of times equal to the width of the R1 register. This process must be repeated that number of times to properly align and shift the R3 and R4 contents to reflect the correct answer to the given problem.

Figures 2.17 and 2.18 show, respectively, the minimum hardware needed and the direct control implementation of such a multiplication algorithm.

To test this scheme out, let us multiply −5 * 15, −27 * −12, and 4 * −3. The solutions to these problems are shown in Figs. 2.19, 2.20, and 2.21, respectively.

Step	Action	R1	R2	R3	R4
		Register Contents			
0 0 0 0	0 → R3, R4	0 0 0 1 0 0	1 1 1 1 0 1	0 0 0 0 0 0	0 0 0 0 0 0
1 1 0 1	$\overline{R1}$ → R1	1 1 1 0 1 1			
1 1 0 0	Shift R1 Right → R1	0 1 1 1 0 1			
0 1 1 0	Shift R4 Right → R4				0 0 0 0 0 0
0 1 1 1	Shift R3 Right → R3			0 0 0 0 0 0	
1 1 0 0	Shift R1 Right → R1	0 0 1 1 1 0			
0 1 1 0	Shift R4 Right → R4				0 0 0 0 0 0
0 1 1 1	Shift R3 Right → R3			0 0 0 0 0 0	
1 1 0 0	Shift R1 Right → R1	0 0 0 1 1 1			
0 1 0 1	R2 + R3 → R3			1 1 1 1 0 1	
0 1 1 0	Shift R4 Right → R4				0 0 0 0 0 0
0 1 1 1	Shift R3 Right → R3			0 1 1 1 1 0	
1 0 0 0	R4 + 100000 → R4				1 0 0 0 0 0
1 0 1 0	R3 + 100000 → R3			1 1 1 1 1 0	
1 1 0 0	Shift R1 Right → R1	0 0 0 0 1 1			
0 1 1 0	Shift R4 Right → R4				0 1 0 0 0 0
0 1 1 1	Shift R3 Right → R3			0 1 1 1 1 1	
1 0 1 0	R3 + 100000 → R3			1 1 1 1 1 1	
1 1 0 0	Shift R1 Right → R1	0 0 0 0 0 1			
0 1 1 0	Shift R4 Right → R4				0 0 1 0 0 0
0 1 1 1	Shift R3 Right → R3			0 1 1 1 1 1	
1 0 0 0	R4 + 100000 → R4				1 0 1 0 0 0
1 0 1 0	R3 + 100000 → R3			1 1 1 1 1 1	
1 1 0 0	Shift R1 Right → R1	0 0 0 0 0 0			
0 1 1 0	Shift R4 Right → R4				0 1 0 1 0 0
0 1 1 1	Shift R3 Right → R3			0 1 1 1 1 1	
1 0 0 0	R4 + 100000 → R4				1 1 0 1 0 0
1 0 1 0	R3 + 100000 → R3			1 1 1 1 1 1	
1 0 1 1	Go To End Execute				
Final Register Contents		0 0 0 0 0 0	1 1 0 0 0 1	1 1 1 1 1 1	1 1 0 1 0 0

Result is − 12

Figure 2.21 Solution to 4 * −3.

2.6 EXAMPLE 3. A SECOND MULTIPLICATION SCHEME WITH ENCODED CONTROL

Another interesting multiplication scheme which is completely memoryless with respect to the initial sign of either the multiplier or the multiplicand will be presented here. The microprogram will be implemented by the encoded ROS word organization scheme, using the exact hardware and data flow used in Example 2 and shown in Fig. 2.17.

Assume that the multiplier and the multiplicand fields are represented as follows:

$$A_{n-1}2^{n-1} + A_{n-2}2^{n-2} + A_{n-3}2^{n-3} + \cdots + A_0 2^0 \qquad \text{(Multiplicand)}$$

$$b_{n-1}2^{n-1} + b_{n-2}2^{n-2} + b_{n-3}2^{n-3} + \cdots + b_0 2^0 \qquad \text{(Multiplier)}$$

Either or both fields are represented by the 2's complement notation if they are negative. Since the a_i, b_j coefficients are either 0 or 1, the product can be represented by:

$$P = \sum_{j=0}^{n-1} \sum_{k=0}^{n-1} \left[b_{j-1} - b_j \right] A_k \cdot b_j(2^{j+k}) \qquad \text{where } b_{-1} \equiv 0$$

Note that, in terms of the familiar hand multiplication method shifting the multiplicand one position to the left before adding it to the accumulator register, this is indicated here by the suffix (2^{j+k}). The prefix multiplier $(b_{j-1} - b_j)$ will be either $+1$, -1, or 0. This says that the multiplicand may either be added in its true binary representation if it is equal to $+1$, or subtracted by adding its 2's complement if $(b_{j-1} - b_j) = -1$, or ignored completely if the prefix $(b_{j-1} - b_j) = 0$. Since the accumulating sum is equal to the sum of the multiplier plus the multiplicand registers ($2n$ bits), the sum will have to be stored into a double-width register.

Figure 2.22 shows the logical structure of the microroutine. The box numbers appear on the top left-hand corner of each box, while the ROS address assigned to each instruction appears on the top right-hand corner of each box. Single-bit branching is assumed. It is also assumed that either the T test (test overflow bit) or the AOB zero-detect test, but not both, may be used for the branch address modification. The desired test will be selected by energizing the *r* or the *n* gate.

The multiplication routine will test both the multiplier and the multiplicand register for zero. If the result of either test is successful, then the microroutine will branch to the instruction fetch routine by passing the multiplication routine.

Since pairs of bits (the b_{j-1} and b_j bits) are needed to determine the actions taken during each iteration of the routine, a zero test on B register is not sufficient to determine the end of multiplication. To terminate this routine at the proper point, a 1-bit is loaded initially in the high-order bit position of the C register. After n iterations (n represents the width of each register), this 1-bit is detected, thereby terminating the multiply operation. Also each b_{j-1} bit must be preserved until it is compared with its adjacent bit, b_j, and the function is performed.

Example

Let the contents of R1 be +29 and let R2 = −23. Multiply R1 * R2 and store the product in registers R3 and R4. R1 is designated as the *multiplicand*, and R2 as the *multiplier*. Figure 2.23 shows the multiplication procedure applied to this example. Note that this figure does not attempt to show the

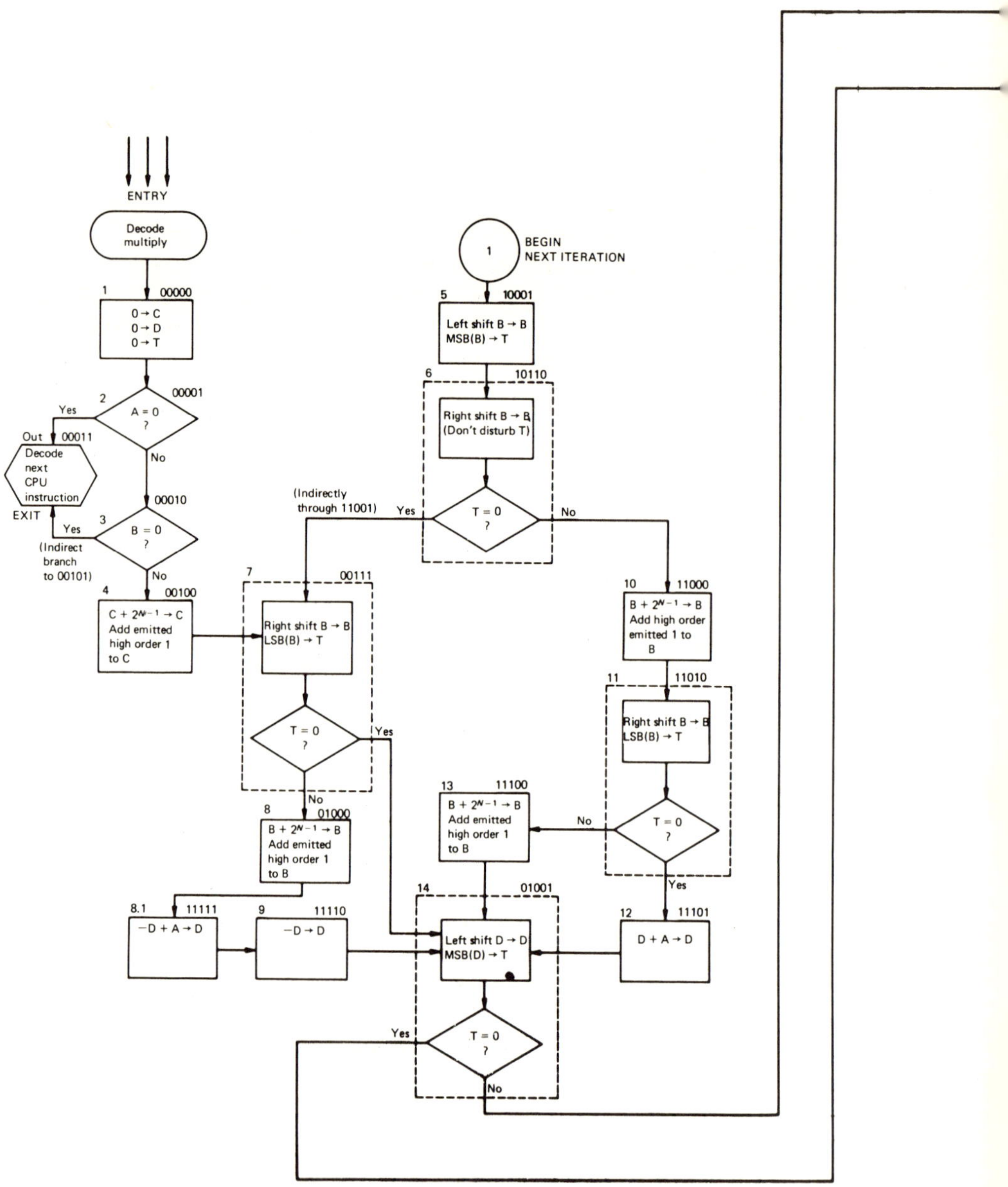

Figure 2.22 Multiply microroutine. *Note:* A = R1, B = R2, C = R3, D = R4.

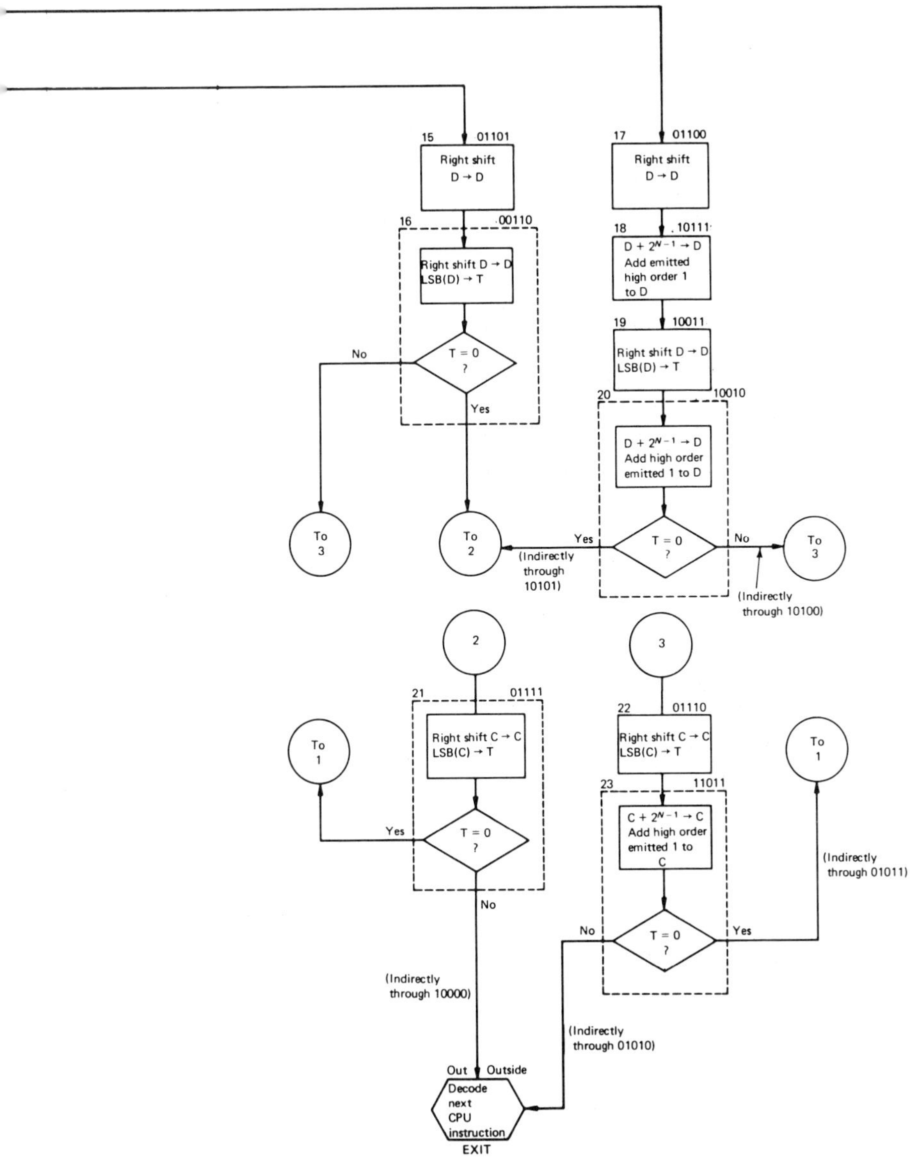

Figure 2.22 (cont.)

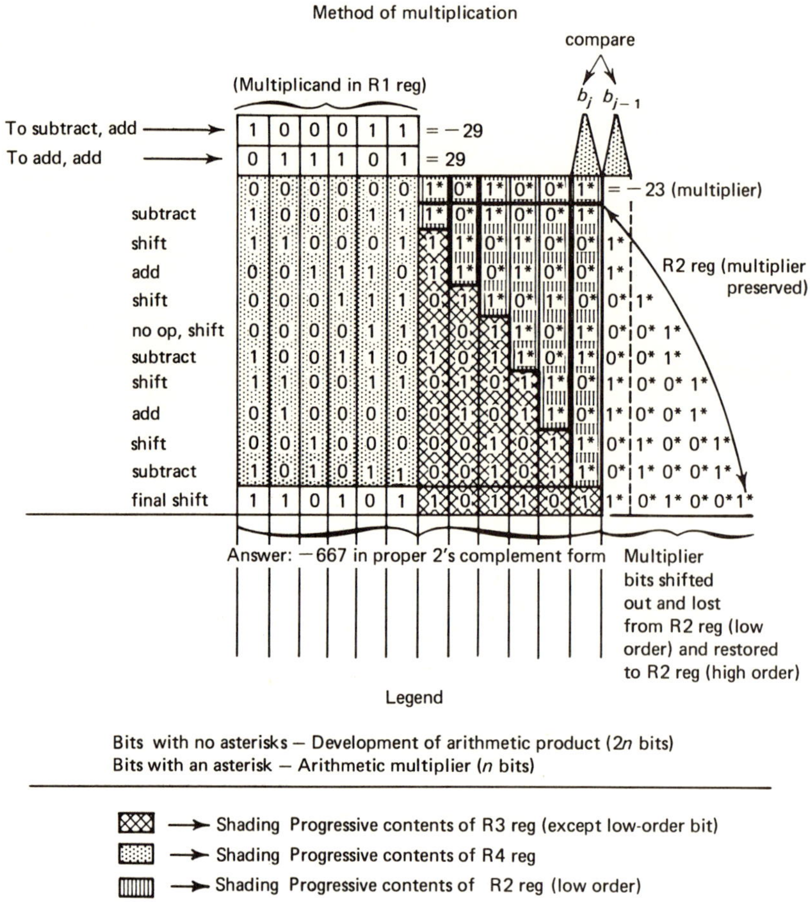

Figure 2.23

intermediate setting of these registers. For a detailed step-by-step application of this scheme to this problem, and for all the intermediate contents of all these registers, another diagram (Fig. 2.24) is included.

Coding the Microroutine

The entire routine (including indirect branches) uses exactly 32 instructions. The routine was hand-"compiled" and optimized so that all intermediate storage locations were utilized. This allowed addressing to be contained between:

$$00000 \rightarrow 11111$$

for the 32 instructions.

The gating operations were organized into the control field decode chart in Fig. 2.25. Therefore, the fields as shown in this chart must be decoded to obtain the actual micro-orders.

Figure 2.17 shows the minimum hardware functions necessary to implement this algorithm. Figure 2.26 shows the actual microroutine words with all fields filled in.

The instruction length using this scheme is 22 bits, including branch address, and a 1-bit emit field which is *decoded* to provide a high-order emitted 1, when needed.

2.7 ROS TIMING

We postulated earlier that a computer system consists of a read/write main storage unit, an arithmetic and logic unit, input and output units, and a control unit in the form of a high-speed nondestructive read-only storage for the control program. The speed of the arithmetic and logical unit (sometimes loosely referred to as the central processing unit or CPU cycle time) is a function of the number of levels of logic decision in the most critical data path in the CPU, times the average delay per logic decision. The average time delay per logic decision is dependent on the circuit technology from which the system is implemented.

Similarly, the main storage cycle time is dependent on the main storage speed plus the memory access delay. Both the CPU cycle time and the main storage cycle time are independent of each other. It is desirable to have the main storage read/write cycle time be equivalent to the CPU cycle time. This would result in an optimum system design where the central processing unit is always operating at its maximum potential without having to wait on the data requested from main storage. Most computer systems cannot afford such an expensive main storage device and usually compromise by using a main storage technology with a cycle time two, four, or more times that of the CPU. In the modern, high-speed, large data processing systems, the CPU main storage cycle gap was made transparent to the total system by including a bank of high-speed buffer memory with a cycle time equal to that of the CPU and with an appropriate paging algorithm. An example would be the 16k-byte minimum size buffer of high-speed memory in the IBM System/360, Model 85.

The speed of the ROS memory which contains the control program has serious implications on the performance of the total system. It must be capable of activating the set of control lines to open both the selected gates for sequencing control and the gates for static control at specific cycle points of every machine cycle. A cycle point is defined as the elemental time interval in the CPU cycle designed for the transfer of information between two specific registers or latches. Examples of this are the register out-gate timing interval,

		STATUS OF REGISTERS DURING MICROROUTINE				
	Microroutine Block No.	(29) R1	(−23) R2	R3	R4	T
	(not address)	0 1 1 1 0 1	1 0 1 0 0 1	0 0 0 0 0 0	0 0 0 0 0 0	0
Initial iteration	4	Not altered		1 0 0 0 0 0		
	7		0 1 0 1 0 0			1
	8		1 1 0 1 0 0			
	8.1, 9				1 0 0 0 1 1	
	14				0 0 0 1 1 0	1
	17				0 0 0 0 1 1	
	18				1 0 0 0 1 1	
	19				0 1 0 0 0 1	1
	20				1 1 0 0 0 1	
	22			0 1 0 0 0 0		0
	23			1 1 0 0 0 0		
2nd iteration	5		1 0 1 0 0 0			1
	6		0 1 0 1 0 0			
	10		1 1 0 1 0 0			
	11		0 1 1 0 1 0			0
	12				0 0 1 1 1 0	
	14				0 1 1 1 0 0	0
	15				0 0 1 1 1 0	
	16				0 0 0 1 1 1	0
	21			0 1 1 0 0 0		0
3rd iteration	5		1 1 0 1 0 0			0
	6		0 1 1 0 1 0			
	7		0 0 1 1 0 1			0
	14				0 0 1 1 1 0	0
	15				0 0 0 1 1 1	
	16				0 0 0 0 1 1	1
	22			0 0 1 1 0 0		0
	23			1 0 1 1 0 0		
4th iteration	5		0 1 1 0 1 0			0
	6		0 0 1 1 0 1			
	7		0 0 0 1 1 0			1
	8		1 0 0 1 1 0			
	8.1, 9				1 0 0 1 1 0	
	14				0 0 1 1 0 0	1
	17				0 0 0 1 1 0	
	18				1 0 0 1 1 0	
	19				0 1 0 0 1 1	0
	20				1 1 0 0 1 1	
	21			0 1 0 1 1 0		0

Figure 2.24

STATUS OF REGISTERS DURING MICROROUTINE

	Microroutine Block No.	(29) R1	(−23) R2	R3	R4	T
	(not address)	0 1 1 1 0 1	1 0 1 0 0 1	0 0 0 0 0 0	0 0 0 0 0 0	0
5th iteration	5		0 0 1 1 0 0			1
	6		0 0 0 1 1 0			
	10		1 0 0 1 1 0			
	11		0 1 0 0 1 1			0
	12				0 1 0 0 0 0	
	14				1 0 0 0 0 0	0
	15				0 1 0 0 0 0	
	16				0 0 1 0 0 0	0
	21			0 0 1 0 1 1		0
6th iteration	5		1 0 0 1 1 0			0
	6		0 1 0 0 1 1			
	7		0 0 1 0 0 1			1
	8		1 0 1 0 0 1			
	8.1, 9				1 0 1 0 1 1	
	14				0 1 0 1 1 0	1
	17				0 0 1 0 1 1	
	18				1 0 1 0 1 1	
	19				0 1 0 1 0 1	1
	20				1 1 0 1 0 1	
	22			0 0 0 1 0 1		1
	23			1 0 0 1 0 1		
		R1	R2	R3	R4	
	Final	0 1 1 1 0 1	1 0 1 0 0 1	1 0 0 1 0 1	1 1 0 1 0 1	

R2 Register Preserved

Notes: Example is given for 6-bit register lengths; however, the microroutine is *independent* of register lengths.

$$\text{Product} = \langle R4 \rangle \langle R3 \rangle = \underbrace{1\,1\,0\,1\,0\,1}_{\substack{\text{high order}\\ (R4)}}\ \underbrace{1\,0\,0\,1\,0\,1}_{\substack{\text{low order}\\ (R3)}}$$

$= -667_{10}$ expressed in proper 2's complement form

Figure 2.24 (cont.)

the adder output bus to register in-gate timing point, etc. This specification would make the timing and access of the ROS device dependent on the CPU cycle time of the system. The best choice would be to have a ROS with a cycle time equal to that of the CPU cycle time, with the proper overlap between the two clocks. The ROS cycle time is dependent on the access time of the ROS memory itself, together with the cycle rate at which it may operate.

CONTROL FIELD DECODE CHART

Field	*Value*	*Description*	*Gates On*
AL	0 0 0	NOP	None
AL	0 0 1	High-order 1 → L (Emit)	*j*
AL	0 1 0	R4 → L	*o*
AL	0 1 1	R1 → L	*d*
AL	1 0 0	R2 → L	*f*
AL	1 0 1	R3 → L	*h*
AR	0 0 0	NOP	None
AR	0 0 1	R1 → R	*e*
AR	0 1 0	R2 → R	*g*
AR	0 1 1	R3 → R	*i*
AR	1 0 0	High-order 1 → R (Emit)	*k*
TR	0 0 0	NOP	None
TR	0 0 1	Gate AOB → R4	*s*
TR	0 1 0	Gate AOB → R3, R4	*c, s*
TR	0 1 1	Gate AOB → R2	*b*
TR	1 0 0	Gate AOB → R3	*c*
T/C	0	NOP	None
T/C	1	1's complement L → LA	*l*
C	0	NOP	None
C	1	Hot 1 → RA	*m*
SH	0 0	NOP	None
SH	0 1	1-bit right shift	β
SH	1 0	Right shift, LSB → T	β, *p*
SH	1 1	Left shift, MSB → T	α, *q*
TS	0 0	NOP	None
TS	0 1	Test T = 0	*r*
TS	1 0	Test AOB = 0	*n*
BZ	0	No branch—use ROSDR bit	N/A
	1	1 → LSB (ROAR), if T = 0	

Figure 2.25

If this ROS cycle time is greater than the CPU cycle time, then the effective machine cycle is the slower one of the two; here, it would be the ROS memory. Hence, we have compromised the speed of the total system for that of the ROS device. On the other hand, if the ROS cycle time is much less than the CPU cycle time, this ROS speed is wasted. As a rule of thumb, the faster the ROS or any storage media is, the higher the cost of storage per bit. Thus, the answer to this problem is to select an ROS technology, with the desired word size and the desired number of words, that can operate at a cycle rate equivalent to that of the CPU.

MICROROUTINE WORDS

Flow Chart Block No.	CROS Word Adder	AL	AR	TR	T/C	C	SH	TS	B Z	EMIT	Next Adder	
1	0 0 0 0 0	0 0 0	0 0 0	0 1 0	0	0	1 0	0 0	0	0	0 0 0 0 1	● = modifiable low-order bit for branching on test
2	0 0 0 0 1	0 1 1	0 0 0	0 0 0	0	0	0 0	1 0	1	0	0 0 0 1 ●	
3	0 0 0 1 0	1 0 0	0 0 0	0 0 0	0	0	0 0	1 0	1	0	0 0 1 0 ●	
OUT	0 0 0 1 1	0 0 0	0 0 0	0 0 0	0	0	0 0	0 0	x	0	x x x x x	Branch outside
4	0 0 1 0 0	0 0 1	0 0 0	1 0 0	0	0	0 0	0 0	0	1	0 0 1 1 1	
INDIRECT BR.	0 0 1 0 1	0 0 0	0 0 0	0 0 0	0	0	0 0	0 0	x	0	x x x x x	Branch outside
16	0 0 1 1 0	0 1 0	0 0 0	0 0 1	0	0	1 0	0 1	1	0	0 1 1 1 ●	
7	0 0 1 1 1	1 0 0	0 0 0	0 1 1	0	0	1 0	0 1	1	0	0 1 0 0 ●	
8	0 1 0 0 0	0 0 1	0 1 0	0 1 1	0	0	0 0	0 0	0	1	1 1 1 1 1	
14	0 1 0 0 1	0 1 0	0 0 0	0 0 1	0	0	1 1	0 1	1	0	0 1 1 0 ●	
INDIRECT BR.	0 1 0 1 0	0 0 0	0 0 0	0 0 0	0	0	0 0	0 0	x	0	x x x x x	Branch outside
INDIRECT BR.	0 1 0 1 1	0 0 0	0 0 0	0 0 0	0	0	0 0	0 0	0	0	1 0 0 0 1	Intraroutine branch
17	0 1 1 0 0	0 1 0	0 0 0	0 0 1	0	0	0 1	0 0	0	0	1 0 1 1 1	
15	0 1 1 0 1	0 1 0	0 0 0	0 0 1	0	0	0 1	0 0	0	0	0 0 1 1 0	
22	0 1 1 1 0	1 0 1	0 0 0	1 0 0	0	0	1 0	0 0	0	0	1 1 0 1 1	
21	0 1 1 1 1	1 0 1	0 0 0	1 0 0	0	0	1 0	0 1	1	0	1 0 0 0 ●	
INDIRECT BR.	1 0 0 0 0	0 0 0	0 0 0	0 0 0	0	0	0 0	0 0	x	0	x x x x x	Branch outside
5	1 0 0 0 1	1 0 0	0 0 0	0 1 1	0	0	1 1	0 0	0	0	1 0 1 1 0	
20	1 0 0 1 0	0 1 0	1 0 0	0 0 1	0	0	0 0	0 1	1	1	1 0 1 0 ●	
19	1 0 0 1 1	0 1 0	0 0 0	0 0 1	0	0	1 0	0 0	0	0	1 0 0 1 0	
INDIRECT BR.	1 0 1 0 0	0 0 0	0 0 0	0 0 0	0	0	0 0	0 0	0	0	0 1 1 1 0	Intraroutine branch
INDIRECT BR.	1 0 1 0 1	0 0 0	0 0 0	0 0 0	0	0	0 0	0 0	0	0	0 1 1 1 1	Intraroutine branch
6	1 0 1 1 0	1 0 0	0 0 0	0 1 1	0	0	0 1	0 1	1	0	1 1 0 0 ●	
18	1 0 1 1 1	0 1 0	1 0 0	0 0 1	0	0	0 0	0 0	0	1	1 0 0 1 1	
10	1 1 0 0 0	0 0 1	0 1 0	0 1 1	0	0	0 0	0 0	0	1	1 1 0 1 0	
INDIRECT BR.	1 1 0 0 1	0 0 0	0 0 0	0 0 0	0	0	0 0	0 0	0	0	0 0 1 1 1	Intraroutine branch
11	1 1 0 1 0	1 0 0	0 0 0	0 1 1	0	0	1 0	0 1	1	0	1 1 1 0 ●	
23	1 1 0 1 1	0 0 1	0 1 1	1 0 0	0	0	0 0	0 1	1	1	0 1 0 1 ●	
13	1 1 1 0 0	0 0 1	0 1 0	0 1 1	0	0	0 0	0 0	0	1	0 1 0 0 1	
12	1 1 1 0 1	0 1 0	0 0 1	0 0 1	0	0	0 0	0 0	0	0	0 1 0 0 1	
9	1 1 1 1 0	0 1 0	0 0 0	0 0 1	1	1	0 0	0 0	0	0	0 1 0 0 1	
8.1	1 1 1 1 1	0 1 0	0 0 1	0 0 1	1	1	0 0	0 0	0	0	1 1 1 1 0	

Figure 2.26

The ROS sequencing and access time are also dependent on the branch logic if the result of a branch test is to immediately determine the address of the next ROS word. In this case, the ROS effective cycle time must also include the data path delay plus the branch logic delay. This means that the CPU cycle time and the ROS cycle time are not overlapped in any way. We can show that the effective cycle time of the CPU is equal to the sum of the CPU cycle time plus the ROS cycle time and can be expressed by the following equation:

Effective CPU cycle = Time to decode the contents of the ROS data registers to enable the control lines
+
The ALU data path delays
+
The branch decision logic delay
+
The ROS access, to gate the next ROS word in the ROS data register.

This total delay and the lack of any overlap between the CPU and the ROS cycle times is the result of an assumption explicitly stated earlier: the result of the branch test is to immediately determine the address of the next ROS word. If an intervening cycle is specified between the branch test and the ROS cycle in which the result of the test is used in defining the address of the accessed ROS word, then the CPU cycle time can be overlapped with the ROS word access time. The address of the ROS word being accessed is independent of any data manipulation going on in the ALU. This overlapped access is shown in Fig. 2.27.

This solution of overlapping of CPU and ROS cycle times is a logical and reasonable one, since most machine cycles executed are sequential and not branch-dependent. Even if there are a number of branch (data-dependent branches) cycles, a large percentage of them would allow for an intervening cycle between executing the data-dependent test and executing the branch.

If the total ROS cycle time is greater than the CPU cycle time, it would be reasonable to design an ROS word which would provide the control for more than one consecutive machine cycle, thus sharing the ROS overhead time among a number of consecutive CPU cycles.

Another method for reducing the effective ROS memory cycle time is to organize the ROS memory bank into two separate stacks (the odd and the even stack) as shown in Fig. 2.3. The two stacks operate on an alternating basis. This organization effectively cuts in half the access time by overlapping the access time of one stack with the recovery time of the other. This ROS organi-

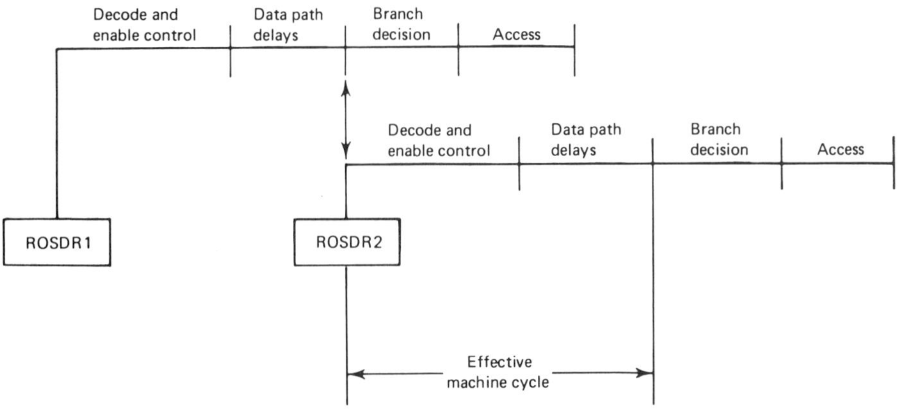

Figure 2.27

zation was successfully used in the RCA Spectra 70/Model 45, as we shall see in Chapter 9.

2.8 ROS STORAGE CAPACITY AND COST

A read-only storage matrix has two basic dimensions (Fig. 2.28)—the number of words, represented by the vertical dimension (h), and the width of the ROS word, represented by the horizontal dimension (w). The number of words is dependent on the functional requirements of the ROS control. The minimum

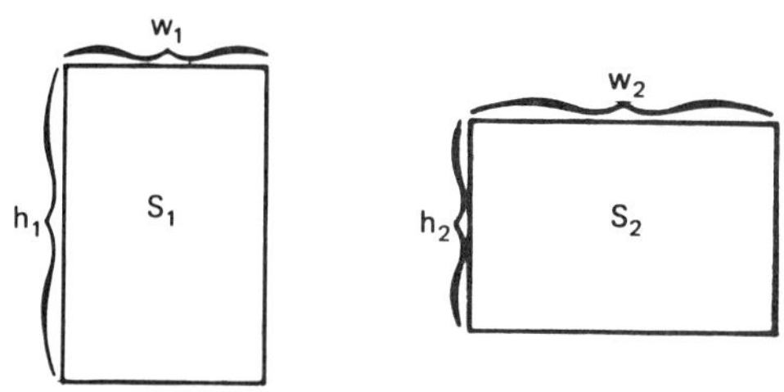

Figure 2.28 ROS storage characteristics.

ROS capacity needed to perform only the static control functions is equal to the number of each possible static control states described by the control routines. This type of ROS control is the easiest to implement, since the microinstruction contains information only to open or close a specific set of gates. However, the sequencing function imposes other control requirements:

1. To link the machine instruction operation code to a specific execution routine.

2. To include the ability to test system indicators and modify ROS sequencing accordingly.

3. Some designs specify that the ROS word contains a field to store the address of the successor ROS word.

In other words, the number of ROS words needed is basically a function of the system architecture, i.e., the number and the complexity of each instruction in the instruction set, the number of different types of instruction formats, the extent of indirect addressing available to the system, the loading procedure, as well as the extent of interrupt handling and priority within the different classes of interruptions.

The largest segment of the ROS storage must be that which contains all the individual instruction execution routines. There is a trend in the small to intermediate systems to share a given hardware unit between two distinct functions or modes of operations. This design imposes new requirements on the ROS control function to handle the bookkeeping and the initialization, as well as the restoration of the system state when it is made to switch between the allowable modes of operations. Among these modes are the CPU mode, the I/O mode, and the emulation mode. These modes of operation will be discussed in more detail in later chapters.

The second dimension, namely, the word size (w), depends on the actual system design implementation. It is dependent on the amount of parallelism which exists in the system. The larger the number of parallel data paths, the more distinct control fields are required to control these data paths. Also, the more complex each data path, the larger the number of control lines, and hence, the wider the corresponding control field assigned to control this mutually exclusive set of control lines. Note that the wider the control word resulting from greater parallelism in the system implies that more functions can be performed in one machine cycle; hence, the number of ROS words is reduced. There is a definite cost/performance relationship that results from this trade-off.

The speed of the ROS is adversely affected by an appreciable increase in either dimension. As the number of ROS words increases, the time required for decoding the ROS address also increases. This increase becomes effective only when the number of additional ROS words necessitates the addition of an extra bit to the ROS address fields. As the size of the ROS word is increased, the ROS access time is increased due to the longer path the word drive pulse is required to propagate and also due to the greater drive signal amplitude level. This increase in the width of the ROS word also has an adverse affect on the signal-to-noise ratio, and thus on the design constraints of the sense amplifier.

The biggest consideration, however, is the cost of the ROS storage. The cost of the control system must include the cost of the ROS storage array plus the costs of the access mechanism, the ROS data register, and ROS field decoders. Each of these is affected as we vary the word size and the number of words per array. Assume for now that we increase the ROS word size. Then the ROS field decoding matrices, as well as the number of sense amplifiers required to activate the control lines will also have to be increased. The cost of such an increase is high. Add to this the increase in the basic cost of the sense amplifier, due to the more stringent design requirements imposed by the aggravated signal-to-noise ratio problem.

On the other hand, an increase in the number of ROS words without an increase in the size of each word would require only a nominal additional cost. To double the number of the ROS words in the array, all that is required is to add a single bit to the ROS addressing scheme, accompanied by modifications to the ROS address decoding hardware.

It can be concluded, then, that the most economic ROS array design would dictate the selection of a simpler, more compact ROS word, with the proportionate increase in the number of distinct words in the ROS array. This criterion must be balanced with the system performance requirement. It would not be difficult to conclude that, the greater the parallelism in the system's data flow, the better the system's performance.

2.9 ECONOMIC MODEL FOR THE MICROPROGRAMMED LOGIC

It is desirable to establish an analytical model to compare the cost of *conventional hardware* control circuits to the cost of the *microprogramming* and the ROS storage required to perform the same data processing function. The purpose of this section is to identify the parameters which must be considered and the trend that will be expected to emerge. For this type of analysis to be meaningful, one has to define system requirements as well as the circuit and ROS technology from which the control system can be implemented.

In conventional control, the cost of the total system depends on the cost of the basic logic blocks, on the complexity of the cycle, and on the number of cycles. For example, doubling the number of machine cycles by increasing the complexity of system architecture would result in almost doubling the cost of the hardware control system. In other words, the cost of increasing the number of machine cycles necessary to implement a given system architecture is directly proportional to this increase.

The cost of the read-only storage includes the cost of the ROS array, the accessing mechanism (which includes the address register, address decoders, and branch logic), and also the cost of the hardware to receive and decode the different fields in the ROS word (including the ROS data register, sense amplifiers, field decoders, etc.). For a given system implementation, this logic which receives the output bits from the ROS array is almost insensitive to an increase in the number of machine cycles. The ROS word organization, ROS field definitions, and the static control mechanism will remain fixed. The only possible change would be in the sequencing field or fields design to specify the address of the next ROS word to be fetched.

The cost of the ROS memory access scheme is not appreciably affected by an increase in the number of machine cycles through which the system will have to sequence. It was pointed out earlier that, if the number of ROS words is doubled, the extent of modification would be the cost of additional bits in the ROS address register and the associated ROS address decoding scheme. Hence, the cost of an ROS control implementation starts out with a fixed value, and, as the number of sequences is increased, the cost of this ROS control implementation begins to increase slowly. This gradual increase tends to be somewhat less than that computed for the hardware logic control implementation. The two lines, representing the ROS control and the hardware control, intersect each other at a point beyond which the ROS implementation begins to be more attractive than a hardware control implementation for the same function. The point of intersection is technology-dependent. This cost analysis is illustrated in Fig. 2.29[23]

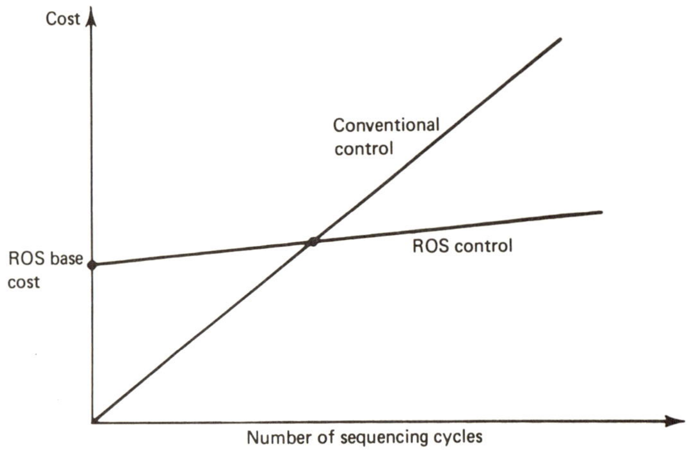

Figure 2.29

The preceding discussion was based mostly on intuition. It is desirable to develop a more accurate, analytical economic model that could be used to compare the cost of conventional hardware control with the cost of the microprogrammed control to perform the same data processing function. It is very difficult to develop a general economic model that is accurate for every data process, every data processing system, and every technology. This task is further complicated by the introduction of the writable control storage technology and the many different options for modifying the contents of the control storage. Another major consideration is the storage technology and the type of accessing mechanism implemented in the control storage. A coincident current selection mechanism, for example, is different from a linear full word drive selection scheme. For these and other major considerations, the only feasible approach is to sketch a simple model for a specific memory technology, assuming a coincident current drive scheme, and to reach some simple preliminary conclusions from this model. This simple model can also help to establish a general direction for the development of a more detailed economic model to be developed for a specific process and a specific circuit and memory technology. The following model was proposed by Beisner and Mills[22] of IBM.

Let us consider a conventional logical device, consisting of circuitry only, which is capable of carrying out a set of ordinary computer instructions. Let the number of logical components be n. Assume that the cost per unit can be expressed as follows

$$C_n = C_C + \frac{C'_C}{N} \tag{1}$$

where C_C is the manufacturing cost per unit and C'_C is the total design and tooling cost, which is spread over the total number of units N. It seems reasonable that the manufacturing and design costs should be proportional to the number of components n:

$$C_C = cn \tag{2}$$

$$C'_C = dn \tag{3}$$

where c is the manufacturing cost per component and d is the design and tooling cost per component.

Now, consider a microprogrammed unit which is capable of performing the same data processing functions as the conventional logical device. The microprogrammed unit consists of two parts: the microprogram memory (read-only storage) and the associated logical circuitry. Let the cost of microprogramming C_p be proportional to the number of microinstructions m:

$$C_p = pm \tag{4}$$

Let the number of components in the logical circuitry be n_L. The logical unit manufacturing cost can then be expressed [by]

$$C_L = cn_L \tag{5}$$

The total design and tooling cost of the logical circuitry is

$$C'_L = dn_L \tag{6}$$

Now, the total cost per microprogrammed unit can be written

$$C_m = \frac{C_p}{N} + C_L + \frac{C'_L}{P} + C_M + \frac{C'_M}{Q} \tag{7}$$

where P is the number of logical units manufactured, which will be considerably greater than N, the number of applications of this particular microprogram; C_M is the manufacturing cost per unit of memory; C'_M is the total design and tooling cost of the memory; and Q is the total number of memory units manufactured, which may be greater than P, the number of logical units.

The memory cost C_M will now be considered as a function of m, the number of microinstructions. It is reasonable to assume that for a coincident current memory, with a given speed and word size, the cost will be roughly proportional to the number of drivers D

$$C_M = \frac{K}{2} D \tag{8}$$

where K is a constant of proportionality.

The number of required address bits s is given by

$$2^s = m \tag{9}$$

For simplicity, we ignore the fact that s must be a whole number. The address bits are halved, and $s/2$ bits are used to select the X-drivers while the other $s/2$ bits are used to select the Y-drivers. This gives a total of

$$D = 2 \cdot 2^{s/2} = 2\sqrt{m} \tag{10}$$

drivers.

The number of drivers for a read-only store of the gate selection type is also given by Equation 10. In this case, one half the address bits select the word drivers and one half the address bits select the gate drivers.

The memory cost can now be approximated

$$C_M = K\sqrt{m} \tag{11}$$

from Equations 8 and 10. Experience tells us that the cost per bit goes down as the size of the memory is increased. Equation 11 gives this general behavior. It, however, is probably not valid for small memory sizes because a certain minimum cost is required independent of the size.

It has been the experience of microprogrammers that, for a given set of computer instructions, there is a proportionality between the numbers of components required and the number of microinstructions required to do the same job. Therefore, let

$$n = km \tag{12}$$

where k is the "power" of a microinstruction. The computing power k of a microinstruction can be increased by increasing n_L, the number of components in the logical portion of the microprogrammed unit. Let these two numbers be proportional with a proportionality constant h:

$$n_L = hk \tag{13}$$

Now assume that the numbers of the three types of units produced N, P, and Q are large in comparison to their respective programming and design costs. Equations 1 and 7 become

$$C_n = cn \tag{14}$$

$$C_m = chk + K\frac{\sqrt{n}}{\sqrt{k}} \tag{15}$$

by virtue of Equations 2, 5, 13, 11, and 12.

In Equation 15, C_m can be minimized by the proper choice of k. Setting the derivative with respect to k equal to zero, one gets

$$\frac{\partial C_m}{\partial k} = ch - \frac{1}{2} K \sqrt{n}\, k^{-3/2} = 0 \tag{16}$$

Solving Equation 16 for k, one obtains

$$k = \left(\frac{K\sqrt{n}}{2\,ch}\right)^{2/3}, \tag{17}$$

which, when substituted into Equation 15, gives

$$C_m = \frac{3}{2}(K\sqrt{2\,ch})^{2/3}\sqrt[3]{n} \tag{18}$$

As can be seen from the form of Equations 14 and 15 or 18 there is always a point, as n increases, where the cost of the microprogrammed unit C_m becomes less than the cost of the conventional logical unit C_n and remains so for greater values of n. Let us determine the cross-over point by setting C_n equal to C_m:

$$cn = \frac{3}{2}(K\sqrt{2\,ch})^{2/3}\sqrt[3]{n} \tag{19}$$

Equation 19 gives

$$cn = \frac{\sqrt{27h}}{2} K = C_n \tag{20}$$

From Equation 12, the power of a single microinstruction in terms of an equivalent number of components may be grossly estimated by setting $m = 1$. Comparing this result with Equation 13 one gets an estimate

$$h = 1 \tag{21}$$

Conservatively, setting

$$h = 10 \tag{22}$$

one obtains from Equation 20

$$C_n = \frac{\sqrt{270}}{2} K = \sqrt{m_1}\, K \tag{23}$$

where m_1 is the size of a memory unit of equivalent cost.

From Equation 23, the number of equivalent memory locations is

$$m_1 \doteq 70 \tag{24}$$

This is a surprisingly small cross-over point. It is so small that the range of validity of the memory equation (Equation 11) has probably been exceeded. Still, the result indicates that small microprogrammed units might be economically applied to data processing functions in competition with logical circuitry only. This would also imply that it might pay to attempt to reduce the cost of small read-only storage elements to conform more closely with Equation 11.

For the case of N not large but P and Q large, consider the microprogramming cost C_p versus the logic design and tooling cost C'_C. Microprogrammers estimate that programming costs and logic design costs are approximately equal for equivalent data processing functions. But tooling costs such as setting up production and testing facilities should make C'_C somewhat greater than C_p. In this case, the above findings are unaffected or even strengthened. This can be seen from examination of Equations 1 and 7.

In summary, a crude economic model has been used to compare the cost of conventional circuits with the cost of stored logic to perform the same data processing functions. The results indicate that there is a point at which, with increasing complexity, the cost of stored logic becomes less than the cost of conventional circuits. This cross-over point may occur at a relatively low degree of complexity. This model can be explored further but the validity of many of the assumptions and approximations does not warrant any further analysis of it.

2.10 MICROPROGRAMMING ADVANTAGES AND DISADVANTAGES

This section considers the advantages and disadvantages of microprogramming in a systematic way. It is difficult to weigh the advantages and disadvantages of this microprogramming concept because what is considered an advantage in one computer system may very well be a disadvantage in another. However, there are a few clear and uncontested advantages which can be stated here.

1. *Flexibility and Tailorability.* One important advantage is that the order code need not be finalized until late in the design cycle. This speeds the machine design cycle, since the microprogramming and system design implementation can proceed in parallel. This also leaves the system architects sufficient room for any architectural modifications deemed necessary at a later time. By changing the microprogramming in the ROS storage, one can make the processor recognize a new and/or richer instruction set or make the processor a slave machine to perform a given task in the most efficient way. Another area of application that is greatly dependent on this capability is emulation which is, in

a way, an extension of one system's architecture to cover another system. Thus, a program written for the emulated system can be successfully recognized and executed by the emulating system.

2. *Changeability.* Related to the first advantage in this list, yet with slightly different implications and emphasis, is the advantage of changeability. Here the ROS matrix may be replaced by a changeable or semichangeable control storage to which information can be easily transferred from other sources when required. This is desirable because the system can be made to switch between a number of different system architectural characteristics, and the programmer would, in fact, be able to select an optimum set of machine instructions for his immediate environment and problems.

3. *Ease of Designing, Maintaining, and Checking.* Microprogramming has been described as a systematic way of designing the machine controls. This is a definite advantage over the ad hoc method which characterizes the conventional control design. In the conventional logic control design, the total effort is predominantly borne by the logic designer. He has the responsibility for the initial conversion of the architecture to the sequencing and timing charts, and for the implementation of such sequencing charts into hardware. He is also responsible for devising the exhaustive test patterns and procedures, and for performing the debugging of the control section. The responsibility for the complete documentations for such a system is also his. This latter responsibility is a difficult task to perform satisfactorily or in a way that can be relied on as an adequate reference document for field debugging, maintenance or modification. All these tasks are indeed difficult, complex, and leave much room for error, because of the designer's inability to execute them with the same proficiency.

Microprogramming, on the other hand, represents a suitable way to rectify these problems. There is a built-in division of labor option that would enhance or improve the process and the quality of the product. For instance, an experienced programmer may be called in to help with the design, flow-charting, and implementing of this section of the design. A programming aid, such as a microprogram simulator, can be implemented to check, debug, and compare different approaches at electronic speeds. A by-product of this simulation is a clear, complete, and standard set of documentations to be distributed to the field. Another difficulty in the conventional hardware control design that is nonexistent in the microprogrammed control design is the packaging problem. Other design steps that can be carried out by the computer are the ROS word address assignment and ROS bit-encoding functions.

The present emphasis on system availability and reliability and the economic penalties for the lack of such in the modern digital computer dictate that all the data paths in the system must be sufficiently checked for any transient, intermittent, or solid failures that can creep into the system. This requirement is very difficult and expensive to effectively design into the hardware control section. Microprogramming, on the other hand, is more adaptable to checking. A parity bit is added to the ROS word which, in effect, permits the sequencing to be checked.

4. *Uniformity of Design.* Related to item 3, yet important enough to be listed separately, is the uniformity of design characteristic. Microprogramming gives order, uniformity, and modularity to the machine control design similar to that of the arithmetic and logical unit functional design. Because of this uniformity, which often spells simplicity, one can expect that it takes less training and experience to produce the necessary skilled microprogrammers than it takes to produce a hardware logic control designer.

5. *Ease of Education.* Because of its simplicity and its methodical and organized approach, microprogramming is much easier to teach to the maintenance engineers in the field. Consider having to trace the execution cycle of an instruction. In microprogramming, one can easily trace the flow charts and programs written in symbolic language that is descriptive of the process going on. In the hardware control design, to trace the same macro, or machine instruction execution, one must resort to the sequencing and timing charts and the hardware logic sheets.

Another area of education is that of computer science, particularly systems architecture and systems organization. Here, microprogramming has also proved to be a successful pedagogical tool.

6. *Microprogramming can extend the useful life* of the system by reprogramming its control function to suit the new demands. This is also true during the whole life of the system. Its ability to respond to new demands and its ability to extend the architecture of one machine to contain another machine are points of extreme importance in a large number of computer installations faced with finding an efficient solution to the data and program conversion problems.

7. *Economy.* We stated earlier that the computer manufacturers have all been striving to give the customer an ideal, comprehensive instruction set, suited for both commercial and scientific programming with equal efficiency. This ideal, general purpose architecture is more expensive to implement by hardware logic than by microprogramming. It has been shown in the section of this chapter devoted to the economic model that this cost saving is dependent on both the technology and the system's architecture.

To be sure, microprogrammed control does have some disadvantages, some of these are as follows.

1. *Operational Loss of Efficiency.* Although microprogramming affords greater flexibility, it can also introduce some difficult problems to the systems programmers, to the operators, and to the management of the computer installation. One can only begin to visualize the difficulty in maintaining system libraries and system monitors in such an environment.

2. *Performance Loss.* This performance loss may result from many reasons. First, the ROS control is sequential and is not adaptable by itself to overlapped operation. This explains, for example, why the IBM System/360, Model 85, resorted to the hardware control in the design of the I-box. Second, performance loss may result from the inability to overlap the ROS access timing with the CPU function. Assume that the ROS memory cycle is 600 nanoseconds wide, and none of that time is overlapped with the 500-nanosecond CPU cycle time. In this case, the effective machine cycle time will be 1100 nanoseconds, the sum of the two cycles.

3. *Cost Disadvantage.* For some systems, where the architecture is simple and fixed, and where the function of the system is slave to the relatively simple and repeated task, one can argue that microprogramming is too expensive.

2.11 DESIRED COMPUTER CHARACTERISTICS FOR MICROPROGRAMMING APPLICATIONS

This discussion has appeared in fragments throughout the first two chapters of the book. The objective here is to summarize the computer characteristics desired for microprogramming applications, and to focus on their implications and trade-offs. Among these desirable characteristics are:

1. The computer should be synchronous, with well-defined timing points.

2. The design implementation should be simple, that is, it should have the fewest possible control lines. This way, we expect to have a large number of sequencing cycles with the narrowest of control word organization. The implications of this can be seen in Fig. 2.29, where the basic fixed cost is minimum, and this basic hardware is utilized over and over.

3. It is desirable to have a significant number of fixed sequences with no branching. That is, in certain implementations, a branching sequence could nullify the gains we realized from the ROS and CPU cycle times overlap. The ROS word fetched to control the next machine cycle may have to be ignored, requiring a correct new ROS word to be fetched.

Another way to dramatize this loss is to review the loss in the machines where the ROS word is designed to control the machine for two or more consecutive cycles. If the branch conditions become excessive, this ROS word will not realize its maximum utility.

4. From the ROS access and timing, we realize that it is desirable to have an ROS technology with timing and access equal to and overlapped with the CPU clock.

5. The performance of the system is partially dependent on how some critical control signals are handled in the system. These conditions (signals) can be handled either by hardware means or by microprogramming stored in

ROS storage whose timing permits activating these control lines early in the machine cycle.

6. In the encoded field organization, it is desirable for the computer to have large sets of mutually exclusive control lines. This utilization of more powerful encoded functions would result in the smallest ROS word width.

REVIEW QUESTIONS

1. Name the two types of ROS word organizations discussed in this chapter, and discuss their implication on the following system design characteristics:
   (a) ROS access time
   (b) Cost of the ROS control
   (c) Flexibility to permit future system architectural modifications.

2. List the functions(s) of each of the following segments of the ROS word:
   (a) Static control lines
   (b) Sequencing control
   (c) Emit field

3. Using the hypothetical computer shown in Fig. 2.7, write a microroutine that compares the contents of register R1 to the contents of Register R2 and branch to the ROS storage address as shown in the following table:

IF R1 = R2	branch to ROS address	0000 (binary)
IF R1 > R2	branch to ROS address	0001
IF R1 < R2	branch to ROS address	0010

   Implement this microroutine in both the direct control matrix and the encoded control.

4. Trace the multiplication algorithm shown in Fig. 2.16 to perform the multiplication of $-19 * 31$.

5. Discuss the impact of:
   (a) Circuit technology, its cost per logic decision, its fan in and fan out capability
   (b) ROS memory access methods such as the coincident current drive and linear word drive

   on the decision to implement the control section of the digital computer from sequential logic blocks or microprogrammed implementation.

6. Discuss the impact of parallism in the data flow of a microprogrammed system on the ROS word and ROS memory organizations.

7. Given the data flow of the hypothetical computer system shown in Fig. 2.17, assume that the dividend resides in registers R3 and R4, and that the divisor resides in R2. Write a generalized binary division microroutine which performs this operation and stores the quotient in R1.

REFERENCES

1. Wilkes, M. V., "The Best Way to Design an Automatic Calculating Machine." Report of Manchester University Computer Inaugural Conference, (July, 1951), p. 16.

2. Wilkes, M. V., Renwick, W., and Weeler, D., "The Design of a Control Unit of an Electronic Digital Computer," *Proc. IEEE*, 105 (1958), p. 121.

3. Wilkes, M. V., "Microprogramming," *Proc. E.J.C.C.* (1958), p. 18.

4. Wilkes, M. V., and Stringer, J. B., "Micro-Programming and the Design of the Control Circuits in an Electronic Digital Computer," *Proc. Camb. Phil. Soc.*, 49 (1953), p. 230.

5. Renwick, W., "BDSACII," *Proc. Inst. Elec. Eng.*, Vol. 102, Pt. B, Supp. 102 (1956), pp. 277–278.

6. Billing, H., and Hopmann, W., "Mikroprogramm-Steuerwerk," *Elektronische Rundschau*, 9 (1955), p. 349.

7. Glantz, H. T., "A Note on Microprogramming," *Journal ACM*, 3 (1956), p. 77.

8. Mercer, R. J., "Microprogramming," *Journal ACM*, 4 (1957), p. 157.

9. Blankenbaker, J. V., "Logically Microprogrammed Computers," *IRE Trans. BC-7* (1957), p. 103.

10. Blankenbaker, J. V., "Logically Microprogrammed Computers—Examples of Small Computers," *Proc. International Conference on Information Processing UNESCO* (1959), p. 430.

11. Dinneen, G. P., Lebow, I. L., and Reed, I. S., "The Logical Design of CG24," *Proc. EJCC* (December, 1958), p. 91.

12. Kampe, T. W., "The Design of a General Purpose Microprogram-Controlled Computer With Elementary Structure," *IRE Trans. BC-9* (1960), p. 208.

13. Gerace, G. B., "Microprogram Control for Computing Systems," *IRE Trans. EC-12* (1963), p. 733.

14. Grasselli, A., "The Design of Program Modifiable Microprogrammed Control Unit," *IRE Trans. EC-11* (1962), p. 336.

15. Hagiware, H., Amo, K., Malsushita, S., and Yamauchi, H., "The KT Pilot Computer—A Microprogrammed Computer With a Phototransistor Fixed Memory," *Proc. IFIP. Congress 62*, North Holland Publishing Co. (1963), p. 684.

16. Emelyanov-Vavoslavsky, L. B., and Timofeer, A. A., "Microprogram Control for Digital Computers," *Proc. IFIP Congress 62*, North Holland Publishing Co. (1963), p. 567.

17. Hawryszkiewycz, Igor T., "Microprogrammed Control in Problem-Oriented Languages," *IRE Trans. EC-16* (1967), p. 652.

18. Harrand, Y., "Evolution of Microprogramming Concepts." *Proc. 3rd AFCALTI Congress of Computing and Information Processing*, Toulouse (1963), p. 187.

19. Meade, R. M., "A Discussion of Machine-Interpreted Macroinstructions." 16th National Meeting of the ACM (1961), p. 6C-1.

20. Tompkins, "Machine Attacks on Problems Whose Variables are Permutations," Proc. for Sixth Symposium on Applied Mathematics. New York, McGraw-Hill Book Co. (1956).

21. McGee, W. C., "The TRW-133 Computer," *Datamation*, 5, No. 2 (February, 1964), pp. 30–32.

22. Beisner, A. M., "Microprogramming Study Report." Department of Computer Mathematics, Center for Exploratory studies, IBM—Federal Systems Division, Rockville, Maryland (September 1, 1966).

23. Pitkowsky, S. H., "Control Storage System." IBM internal document.

24. Campbell, C. R., and Neilson, D. A., "Microprogramming the Spectra 70/35," *Datamation*, 7, No. 9 (September, 1966), pp. 64–67.

25. van der Poel, W. L., "Zebra, a Simple Binary Computer," *Information Processing, Proceedings of International Conference on Information Processing, UNESCO, 1959* London: Butterworths, (1960), p. 361.

26. van der Poel, W. L., "Microprogramming and Trickology," in *Digital Information Processing*, W. Hoffman, ed. (New York: Interscience, 1962), p. 269.

MICROPROGRAMMING APPLICATIONS

3

3.1 INTRODUCTION

The potential of microprogramming as a means to alter the architecture of a given data processing system has presented subject matter for researchers, engineers, and programmers for a good number of years. The field has come a long way from the original concept, annunciated by the Wilkes group, where microprogramming was thought of as a means for designing the control unit of an otherwise conventional Von Neumann machine. Since then, many of these concepts have been pursued by computerologists to various levels of intensity and in many various ways. Most of these activities, however, were directed in the area of control memories, their organization and sequencing. Little has been written about the merits of microprogramming and its potential application, other than as another way to design the control section which would make the process of changing the computer instruction set a fairly simple one. It is this ability to alter the complete instruction set without affecting the hardware design which presented the computer designer and the user with a most intriguing yet challenging problem. Wilkes did not feel that the user should be permitted to make such changes. He felt that chaos would result within the computer center if users were permitted to have their private order code[1,2]. Clearly, Wilkes felt that the microprogram should remain the province of the computer designer rather than the programmer.

While some continued to pursue this concept of specialized application to a given data processing requirement, other researchers began to look at microprogramming as an economic way of standardization, of implementing a rich instruction set for a whole line of compatible processors with variable speed capacity and cost. This new trend, in which the computer designer hopes to achieve a truly general purpose computer through the complex architecture and rich intricate set of machine instruction now feasible due to microprogramming, was the guiding principle in the design of two such lines of computer processors, namely, the IBM System/360 and the RCA Spectra/70. These two systems presented the first concerted use of microprogramming. Yet, the lack of microprogramming education, the lack of reference information, and the difficulty the user would encounter from the designer in order to effect any changes have kept microprogramming applications largely in the domain of the designer.

Nevertheless, there is definitely a growing interest in microprogramming, and the trend toward its more efficient use is unmistakably clear. More and more military aerospace system designers, and commercial users are turning to microprogramming in designing data processing systems with an efficient operating characteristic. An example of this trend for more systems efficiency achieved by microprogramming new instructions particularly suited to a specific application is found in the area of sensor data processing, such as the large-aperture seismic-array processing. Another example is in the area of formatting and processing radar data. A third example can be found in the extent of microprogramming research activity in optical data processing which has a wide range of applications including pattern recognition and image processing.

Microprogramming has been advantageously used in computers which control communication switching, as well as in the so-called intelligent channels for message processing, telemetry decoding and routing, data manipulation, input/output data formatting[3,4]. Other more advanced types of applications being experimented with are in the areas of process control, space communications, and satellite data management. These future applications will be evaluated later in the chapter.

The application of microprogramming since its inception has been oscillating between designing the controls to implement some architectural specification and accelerating the existing system's performance by the addition of a few new operation codes to the existing ones. The new operation codes are selected for the mission and the environment for which the system is designed.

The first phase in this application cycle during the early 1950s was characterized by the emphasis microprogramming received as a means to implement the control function of the conventional von Neumann System. However, due to the lack of fast, efficient, inexpensive ROS devices, and due to the lack of familiarity with this concept, microprogramming lay almost dormant for the duration of the 1950s. Reference 19 described the control system of the

EDSAC-2 computer built at the University of Cambridge. The read-only memory was composed of a matrix of ferrite cores through which wires were threaded. Early in 1960s, we recognized a new phase in the development and application of this concept. It is characterized by the specialized system design for military applications and radar communications. One example of this is the number of "stored logic" computers. The trend shifted again during 1963–1970 in favor of standardization. This is evident from the number of product lines with the upwards and downwards compatibility. A quick examination of the future applications of microprogramming points to appreciable growth in both standardization and specialization of microprogramming.

In general, microprogramming applications are classified into the following classes:

1. Design of an efficient control for a well-specified system architecture.
2. Modification of an existing computer architecture with or without modifications to the existing data flow.
3. Emulation of some existing hardware capabilities.
4. Integration of the hardware and different levels of software (operating system through the application programs) into one integrated and total system design.
5. Real-time applications for process control.
6. Future applications in the areas of communications, control, and space.

These classes are by no means mutually exclusive. On the contrary, the areas of overlap among them are significant. Only the emphasis and the specialized problems that have to be addressed suggested such a classification.

An example of the first class is the implementation of the IBM System/360 in which many processors with radically different internal organization and data flow were made through microprogramming to recognize the same architectural specifications. This is really the essence of the upwards and downwards compatibility. This class of application will be explored in Chapter 6.

Class two is more user-oriented than all remaining ones. The concepts of tailored instruction sets, and additional macros for system architectural extensions (with or without some modifications to the data flow), do unquestionably fit in this class. Substantial segments of the application programming such as process control, real-time sharing, inventory maintenance and control, and traffic control and reservations are all potential and suspected fields of microprogramming application.

This class is often described as interpretive class or interpretive microprogramming in which existing data flow is made to interpret and execute another architecture or higher-level user language. Emulators, one can argue, are also interpretive; this is true. The reasons, then, for placing them into separate classes of application are twofold. First, it reflects the large extent of emulator design activity. (Among the approximately 12,000 general processing systems installed, e.g., the IBM System/360, RCA Spectra/70, Honeywell, GE 400 and GE 600, IC-6000E, and others, emulation has been, at least initially, the overwhelming choice for program conversion. Of those users who have installed emulators, 75% are operating in "emulator mode" more than 50% of their total running time.)

The second reason is that emulators have many similar design problems and characteristics that should be discussed separately from all other applications. There are numerous examples of the second and third classes of applications. In the second class, an important example which will be discussed in detail is the Conversational Programming System (CPS-System) developed at Allen-Babcock Computing, Inc. under a contract to the International Business Machine Corporation.

As indicated earlier, the emulator class can be considered as a subclass of the second class. Emulators have the added implication of augmenting the data flow in the emulating system in order for it to recognize and execute the emulated system instruction code efficiently. There are many examples of this: the IBM 7090 emulated on the IBM System/360, Model 65; the 1620 System emulated on the IBM System/360, Model 30; etc.

Very little was done in the fourth class of application. However, programming systems and operating systems, particularly in the fields of IOS macros, control programming, schedular, dispatcher, interrupt handlers, remain fertile areas for performance improvement through microprogramming. Opler[20] has suggested the term *firmware* for microprogramming written to support software. He suggested that firmware will take its place along with software and hardware as a main design option.

Another area of growing interest in microprogramming application is that of process control. The main problem here is that the process control computer must be capable of responding to real-time demands of a number of either distinct or interdependent functions. It must have sufficient computing power to meet the peak demand placed upon it. It is therefore desirable to include background programs and tasks that are not time-dependent to utilize the reserve computing power necessitated by the peak demand intervals. The control program in such an environment must interrupt the fill-in tasks to handle the real-time request. After such a request is handled, and if no other real-time control function is required, the system is then returned to the interrupted background task for completion.

The last class of microprogramming is that of the future applications. This area points to the extent and direction in which microprogramming will be applied in the late 1970s and beyond. We have given as examples of this class applications such as signal processing and communication switching control, and applications in the so-called intelligent channel design for message processing and switching, input/output formatting and routing, and telemetry decoding and routing. We also speculated on the microprogramming application for radar data processing for air traffic control and for the wide field generally referred to as image processing.

A particularly interesting application is in the basic microprogrammed controller area. The system here consists basically and principally of the read-only storage. Such a machine may have a wide variety of military and space applications in the control of missile interceptors, spacecraft, aircraft, anti-aircraft guns, locomotion, and the control of undersea vehicles over rough terrain on the basis of sensor information.

One example is how this microprogrammed controller might be used to help control a manned spacecraft. The microprogramming sequencing is made to depend partially on the result of an environmental test specified by the microinstruction. The variables which might be tested in the spacecraft are cabin pressure, cabin temperature, spacesuit temperature, roll rate, yaw rate, pitch rate, etc. The idea is to use the microprogram to control and issue the proper indications and warnings to the astronauts and ground station whenever such a dangerous condition occurs. For example, binary numbers representing upper and lower safe limits for the cabin pressure could be stored in the microinstructions. These numbers can be fed by the microprogram to a binary-to-analog conversion system which produces voltage against which the cabin-pressure sensor output is tested. The testing can be performed by means of a simple threshold device which communicates back to the microprogram by forming a bit (or more) in the low-order bit(s) which defines the address of the next microinstruction sequence. Thus the microprogram is forced to branch to the proper microroutine to take the proper control steps.

Another area of microprogramming activity in which substantial growth can be anticipated is aerospace navigation and control. This implies microprogramming a system in which the computer is but a component of the total system. The integrated system control is difficult to implement with conventional hardware due to the lack of standardization of peripheral equipment, interfaces, or formats from one application to another. Such a system may encompass such diverse applications as process control, command and control, navigation and guidance, message switching, fire control, reconnaissance and surveillance, time sharing, image processing, and tracking. In these and other applications, the computer system is operating on-line, performing highly specialized tasks such as scanning, polling, function generation, interrupt handling, and format conversion for a significant portion of the operating time. Therefore, a convenient and inexpensive way to customize the system features is needed. Microprogramming is a natural media for this type of specialization[12].

Another application is building experimental processing systems for algorithmic languages. This type of application opens a new dimension in the usage of microprogramming for processing of interpretive code[13,14,15]. An example is the microprogrammed implementation of Euler on the IBM System/360, Model 30, by Helmut Weber of IBM. Since his article* on the

*Helmut Weber, "A Microprogrammed Implementation of Euler on the IBM System/360 Model 30," *Communications of ACM*, Vol. 10, No. 9 (Sept., 1967), pp. 549–558.

subject clearly describes this system, it is sufficient to reference it without duplicating his effort.

3.2 POSSIBLE PERFORMANCE IMPROVEMENTS THROUGH MICROPROGRAMMING

If one accepts the following simple premises:

1. A true general purpose computer that suits every programmer's needs in every environment with the same simplicity of programming and efficiency of executing the code does not exist.
2. Today's general purpose computers are the product of a series of compromises that result in a "balanced" performance across a wide range of applications, but certainly not optimum with regard to any one application.
3. No two computer installations are alike.

then it is easy to conclude that every installation, especially those dedicated to certain types of long, repeated production runs or to specialized applications can profit considerably from hand-tailored instruction sets and other architectural characteristics added to the system. A capable programmer could identify an unlimited number of analytical functions which would substantially improve the performance of that system.

Implicit in the original concept of microprogramming is the idea that the logic designer could and should employ the talents and experience of the programmer who would be more qualified in selecting and even microprogramming the instruction set. This has been realized to a limited extent as evidenced by the birth of the *new discipline* in computer design known as *computer architecture.* This development represents a step in the right direction, namely, to produce a balanced system based on what the application programmer sees, or would like to see, in the system. But since the programmer's needs vary drastically from one application to the next, it would be profitable if some means for augmenting the instruction code is made available to him.

Almost all existing general purpose computers that are used for commercial processing can profit considerably from the addition of a few macros for bit and byte manipulation in microprogramming. Similarly, in scientific applications, additional instructions such as matrix operations and inverse divide would reduce compile time and would lead to a fuller realization of the computer potential. In the area of system monitors, compilers, interrupt handling, schedulers, interpreters, resource allocation schemes, etc., where these routines are used over and over, and where no changes are expected to be made, it is imperative that they be seriously considered for microprogramming.

A study conducted by Hans Jeans of IBM on the possible performance improvements by microprogramming for IBM System/360, Models 40, 50, and 65, indicated[5]:

1. "The mathematical inner loop subroutines such as polynomial evaluation and vector scalar multiplication will not be greatly improved by microprogramming. The inherently slow arithmetic operations are not appreciably expedited by the microprogrammed composite operations, since the instruction fetch time saved remains a small fraction of the overall execution time. However, performance improvements of 1.3 to 1.6 times for polynomical multiplications are possible."

2. "A table search operation can be microprogrammed to execute at nearly the cycle time of the main storage, with a resultant performance improvement of 10 to 15 times, while a character scan operation can be improved 5 to 11 times through microprogramming. The fact indicates that the payoff of microprogramming lies in logical operations, essential to compiling, list processing, and pattern recognition."

3. "It is not inconceivable that a compiler with many of its frequently used routines microcoded may improve its performance substantially. The addition of special instructions will increase the efficiency of compiled programs. A somewhat different approach would be to implement a compiler (one pass) and an interpreter with extensive use of microprograms so that total time for the "compile and execute" process can be greatly reduced."

These conclusions by Jeans are not limited to any particular system or product; rather they can apply to most general purpose computers with microprogrammed control.

The extent of the improvement is a function of the amount of parallel data paths and the extent of parallel operations that can be performed in one machine cycle. In the IBM System/360, Model 50, for example, in one machine cycle the system can add two operands through the adder data path, perform logic operation on two bytes of data through the parallel mover data path, read one word (four bytes) from local storage to the CPU and store another word in the same location in local storage, decrement a counter, and initiate a main storage request. Another feature of microprogramming is the powerful built-in decision logic capability. Again, Model 50 is capable of performing a 64-way branch in one microinstruction, while the Model 40 is limited to a 32-way branch. This fact accounts for the tremendous performance improvement realized in table search and sorting routine when implemented in microcode.

A third factor contributing to the performance improvement is the reduction of instruction fetch and excessive memory references. The first is dependent, of course, on the number of the different instruction formats, the complexity of the address computation algorithms, and the word boundary alignment constraints. For a medium range, general purpose computer, to fetch an instruction may require 5 to 30 machine cycles. If a given subroutine is microprogrammed and made accessible by a macro, only one instruction fetch for that macro is needed. Thus, a substantial saving in the total instruction fetch time is realized. Another saving results from the reduction of excessive memory references. This saving is due to the discrepancy that exists between the main storage cycle time and the CPU cycle time. Thus, a microprogrammer would enhance his microprogram performance improvement by utilizing internal CPU storage capabilities (hardware registers) for intermediate results and limit the main storage references to fetching the operands and storing the final results.

It must be realized that the variation of performance improvement from one system to the next is a function of the number of parallel data paths that can be activated simultaneously, the width of the hardware facilities, the width of the data paths in relationship to the width of the operands, the number of hardware registers, and the size of the local storage section that the microprogrammer can use as work space for storing intermediate results.

The performance differential between the two IBM System/360 models, 40 and 50, is primarily due to the differences in their internal structure. The Model 50 main storage data register is 32 bits wide (equal to the System/360 word length), whereas the Model 40 storage data register is only 16 bits wide. The Model 50 adder is also 32 bits wide, while the Model 40 is only 8 bits. Hence, there are more microprogramming constraints imposed on the Model 40 microprogram than on the Model 50. These constraints and others substantially affect the performance improvement that one can expect from a given system.

Table 3.1[3] gives the performance improvement ratio of the execute time of the microroutines to the regular subroutines for three different IBM System/360 models—40, 50, and 65—utilizing standard existing equipment without any hardware changes such as faster-access storage, more hardware registers, or modifications to the data paths which would result in further performance improvement. It will be a valuable exercise, after studying the internal data flow and the microprogramming of the IBM System/360 Models 40 and 50, to refer back to this table and attempt to justify their relative gains.

Table 3.1 PERFORMANCE IMPROVEMENT

	Model 40	*Model 50*	*Model 65*
Matrix multiplication	1.47	1.59	1.49
Polynomial evaluation	1.64	1.45	1.34
Table search	11.8	13.8	10.0
Byte test	10.8	6.2	5.0

In the mathematical subroutine, single precision floating-point arithmetic was assumed. The polynomial

$$P(x) = A_0 + A_1X + \cdots + A_nX^n$$

was evaluated by nested multiplication and additions.

$$P(x) = (\cdots((A_nX + A_{n-1})X + A_{n-2})X \cdots)X + A_0$$

A special System/360 RX format instruction to evaluate the summation of product was devised with suitable pointers to the coefficients, variable X, and the order of the polynomial N, stored in the general and floating-point registers in local storage. A minor point here is that the performance improvements that would be expected in both of these mathematical subroutines are functions of the sizes of the matrices and the order of the polynomial. This improvement is larger for larger matrices and polynomials.

For the nonmathematical routines which rely extensively on the microprogram's capability for decision logic and reduction of memory fetches, plus overlapping the main storage references with CPU operation, the performance improvement was found to be substantial. For subroutines without floating-point arithmetic, the performance improvement becomes even more appreciable.

The table-search subroutine searches selectively a table of 32-bit words, W_i, in main storage with a given argument, A, until it finds the first word satisfying a predefined relationship to A.

The byte-test operation is even more dependent on the internal logic decision and multiple branch capability of microprogramming. However, since only one memory reference is made, the performance improvement due to the overlapped memory references did not contribute in the byte test as it did in the table search.

3.3 TRANSLATION, SIMULATION, AND EMULATION

There are several approaches to converting a program written for machine X to run on machine Y. These approaches cover the whole range from reprogramming, code-to-code translation, simulation, emulation, and, finally, system compatibility. These five techniques differ in the effort needed to perform this conversion, the efficiency with which machine Y is utilized, and the amount of hardware needed to augment machine Y in order to efficiently recognize and execute programs coded for machine X. Each point on this conversion scale (Fig. 3.1) has its advantages and disadvantages and must be justified on the cost/performance gains realized in a given environment. These conversion aids fall into two distinct categories:

1. Those which allow the user to continue running of the old programs on the new system, which includes emulators and simulators.

2. Those conversion aids which help the user to create new programs and data sets from the old ones. This class includes translators and language-conversion and data-conversion programs.

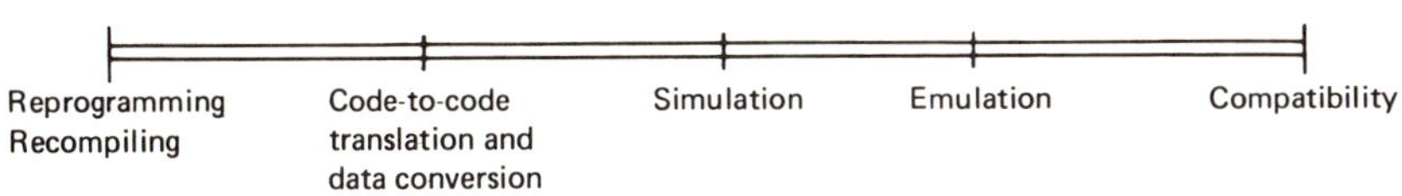

Figure 3.1

At one end of the scale, the user must reprogram his problems in machine *Y* language. This would permit him to take any advantages of the additional or favorable features of machine *Y*. The cost involved comprises human effort, computer time, and debugging time, and can be very prohibitive.

If the user programs are all written in a higher-level language such as FORTRAN and COBOL, the task of conversion is reduced to recompiling these programs for the new system *Y*.

Code-to-code translation is perhaps the least efficient step on the ladder of conversion. It is often intended only as an intermediate step. Here each machine instruction in the source program for system *X* is replaced by essentially one or more instructions of machine *Y*, which would execute the same operation intended by machine *X* instructions. The resulting source deck for machine *Y* will then be assembled and executed. The two factors that would influence the performance loss in machine *Y* are:

1. The architectural similarities or differences between the two systems.
2. The frequency with which the resultant object code is executed.

The three major classes of problems in this type of conversion by translator are:

1. *Program Segments That Are Completely Untranslatable.* An example of this is a program written to test the core memory of the CDC 1604. This program depends on the way this particular memory is wired and driven. Here the test is completely dependent on the physical structure rather than the architectural characteristics of the system itself. Other untranslatable routines are those which are peculiar to how the I/O instructions and program segments would purposely cause interrupt program segments to turn a console light on.

2. *Idiomatic Expressions.* These are program segments of one or more instructions which have a meaning as a whole that is not a direct result of the implied meaning of each individual instruction[21]. An idiomatic expression must be translated by identifying its meaning in terms of the function it performs rather than the sequence of codes of the target machine.

3. *Program Time-Varying Structure—Self-Modification.* This third, general problem of self-modification is an obvious one and is more difficult to solve. The time-varying structure may depend on the data or on machine states that are unpredictable. A partial solution to all three classes of program translation calls for programmer assistance in defining these problems and programming them by hand. Dellert[22] proposed the use of many levels of macros in the translation of symbolic assembly language of one computer to another. This set of macros can be inserted at the beginning of the target machine source deck. These macros will treat incompatible codes as macros to produce new codes compatible with those of the host machine.

3.3.1 Simulation

Simulation is defined as the process of representing the functioning of one system by another. In the data processing field, it is a process of modeling the physical behavior of one machine on another. It differs from translation in that each instruction or line of code is interpreted in sequence at the time of execution. That is, the image of the simulated machine is mapped, bit by bit, character by character, on the host machine.

The most obvious objection to simulation is that its program may run more slowly than on the original machine. The degree of inefficiency to which the host machine is run depends on the depth to which the source machine is simulated each time a program is executed, rather than once.

Simulators consist of two major parts:

1. The internal CPU program, tailored to each target machine.
2. A supervisory program to handle I/O and all other exceptions.

A simulator consists of several well-defined interacting functions. The most fundamental of these is the interpretive routine which fetches the source machine instruction, converts addresses into the host machine, manipulates operation codes to get addresses of the routines programmed to execute each of the target machine operation codes.

A second important section of the simulator routine contains the execution routines not related to the input/output. The most important feature of these routines is their speed, which is necessary to reach an acceptable efficiency in the simulation environment.

Another segment of the simulation is the control program which is designed to handle any exception to the fetch, address computation, and execution of the simulated operation code.

A fourth section is designed to perform an I/O interface between the two systems. The greater the differences between the I/O devices and the data characteristics, the greater the complexity of this part. This section must also consider the differences in data formats, record lengths, data conversion, and transmission.

The host's I/O control system and the linkage between the CPU and I/O simulation routine form an important section of the supervisory program.

Other routines normally found in the simulator are core storage mapping, radix conversion, and interrupt handling.

The advantage of this purely software approach to program conversion, called simulation, is that it does not require any hardware and it is conceptually simple to design due to its strict adherence to the interpretive mode of operation. The main objection, mentioned earlier, is that simulators are slow. Their acceptability is fundamentally an economic compromise based on how well they perform on the host machine and on the depth to which the source machine is to be simulated.

The lack of efficiency here may be a result of one or more factors. Among these is the inability of the simulation program to take advantage of many hardware features on the host machine. Another factor is the failure of the simulation program to effectively preserve the parallelism that exists in the target machine and to interpret the parallel operations in an efficient sequential manner.

3.3.2 Emulation

A higher point on the program conversion scale, shown in Fig. 3.1, is the emulation step. Emulation is defined here as a combined hardware-software approach to simulation. The hardware portion may consist of augmenting the data flow of the host system as well as microprogrammed modifications and extension. Thus an emulator is basically an extension of the host machine's architecture, hardware, and software to include the image of the target machine. This method of conversion is intended to emphasize the similarities of the two systems, and the emulation is thus performed at the point of minimum difference. Therefore, as one would expect, emulators are normally five to ten times faster then the corresponding pure simulator written for the same machine. Generally, emulators offer at least a one-to-one throughput as compared to the target system.

A successful design of an emulator requires an efficient solution to four highly related problems:

1. The first involves the mapping of the image of the target machine into the host machine. This consists of:

(a) Main storage mapping.
(b) I/O system mapping.
(c) Registers, triggers, counters, accumulators, etc.
(d) All other resources in the target machines that are addressable in the source program.

The memory mapping function must ensure that a minimum addressable storage cell of the target machine will be contained in the smallest possible host machine storage cell. This results in a more efficient address conversion, as well as in conserving of the host memory storage. Mapping of I/O devices is considerably simpler than storage mapping, since a static one-to-one correspondence is maintained.

Existing hardware registers, counters, etc., in the host machine can now be used in every conceivable way to perform this simulation. This freedom can be exercised, for example, in using the hardware registers to implement a highly efficient addressing scheme.

2. The main problem in designing an emulator is the selection of a small group of machine instructions to augment the host machine instruction set. In a microprogrammed machine, the implementation of this step, once the instructions are selected, is a relatively simple task.

These special instructions are selected to speed up the software simulator, and they are based on the length of time the simulator spends in performing specific functions. A common task performed by all emulators is the instruction fetch and address translation simulation. This process has to be repeated for every instruction in the program to be converted, before branching off to another special routine for executing it. There are undoubtedly other interloops that enjoy high utilization in executing a wide range of application programs on the target machine. These must also be evaluated as candidates for this special instruction class. Among these, one can identify some interrupt handling routines, floating-decimal operation, storage boundary alignment, branching, character conversion, etc.

3. The third problem is that of designing efficient linkage from the host machine mode of operation to the emulation or interpretive mode of operation. It is necessary to maintain a complete separation between the two modes in order to maintain the architectural identity and system compatibility between this host machine and other models in a compatible product line. Since the architectural characteristics have been defined as those characteristics seen by the programmer, the system designer can, in fact, anticipate other functions which are helpful in servicing, maintaining, and adding any experimental features that are deemed desirable, and which are not available to the user. In the IBM System/360 architecture, a special privileged *diagnose* feature or instruction was defined primarily for the field engineer to transfer to one of many built-in diagnostic routines to verify the proper functioning of the equipment and locate faulty components. This special instruction is executed in a supervisory (restricted) mode, and the application programmer would have no means to get to it. Since an emulator is designed by the systems designer as a total efficient hardware-software package, the manufacturer can then use the diagnose instructions and other privileged operations as part of the total systems implementation that is transparent to the architecture. This diagnose instruction is, in fact, used quite extensively by IBM engineers as a means to implement a number of special software macros into microroutines. Since it is a privileged operation, the diagnostic instruction is implemented differently for the different IBM System/360 models. Each model designer in the IBM System/360 has the responsibility of defining the linkage and the interface between the CPU and the diagnostic modes of operation.

4. Since each simulator and emulator is developed as a stand-alone programming package, it is necessary to develop an operating structure under which each simulator may be run. Such an operating structure must recognize the extensive similarities between the simulators and must use these similarities to enact efficient operating systems.

The problems of I/O simulation and emulation are divided into four main classes:

1. The static one-to-one mapping of the subject I/O devices into the I/O system, and the treatment of exceptional conditions. This one-to-one mapping

must be dynamically modifiable by the operator so that the conversion does not become too expensive.

A number of exception conditions can occur in the I/O process. These conditions are communicated to the control program in order to interrupt the instruction fetch to service them or take other appropriate action. An exception is the target machine *halt* instruction.

2. Data conversion and transmission.
3. Linkage between the CPU and I/O.

These last two problems are greatly influenced by the extent of the required interface between the target I/O instructions, data formats, record length and status data, and the host I/O control system.

4. *How to Maximize the Overlap of I/O With CPU Emulation.* Tucker[23] estimated this nonoverlapped portion of I/O and CPU for the operation of a

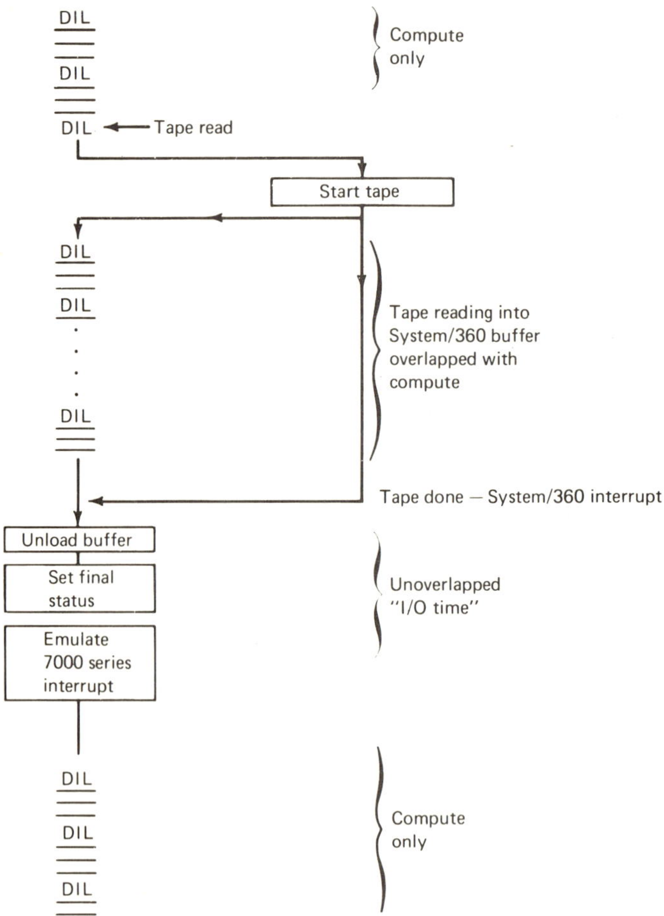

Figure 3.2 Emulation of input/output.

90,000 character per second tape, with a host system buffer unload operation faster than a million characters per second, to be less than 10%. Figure 3.2 shows this partial overlap of I/O and CPU emulation. Note how the DIL (Do Interpretive Loop) microroutine continues to pick up the target machine CPU instructions, emulates and ends up transferring back to DIL. This continues till a tape-read operation occurs. Note how the emulation control program interrupts the CPU emulation to issue the tape start operation, and then continues with the CPU emulation by transferring control back to the DIL routine. This overlapped CPU emulation and the tape reading into the host machine buffer continues until this tape operation is satisfied; then it interrupts the CPU emulation and takes over the system to perform the following unoverlapped operations: unload buffer, set final status, and emulate the target systems interrupt.

3.3.3 Simulation of Second Generation Systems on the IBM System/360

A large number of emulators for the IBM second generation systems have been implemented and are being used widely to ease the program conversion problem. Each simulator and emulator consists of four different parts:

1. An internal CPU program tailored to each target machine.
2. Special hardware and data flow augmentation of the host machine to enhance the simulator performance.
3. Additional interpretive routines in microcode designed to further speed up the simulator.
4. The supervisory package that would handle the linkage to the internal CPU program, the I/O, interrupts, and other common functions.

Parts 2 and 3, if present to any extent, would distinguish that conversion package as an emulator. Otherwise, it will be called a simulator.

A partial list of software simulators developed at IBM as conversion aids to System/360 include: 1410/7010, 1620, 7070/7074, 7080, 7090/7094. Note that these simulators are System/360 model-independent since they are not optimized by the use of either microprogramming or data flow modifications.

The most widely accepted means of conversion in industry today is the emulator. Since each emulator is designed as a stand-alone package capable of meeting a certain prescribed cost/performance criteria, their design varies from one system to another. In some cases, the emulators rely heavily on augmentation of data flow with relatively small microprogrammed functions, and in other cases, the emulator is heavily implemented by microcode with little or no data flow modifications to the host system. The balance of these ingredients of software, hardware, and microprogramming is left to the system designer and is subject to cost/performance justification. For example, the IBM 1620 emulator on the IBM System/360, Model 30, is composed of a large software simulator package with up to 4000 words of microprogramming, but with no hardware modification or augmentation of data flow. On the other hand, the IBM 7080 emulator on the IBM System/360, Model 65, is heavily dependent on the hardware modification to the Model 65 data flow. It uses up to 800 words of microcode, plus a large software simulator.

Figure 3.3 shows a comparative analysis of the internal performance of some emulators and simulators written in IBM[8]. The tail of the arrow shows

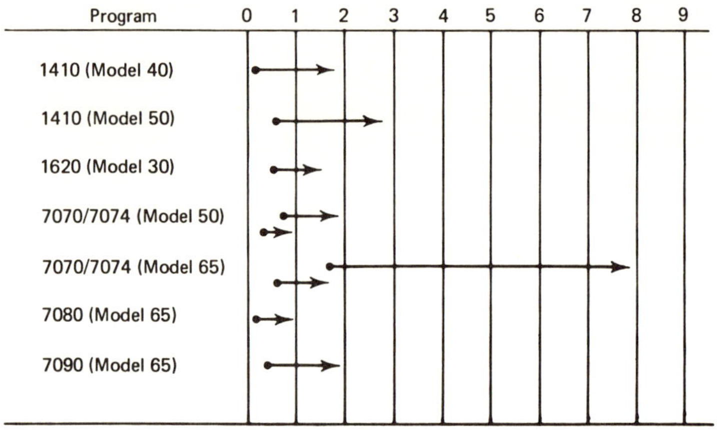

Figure 3.3

the performance of a strict software simulation package in relation to the target system's execution time; the head of the arrow shows the achieved performance gains realized from a specific emulator implementation. Generally, the emulator offers at least a one-to-one throughput as compared to the target system; a typical simulator performs at about one-half the efficiency of the target system.

At this point in the discussion of emulators and their reliance on microprogramming, it would be fruitful to select a specific emulator and treat it as a case study, pointing out some of the compatibility features that were incorporated in the data flow, the selection of a subset of special instructions operative only in the emulator mode, and other problems that were encountered in the design of the emulator. The IBM 7080 emulator on the IBM System/360, Model 65, is selected for that purpose.

The main features of the 7080 emulator on the Model 65 are:

1. The 7080 memory and storage mapping scheme.
2. The instruction interpretive routine (the DIL instruction).
3. The technique developed for internal tape record conversion (from EBCDIC to BCD, and vice versa).
4. The design of over fifty special System/360 instructions intended to emulate 7080 instructions directly.

The obvious problems in designing this emulator are:

1. The difference in instruction set and other architectural characteristics between the target and host systems.
2. The difference in the internal code of the two systems: binary, binary-coded decimal (BCD), hexadecimal, etc.
3. The reconciliation of address or instruction modification, especially when the internal codes of the two systems are so different.
4. The question of how to distinguish between data and object codes in the object code translation.

The 7080 machine is a binary-coded decimal character machine with a fixed instruction format. The 7080 operands are of variable field length. Each data field occupies the necessary number of contiguous BCD digits in the 7080 memory, and the data field is terminated by either a record mark or a group mark, and it is addressed by the decimal address of the rightmost digit of the data field. A binary-coded decimal digit occupies six bits of core storage in the 7080 memory (BA8421). These digits (six bits) are mapped into the six right-most bits of the System/360 eight-bit bytes, as shown in Fig. 3.4.

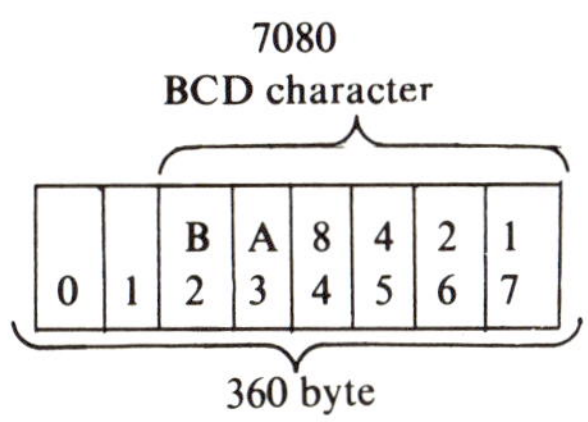

Figure 3.4

The five-character, System 7080 instruction format is mapped into System/360 double words, where the 7080 word is mapped into bytes 3, 4, 5, 6, and 7 of the double word. Similarly, a series of ten contiguous 7080 characters with addresses ending between zero and nine are mapped onto 16 contiguous bytes of 360 storage, as shown in Fig. 3.5[11]. This address mapping is not unique, but it is the simplest address translation possible.

Emulated 7080 address	0	0	0	0 A	1 W	2 O	3 R	4 D	0	0	0	5 A	6 W	7 O	8 R	9 D
System/360 address	0	1	2	3	4	5	6	7	8	9	A	B	C	D	E	F

Figure 3.5 7080 word mapping.

In addition to memory and character mapping schemes, a powerful instruction called *Do interpretive loop*, or simply DIL, had to be implemented. This most important of all the special instructions developed replaces a whole subroutine that a pure software simulator would use to:

1. Fetch the simulated instruction counter (IC).
2. Convert the IC value to the address of the storage cell which continues the next instruction to be interpreted.
3. Fetch this next subject instruction.
4. Update and restore the simulated IC.
5. Perform any indexing required for the subject instruction.
6. Convert the effective address obtained to the address of the storage cell containing the subject operand.
7. Interpret the subject operation code and branch to the correct E-type emulation routine for the decoded instruction.

All this is done in two microseconds and is repeated for every machine instruction simulated.

Two other terminations of the DIL instructions can occur. Both terminations are a result of setting a program-controlled hardware switch, and they occur prior to fetching the next 7080 instruction and associated functions. If the DIL *interrupt trigger* is on, the DIL instruction replaces the System/360 IC with the contents of the floating-point register 2. If the DIL *count trigger* is on and the DIL count (byte zero of general purpose register) is zero, the DIL instruction replaces the System/360 IC with the contents of the floating-point register 2 plus eight.

The 7080 branch instruction simulation is aided by the addition of special System/360 instruction called *branch if.* This instruction tests the condition specified by the 7080. If the test is successful, it will move the effective address into the emulated IC and then do the DIL interpretive routine. Otherwise, it will ignore the specified condition and branch immediately to the DIL routine.

A number of other instructions have been implemented to emulate some 7080 instructions that have high usage and are difficult to simulate, or which take a large portion of time when simulated by software. Among these special instructions are the I/O conversion of information transferred between the I/O unit and the buffer to the emulated machine core.

Basically, then, the structure of the CPU simulator is very simple. The central component is a basic interpretive routine which extracts the next instruction to be simulated, updates the simulated instruction counter, decides the operation code of the simulated instruction, performs all address computations according to a well-defined mapping algorithm, and then branches to a particular operation code simulation routine. Each operation code simulation routine manipulates the simulated storage, registers, and status triggers in the same manner as for the target machine. At the end of each operation code simulation, if no exceptional conditions or branch conditions are present, a transfer to the beginning of the basic interpretive routine is executed.

There are a large number of additional macros that can be identified in each case; if implemented in microcode, these macros would speed the simulation process and minimize the core required by the simulator program. Each of these instructions and data-flow modifications are evaluated separately and are justified on the basis of their performance/cost contributions.

Similarly, the design of the I/O simulator is greatly influenced by the fundamental design considerations and by the technological differences between the host machine's and the target machine's I/O devices. This simulation activity is also affected by the data formatting, radix, and also by the extent of any information code translation. Generally, it is in this area of the simulation activity that hardware augmentation and simulation functions pay the greatest dividends.

The physical arrangement of the CPU interpretive routine and other microprograms that are part of the simulators, as well as of the physical I/O interface, code translation, and address translation paths, are illustrated in Figs. 3.6 and 3.7.[9]

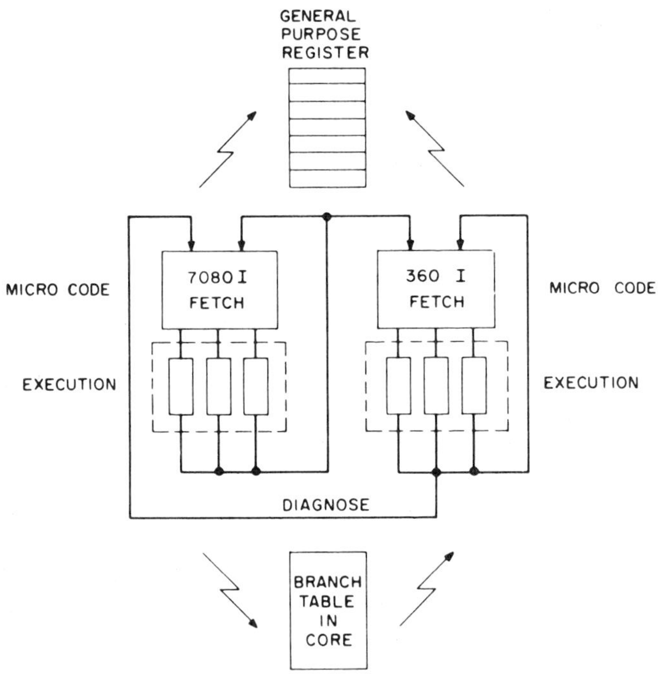

Figure 3.6 Communication relationship between System/360 instruction set microcode and emulator instruction set microcode.

The linkage between the emulator software package to the microprogrammed portion of the simulation is accomplished by the privileged operation, Diagnose (op code 83), which contains the starting address of the microroutine plus some switch setting which provides additional functions specified by the programmer. Linkage data are stored in the general purpose registers. The

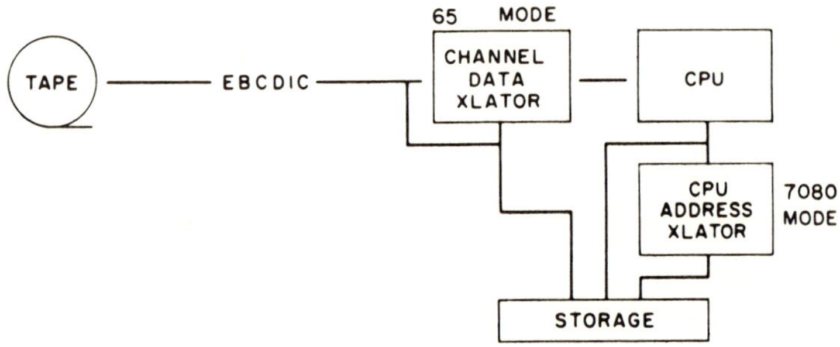

Figure 3.7 Data and address translation paths in Model 65, 7080 emulator, implemented in hardware.

transfer of control from the microprogram back to the software package of the simulator is accomplished by picking the address of the next System/360 instruction from the PSW (program status word) and then executing System/360 instruction fetch.

3.4 CONVERSATIONAL PROGRAMMING SYSTEM

The conversational programming system (CPS) is a time-sharing system that was developed for IBM by Allen-Babcock Computing, Inc. and runs under OS/360, option 2. The system can be described as an interpretive processor using an extended PL/1 language for both programming and control. The system was designed to run on the IBM System/360, Model 50, and it uses microprogramming extensively.

IBM's primary objectives from this project were: (1) to advance the understanding of how to employ PL/1 in-time sharing environment, (2) to advance the art of using microprogramming, and (3) to appraise the importance of specialized microprograms to enhance particular programming systems.

One of the basic concepts of CPS is the use of microprogramming to improve the performance of the interpreter in an attempt to regain some of the throughput efficiency and to maintain some of the flexibility and convenience of the conversational environment.

Figure 3.8 shows the hardware configuration in the system. It consists of the System/360, Model 50, with 128k bytes of the 2-microsecond storage, and 1024k bytes (box 2381-1) of core storage with the 8-microsecond speeds. It also contains disk storage and a variety of I/O devices, including 62 terminals of the Model 2741 type.

Storage is allocated to the users dynamically, one page at a time. Each page contains 4096 bytes of storage; 8000 bytes are reserved for the CPS system program itself.

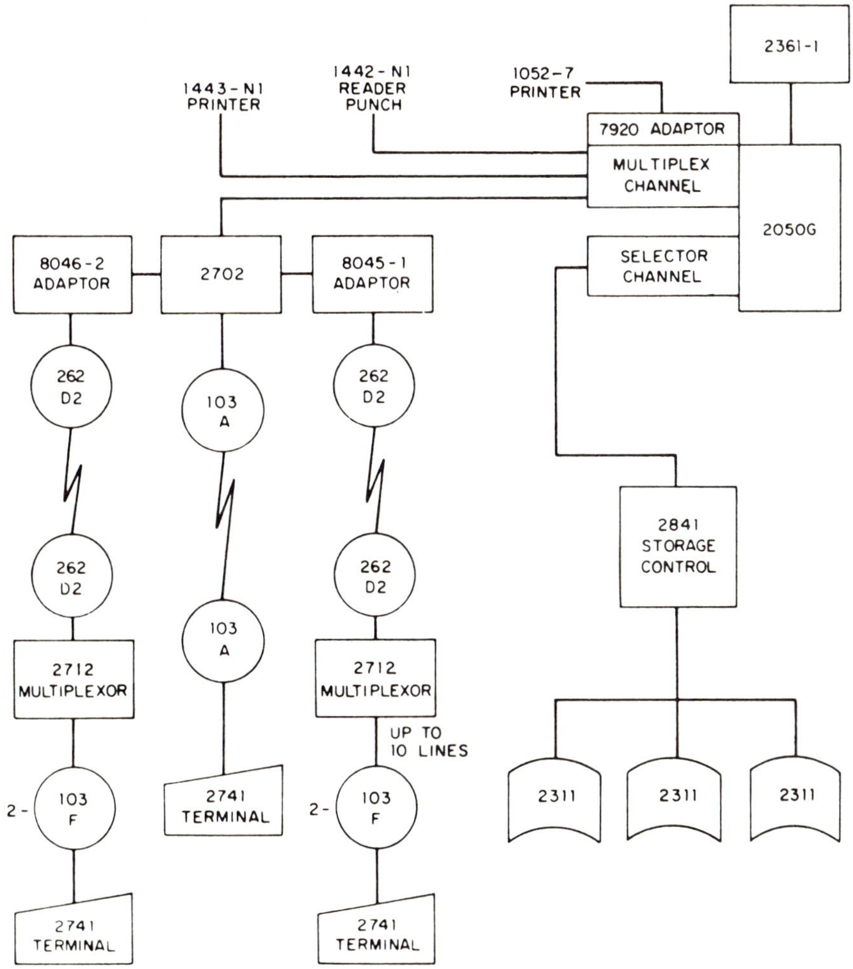

Figure 3.8 System/360 configuration required for CPS.

The novel feature of this system and the area in which most of the inventive efforts have occured is the way in which the system accepts, preprocesses, and interprets the user's PL/1 statements and real-time commands. To accomplish this, seventeen special machine instructions were microprogrammed to augment the standard System/360 instruction set. These additional instructions can be divided into three categories:

1. Instructions to reduce red tape.
2. The decimal floating-point instructions which would produce the results of the operations in a meaningful decimal form immediately after they are calculated.
3. A powerful superinstruction, EVAL, which would accept a PL/1 state-

ment which has been translated from a normal PL/1 form into polish notation, consisting of a string of variable names and operations appearing in the order in which they are to be executed. Once the decomposition of the statement is finished, control is transferred to the EVAL microroutine for execution.

Table 3.2 gives the list of the seventeen new special instructions, their functions, and the number of ROS words required for their execution. These are grouped into the following classes:

Table 3.2 NEW SET OF OPERATION CODES FOR THE CPS SYSTEM

Code Name	*Name or Function*			*Words of ROS*
EVAL	Evaluate an arithmetic expression			184
CHB	One-byte	List search		
CHBE	item	operations		
CHH	Two-byte			
CHHE	item			43
LDM	Load multiple floating-point			
STDM	Store multiple floating-point			12
LUM	Load under mask			13
STUM	Store under mask			12
TAC	Table-ANDed characters		Extensions of	
TOC	Table-ORed characters		translate and test	41
BS	Binary search			20
ADD, SUB, COMP, MULT, DIV	Floating-decimal instructions			320
			Total ROS words	645

1. The five floating-decimal operations are ADD, SUB, COMP, MULT, and DIV. They have the same format and specifications as their comparable IBM System/360 floating-point Hex (hexadecimal) double precision operations, except that the mantissa is made up of decimal digits rather than Hex digits. One problem that exists here is that the operands are not checked for valid decimal digits in the fractions. If invalid digits are used as input or operands, the operation is executed with no interrupts or machine check, and the results are unpredictable. These five instructions require 320 words of ROS storage space.

2. Four list-search microroutines were implemented to execute the four new added instructions in this area (CHB, CHBE, CHH, CHHE). Two of these instructions are specified for equal or greater comparison on ordered lists, and the

remaining two instructions are specified for strict equality on nonordered lists. These operations search a list of chained entries and compare a specified field against a one- or two-byte argument. Search continues until the end of the list is reached, if the comparison criterion is not reached after a specified number of items have been searched or when the requested comparison is satisfied.

3. The two instructions, LDM (load multiple floating-point) and STDM (store multiple floating-point), are the floating-point equivalents of the fixed-point instructions, load multiple and store multiple, respectively. The format used is the System/360 RX format shown in Fig. 3.9.

LDM (96)

	R1	R3	B2	D2

STDM (95)

Figure 3.9

R1 must be the first legal floating-point register (0, 2, 4, 6) to be stored, and R3 is the last floating-point register (with wrap-around from 6 to 0) to be loaded. If R1 = R3, then the contents of one register are transmitted. D2 (B2) is the address in main storage in which or from which the first word is transmitted.

4. The two new instructions, load under mask, LUM (A8), and store under mask, STUM (A9), were added to allow the transfer of bytes of storage between the main storage and byte positions in a general register under the control of a mask. In the store under mask, for example, if the mask is 1011, three bytes will be stored from R*i* (0–7), R*i* (16–23), and R*i* (24–31) into storage locations D2(B2), D2(B2) + 1, and D2(H2) + 2, respectively. The instruction format is shown in Fig. 3.10, where the M is the bit mask; R1 is the general register from which or to which the byte(s) is stored. D2 (B2) gives the address of the first position in storage from which or to which the corresponding byte transfer occurs.

A8 or A9	M	R1	B2	D2

Figure 3.10

5. The new added binary-search instruction searches on an order table of full words until the specified conditions for the search are satisfied. The search will stop when it has found a match or when it exhausts the search in the allowable range.

6. Two other extensions to the translate and test instruction, namely, the table-And'ed characters (TAC) and the table-(EXCLUSIVE)-Or'ed characters (TOC), were also implemented. They both use the eight bits of the characters in operand 1 as arguments to reference function bytes in the function table

addressed by operand 2. However, when they find an "interesting" character, they do not stop as the System/360 translate and test instruction does. They do the following.

The TAC instruction ANDs the function byte with the mask in GPR2 (0–7). If any of the bits that are 1 in the mask are also 1 in the function byte, then that character in the string is called "interesting." Similarly in the TOC instruction, the mask bits are EXCLUSIVE-ORed with the function byte bits. If the two match perfectly the character is the string is called "interesting." When an "interesting" character is found, the byte is interred in bits (0–7) with its address in bits (8–31) of the first available word in the table of "interesting" functions. Figure 3.11 shows the instruction format used for the two instructions, where:

OP	L	B1	D1	B2	D2

Figure 3.11

L	=	number of characteristics to examine *after* the first;
B1 + D1	=	address of the character string—operand 1;
B2 + D2	=	address of the function table—operand 2;
GPR2—bits (0–7)	=	mask for determining which character is interesting;
bits (8–31)	=	address of full word table of interesting characters stored into by the instruction.

7. The last and most powerful of the seventeen new instructions is the EVAL instruction. It performs interpretive evaluation of expressions which are written in polish string form. EVAL performs all necessary housekeeping for the operation of the push down stack. It also maintains control of the execution of the statement, calling upon other microprograms in the ROS memory to multiply, compare, add, etc., and upon some software routines to evaluate special functions, until a certain termination character appears in the polish string. When the EVAL operation is terminated, control is transferred to the next machine instruction following EVAL.

The EVAL instruction can be illustrated by following the execution of the following PL/1 statement (Fig. 3.12):

$$X = A + B + C * D **(E + F)$$

The statement is first decomposed to the internal polish parenthesis-free notation. The operators appear in the order in which they are to be executed, and the operands are implied by the order of preceding the symbol. In the CPS system, variables are designated by single letters, so that the final polish form of this expression is not ambiguous. The preceding statement is decomposed to:

$$XAB + CDEF + \uparrow * + = ;$$

After this decomposition is finished, control is shifted to EVAL for its execution. EVAL scans this statement from left to right, and when it comes to an operator, it executes it on the previous two entities. When it comes to a variable, it adds it to a push-down list. The push-down list is headed by the four floating-point registers of the Model 50. Any overflow from these four goes to memory.

The execution is explained by showing where each item of data is stored and what instructions are invoked at what time. Only the evaluation of the expression on the right will be explained because this illustrates the concept without getting into needless detail. In the action of the push-down list, it is sometimes pushed down and sometimes popped up; and sometimes, an item of data is merely stored at the top of the list.

			Floating-Point Registers of Model 50			
Step	*Character Read*	*Action*	*1*	*2*	*3*	*4*
1	A	push A	A	empty	empty	empty
2	B	push B	A	B	empty	empty
3	+	call ADD:pop:store	A + B	empty	empty	empty
4	C	push C	A + B	C	empty	empty
5	D	push D	A + B	C	D	empty
6	E	push E	A + B	C	D	E
7	F	push F	F	C	D	E A
8	+	call ADD:pop:store	empty	C	D	E + F A
9	↑	call EXP:pop:store	empty	C	D (E + F)	empty A
10	*	call MULT:pop:store	empty	C * D (E + F)	empty	empty A
11	+	restore A + B	A + B	C * D (E + F)	empty	empty
12		call ADD:pop:store	A + B + C * D (E + F)	empty	empty	empty

Note: At step 3, EVAL involves another microprogrammed instruction ADD. At Step 8, the overflow A + B is not restored. It has been found to be more efficient to flag the register as empty and restore the contents only when needed—in this case, in step 11.

Figure 3.12 Evaluation of AB + CDEF + ↑ * + by EVAL.

In step 9 of Fig. 3-12, EVAL calls a program to carry out exponentiation, and this program, in turn, invokes the microprogram MULT. The facility of being able to call a program from EVAL is used to evaluate expressions involving subscripted variables and functions defined by the user. EVAL can be used recursively so that it can call a program that uses EVAL.

Provision is made for the unary minus which is needed in $-A + B$, although not in $B - A$. This is treated as a different operator from the binary minus. The unary operators operate on the top member of the push-down list. Binary operators operate on the top two members, and subroutines with arguments operate on the top n members of the push-down list.

The addresses used by EVAL are compressed when the statements are put in internal form. Where needed, they are expanded to full-fledged addresses by a scheme much like the dynamic relocate of the Model 67. This improves the speed of EVAL and allows the data in memory to be dynamically relocated in order to tidy up the chaotic memory allocation that conversational operation creates.

The overall performance gain that microprogramming will bring to CPS has not been released. A factor of five has been estimated by Babcock. Eventually, a good assessment of this will be available because all the microprogramming is being paralleled by ordinary subroutines to allow checkout of code on a Model 40. Therefore, the system can and will be run with and without microprogramming to get a comparison.

Figure 3.13 shows the way the EVAL microroutine links with the System/360 microprogram. It is important to recognize that EVAL can be nested so that it can call a program which also uses EVAL[9].

Another application similar to CPS is that of building experimental processing systems for algorithmic languages. This type of application opens a new dimension in the usage of microprogramming for processing of interpretive codes.

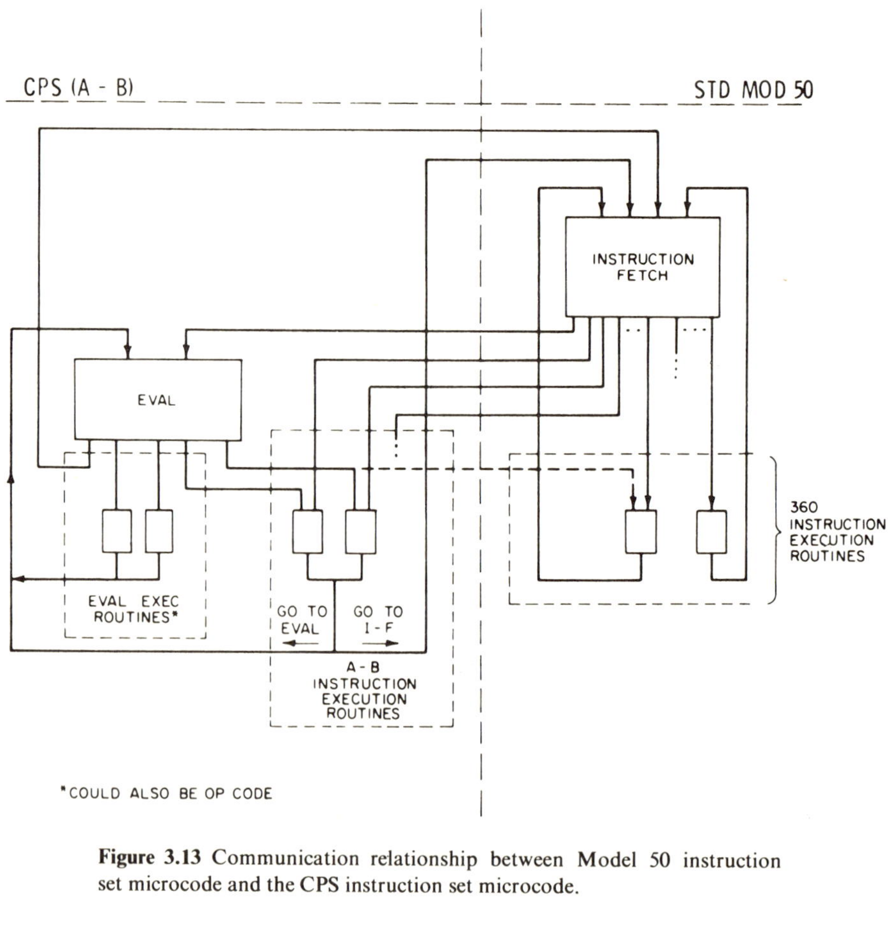

Figure 3.13 Communication relationship between Model 50 instruction set microcode and the CPS instruction set microcode.

Hawryszkiewycz[24] has written on the use of microprogramming support for problem-oriented languages. He experimented with a special set of analog computer functions and reported a three-to-one speedup in this simulation as a result of microprogramming this special set.

Lawson[25] discussed the implications of the inclusion of a writable control storage in the fourth generation hardware. He also discussed some of the considerations required in evaluating various forms of instruction streams. Six various types of instruction sets were discussed and compared with respect to their conciseness, complexity, dynamic capability, and flexibility.

3.5 LARGE-APERTURE SEISMIC-ARRAY SIGNAL PROCESSING—LASA

An area of application which is presently receiving considerable attention is the sensor data processing. An example of this is the signal-array processing for the large-aperture seismic analysis (LASA). In this section we will examine the system's requirements and the programming requirements, and we will review how microprogramming could be used and was used in a specific array processing system[16].

The primary functions of LASA signal processing are:

1. To detect, edit, and record the seismic events such as earthquakes, tremors, shockwaves, or underground nuclear explosions occurring within a specified azimuthal sector scanning and within a specified radius from LASA.

This function must be executed in real time, through repeated and periodic sampling of the seismometer outputs and analyzing this data. This real-time analysis will produce an event-tape recording which will be analyzed for the purpose of classification and warning.

2. The second function, then, is the classification of these events. While this function is allowed to be done off-line, it must be done fast to retain its value and to help avoid disastrous consequences.

The array processing system consists of 21 subarrays covering a surveillance area 200 km in diameter. Each subarray contains 25 seismometers which are distributed over an area 7 km in diameter. Most of the subarrays are clustered in the center of this surveillance region. The LASA systems receive the output of every seismometer of every subarray on a real-time basis, once every 50 millisecond, to be processed.

Figure 3.14 shows a block diagram of the LASA processing system. It consists of a subarray beam former (SABF) located at the LASA site. The SABF receives its inputs from the 21 subarrays in a multiplexed serial format. Each subarray transmits 25 words of 13 data bits, one sign bit, and one parity bit. Actually, each subarray has one padded seismometer which is not really in-

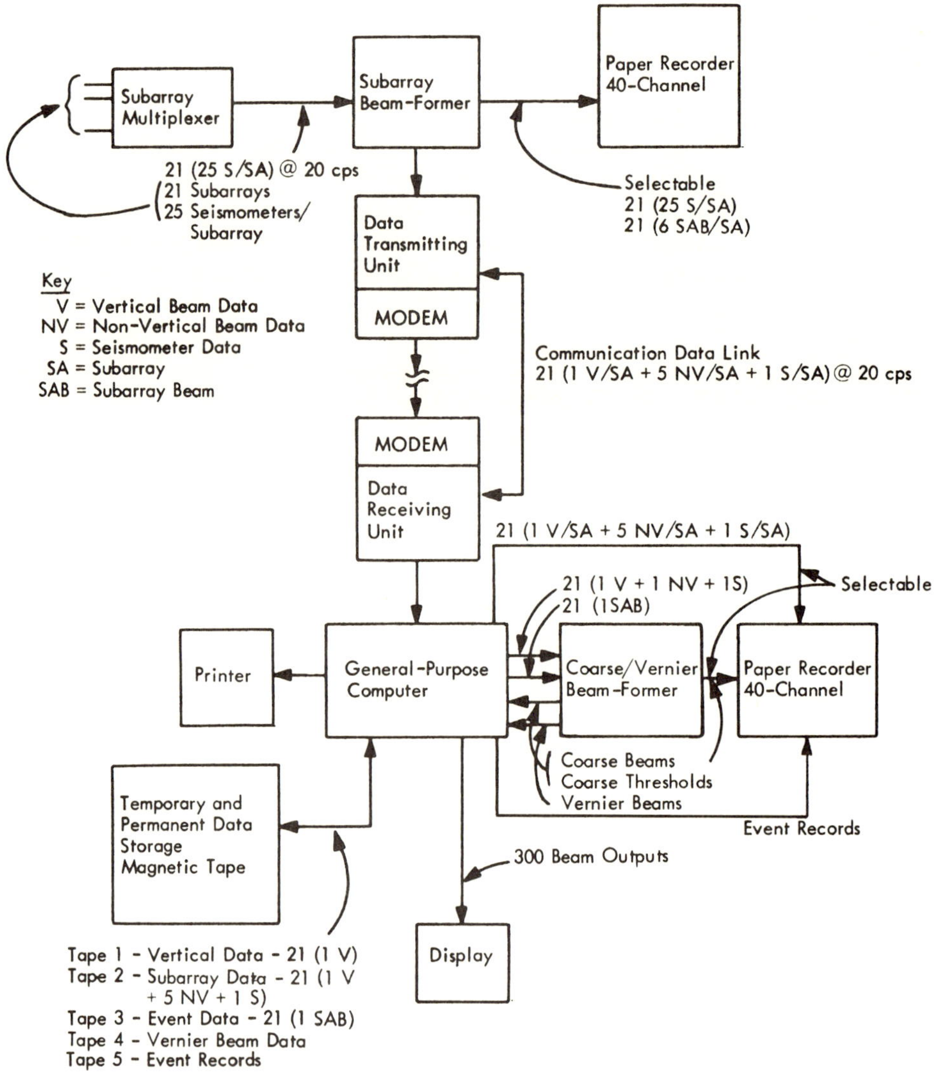

Figure 3.14 LASA processing system functional block diagram.

volved in this beam-forming process. Rather, this padded seismometer output is used to record events with very high amplitudes. The subarray beam former reduces the 21 × 25 inputs to 6 beams for each of the 21 subarrays, and it transmits the output to the main signal-processing site.

The general purpose computer receives this information on a real-time basis and begins its analysis on the data for the purpose of detecting, editing, and finally recording the location and the intensity of the seismic event. The input to the general purpose, data processing system is filtered, rectified, and integrated. Also, the set of 21 padded seismometer outputs is used to localize events which are strong enough to overload the remaining unpadded 24 seismometers in each subarray. This localization is also time dependent. By recording the arrival times of such an event at the various subarrays and fitting a "least square" planar wavefront to this data, the location and the time of the event can be estimated.

Once an event is detected and localized, it must be recorded on the event tape for further classification by the vernier beam former (VBF) which allows more accurate event location by finer signal waveform model fitting and examination.

The data processing system is slave to this process; it is required to repeat the same analysis once every 50 milliseconds. This type of data processing, could, of course, be done by conventional machine programming, provided that the chosen system is fast enough to accommodate the real-time demands. However, such a general purpose system is too expensive to handle this function. The other option here is to design a special data processing system to carry out this specified and highly specialized task. This option is also too expensive, since the whole cost of such a development must be borne by the few systems needed to accomplish this mission.

The third and most practical alternative was to select a microprogrammable, general purpose, data processing system, and microprogram the critical time-consuming loops of the different data processing subroutines.

Once the decision to select this third and most practical alternative is made, a new, well-defined set of problems is created. Among these are the following:

1. Select the general purpose microprogrammable systems. These candidate systems must meet the general storage, I/O, reliability, and cost requirements. The systems control storage must have sufficient vacant locations to permit the storage of some additional functions.

2. Analyze the function to be processed. This includes flow-charting each process to identify the most frequently utilized loops and subroutines in each process. This analysis must also include computing the total executing time for each function and the executing time for the total process when performed by the conventional programming steps.

3. Select the subroutines in the process that will be microprogrammed on each candidate data processing system. The microroutines are simulated to check their accuracy and timing. This step must be based on a thorough knowledge of the process and the internal structure of each of the proposed data processing systems. This is because any performance improvement, as was discussed earlier in the chapter, is dependent on the internal data flow, the parallelism, and the width and timing of the CPU-main storage access.

4. The final selection of the microroutines that will be incorporated in the control section of any data processor must favor the ones with maximum performance improvement over the conventional programming requirements. These new microroutine storage requirements must not exceed the available

ROS vacancies; otherwise, an appreciable cost can be anticipated if the ROS storage is to be increased.

5. The total process, including the microroutines, will have to be simulated for each candidate system. This step will test the integrated system's capability to handle the real time processing requirements.

6. The final selection is based on the cost of the system, its performance, its modularity, and system availability. The latter characteristic is to ensure that the system can continue to operate successfully despite the presence of some malfunction.

The System/360 Models 40 and 50 were carefully evaluated in one feasibility study conducted by the IBM Federal Systems Division. The complete process was programmed and timed for both models. The Model 40 took 545 milliseconds to execute this program, while the Model 50 executed it in 199 milliseconds. Both systems failed to meet the 50 millisecond time requirement. Microprogramming was therefore proposed as a means to achieve the required real time processing. The Model 50 was soon eliminated from this study since cost was prohibitive, and the remainder of the project was concentrated on bringing the Model 40 to perform the function within the specified cost/performance limits. The final recommendation was to microprogram some care-

Loop					Time M50	Time M40
SAB Loop →	1.	BBA	LR	R14, R4	2.50	7.50
	2.		SR	R0, R0	3.25	7.50
S Loop →	3.	BBB	L	R15, 0(R8, R14)	4.50	13.13
→	4.		BXH	R15, R2, BBC	5.00	16.26
50% time	5.		AR	R15, R6	1.63[a]	3.75[a]
→	6.	BBC	A	R0, 0(0, R15)	4.00	11.88
→	7.		BXH	R14, R10, BBB	5.50[b]	16.26
	8.		ST	R0, 0(0, R1)	4.00	12.50
	9.		LA	R1, 4(0, R1)	2.75	10.00
	10.		AR	R2, R4	3.25	7.50
	11.		AR	R8, R4	3.25	7.50
→	12.		BCT	R7, BBA	4.50	10.63
			Done			

1. SAB loop—instructions 1, 2, 8–12 inclusive.
2. S loop—instructions 3–7 inclusive.

$$\text{Execution time} = \begin{matrix} \text{SAB} * (22.50 + \text{S} * 20.63) \text{ microseconds for Model 50} \\ \text{SAB} * (63.13 + \text{S} * 56.28) \text{ microseconds for Model 40} \end{matrix}$$

where

SAB = total number of beams formed
S = number of seismometers (or other inputs) for each beam

a. AR instruction executed approximately 50% of the time.
b. When branch does not occur, 1.0 microseconds are subtracted from the SAB loop for the Model 50.

Figure 3.15 Beam form filtered data.

fully selected loops in the LASA process and to augment the Model 40 data flow if the cost/performance gains could justify such a step.

The process of selection of the specialized instructions and architectural extensions in the host system is similar to emulator design. It is based on the designer's ability to identify the critical time-consuming loops in the process. Figure 3.15 illustrates this evaluation process for the form filtered data routine in the larger-aperture, seismic-array analysis.

Three options were considered:

1. A minimum strategy would be to microprogram the S loop (instructions 3–7) which is executed twelve times for each beam.

2. Another option was to microprogram the SAB loop (instructions 1, 2, 8–12 inclusive).

3. A third solution would be to take this whole routine (instructions 1–12) and code it into the ROS control.

One area where considerable time was being consumed was in the filtering of the signals received from the seismometers. This involves a large number of multiplications and summations. To do this by normal computer instructions becomes a time-consuming task. It can be speeded up considerably by microprogramming. However, the LASA data processing function was further speeded up by augmenting the data flow of the general purpose processor with a special purpose hardware function named the *multiplier-summation processor* (SUMP). This is equivalent to adding a new data path in the system which can operate in parallel with the base processor. The SUMP hardware was designed to perform the function:

$$\sum_{i=1}^{n} X_i \cdot Y_i$$

This specialized function, coupled with the microprogramming, resulted in a system performance improvement of ten to fifteen times. The final system was capable of executing the total LASA process in 49 milliseconds, slightly less than the process requirements.

3.6 MICRODIAGNOSTICS

As the complexity of the data processing systems increases, the problem of servicing and maintaining them increases also. This situation is further aggravated by the ever increasing number of data processing systems finding their way to the market place. We are rapidly reaching the point when the computer industry will be unable to hire and train the needed number of technically qualified people to fill the demand. Even if such a field force is made available, the cost of training them to any level of proficiency will be economically prohibitive.

The data processing system today is used in the midst of the decision-making process; it is an integral part of the management and control tools which affect the whole decision- and policy-making process. As such, any undetected error or failure due to the system could have catastrophic or grave economic consequences. Hence, we find that the availability of the system becomes an important characteristic—as important as speed, throughput, utilization factor, etc. Therefore, it is imperative that the user and designer formulate a sound approach to the development of effective tools for servicing and maintaining the system in order to enhance its effectiveness. Any such approach must include all levels and stages of computer development, from the component design specification, stress analysis, and life testing, to the functional design and, finally, to the system design in the field. Yet, regardless of how effective a job the designer does, the system will still fail. Therefore, an effective diagnostic philosophy for checking, locating, and isolating the errors in the field must be developed.

The diagnostic package must have the following characteristics:

1. It must have a high resolution for detecting an error and isolating it to the smallest replaceable unit in the field.
2. It must be simple to apply so that fewer and less technically trained field engineers can apply it.
3. It must require the minimum operating portion of the system in order to be executed.
4. The error detection and isolation must be performed as rapidly as possible, and their findings must be communicated to the maintenance engineers in a precise, concise, and unambiguous way.
5. The diagnostic package should be complete and selfmanaged to the extent that the maintenance engineer should not have to make a decision as to what package to use and what data path to check.
6. The diagnostic routines must be independent of the system configuration.
7. They must also be independent of the architecture to the extent that any future architectural extensions and modifications could be covered.
8. It must be economical to design and run.
9. It must not impact the resources available to the user.

While it is feasible to design a diagnostic technology which will meet all the preceding requirements, the cost of such a project can be prohibitive. Hence, depending on the machine project, some cost/performance trade-off can be found. For example, a diagnostic routine could be designed to isolate the error to a few replaceable parts rather than to a single replaceable part.

In the past, software routines were written for each specific system. They were found to be unacceptable because such a maintenance package could contain over 100 diagnostic routines, each requiring possibly more than 4k bytes of storage. They are difficult to use since the maintenance engineer would have to be familiar with each section and select the ones he thinks he needs to diagnose a specific situation. This type of diagnostic, for example, was implemented on the IBM 7074. The CPU coverage alone takes over 45,000 machine instructions. It took 7.5 man-years of effort to develop, yet its error detection capability is estimated at 70% and its error diagnostic capability is less than that—this despite the fact that the 7074 had an extensive hardware checking capability and additional 1.5% diagnostic hardware was added to facilitate the diagnostic package.

On some of the System/360 models, the CPUs were fitted with hardware fault locating tests (FLT). However, these FLTs were difficult to use and expensive to generate and implement.

The solution to this problem can be found in microdiagnostics. Here a given data path can be diagnosed by forcing a predetermined sequence of bits (test patterns), which can be stored in core, through the different circuits in the data path. The output from the data path is analyzed in order to select a finer set of test patterns to isolate the error to the smallest replaceable unit. These microdiagnostics could reside in the read-only storage unit or could reside in a writable control storage unit. In the latter case, a small resident microdiagnostic routine must reside in the CPU (in a small ROS) to ensure that the circuitry needed for loading the writable control storage unit is error-free. It may be sufficient for this resident microdiagnostic to detect the error without attempting to isolate it. The main body of the microdiagnostics could be loaded into the writable control storage as independent segments. Each section addresses a logical portion of the hardware such as the adder data path, the shift data path (if it is separate from the adder data path), local storage, and miscellaneous stats.

The superiority of microdiagnostics over software diagnostics is due to many factors; among these are:

1. Microdiagnostics is much closer to the actual hardware; hence, it has a finer resolution in detecting and isolating errors to the smallest replaceable unit. It is feasible to design microdiagnostics with estimated error detection and error isolation capability to two or more replaceable units close to 90%, and an error isolation capability to a single replaceable unit of greater than 80%. This, of course, is dependent upon the circuit technology and packaging density.

2. Microdiagnostics are faster and less costly to produce and store. These advantages again stem from the power of the microinstruction to test the largest data path and thus avoid the great deal of redundant testing made by the software diagnostics. This accounts for the reduced storage space required by the microdiagnostics. It is estimated that the complete microdiagnostic package for the IBM System/360, Model 85 (which has ten times the hardware found in the IBM 7074), could be fitted in an 8k (100 bits wide) ROS storage. As far as speed is concerned, we have shown theoretically in earlier chapters that microroutines outperform their computable software routines by more than a 10 to 1 ratio. While a cost comparison between software and microdiagnostics generation has not been made, it is believed that microdiagnostics can also be less expensive to generate and implement.

3. Microdiagnostics require much less hard core to be operational than do the software diagnostics. The latter may require much in terms of channels, main memory, I/O devices, ROS storage, etc; while in microdiagnostics, all that is required is a trouble-free data path to gate the test pattern into the ROS data register.

4. Because of their speed and their finer control in the selection of the data path to be tested, microdiagnostics could be interleaved with the user program to check out a functional unit that is found to be nonoperational during any machine cycle. Even in the environment of failure, microdiagnostics could aid the instruction re-try capability.

There are several disadvantages to microdiagnostics. For instance, they are unable to catch intermittent failures and cannot check to see if the machine meets its architectural specifications. This latter disadvantage could be considered an advantage also, depending on the specific system requirements.

Every indication shows a great potential and bright future for microdiagnostics. This prediction and this concept are aided with the increasing interest in writable control storage. The WCS could further reduce the cost of storage of the microdiagnostics and thus become an even more attractive diagnostic tool to apply.

3.7 AN EXPERIMENTAL TIME-SHARING SYSTEM

The main features of this experimental System/360, Model 40, system have been described by Hellerman and Hoernes[17] and by Lindquist, Seeber, and Comeau[18]. The main objective of this experiment is to modify an existing System/360, Model 40, to provide the essential functions required by time sharing.

The principal area of modification was to incorporate an associative storage as a device to translate from the user's virtual page number to a physical page. This would hopefully relieve an otherwise big overhead factor. Through this process of segmenting the main storage into pages and swapping pages between available main storage and disk storage, and through time segmentation, any one of the maximum allowable number of users would be unaware that somebody else is using the system. Indeed, he would think that he has virtually the whole system to himself. This type of system was therefore given the name *virtual system* or *virtual machine.* This particular system was designed to service up to fifteen users at one time.

The inclusion of the associative storage in the system resulted in a system that was architecturally modified to function in one of three modes of operation selected by a special mode switch setting. These three modes are: the standard System/360, Model 40, original mode; the time-sharing mode; and a standard System/360, Model 40, with a 64-word by 16-bit associative memory box that can be used in solving any application problem. To accomplish all this, the hardware extensions were accompanied by compatible microprogram modifications and additions which will do the following:

1. Facilitate the transfer of data between the CPU and the associative memory box. This is accomplished by adding new SI type of instruction to the System/360 repertoire.
2. This AM instruction took approximately 30 microinstructions to implement. The modified SI instruction is shown in Fig. 3.16, where the A0, a

0	4	7 8	15 16	20	24	28 31
A	0	Operation	R1	R2	R3	

Figure 3.16

standard System/360 invalid operation code, has been modified to imply the new associative memory operation. However, the actual function to be performed by the association's memory is indicated by the proper setting of some bits in the AM operation field bits 8–15.

Bit 8:	Would be interpreted as a command to clear all match indicators.
Bit 11:	Multiple.
Bit 12:	Read physical address.
Bit 15:	Clear match indicator bit of selected word.
Bits 9, 10, and 13:	Imply interrogate for occupied words, interrogate for vacant word, or write, in that order.

Bits 9, 10, and 13 are further decoded to determine the condition under which the match indicators of the particular associative memory word will be set. This new validated (A0) operation has, in reality, provided a host of machine operation codes or commands which are made available by encoding them into another part of the modified SI instruction format.

The R1, R2, and R3 fields in the associative memory instruction format (Fig. 3.16) refer to three general purpose registers used for the AM operations. The last bits, 29–31, are not used.

The microprogram modification must obviously start from the point in the standard, Model 40, I-fetch routine where the (A0) operation code was signaled to be an invalid operation trap and would lead to the program interrupt handling if and when it is encountered. This has to be changed now so that it will be recognized as a valid AM operation if the mode switch is set in the AM or in the time-sharing mode, and be recognized as an invalid operation if the switch is set in the standard Model 40 mode. In the time-sharing mode, this AM instruction is privileged and can only be used when the CPU is in the supervisor state.

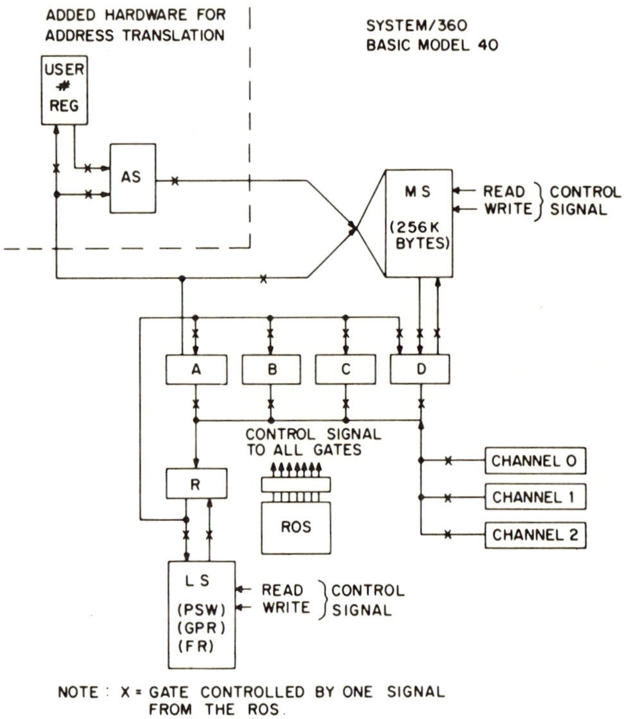

Figure 3.17 Flow diagram of the experimental system.

Other changes to the microroutines include the addition of about 30 microinstructions for interrupt processing and about 82 for operand pretesting. The interrupt processing involves saving the PSW, the sixteen general purpose registers, and the four hardware registers. The major modifications were made to ensure that an instruction which has been started must be completed. To ensure this completion, especially in the case of variable-field-length instructions crossing a one-page boundary and continuing on another page assigned to the same user, these operands are later pretested to ascertain that the pages are all in core to ensure successful conclusion of the variable-field-length instruction execution. Figure 3.17 shows a simplified diagram of the Model 40 hardware modification.

REVIEW QUESTIONS

1. Discuss the concept of upward and downward compatibility.
2. Discuss the different classes of microprogramming application and the justification for each class.

3. (a) Define the terms *simulation* and *emulation*.
   (b) What are their similarities and differences?
   (c) Discuss the basis for their justification.

4. What are the main difficulties in code-to-code translation for program conversion?

5. What are the four main functions in simulation and emulation routines?

6. List all possible cost/performance improvements that can result from microprogramming an algorithm. Assume that the microroutine can be accommodated in existing ROS and that no data flow augmentation is required.

7. What is the contribution of each of the following machine functions to the overall performance realized from microprogramming a mathematical routine (such as matrix multiplication)?
   (a) Multi-data path structure
   (b) Main storage to CPU cycle time ratio
   (c) The ratio of the width of main storage interface bus to the width of the main arithmetic and logic functions in the CPU

8. Same as Question 7, except the microprogramming application involves bit and byte manipulations.

REFERENCES

1. Wilkes, M. V., "The Best Way to Design an Automatic Calculating Machine," Report of Manchester University Computer Inaugural Conference, Manchester, England (July, 1951).

2. Wilkes, M. V., "Microprogramming," *Proc. Eastern Joint Computer Conference* (1958) pp. 18–20.

3. Beck, L., and Koeler, F., "The C-8401 Data Processor," *Datamation 5*, No. 2 (February, 1964), pp. 33–35.

4. "A Unified Information Processing Telemetry System," Adaptable Telemetry Conference, Goddard Space Flight Center (Feb. 15–16, 1966).

5. Jeans, Hans, "Possible Performance Improvement by Microprogramming for System/360 Models 40, 50, and 65," IBM internal report.

6. Martin, H. M., "System/360 Conversion Aide," IBM System Development Division, Kingston, New York.

7. Baron, R. G., Vanderkulk, W., and Lorenz, S. D., "A LASA Signal Processing System," IBM Federal Systems Division, internal report, Bethesda, Maryland (November, 1965).

8. Owens, J. T., "Designing An Emulator," *IBM Technical Report No. TR21-232.* SDD Division, Kingston, New York.

9. Bush, G. H., "Microprogramming," *IBM Technical Report No. TR00-1581.* IBM-SDD Division, Poughkeepsie, New York (March 7, 1967).

10. Rochester, Nathaniel, "Conversational Programming System, Progress Report."

11. Boeving, R. W., "System/360-7080 Emulator Program Development," *IBM Technical Report No. TR21-213.* SDD Division, Kingston, New York.

12. Patzer, W. J., and Vandling, G. C., "Aerospace Systems Implications of Microprogramming." IBM internal report.

13. Melbourne, A. J., and Pugmire, J. M., "A Small Computer for the Direct Processing of Fortran Statements," *Comput. J.*, Vol. 8 (April, 1965), pp. 24–27.

14. Lawson, H. W., Jr., "Programming Language-Oriented Instruction Streams," *IEEE Transactions on Computers*, Vol. C17, No. 5 (May, 1968), pp. 476–485.

15. Weber, H. A., "Microprogrammed Implementation of Euler on IBM System/360 Model 30," *Commun. ACM*, Vol. 10 (September, 1967), pp. 549–558.

16. "Large Aperture Seismic Array Signal Processing Study," *IBM Final Report*—prepared for the Advanced Research Projects Agency, Washington, D. C., Contract No. SD-196 (July 15, 1965).

17. Hoernes, G. E., and Hellerman, L., "An Experimental 360/40 For Time Sharing," *Datamation* (April, 1968).

18. Lindquist, A. B., Seeber, R. R., and Comcau, L. W., "A Time Sharing System Using an Associative Memory," *Proc. IEEE* (December, 1966), pp. 1774–1779.

19. Wilkes, M. V., Renwick, W., and Wheeler, D. J., "The Design of a Control Unit of an Electronic Digital Computer," *Proc. IEEE*, Vol. 105 (1958), p. 121.

20. Opler, A., "Fourth Generation Software," *Datamation 13*, No. 1 (1967), p. 22.

21. Gaines, R. S., "On the Translation of Machine Language Programs," *Communications of the ACM*, Vol. 8, No. 12 (December, 1965).

22. Dellert, G. T., Jr., "A Use of Macros in Translation of Symbolic Assembly Language of One Computer to Another," *Communications of ACM*, Vol. 8, No. 12 (December 1965).

23. Tucker, S. G., "Emulation of Large Systems," *Communications of ACM*, Vol. 8, No. 12 (December, 1965).

24. Hawryszkiewycz, Igor T., "Microprogrammed Control in Problem-Oriented Languages," *IRE Trans. EC-16* (1967), p. 652.

25. Lawson, H. W., Jr., "Programming—Language-Oriented Instruction Streams," *IEEE Transactions on Computers*, Vol. C-17 (1968), p. 476.

WRITABLE CONTROL STORAGE

4

4.1 JUSTIFICATION

Many computer applications require a control storage structure which is easily modifiable at electronic speeds to achieve optimum processing capability. This requirement stems from these three basic facts:

1. The requirements for data processing capability vary from one user to another. Even in the same customer's office, the requirements may vary from one application to another. These individual requirements add up to the most basic definition of a general purpose computer. A computer that suits every programming need with the same simplicity and processing efficiency does not exist.
2. Today's general purpose computers are the product of a series of compromises which result in a balanced performance across the widest range of application within a given price range, but certainly not optimum to any given application.
3. As stated earlier, no two computer installations are alike.

These facts lead to the conclusion that most installations with special applications where the bulk of the data processing activity is dedicated to a repeated application of few programs can greatly benefit from this writable control feature. In the first three computer system generations, such a change was provided by the manufacturer (in the form of ***request for proposals***, RFPs) at a great expense and effort. Today's circuit and memory technology has altered the picture to the extent that a computer system can be designed without any architectural identity of its own; rather it can assume any computer architectural specification by reading the new architecture in the form of microroutines into the writable control storage device. We are referring here to the large-scale integration technology which promises to produce a read/write memory in the desired speed range for a fraction of the cost of any other ROS or read/write technology in the same speed and capacity.

Since the inception of microprogramming, these concepts and this potential were the subjects of much speculation. The proponents would present the advantage, to be challenged by the opponents. Most arguments against writable control storage are based on the belief that such a flexibility would create chaos in the computer installation operation and management.

In his first article on the subject, in 1951, Wilkes astutely observed then that there was probably no real requirement for a variable code machine when he stated:

> "... The matrices may be regarded as very high speed stores holding fixed information. If they could be replaced by an erasable store to which information could be transferred from the main store of the machine when required, we should have a machine with no fixed order code; the programmer would, in fact, be able to choose his order code to suit his own requirements and to change it during the course of the program if he considered it desirable. Such a machine would have a number of fascinating possibilities, but I doubt whether, in view of the amount of equipment it would doubtless involve, its construction could be justified."

Seven years later, Wilkes reiterated his original assessment of the lack of need for such a flexibility: "... the many problems involved in running a computing laboratory are bad enough as it is, without the additional license which would be created by a system of private order code." The concern about the potential chaos that such a flexibility can create is realistic and is shared by most computer designers today. All of these concerns, however, deal with the operations and management of an average, conventional computer center.

Nevertheless, a large number of computer applications exist in which disciplined use of this WCS (writable control storage) feature is justified. A valid argument can be made here for a number of research projects which seek to compare different architectures and evaluate architectural extensions and modifications. A valid argument can also be made for research projects in other areas of the scientific applications, and compilers in which such a feature would allow smooth convergence toward optimum processing capability. The other alternative to this flexibility is simulation; however, simulation tends to be slower and would require data on the parameters involved in the simulation model that are not readily available.

The use of a dynamic microprogram storage device whose contents can be changed at electronic speeds seems to offer quite a number of advantages over present static read-only storage devices. These advantages can be summed up in few words: cost, tailorability, serviceability, engineering changes, reduction of storage requirements, and additional functional capabilities. These are examined in more detail in the following paragraphs.

4.1.1 Cost

The cost-saving argument can be viewed from a number of points. Cost saving may result from the reduction of storage requirements for the system architecture, emulators and other features such as microdiagnostic routines by overlaying portions of the normal control functions with the emulator or microdiagnostic routines. Writable control storage can be justified on the basis of the reduced hardware cost in making engineering changes. It may be justified on the basis of the saving of time and manpower needed for responding to a large percentage of requests for proposals which can be implemented by microcode. Add to this the cost of the special hardware needed for making such changes in the read-only storage control. The trade-offs implied here will continue to provide a challenge for future system designers and implementors.

Another big saving factor in smaller machines is derived by further reducing its development cost through the use of a single memory box for both the information storage (main storage) and control storage. This was wisely exploited in the design of the IBM System/360, Model 25.

4.1.2 Tailorability

The main benefit derived from such an added feature is the increased performance for certain specialized functions. This is especially true in environments where the computer spends the major portion of its processing time executing a given function. It is generally accepted that a saving of five to fifteen times can result from microprogramming this function. This speed advantage tends to favor the logic operations and bit and byte manipulations over straight arithmetic operations such as matrix multiplication. If such special functions are needed for the different missions assigned to the system, and if such a system has to retain its ability to perform its normal functions, then a writable control storage is clearly justifiable.

This is also true for other systems that can be controlled by microcode. An example would be specializing a channel to specific I/O devices by including the control unit functions in the channel. The same argument can be made for the design of remote terminals that are used to communicate with a geographically dispersed large data base system. The architectural capability of that terminal must be specialized to suit the particular terminal user's needs. The terminal can be specialized to accept input data in the available format and convert it to a format acceptable to the central system. A writable control storage device may prove to be the best means to achieve this end. A terminal with a WCS can be made to perform certain edit, format and compute functions too expensive to provide otherwise.

Another area of interest today is building efficient language processors. It has been shown by Weber (6) that these types of programs can achieve an improved performance of one order of magnitude over a software compiler when such a compiler is completely microprogrammed.

Several proposals have been made recently for FORTRAN machines (7, 8). The trend in this area shows that it would be easier and more advantageous to design a microprogram FORTRAN interpreter than designing a FORTRAN compiler or building a FORTRAN hardware system.

4.1.3 Increased System Availability

Availability is defined as the probability that a specific function(s) will be adequately performed within a specified interval of time under specified operating conditions. Availability is a measure of the system's uptime. This is the time during which the system is useful to the user to the maximum time that it is possible for it to be under the user's control. A few factors that degrade the availability characteristics are:

1. The time for diagnosing, locating, and correcting the error. As computer systems become more complex, the present software diagnostic techniques become inadequate. Their ability to locate and isolate errors to the smallest replaceable units is extremely limited. Also, the design of such a function in small processors may be uneconomical. The application of microdiagnostic techniques in which the test patterns are programmed to propagate through the selected data paths has been shown to be superior to and more economical than the conventional hardware fault location technology (FLT) tests. Microdiagnostics has the added capability of isolating the error (through more specialized microdiagnostics) to the smallest replaceable unit possible. If such a capability is added to the system, as was done in the IBM System/360, Model 85, then the time required by the field engineer (an unproductive time to the customer) can be reduced.

2. Another factor that degrades the system availability is the system time required by the designer to make the ever-increasing number of engineering changes. These changes are made faster, with less downtime, if a writable control storage is available to the system.

3. In a multiprocessing environment, the error detection, error isolation, and reconfiguration can be performed automatically at electronic speeds, rather than through the manual reconfiguration procedure.

4. System availability can be further improved by an added architectural feature called self-repairability. This implies that the system has the ability to detect, re-try, isolate, and correct any isolated or clustered transient, inter-

mittent, or solid failures in the system. This reconfiguration implies that the system is partitioned in such a way that any partition could diagnose itself and one or more other subsections. It also implies that, once the error is isolated to a particular function or subfunction, this failing part is reconfigured out of the system and is replaced by an identical function which has been held in reserve. This means that the system must be capable of changing the machine control section to execute the desired function by an alternate approach which utilizes the remaining trouble-free data paths and functions if such is feasible. An example of this is the telephone network. A call from New York City to San Francisco, for example, would be directed through the shortest and most available central exchanges between the two cities. If, however, the Chicago central exchange is out of order, a next least critical path between the two cities will be attempted.

This example suggests a critical and complex control and priority algorithms that may vary as new features are added to the system. A dictionary of the options and available data paths must be maintained. A detailed reconfiguration strategy which takes all these environmental parameters into consideration must be developed and microprogrammed. Finally, complete and concise information related to the system's failure and the maintenance procedure required must be communicated to the operation management and field engineer. It is obvious that this flexibility and its complexity cannot be economically achieved except with a writable control feature.

4.1.4 Extended Capability

Another possible advantage can be realized from the inherent capability of the WCS to support a multiplicity of architectural definitions within a single processor. This may involve the mapping of specific features of another architecture onto the system, such as in the design of emulations. Or it may involve providing a complete system identity by replacing its old identity with a new one. The Standard Computer Corporation of Los Angeles announced the IC4000 system which encourages the user to change its microprograms. The system comes with a number of different machine codes. It also includes a specialized FORTRAN system supported by a microprogramming system called *miniflow*.

In general, one can see no limitation on the architectural extensions and on the applications for which such a capability can be used.

4.2 DISADVANTAGES

This panacea, if not properly used and closely checked, can lead to panic. And it is this type of undisciplined use of the facility that concerns the designer most. The majority of computer installations function under an operating system of some type and with a system that provides room for growth without the expensive and time-consuming conversion to the new system. This extensive support by the manufacturer can be provided only when his product line, which he hopes will satisfy the broadest market spectrum, shares the same architecture and the same operating system. Once the undisciplined use of the writable control storage to satisfy any special function requirements outside the scope of the fixed system architecture is introduced, it will create problems for both the user and the manufacturer.

4.2.1 Disadvantages to the User

For the user, the first disadvantage is the loss of compatibility. The new modified system can no longer run under the manufacturer-provided operating systems. Some modifications, to varying degrees, have to be made to such an operating system. Such modifications are expensive, even when made by those closely affiliated with the original design of the system. A few modifications and extensions to the instruction set could possibly be made without this loss of compatibility, if they were made under the supervision and with the consent of the manufacturer. Another type of loss of compatibility occurs between the modified processor and other processors in the same product line. That is, the upward and downward compatibility, which was the original intent of the manufacturer, is now sacrificed for only upward compatibility. This means that, if the architectural changes made to the system were extensions to the existing instruction set which are suspected to be the most common type of modifications, then a program written for this modified machine cannot run on another machine in the compatible product line. A program which complies with the basic system architecture can run on this modified machine.

At a time when the proliferation in programming languages and computer architecture has caused considerable concern to the many newcomers needed in this fast growing field, the most rational course of action is for the designer to guarantee the typical user a standard total system. A total system is defined here as the sum of all the computing functions that can economically and efficiently meet the user's requirements. It consists of the programming languages, compilers, assemblers, an executive system, and an executing system. The designer would, in fact, be left with the more challenging problem of utilizing any technological advancement for improved performance and, thus, the reduction of computing cost.

A second major disadvantage from the user's point of view is that any gains in the internal speed of the processor may not result in any appreciable gains in throughput. The system throughput is a complex function of many factors which are interwoven and carefully balanced against each other. The internal speed, the main storage and storage hierarchy, the operating systems, and many other functions contribute to this all-encompassing factor named throughput. The decision to undertake such an architectural change must be carefully based on an analytical evaluation and simulation of its contribution to throughput.

A third disadvantage concerns the difficulty encountered and the special training required by the microprogrammer to produce an optimum microcode. This problem is compounded by the fact that microprogramming languages are not compatible even between two adjacent models of the same architecturally compatible product line. These languages and their supporting microcode translators and simulators tend to be difficult to learn and are scarcely documented. The net result would be for semi-experts to produce a microcode which is inefficient and which requires more extensive storage space than can be produced by the expert microprogrammer in the designer's house. This is not to suggest that any one manufacturer or group of designers has a monopoly on talented microprogrammers. Rather, it is to point out the need for a suitable and standard higher-level microprogramming language and the associated compilers which the user can employ in producing an efficient microcode, without concern about the individual data flow and microprogramming language for every system. The need and a specific proposal for such a higher level microprogramming language will be discussed later in the chapter.

A fourth disadvantage is that microprogramming debugging can be more difficult and more time-consuming than in conventional programming. The microprogrammer may resort to the maintenance engineer's facilities for debugging the system. This exposure, besides the higher cost of debugging, may also cause serious disruption to the user's machine room operations.

4.2.2 The Designer's Point of View

The designer's major problems result from his concern for protecting the average unsuspecting user from the pitfalls of this outwardly most glamorous added feature to his system. His main problem is to ensure that compatibility with the original architecture is not lost. The designer is also interested in minimizing the potentially great expense resulting from greater field force needed to install and cope with the problems of both hardware maintenance and diagnostics. He must also contend with the user's dissatisfaction which may result from an unchecked use of a writable control storage unit.

Another problem the designer must face is that, for this feature to be useful to the customer, it would require considerable and costly support for diagnostics and simulation, higher-level microprogramming languages and compilers, and a force to educate the potential microprogrammers in the field. This added expense will undoubtedly have to be reflected in the rental and sales cost to the user.

On the architectural side, the designer must answer a number of questions:

1. How to allow the customer the microprogramming flexibility without the loss of compatibility?

2. What microprogramming language is the best compromise? If the user's use for microprogramming is very limited, how may this function be best provided?

3. What debugging aids should the microprogrammer in the user's office have?

4. What types of problems can be anticipated when running such a system in a multiprogramming or multiprocessing environment?

5. What impact will the engineering changes have on this added capability? Would it be sufficient to recompile the microroutines affected by such engineering changes?

6. What loading procedure is best suited for this type of computer installation? Also, how could the unsuspecting user know what architectural characteristics are mapped onto the system at any point in time?

These are but a few of the many technology and architectural questions the user and the designer must answer to their mutual satisfaction. One suspects even a longer list of policy and marketing questions for which the designer must formulate answers to suit his interests and operations. These marketing and policy questions are very much dependent on the individual manufacturer, and we will not attempt to list any of them in this chapter.

Complete answers to these architectural questions depend on the extent to which any installation is permitted to control the use of such a writable control storage. Yet, even in the most limited case, once the user is given any freedom to modify the architecture by loading a new control routine into the writable control section, we begin to see a list of physical and support requirements that the designer must provide.

The writable control storage must be loadable by the user from main storage (which can be loaded from any I/O device prior to this time). This loading procedure may be provided through a new machine instruction such as load microprogram, or via a special privileged operation in the supervisor mode or a special system mode. In any case, a manual procedure for initial microprogram load is advisable in the general sense, and it is a must if the system is operating at least partially in a multiprogramming or multiprocessing environment. In the latter case, a means for identifying the status (architectural characteristics) of the machine must be provided.

The loading procedure requires a clear strategy for reloading in case of a memory failure. It should also be part of the total system architecture, independent of the model or the size of the microroutine to be loaded. This latter requirement suggests a fixed format that the microprogrammer can comply with in the preparation of his input cards prior to the loading time. The microprogram control input cards may contain two types of information.

One type is the model-dependent information, while the second type contains information related to the microprogram.

4.3 DESIGNER SUPPORT NEEDED

In the support area, the chief requirement is that the manufacturer provide a high-level microprogramming language and the necessary compilers which the user can apply in the performance of this function. Such a language must be procedural, descriptive, machine-independent (model-independent also), and familiar to FORTRAN and PL/1 users. This is indeed a high and difficult goal to attain. A more realistic one perhaps would be for the designer to specify a standard set of macros which would permit the same microroutine to be assembled for the different models sharing this standard set of macros. It would also permit the microprogrammer to microprogram for a computer different from the one he is familiar with, taking advantage of the new system organization and data flow without having to learn an entirely new microprogramming assembly language and procedure. The remainder of this chapter summarizes a project which was initiated by the author in which both options were examined.

4.4 HIGH-LEVEL MICROPROGRAMMING LANGUAGE DESIGN SPECIFICATIONS

Any such language must meet the following requirements:

1. *Descriptive.* The language must be capable of describing the structural properties and the relevant behavior of the system. That is, the system designer must be able to declare every facility such as register, latch, data bus, memory, and local storage. He must be able to declare the timing points which will affect the execution of the microprograms and the facilities affected by these timing points. The system designer must be able to express with ease and simplicity the action implied by every micro-order recognized by the system. Finally, he must be able to describe the makeup of the ROS control word designed for his system.

2. *Procedural.* Most existing higher-level languages such as FORTRAN, PL/1, and COBOL possess this characteristic. The language must have a well-defined hierarchy, syntax, and semantics.

3. The language must be as similar as possible to natural language so that it can serve as an easily understood and meaningful design documentation. In our proposed language, we endeavored to remain as close as possible to PL/1 and FORTRAN.

4. *Timing and Concurrency of Operations.* The language must be capable of recognizing the parallel data paths in the system in order to improve the efficiency of the resultant microcode. It must also recognize any parallel data transfer within a given data path.

5. It must not be ambiguous; it must be precise and concise.

6. It must be sufficiently flexible to adapt to future machine needs. It must be flexible enough to handle various word sizes and number representations without major inefficiencies.

7. It must permit hierarchical description of the machine. That is, the user must be able to describe the system exactly or conceptually.

In recent years, many languages have been developed which permit the simulation of a proposed digital computer and its control section. Examples are the LOCS developed by M. S. Zucker,[1] the LOTIS (a formal language for describing machine logic, timing, and sequencing) developed by Schlaeppi,[2] and a register transfer language developed by M. H. Schorr.[3] Other languages are discussed in articles by Metze and Seshu[4] of the Coordinate Science Lab and Department of Electrical Engineering at the University of Illinois, and by Iverson[5] of IBM. With the exception of the last reference, all other languages seriously fall short in more than one of the aforementioned requirements. Iverson's language, while it seems to be the most adaptable, is based on notations somewhat difficult, cumbersome, and unfamiliar to the average programmer.

In the remainder of this chapter, we will examine a first attempt by the author to design such a language and to identify the major elements of the associated compiler. This discussion is based on three months' effort primarily by the author assisted by two colleagues, Dr. I. Miller and Mr. B. Clark, from the IBM Data Processing Divisions Advanced Technology Group.

At this point, prior to the language description, a few qualifications and disclaimers for this proposed language are in order. While the language seems to satisfy all the requirements stated earlier, the language is incomplete since the I/O statements were intentionally left out to simplify the initial task of proving the feasibility of such a language. Also, as of the time this chapter was written, no compilers have been produced. All the work done thus far to prove the feasibility of this language, which is still in its embryonic stages, has been done manually.

4.4.1 Available Options

Two options that can yield the desired compatibility in microprogramming different systems with radically different internal organizations and capabilities are:

1. The high-level language approach.
2. A compatible microcode approach.

Both options are difficult due to radical differences in internal organizations such as:

1. The number of independent data paths that can operate in parallel.
2. The differences in the widths of each of these data paths and the width of the most frequently interrogated facilities in the system, e.g., the adder and the shifter.
3. The presence or absence of counters to perform index arithmetic without tying up the main adder data path.

These three factors and others can present a varied demand on either of the two options. Of these options, the high-level language was chosen because it represents greater potential. It was recognized early in the project that some serious problems have to be solved in order to produce a sufficiently efficient microcode. Hence, building an optimization capability into the compiler will continue to be an important advanced technology project.

Another important consideration is that, depending on the system organization, it would be more efficient to describe the control programs at an architectural level. For IBM System/360, this would be in terms of the general purpose registers, main storage, core storage, and instruction formats and lengths. The action here could be described in terms of a vector transformation and operations. However, in certain processors, it may be more desirable to give a step-by-step description of the task to be performed by the processor for which the microcode is to be prepared. It will be shown that the compiler is designed to handle both of these varied levels of application.

A third problem in designing the compiler is that, although the language is machine-independent, it must have available to it the description of the processor and the necessary capability to take these microprograms written in a machine-independent, higher-level language and produce an efficient, very

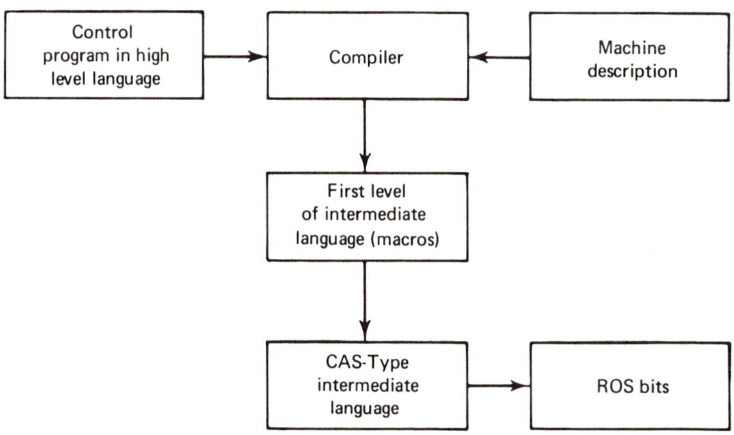

Figure 4.1 Description of compiler process.

machine-dependent microcode. This problem was solved by designing and utilizing two levels of intermediate languages—the macro level, and an assembly level language similar to the IBM CAS (control automation system). The higher-level language was first compiled to the macro level while building the proper dictionaries. Then, this set of macros was translated to a CAS-type, second-level, intermediate language which could then be translated into the ROS bit pattern. It was reasonable at that point to anticipate some optimization activity after each level of translation. Figure 4.1 gives a description of the process involved.

4.4.2 User's Hints

A compiler cannot detect logically consistent errors. It is the user's responsibility to describe the machine correctly and efficiently. The purpose of this section is to discuss the user's responsibility and the reaction of the compiler to it.

Since the user could potentially describe an unlimited number of machines and the concepts of each in an unlimited number of ways, it would be expected that, in a real situation, he would try to describe a machine that meets a particular need and is efficiently designed. Since it is expected that this program will be used primarily for describing the functions of a CPU, we shall prescribe a recommended procedure.

The basic outline of any CPU is shown in Fig. 4.2, which is a graphic representation of how the functions are related, where each branch represents a call or transfer and each loop is a complete function. (The user is not expected to work with such a diagram, but it is obvious that, the simpler the diagram, the better the design.)

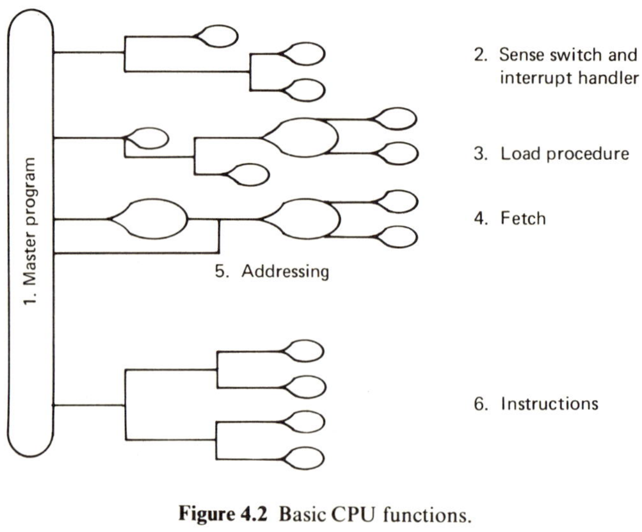

Figure 4.2 Basic CPU functions.

The routines have been specified as above because experience has indicated that this is a simple, logical, and efficient manner for subdividing the CPU. The master program obviously keeps the CPU running continuously, delegating authority to whichever routine is needed. The interrupt and load routines respond to pushbutton operations on the console. The fetch routine gets the data from core at the end of the execution of each instruction. The addressing routine is the most important routine since it is referenced in the fetch of every instruction and is referenced by every instruction routine. Heavy usage of the addressing routine justifies its being defined explicitly as a subroutine. Finally, each of the machine instructions is defined as a separate subroutine.

Simplicity in design has been stressed in the previous paragraphs; however, conceptuality of design is also to be emphasized. The natural tendency is to describe a machine in accordance with the logic of its operation. However, if the description of the machine were used on another machine with different architecture, full advantage might not be taken of the new machine. For example, the IBM 1401 could be described as fetching one byte at a time until it finds a byte with a word mark. Conceptually, it could be described as fetching several bytes, looking at all simultaneously to see which is the first with a word mark. The second description allows one to take advantage of a multiprocessing CPU with a several-byte fetch capability. Thus we describe the result of the fetch rather than the mechanism.

In summary, in using the microprogram compiler to describe a machine, simplicity and generality are good objectives by which to guide ourselves.

4.5 LANGUAGE DESCRIPTION

It was stated earlier that the language must have two important ingredients: the machine description and the control program description. The first is handled through a set of facilities, subfacilities, connected facilities, declarations, and definition statements.

4.5.1 Facilities Declaration

The facilities declaration is similar to a FORTRAN dimension statement. The user can declare arrays, vectors, and lengths of facilities. He can also connect a number of facilities to form structured facilities. Consider the following examples:

1. MS(BT5, 32)—declares MS has a two-dimensional array (main storage) containing 2^5 elements, each of which is 32 bits wide. The address of the first element in the array is assumed to be zero unless otherwise specified.
2. RM = R[0, 15], M[16, 31] is a vector which consists of the ordered set

of binary bits formed by the leftmost 16 bits of register R and the rightmost 16 bits of register M.

4.5.2 Timing

Every system is assumed to have a basic timing ring split into a number of overlapping pulses of the same or different lengths. The timing ring and timing points must be declared in the machine description section by assigning a name to the timing ring, giving its length, and assigning a name to every timing point. The set of declared facilities is followed by the group of timing points in the order in which they affect the facilities. The compilation, simulation, and optimization phases are dependent on this timing.

4.5.3 Statements of Language

The basic language is an adaptation of the FORTRAN and PL/1 languages, with additional statements and declarations to accomplish the proposed task. The following statements analogous to this base are:

1. ASSIGN statement
2. IF statement
3. GOTO
4. DO
5. CALL

The ASSIGN statement has additional features which allow one to specify the length of the data to be manipulated and one of three modes of operation:

1. Binary
2. Logical
3. Decimal

A DECODE statement has been introduced which is similar to the computed GOTO. An example is:

DECODE (M[0, 1]) RR, RX, RSS1, SS

In this example, the four combinations of the two bits of M are used to indicate which branch is to be chosen, where 00 implies RR, and 11 implies SS.

Examples of how the language can be used are given in the appendix. The following data processing functions were described for the IBM 1401 system:

1. Instruction fetch
2. Addressing
3. A typical instruction

In the next section, we shall consider the problems peculiar to the compiler for this language and application.

4.6 THE COMPILER

The compiler has been designed with simplicity and generality in mind. Essentially, it breaks down the high-level language statements into a more basic form of the language. The immediate output is a macro language which is intermediate between the high-level language and assembly language. This macro language will be described later in more detail. A macro is essentially an in-line subroutine repeated many times throughout the program. The macros for this compiler are, in essence, calling sequences to the compiler's library. This design has been chosen because of the two modes of operation of the compiler-simulation interpretation. In the simulation mode, the library subroutines will execute the commands of the high-level language, whereas in the interpretation mode, the commands will be translated to a microprogram language completing the compiler function. Let us now consider the first mode, that of simulation, which appears in Fig. 4.3.

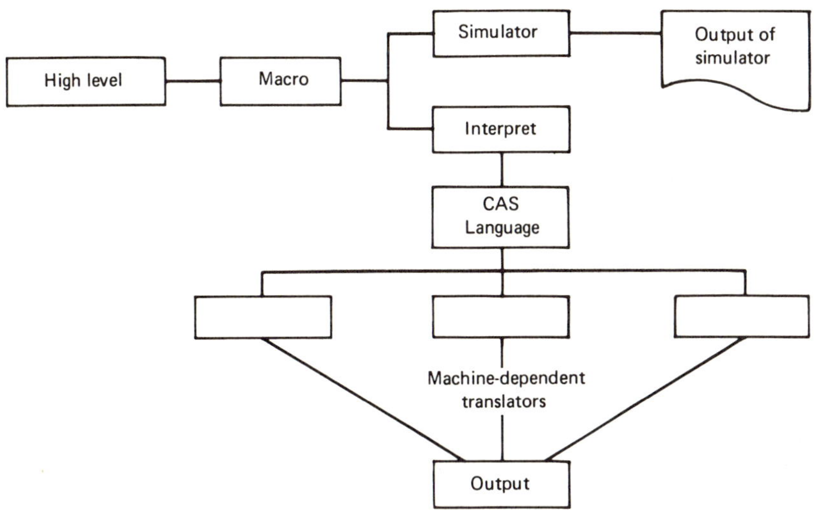

Figure 4.3 Compiler modes.

When the user describes his machine in the high-level language, it is probable that he will make several errors. In this situation, the final output from the compiler will not be in agreement with the conceived design. The user would then be forced to debug his design on the new machine. To simplify the design task, the simulator will describe the flow and execution of his logic before construction of the actual hardware. Of course, in order for the simulator to test the logic, it is necessary for the user to write the test and diagnostic cases.

The user receives a number of benefits from the compiler. First, he is forced to organize his design and specify test cases, thus completely planning the entire project before making any hardware. Secondly, he has a means of controlling and checking his diagnostic routines, thus saving much time during the debugging phase of his work. In summary, the design can be organized and planned completely in the first phase of the project.

Whereas simulation is not necessary for this compiler, interpretation is. During the interpretation, the compiler creates a branching tree and converts the macro language instructions into microcode using the dictionary. To simplify the task, we define sets of microcode instructions to which the macro language can be directly translated. The dictionary is used essentially as a monitor of the length of the data to ensure that they are moved properly. A usage table for the microprogram registers is used in conjunction with the dictionary so that the change of location of the data can be used during the optimization phase. The completion of this phase represents completion of the compiler's function. To improve the diagnostic performance, another simulator will be written to test the microcode.

4.6.1 The Dictionary

The dictionary is built up as declarations are made and as constants are used. An entry will be made for every facility, subfacility, or connected facility that is declared, together with entries for all of the self-defining constants used during statements. Each entry will have a set of attribute bits describing the nature of the facility. These bits will describe the following:

1. The type of facility, e.g., subfacility, connected facility.
2. Whether the facility is permanent or temporary.
3. Whether the facility is a hardware register, an architecture register, or a constant.

For any facility, the dictionary entry will be a pointer to a memory area containing:

1. The address of the first bit of the facility.
2. The dimensions of the facility.
3. Attributes of the facility.
4. The address of the storage area representing this facility for simulation.
5. A pointer to a reservation list used during compilation.
6. The timing framework for compilation.

For a subfacility or a connected facility, the information will be the same, but the alias will be given explicitly, i.e., a pointer will be added to the facility dictionary entry.

Constants will be entered into the dictionary as the program encounters them in the input. Since the length of a constant is implied by the length of other operands in the defining statement, the length is not given as a parameter; nor is the starting position. It has no timing, and so its parameter list consists of its value and its attributes.

4.6.2 Example of Dictionary Entry

The dictionary entry for the PSW or System/360 (DECL PSW(64)) would look like this:

In the dictionary area		PSW = AREAN
AREAN	0	The address of first bit of MAR
AREAN + 4	1	Dimension
AREAN + 8	64	Length in bits
AREAN + 12	X'80000000'	Attributes
AREAN + 26	A(PSWAREA)	Simulation address
AREAN + 20	A(RESAREA)	Compilation reserved area
AREAN + 24 on timing information		

For a constant 2 used in the statement A(0, 31) = B(0, 31) + 2, the dictionary would contain:

C2 = AREAX	
AREAX = 0	
AREAX + 4 = 1	
AREAX + 3 = 32	
AREAX + 12 = X'40000000'	Attributes definite as constant
AREAX + 16 = 2	The explicit value of the constant

4.6.3 Uses of the Dictionary

The dictionary is set up mostly during compilation and at this point is used to check for validity of declares and for duplication of names. Since it will be required that the same name, e.g., TEMP, be used repeatedly for temporary storage, it will be necessary to have a means of indicating when an item is to be removed from the dictionary. This can be done either by an ERASE statement or by a DECLT, which indicates that the item is to be kept in the dictionary until another entry with the same facility name is found. Then the new one will replace the old. The dictionary is used frequently to help in macro compilation. It is necessary here to define implicit lengths and starting points.

During simulation, the dictionary is used primarily as a table of reference to the memory locations used to represent the facilities being simulated, and also as a source of constants.

During the interpretation process, the dictionary is used in conjunction with list-processing techniques to keep track of both the contents of the facilities and the cross-referencing of hardware and software facilities. For each architecture facility in the dictionary, the reservation list pointer will indicate an entry in a list, showing in which hardware facility it is contained. This hardware facility will also point indirectly to the architecture facility it contains. Since each hardware facility entry contains all related timing information, this, too, can be accessed by the interpreter.

4.7 MACROS

Because of the objectives and ease of design, it was decided to translate the high-level language into a macro language. By using a macro language, one can either use the existing assembler or write one which is a simple subset of the existing assembler. The following macros have been defined:

MOVE
ADD, SUBTRACT, MULTIPLY, DIVIDE
COMP
GOTO
CALL
RETURN
DO

Each one has a parameter list, where the parameter is defined as follows:

PARM ≡ LABEL, DIMEN, START, COUNT

LABEL is the address of the parameter, DIMEN is the referenced dimension (it equals 1 when the parameter is not dimension), START is the first referenced bit of the parameter, and COUNT is the number of bits to be processed.

Let us now consider the function and format of each of the macros. The move macro is of the form:

MOVE PARM 1, PARM 2

where PARM is an abbreviation for the four items in the parameter list. This instruction causes the contents of PARM 2 to be moved to PARM 1.

The ADD, SUBTRACT, MULTIPLY, and DIVIDE macros are of the form:

ADD PARM 1, PARM 2, PARM 3

where the operation of PARM 2 and PARM 3 is stored in PARM 1.

The COMPARE macro is of the form:

COMP PARM 1, PARM 2

where PARM 2 is compared with PARM 1.

The GOTO macro is of the form:

GOTO CC, STATE

where CC is a condition code, and STATE is the address to which the branch refers.

CC has the following meanings:

CC = 2 greater than
4 less than
8 equal to
15 unconditional

The CALL macro is of the form:

CALL SUB, LIST

where SUB is the name of the subroutine, and LIST is a pointer to the list of parameters. The return statement has no parameters.

The DO macro is of the form:

DO STATE, VAR, INIT, FINAL, STEP

where STATE is the statement at the end of the DO, VAR is the running variable, INIT is the initial value of VAR, FINAL is the last value, and STEP is the increment size for VAR.

There are two modes of operation for these macros. The first is one of simulation, and the second, interpretation. In either mode, the macros call a subroutine for handling their operation. A flag is set which causes this subroutine to operate in the selected mode. In the simulation mode, the subroutine performs the operation indicated by the macro, while in the interpretation mode, it generates the microcode for executing these instructions.

The prime purpose of the simulation mode is to provide a debugging tool for the user. In this mode, the user has the opportunity to see if the execution of the description and his design concept are the same. He can also use the simulation mode to test his diagnostic programs for the machine, thus providing a means for evaluating their effectiveness. The simulation mode also provides a consistency check in the compiler, for by performing the simulation before and after the microcode is generated, the compiler can indicate any problems that exist.

During the interpretation mode, the compiler continues the compiling process by converting the macros into microcode. Again, the macro concept, in which a set of microcode macros is used to correspond to the assembly language macros, will be used. Thus the interpretation mode will be very similar to the first compilation from high-level language to assembly language. The key difference will be the trace on the tree which will be built for use during conversion to microcode.

4.7.1 Assembler Macros

Until this phase in our compiler design, we have managed to remain machine-independent. We have simply taken the control routines written in the proposed high-level language, translated them into an intermediate set of macros, and built the necessary dictionaries and bookkeeping facilities for the internal use of the compiler. This intermediate macro listing under the programmer's option could be simulated, or it could be translated into another intermediate language similar to CAS.

The SDD Intermediate System Development Group in Poughkeepsie, working with the SDD Endicott Small System Development Group, has been designing and coding a microprogramming assembler in System/360 language. It is hoped that this program will eliminate some of the serious problems one experiences when using the CAS system.

The success of this microprogram assembler effort and the degree of its generality will eventually dictate whether the second-level translation in our compiler will be in terms of the CAS-type language or the assembly language.

In the next section three examples are given of this high-level language compiled to the first-level intermediate macro language. The first example will show the high-level listing of a 1401 I-fetch, instruction, and addressing routine expressed on the architectural level. The second example will show translation of the 1401 system control program expressed in the high-level language to the macro language; and in the third example, to the microcode (CAS) output.

For the second option, that of translating the macro level to the control program bit pattern which makes up the microprogram, one can envision a number of microinstruction boxes (microroutines) written with variable facilities that will be defined by the microprogram compiler at the compilation. This translation of the macros, along with the machine description, forms the bridge between the machine-independent compiler and the machine-dependent microprogram.

4.8 EXAMPLE

In this section, examples of an application of the language, a hand translation to the intermediate language and then to the microcode language (CAS), are presented.

The fetch routine, addressing routine, and one instruction routine in Section 4.9.1 were chosen for the IBM 1401 to illustrate the application of the language and to give the reader an appreciation of the power and simplicity of the language. The load sequence and other routines were coded for testing the completeness of the language, but they are not presented here.

The fetch routine retrieves data from main core and stores the data in the OP, A, and B registers of the 1401. The addressing routine illustrates the calculation of an address for a 1401 without the indexing feature. Finally, we chose the load characters to a word mark as a representative instruction.

In Section 4.9.2, we have translated the high-level language into the intermediate language. It should be noted that for LCA, one line of code was optimized. In Section 4.9.3, we translated the macro language for the fetch routine only through statement 1—the end of the DO.

Though these examples are limited, IPL, fetch, addressing, and the instruction set for the 1401, 7090, and the 360 have been described by the authors in the high-level language as an exercise; they are not presented here.

4.8.1 1401 Description in High-Level Language

```
    DECL    A (6)
            B (6)
            OP (6)
            OA (3, 6)
            OB (3, 6)
            ILC (4)
            IC (14)
            MS (BT 14, 7)
            CC (2)
            AA (14)
            BA (14)
SUBFETCH
            OP, B = MS (1C) [0, 5]
            ILC = 0
            DO 1 K = 1, 3
            IC = IC + 1
            ILC = ILC + 1
            B = MC (1C) [0, 5]
            IF (MS (1C) [6] = 1) THEN GO TO 3
            ELSE A = B
            OA (K) = A
            OB (K) = A
            DO 2 K = 1, 5
            ILC = ILC + 1
            IC = 1C + 1
            B = MC (IC) [0, 5]
            IF (MS (IC) [6] = 1) THEN GO TO 3
            ELSE A = B
       2    OB (K) = A
            CALL CHECK
       3    IF ((ILC = 3)/ + (ILC = 6)) THEN CALL CHECK
            ELSE RETURN
```

The following routine takes a decimal address in ADDR (either OA or OB), and converts it to binary. The conversion is stored in ADDC (IC, AA, or BA).

```
SUBAADDR (ADDR, ADDC)
   DECL     ADDR (3, 6)
            ADDC (14)
ADDC = 10 * (10 * ADDR (0) [2, 5] + ADDR (1) [2, 5]) + ADDR (2)
         [2, 5] + 1000 * (ADDR (0) [0, 1] + 4 * ADDR (2) [0, 1])
RETURN
LCA
   1. MS (BA) = MS (AA)
      BA = BA - 1
      AA = AA - 1
      IF (MS (AA + 1) [6] = 0)
      THEN GO TO 1
      ELSE RETURN
```

4.8.2 1401 Translated to Macro Language

```
SUBFETCH
      MOVE    A(B), 1, 0, 6, A(MS), A(IC), 0, 6
      MOVE    A(OP), 1, 0, 6, A(B), 1, 0, 6
      MOVE    A(ILC), 1, 0, 4, A($0), 1, 0, 4
      DO      S1, K, 1, 3, 1
      ADD     A(TEMP), 1, 0, 14, A(IC), 1, 0, 14, A($1), 1, 0, 14
      MOVE    A(IC), 1, 0, 14, A(TEMP), 1, 0, 14
      ADD     A(TEMP), 1, 0, 4, A(ILC), 1, 0, 4, A($1), 1, 0, 4
      MOVE    A(ILC), 1, 0, 4, A(TEMP), 1, 0, 4
      MOVE    A(B), 1, 0, 6, A(MS), A(IC), 0, 6
      COMP    A(MS), A(IC), 6, 1, A($1), 1, 0, 1
      GOTO    8, S3
      MOVE    A(OA), K, 0, 6, A(B), 1, 0, 6
S1    MOVE    A(OB), K, 0, 6, A(B), 1, 0, 6
      DO      S2, K, 1, 5, 1
      ADD     A(TEMP), 1, 0, 4, A(ILC), 1, 0, 4, A($1), 1, 0, 4
      MOVE    A(ILC), 1, 0, 4, A(TEMP), 2, 0, 4
      ADD     A(TEMP), 1, 0, 14, A(IC), 1, 0, 14, A($1), 1, 0, 14
      MOVE    A(IC), 1, 0, 14, A(TEMP), 1, 0, 14
      MOVE    A(B), 1, 0, 6, A(MS), A(IC), 0, 6
      COMP    A(MS), A(IC), 6, 1, A($1), 1, 0, 1
      GOTO    8, S3
      MOVE    A(A), 1, 0, 6, A(B), 1, 0, 6
S2    MOVE    A(OB), K, 0, 6, A(A), 1, 0, 6
      CALL    CHECK
```

```
S3      COMP    A(ILC), 1, 0, 4, A($3), 1, 0, 4
        GOTO    8, CHECK
        COMP    A(ILC), 1, 0, 4, A($6), 1, 0, 4
        GOTO    8, CHECK
        L       15, RETURN
        BCR     15, 15

SUBAADDR (ADDR, ADDC)
        MULT    A(TEMP), 1, 0, 14, A(ADDR), 1, 2, 4, A($10), 1, 0, 4
        ADD     A(TEMQ) 1, 0, 14, A(TEMP), 1, 0, 14, A(ADDR), 2,
                2, 4
        MULT    A(TEMR) 1, 0, 14, A(TEMQ) 1, 0, 14, A($10), 1, 0, 4
        ADD     A(TEMS) 1, 0, 14, A(TEMR) 1, 0, 14, A(ADDR), 3,
                2, 4
        MULT    A(TEMT), 1, 0, 14, A(ADDR) 3, 0, 2, A($4), 1, 0, 3
        ADD     A(TEMU) 1, 0, 14, A(TEMT), 1, 0, 14, A(ADDR) 1,
                0, 2
        MULT    A(TEMV) 1, 0, 14, A(TEMU) 1, 0, 14, A($1000) 1, 0,
                14
        ADD     A(ADDC) 1, 0, 14, A(TEMV) 1, 0, 14, A(TEMS) 1, 0,
                14
        L       15, RETURN
        BCR     15, 15
LCA

S1      MOVE    A(MS), A(BA), 0, 7, A(MS), A(AA), 0, 7
        MOVE    A(TEMS), 1, 0, 14, A(AA) 1, 0, 14
        SUB     A(TEMP), 1, 0, 14, A(BA), 1, 0, 14, A($1), 1, 0, 14
        MOVE    A(BA), 1, 0, 14, A(TEMP), 1, 0, 14
        SUB     A(TEMR) 1, 0, 14, A(AA), 1, 0, 14, A($1), 1, 0, 14
        MOVE    A(AA), 1, 0, 14, A(TEMR), 1, 0, 14
        COMP    A(MS), A(TEMS), 6, 1, A($0), 1, 0, 1
        BCE     S1
        L       15, RETURN
        BCR     15, 15
```

4.8.3 1401 Description in CAS

STORAGE ASSIGNMENT

1401 Labels	360 Local Store	1401 Labels	360 Local Store
AA	WS3	A	H
BA	WS2	OP	WS4
IC	WS1	OA	WS5
ILC	MD	OB	WS6
CC	CC	TEMP1	WS7
B	M	TEMP2	WS8

SUBFETCH IN MICROCODE

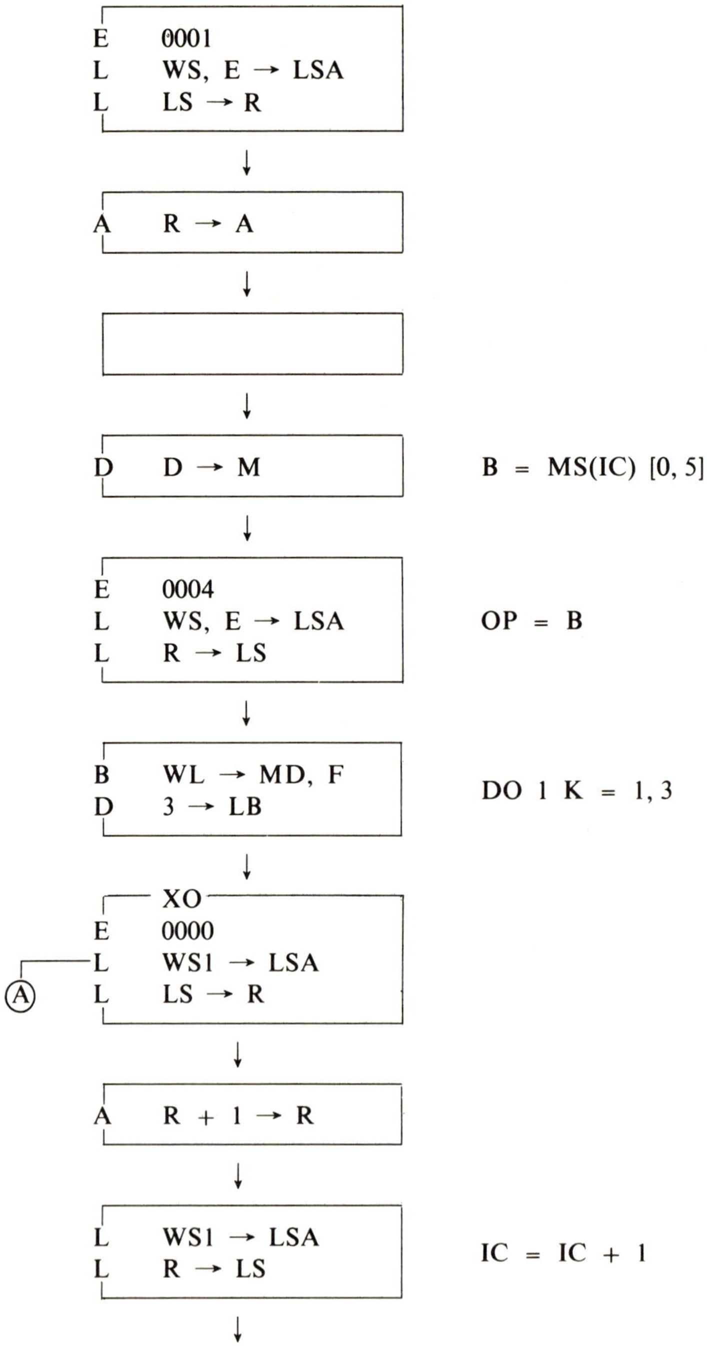

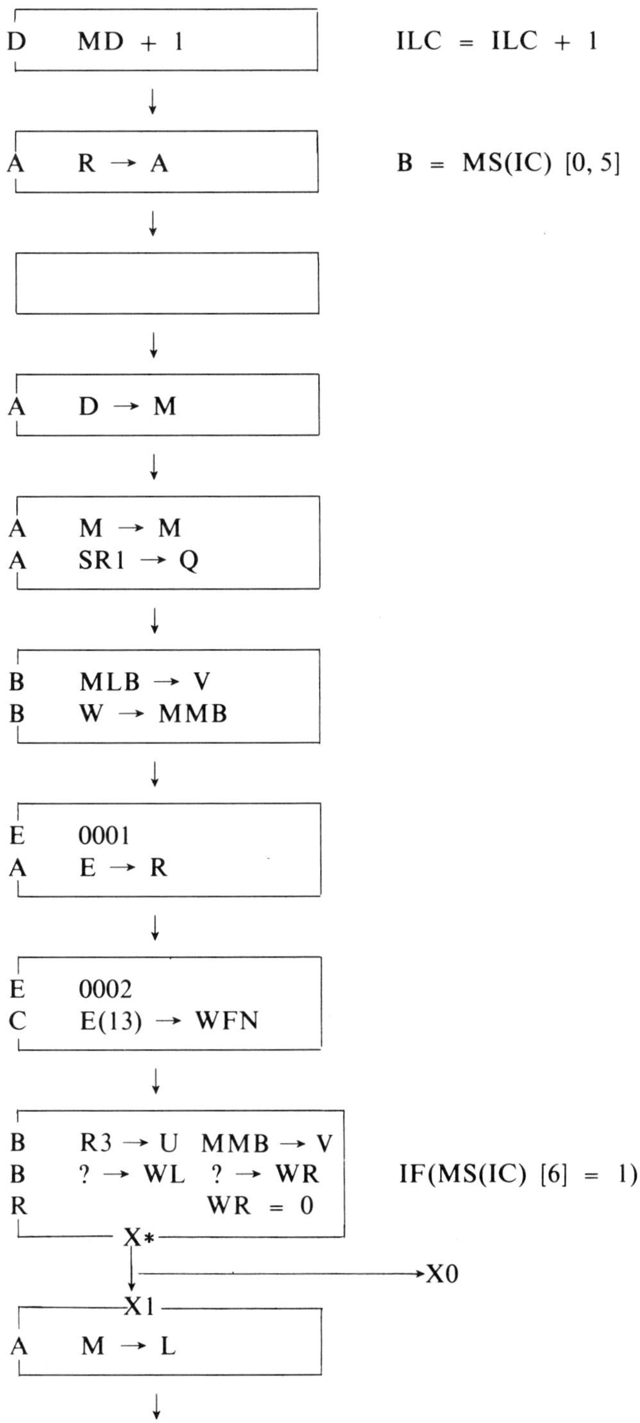
D MD + 1
ILC = ILC + 1
A R → A
B = MS(IC) [0, 5]
A D → M
A M → M
A SR1 → Q
B MLB → V
B W → MMB
E 0001
A E → R
E 0002
C E(13) → WFN
B R3 → U MMB → V
B ? → WL ? → WR
R WR = 0
IF(MS(IC) [6] = 1)
X*
X0
X1
A M → L

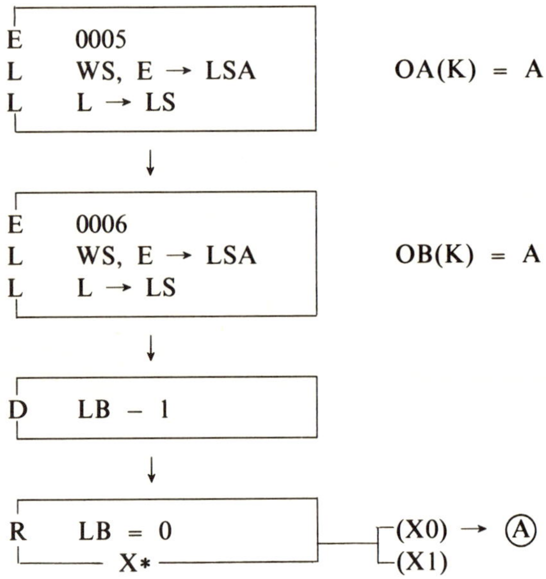

4.9 SUMMARY AND CONCLUSIONS

The compiler has been considered from the high-level language to the production of a microcode. To prove the feasibility of the compiler, we traced the translation of several high-level statements through various phases of the compiler. Tables were constructed and used during this trace process to explain the functions of the compiler. See the appendix for examples.

The key to the proposal of feasibility was the simplicity of the compiler's design. We realized that speed of compilation and optimization of final code might be desirable; however, for the present discussion, they were not considered to be of prime importance, whereas feasibility was. By classifying the organization of the compiler through the specification of macros, we were able to simplify the description of the compiler, and we have also simplified the task of implementing it.

The main task of a compiler is that of bookkeeping, where the bookkeeping is done through the construction and use of the dictionary. We use the dictionary in the first compilation phase to store labels and their attributes. The dictionary is created by the declaration statement and default characteristics derived from executable statements. During the other phases, the dictionary is referenced many times to obtain information for use during compilation, simulation, or interpretation.

An important question to ask about this project is what is its value to the computer industry. We feel that there are three important uses for this program:

1. Designing
2. Debugging
3. Translating

The high-level language provides a documentation tool for the designer in that he has a uniform language in which to describe his work. Simply being forced to organize is of value in itself. Then, the designer is forced to write test cases and diagnostics before implementation of the design; thus he is forced to expose potential design difficulties. When the actual machine is checked out, the results of the simulation runs will be directly applicable. When the next generation machine is produced, making the present one obsolete, the compiler can be used to translate the description into microcode for the new machine. In this way, the user will not be led into a dead end where the next generation machine will not support his existing system.

In Sections 4.4 through 4.7 we have tried to prove feasibility, as illustrated by examples in Section 4.8, by tracing representative programs through different phases of the system. The biggest shortcoming has been that of not considering I/O. First, the problem of conversion must be solved, and second, the problem of channel commands. It is felt that these problems can be handled easily based upon the experience obtained in describing CPU.

To continue this project, the next step will be to begin the implementation: first, by writing a compiler for converting to the macro language (with options allowed for the implementation of I/O), and then designing the macro subroutine support for the simulation mode. Then the macros will be modified for the interpretation mode. The concluding step of this phase will be the translation to microcode tapes. Each of these phases will represent minor checkpoints, and the last phase will represent an evaluation checkpoint. Next, I/O will be implemented. Finally, the mechanism for optimization of the microcode will be generated.

In summary, the feasibility and plan of implementation have been considered in some detail. Checkpoints have been chosen so that the success of the project can be evaluated easily during the entire development of the project.

REFERENCES

1. Zucker, M. S., "LOCS: An EDP Machine Language and Control Simulator," *IEEE Trans. on Electronic Computers*, Vol. EC-14 (June, 1965), pp. 403–416.

2. Schlaeppi, H. P., "A Formal Language for Describing Machine Logic, Timing, and Sequencing (LOTIS)," *IBM Research—Technical Report RZ 125* (December 24, 1963).

3. Schorr, H., "Computer-Aided Digital System Design and Analysis Using a Register Transfer Language," *IEEE Trans. on Electronic Computers* (December, 1964), pp. 730–737.

4. Metze, G., and Seshu, S., "A Proposal for Computer Compiler," *Proc. Spring Joint Computer Conference* (1966).

5. Iverson, K. E., *A Programming Language*. New York, John Wiley & Sons, 1962.

6. Weber, H. A., "Microprogrammed Implementation of Euler on IBM System/360 Model 30," *Commun. ACM*, Vol. 10 (September, 1967), pp. 549–558.

7. Bashkow, T. R., Sasson, A., and Kronfeld, A., "System Design of a FORTRAN Machine," *IEEE Transactions on Electronic Computers*, Vol. EC-16, No. 4 (August, 1967), pp. 485–499.

8. Melbourne, A. J., and Pugmire, J. M., "A Small Computer for the Direct Processing of Fortran Statements," *Comput. J.*, Vol. 8 (April, 1965), pp. 24–27.

CONTROL STORAGE TECHNOLOGY

5

5.1 INTRODUCTION

In the first four chapters the distinction was made between the static ROS control systems and the dynamic microprogrammed systems with a writable control storage. In this chapter we will review the memory technologies available for both types of control implementations. Except for a few systems such as the IBM/360 Models 85 and 25, the Standard Computer Corporation's IC 4000 and IC 6000E, and Interdata Systems 2, 3, and 4, all microprogrammable systems today are of the static type that utilize a ROS storage of some type. The main reason a read/write storage has not been used in the past was the lack of a read/write storage technology with the desired cost/speed tradeoff. However, the advent of the large scale integration technology and the significant cost reduction projected for a high speed read/write memory may alter this balance between the ROS and read/write storage implementation. Since the balance is in favor of a ROS control system implementation, we will start with and devote considerable space to the ROS technology.

5.2 READ-ONLY STORAGE (ROS)

The read-only storage (ROS) is a special type of storage device in which information (zeros and ones) is permanently or semipermanently stored. The name implies a nondestructive readout cycle as well as the absence of any electronic means for modifying the information stored therein. The information in a ROS can be altered or replaced by changing some physical connections or by replacing or modifying some physical part in the system.

The read-only storage devices have been used primarily as an alternative for the transistor logic control sections of many data processing systems. They are used to store the basic microroutines such as instruction fetch and decode, instruction execution routines, address arithmetic, interrupt handling, and other numerous microroutines that define the basic computer architecture as it is implemented on a given data processing system. The ROS can also be used to store many other types of permanent or semipermanent information such as in telephone code translation switches, standard mathematical and trigonometric functions, supervisory routines in operating systems or industrial control, dictionaries, code translator tables, mathematical constants, sometimes entire compilers, business tariffs, rates, and time schedules.

Another ROS application was proposed as an addendum or erratum to the read-write main storage device. The ROS interprets an address and shifts the incoming information to another memory module or another unused area of the bulk storage if a failure of a bit in the selected word in main storage is found. As new failures are detected during the life of the system, a new card may be punched and inserted to circumvent them and mask them out.

The main advantage of ROS storage over a read-write ferrite core storage is the operating speed, since the information is retrieved from ROS storage without altering it and, hence, no rewrite cycle is necessary. Another advantage is its high reliability characteristics due to the simplicity of its construction and the undestructibility of information by stray pulses caused by intermittent errors external to the ROS memory or by accidental breaks in the electrical circuits. A third advantage is that the information stored therein is protected from being damaged or destroyed, erased or altered accidentally, even when the hardware itself remains trouble free. Such destruction can be caused in conventional read-write memories by software errors or unresolved contention for storage space causing stored data to be inadvertently overwritten. A fourth advantage is its high density, but low power dissipation. A fifth advantage is that ROS units tend to be cheaper than their read-write counterparts at a given speed.

Control storage devices can be characterized and classified in many ways. One classification is based on the ease of modifying the information as follows:

1. *Permanent Storage*. The stored information is fixed by the manufacturers so that any changes in the contents of this ROS device would require the physical reconstruction of the device at least partially, if not completely. Generally, these modifications require extensive laboratory equipment and facilities and cannot be made in the field or in the user's office. Examples of this type are the capacitor read-only storage, the transformer read-only storage, etc.

2. *Semichangeable or Semipermanent ROS.* Data stored in devices of this class can be simply, cheaply, and conveniently modified by removing easily accessible words, cards, plans, etc., and then inserting new ones containing the new modified information. Such a storage device permits changes to be made in the user's office where specialized equipment and tools are not usually available. An example is the card capacitor read-only storage (CCROS) where information is contained in a removable mylar or plastic card of the size and format of a regular IBM programming card (Fig. 5.1). The new information

Figure 5.1 Read-only unit utilizing a semipermanent card capacitor read-only storage technology (CCROS).

(zeros and ones) are coded into the card by punching the card on a general card punch equipment found in the user's office. The oldest example of this type of ROS device is the back panel wiring on data processing equipment. It is desirable to limit the density of information stored per card to avoid excessive waste of time in punching all the information needed to make a minor change in one word.

3. *Electronically Changeable Control Storage Device.* The dominant feature of this class is that the control (the microroutines) is now stored in a read-

write storage device, which is capable of being changed by the user electronically, by loading it through the main storage (which has been previously loaded from an I/O storage). The advantages and the possibilities that an electronically changeable control storage device can provide are considerable.

A few of the advantages are listed below; some of them were discussed in Chapter 4 on writable control storage.

1. Ease, speed, and the reduction in cost of making engineering changes.
2. Ease in tailoring the general processing system to special applications. This can also be interpreted as an efficient way to provide some special features for specific applications.
3. Reduction of storage requirements for the system's architecture, microroutines, and emulators by overlaying portions of the microroutines required for a given application.
4. Availability, maintainability, diagnostic, and serviceability features can be provided economically.
5. Rapid conversion among different sets of microroutines and different modes of operation.

These advantages must be weighed against the many possible disadvantages that such a writable control storage option can create. Among these are performance loss and confusion which can be created in the user's computing center if undisciplined access to the writable control storage is permitted. Another area of concern is in the difficult management decisions regarding the legal responsibilities of the manufacturer to the user under an environment of indiscriminate changes in the system control section.

Another type of classification of the read-only storage device is based on the physical concepts and the technology employed. Among these classes are:

1. Mechanical storage with mechanical, electrostatic, electrodynamic, electromagnetic, magnetic, optical, and other read-out principles including readout by means of radioactivity.
2. Electronic storage such as the two cores per bit with nondestructive readout which is electronically changeable.
3. Read-only storage with optical storage methods such as holographic wave interference pattern to store the image of the control information matrix, light-coupled diode read-only storage, and others.
4. Read-only storage with other storage and readout principles not included in the first three classes. Among these principles is the use of thermoelectricity to sense a position in the control information matrix.

Examples of these different classes will be discussed with more emphasis on the practical applications, electrical problems, and on the future of the read-only storage technology in general.

Among the design features that will be used in analyzing the different technologies and in comparing the different ROS matrices are:

1. The signal-to-noise ratio which is required to relax the design specifications of the sense amplifier circuitry.
2. Low power dissipation characteristics to relax the requirements for cooling the system.
3. In many applications, ease of changeability.
4. Ease and economy of construction.
5. Access and rate of retrieving information from each specific ROS technology.
6. Capacity of storage as well as storage density per unit area or volume.
7. Cost per bit of stored information.

5.2.1 Principles of Operations

Prior to discussing any specific implementation or technology, a more general discussion of the operating principles is in order.

Generally, and because of the desired speed, the read-only storage device is linearly organized on a word basis. Each word can be accessed by a linear

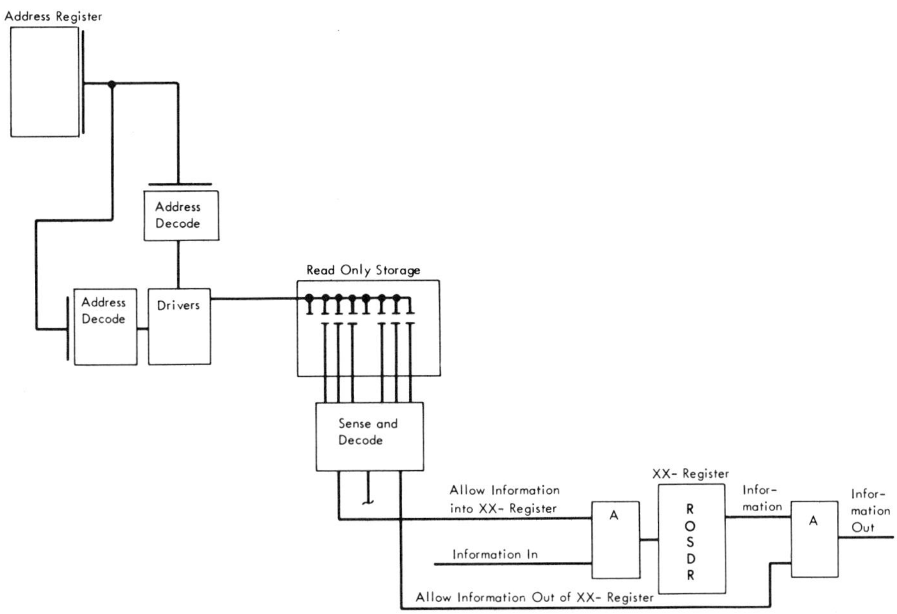

Figure 5.2 General operation of the ROS.

or nonlinear access mechanism that will decode the bits placed in a ROS address register (ROSAR) at a specified point in the ROS cycle. The selected ROS word is driven by a full-word drive, and the ROS word contents are sensed by a sense amplifier which sets the corresponding bits in an ROS data register (ROSDR) (Fig. 5.2). The output of the ROSDR is either used directly to activate the control signals or is grouped and decoded as a set of control fields, each of which is assigned a specific function and specific control gates.

Essentially, a read-only storage can be considered as a black box with a number of input lines used to select a unique word of storage in ROS. The output bit lines specify the binary state of the corresponding binary bits in that word. An impedance coupling of some type is provided between each word drive and a certain combination of bit lines according to the information stored in the specific ROS word. Storage arrays are designed in a matrix form where all bits in an ROS word are selected by unique word drive lines, and corresponding bits of each word in the storage array are sensed by a common ROS bit sense line. Figure 5.3 shows a simple ROS storage matrix. A given

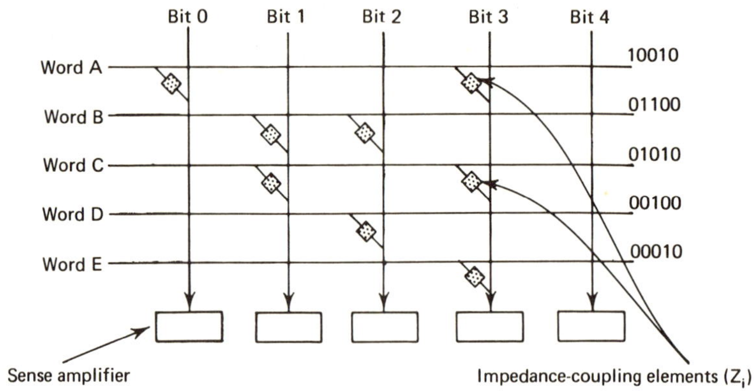

Figure 5.3 ROS matrix.

word-line drive is permitted to couple through to any bit line by connecting the word line and the bit line through some sort of impedance coupling, Z.

Each intersection between the word drive and bit sense line represents a storage bit. The locations of such intersections where an impedance coupling is shown represent a storage of 1; if an impedance coupling is not present, a 0 storage bit is implied.

Word A in Fig. 5.3 represents a binary field of five bits. It is permanently wired to store 10010; word B has 01100; etc.

The existing read-only storage devices differ from one another mainly in the technology and method used to implement impedance coupling. The technology used for implementing this impedance coupling yields the following classification of the ROS arrays:

1. *Linear arrays* use resistive, simple inductive, simple capacitive, and transformer types of devices.

2. *Nonlinear arrays* in which permanent magnet, twistor store, plated wire store, eddy current twistor store, and biased thick-film storage media are used in the implementation of such a device.

3. *Optical systems*, including fiber-optics storage, phototransistor matrix, integrated photodiodes, and holographic storage technology.

5.3 LINEAR ARRAYS

Resistive, capacitive, and inductive arrays are generally organized in a matrix form similar to that shown in Fig. 5.3. The coupling impedance Z may be a resistor, a capacitor, or an inductor, and it is used to couple the signal from the word drive line to the bit sense line. Since a selected bit sense line in this matrix may be coupled to an unselected word line because of the available impedance path in the matrix, each bit sense line is terminated in a low input impedance sense amplifier, and the ROS matrix will act as an attenuator. For example, if word drive A in Fig. 5.3 is made active, the voltage shift (or current) on the drive line is coupled to bit sense lines 0 and 3. However, the impedance coupling between bit sense line 3 and word C couples the voltage shift on bit line 3 to word drive C which, in turn, couples that voltage shift to bit sense line 1. Because any unwanted signal as that between bit line 3 and word drive C and then word drive C and bit line 1 must pass through that many elements in that sequence, this unwanted signal is greatly reduced. The sense amplifier must be designed to reject this low-amplitude signal as noise.

5.3.1 Resistive Read-Only Storage Arrays (RROS)

Resistor matrices were used as early as 1943 in the first electronic computer, ENIAC, to store functional tables. Subsequently, they fell out of favor, perhaps because of their high cost and high power dissipation; however, a revival in interest in this storage has been shown recently. A block diagram of an IBM RROS model is shown in Fig 5.6(a). Figure 5.4 shows the impedance matrix presentation similar to that in Fig. 5.3, except that the impedance element, Z, has now been replaced by a resistive coupling element, R. The unwanted current paths in this RROS impose a serious limitation on its application; the attenuation capability of this matrix depends largely on the number of resistors, their values, and the values of the terminating resistors for both the bit and word lines.

RROSs are being developed and may be produced by silk screened, rotogravure printing or by vacuum deposition. Actually, several fabrication

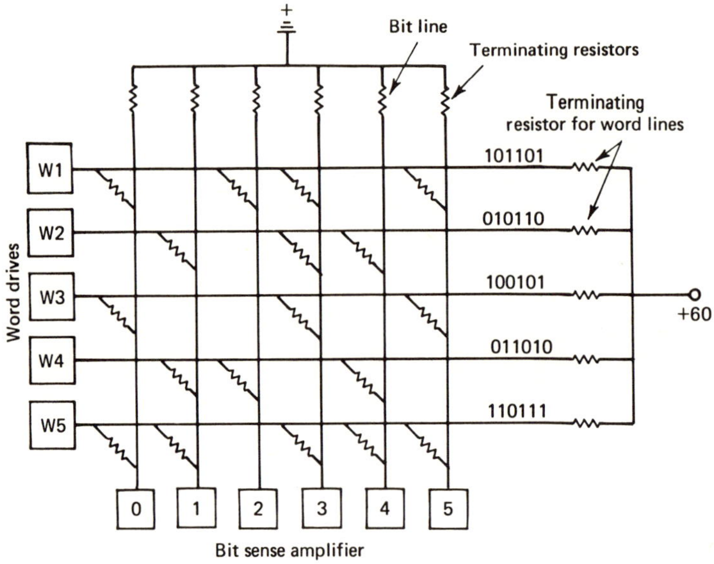

Figure 5.4 Resistive read-only storage matrix.

methods were used in realizing the resistor matrix array. Some methods proved unstable. Others proved to be more stable; among these are:

1. Paper stack card with printed and/or vapor-deposited resistor and lines (Fig. 5.5). The cards contain a set of silver-painted word drive lines on one

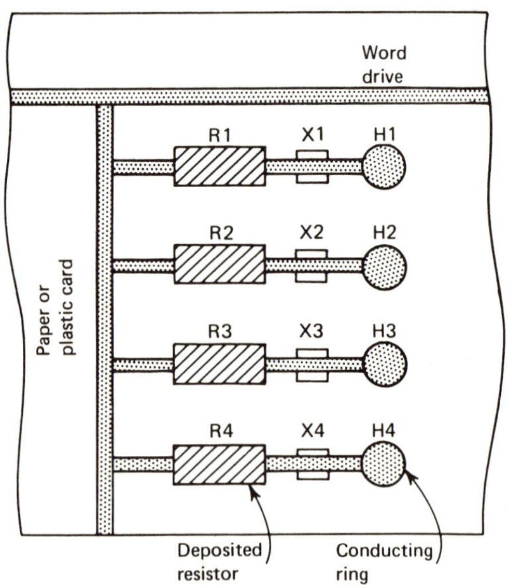

Figure 5.5 Resistor punch card.

surface. Figure 5.5 shows one word drive line coupled to a set of conductive rings, H_i, by a resistive element, R_i. In this specific method of fabrication, each card is programmed prior to assembly of the RROS array. Programming a card involves breaking the connection between the resistive body R_i and its conducting hole H_i by punching a hole in the position X_i. Thus, if the X_i tap is removed, the bit is said to be storing 0; otherwise, it is said to store a 1 bit. After all the cards have been programmed, they are assembled in a stack, and the open channels formed by the holes in the conducting ring, H_i, are filled with a low-temperature, molten alloy and are set aside to solidify. This alloy going through the stack now connects the corresponding bits of all the words in the array and forms the bit sense lines.

This method of fabrication is desirable because of its simplicity and the low cost of the paper stack. However, attempts to print on the card to produce uniform resistors and reliable insulation were found to be a prohibiting factor in continuing this manufacturing effort. The main reasons were the card porosity and surface roughness. A number of alternate approaches can be taken to solve this problem. Each new approach, while it can solve an immediate problem, will also present a new set of problems. For instance, the paper cards can be replaced by a stack of etched copper drive lines with vapor-deposited resistors. This, too, was found unworkable because of the thickness differential between the etched copper and vapor-deposited resistors. Any attempt that is made to press the copper into the card and make it flush with the surface is found to be unsuccessful due to the poor bonding between the card and copper after such an operation.

2. The IBM resistor matrix is made on a standard size IBM punch card made of Mylar* film. The card contains a set of eleven horizontal word drive lines which coincide with eleven of the twelve row positions of the standard IBM cards and sixty vertical conducting lines coinciding with sixty of the eighty columns on the IBM punch card. The conducting lines form a grid structure but are electrically insulated at the crossover points, and the resistors are printed as shown in Fig. 5.6(b). The resistors are placed to form a resistive coupling which can be punched out with standard card punch equipment.

Resistor cards are called *information documents* and are packaged between the folds of an etched Mylar tape, shown in Fig. 5.6(c), which has land patterns to make contact with the drive line tapes and the sense line tapes on the information document. This pattern is repeated on both sides of the tape so that contact can be made with two cards per tape. The assembly of the folded tapes and information document is housed between metal plates which are brought together to provide pressure over the area of contact.

The RROS storage was initially conceived to satisfy applications that require low storage capacity of 500 words or less. The objectives were for variable storage capacity (up to 500 words) of 10 to 60 bits per word, where speeds of 5 to 10 microseconds can be tolerated, and where information modi-

*Registered trademark, E. I. Dupont deNemours Co., Inc., Wilmington, Delaware.

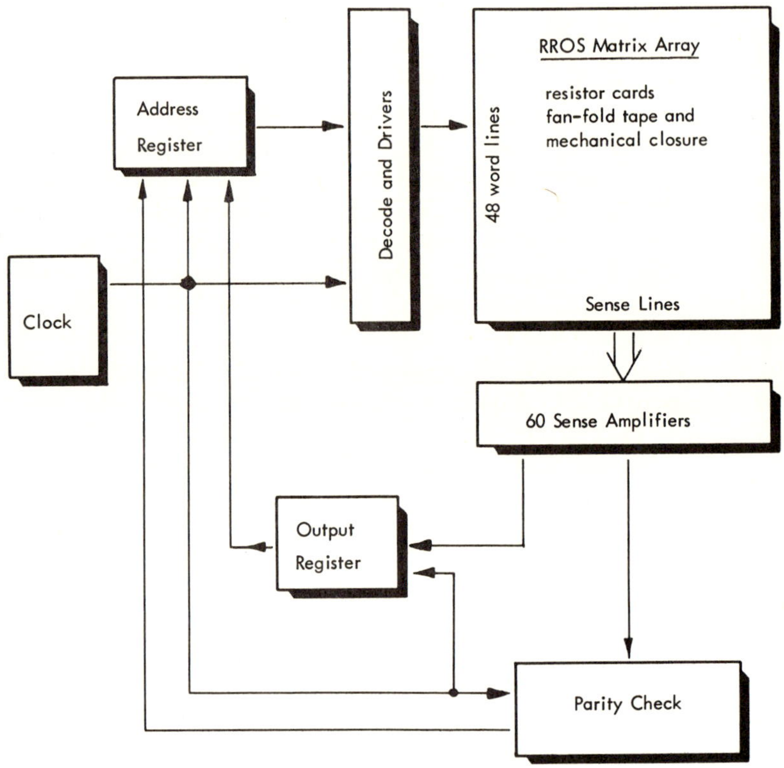

Figure 5.6(a) Block diagram of RROS model.

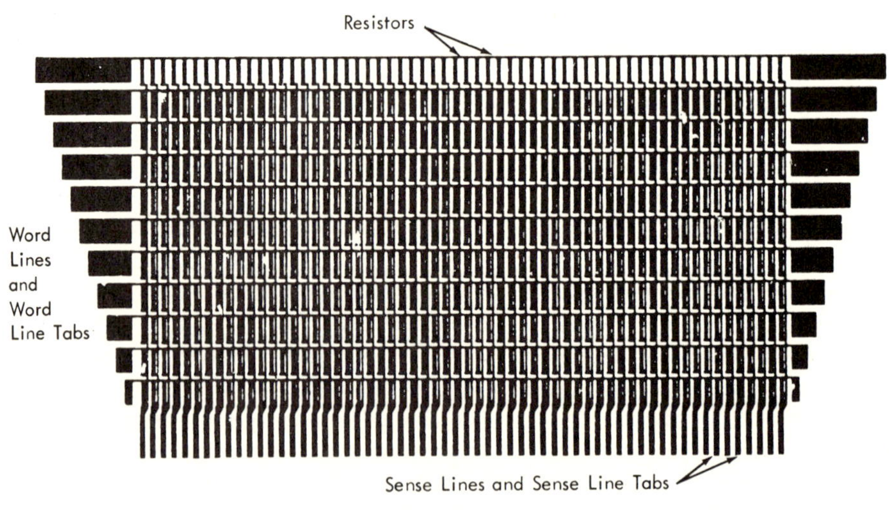

Figure 5.6(b) Composite resistor matrix card.

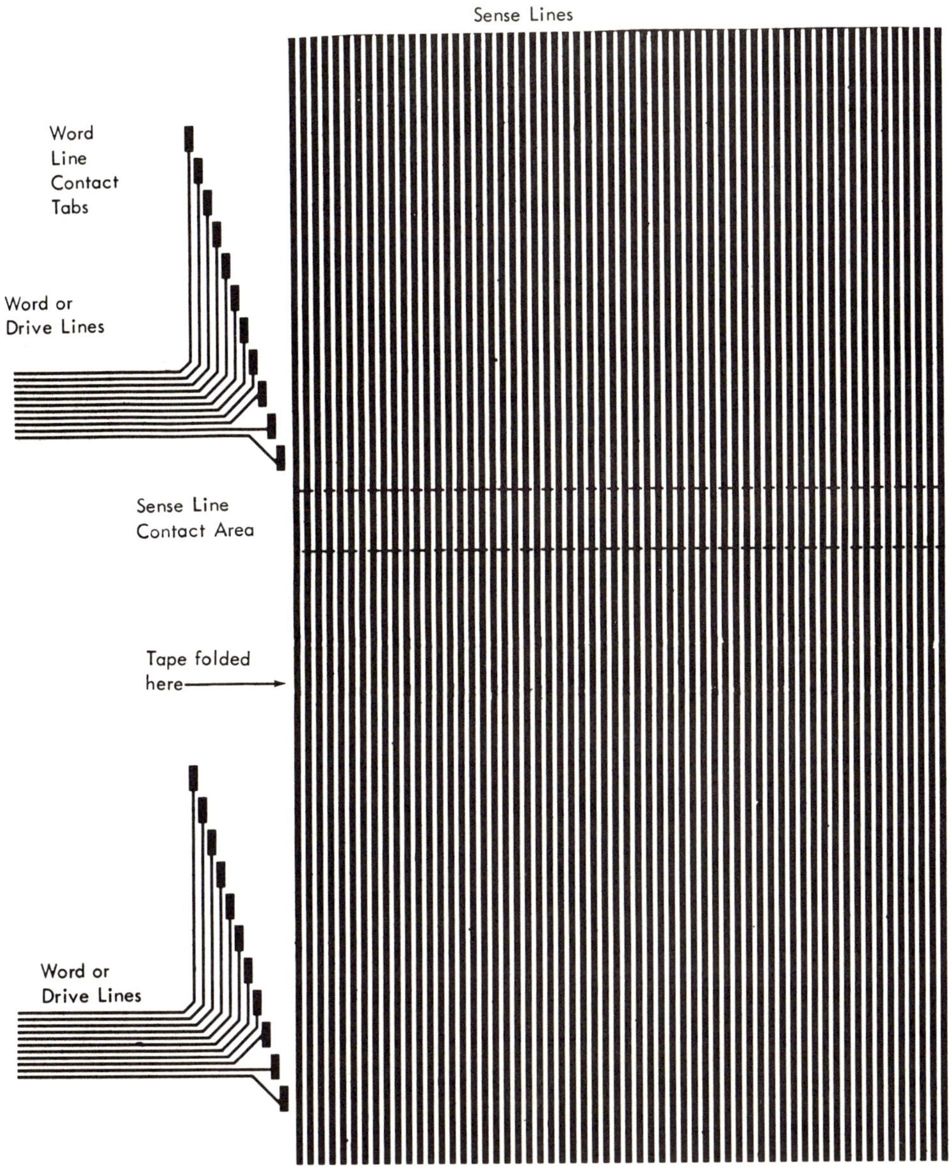

Figure 5.6(c) RROS etched tape pattern.

fication is a simple and inexpensive process. The main disadvantages of any RROS device of sufficiently large storage capacity is the large power dissipation and slow operation.

5.3.2 Capacitive Matrix

If the impedance coupling Z_i in Fig. 5.3 is replaced by a capacitor, then we will have a simple capacitive read-only storage. However, the manner in which this capacitive coupling is implanted over the orthogonal wire mesh and the ease in which the information is modified gave us two types of capacitive read-only storage systems: the semipermanent card capacitor read-only storage, CCROS, and the more permanent assembly represented by the capacitor read-only storage, CROS. The principle of operation of the first type is shown in Fig. 5.7, where two cards are arranged, face to face, with two sets of lines

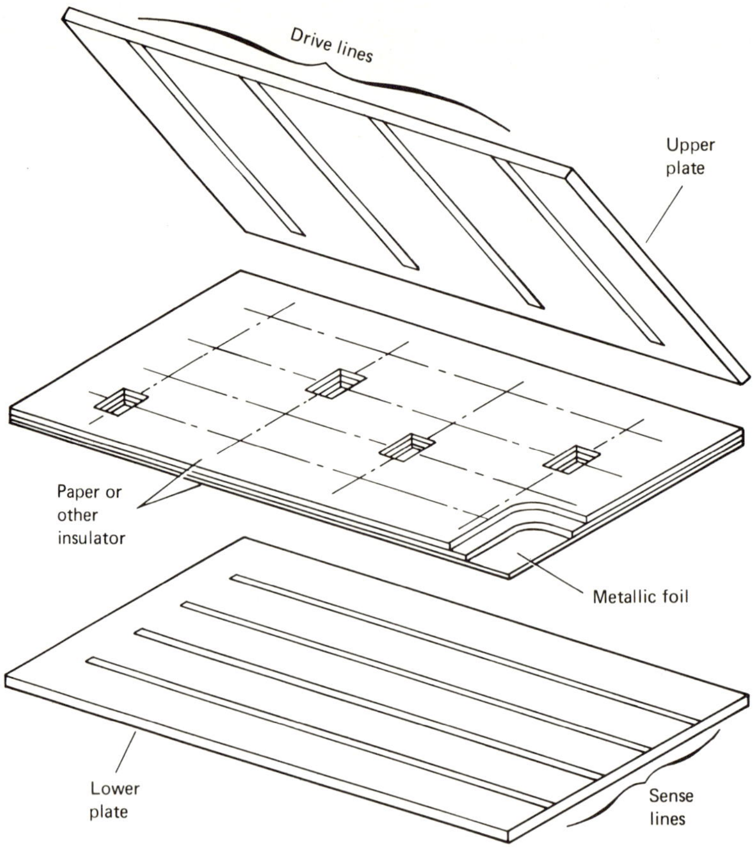

Figure 5.7 Grounded electrostatic shield memory.

crossing orthogonally, and where an aluminized Mylar grounded card is sandwiched between them to act as a shield. The principal physical concept involved here is that, when the horizontal set of lines is placed over the vertical set of lines in proper registration and separated by the proper dielectric media, a capacitive coupling is formed which is directly proportional to the area common to the two wires and inversely proportional to the separation between them:

$$C = \epsilon \frac{A}{l}$$

where C = the capacitance in farads;
ϵ = the permittivity constant (which is the product of $8.85 \times 10^{-12} \times K$, where K is the dielectric constant of the material);
A = the common area between the parallel plates;
l = the distance that separates the two plates in meters.

Hence, by separating the two plates by thin insulating film (which forms the dielectric), a matrix of capacitive coupling is formed. Any voltage induced on any word line will therefore be capacitively coupled to every sense line in the lower plates. By sandwiching the grounded metal plates between the upper and lower plates and then punching holes in the proper spots, we will be effectively permitting capacitive coupling to exist between some selected word lines and sense lines.

This grounded metal plate is the information document which can be removed, modified, or replaced by another card carrying the correct information. The hole in the information document represents a 1, while the absence of a hole at a potential ROS bit position defined by the intersection of any word drive line and bit sense line represents a 0 bit. During the past few years, stores using this principle of voltage coupling have been described by Foglia, McDermid, Petersen, and Garcia,[1] and by Takahashi and Watanabe.[2] The

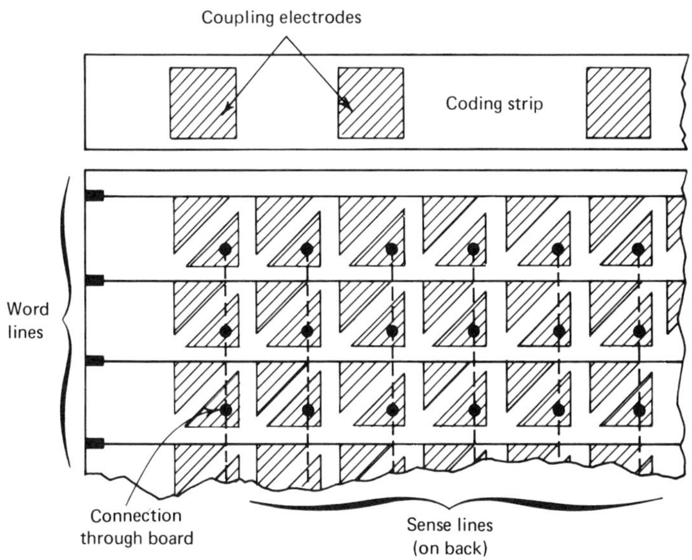

Figure 5.8 Van Goethem's capacitor store—original version.

one described by Takahashi and Watanabe has a capacity of 4096 words with 50 bits per word and a cycle time of 200 microseconds.

Another capacitive coupling matrix is the Van Goethem's storage shown in Fig. 5.8.[3] Here, the word drive lines and bit sense lines which form capacitor plates do not face each other; rather they are etched on the same surface. If a card strip containing a pattern of insulated metallized areas is placed directly over the capacitor bit pattern, coupling capacitance exists only in those bit positions where the coupling electrodes are present. Note that no connections are made to the insulated coupling strip which serves as an information document. Note also that the word drive lines and bit sense lines are made (printed or vapor-deposited) at the opposite sides of the capacitor plate with plated-through holes to the bit sense lines.

Improvements on this capacitive coupling were made by Van Goethem as shown in Fig. 5.9, where the sense lines themselves form one electrode of the coupling capacitor.

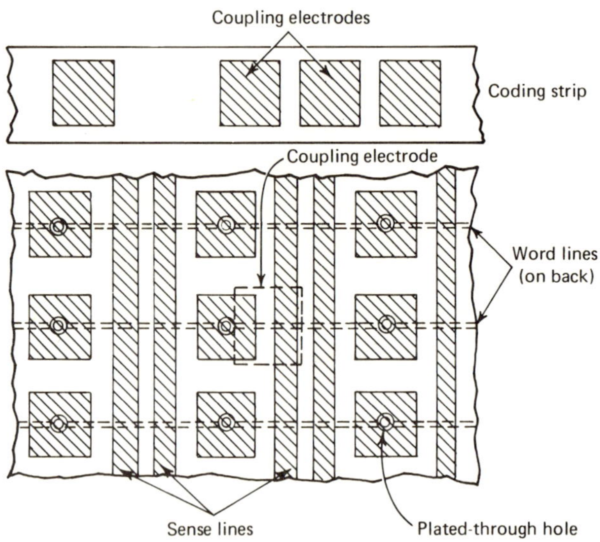

Figure 5.9 Van Goethem's capacitor store—recent version.

Another card capacitor ROS matrix form of construction was implemented by IBM for the IBM System/360, Model 30, computer. The word drive lines and the bit sense lines for this matrix are shown separately in Figs. 5.10(a) and (b). All line conductors have rectangular enlargements or spots at regular intervals corresponding to the hole spacing in a standard IBM punch card. Figure 5.10(a) shows a copper-clad epoxy glass sheet of standard SLT (solid logic technology) board material on which a set of vertical lines is etched. This board is fixed in position by a card locator frame. A layer of 5-mil

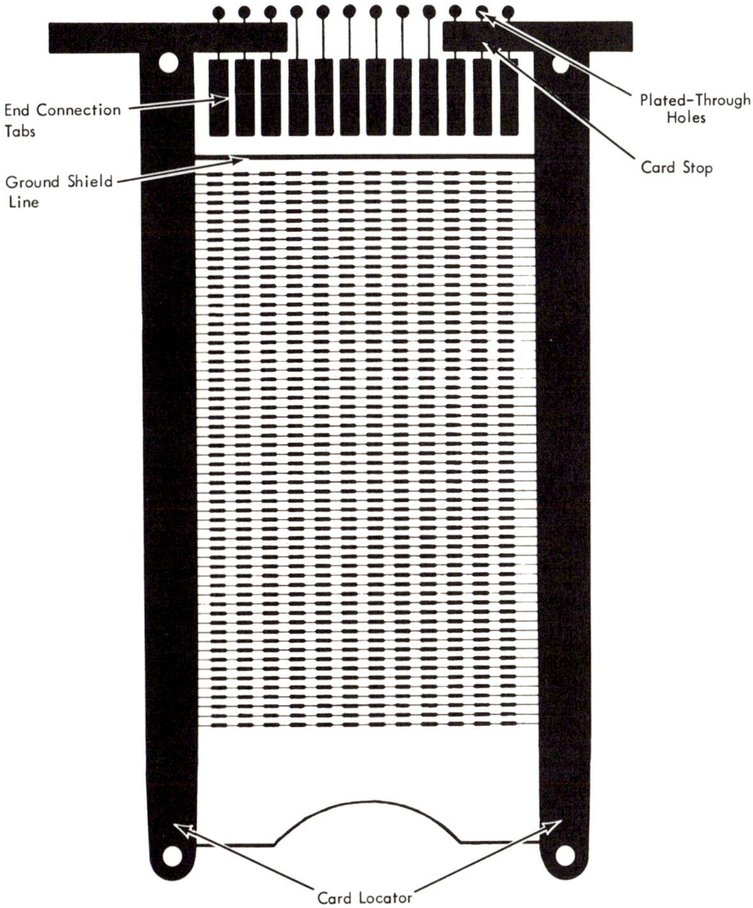

Figure 5.10(a) Storage board vertical bit lines.

Mylar dielectric material is bonded to the surface. Figure 5.10(b) shows the removable information document which contains the horizontal lines; it is printed with silver ink on an ordinary IBM paper or else etched on 6-mil copper-clad Mylar film.

The dimensions and spacing of the pattern printed on the information document are such that, if it is placed over the vertical set of lines in proper registration, the spots and the film on the information document form a capacitive coupling matrix. This information document is to be fed through regular IBM card punch equipment to punch out the selected spots in order to remove the capacitive coupling element between some vertical and horizontal lines. The CCROS information document used in the IBM System/360, Model 30, contains twelve words or twelve horizontal lines that terminate in large tabs. These tabs are often gold plated to improve contact reliability when pressed

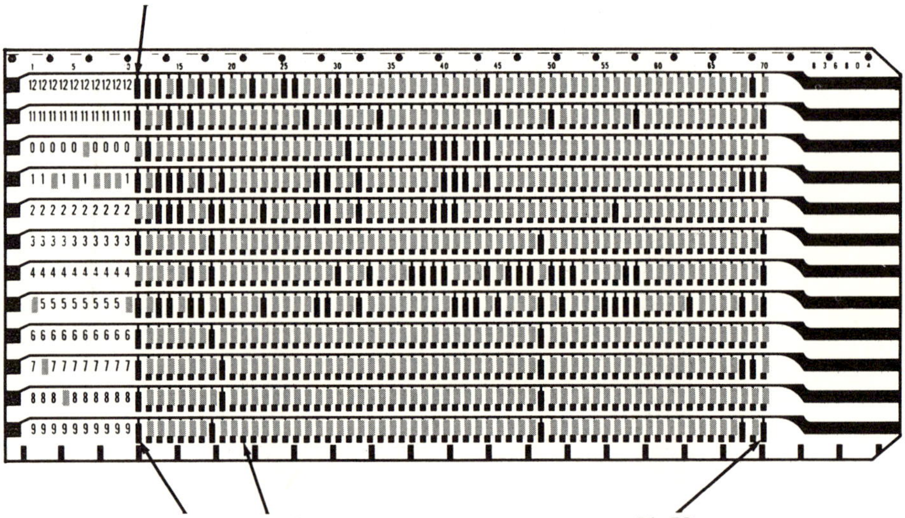

Figure 5.10(b) IBM CCROS information document.

against the corresponding end connection tabs in the vertical bit storage board, Fig. 5.10(a). Each word is made up of 60 bits which occupy columns 11 through 70 on the standard IBM punch card. Actually, the vertical bit storage boards are packaged on a standard large SLT board with four information documents on each side of the board and with the drive adaptor sockets and output sense line terminals as shown in Fig. 5.11.

Forty-eight drive lines per storage board side are connected to the opposite side of the board by plated-through holes inserted at the top of the board. This will accommodate 48 inputs to provide either 48 words of 120 bits each, or 96 words of 60 bits each. The Model 30 uses the latter arrangement, where two words (odd and even) with consecutive address words are sensed by 120 sense amplifiers arranged in 60 pairs so that either the odd or the even word is finally gated to a 60-bit ROS data register. The Model 30 uses 43 such boards for a total of 4032 words of 60 bits each. The CCROS operates on a cycle time of one microsecond, although it is capable of much higher speeds. The input drive is normally 10 volts of 75 milliampere pulse and of approximately 150 microseconds duration. The matrix 1-output signal is measured at an amplitude of 1–7 millivolts, 140 microamperes. The outstanding feature of this CCROS is the ease and rapidity of changing stored information.

It is essential that the difference between a punched and unpunched intersection coupling capacitance be at least an order of magnitude for the system to be of practical size. It is also essential that a ground reference exists to shield any unwanted stray capacitance between the bit sense lines and word drive lines. To see the implication of this, let us assume that each card has 12 words of 60 bits, or 720 bits per card. To build up a memory model,

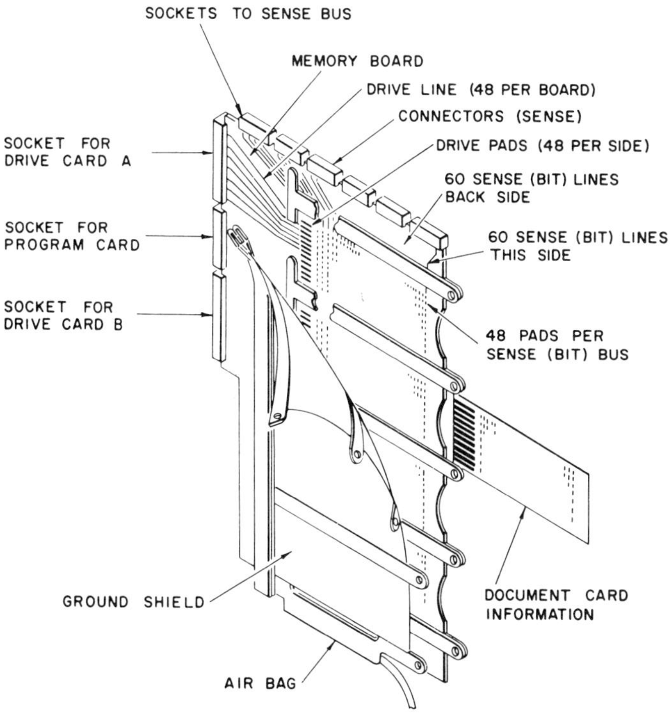

Figure 5.11 CCROS memory board.

assume that we take 50 boards, as shown in Fig. 5.11, where the sense lines of the corresponding bit positions of 50 words are connected in parallel. Let V_{in} represent the drive voltage, and let V_o represent the output voltage at the input of the sense amplifier. Let C_n represent the value of the capacitive coupling, and let C_x represent the sum of the stray capacitance that exists between the sense line and ground at the unpunched intersection plus the residual coupling capacitance due to fringing effects. Out of the 50 bits sharing the same sense line, only one is selected at any time; hence, the output voltage is taken across a capacitive voltage divider as shown below and can be expressed mathematically as:

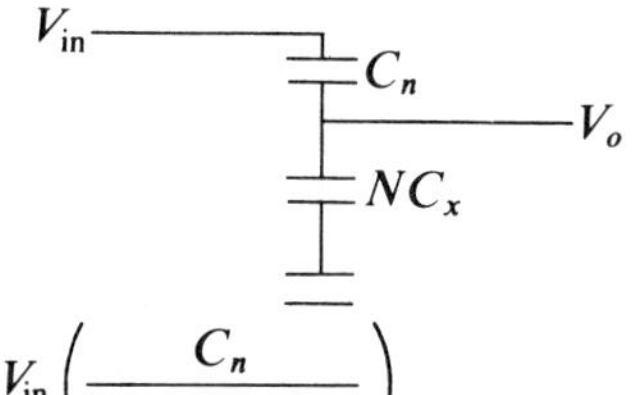

$$V_o = V_{in}\left(\frac{C_n}{C_n + N \cdot C_x}\right)$$

$$V_o = \frac{C_n}{NC_x} \qquad \text{since } C_x \gg C_n \qquad \left(\frac{C_x}{C_n}\right) = 200$$

Then for $N = 50$ (50 sense line boards in the memory module), and for an input of 100 volts, the output signal is:

$$V_o = \frac{100}{(50)(200)} = 0.1 \text{ volt}$$

This great loss in the signal amplitude makes the design of large capacitive memories difficult.

Many attempts were made to alleviate this high attenuation factor. One method, described by MacPherson and Fork, is called the *cap-store*. The basic principles of operation in the cap-store array shown in Fig. 5.12 are similar

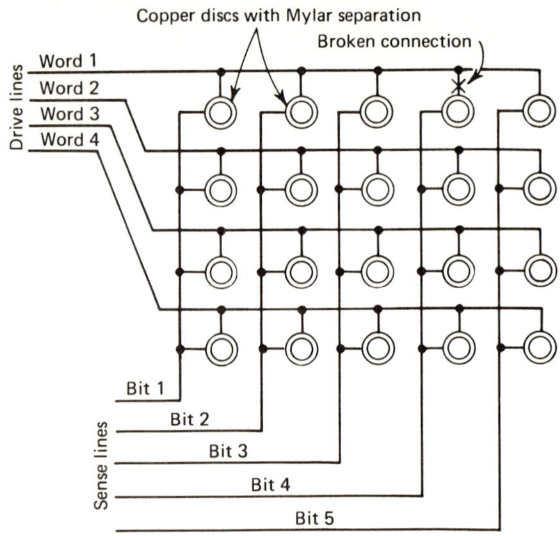

Figure 5.12 Cap-store card layout.

to those for the simple card capacitor ROS array. The main difference is that in the cap-store, if a capacitive coupling in any bit position in the array is not needed, the drive connection to the individual capacitor is removed. Hence, 1 is stored in cap-store by leaving the drive line connected to one plate and the bit line connected to the other plate of the capacitor. Removing the drive connection to any individual plate would imply that a 0 is stored in that position.

The construction of the cap-store array is simple. Both sets of lines and capacitor plates are printed on both sides of a 5-mil Mylar sheet or etched on both sides of a two-surface copper-clad epoxy glass sheet. The resulting board now has all 1's in every bit storage position. The two sizes of discs are constructed to make the indexing easier. In this storage assembly, it is essential that all except the selected drive line be grounded. The main advantage of this technology is an order of magnitude improvement in the amplitude of the output signal. Here the output voltage is now independent of the value of the capacitor and is dependent only on the number of bits sharing the same sense line. In our example,

$$V_o = V_{in}\frac{1}{N} = \frac{100}{50} = 2 \text{ volts}$$

In the IBM CCROS assembly described earlier, to minimize the unwanted line-to-line coupling that exists between the vertical and horizontal lines, the two sets of lines are closely sandwiched between two ground planes. These ground planes also present the shunt capacitive load to both sets of lines. This additional shunt loading results in a better signal discrimination capability between wanted and unwanted outputs. The main disadvantage is that the shunting tends to limit the operating speed.

5.3.3 Balanced Capacitor ROS Matrix

Earlier in this chapter, we alluded to a most serious difficulty which makes a large-capacity, high-speed capacitor (and other types of impedance coupling arrays) impractical. This was the unwanted capacitive coupling in a simple capacitor storage matrix.

In Fig. 5.13, when driver 1 is made active, capacitor C1 couples the voltage shift to the A sense line, capacitor C2 couples the voltage shift on the A sense

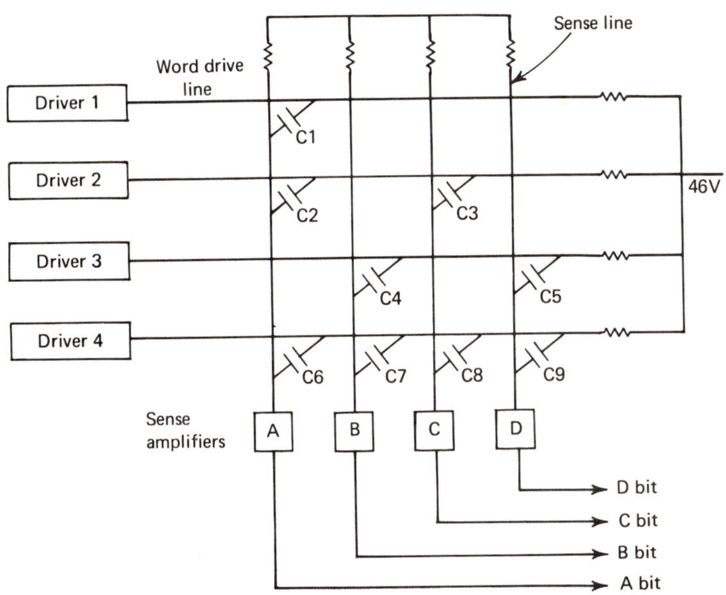

Figure 5.13 A CROS matrix.

line to the driver line 2, and capacitor C3 couples the voltage shift on word drive line 2 to the sense line C. This unwanted signal is called noise. The difficulty here is how to shield the sense lines against such noise. Because of the nature of this array, this means that the word lines themselves must also be shielded against noise. One way to overcome this difficulty was suggested earlier, i.e., design the sense amplifier with a threshold that will reject the weak signal (noise) and amplify only the desired strong signal. To further reduce the unwanted signals, each sense line is divided into a pair of sense lines. These lines feed a differential amplifier which amplifies only the difference between the two input lines. This configuration is shown in Fig. 5.14.

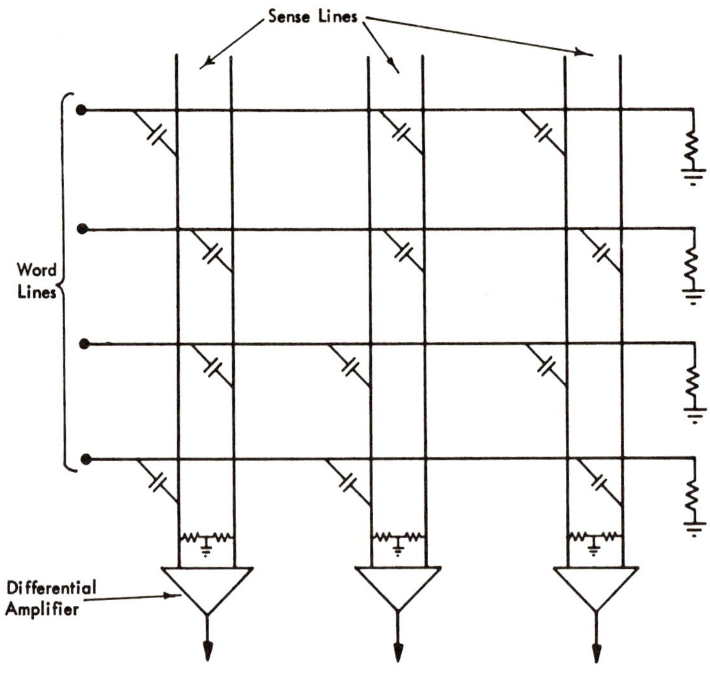

Figure 5.14 Balanced capacitor matrix (Cubbage's system).

Each word line couples with one sense line or the other, depending on whether the information stored is 0 or 1. Since most of the noise on the sense lines exists as a common-mode voltage between the conductor and ground rather than between the conductors themselves, the differential sense amplifier does not register this noise. A capacitor that causes a positive output from the differential amplifier represents a zero bit, while a capacitor that causes a negative output represents a one bit. Thus each drive line has the same capacitive load regardless of the bit configuration. This limited solution was proposed by Cubbage in his U. S. patent No. 3,183,490.[8]

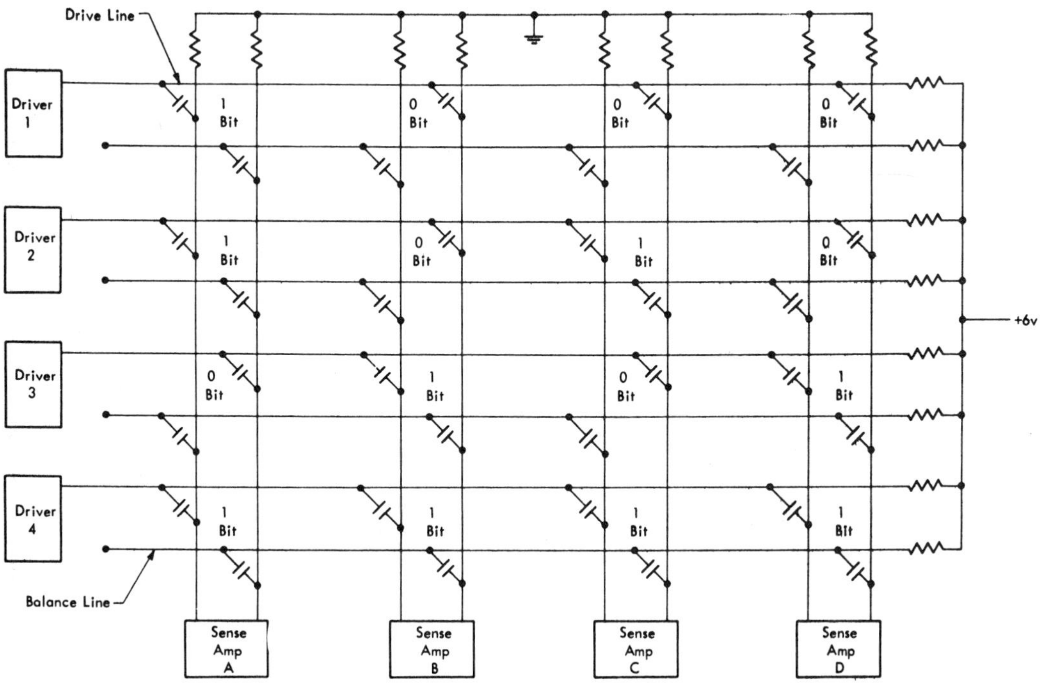

Figure 5.15

An extra line, called a balance line, was later added to each drive line by Owen and Proudman.[5] This balance line, shown in Fig. 5.15, is not connected to a driver; instead it is connected to the drive line voltage source at one end and allowed to float at the other. Where a word line couples with one conductor of the balance sense lines, the drive balancing line couples with the other. Note that the balanced lines are terminated in exactly the same way as the word lines. This arrangement yields a balanced system with uniform transmission characteristics at all frequencies, independent of the bit configuration stored in the array.

A ROS memory of this type is used in the IBM System/360, Models 50, 65, and 67.[6] The IBM balanced capacitor read-only storage system has a 200-nanosecond cycle time and a 90-nanosecond access time. Its total capacity is 2816 words of 100 bits each, organized into 16 ROS planes with 176 words per plane. Each address line can select two words of 100 bits each, designated as upper and lower words. The number of address lines in the system is therefore 1408, and those are divided into two gates as shown in Fig. 5.16. Each gate is made up of eight planes, four on each side of the gate. Two of them, A and B, are mounted back-to-back on one frame, and another two, C and D, are mounted back-to-back on the other frame. One concludes that there are 704 address lines per gate, whose individual selection is achieved by a 22 × 32 transistor matrix.[8]

1. Address lines and terminating resistances
2. Bit planes
3. Drive circuits
4. Sense lines terminating resistances
5. Sense amplifiers
6. Latches

Figure 5.16

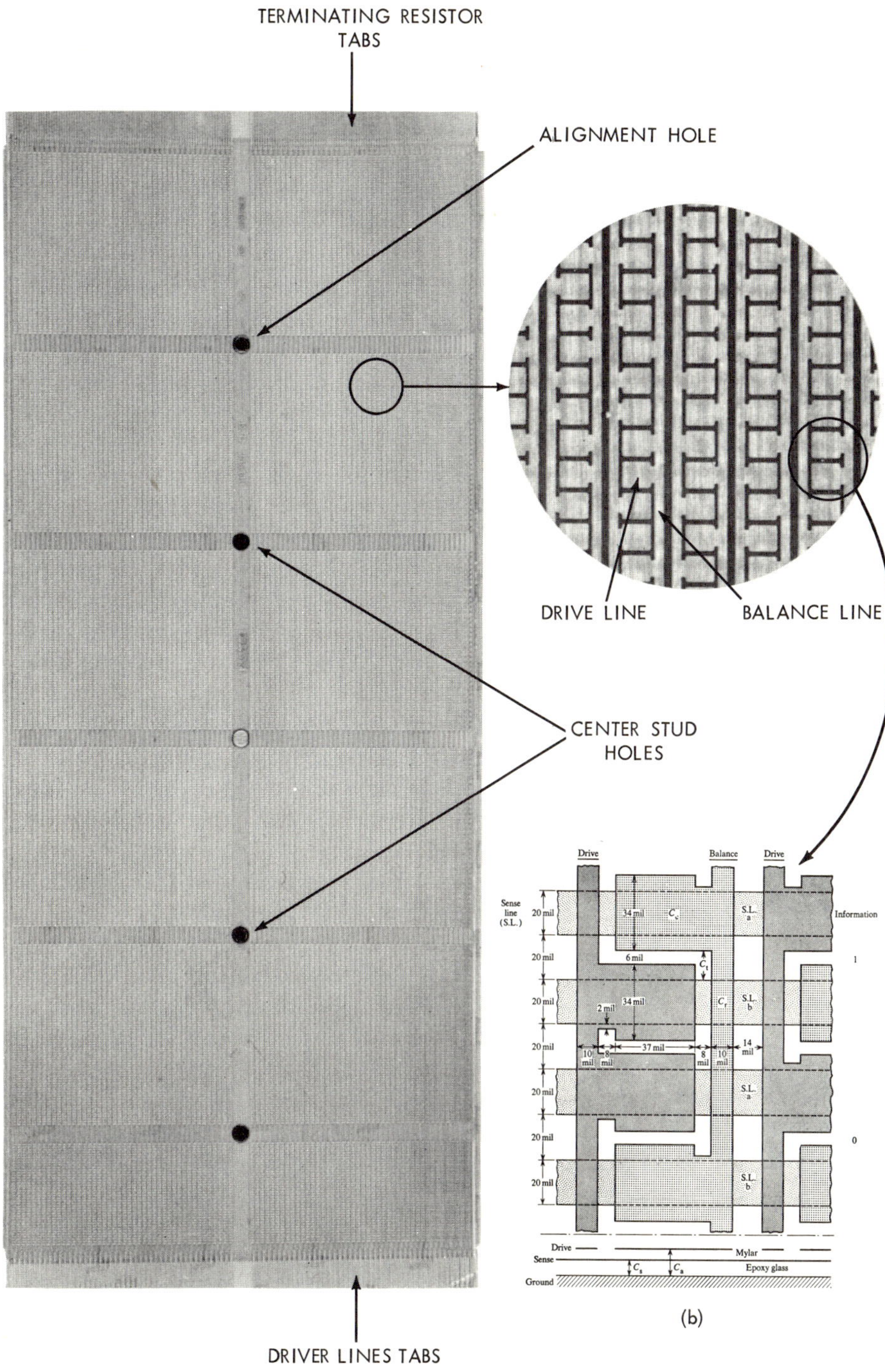
TERMINATING RESISTOR TABS
ALIGNMENT HOLE
DRIVE LINE
BALANCE LINE
CENTER STUD HOLES
DRIVER LINES TABS
Drive
Balance
Drive
Sense line (S.L.)
20 mil
34 mil
C_c
S.L. a
Information
6 mil
C_t
1
34 mil
2 mil
C_r
S.L. b
37 mil
10 mil
8 mil
8 mil
10 mil
14 mil
S.L. a
0
S.L. b
Drive
Mylar
Sense
C_s
C_a
Epoxy glass
Ground
(b)

(a)

Figure 5.17

Below the storage arrays in each gate are four laminated printed circuit boards which contain the selection matrix and the word drive circuits. To the left of these arrays are more laminated boards which contain the sense amplifiers, latches, power distribution, and timing control circuitry associated with this ROS array.

Each of the sixteen bit planes is made of a copper-clad epoxy glass laminate, approximately 8 × 18 inches, with the bit pattern etched according to the desired bit configuration; these are often referred to as the information documents. The capacitive storage is achieved by sandwiching a 1-mil Mylar dielectric between the drive and balance and the sense plane. The drive and balance lines are photo-etched from a sheet of copper-clad epoxy glass, shown in Fig. 5.17(a). There are 88 driver line tabs at the bottom, and twice that many at the top. The top tabs are used to connect the drive and balance lines to terminating resistors, and the bottom tabs connect the drive lines to the driver circuits. The four holes in the bit plate are used to align the bit plane to the sense plane. A pressure plate system is applied to hold the assembly together and maintain an accurate registration between the bit and sense plane.

Figure 5.17(b) shows an exploded picture of a small area of the bit plate; it shows the protrusions of the drive and balance lines of two bits of information. These protrusions increase the size of the capacitors to form the bit configurations which, because of their shapes, are called *flags*. The flags are positioned over either sense line of the pair of sense lines. This pair of sense lines is connected to one sense amplifier which amplifies the difference current. The positioning of the flag (the protrusions from the drive line and the balance line) depends upon the desired stored information. In Fig. 5.17(b), the top bit shows the drive line flag over sense line b and a balance line flag over sense line a. This represents the storage of a binary 1 bit in that location; while the bottom bit, where the positioning of flags is reversed, represents a binary 0. The effective capacitor coupling here is approximately 0.5 pF.

The drive line on the bit plane connects to 200 bits serially. This is equivalent to two 100-bit ROS words, and it is terminated at one end with 100 ohms to +6.0 volts. This means that the sense plane contains 400 sense lines in each CROS plane. Two sense lines are required to read one bit. These sense lines are photo-etched into copper-clad epoxy glass plates and are permanently mounted to the array gates.

The IBM Models 65 and 67 use all 100 bits per ROS word, while Model 50 uses only the top 90 bit positions, leaving 10 bits of each word unused.

5.3.4 Inductive Read-Only Storage Arrays

Linear inductive arrays are similar to the capacitive arrays except that, in the inductive ROS storage, a drive current is pulsed through the word line. The pulse is inductively coupled to the sense lines if an inductive coupling is present. The presence of an inductive element is, in turn, a function of the stored information pattern. The inductive matrices may be divided into three classes: the air coupled which can be made either fixed or semipermanent; the open magnetic paths similar to the one used in the Atlas 1 computer; and finally the closed magnetic path. Examples of this last type are the transformer read-only storage memory used in the IBM System/360, Models 20 and 40 and the E core transformer memory used in the RCA Spectra 70/45. It is this latter type that will be emphasized here; but first we will illustrate the first type.

An example of the air coupling inductive array which is of the permanent type is the alternate current path inductive store shown in Fig. 5.18. In this

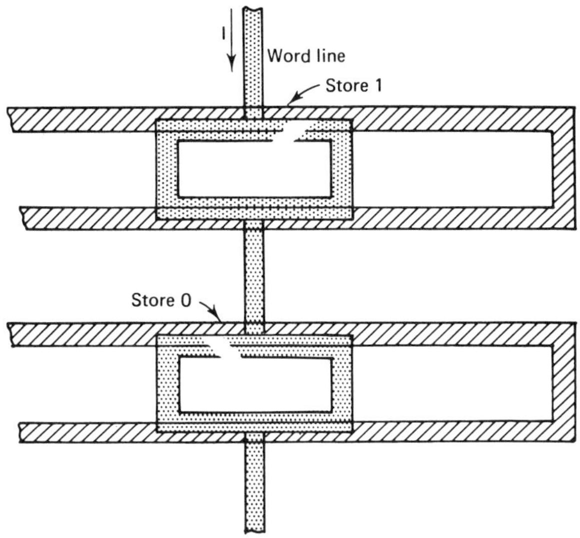

Figure 5.18 The inductive coupling array.

array, the word drive in the form of a current pulse is induced in the selected word line. Note that the sense lines in the capacitive arrays are replaced by sense loops, each terminated by a sense amplifier. The word line also contains loops that are designed to overlap the sense loops. These two sets of lines are etched on either side of a thin insulator sheet. At each word bit loop position, the drive current is forced to go in either a clockwise or counterclockwise direction, depending on the cut made in this loop. The net result is to force the drive current and the resulting magnetic flux in such a way that the polarity of the output voltage measured at the input of the sense amplifier is positive or negative, depending on the location of that cut. If the cut is made to the right of the word drive line (by convention only), this bit is said to store 1. This condition then forces a negative voltage to be measured between the upper and the lower terminals of the bit loop. If the slit is made to the left, the bit is designated as 0, and the polarity of the output voltage is reversed.

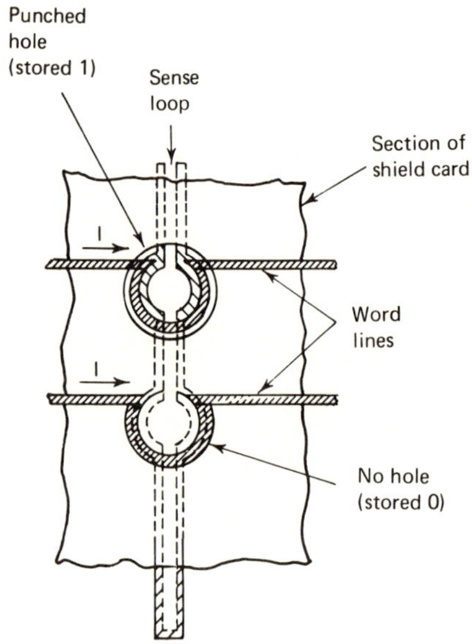

Figure 5.19 Card changeable inductive array.

Another example of the air coupling induction memory which is changeable is the shielded induction array shown in Fig. 5.19. The current induced in the word lines (from left to right) is made to travel in a counterclockwise direction; thus the measured output voltage at the sense loop terminals is always of the same polarity. Therefore, another scheme for discriminating between 0 and 1 bits in storage is required. If the word loops and the sense loops are formed on two separate boards and a thin insulating material is sandwiched between the two sets of lines facing each other, each bit position will have a 1 stored in it. On the other hand, if a ground metallic foil is sprayed with insulating material on both sides and is sandwiched between the two planes, the inductive coupling is virtually eliminated. Therefore, by punching the holes in the proper locations in this metal shield card, any bit pattern can be stored. By simply replacing the information document in a manner similar to that for the card capacitor read-only storage array discussed earlier, we obtain a semichangeable storage matrix.

The final class of the inductive array coupling to be discussed is closed magnetic path arrays. Among these is the transformer read-only storage array. First described by T. L. Dimond in 1951, it has become known as the "Dimond Ring." This type of storage was first used for code translation in the American No. 5 crossbar telephone system and for subroutine storage in the Bell Laboratories Model 6 Computer. More recent uses include the Atlas ground guidance computer system and the IBM System/360, Models 20 and 40, for storing the microroutines.

The principle of operation is illustrated in Fig. 5.20. One input wire is provided for each word stored; this is called the word line drive. One transformer is provided for each bit in the output word. The word line is either threaded through or bypasses a ferrite core, depending on whether a 1 or a 0 is to be stored in the corresponding bit location. Thus, in Fig. 5.20, word A stores 1011, word B stores 0101, and word C stores 1100.

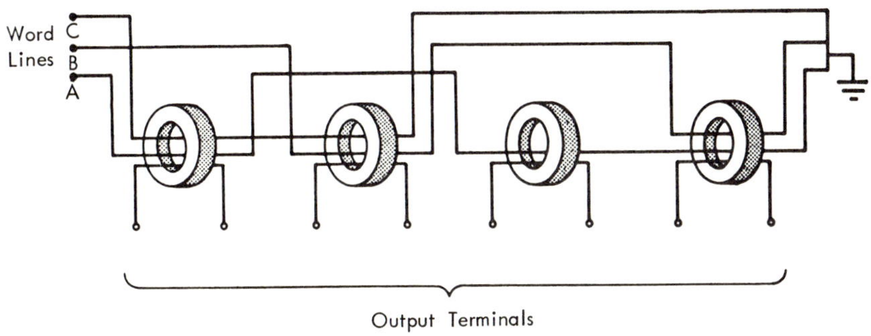

Figure 5.20 Transformer read-only store.

To read out a word, a current pulse is passed through the corresponding word drive line. This causes an output current to be measured at the sense amplifier of the bit through which the word drive line is threaded. This output current represents 1. Ideally, the transformer through which the selected word drive is not threaded should have no output; but in practice, a small output current is sensed because of the impedance of the word lines and the capacitance through them. This unwanted output must be kept small compared to the 1 current output.

TROS CONSTRUCTION. The main advantages of the TROS memory are its simplicity of construction and the large output signal available. This advantage must be weighed against the speed disadvantage which can be the most critical requirement in the system.

Earlier TROS memories used ordinary insulated wires as the word lines. They were threaded by hand or by machines through cores made of soft ferrite, arranged in a row or around the circumference of a circle. This construction gave way to a more durable and more reliable construction which will be described next. This assembly is used by IBM for the microprogram storage in the IBM System/360, Models 20 and 40.[8]

The principle of operation of the new assembly is similar to the one just described. Soft ferrite cores made into a U-shape with a second I-piece were used as current transformers for the storage of data. Each current transformer represents one bit position in the output word which may vary in length up to

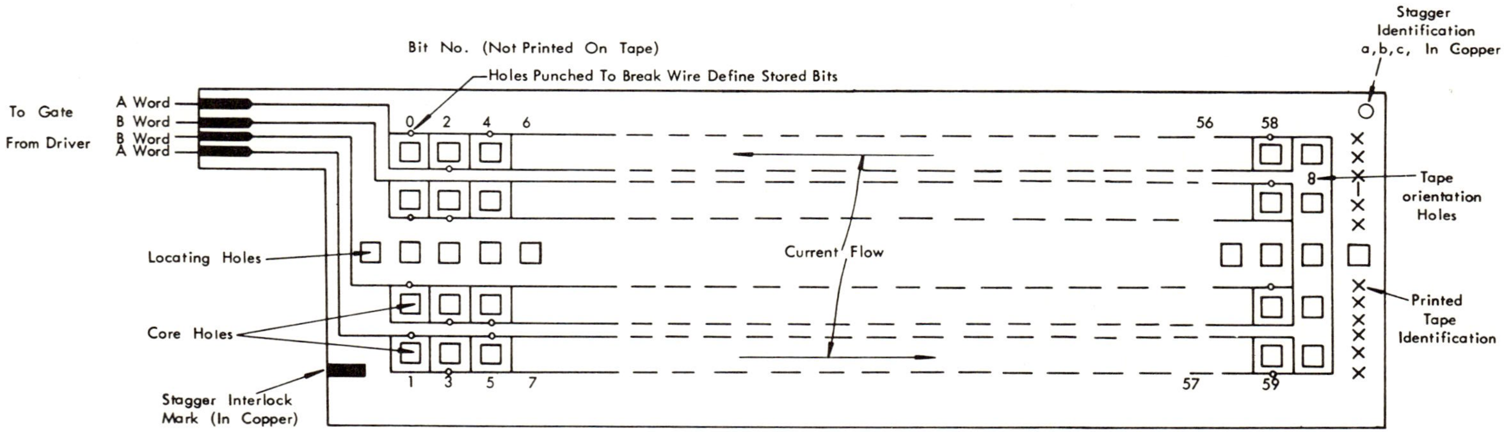

Figure 5.21 Printed circuit on plastic tape for transformer ROS.

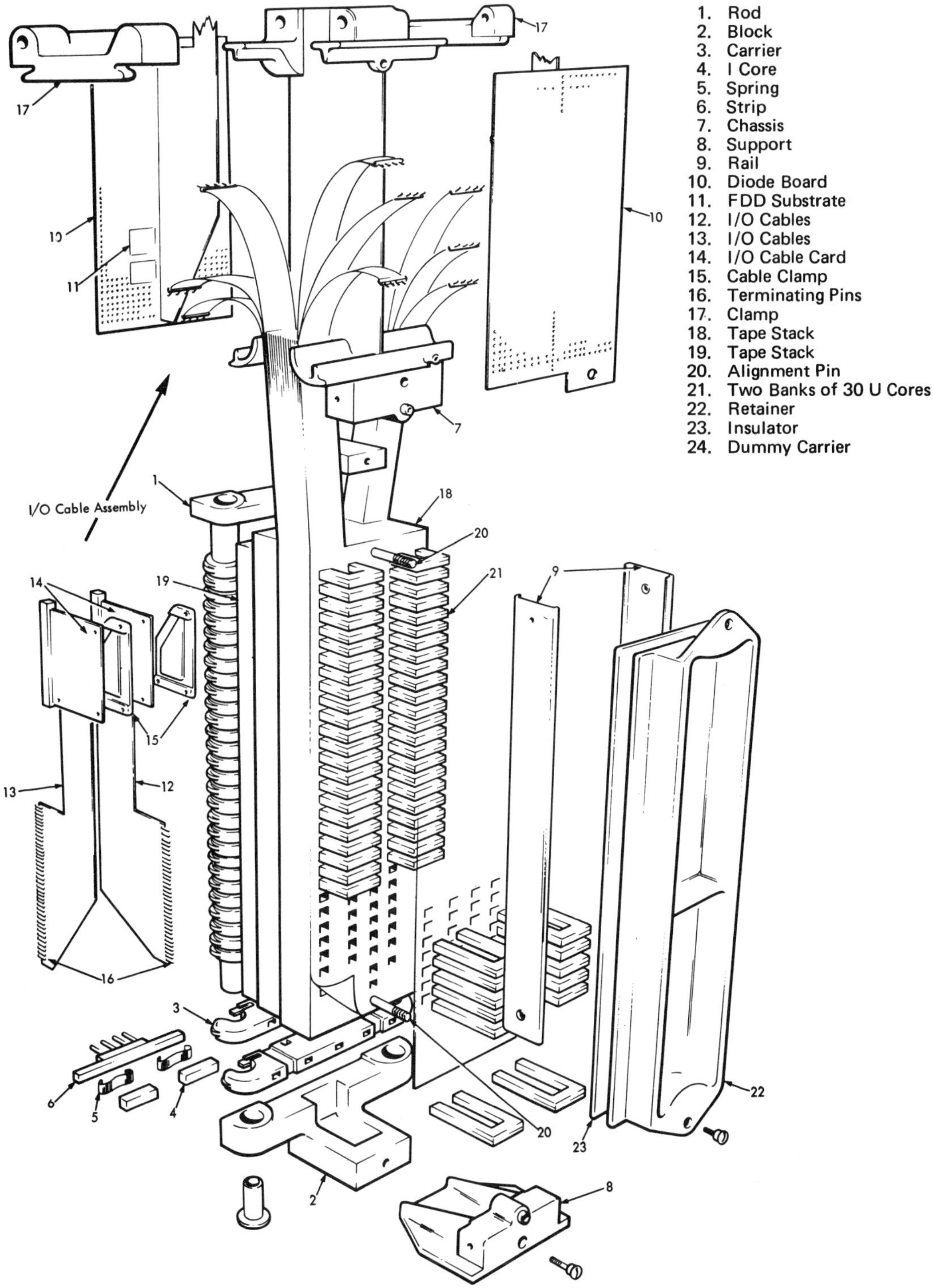

Figure 5.22 TROS module (exploded view).

a maximum of sixty bits. The word lines are photo-etched on flexible Mylar or plastic tapes with copper sheet bonded to them. Two word drive lines are etched on each tape. Figure 5.21 shows the layout of the TROS tapes. The two conductors (word drives) are made into a ladder design to encircle one leg of the U-shaped transformers. There are a maximum of sixty bits per word; hence, there can be a maximum of sixty transformers going through the TROS tapes. Each bit position stores either 1 or 0 simply by punching a hole in the inner or outer side of the ladder conductor. When a hole is punched (breaking the contact) in the inner side, the drive line current is forced to bypass that bit position. The absence of primary drive current in that bit position results in no appreciable output pulse normally sensed in the thirty-five-turn secondary winding around the I-piece. The two conductors per tape share the sixty transformers going through the tape.

The transformers are arranged in modular units; an exploded view is shown in Fig. 5.22. Each module contains 128 TROS tapes, or 256 words of 60 bits per word. A number of these modules may be interconnected to increase the capacity of the system from 256 to 6144 words. In the IBM System/360, Model 40, they are mounted eight to a subframe. There are two such subframes arranged on the two sides of a gate which is arranged to swing to facilitate easy access. The address decoding, drivers, timing, and sensing circuits are

Figure 5.23 Complete store for the IBM System/360, Model 40. A second set of eight modules is mounted behind the eight shown.

housed on two large SLT cards. The gate circuit cards used in the drive word scheme are mounted above the TROS modules on a special printed board. Figure 5.23 shows the total TROS assembly for the Model 40.

To reduce the capacitance between the drive lines on adjacent tapes, three different types of tapes are used. On each type, the conductor pattern is displaced from the holes by different amounts. The three types are labeled A, B, and C, and they are arranged sequentially throughout the deck of tapes.

5.4 NONLINEAR ARRAYS

Having discussed linear resistive, capacitive, and inductive arrays, we will now proceed with the nonlinear class of read-only storage arrays. Among the best known are the diode arrays, the permanent magnet twistor store arrays, the eddy current twistor store arrays, and the biased flat film arrays. The main advantage of nonlinear arrays over the linear ones is their immunity to many of the sneak signal paths and to the high signal-to-noise ratios that plague the linear arrays. This advantage can be translated into better and less-stringent sense amplifier design characteristics. Because of their improved input-output signal ratio, nonlinear arrays require less input word drive than do the linear ones; thus, the power dissipation problem is reduced substantially.

5.4.1 Diode Arrays

Diode matrices (Fig. 5.24) as an encoding medium have been under review since 1949; however, since the cost of such an array was prohibitive, little interest was shown in them. The recent advances in semiconductor technology and the large-scale integration and fabrication technology have pushed semiconductor diode arrays to the forefront, where they are receiving considerable attention. A number of experimental ROS arrays have been fabricated. Earlier matrices were essentially permanent in construction, but later developments resulted in punchable printed circuit cards. One such development uses vacuum deposition of semiconductor material to form diode arrays on these printed circuit cards. Information (0's and 1's) can be encoded by breaking the input connection to the diode where a zero is required. The array shown in Fig. 5.24 is similar to linear arrays, except that the impedance coupling element is a diode rather than a resistor, capacitor, etc.

The principle of operation is simple: in the quiescent state, all the diodes are reverse-biased. In a readout cycle, the selected word drive line is driven with a positive-going pulse of sufficient magnitude to forward-bias the diodes connecting this word drive and the selected bit sense lines, causing them to conduct. The outputs of the diodes are sensed by the corresponding sense amplifiers. The input to each sense amplifier is in reality an n-input OR gate,

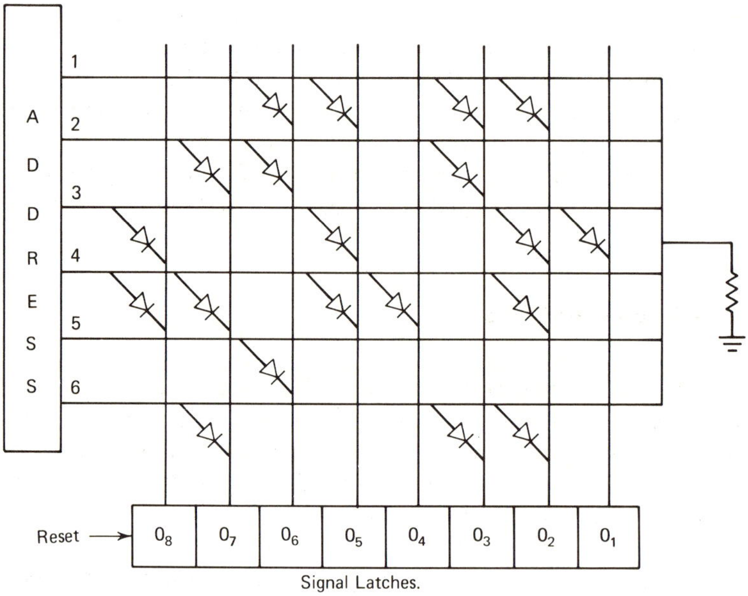

Figure 5.24 The diode matrix.

where n can be equal to the number of words in the matrix. This imposes some design construction if n is large and if the diodes have wide variations in their switching transient characteristics. One positive characteristic that promises greater economics and better packaging density is the possibility of fabricating the ROS diode arrays and the ROS address decoders on one substrate.

5.4.2 Permanent Magnet Twistor Store Arrays

The magnetic matrix was developed in 1955, by Bell Telephone Laboratories, as a mean for storing semipermanent digital information. This memory was designed for computers which require random access to stored information which is changed very infrequently. The information is stored in an array of small permanent magnets. The presence of this magnet at a given storage location in the array represents a binary 0; it is sensed by a wire wrapped with magnetic tape placed close to the magnet.

The basic hardware needed to understand the underlying principle of operation in this ROS technology is shown in Fig. 5.25. Each word of storage consists of word access core and a word solenoid associated with it. A twistor wire is inserted at each bit position as shown in the figure. The twistor wire is formed by winding a 4-79 permalloy tape in a helix around a 0.003-inch copper wire at an angle of about 45°. All the twistors in the array are

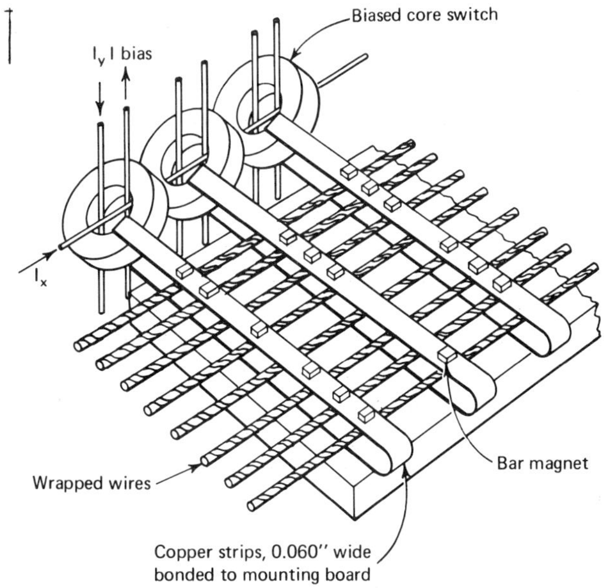

Figure 5.25 Permanent magnet twistor store.

magnetized in the same direction. Each intersection of a twistor with a copper strip represents a bit position.

Coincident currents I_x and I_y select a biased core switch which, in turn, selects a word solenoid. This drive current induced in the selected word solenoid produces a magnetic field in the direction which reverses the magnetization of the twistors and produces a voltage output. The twistor magnetic field reversal can, however, be inhibited locally by placing a permanent magnet at the appropriate bit positions in the array where a stored 0 is desired. The magnitude of the field in the solenoid can be high to achieve fast switching. The field of the bar magnet must therefore be sufficient to inhibit the switching of the wrapped wire. The interrogation is nondestructive since, on removal of the core switch drive current, a reset (reverse) current is passed through the word solenoid. This current switches the twistors to their original states.

The earliest operating twistor memory model consists of sixteen cards. Each card stores thirty-two 26-bit words, for a total of 512 words. Several improvements have since been discussed. C. F. Ault et al., of the Bell Telephone Laboratories, described an operating 65k-word, 88 bits per word model designed for the ESS program store. Most of these modifications were aimed at achieving the desired density by solving the stray field interactions which have in the past dictated that the spacing between the various elements be large. The other problem which had to be overcome was that the size of the memory module was limited by the attenuation and delay along the twistor wire, both being greater than in ordinary wire.

One variation which yields a bipolar output signal was achieved by eliminating the return wire. The sense loop now consists of a single twistor wire. Two magnets per bit are required, but only one of the pair is magnetized. For storage of 1, the left magnet is magnetized, and for a 0, the right one is magnetized. Hence, one or the other is able to switch, and a positive or negative signal is sensed. This has the effect of increasing the output signal and giving the sense lines more uniform characteristics. Another modification which can produce the desired density improvement is achieved by placing the return wire outside the solenoid, thus reducing the spacing between the twistors and the permanent magnets. The eddy currents induced in the conducting cards which hold the permanent magnets enhanced the field produced by the word solenoids, causing the twistors to switch faster. Many other variations have been proposed and discussed. However, a complete discussion of these variations in any technological depth is outside the scope and objectives for this book. The reader is referred to a number of references cited at the end of this chapter.

5.4.3 Eddy Current Memory Arrays

Eddy current memories make use of the eddy current as the coupling media between the drive wire and sensing wire. The basic elements are illustrated in Fig. 5.26. The principle of operation is a simple one. If a pair of wire loops, as shown in Fig. 5.26(a), are orthogonal, they have no mutual flux linkage. A drive current in one loop will have no effect on the voltage output of the other. If a closed loop, Fig. 5.26(b), is placed over the two orthogonal loops, eddy current will be induced in the loop which, in turn, gives rise to the mutual flux linkage. If the loop is punched open as shown in Fig. 5-26(c), the mutual flux linkage is removed, thus removing the eddy current coupling. The information, 0, can therefore be stored by punching a small slot at the one side of the conductor ring.

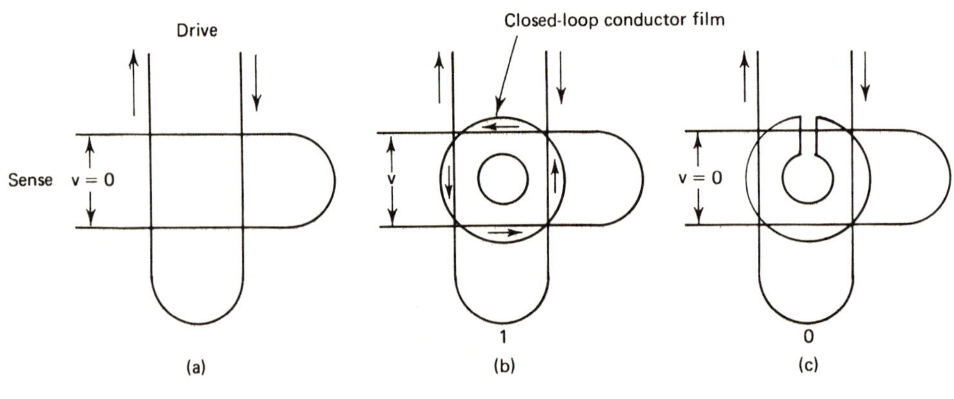

Figure 5.26 Eddy current memory element.

An equivalent circuit for the closed-loop conductor is shown in Fig. 5.27. The circuit consists of a driver coil L_1; the eddy current closed-loop conductor with inductance L_2 and resistance value R_2; and a sense coil L_3. M_{12} is the mutual inductance between L_1 and L_2. M_{23} is the mutual inductance between L_2 and L_3. We will assume that the mutual inductance M_{13} is zero. Analyzing the circuit by the Laplace transform method yields the final expression:

$$e_3(S) = i_1(S)\left[S^2 M_{12}\frac{M_{23}}{L_2}\left[S + \frac{R_2}{L_2}\right]\right]$$

If the current i_1 is chosen to be a ramp ($i = mt$), the output voltage $e_3(t)$ will be:

$$e_3(t) = \left(mM_{12}\frac{M_{23}}{L_2}\right)\exp -\left(\frac{t}{L_2/R_2}\right)$$

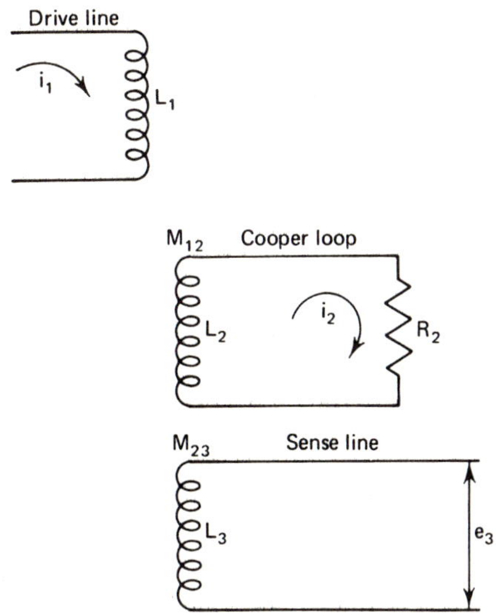

Note: $M_{13} = 0$

Figure 5.27 Eddycard equivalent circuit for 1.

For an alternating current with an angular frequency of ω and an amplitude i_1 applied to the drive coil, the output voltage is expressed as:

$$e_3 = \frac{\omega M_{12} M_{23}}{R_2 + j\omega L_2}\, i_1$$

Hence the L_2/R_2 ratio can be adjusted to give any pulse width. The current in the loop is proportional to M_{12}, and the output voltage is proportional to the product $M_{12}M_{23}$. The loop current can therefore be reduced by reducing M_{12} without affecting the output voltage by a proportionate increase in M_{23}.

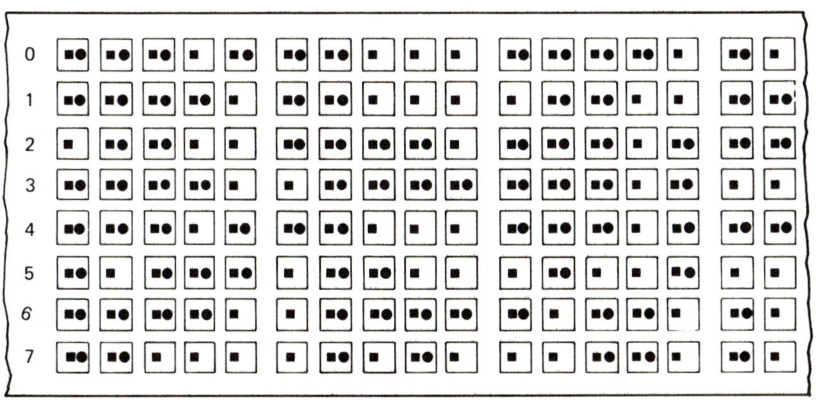

Figure 5.28 Close-up view of punched eddycard.

Figure 5.28 shows a simplified model of an experimental eddy current memory with a capacity of 1024 words, each having 45 bits.

Cards used in the eddy current memories are fabricated by chemical etching of the silver-coated copper conductors on an epoxy glass sheet. The complete card has eight 45-bit words. Etched on another board large enough to accommodate 16 eddy cards are 128 drive wire pairs. The whole memory is made up of eight such drive boards, for a total of 1024 words. A sponge cover is used to hold the eddy cards firmly and evenly against the drive boards.

5.4.4 Eddy Current Twistor Store

We have already discussed a permanent twistor memory store in which the twistor is taken outside the word solenoids to increase the switching field. This field is further increased by placing a conducting sheet above it, to which the magnets are bonded. Eddy currents produced in this sheet help to switch appropriate twistor bits. Krylow, Perry, and Reimer proposed a scheme in which the magnet bars are found to be unnecessary. Instead, holes are punched in the sheet in locations where a 0 is to be written. These holes lower the switching field at that twistor position below the threshold value; hence, no flux reversal takes place. Figure 5.29 shows the basic eddy current

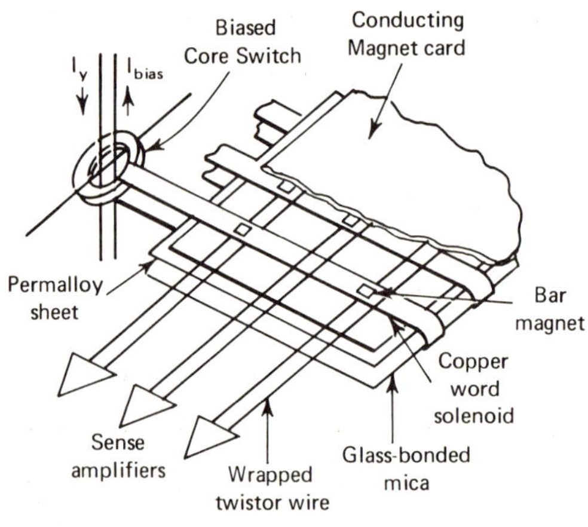

Figure 5.29 Eddy current twistor memory elements.

twistor memory element. The twistor is used to sense the presence of eddy currents in the sheet. If eddy currents are present, their fields will reinforce the solenoid field and switch the twistor; otherwise, the twistor is left in its original state.

5.4.5 Magnetic Film Read-Only Memories

The EDSAC II read-only storage utilizes a conventional core selection array with one core per word and a coincident x and y drive and a third line for current bias. Using coincident current technique, only the selected core is switched during a memory cycle. The bit sense winding is threaded through the cores of the corresponding word bits only if a 1 is required in that bit position. Bypassing a core is then equivalent to having a 0 stored in that position.

Another magnetic film read-only memory which was described by Hart in 1961 uses a pair of thin vacuum-deposited films as memory cells. This memory operates on the general principle of storing a binary digit in one film and sensing its remanent state by switching a second film in a manner which leaves the magnetic state of the first unchanged.

5.5 OPTICAL SYSTEMS

The basic characteristic of this class of ROS memories is that the information is stored as a pattern of opaque or transparent spots on a normally flat surface, such as card, plate, or disk, or in the form of holes punched in a specific pattern on an opaque card. This means of photographic storage has a high storage density unmatched by any other class. Initial optical storage arrays utilized a photoelectric device to sense a light beam emerging from holes on the mechanically selected punch cards.

Among the earliest of the optical systems is the photographic disk store. Information, here, is stored on a glass disk store. A light source generated by a cathode ray tube is directed along two paths; one path leads directly through a photomultiplier tube (PMT); the second path is directed through the selected information plate to be scanned before going to another PMT. The intensity of the light passing through the second PMT is compared to the light intensity passing through the first. If this ratio exceeds a preselected threshold, the system registers a hit or a 1 in that storage array position; otherwise, it registers a 0.

Another optical system which has received considerable attention is the flying spot store system. Information is permanently stored on photographic plates as a pattern of opaque and transparent spots. The CRT is again used as a light source whose output is projected and optically focused on the photographic plate. A PMT detector is situated behind the plate to sense the intensity of the output light beam which decodes the presence or absence of an opaque spot, thus indicating a 1 or 0 at a particular storage spot on the storage plate. A closed-loop servo system is used to position the CRT beam on the small area of the information plate. The address of the selected word or

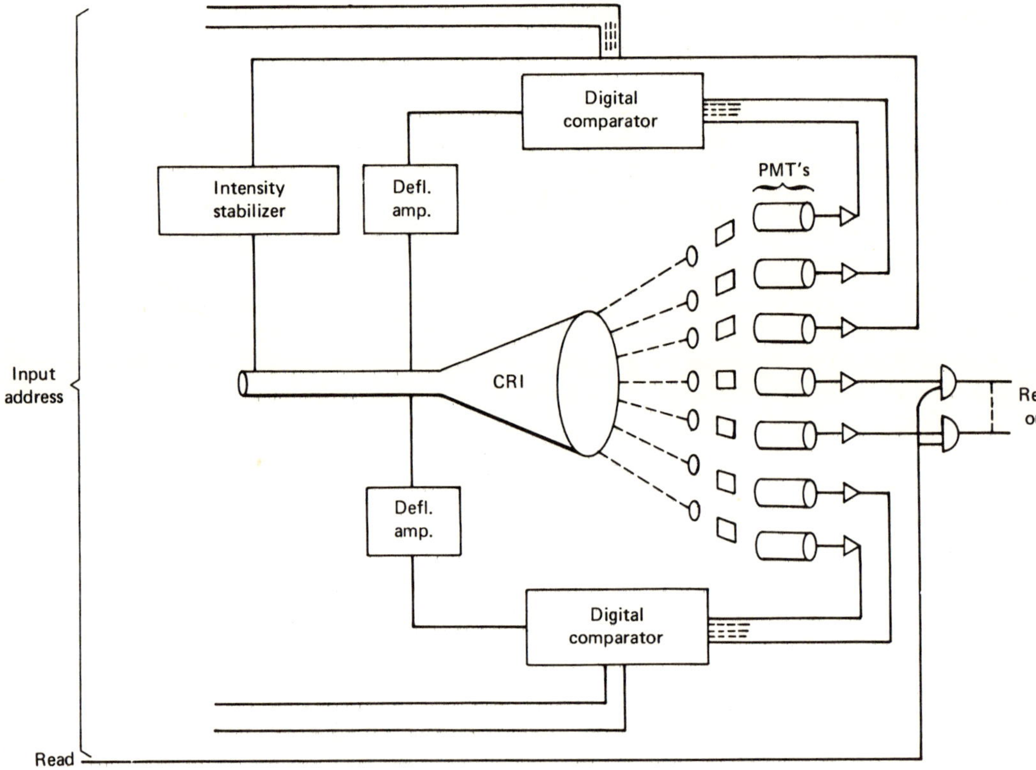

(a) Flying spot store circuit organization

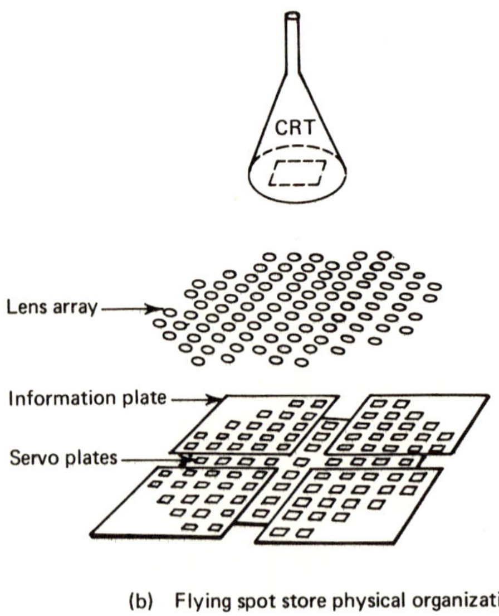

(b) Flying spot store physical organization

Figure 5.30

selected submatrix is located on address code plates which are located in the same plane as the stored information plane.

Figures 5.30(a) and (b) show a simplified flying spot storage. The A area represents a substorage matrix. The system shown in these figures consists of one CRT, B focusing lenses, the changeable photostorage plate, and finally, the condensing lens and photomultiplier tube array.

5.5.1 Fiber-Optics Store

Another system proposed by Hoffman and Jeffreys uses a fiber-optics system to replace the focusing and condensing lens arrangement. The light source can be a CRT or a gallium arsenic diode whose light output is transmitted through the glass fibers ("light pipes") to a photodetector or PMT tube.

5.5.2 Holographic Store

A relatively new photographic storage process which has been developed recently is called *holography*. Its primary advantage is that it is capable of storing a three-dimensional image. This has many advantages and many interesting applications; for example, it facilitates the three-dimensional array processing economically and fast.

The basic concept here is a means of recording the nature of the light wave emanating from any source—such as the information array. The recorded image is really the wave pattern resulting from the interference of the information-carrying beam emanating from the transparent and opaque spots (representing 1's and 0's on the information document) with another reference light beam. These two beams are coherent with each other. Hence, the interference between them is stationary in a plane in space, the position of which can be calculated. This interference can be recorded on a photosensitive plate, developed, and stored for later regeneration or reading.

The reading is carried out by illuminating the developed photographic plate with the reference beam. This beam is diffracted by the wave structure recorded on the plate. This diffraction causes another beam of light to form which is an accurate reproduction of the information document.

5.6 SEMICONDUCTOR READ-ONLY STORAGE

The semiconductor flip-flop has been introduced by a number of semiconductor manufacturers as a practical contender for a significant portion of the memory in a computing system. This reality can be achieved only by the mass production of similar items—a process which is in exact accordance with the present production methods in the semiconductor industry and is far more attainable than that of the LSI logic field. For a number of reasons the memory system, both NDRO and writable cells, forms a useful vehicle for proving out the concept of large-scale integration in general, and the techniques of interconnecting the cells such as the discretionary wiring. The LSI process is even more feasible in the form of a linear array of x words containing y bits, than in the more general, nonlinear, coincident type of storage. This linear array can be used in many functions, ranging from the control and scratch pad memories to shift registers and other functional memory applications. The regularity of the interconnections of these memory cells also permits maximizing the number of circuits per unit area. By including the word driver and address decoder on the same LSI slice, a drastic reduction in the number of pin connections can be achieved. This pin-to-circuit (cell) ratio is perhaps the most difficult problem to be solved in the LSI logic field.

A memory cell designed for larger-scale integration must meet several criteria.[8] It must be small to allow a maximum number of cells per slice and, thus, reduced cost per bit. Second, it must be capable of operating with the maximum component tolerances. This implies greater yield and minimum cost per bit. Third, it must have low power dissipation for improved speed, reduced packaging density, and increasing density. Fourth, the cell should be fast enough to ensure its utility in the maximum number of applications. Finally, its accessing and driving circuitry must be compatible with the other circuits in the system.

Figure 5.31 shows a basic memory cell designed by the Autonetics Microelectronics Laboratories. It is a flip-flop with multiemitter transistors arranged to permit reading and writing of the contents of the cell. More than 3800 such cells are placed on a slice with the word drivers required.

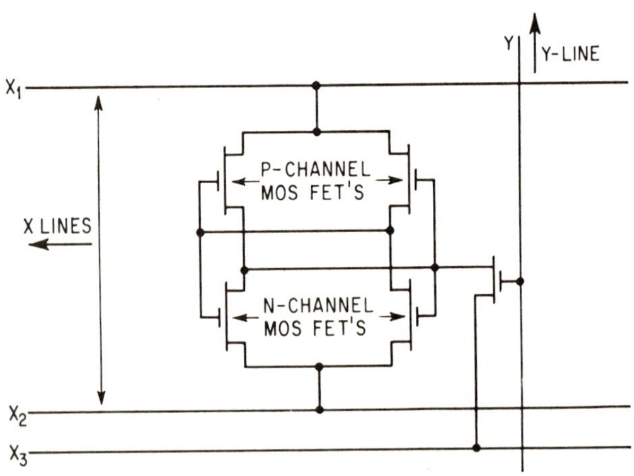

Figure 5.31 Simple MOS memory cell for an alterable NDRO array consists of two inverter circuits connected back-to-back to form a flip-flop, with a fifth MOS transistor added for cell readout. The basic flip-flop occupies 225 square mils.

At the present time, three distinct chip technologies are competing for the LSI sweepstakes. One uses conventional bipolar integrated circuit techniques; the second is the P-channel metal oxide semiconductor (MOS) field-effect technique, and the third is the complementary MOS. The last is distinguished by its integration of the P-channel and N-channel MOS devices. In recent years, MOS was bypassed in favor of the bipolar technique for two reasons:

1. *Speed.* The MOS device with its high impedance can be slower in the presence of stray capacitance.
2. MOS device parameters are unstable over a long term. This instability is due to the ion migration in the silicon oxide.

These disadvantages of the MOS must be weighted against the list of advantages, especially when it is used for memory array production. Among the advantages are:

1. MOS provides greater bit density. This is because the bipolar integrated circuit requires an isolation around each transistor.
2. Higher yield is obtained because of the simpler and fewer manufacturing steps.
3. MOS arrays have been in production for some time, while the bipolar is largely experimental.

REFERENCES

1. Foglia, H. R., McDermid, W. L., and Petersen, H. E., "Card Capacitor—A Semi-permanent Read Only Memory," *IBM J. Res. Development*, Vol. 5, No. 1 (1961), p. 67.
2. Takahashi, S., and Watanabe, S., "Capacitor-Type Fixed Memory." In Yovits, M. C. (Ed.), *Large-Capacity Memory Techniques For Computing Systems.* New York, The Macmillan Company, 1962, p. 53.
3. Van Goethem, J., "The Capacitive Semi-Permanent Information Store and Its Uses in Telephone Exchanges," *Proc. IEEE*, Vol. 107B, Supplement 20 (1960), p. 346.
4. MacPherson, D. H., and York, R. K., "Semi-Permanent Storage by Capacitive Coupling," *IRE Trans. on Elec. Comp.* (September, 1961).
5. Taub, M. D., "Analysis of Sneak Paths and Large Line Distortions in an Improved Capacitor Read Only Memory," *Proc. IEEE*, Vol. 51, No. 11 (November, 1963), p. 1554.
6. Fugg, P., Brown, J. L., and Doody, D. T., "IBM System/360 Engineering," *AFIPS Conf. Proceedings*, *26*, Pt. 1 (1964), p. 205.
7. Cubbage, J. F., U.S. Patent No. 3,183,490.
8. Canning, M., Dun, R. S., and Jeansonne, G., "Active Memory Cells for Discretion." *Electronics* (February 20, 1967).

MICROPROGRAMMING THE IBM SYSTEM/360

6

6.1 INTRODUCTION

The upward and downward architectural compatibility in the IBM System/360, coupled with the demands for inexpensive, efficient, and fast conversion tools for the user application programs, resulted in a fundamental design decision to implement the control section where possible in microcode. The primary targets were the small- and medium-size systems since they would be the prime beneficiaries in that they would have a rich, comprehensive instruction set through microprogramming at a development cost much lower than that for a compatible hardware control implementation. Many other factors contributed to favoring microprogramming implementation over conventional hardware control. These advantages were presented in Chapter 2. Among the most important ones in this case were:

1. The simplicity and efficiency in designing any number of emulators and other compatibility features. The Model 30 and the Model 40 have two different emulators each. Because of this, these systems accepted an application source deck or binary deck written for another machine and performed the job at an even better cost per job than was possible on the old system.

2. Engineering changes can be implemented in the field much faster than in a hardware control design. This resulted in more system availability for the customer.

3. Faster implementations for new and optional features requested by the user. These features are frequently added to optimize the system's performance in a particular application for which it is designed to spend most or all its running time.

4. A major consideration in the design of the System/360 was to give the user room for growth into a larger system at a later time without the heavy penalty in time and manpower and the large investment in retraining and reprogramming. This means an architectural compatibility.

Since the IBM System/360 announcement in 1964, a number of other systems which share the same design philosophy of a microprogrammed and architecturally compatible produt line have appeared in the market place. Among these systems are the RCA Spectra/70 and the Honeywell H4200/H8200. Another IBM product line which utilizes the stored program control and which is architecturally compatible with System/360 was implemented. This military version of the commercial System/360 is named the 4PI System.

There are several important features shared by the IBM System/360 processors. These features include:

1. An 8-bit (one byte) alphanumeric coding—two digits per byte.
2. Signed 2's complement, fixed-point binary arithmetic.
3. Sixteen full-word (four bytes per word) general purpose registers that can assume one of many functions.
4. Four double-word floating-point registers.
5. Two states of operation: the programmer state and the supervisory state.
6. Interrupt control and the priority in handling five types of interrupts.
7. Up to two levels of indexing.
8. Multiplexed I/O as well as two to six selector channels.
9. Rich instruction set of different width, different addressing, and a different number of main storage accesses implied by each instruction format.
10. SLT (solid logic technology) circuitry.
11. Microprogrammed control where feasible.

On the conceptual level, all models share the same basic data flow model shown in Fig. 6.1. Within this framework, major design decisions were left to be made by the system developers and engineers responsible for implementing each model. These decisions provided the point of departure from one processor to the next, and they include:

1. The speed of the basic SLT circuitry.
2. The width and speed of the buses and registers, in particular, the CPU main storage and CPU buses.
3. The width, speed, and number of separate adders in the arithmetic and logic section; also, the extent of parallelism between the data paths and the extent of gating within each data path.
4. The width, speed, and maximum addressable size of the main storage unit.

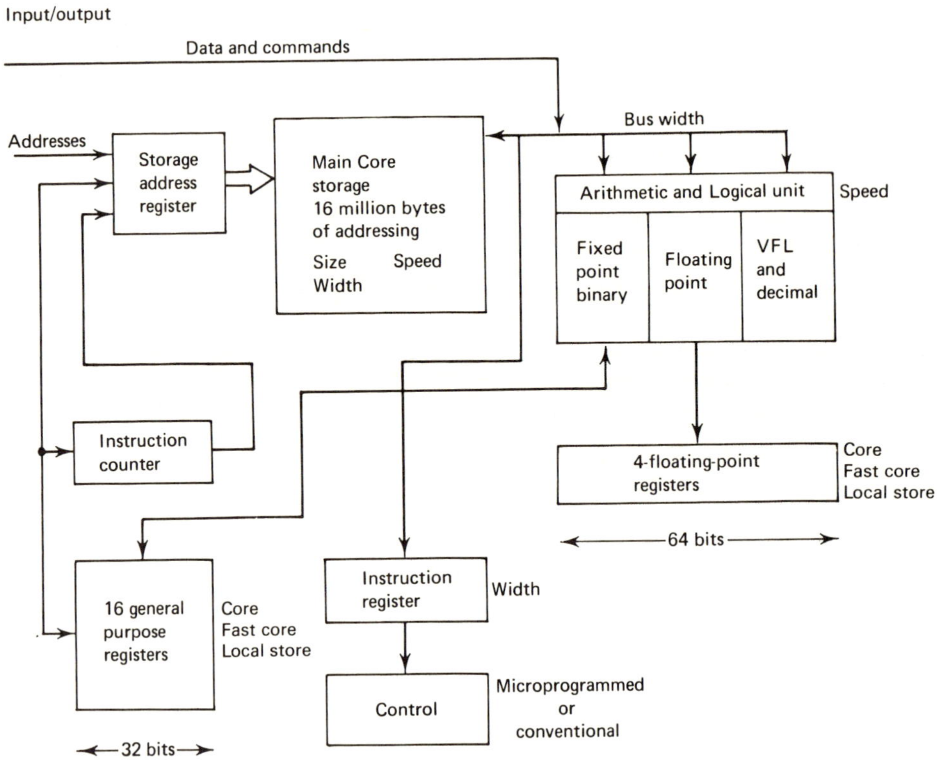

Figure 6.1 Data flow diagram indicating design decisions for all System/360 processors.

5. The speed, width, and size of the high-speed local storage. The type of information to be stored in the local storage, the extent of local storage, CPU clock overlapping, and the type of interface to be provided between the CPU and local storage.

6. The method of the data flow control. If a microprogrammed control storage is chosen, then the type, capacity, speed, and organization of the ROS memory must be decided.

7. The method of handling the decimal arithmetic on a basically binary arithmetic functional unit. In particular, a decision had to be made as to whether the six-correction method (to be discussed later) was to be implemented by microcode in two cycles (as in Model 50) or in one machine cycle and by a hardware decimal file and correct box (Model 40).

8. Another point of departure was the size of the instruction register and the extent of instruction look-ahead applied in the design.

9. The foregoing eight decisions are by no means the only points of departure between the models. Another important point prevalent in the models on the upper end of the scale is the extent of buffering from a relatively

slow main storage to a very fast high-speed buffer storage. This memory hierarchical organization was intended to close the access gap between the CPU cycle time and the main storage cycle time, making the low-speed large-capacity main storage almost transparent to the CPU.

At the time of the System/360 development cycle, the read-only storage memory technology was lagging. The best possible ROS technology with the cost and access characteristics compatible with the goals of the project was the 200-nanosecond capacitor read-only storage with a density of 100 bits per word. The lack of the proper ROS technology with an access time equal to that of the CPU time for the Model 75 necessitated that the Model 75 be designed with a conventional hardware control (Table 6.1).

Figure 6.2 shows how the System/360 control system operates. It is based on the encoded ROS word in which the selected ROS word is gated into a permanent ROS data register. The outputs of the ROSDR storage bits which make up a field are decoded to provide the controls to the mutually exclusive subset of the static control lines. A single word in the ROS device is called

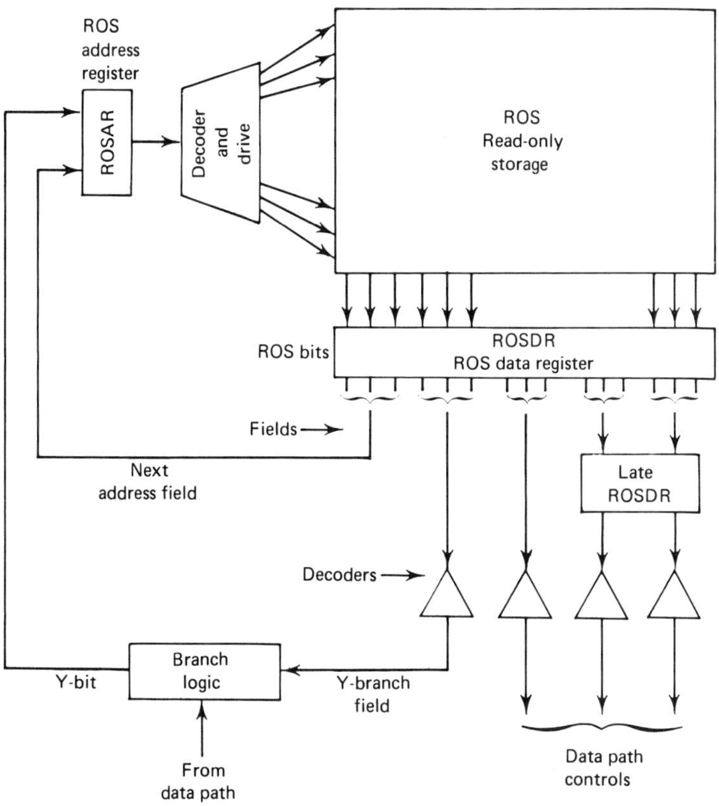

Figure 6.2 System/360 read-only storage microprogram control.

Table 6.1 SYSTEM/360 STORAGE CHARACTERISTICS

IBM System/360 Model	CPU Cycle (μsec)	Main Storage (MS) Cycle (μsec)	Width of MS Data Bus	ROS Storage: ROS Speed (μsec)	ROS Storage: ROS Type	ROS Storage: Capacity in Kilo Words	ROS Storage: Bits/Words	Local Storage: Words	Local Storage: Bits/Words
20	.600	3.6 R—W	8	.625	TROS	4.096	60*	256	8
25	.900	0.9 R/W	16	0.9	Main Store R/W	16	16	64	9
30	1.00	2 R/W	8	1.0	CCROS	8.064	50	512	8
40	.625	2 R—W	16	.625	TROS	4.096	56	144	22
50	.500	2 R/W	32	.500	CROS	2.816	88	64	32
65	.200	0.8	64	.200	CROS	2.816	100	24	32
85	.080	0.96 for MS; 0.08 for buffer R/W	144 bits	.080	CROS	2.048 w/cont. 512 108 bits	108	24	32

*Model 20 stored three microinstructions in each 60-bit TROS word.
R/W implies continuous read-write cycle.
R—W implies split read-write cycle.

a microinstruction, and the encoded information into each ROS control field is called micro-orders. Thus a micro-order can affect the control of one gate such as the true/complement gate setting, or the control of a subset of gates simultaneously such as the gating of the output of the adder to one or more of the hardware registers. As such, the micro-orders in the System/360 microprogramming are more powerful than those in the original definition of micro-orders. For example, one micro-order may involve shifting a number right or left one or four places, storing the overflow bits in a designated register, and filling the vacant bits from still another register. The set of micro-orders in a single microinstruction can be executed in one machine cycle. Therefore, as an example, in a single machine cycle, one can gate various fields from various registers to the adder, perform a decimal or binary addition, shift the resulting adder output four bits right or left, and store the result into one or more registers.

6.2 SYSTEM/360 MICROPROGRAMMING CONCEPTS

The microprogramming controls used in the System/360 are similar to those described in Chapter 1. The ROS dimensions and organization vary from one system to another in the total System/360 product line. This variation is due to the number of parallel data paths, the extent of gating, and the extent of conditional and functional branching in each model. The main guiding principles in the design of the System/360 microprogrammed control are described in the following sections.

6.2.1 The System/360 ROS Word Organization

The ROS word is divided into three types of fields of information:

1. *The Static Control Field.* Each field in this group is designed to control a mutually exclusive subset of the total set of control gates in the system. The grouping is based on the input and output gates to a functional unit or on the number of parallel data paths in the system. An example of this is to specify the left and right inputs to the adder in two separate fields, the operation to be performed by the adder in a third, and the destination of the final results in a fourth field. The guidelines for achieving this classification of the control gates into the mutually exclusive subset and encoding them into fields were discussed in Chapter 1.

2. *Emit Field.* In all the microprogrammed models of the System/360 processors, the ROS word has a four-bit field which contains constant bits of information that are used to modify the statistizer's setting; set the condition code bits in the PSW; set certain bits of ROSAR, indirect function register,

and local storage address register; as well as forcing four bits onto the arithmetic and logic data paths in the system. The latter option is very useful in the bit and byte manipulation functions.

3. *The Sequencing Control.* In the System/360, the basic ROS addressing mechanism is to store the maximum number of high-order address bits for the next ROS word to be interrogated into the one that is currently in control. Thus, each ROS word has all the address bits or the high-order address base plus all additional branch conditions and sources that will determine the remaining low-order base. Sequencing the ROS memory in System/360 environment is accomplished by number of control fields. Some fields are fixed (high-order base bits) and are not decoded, and some fields specify only machine-dependent and/or data-dependent conditions used in resolving the few remaining low-order bits of the ROS address.

The different System/360 models vary in the extent of their branching capability. The Model 50 processor, for example, can execute a 64-way branch in one machine cycle, which is equivalent to saying the microinstruction can specify one of 64 potential successor microinstructions. This capability is limited to one of a 32-way branch in the Model 40. All these systems have the capacity to execute 2-way, 4-way, 8-way, 16-way, or 32-way branches, or a 64-way branch as in the Models 50 and 65. This branching is dependent primarily on the binary setting or some hardware statistizers or conditional flip-flops set to remember a given occurrence of a data or machine condition in a previous cycle. There are few data-dependent and branching capabilities in which the branch-executed next cycle depends on the result of some arithmetic or logical operation performed in the current cycle. This limitation is imposed by the ROS timing and access conditions which permit maximum overlap between ROS access time and ALU operations that can affect ROS addressing.

The complexity of the ROS sequencing logic in the System/360 environment is a direct result of the complex System/360 architecture. (See a copy of the IBM System/360 principles of operation manual.)

The 2- or 4-way branch sequencing in the System/360 is accomplished by hardware logic that sets each of the two low-order address bits separately, based on the outcome of one or two separate tests specified by one or two separate branch test fields. Sequencing is also accomplished by forcing the whole ROS address register bits from either the operator's console or other hardware source in the case of an abnormal operation or interrupt condition.

To speed up the ROS access timing and to reduce the amount of address drive circuitry and gating circuitry, a technique which is generally applied is to access the two adjacent ROS words: the even address with the low-order bit, 0, and the next higher address odd word with the low-order bit, 1. This organization is based on the premise that most branch testing involves one of two successor microinstructions. If those two successor microinstructions are assigned the two even/odd consecutive addresses in ROS storage, then the two words can be fetched and gated to two ROS data registers. This permits delaying the testing and final selection between the two successor microinstructions until the last possible moment in the machine cycle.

6.2.2 System/360 Instruction Formats—Fetch and Decode

The System/360 instructions are organized into five basic formats (Fig. 6.3) varying from two to six bytes in width. These are the RR, RX, RS, and SI formats, and finally the SS format which is six bytes wide. All these formats use the first byte (leftmost) to specify one of the 256 possible unique operation codes. The remaining one to five bytes specify the source and the address of the operands involved in these instructions. There are 142 valid System/360 operation codes. The remaining 112 binary codes of the 256 are considered invalid for the time being and can be assigned some valid operation codes if a decision is made to expand the System/360 architecture.

Table 6.2 lists these valid operation codes and their classifications. The underlying philosophy in this operation code assignment is that each operation code can specify its instruction format, its class of operation within the given instruction format, and the specific instruction to be executed. Furthermore, this code was to be used during the instruction decoding cycle for sequencing the ROS memory. This implementation can be seen in Fig. 6.4.

The two leftmost bits of the Op-Code field specify the instruction format and, thus, the number of bytes to be fetched and the source from which the addresses for the operands or the operands themselves are to be retrieved. If the two bits are:

Bits (0, 1)	*Instruction Length*	*Instruction Type*	*Number of Main Storage References During Execution*
00	two byte	RR	None
01	four byte	RX	1
10	four byte	RSSI	1
11	six byte	SS	2

then, as shown in Table 6.3, each type of instruction format contains four classes of operations. The instructions in each class share maximum commonality of the processing steps to be carried out during the instruction fetch; hence, the type of operation implied within a class is specified by bits 2 and 3 of the operation code. The final four bits of the operation code are used to specify the exact instruction within the group of sixteen possible operations, each of the same type.

These groups of bits are used gradually and systematically to sequence the ROS memory into the instruction fetch and decode microroutines which would

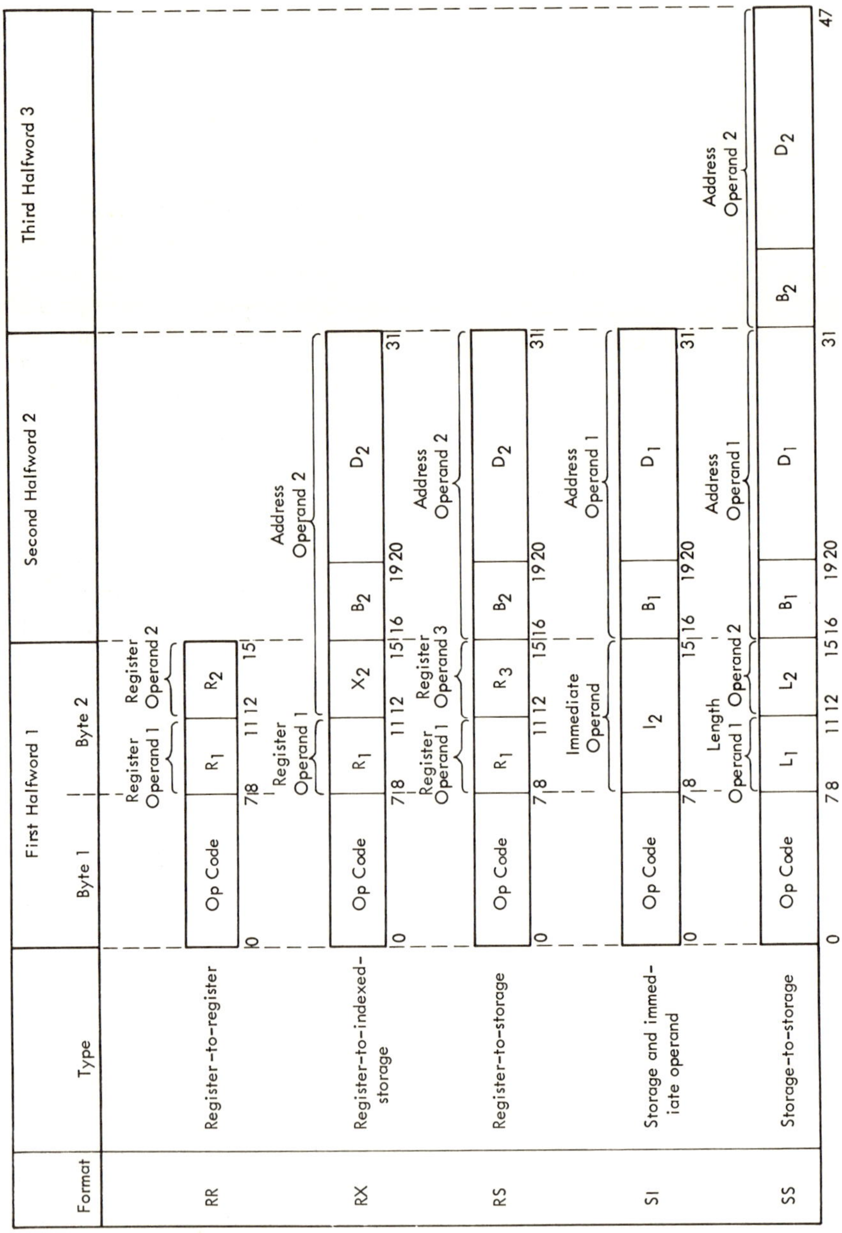

Figure 6.3 Five basic instruction formats.

Table 6.2 SYSTEM/360 VALID OPERATION CODES

	LH. DIG. (HEX)	0	1	2	3	4	5	6	7	8	9	A	B	C	D	E	F
		RH. DIG. (HEX)															
RR	0					SPM	BALR	BCTR	BCR	SSK	ISK	SVC					
RR	1	LPR	LNR	LTR	LCR	NR	CLR	OR	XR	LR	CR	AR	SR	MR	DR	ALR	SLR
RR	2	LPDR	LNDR	LTDR	LCDR	HDR	LRDR	MXR	MXDR	LDR	CDR	ADR	SDR	MDR	DDR	AWR	SWR
RR	3	LPER	LNER	LTER	LCER	HER	LRER	AXR	SXR	LER	CER	AER	SER	MER	DER	AUR	SUR
RX	4	STH	LA	STC	IC	EX	BAL	BCT	BC	LH	CH	AH	SH	MH		CVD	CVB
RX	5	ST				N	CL	O	X	L	C	A	S	M	D	AL	SL
RX	6	STD							MXD	LD	CD	AD	SD	MD	DD	AW	SW
RX	7	STE								LE	CE	AE	SE	ME	DE	AU	SU
RS	8	SSM		LRSW	DIAG	WRD	RDD	BXH	BXLE	SRL	SLL	SRA	SLA	SRDL	SLDL	SRDA	SLDA
RS	9	STM	TM	MVI		NI	CLI	OI	XI	LM				SIO	TIO	HIO	TCH
SI	A																
SI	B																
SS	C																
SS	D		MVN	MVC	MVZ	NC	CLC	OC	XC					TR	TRT	ED	EDMK
SS	E																
SS	F		MVO	PACK	UNPK					ZAP	CP	AP	SP	MP	DP		

0 ↓ 1	2 ↓ 3	4 ↓ 5 ↓ 6 ↓ 7
Type of Instruction	Type of Operation	The Exact Operation to be Executed

Figure 6.4 Instruction—operation code fields.

do the common work implied in each class, e.g., fetch the right number of bytes, compute the addresses, fetch the operands, and finally branch to the execution microroutine for the specific instruction designated by the operation code. The instruction fetch and decode cycles will be traced on both Models 40 and 50 in later chapters.

Table 6.3

Type of Instruction	*Op Code First Four Bits*				*Type of Operation*	*Number of Instructions*
	0	*1*	*2*	*3*		
RR Format	0	0	0	0	Branch on status switching	7
	0	0	0	1	Fixed-point fullword and logical	16
	0	0	1	0	Floating-point long	13
	0	0	1	1	Floating-point short	13
RX Format	0	1	0	0	Fixed-point halfword and branching	15
	0	1	0	1	Fixed-point fullword and logical	13
	0	1	1	0	Floating-point long	9
	0	1	1	1	Floating-point short	9
RS/SI Format	1	0	0	0	Branching status switching and shifting	15
	1	0	0	1	Fixed-point logical and I/O	12
	1	0	1	0	Invalid Format	0
	1	0	1	1	Invalid Format	0
SS Format	1	1	0	0	Invalid format	0
	1	1	0	1	Logical	11
	1	1	1	0	Invalid format	0
	1	1	1	1	Decimal	9
					Total	142

6.2.3 Decoding of the ROS Fields

Most System/360 processors share at least part of their data paths with the multiplexor and selector channels. This is more prevalent in the small and intermediate processors. The multiplexor channel often has little circuitry of its own. It relies on the processor for updating its byte count and for providing the data path to main storage as well as the intermediate storage facilities for assembling the bytes of information to be transmitted to main storage. The selector channel, on the other hand, often shares the processor's main storage data path, but it has the hardware control built into the channel to handle all other I/O functions.

The timing, functional requirements, conditional tests, and other constraints imposed on the CPU data paths are different when it is operated as part of the I/O channel than when it is operated as an arithmetic and logic unit. The organization of the ROS word and the decoding of the ROS fields vary in both requirements. For this reason, an attempt was made to minimize these differences between the CPU and the I/O ROS word organization, thus using the minimum modifications in the decoder circuitry. This resulted in defining two modes of operation, the I/O mode and the CPU mode, with a clear procedure for switching and restoring the machine to its normal CPU mode when an I/O request is reorganized and handled successfully.

Most models in the System/360 processors have the capability of emulating other second generation IBM systems. Often, this emulation capability is achieved by microprogramming.

Sometimes, to achieve the efficiency and the desirable performance, some minimum hardware is added to the host machine to close the hardware and architectural compatibility gap between the target and the host systems. This procedure of emulation may lead to yet another mode of operation and ROS word interpretation called the *emulation mode*.

6.2.4 Stat* Control

The constraints imposed on the ROS access and timing prohibit the dependency of the selection of the next ROS control word on the result of an operation performed in this cycle. Hence, numbers of one-bit storage registers called *stats* are used to store the machine conditions. The stats can then be interrogated in a later cycle to effect a branch based on this condition. Stats can also be set and reset by the microprogrammer as well as by some external machine correlations.

6.2.5 Hardware Microprogrammed Control

While most of the control functions in the System/360 environment are implemented in the ROS storage, there is still a number of hardware control circuits in few critical data paths. The main reason for this sharing of control is to improve the efficiency of the system. An example is the handling of an I/O request. In the intermediate systems, there are several I/O microroutines associated with the channel operations; however, the break-in and break-out

*A stat (statistizer) is a one-bit indicator (trigger) used to memorize the occurrence of a specified condition.

cycles are initiated by the channel by requesting the appropriate action to be taken by the CPU. Another example is the handling of the different priorities of interrupts, specifically those that result from machine errors signalled to the microprogram by forcing a special ROS address in the ROS address register after carrying the necessary initialization steps.

6.2.6 Indirect Function Control

An indirect function control is a kind of second-level control. The action specified for a particular data path in the system is decoded from another source (indirect function register) rather than from the normal ROS control field assigned to the specific data path. In the indirect function cycle, the contents of the ROS field, when decoded, point to the indirect function register. In the Model 50, this concept was used to specify the action of the logical adder (mover) function, while in the Model 40, this was used to indicate the ALU action or operation to be performed.

The main justification for the use of the indirect control register is the potential saving of ROS storage size. In routines where similar machine instructions are executed, a number of microinstructions may be identical except for that function. In this case, a common routine can be shared by both instructions, the only difference being the specification of the direct or indirect function in the proper microinstruction.

The indirect control function registers are normally under microprogram control. They can be set by specifying the bits in the emit field and transferring its contents to the indirect control register.

6.2.7 Decimal Arithmetic

System/360 functional units are designed to execute the arithmetic and logic operations in the hexadecimal code. Therefore, if the decimal numbers are corrected properly, the same binary adders can be used to perform the decimal operations.

To add two decimal numbers, the following steps are taken:

1. A correction of 6 (decimal) or 0110 binary is always added to each digit of the first operand.
2. Binary add the second
3. If the high-order bit position of a *packed decimal digit* generates a carry, the result is correct.
4. If this digit does not generate a carry the result has to be decreased by 6. This is accomplished by adding the 2's complement of 6 (which is 1010), ignoring the carry that can occur.

Example

		Bits 0 1 2 3	4 5 6 7
38		0 0 1 1	1 0 0 0
+47		0 1 0 0	0 1 1 1
85			
Add binary 6 to the first operand digit		0 0 1 1	1 0 0 0
		0 1 1 0	0 1 1 0
		1 0 0 1	1 1 1 0
Add second operand		0 1 0 0	0 1 1 1
	nc		c
Add 1010 to decimal digit without a carry		1 1 1 0	0 1 0 1
		1 0 1 0	0 0 0 0
	c	1 0 0 0	0 1 0 1

Ignore the carry. The result is 85, which is the correct answer.

In the IBM System/360, negative numbers are expressed in their 2's complement form. The complement of any number in any number system is obtained by subtracting the number to be complemented from the highest number in the system. Then, a 1 is added to this intermediate result.

Assuming a decimal number system of three positions, then the highest number here would be 999. The 10's complement of 115 is obtained by subtracting 115 from 999, leaving 884. Then add 001 to this intermediate result, and obtain 885 as an answer. This is said to be the 10's complement of 115.

In the binary number code, the negative numbers are expressed in their 2's complement form. The rules for complement add are:

1. No 6 correction is made to the first operand.
2. Add the 2's complement of the second operand.
3. If the high-order bit of the packed decimal digit generates a carry, the result is correct.
4. If this bit position does not generate a carry, the result has to be decreased by 6, adding the 2's complement of 6 (or 1010).
5. If a sign change occurs, the result has to be recomplemented (subtracting the results from zero).

Example

		Bits 0 1 2 3	4 5 6 7
47		0 1 0 0	0 1 1 1
−75		0 1 1 1	0 1 0 1
−28		0 1 0 0	0 1 1 1
Add 2's complement of second operand		1 0 0 0	1 0 1 1
		1 1 0 1	0 0 1 0
			c
Add 1010 to digit without a carry		1 0 1 0	0 0 0 0
		0 1 1 1	0 0 1 0
The carry out indicates the sign change; hence, recomplement		c	
		1 0 0 0	1 1 1 0
		nc	nc
		1 0 1 0	1 0 1 0
Add 1010 to digits without carry		0 0 1 0	1 0 0 0
	c		c

Ignore the digit carries. The answer is −28.

The different System/360 processors all handle decimal operation in this fashion; however, the actual implementation varies from model to model. In Model 40, the 6-decimal fill, decimal operation, and decimal corrections are all performed by hardware in one machine cycle. The Model 50 requires a second machine cycle to do the correction step. In all cases, the micro-orders indicate the exact decimal operation implied.

6.2.8 General and Floating-Point Registers in High-Speed Local Storage

Another architectural and implementation similarity among most System/360 processors is the location of both sets of general and floating-point registers in the high-speed local storage. These local storage devices vary from one machine to another in their addressing mechanism, partitioning, cycling through the whole read-write cycle or split cycle, the extent of their interface mechanism with the CPU, and the technology from which they are implemented. In all cases, the accessing of local storage was put under the control of the microcode. In the Model 85, the local storage control which is in the E box is also shared by the hardware of the instruction box and has an access to it during the instruction fetch cycle.

6.3 MICROPROGRAMMING SUPPORT

The micro-order mnemonics in which the different microroutines are written vary from one machine to another in the System/360 product line. However, the basic structure of the microroutines and the mechanics of translating them into the ROS bits are the same.

Microprogramming the System/360 is similar to flow-charting the action implied in a given operation at microinstruction and micro-order level, in which the microinstructions are represented by boxes connected by a network of lines showing the order and the branching that would result during execution of a linear net of microinstructions. This flow-chart-type language, which has the appearance of a higher-level language, is designed by IBM to unify and standardize the microprogramming steps and automate the translation, simulation, documentation, address assignment, and bit-encoding steps that will finally lead to the manufacturing tape that contains the bit pattern of each ROS word. This language then forms one input to a more general program called the control automation system, CAS.

The objective of the CAS system is to aid the microprogrammer in designing the system control section, recording the microroutines in their symbolic flow-chart form, testing and simulating their actions on a given machine, translating the microinstruction into an intermediate language used by the simulation and physical assignment routine, and finally, performing the ROS bit-encoding function.

A second input to the 7090 CAS program is the machine description. This is prepared by the hardware designer. It includes all data flow paths, showing the registers and their in-gate and out-gate position, the functional units such as the adders, shifters, movers, etc., and all the interfaces, communications, and data paths that exist in the system. This section of the CAS input must include the timing, the control signal lines, and all the parallelism in the system.

The third input to the CAS system is the set of rules which governs the translation simulation and encoding processes.

Figure 6.5 shows the central flow of the CAS system. The microprogrammer submits his microprogram for a specific machine whose physical

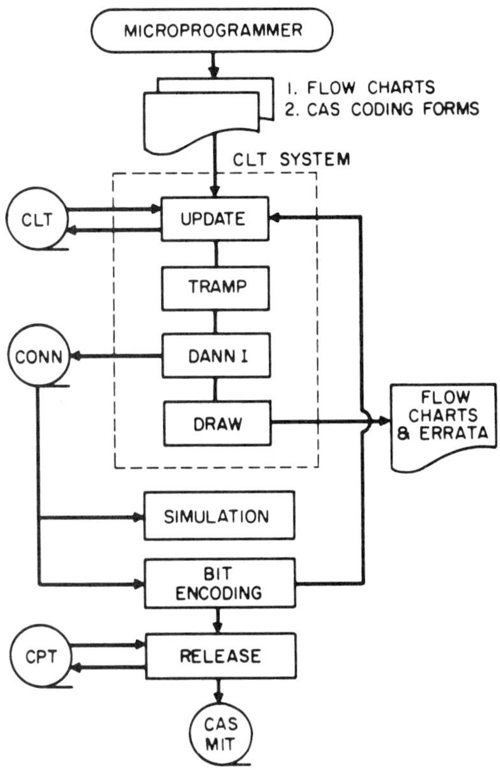

Figure 6.5 General flow of a CAS system.

description is contained in the master control logic tape, CLT. This input is submitted on a standard logic sketch sheet, shown in Fig. 6.6, which is then punched into a deck of IBM cards and put on tape off-line. The tape is called the new CLT tape. The old CLT, the new CLT, a history file tape called the

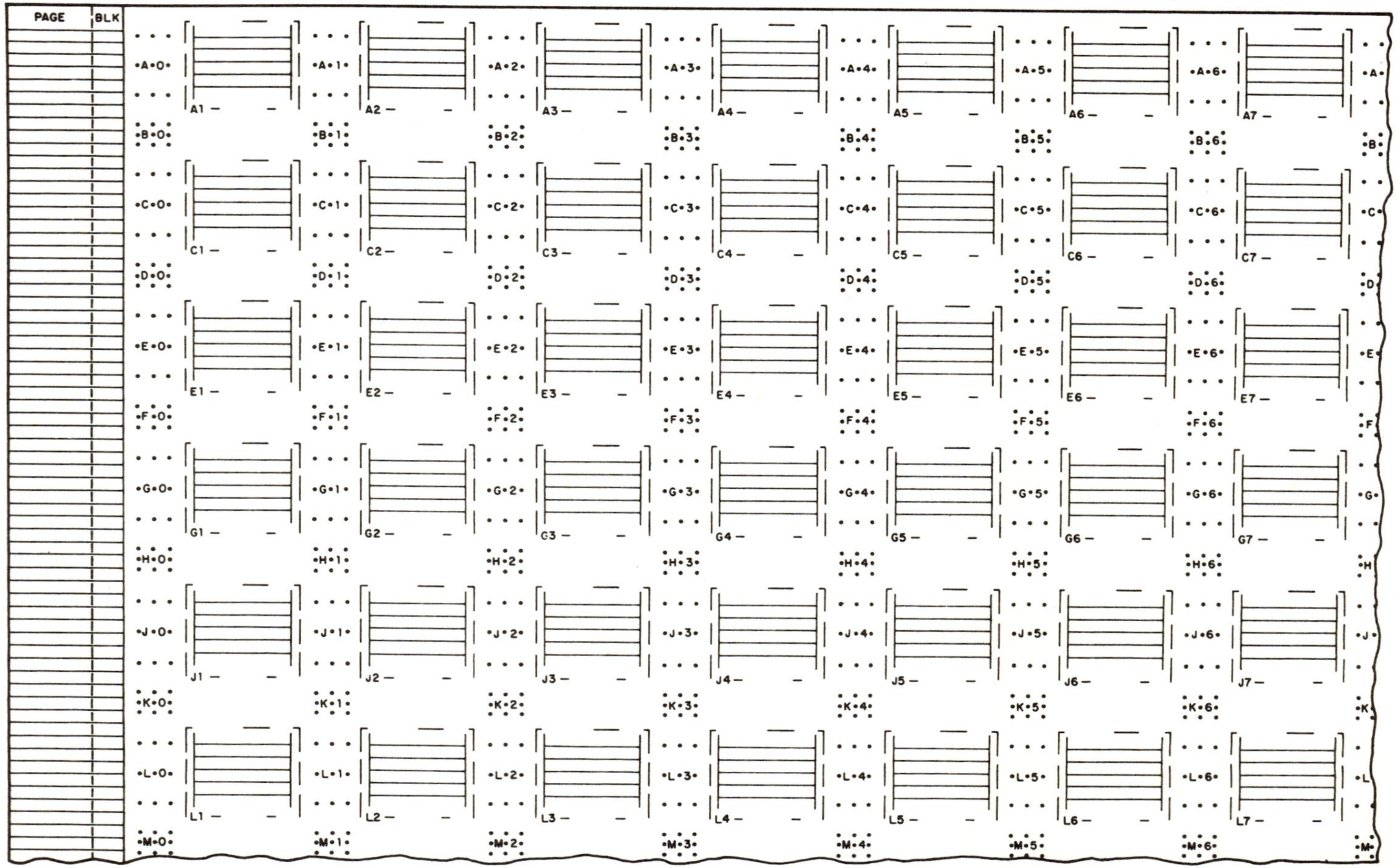

Figure 6.6 Portion of a logic sketch sheet.

CPT, and the CAS system tape are mounted on the 7090 system. The output tapes are: an updated CLT, a microprogram logic sheet, a ROS bit tape (CASMIT, CAS manufacturing interface tape), and an errata and simulation information tape.

The main subroutines in the CAS system are shown in Fig. 6.5; they are:

1. *The UPDATE Routine.* This is responsible for maintaining the design information of a specific system on the CLT tape. It is also responsible for updating the microprograms in their symbolic form. The microprograms are maintained on a logic sheet basis. Each of the various blocks and nets on a sheet can be updated. The functions available for update are:
(a) Addition of a microinstruction or a micro-order within a microinstruction.
(b) Deletion and/or replacement of such items.
(c) Deletion of a whole logic sheet or a group of sheets within a particular microprogram.
This program, therefore, keeps the current status of the system design, including the engineering changes identified by an engineering change-level number. Only one engineering change level of each sheet is maintained on the CLT tape.

2. *The TRAMP Program.* The main objective of this program is to check and translate the microprogram and microinstructions to a standard format which is used by the simulation and address assignment routines. The TRAMP rules are defined by the microprogrammer. These rules include the details of the ROS word organization, a comprehensive micro-order mnemonics decoding chart for each ROS field. TRAMP checks each microinstruction to see if it contains sufficient information to identify its set of successor microinstructions. TRAMP must also know the microinstruction format, edge characters, and the meaning of all special symbols used in this microinstruction template.

This program then translates each microinstruction into some intermediate language called the *decimal orders.* This consists of two leading alphabetic characters which identify the ROS field, followed by one or more decimal characters which represent the decimal equivalent of the binary bit encoding of the specific micro-order in the field. The CAS system has a unified set of symbols that are used in microprogramming the different models of the System/360. Table 6.4 gives this unified list.

3. *DANNI.* This is a program which checks that each net on a logic sheet contains a group of compatible microinstructions such that every successor microinstruction has a predecessor connected to it by a line. In other words, DANNI checks the logical flow of the microroutine, testing to ensure that no connecting lines are left out, that every branch micro-order is connected to all possible successor microinstructions, and that every logic page is identified properly to facilitate tracing of the microroutines.

DANNI is heavily dependent on the leg identifier and leg selector concept which is built into the microprogramming CAS language. The selector-identifier code is written by the microprogrammer in order to identify both the extent

Table 6.4 CLD BLOCK SYMBOLS

Symbol	*Definition*	*Example*
+	True Add/Positive	A + B: B is added (True) to A
−	Complement Add/Subtract, Negative	A − B: B is Complement Added to A
=	Equal	A = B: A is Equal to B
≠	Unequal	A ≠ B: A is Unequal to B
→	Is Set Into	A → B: A is Set Into B (Destructive Read-In is Implied)
·	Is ANDed With (Logical)	A · B: A is ANDed with B
,	AND (Nonlogical)	A → B, C: A is Set Into B and C
Ω	Is ORed With (Logical)	A Ω B → C: A is ORed with B and the Result is Set Into C
/	OR (Nonlogical)	A/B → C: A or B is Set Into C
⊻	Is EXCLUSIVE ORed With	A ⊻ B → C: A is EXCLUSIVE ORed with B and Set Into C
±	True or Complement Add/Positive or Negative	A ± B → C: B is True or Complement Added to A and the Result is Set Into C
±	Binary Add Under T/C Control	A ± B → C: B is True or Complement Added to A and the Result is Set Into C
±	Decimal Add Under T/C Control	A ± C → C: B is True or Complement Decimal Added to A and the Result is Set Into C
<	Is Less Than	A < B: A is Less than B
¬	Not (Boolean—Used as the Not Function on CLD's)	
:	Is Compared to	A:B: A is Compared to B
( )	Used for Normal English Punctuation or to Enclose an Expression Within a Statement	
*	Special. 2030 Uses the * for One Mnemonic.	*aa: Explained in Section on Mnemonics
?	Indeterminate Function (This Describes a Function which is Hardware Controlled Rather than Under the Direct Control of the Microprogram).	A ? B → C: A and B are Logically Combined (Under Hardware Control) and the Result is Set Into C

of branching and all the successor microinstructions. This leg selector-identifier concept is also used as the input to both the simulation and the physical address assignment routines. The basic concept is simply to leave some bits corresponding to ROS low address bits to be set to 1 or 0 based on the outcome of the specified test in the microinstruction. This leg selector field is normally five to six bits wide and appears at the bottom line in the microinstruction box (see Fig. 6.10). The branch bit or bits are indicated by an asterisk in the corresponding print position. If one asterisk appears in the leg selector microinstruction, then two successor microinstructions must be identified by having a 0 in one and a 1 in the corresponding bit of the leg identifier field of the two successor microinstructions. Figure 6.7 shows a two-way branch

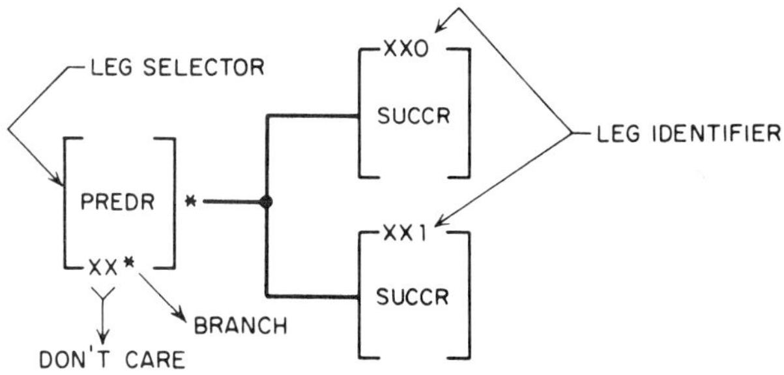

Figure 6.7 CAS branch bit.

with the leg selector in the predecessor microinstruction and the leg identifier in the two successor ones.

Leg selector will be two to six characters right-justified in the print positions of the bottom line of the microinstruction box. It represents the bits of the ROS next address which are controlled by branch conditions in that block. The characters will be:

*	There is a branch condition on the bit.
X	No branch, and the microprogram does not care whether the corresponding bit of the ROS next address is assigned 0 or 1.
0	No branch, but microprogram needs the corresponding bit of the ROS next address assigned 0.
1	No branch, but microprogram needs the corresponding bit of the ROS next address assigned 1.

Leg identifier will also have two to six characters right-justified in positions 1 through 6 in the top line of the block. It will represent the bits of the ROS current address which are controlled by branch conditions in the preceding block or blocks. The characters will be:

X	There was no branch condition on the corresponding bit in any of the preceding blocks; the microprogram does not care whether this bit is assigned 0 or 1.
0	Microprogram needs this bit assigned 0.
1	Microprogram needs this bit assigned 1.

4. *SIMULATION Routine.* This program, when called to action by the microprogrammer, attempts to simulate the execution of the selected microroutine on the machines whose full description in terms of the facilities, data paths, timing points, etc., has been fed to the CAS system. This SIMULATION routine is also useful in debugging the initial machine description itself, as well as for simulating microprogram signals chained in real time.

The program can also simulate a few CAS logic sheets that form a section of a microroutine. In any case, the microprogrammer is responsible for initiating the action by specifying the state of the machine at the point just prior to the required simulation. This initialization includes setting the registers and all other data facilities and machine functions to any predetermined values. The simulation routine gives the microprogrammer considerable control over the total running time as well as over the extent of the monitoring he wishes to exercise. Figure 6.8 shows the general organization of the CAS simulator.[7]

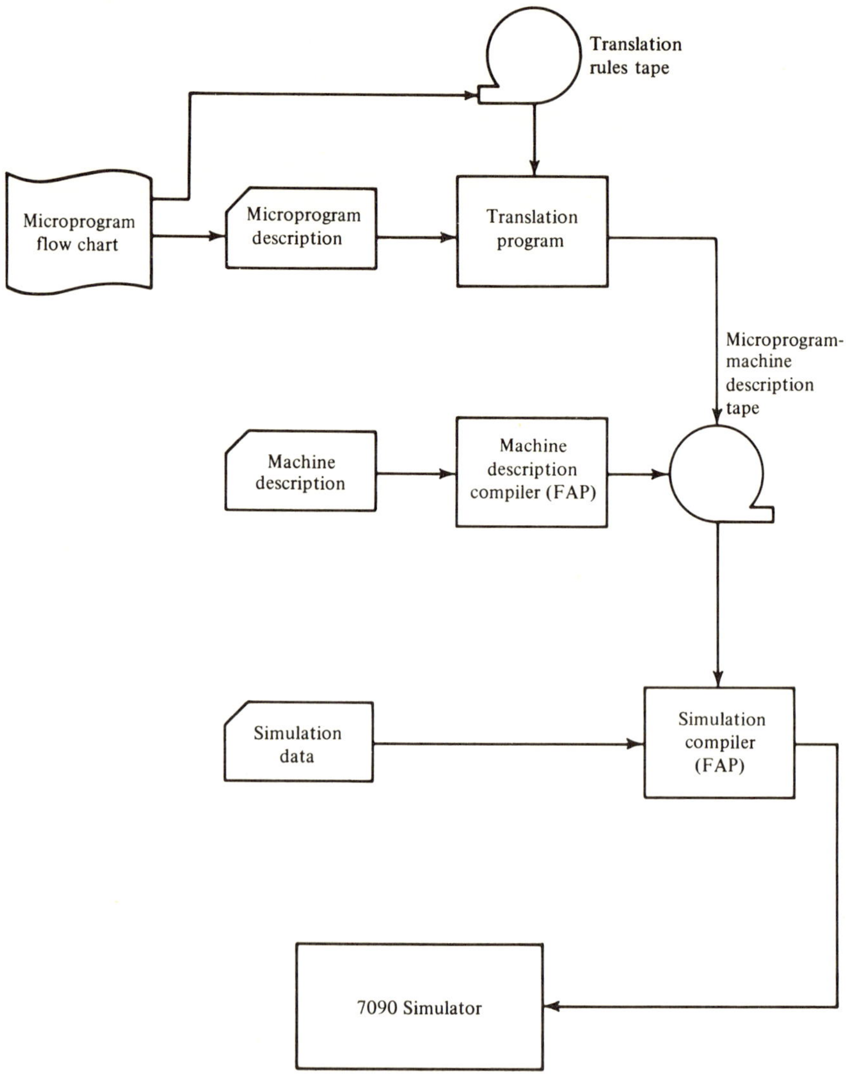

Figure 6.8 Organization of CAS simulator.

5. *DRAW*. This is a subroutine which produces the final microprogramming document in the familiar flow-chart form with the physical addresses assigned to each box in the microroutine. The output is printed to reflect the

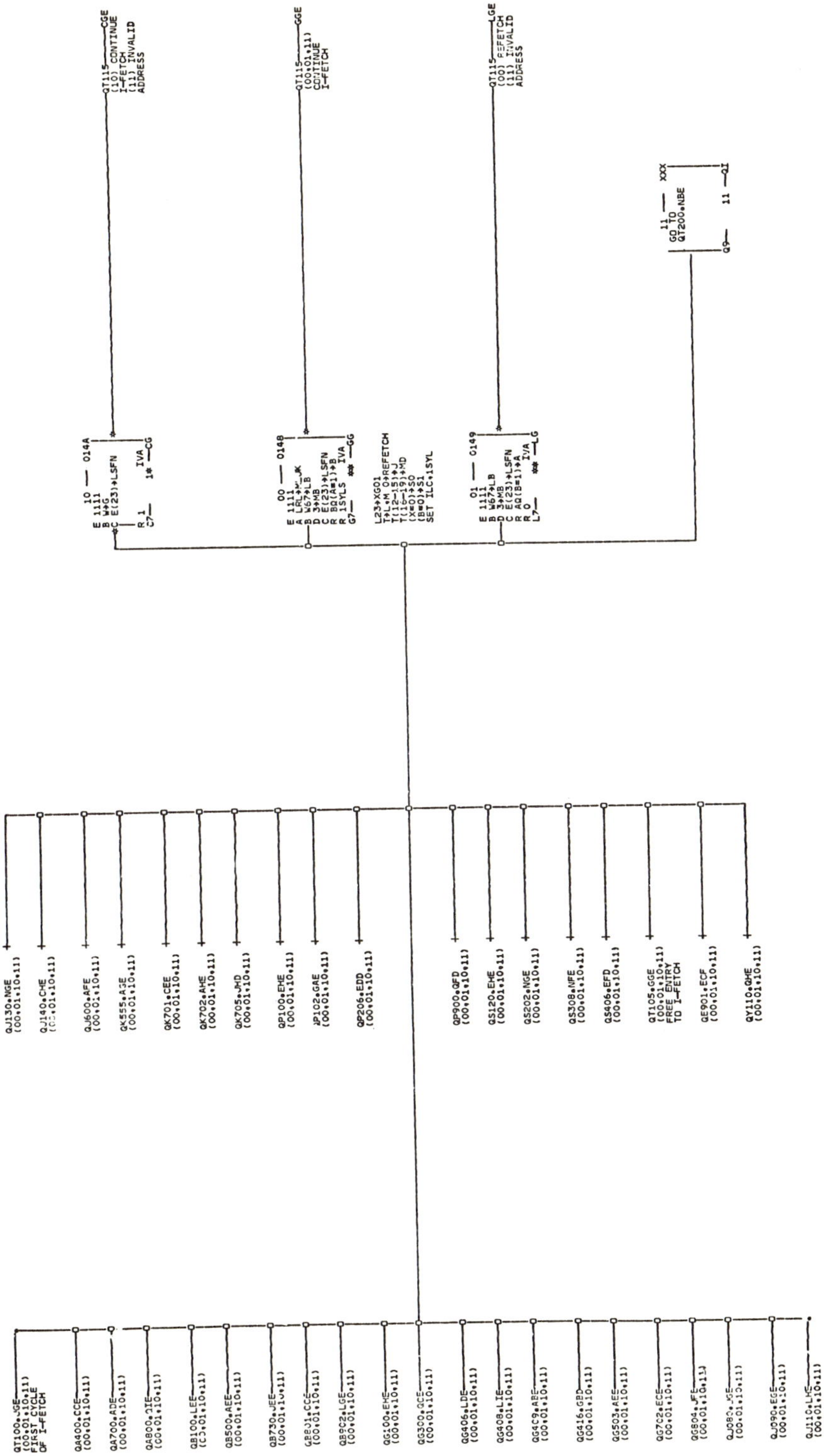

Figure 6.9 CLD sample page.

HEX ADD	BINARY ADDRESS	CLD BOX	MODE	BINARY BIT PATTERN
				P LU MV ZP ZF ZN TR ZR WS SF P IV AL WM UP MD LB MB DG UL UR P CE LX TC RY AD AB BB UX SS 92—97
09BE	0 1001 1011 1110	A8	-	0 000 00 100101 0001 100 00011 0 100 111 0 000 11110 0111 10 0 0 0 000 01 00 1 0001 000 1 000 0001 000001 00000 0 000000
09BF	0 1001 1011 1111	J8	-	0 000 00 100101 0001 100 00011 0 100 111 0 000 11110 0111 10 0 0 0 000 00 01 1 0001 000 1 000 0001 000001 00000 0 000000
09C0	0 1001 1100 0000	E3	-	0 000 00 100101 0001 100 00001 0 001 000 0 000 00000 0001 10 0 0 0 000 01 00 1 0100 000 1 000 0001 000000 00001 0 000000
09C1	0 1001 1100 0001	N3	-	0 000 00 000110 0101 100 00000 0 100 111 0 000 00000 0000 10 0 0 0 000 01 01 1 0011 000 1 000 0001 000001 00001 0 101000 00 000000
09C2	0 1001 1100 0010		-	0 000 00 100111 0000 100 00000 0 000 000 0 000 10111 0000 00 0 0 0 010 00 00 0 0000 000 0 000 0000 000001 00000 0 000000
09C3	0 1001 1100 0011		10	0 000 00 100111 0000 100 00000 0 000 000 0 000 10111 0000 00 0 0 0 010 00 00 1 0000 000 0 000 0000 000001 00001 0 000000
09C4	0 1001 1100 0100		10	1 000 00 100111 0001 100 00000 0 000 000 0 000 10111 0000 00 0 0 0 010 00 00 1 0000 000 0 000 0000 000000 00000 0 000000
09C5	0 1001 1100 0101		10	1 000 00 100111 0001 100 00000 0 000 000 0 000 10111 0000 00 0 0 0 010 00 00 0 0000 000 0 000 0000 000000 00001 0 000000
09C6	0 1001 1100 0110		10	1 000 00 100111 0001 100 00000 0 000 000 0 000 10111 0000 00 0 0 0 010 00 00 0 0000 000 0 000 0000 000001 00000 0 000000
09C7	0 1001 1100 0111		10	1 000 00 100111 0001 100 00000 0 000 000 0 000 10111 0000 00 0 0 0 010 00 00 1 0000 000 0 000 0000 000001 00001 0 000000
09C8	0 1001 1100 1000		-	1 000 00 100111 0010 100 00000 0 000 000 0 000 10111 0000 00 0 0 0 010 00 00 1 0000 000 0 000 0000 000000 00000 0 000000
09C9	0 1001 1100 1001		-	1 000 00 100111 0010 100 00000 0 000 000 0 000 10111 0000 00 0 0 0 010 00 00 0 0000 000 0 000 0000 000000 00001 0 000000
09CA	0 1001 1100 1010		-	1 000 00 100111 0010 100 00000 0 000 000 0 000 10111 0000 00 0 0 0 010 00 00 0 0000 000 0 000 0000 000001 00000 0 000000
09CB	0 1001 1100 1011		-	1 000 00 100111 0010 100 00000 0 000 000 0 000 10111 0000 00 0 0 0 010 00 00 1 0000 000 0 000 0000 000001 00001 0 000000
09CC	0 1001 1100 1100		-	0 000 00 100111 0011 100 00000 0 000 000 0 000 10111 0000 00 0 0 0 010 00 00 1 0000 000 0 000 0000 000000 00000 0 000000
09CD	0 1001 1100 1101		-	0 000 00 100111 0011 100 00000 0 000 000 0 000 10111 0000 00 0 0 0 010 00 00 0 0000 000 0 000 0000 000000 00001 0 000000
09CE	0 1001 1100 1110		-	0 000 00 100111 0011 100 00000 0 000 000 0 000 10111 0000 00 0 0 0 010 00 00 0 0000 000 0 000 0000 000001 00000 0 000000
09CF	0 1001 1100 1111		-	0 000 00 100111 0011 100 00000 0 000 000 0 000 10111 0000 00 0 0 0 010 00 00 1 0000 000 0 000 0000 000001 00001 0 000000
09D0	0 1001 1101 0000		-	1 000 00 100111 0100 100 00000 0 000 000 0 000 10111 0000 00 0 0 0 010 00 00 1 0000 000 0 000 0000 000000 00000 0 000000
09D1	0 1001 1101 0001		-	1 000 00 100111 0100 100 00000 0 000 000 0 000 10111 0000 00 0 0 0 010 00 00 0 0000 000 0 000 0000 000000 00001 0 000000
09D2	0 1001 1101 0010		-	1 000 00 100111 0100 100 00000 0 000 000 0 000 10111 0000 00 0 0 0 010 00 00 0 0000 000 0 000 0000 000001 00000 0 000000
09D3	0 1001 1101 0011		-	1 000 00 100111 0100 100 00000 0 000 000 0 000 10111 0000 00 0 0 0 010 00 00 1 0000 000 0 000 0000 000001 00001 0 000000
09D4	0 1001 1101 0100		-	0 000 00 100111 0101 100 00000 0 000 000 0 000 10111 0000 00 0 0 0 010 00 00 1 0000 000 0 000 0000 000000 00000 0 000000
09D5	0 1001 1101 0101		-	0 000 00 100111 0101 100 00000 0 000 000 0 000 10111 0000 00 0 0 0 010 00 00 0 0000 000 0 000 0000 000000 00001 0 000000
09D6	0 1001 1101 0110		-	0 000 00 100111 0101 100 00000 0 000 000 0 000 10111 0000 00 0 0 0 010 00 00 0 0000 000 0 000 0000 000001 00000 0 000000
09D7	0 1001 1101 0111		-	0 000 00 100111 0101 100 00000 0 000 000 0 000 10111 0000 00 0 0 0 010 00 00 1 0000 000 0 000 0000 000001 00001 0 000000

Figure 6.10

control logic tape (CLT) organization which is on a sheet basis. Each sheet contains the machine identification, a history of all engineering changes, the function of the particular microroutine, a unique CLD page identification number, and all the left and right net identification numbers. The microinstruction boxes and the lines which connect them are also printed. Figure 6.9 shows a sample CLD page.

6. *Bit Encoding*. This program converts the symbolic microprogram on the CLT to the actual bit patterns in the ROS memory. Under the designer's control, assignment of microinstructions to the ROS words is also done. Figure 6.10 shows a sample ROS encoded output.

7. *RELEASE*. This program records the ROS word bit pattern on a history file known as the CPT (CAS physical tape). Any engineering change from this history tape level can be extracted. The extracted information is transmitted to the manufacturing group in a special manufacturing interface tape called CASMIT.

6.3.1 The Microinstruction Box

The main information contained in the logic sheet is that which appears in and in the vicinity of the microinstruction box (Fig. 6.11). Each box

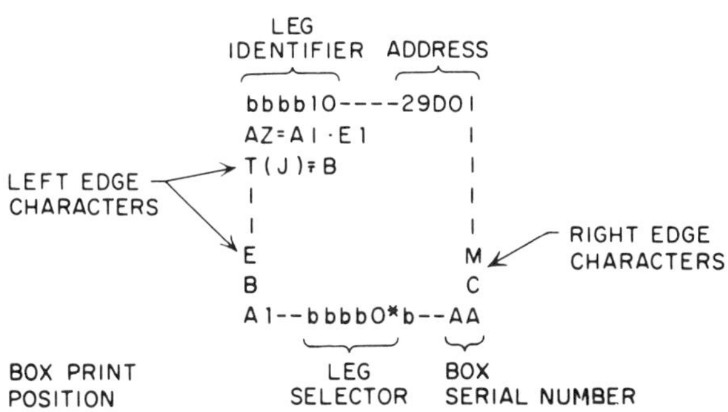

Figure 6.11 Box character.

identifies one microinstruction. The print area within the box is 13 print positions wide and can contain a maximum of 18 print lines, including the leg identifier and leg selector.

The left side of the box contains the left edge character. Each edge character identifies a particular data path in the system. Normally, the top edge character on print line 2 is E, which specifies the binary bits stored in the emit field of that specific ROS word. The number and the meaning of such edge characters are machine-dependent; however, in each machine, the order in which these characters appear is well-defined and must be strictly adhered to. For example, if the adder data path edge character, say A, appears on print line 3 after E, and before the local storage addressing on print line 4, the order must always be preserved without exception. In few cases, a number of print lines share the same edge characters. In this case, the order in which the specific groups of micro-orders occur in the given data path must also be preserved. This order and the left and right edge characters are specified in the microinstruction template which forms part of the machine description.

The following rules must be obeyed:

1. One blank space must be provided between a micro-order and its respective edge character.
2. A line of data which can be identified by a single edge character must always be identified by the left edge character only.
3. A line can contain a maximum of two micro-orders or statements. The two statements must be separated by at least one blank.
4. No micro-order or statement may overflow to another print line.

The leg identifier and leg selector fields which appear at the top left corner and the middle of the bottom print lines of the instruction box are used to define the addressing and branching of the microinstructions within the logic note as we have discussed earlier. The hexadecimal number at the top right corner of the box is the physical address of the ROS word containing that microinstruction. This address can be specified by the microprogrammer completely, or the actual address assignment can be left up to the CAS system itself.

The lower right corner of the box contains two alphabetic characters called the block serial number. These are used to indicate the flow of the microprogram on the CLD page. Initially, the block serial number will be generated from the block coordinate (retaining the alphabetic row character, and converting the number column characters 1 to 9 into corresponding alphabetic characters A through I, e.g., N3 becomes NC). That block serial number will then remain unchanged even if the block coordinate subsequently changes.

The final microinstruction box identification field is the box print position which is a two-character field in the bottom left corner of the box. This two-character coordinate (row, column) is used to indicate the block's position on the sheet. The first is an alphabetic character to place the block in one of nine possible rows, A–I; the second is numeric and places the block into one of nine columns, 1–9.

The micro-order and all other machine-dependent microprogramming steps will be illustrated in the chapters on Models 40 and 50.

6.4 MODEL-DEPENDENT FEATURES

In this section, we will review the unique features of System/360 models. Models 40 and 50 will be presented and their microprogramming will be discussed in detail in the next two chapters.

The two more recent additions to the IBM System/360 product line were announced early in 1968: the Model 85 on the higher end of the product line and, at the lower end, the Model 25. Both models use microprogramming to different extents and in unique ways. We will examine the Model 85 first.

6.4.1 The IBM System/360, Model 85

This model was intended primarily as a growth system for the Model 65 and Model 75 users. As such, it is architecturally upwards compatible with the System/360 product line. It is intended to provide high-speed performance across the entire spectrum of editing, file maintenance, and computational work loads.

The main design objectives were:

1. To provide efficient performance in terms of more jobs per hour while the system is running.
2. To provide balanced performance across a wide range of job streams in both commercial and scientific applications, fixed-point, floating-point, decimal, logical, and input/output operations.
3. Improve availability by:
   (a) reducing the duration of unscheduled maintenance.
   (b) providing efficient diagnostics for quick error isolation.
   (c) instruction re-try and log-out.

The first two objectives were partially met by implementing an instruction fetch and decode unit in the CPU, separate from the executing unit. The operations of the I-unit and E-unit are overlapped. The I-unit is capable of fetching, decoding, and transmitting new instruction for execution by the E-unit every machine cycle. This performance is achieved by overlapping the instruction fetch while keeping the execution of the instruction in sequence. This parallelism in the instruction fetch and decode cycle made it necessary to design the I-unit with hardware control since the read-only storage execution is sequential. The I-unit may be preparing several instructions simultaneously while the E-unit is executing a single instruction. Because of its sequential nature, the E-unit can benefit from and does use a microprogrammed control.

Microprogram control in the Model 85 operates out of two dual control units: the capacitive read-only storage and the writable control storage unit. The first holds the System/360 instruction execution and other specialized routines, while the writable control storage is used only for microdiagnostics and other microroutines which are not part of the basic Model 85.

The basic ROS word is 108 bits in width and is accessible to the E-box at a CPU cycle rate (80 nanoseconds per ROS word). The ROS storage consists of 2048 addressable words. Each word is divided into 27 fields which are decoded to control individual machine functions. Each ROS word contains all the control information needed to activate the E-box for one machine cycle. It also contains all the information necessary for setting the ROS address register. Bits 37–43 of this ROS word are set into the high-order seven bits of the ROSAR. The remaining five bits of the ROSAR are determined by a number of hardware and other data-dependent conditions indicated by the current microinstruction.

The writable control storage contains 512 words of 108 bits each. It is built from monolithic volatile circuits which can be accessed at an 80-nanosecond cycle time. The writable control storage unit is loaded from main storage via a special System/360 SI format instruction called *load microprogram* (LMP). The effective address points to the beginning of an 80-byte record in main storage. The leading double words of this 80-byte record contain the record count, a nonzero byte record indicator, and the address in the writable control storage of the first word to be loaded.

The writable control storage in the Model 85 was originally intended for microdiagnostics of the E-box and, indirectly, the I-box. Since it controls the data flow and the gating of the E-box, it can check this box by forcing test patterns to flow from a functional unit to the next functional unit; thus, it isolates the error to a very small section of this data path. The E-box communicates with the I-box over a fixed number of data interface buses; therefore, the E-box can view the I-box as a black box with limited input/output lines. The errors in the I-box must be synthesized from its response to the specific data and from signals sent to it by the microdiagnostic routine in the E-box. The Model 85 is therefore a semimicroprogrammable system. In their attempt to achieve performance through parallelism and pipe-lining, the designers of the system have used hardware control, microprogrammed control, and microdiagnostics in a writable control storage.

6.4.2 The IBM System/360, Model 25

This model is microprogram controlled. The microroutines used for system control are contained in the control storage area of core storage. Control storage is always the high-order 16k bytes of the composite storage array.

The functions of the microprogram in the Model 25 are:

1. Fetch and decode machine language instructions and data located in the main storage area in the core storage.
2. Control the execution sequences for the machine instructions.
3. Handle all I/O operations for the integrated attachment and I/O channels.
4. Perform diagnostic operations on the system.
5. Handle console operations.
6. Formatting the information on an output device to produce a meaningful record.

Since the microroutines reside in core storage, new and different information can easily be transmitted. The Model 25 has a special switch at the operator's console called *control storage load* (CSL). Operator's activation of the CSL switch causes a deck of cards to be read from a selected I/O device; and it causes their contents to be transferred to the core control storage under the control of a permanently resident microprogram. Thus an operator may switch from System/360 mode to 1401 mode by loading a 1401 emulator deck. Other microroutines, indeed, an entirely different set of instructions or additional machine instructions, can be added with minimum effort. Hence, this Model 25 must be viewed as the first machine with a writable control storage introduced by a major computer manufacturing company.

Like the Model 85, Model 25 uses microprogramming in a unique way. The control word is two bytes wide, divided into specific fields defined for controlling the data flow of the system. Seven control words are defined for the Model 25. A brief description of each of these seven words is as follows:

1. *Set-Reset Word.* This control word type is used to set or reset bits of a designated external facility, to store the link address into local storage when used as link word, and to retrieve the same link address when used as return word.

2. *Arithmetic Constant Word.* This is decoded to determine the ALU operation, the ALU input gating, and the ALU destination after the operation is performed. This type of control word is defined as word type 1, with bits 0, 1, and 15 of the ROS word = 001.

3. *Storage Word.* The functions of this type of control word are:
   (a) Read data (byte or halfword) from any section in the main storage (program, control, or auxiliary section).
   (b) Store data into any section of main storage.
   (c) Do halfword update of storage addresses or data count.
   (d) Move halfword from one local register to another.

Bits 0, 1, and 15 in the control word are equal to 010. This storage control word is designated as type 2. Figure 6.12 shows the subdivision of this two-byte control word into the significant bit fields, and it shows how the bits are decoded. This division into fields of operation is typical of the type of division in all other types of control words in Model 25.

4. *Move-Arithmetic Word.* This fourth class of ROS word designated as type 3 by the setting of bits 0, 1, and 15 to 011 is known as move-arithmetic word. It performs arithmetic and move operations on local storage data. It moves data from local storage to external facilities and from external facilities into local storage.

5. *Branch Unconditional Word, Type 4, with Bits 0, 1, and 15 = 100.* This word type, when executed, will perform an unconditional branch to some designated address and will store the address of the next sequential control word in a special CPU backup register in local storage zone 4. The unconditional branch control word appears in different forms: the BR form when a return to

Word Type	Storage Control	Data Register Address	Address Register Address or K-Field	Modifier Control	Word Type
0 1	2 3	4 5 6 7	8 9 10 11	12 13 14	15
0	00=Read Control	This field may address Local Storage for a byte or halfword access.	For Indirect Addressing	000=No update	0
1	01=Read Aux/Prog	Byte selection of Local Storage is limited to the odd address, where control word bit 7=1.	The Address Register selected by this field is a halfword Local-Storage Register	001=Direct Address	
	10=Store Control	Halfword selection is limited to even addresses, where bit 7=0	Bit 11 does not enter into the addressing of the address register.	010=No access, + update	
	11=Store Aux/Prog	External facility addressing is always done in byte mode, where control word bit 7=1.	Bit 11 designates the following: Bit 11=0, Bits 2,3=01: Read Aux Bit 11=1, Bits 2,3=01: Read Prog Bit 11=0, Bits 2,3=11: Stor Aux Bit 11=1, Bits 2,3=11: Store Prog	011=No access, − update 100=LS Data Reg,Access, + update 101=LS Data Reg,Access, − update 110=Ext Data Reg,Access, + update 111=Ext Data Reg,Access, − update	
When no access to storage is designated, modifier control decode of 010 or 011, Bits 2 and 3 determine the value to update by: Bits 2 and 3 = Value 00 = 0 01 = ±1 10 = ±2 11 = ±3		Local Storage * 0000 U0 0001 U1 * 0010 V0 0011 V1 * 0100 G0 0101 G1 * 0110 D0 0111 D1 * 1000 I0 1001 I1 * 1010 T0 1011 T1 * 1100 P0 1101 P1 * 1110 H0 1111 H1 * Even registers can be used as address registers (ctrl bits, 8, 9, 10, 11)	For Direct Address Bits 8, 9, 10, and 11 are forced to values that will result in the K-address for the M1-Register. The M1-Register is forced to the following values for the K-field values indicated: Bits 8, 9, 10, 11 — M1-Register in Hex 0000 88 0001 8A 0010 8C 0011 8E 0100 98 0101 9A 0110 9C 0111 9E 1000 A8 1001 AA 1010 AC 1011 AE 1100 B8 1101 BA 1110 BC 1111 BE M0 = X0 for direct access of Auxiliary Stoarage M0 = X3 for direct access of Control Storage		

Fig. 6.12 Bit-significance chart; storage word (word type 2).

the next sequential instruction is not anticipated, and the BAL form which implies that a return to the next sequential instruction will occur when some subroutine starting at the branched-to address is finished.

6. *Branch on Mask Word, Type 5, as Designated by Bits 0, 1, and 15 to be Equal to 101.* This type of microinstruction performs a 2-, 4-, 8-, or 16-way branching. There are over forty different control word settings and system machine combinations that will effect this branching.

7. *Branch-on-Condition Word, Type 6 or 7, with Bits 0, 1, and 15 set to 110 or 111.* This microinstruction type tests one of eight bits of a designated

byte source for a 0 or 1 condition. The branch option will be executed if the condition of the bit tested compares favorably to the condition indicated by the branch-on-condition statement.

One can see from these seven types of microinstructions that the limited machine activity specified by each machine action would permit a simple microprogramming assembly language buildup with a special assembler routine to set the type of instruction as well as the ROS bits to indicate the action performed in a given ROS word. With this language, loading and overloading some segments of this control program would be as simple as loading the user's source deck from the card reader into the program section of main storage.

The main physical precaution that was taken in the Model 25 was to provide a control storage load (CSL) switch on the system control panel. Pressing the CSL switch starts the control storage load routine. The control storage load switch must also be pressed during a store into the control storage area of core storage. This loading of the control storage is more fully described in the IBM Model 25 functional characteristics, Form #A24-3510-0 manual.

The key new microprogramming concepts used in microprogramming the Model 25 are:

1. The microcontrol resides in a conventional core read-write storage and can be modified. This, of course, involves a core load procedure.

2. The seven different control word formats. Each word is a bit significant that has specific fields defined for controlling the data flow of the system.

3. The microlisting, which contains the actual contents of the control

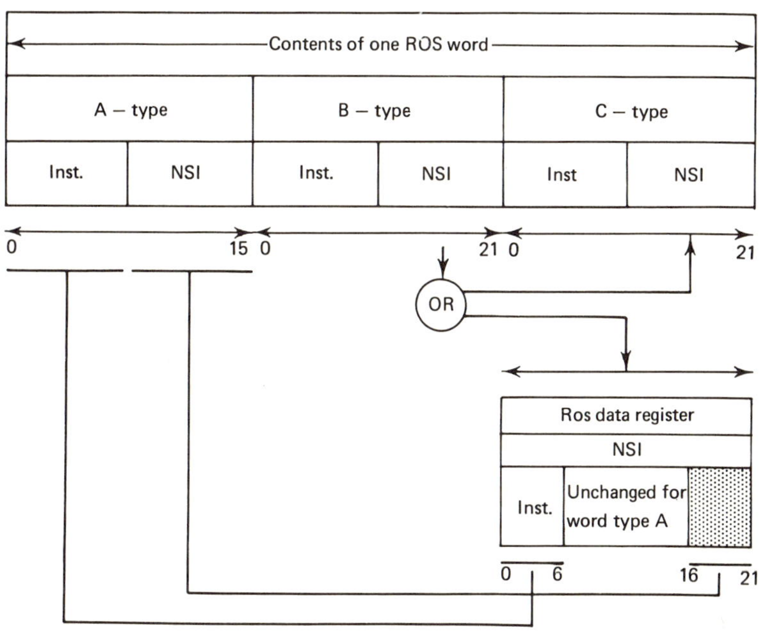

Figure 6.13

SLDA Repr	CAS Repr	0	1	2	3	4	5	6	7	8	9	10 - - - - - - 17	18 - - 21	Remarks		Figure	
MX*X	X → X	P	0	1	X*	X*	X*	0	X	X	X	← NSI Address →		Transmit the content of a selected data register (X) into another selected data register (X*)		3-3	
MX*N	N → X	P	1	1	X*	X*	X*	N	N	N	N	← NSI Address →		Transmit the N part of the micro instruction, a binary value between 0 and 15, into a selected data register (X*)		3-4	
UOX	UOX	P	0	0	0	1	1	1	X	X	X	← NSI Address →	Mask	Use the content of a selected data register (X) and OR it with the NSI address to determine if a branch in the micro-program is required		3-5	
UAX	UAX	P	0	0	0	1	0	1	X	X	X	← NSI Address →	Mask	Use the content of a selected data register (X) and AND it with the NSI address to determine if a branch in the micro-program is required		3-6	
UXX	UXX	P	0	0	0	0	1	1	X	X	X	← NSI Address →	Mask	Use the content of a selected data register (X) and Exclusive OR it with the NSI address to determine if a branch in the micro-program is required		3-7	
INCR X	X + 1	P	0	0	0	1	0	0	X	X	X	← NSI Address →	Carry	Increment the selected data register (X) by 1		3-8	
DECR X	X - 1	P	0	0	0	0	1	0	X	X	X	← NSI Address →		Decrement the selected data register (X) by 1		3-9	
F N	FN	P	1	0	0	1	Y	N	N	N	N	← NSI Address →		Read one byte out of AUX storage into the U and L data registers. The AUX storage is addressed by the N part of the micro instruction		3-13	
F PN	FPN	P	1	0	1	1	Y	N	N	N	N	← NSI Address →		Read one byte out of AUX storage into the U and L data registers. The AUX storage is addressed by the content of data reg P and N part of the micro instruction		3-12	
F STR	FSTR	P	0	0	1	1	Y	0	0	1	1	← NSI Address →		Read one byte out of main storage into U and L data registers. The main storage is addressed by the content of the data registers A, S, T, R		3-10	
S N	SN	P	1	C	0	0	Y	N	N	N	N	← NSI Address →		Write one byte into AUX storage from U and L data registers. The AUX storage is addressed by the N part of the micro instruction		3-13	
S PN	SPN	P	1	0	1	0	Y	N	N	N	N	← NSI Address →		Write one byte into AUX storage from U and L data registers. The AUX storage is addressed by the content of data register P and the N part of the micro instruction		3-12	
S STR	SSTR	P	0	0	1	0	Y	0	0	1	1	← NSI Address →		Write one byte into main storage from U and L data registers. The main storage is addressed by the content of the data registers A, S, T, R		3-11	
SENSE	SNS n	P	0	1	n	n	n	1	n	n	n	← NSI Address →		Sense the condition of various CPU and I/O indicators and set the sensed condition into data registers U and/or L (Sense 0-63 into L, Sense 31-63 into UL)		3-14/15	Sense Table
CTL	CTL n	P	0	0	1	n	n	1	n	n	n	← NSI Address →		Control any CPU or I/O operations such as stop the CPU or release the card reader clutch	0-31	3-16	Control Table
CTL	CTL n	P	0	0	0	0	0	1	n	n	n	← NSI Address →		Control any CPU or I/O operations such as stop the CPU or release the card reader clutch	32-39		
CTL	CTL n	P	0	0	0	1	1	0	n	n	n	← NSI Address →		Control any CPU or I/O operations such as stop the CPU or release the card reader clutch	40-47		
	X 0-7				X* 4	X* 2	X* 1		X 4	X 2	X 1						
	N 0-15							N 8	N 4	N 2	N 1						
	n 0-63				n 32	n 16	n 8		n 4	n 2	n 1						
							Y 0-1	0 = Long Cycle 1 = Short cycle									
		P 0-1	Parity bit for address (14 bits) and Micro Instruction (22 bits)														

Figure 6.14 Model 20 microinstruction summary.

storage area and descriptive text. This appears more like an assembly listing than the CAS output.

6.4.3 IBM System/360, Model 20

The 2020 processing unit of the IBM Model 20 at the lower end of the System/360 product line is a microprogrammed machine where 36 machine instruction execution routines are stored in a transformer read-only storage, TROS. The electrical characteristics and the principles of operation governing such a ROS device were discussed in Chapter 5. The TROS used in the 2020 is identical to that used in the IBM 2040 processor. It consists of 16 TROS modules. Each module contains 128 plastic tapes which form the information document. Each tape contains two 60-bit words.

The 60-bit word from the TROS is divided into one 16-bit microinstruction and two 22-bit microinstructions. The selected TROS word is gated to the 22-bit ROSDR. Combinational logic is used to select the proper one of the three microinstructions. Thus there is a total of 12,288 22-bit ROS words (Fig. 6.13).

When the 16-bit, type-A microinstruction is read from the ROSDR, bits 10–15 of the ROSDR remain unchanged from the previous microinstruction, and bits 10–15 of the TROS are gated into bit positions 16–21 of the ROSDR. This 16-bit microinstruction is used when only the low-order 6 bits of the ROSDR bit positions 10–21, which contain the address of the next TROS word, are changed.

Every microinstruction consists of a 10-bit instruction part followed by a 12-bit or a 6-bit next ROS address part. Branch type microinstructions change the last four bits of the read-only storage address register bits that contain the next microinstruction address.

Fifteen different micro-orders are encoded into the low-order ten bits of the ROS word. These micro-orders form the building blocks of all the microroutines in the 2020 processor. Figure 6.14 gives the complete Model 20 microinstruction listing.

REFERENCES

1. Tucker, S. G., "Microprogram Control for System/360," *IBM Systems Research Journal*, Vol. 6, No. 4 (1967).

2. Stevens, W. S., "The Structure of System/360; Part II—Systems Implementation," *IBM Systems Research Journal*, Vol. 3, Nos. 2 and 3 (1964).

3. Fagy, P., Hipp, J. A., et al., "IBM System/360 Engineering," *AFIPS Conference Proceedings*, Vol. 26 (1964), pp. 205–231.

4. Carter, W. C., Montgomery, H. C., et al., "Design of Serviceability Features for the IBM System/360," *IBM Journal of Research and Development,* Vol. 8, No. 2 (1964).

5. Weber, Helmut, "A Microprogrammed Implementation of Euler on IBM System/360 Model 30," *Communications of the ACM*, Vol. 10, No. 9 (September 1967), pp. 549–558.

6. Conti, C. J., Gibson, D. H., and Pitkowsky, H. S., "Structural Aspects of the System/360, Model 85, Part 1 General Organization," *IBM Systems Research Journal*, Vol. 7, No. 1 (1968).

7. Hellerman, L., "An Introduction to IBM System/360 Model 40 Microprogramming," *IBM Technical Report TR-00.1766* (August 16, 1968).

8. "IBM Field Engineering—System/360 Model 20; 2020 Processing Unit," *Document No. Y26-5909-4.*

9. "System/360 Model 20 CPU IBM Field Engineering Education Instruction Guide," *Document No. Y25-3027-0.*

MICROPROGRAMMING THE IBM SYSTEM/360, MODEL 40

7

7.1 INTRODUCTION

Model 40 is a medium-to-small member of the IBM System/360. It combines some features of the 1400 and the 7000 series. Its comprehensive binary operations combined with its character-handling ability make it attractive as a commercial, scientific, data-acquisition, and general purpose system.

The commercial performance of the Model 40, in comparison with the 7070, is:

3.34 × 7070 internal
2.95 × 7070 throughput

The scientific performance of the Model 40 system is:

2.11 × 709 scientific internal (short)
1.14 × 709 scientific internal (long)

Model 40 operates in both binary and decimal, and it handles both fixed- and variable-length fields. The philosophy in the System/360, Model 40 physical implementation is to design the arithmetic and logic circuitry as the center of all information handling.

The I/O information flows through channels which are regarded as independent small processors to handle the input/output operations concurrently with the CPU operations. The information flowing through these channels does not flow directly to and from main storage; it passes through the common data-handling bus in the CPU and, subsequently, through the only main storage access path. This last restriction puts some constraints on the microprogramming and on the system's performance, as we shall see later.

The basic Model 40 CPU cycle time is 625 nanoseconds. It has three memories: the main storage (MS), the local storage (LS)—this is sometimes called scratch pad memory (SPM)—and the read-only storage (ROS).

The main storage is a split cycle read/write memory. The read operation is initiated by transferring an address into the main storage address register (MAR) and initiating a read signal from the read-only storage data register (ROSDR). The read operation takes two machine cycles before the two-byte wide word from MS is accessible to the CPU registers. This read cycle is followed by any number of CPU cycles before a write operation is executed. During the intervening cycles, the data read from MS can be operated upon and the result is brought back to the main storage data register (MDR) to be stored into MS. The read cycle is a destructive read, hence, before writing into any location in MS, the write command must be preceded by a read command to zero that location in MS. The Model 40 MS is expandable to 385 kilobytes.

The local storage is considered as an extension of the CPU hardware registers. It operates at the CPU cycle time (625 nanoseconds) and contains the sixteen general purpose registers (GPR), the four floating point registers (FPR), the current program status word (PSW), the I/O dump area, the instruction register (IR), and a number of work areas used by the microprogrammer for intermediate storage.

The ROS is a part of the control system and stores the microroutines that define the basic characteristics of the system. In this case it contains the System/360 architecture. It uses a transformer read only storage (TROS) device with a cycle time of 625 nanoseconds. The CPU logic is organized around a one-byte arithmetic unit capable of performing all arithmetic and logic operations in binary, decimal, and hexadecimal modes.

Figure 7.1 shows a simplified data flow of the Model 40 CPU (generally referred to as the "2040 box"). There are a number of hardware registers that are interconnected by eight- and sixteen-bit data buses.

The Model 40 central processing unit contains the following major functional units:

Arithmetic unit
High-speed local storage
Main storage
Read-only storage
Timing chain and clock
I/O interface
Selector channels

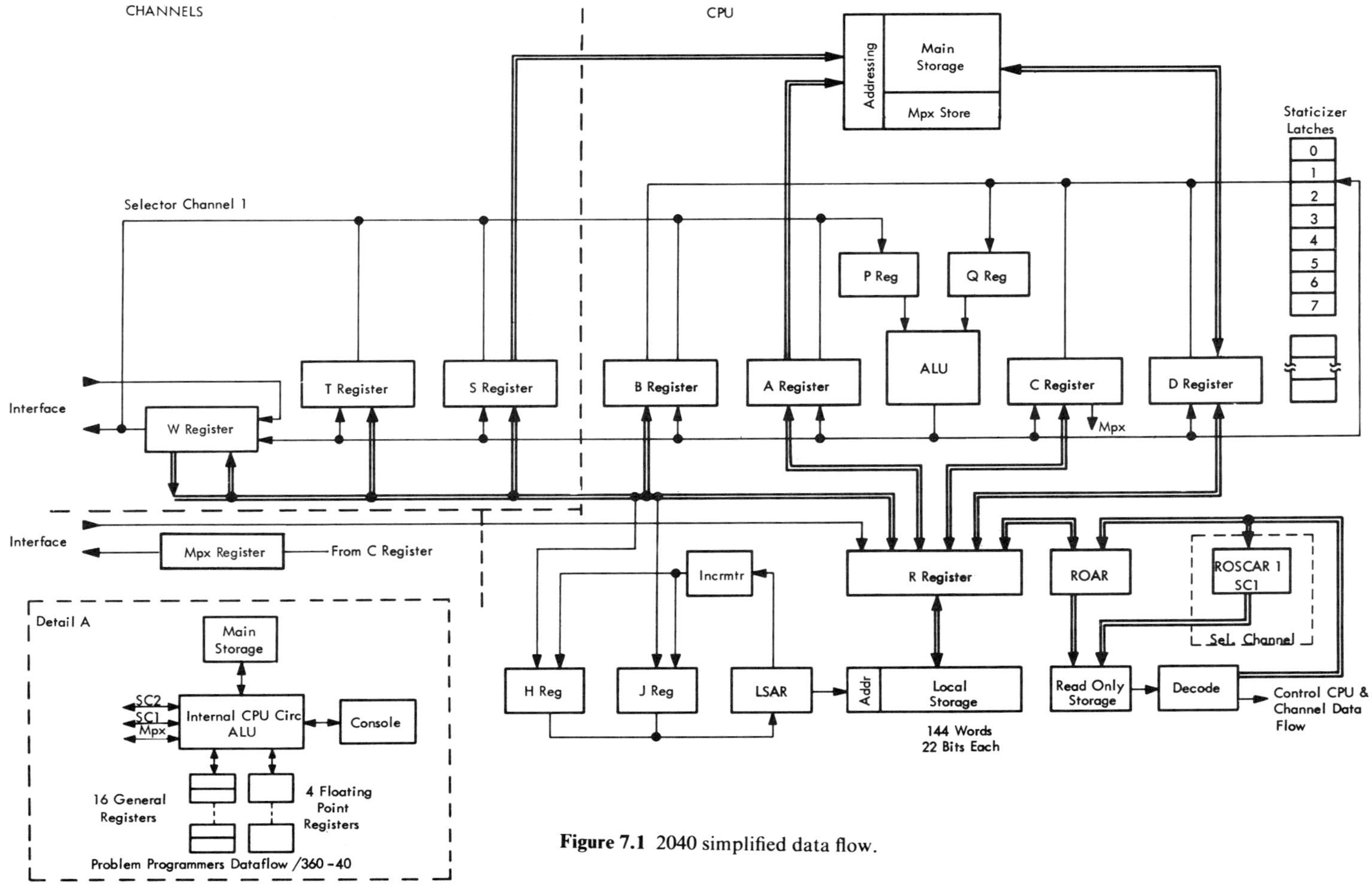

Figure 7.1 2040 simplified data flow.

Multiplex channel
Direct control channel (optional)
Power supplies
Interval timer

The first five functional sections of this list are the primary elements in the data-handling computation, instruction fetch, decode, execute, and sequencing. The next three, namely the I/O interface, the multiplex channel, and the selector channel, permit the attachment and data transfer to and from I/O devices of different rates through the appropriate channel, over the common I/O interface, and through the processor circuitry to the main storage. The direct control attachment is optional and permits the connection of specialized user equipment. Finally, the interval timer is a real-time clock used by the programmer to time out certain segments of his program.

The Model 40 also contains a number of hardware working registers, each of which is assigned specific priorities on the information that it can handle and the interfaces that it can provide. Also, a number of general and special purpose stats (indicators) are provided for microprogram status storage.

The problem program is a sequence of instructions which is fetched, decoded, and executed from main storage, one instruction at a time, under the control of the microroutines stored in the read-only storage. The data flow control circuits receive the control signals from the ROS unit and, in turn, sequence the execution of these instructions synchronized by the clock distribution system.

The basic data unit in the Model 40 CPU is the byte (8 bits plus a parity bit). This allows an extended character set of up to 256 characters. It also permits packing of two decimal characters per byte. All data flow sections are therefore integer multiples of the byte.

7.2 MODEL 40 DATA FLOW

The Model 40 data flow may be divided into five sections.

1. One-byte arithmetic and logic section.
2. Two bytes of data handling to or from local storage.
3. Two-byte handling of main storage address and data fields.
4. Two-byte service to channels, where one byte contains data and the other contains the channel interface.
5. One-byte local-storage address servicing.

Figure 7.2 shows a detailed data flow of the Model 40 with the multiplex channel interface. Figure 7.3 shows the selector channel data flow—the interface lines and hardware registers that connect it to the CPU. However, Fig. 7.1 gives a more simplified data flow showing all the main hardware

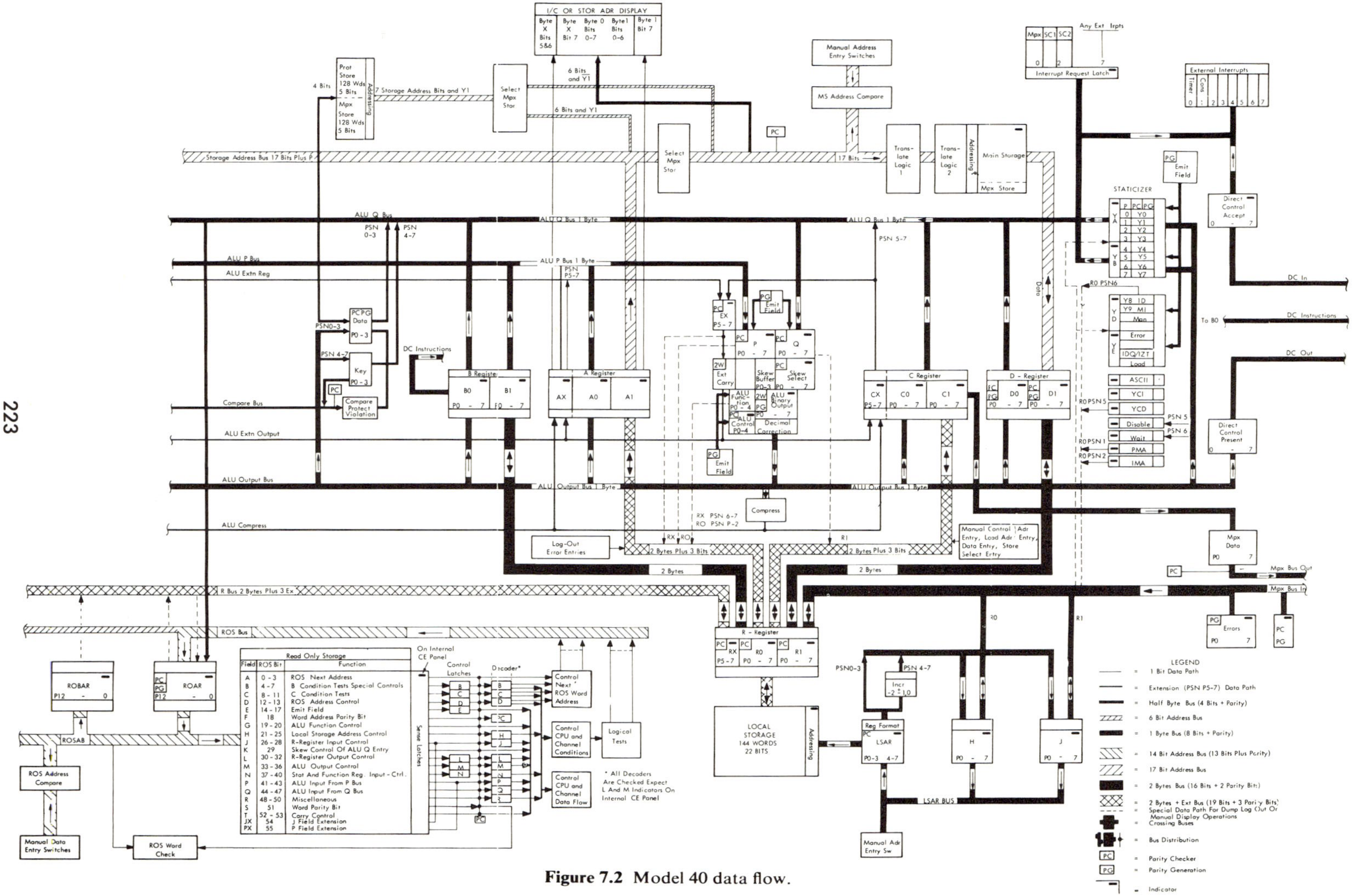

Figure 7.2 Model 40 data flow.

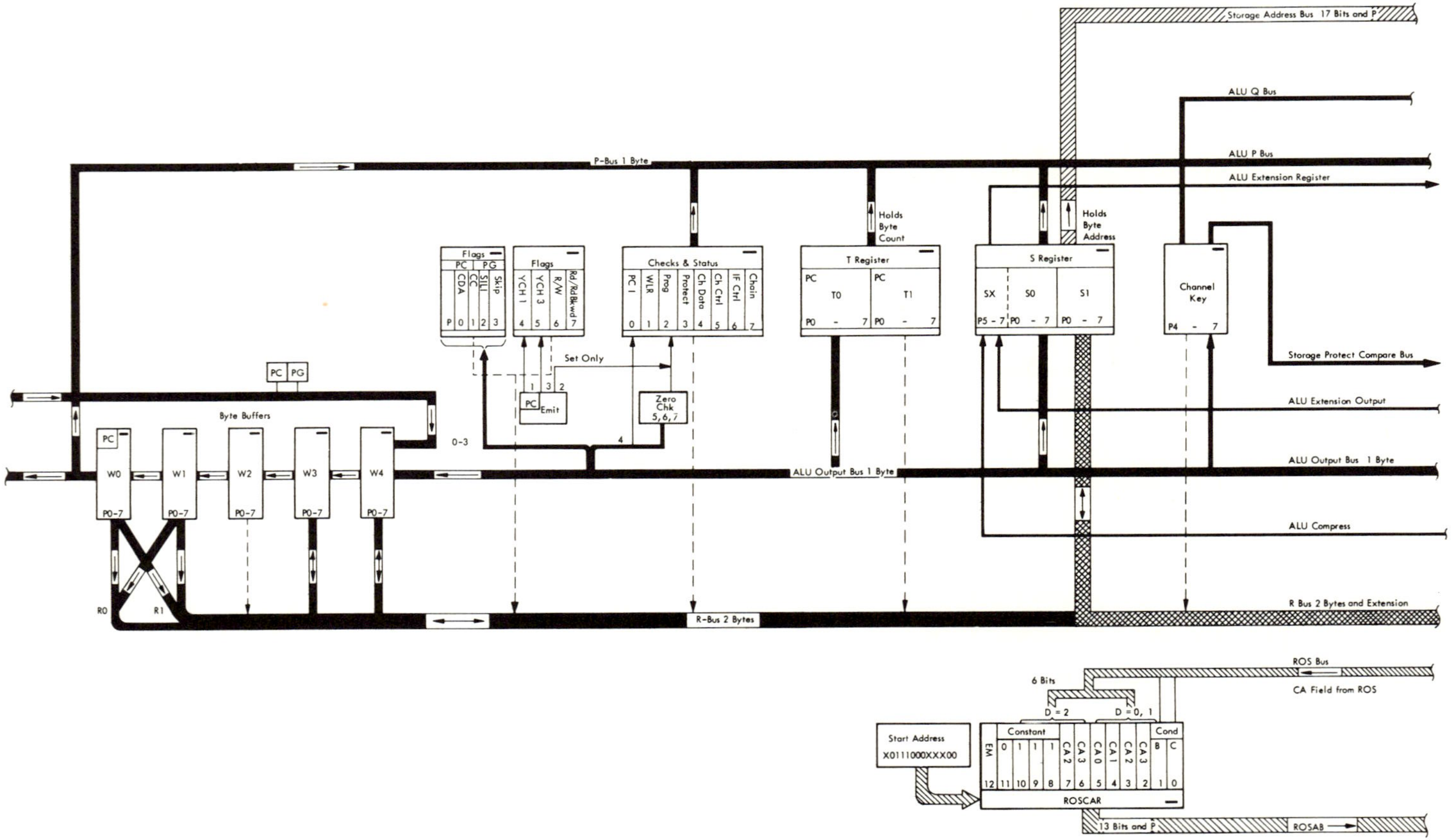

Figure 7.3 2040 selector channel data flow.

functions, the staticizer latches, local storage, main storage, the selector and multiplex registers, the local storage, the control storage, and all the data buses which connect these hardware functions into one total system. Figure 7.1 does not show the widths of these different facilities and other valuable information. Therefore, when reference is made to Fig. 7.1, it would be advisable for the reader to also refer to Fig. 7.2 to get the additional information.

7.2.1 The Main Hardware Registers[1]

These registers, labeled A, B, C, D, and R in Figs. 7.1 and 7.2, are the five main registers provided in the CPU which can be used by the microprogrammer for temporary storage of data and control information. These registers form the basic "skeleton" of the system. They are divided into two groups, based upon their width of data handling: registers B and D are in one class, and A, C, and R are in another class. The latter class has the three extender bits in the high-order positions (the use of the extender bits will be discussed later). The transfer of information between the first four registers is handled over a two-byte data bus which connects each register to the R register. Thus, any transfer of data between any two registers has to be routed through the R register. For example, a transfer of the contents of register A to register C implies that the contents of register A are routed over a two-byte plus three-bit data bus to register R, and then from register R over another data bus also of two bytes plus three bits to register C. A transfer from register D to register A would entail transferring the contents of register D over a two-byte data bus to the two low-order bytes (R_0, R_1) of R, and then setting the R_x portions (R-extended bits) to zero prior to transmitting the full contents of R over the two-byte plus three-bit bus to register A.

Each of these registers is assigned certain functions which have priority on its services over all other functions in case of any such contention. Hence we shall review each register separately.

A Register. The A register is two bytes (A_0, A_1) plus the three-bit extension section (A_x). It serves the microprogrammer as a temporary storage; however, its main purpose is to address main storage during a CPU or multiplex channel operation. The A register is set from the R register, or a selected byte of the A register is set from the arithmetic and logic unit (ALU) output over the eight-bit plus parity ALU bus. Its output can be gated into one or more of the following facilities: the storage address bus, R register bus, and the P bus. Since the source of the A register data can be specified in the same microinstruction from the R bus and the ALU output, the priority in such a conflict is given to the ALU output by blocking the R bus gating for the A register part which is designated to receive the ALU output.

C Register. This is identical to the A register in width and composition. The C register extension bits, C_x, are provided for more efficient access to main storage. During a main storage cycle, the A register which contains the address of the main storage location being interrupted may not change. Hence, with the extension bits C_x available, any address updating can be carried out in the C register during the main storage cycle.

The C register is set from the R bus and the output of the ALU. Its output can be microprogrammed to set into the R bus, or any selected byte from the C register can be gated to the Q input of the ALU section.

Byte C_1 is the only source which is fed to the multiplex data register. All output information of the multiplex channel must therefore be transferred to C_1.

B Register. This register is only two bytes wide (B_0, B_1), each with a parity bit. It is microprogrammed-controlled (as are all other registers) to serve as a work area and temporary storage. The B register is set from the R register bus or the ALU output. Its output can be gated into the R register, and the P or Q bus inputs to the ALU unit. It is the only register that is not assigned any specific function, and it is the only register that can be gated to both the P and the Q buses.

D Register. Except for its function and output gating, the D register is identical in width and circuitry to the B register. It is used for temporary storage for two bytes of data; however, its principal function is to act as the main storage data register. In a main storage read operation, the contents of the selected two consecutive bytes of main storage are gated into the D register during the second main storage read cycle. If an invalid read main storage address occurs, or if storage protection is violated, the ALU and R gating to the D register input are blocked to ensure that the storage data are not destroyed.

Since the microprogrammer may inadvertently specify the inputs to the D register from the ALU, MS, or R register in the same microinstruction, a hardware priority scheme is incorporated to resolve any such contention to the D register. In such a conflict, the highest priority is given to the ALU, the next lower to the R register, and the lowest to the main storage interface. No provision is made to override one byte of main storage by the specified ALU byte; in such a case, an invalid microprogrammer specification is given.

R Register. Since the previous four registers and many other facilities in the Model 40 literally use the R register for the central distribution of data, its width must be sufficient to accommodate the widest of all these facilities, namely the A and C registers. Hence, the R register is a two-byte plus three-bit extension (R_x, R_0, R_1), with a separate parity bit for each segment.

The special purpose of the R register is to provide a buffer for the local storage and also a display buffer for the console lights associated with the Model 40 console roller switches. (Incidentally, in Figs. 7.2 and 7.3, the hardware facilities with a dash (—) in the upper right corner have unique indicator lights on the customer engineer's indicator panel.)

The input and output gating to and from the R register is microprogram-controlled. In one CPU cycle, information from any selected source is gated into the R register. It is then transferred back into the same source or into one of several other available sinks or destinations specified by the microprogrammer. The R register (bus) consists of a group of latches which can hold information for a maximum of one machine cycle, and which are reset at the same timing point of every machine cycle. Therefore, to preserve the information buffered into the R latches, the microprogrammer must transfer it to a more permanent storage media prior to the R register reset pulse. During a hardware log-out cycle, the contents of R register are stored into the local storage. Also, in an undump hardware cycle, the R register contents are gated to the read-only storage address register (ROAR).

7.2.2 One-Byte Arithmetic and Logic Unit

The ALU is a one-byte adder/subtractor which operates in either hexadecimal or decimal mode; it is capable of performing arithmetic and logical operations on the input data. The ALU schematic[1] is shown in Fig. 7.4.

Data are directed onto the P and Q buses in one-byte increments from the A, B, C, and D registers or, alternatively, from the skew function in response to the current microinstruction. The ALU then proceeds to manipulate and combine the two bytes of data as directed by the microprogram. The result appears on the ALU output bus to be returned to the location specified by the current microinstruction being executed. An ALU operation is attempted every machine cycle. If no operation is required by the microinstruction being executed, zeros will be gated to the ALU and will be ANDed.

The actual arithmetic or logical operation that occurs is either a direct or an indirect function of the ROS word. For the direct functions, the ALU control circuits are determined by the contents of the ROS word. An indirect function is one for which the required function is preset in a five-bit *function register* and then used as required by the microprogram. This means that several instructions may have a common microprogram subroutine when the difference is only in the ALU function. Direct and indirect functions have separate carry latches, YCD and YCI, respectively.

On the left side of Fig. 7.5, the box labeled ALU control illustrates the direct and indirect operations. It has three inputs: the indirect control signal comes from the function register; the signal for direct operation comes from the CG field in the ROS word; and the third input determines which of the two types of operations will be executed. This choice depends on the setting of the CG field and on the status of the inhibit dump stat, ID (Y8), and the

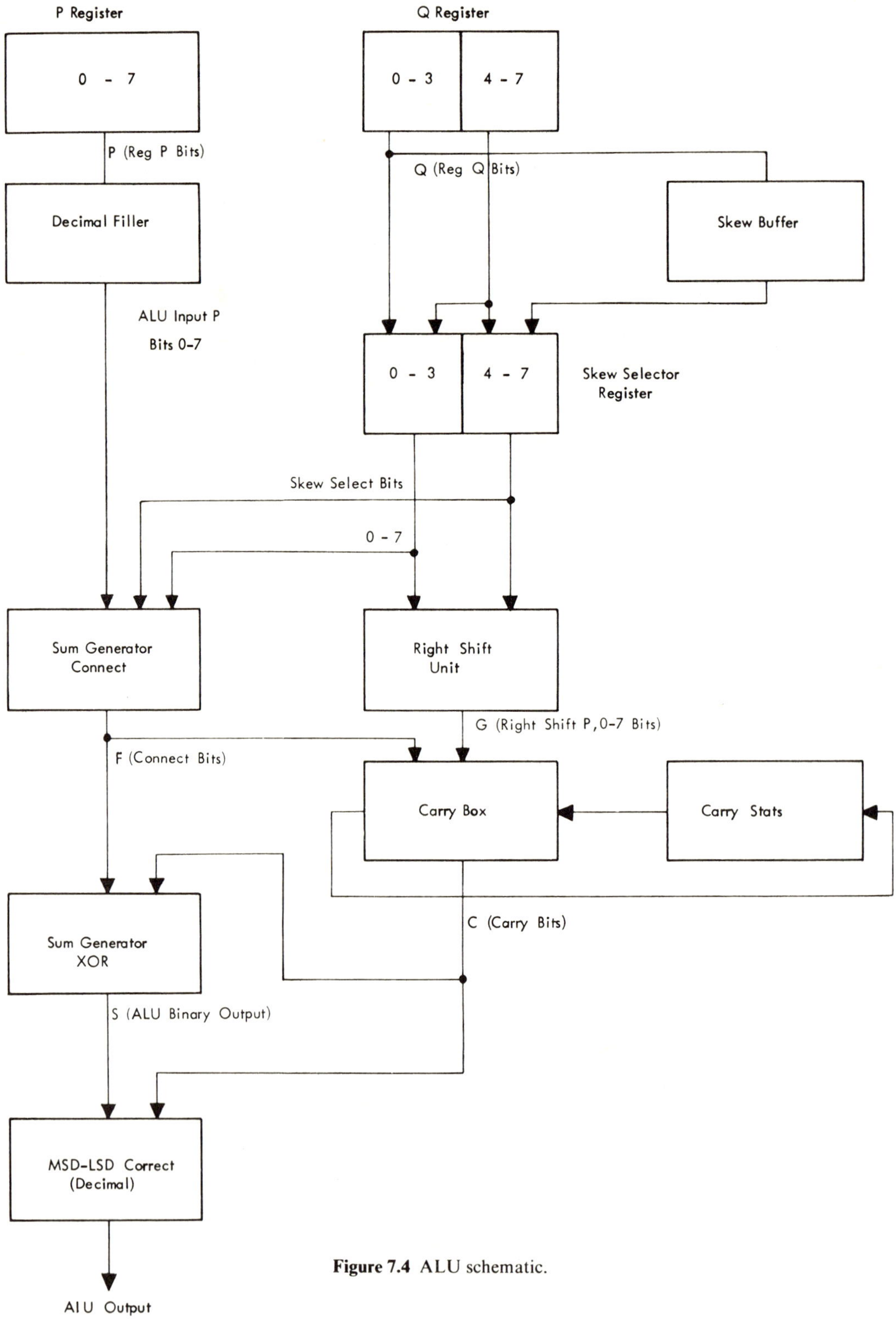

Figure 7.4 ALU schematic.

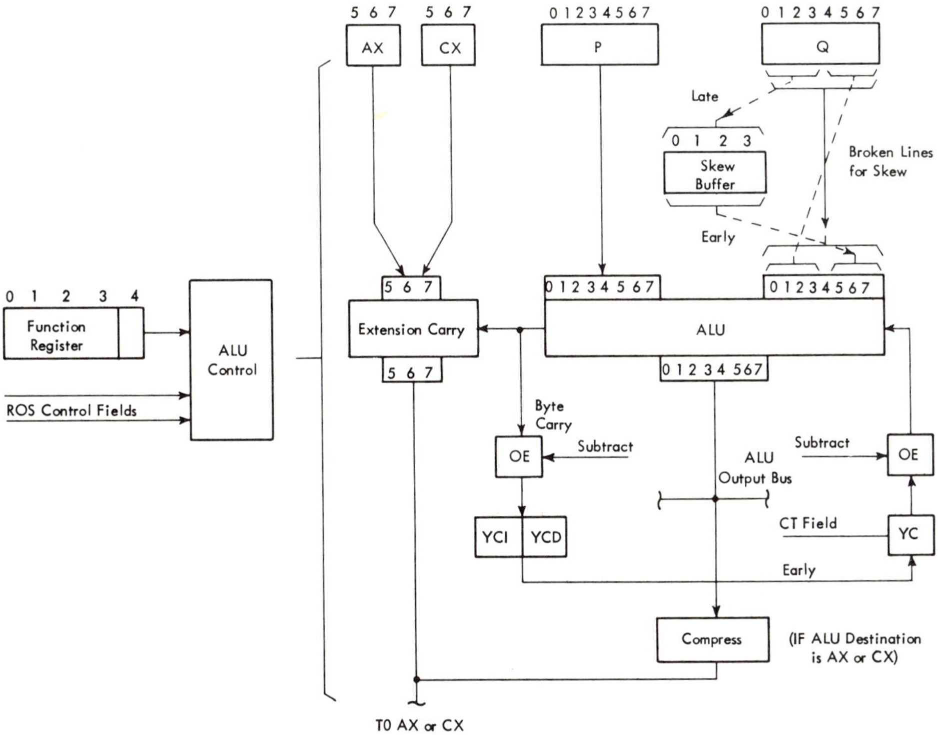

Figure 7.5 ALU principle.

reentrant control stat. Both of these stats are under the control of the microprogrammer.

The indirect function register is also under the control of the microprogram; but once it is set (from the emit field), it remains until reset by the microprogram.

The ALU essentially treats each digit as binary data, and a carry from each decimal digit is guaranteed after the count of 15. Additional logic is added to process decimal arithmetic. This is shown on the ALU schematic diagram (Fig. 7.4) by the Decimal Filler and Decimal Correct (MSD-LSD) boxes (LSD = least significant decimal digit; MSD = most significant decimal digit).

Decimal arithmetic, then, is performed by effectively adding Hex 66 to the input byte from the P latches. This is then added to the second operand from the Q latches. If the high order of the packed digit generates a carry, the result is correct; otherwise, the result must be decimal-corrected. For both (0–3) MSD and (4–7) LSD, all sums falling between 9 and 15 are corrected by subtracting 6 (effectively adding 1010 1010, the 2's complement of 66).

The Skew Box affects the output of the Q bus only. It is effectively a four-bit left shift with the contents of the four-bit skew buffer moved to fill the low-order four bit positions. The leftmost four bits which are shifted out are entered into the skew buffer and are available for a later skew cycle. The order in which the half bytes are transferred is as follows. The skew-buffer bits (0–3) are transferred to skew selector register (4–7). Q (4–7) is then set into the skew selector register bits (0–3). Finally, the Q-register bits (0–3) are transferred into the skew buffer (0–3). If the skew buffer is reset prior to an ALU skew cycle, zeros are gated into ALU bits (4–7).

The ALU Carry Output is stored in either the YCD or YCI stat, depending on whether the current ALU function is direct or indirect (Fig. 7.6). Early in

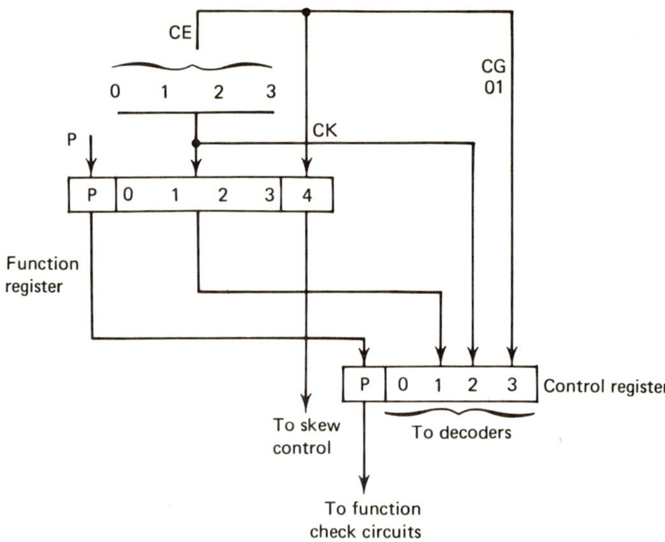

Figure 7.6 Function and control registers.[1]

the following cycle, the YCD or YCI contents are transferred into the YC stat (the choice between YCI or YCD depends on the ALU function in the cycle in which the transfer does occur). The contents of YC are then EXCLUSIVE ORed with a subtract control signal. In other words, if an indirect subtract operation results in a carry-out, the YCI will be set to zero.

Similarly, the carry-out signal which is gated into the extended-carry control logic is EXCLUSIVE ORed with this subtract signal. In a true add, therefore, the output of YC is gated directly into the low-order bit of the next byte. In any case, these steps are merely logic tricks, and the carry is effectively propagated to the next byte.

The YCI, YCD, and YC are available for the microprogrammer to test and branch to one of two successor microinstructions based on their setting. The results of these tests force a zero or one in the ROAR bit zero, ROAR(0), to reflect the status of the specified carry stat.

Extended-Carry Box. Whenever the high-order byte of the A or C register (A_0, C_0) is specified as an input to ALU, the corresponding extension bits, A_x or C_x, are automatically gated into the ALU extended-carry circuits. If both A_0 and C_0 are specified ($A_0 \rightarrow P$, $C_0 \rightarrow Q$) as inputs to the ALU, A_x overrides C_x. The extended-carry circuits are updated by any carry signal emanating from the byte logic; and the updated extended-carry is gated into the A_x or C_x, depending on the specified destination of the ALU output.

Compress Circuit

In all ALU operations, whenever the ALU output byte is gated into A_x or C_x, the byte must be compressed into three or two bits. If the specific Model 40 has less than 256K storage, then the bit resulting from ORing ALU (0–6) bits is set in A_x or C_x bit 6, and A_x or C_x bit 7 is 0. However, if the Model 40 has 256K storage or greater, resulting in a full 19-significant-bit address, then ALU bits (6, 7) are set in the A_x or C_x (6, 7), and bits ALU (0–5) are ORed and transferred into A_x or C_x bit 5. Since the maximum standard Model 40 storage is 256K byte, an 18-bit address bus would be sufficient; however, since the main storage to the CPU data bus is 2 bytes wide, the low-order bit of the 18-bit address is not needed. Hence, in Fig. 7.2, the storage address bus is shown to be 18 bits wide.

ALU—Function Control. The ALU control circuits are either specified directly from the ROS word, in which case the control is said to be direct; or it can perform an operation specified by a special function register. The control of the ALU in this latter case is called indirect. Once the function register is set (under the control of the ROS word), it remains set until modified by the microprogrammer.

In routines where similar operations are performed, except for the ALU function, this indirect operation capability can save a number of ROS words. The indirect function register is set from the CE field. The low-order bit is set by the CK field for skew control. The high-order four bits of the function register are transmitted to the control register to be decoded. The control register issues up to eleven control signals to the different microscopic functions in the ALU. Figure 7.6 shows the function and control registers.

7.3 LOCAL STORAGE

A small high-speed storage unit is incorporated into the Model 40 to provide storage for fixed- and floating-point registers, for channel operators, for dumping of CPU registers in case of error condition or I/O interrupt, and to provide a few working registers which can be used by the microprogrammer as a scratch pad.

The local storage unit contains 144 words of 22 bits each (19 bits plus 3 parity bits). The local storage addressing is completely random and operates on a split cycle. A read or write cycle takes 625 nanoseconds. For high-speed access to several locations, local store is cycled continuously in the read mode, and the contents of each local-store location designated every cycle will be transferred through the R register and then on into its specified destination. The local store is essentially a destructive read memory; hence, a read call must always precede a write to clear the specified storage location.

7.3.1 Local-Storage Address Loop

Local storage is addressed by an 8-bit LSAR register (local-storage address register). Two other 8-bit registers, H and J, the incrementor logic, and the LSAR form a local-store address loop. The incrementor can increment the LS address by +1, −2, or 0. The contents of LSAR can be routed through the incrementor to the H or J register later in the cycle so that the proper address can be gated into LSAR early in the next machine cycle.

Figure 7.7 shows the contents of the local storage. Note the invalid address locations which have addresses of the form XX01XXXX. It must be stated that the 10XXXXXX and 11XXXXXX addresses are equivalent. Local-storage address specification and the ROS fields involved will be discussed later.

7.4 MAIN STORAGE

The main storage used for the Model 40 is a 2.5-microsecond split cycle (read or write) memory. It consists of two parts, the true main storage and an additional segment called the *bump* stores. The later is accessible only by the microprogrammer and is used to hold channel control words. The main storage is available in five sizes, 16k, 32k, 65k, 131k, and 262k bytes, with a bump storage of 256, 512, 1024, 2048, and 2048 bytes, respectively, associated with the MS.

Main storage data are transferred to and from the CPU via a 16-bit data bus. Register D is the main storage data register. Main storage address circuits are fed from the A register by a 17-bit main storage address bus. In the selector channel mode, the S register acts as the main storage address register. Main storage accesses for read or write operations are completely independent and must be called individually by the microprogrammer. There may be a gap of an indefinite period following either a read or a write cycle; but a write cycle must always be preceded by a read cycle to zero the selected two-byte locations in main storage.

The main storage clock, once started, proceeds under control of the storage clock independently of the system clock.

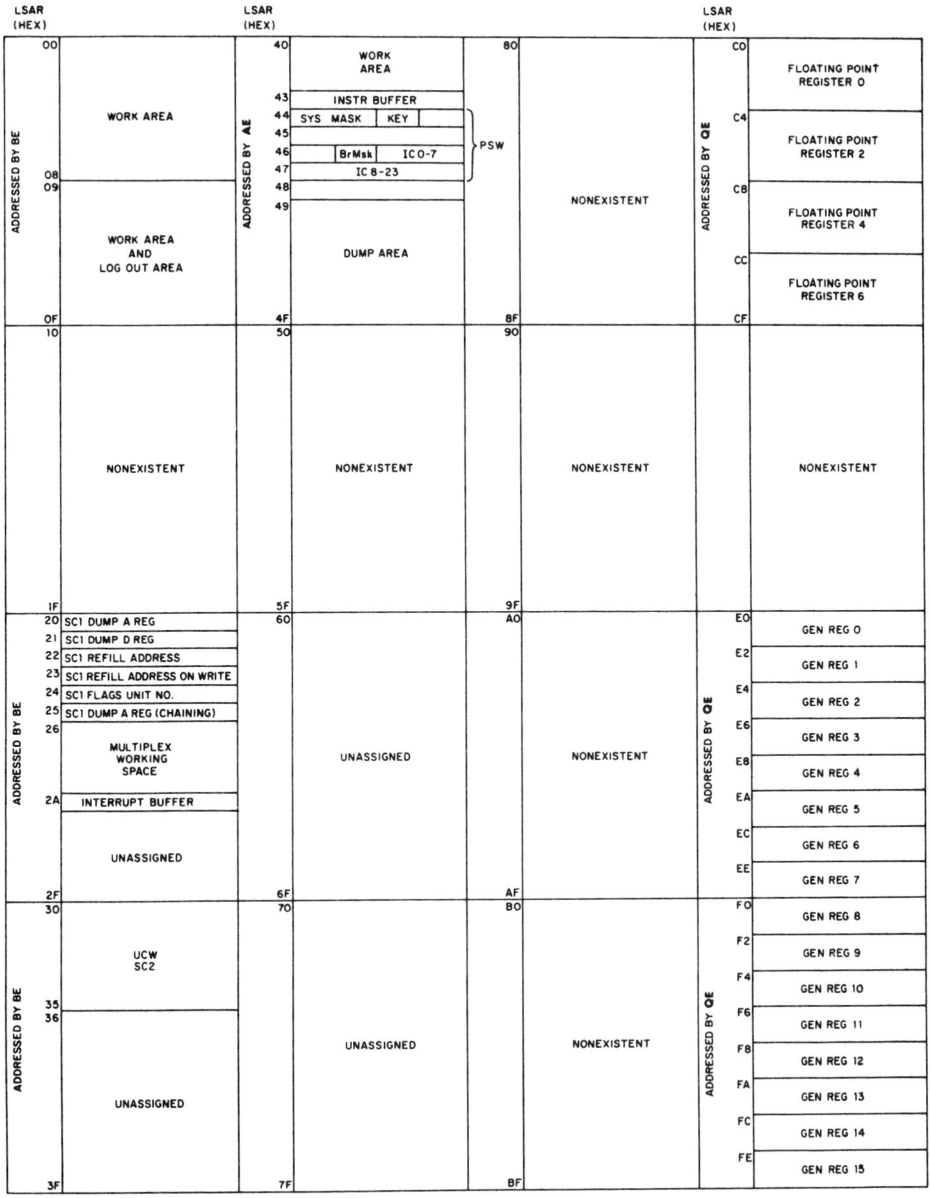

Figure 7.7 Local storage contents for System/360, Model 40.

To initiate a main storage read cycle, the address is set in the A register in the second half of the previous cycle. During the second storage read cycle, data from the MS are gated into the D register. The timing of MS read and write operations and the ROS control fields that affect the memory storage cycle will be reviewed later.

7.5 STATICIZER LATCHES (STATS)

There are sixteen multipurpose, single-bit registers, called stats, which are set to indicate the status of the machine. They can also be grouped into registers for intermediate temporary storage. Some stats are set by hardware to indicate that a specific condition has arisen, others are set by the microprogram individually or in groups. Figures 7.1 and 7.2 show these stats and some possible ways in which they can be grouped. For example, the top eight stats (0–7) can be grouped into one 8-bit register that can be gated onto the ALU Q-input bus. Stats can be displayed on the console or tested by the microprogrammer to alter the microprogram sequencing.

The sixteen general purpose stats are also grouped into YA, YB, YD, and YE, with four stats in each group. Other stats that are not called Y stats are: HALT, WAITE, ENABLE, ASCII, CPU, PSA (protected storage address—set when a protected storage location is addressed), and ISA (invalid storage address is set when an invalid storage is addressed). The YCD or YCI stats were discussed earlier in conjunction with the ALU direct carry or indirect carry operation. Again, the micro-orders and ROS fields which control their input, output, and function will be discussed later.

7.6 SYSTEM TIMING[1]

The basic clock time is derived from a crystal-controlled oscillator of 3.2 megahertz (3.2×10^6 Hz) fed into a paraphase amplifier and emitter-follower logic to produce two separate timing pulses. The odd and even clock pulses produced are 180° out of phase. Two sets of pulses, the P and T pulses, are generated. One continuously available set of pulses, called the P (P1–P4) pulses, control the main storage and the ROS storage, while the T (T1–T4) pulses can be started and stopped by CPU control. The P and T pulses are shown in Fig. 7.8.

Figure 7.9 shows a few basic timing pulses and the order in which the main hardware facilities are affected. LSAR and the R, P, and Q registers are set early in the machine cycle, while registers A, B, C, and D are set late in the cycle. The contents of local store are available to the CPU (with R register) during the first half of the cycle and can be gated into any register in the second half of the cycle. Accordingly, the contents of any hardware register can be gated into another hardware register via the R register in one cycle. Also, the address of a selected word in LS can be gated into the LSAR, and a transfer between LS and any selected hardware register can be executed, all in one cycle. In the same cycle, one can gate two bytes to the ALU P and Q buses to perform any arithmetic or logic operation, and then transfer the resulting byte into any hardware register specified. This is possible because the ALU 8-bit data flow and the register-to-register 16-bit data flow systems can operate in parallel.

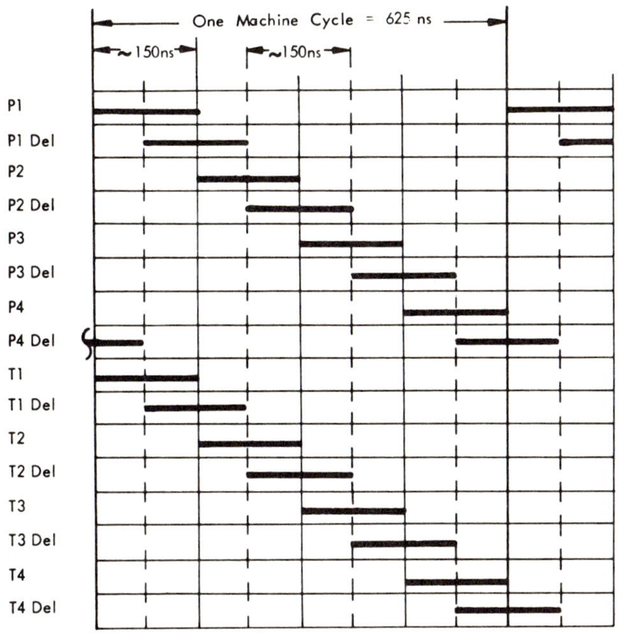

Figure 7.8 Basic clock pulses.

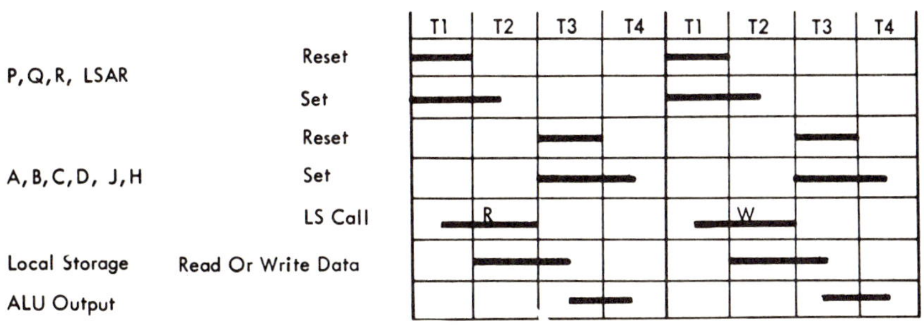

Figure 7.9 Basic data flow timings.

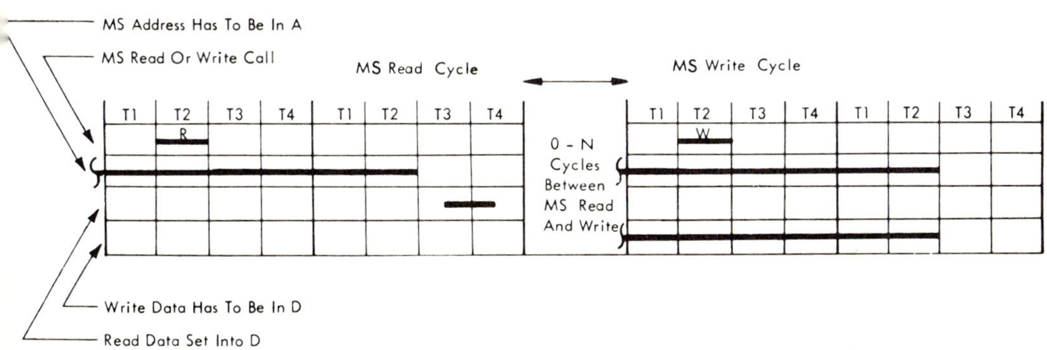

Figure 7.10 CPU and main storage timing relationship.

Figure 7.10 shows the CPU and main storage timing relationship. Basically, the main storage clock is 2.5 microseconds, which is equal to four CPU clocks, and it is split in two halves: the read half and a write half. Between the read half and the write half, any number of basic machine cycles can be taken.

7.7 READ-ONLY STORAGE CONTROL

As in most microprogrammed control machines, the Model 40 receives its control from a nondestructive transformer read-only storage (TROS) unit every machine cycle. Hence, the total cycle time of the read-only storage unit is 625 nanoseconds.

TROS is built of modular units, each containing 256 addressable words, stored on plastic tapes, two words per tape. Each word has 56 bits, including two parity bits. The standard Model 40 has 16 TROS modules. This capacity can be expanded to 32 if needed.

TROS is addressed by a 12-bit read-only storage address register (ROAR) or by an ROSCAR register if in a selector channel mode of operation. There is a unique ROSCAR for each of the two selector channels permissible in the Model 40 system configuration.

7.7.1 ROS Control Word

The TROS word is 56 bits wide and is divided into 18 separate control fields. The number of bits in any field is dependent on the number of functions the field is intended to perform.

The field can be grouped into four general classes: the parity bits (18, 51); the ROS sequencing fields consisting of CA, CB, CC, and CD and occupying the first 14 bits of the ROS word; the emit field, CE, which is a 4-bit field that can supply a maximum of four bits of fixed data directly into various locations in the CPU; and the status control fields, which control the CPU, multiplex channel and some selector channel operations.

The emit field can gate into one or more of the following:

1. Bits 0–3 or 4–7 of the P bus.
2. Bits 0–3 or 4–7 of the Q bus.
3. Bits 0–3 of the ALU function register.
4. Bits 0–3 or 4–7 of the LSAR.
5. YA, YB, YD, or YE stat groupings.
6. Selector channel flag registers.

Before reviewing the catalog of micro-orders for the IBM Model 40 processing unit and introducing any microprogramming examples, an important area in the system's data flow remains to be discussed. This area, presented in the next section, is the input/output and channel controls.

7.8 CHANNELS

A large number of I/O devices can be attached to the Model 40 through the multiplex or the selector channel A or B. Channels and control units are connected to the processor via a standard interface which is a data path that permits the transfer of one byte of information between the channel and control unit.

During the actual transfer of information, the whole data path between main storage and the selected I/O device is dedicated to this operation. Data transfers are initiated by the CPU program which, in turn, sends commands to the proper control unit in the I/O chain. Examples of these commands are READ, WRITE, READ BACKWARDS, etc. Once these commands are accepted by the unit, it releases the channel and interface facilities and resumes the execution of these commands on its own.

Channels are controlled by channel programs. They may be physically separated and regarded as independent computers for processing I/O instructions; or, as the case in the Model 40 design, the channels may share CPU facilities. While it is economical, the latter case can present a great deal of interference with the CPU and thus affect the system's performance. A channel is unbuffered if it shares all the facilities with the CPU. A buffered channel needs to share only the main storage access path. In either case, when the channel requests any CPU facilities, the Model 40 is switched from the CPU to the I/O mode; however, the two operations can be interleaved to give maximum performance efficiency.

7.8.1 Multiplex Channel

The multiplex channel in the Model 40 has no data flow of its own but shares almost all its facilities (except the circuits required to communicate with the interface) with the CPU data flow. In a read operation, information is gated into the R bus; for a write, the contents of byte C1 are gated to the interface lines. The multiplex channel is controlled by the ROS memory.

In accepting one byte of information from an I/O device, the device would first request a program interrupt. This interrupt causes the CPU to dump the contents of its data flow in local storage to preserve it. Now the channel routines provide the necessary channel control and update till the byte is safely tucked away in main storage. At the end of operation signal, the CPU is undumped and the operation resumes from where it left off prior to the interrupt.

7.8.2 Selector Channels

The Model 40 selector channel is almost completely independent of the CPU circuitry, except for the D register. A 5-byte buffer (Fig. 7.2) and other independent data and control registers are provided in each selector channel.

Among the CPU facilities duplicated in the channel are the main storage address register, S, and the ROSCAR. Thus the CPU channel interference is kept to a minimum. During the data transmission, the only CPU facility that has to be dumped and undumped at the beginning and end of the data transfer operation is the D register. Hence, when the channel is ready with two bytes of data, the D register is undumped, and the machine is shifted from the CPU mode to the I/O mode by simply shifting control from ROAR to ROSCAR. When the I/O operation is completed, the D register is undumped, and control is returned to the ROAR register.

PROBLEMS

*Data Flow**

The following are exercises in reading Figs. 7.1, 7.2, and 7.3. Although they are mostly trivial, they should be gone through in order to gain a working knowledge of the Model 40 data flow.

1. What are the bit capacities of each of the following?
   (a) B register
   (b) H register
   (c) Bus between the A and R registers
   (d) Bus between A register and MS (main storage)
   (e) Bus between MS and the D register
   (f) Bus between Compress and CX (C register extension)
   (g) Bus between LS (local storage) and the R register
   (h) A register

2. Which of the following is *not* displayed by console indicators?
   A, B, LSAR, R0, LS, MS, E field of ROS word

3. True or false: information may be gated directly from:
   (a) B0 to B1
   (b) A0 to P
   (c) A1 to P
   (d) A1 to Q
   (e) CX to EX
   (f) C to R
   (g) D0 to P
   (h) D0 to Q
   (i) B0 to P
   (j) B0 to Q
   (k) D to C
   (l) D to R
   (m) R to C
   (n) MS to C
   (o) B to MS
   (p) R to LS

4. True or false: the output of the ALU may be gated to:
   (a) R
   (b) C
   (c) C0
   (d) AX
   (e) T1
   (f) S0

*Many of the practice problems in this chapter were originated by Dr. Leo Hellerman and are used here with his permission.

(g) Channel flags
(h) Buffer W4
(i) Storage protect keys
(j) ROAR

5. How is ROS addressed? That is, what are its address registers? How many bits do they contain?

6. Concerning the ROS control storage:
   (a) How many parity bits are there in each ROS word? What do they check?
   (b) What are four ROS word fields that determine the next ROS word to be executed?
   (c) What ROS word fields control the inputs to the P bus and the Q bus?
   (d) What ROS field determines carry control for the ALU? How many bits does it have?
   (e) Which ROS word field has the most bits? Which fields have only one bit?

7. Why is there no YC group of stats? (*Hint:* If there were, confusion might arise with other specific stats. Which ones?)

8. Note that the capacity of the A register is 19 data bits, but the address bus from the A register to MS addressing is only 17 bits. Why?

7.9 THE MECHANICS OF MICROPROGRAMMING THE MODEL 40

The general mechanics of writing a Model 40 microprogram are similar to those discussed in the System/360 general concepts. Microinstructions that are intended to control the machine cycles on a one-to-one basis are documented on the ALD (Fig. 6.6) logic sheets in a specific order and format. The concepts of edge character, leg selector, leg identifier, Hex address, box print position, and box serial number have the same meaning here. Only the contents of the box and the decoding of these contents are different from any other System/360 microprogram.

Figure 7.11 shows a Model 40 microinstruction box with all the allowable edge characters and their functions.

Print Line				*ROS Word Address in Hex*
1	*Leg Identifier*			XXX
2	E	b	Emit field, carry C	
3	A	b	Arithmetic statements	
4	L	b	Local storage—loading and addressing	
5	D	b	Data transfer (the 16-bit transfer)	
6	S/C	b	Main storage and miscellaneous	
7	R	b	ROS addressing—tests	
8	XXX		*Leg Identifier*	XXX
	Box Position			*Box Serial Number*

Figure 7.11 Microinstruction box.

Note that, when writing micro-orders, the order of these edge characters must be preserved. The edge character must be specified, followed by a blank position before the first character of the micro-orders on that line. Also, no micro-orders or statements may be permitted to overflow to another print line. Finally, if no statement appears on a particular line, then no edge character will be printed for that line.

When the microprogram has been written on the ALD sheets, the microroutines can then be assembled, simulated, edited, and translated to the final 0's and 1's to be punched in the TROS tapes.

An example of a Model 40 microinstruction is shown in Fig. 7.12.

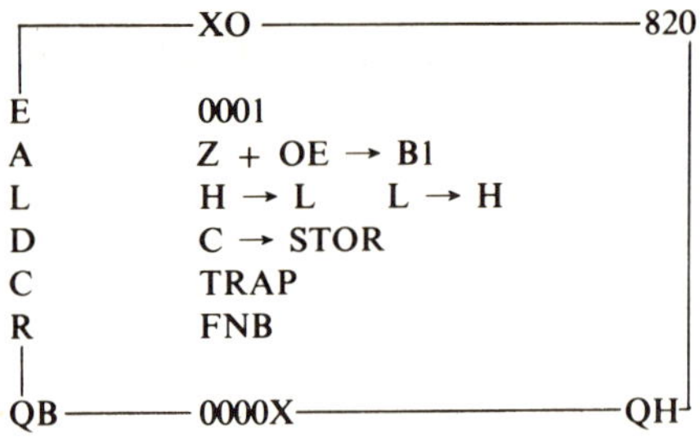

Figure 7.12 An example of a microinstruction box.

7.9.1 Control Field Specifications

Figure 7.13 shows the ROS control fields. It also gives tabulated ROS field decoding charts, the numeric codes, the decimal orders, and the conditions (machine stats and modes of operations) under which the same contents of the fields are interpreted.

CA, the first field, occupies bits 0–3 of every ROS word, forms part of the next ROS address and is set into ROAR or ROSCAR. It is dependent on CB and CD fields in determining where in the ROAR the CA bits are to be set. For example:

If CB = 0–13, and CD = 2, then CA is set into ROAR bit positions 5–2.
If CB = 15, then CA is set into ROAR bit positions 9–6.
If CB = 14, and CD = 3, then CA not used (undump format).

CB field is a 4-bit field occupying bits 4–7 of the ROS word. It is used to specify the B condition tests which set the ROAR or ROSCAR bit 1 according to the outcome of the specified test and if the CD field = 0 or 2. The CB interpretation is controlled by the CPU I/O latches. If CD field is equal to 3, special branch conditions are generated. If, on the other hand, CB is 15, it will override all other conditions and will specify a functional branch.

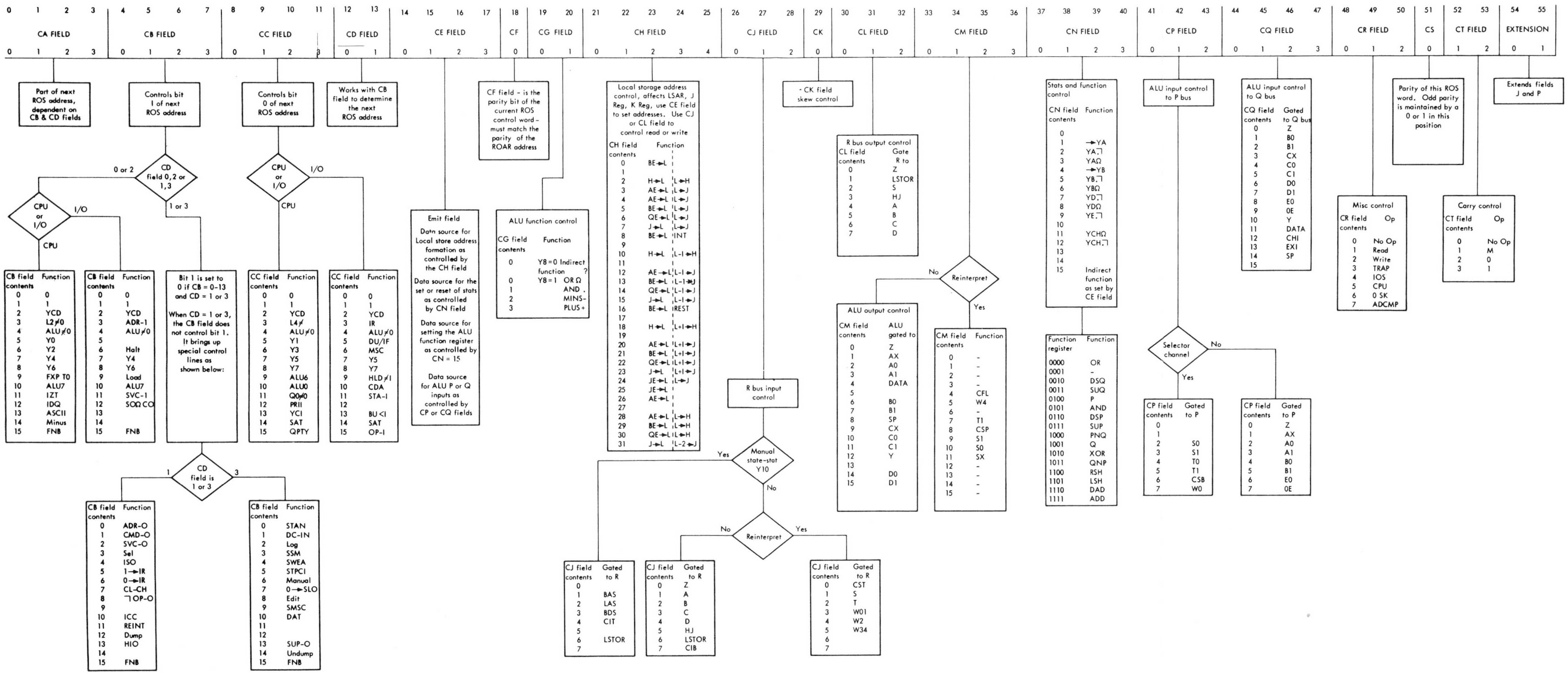

Fig. 7.13 ROS control word.

Examples

Assume that the CD field is 0 or 2; assume also that the machine is in the CPU state. Then if CB = 0, a zero is forced into ROS address bit 1. If CB = 4 and the specific test is (ALU $\stackrel{?}{=}$ 0), then a 1 is forced into ROS address bit 1 if the ALU output is nonzero.

If CB = 15 (symbolic FNB), then a functional branch is taken.

If none of the allowable specifications are present or if CD = 1, ROS address bit 1 is set to zero.

CC field is also a 4-bit field; it is used to specify the C condition test which sets ROAR and ROSCAR bit 0 accordingly. Depending on the mode of operation, some values of CE will be interpreted differently. For example, if CC = 5, in the CPU mode, and the Y1 micro-order is executed, a 1 is forced if Y1 is on. However, if the system is in the I/O mode, CC = 5 will have altogether a different mnemonic and will perform a different function. It will be represented by DU/IF which implies a 1 is forced if the device is unobtainable or if the interface is not free.

CD field is 2 bits wide and occupies bits 12 and 13 of the ROS word. It controls the decoding of the CB field and the generation of the ROS address field. For example, if CB = FNB, the CD field is set into ROS address bits 11 and 10.

CE field is the 4-bit emit field which can be gated into a number of facilities in the system. Three types of emit statements are possible to control the CE field:

1. A single statement to set 4 binary bits in the CE field.
2. A statement which combines the CE and CN fields to set and reset the stats.
3. A statement which controls the setting of the ALU function register when CN = 15.

CF field is the parity bit of the current ROS control word. It must match the parity of the ROAR address. If the two do not match, a machine check is indicated.

CG field occupies the next 2 bits in ROS and provides the ALU function control.

CH field controls local store addressing. Its contents affect the LSAR, and the J and K registers. It uses CE to set addresses, and CJ and CL to control the read and write operations.

CJ field is a 3-bit field which provides the R-bus input control.

CK field is used to indicate the skew option on the Q-bus input.

CL field occupies bits 30–32 of the ROS word and is used to designate the R-bus output control. It will affect the LSTOR, S, H, J, A, B, C, and D registers.

CM field is used primarily for ALU output control. It occupies bits 33–36 in ROS. More detailed information about this field will be available from the catalogs and examples that will be presented later.

CN field specifies a number of stats and function control micro-orders. It is used in conjunction with the CE field to set and reset stats. Examples are:

1100 → YA	→ means transfer
YA Ω 1010	Ω means Logical OR
YA ·⏋ 70010	·⏋ means Logical AND

CP and *CQ* fields provide the necessary input control for the P bus and Q bus, respectively. Since the P and Q buses are one byte wide, the mnemonics stated in the symbolic micro-order specification catalog do relate to the designations we have used for the one byte of hardware register in the CPU.

CR field specifies main storage functions such as read and write, along with other miscellaneous controls.

CS field, in bit 51, contains the odd parity of the whole ROS word.

CT field is a 2-bit field designated for carry control, to set and reset the ALU carry latches.

Figure 7.14 shows the locations of the ROS field on the CAS block template.

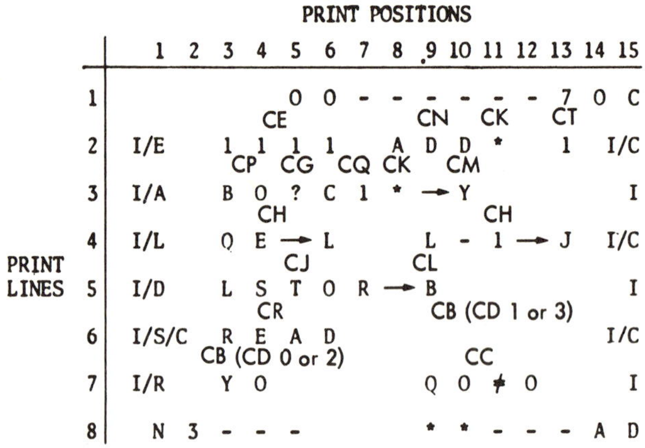

Figure 7.14 Microinstruction box.

7.10 MODEL 40 MICRO-ORDERS, CATALOG, AND DECODING CHARTS

This section gives a complete catalog for the Model 40 micro-order specifications and field decoding. The reader should look through the catalog to become familiar with the contents and language. In the final section of this chapter, which follows the catalog, we will begin to use these micro-orders in microprogramming the different data paths in the Model 40. Thus, through examples and planned exercises, the process of learning how to microprogram this machine will be completed.

The catalog consists of a number of separate decoding charts, each representing the decoding of a specific ROS field under specific machine conditions. Each decoding chart is identified by the name of the field and the ROS bits the field occupies. The top line also gives the machine state(s) in which this chart is valid. The table is divided into five columns which show the edge character (in the microinstruction box), the micro-order mnemonics, the binary ROS bits that will be stored in the ROS field to represent this micro-order, the decimal order and the function(s) performed by this field, and each micro-order therein. The decimal order always contains the two alphabetic characters that specify this field, followed by one or more decimal digits which represent the binary-to-decimal encoding of the ROS binary bits assigned to the micro-order.

CA Field | ROS Bits 0-3 | Part of Next ROS Word Address Under CB and CD Field Control (ROS Address is Shown Under CD Field) | Sense Latches EB181 A-Field Entry RX051

Edge Char.	Micro-Order	Bits	Dec. Order	Function	ALD
Not applicable		XXXX	-	CA ROAR PSN 5-2 when CB = 0-13 AND CD = Not 2	RX091
				CA ROAR PSN 9-6 when CB = 0-13 AND CD = 2, or CB = 15	RX011
				CA not used when CB = 14	-

CB Field | ROS Bits 4-7 | Forms PSN1 of Next ROS Address by Testing Specified Conditions With CD = 0 or 2 in CPU or I-O State. (ROS Address is Shown Under CD Field). Issue Special Controls when CD = 1 or 3

CPU State, CD Field = 0 or 2

Edge Char.	Micro-Order	Bits	Dec. Order	Function - Branch Condition	ALD
R	0	0000	CB0	No Output (ROAR PSN1 is Reset to 0)	DR751
R	1	0001	CB1	Set PSN1 of ROAR to 1	DR751
R	YCD	0010	CB2	Direct Carry	RX111
R	L2 ≠ 0	0011	CB3	LSAR PSN6 or 7 non zero (Test at T2 Del. same cycle)	DR751
R	ALU ≠ 0	0100	CB4	ALU output non zero (Test at T2 Del. same cycle)	RX111
R	Y0	0101	CB5	Storage Protect Primer Stat = 1	DR751
R	Y2	0110	CB6	Condition Register PSN0 = 1	DR751
R	Y4	0111	CB7	General-Purpose Stat Y4 = 1 (Not REINT or ROSCAR in control)	DR751
R	YCH3			(REINT or ROSCAR in control) channel general stat YCH3	GE571 HE571
R	Y6	1000	CB8	General-Purpose Stat Y6 = 1 (Not REINT or ROSCAR in control)	DR751
R	SI (7)			(REINT or ROSCAR in control) Bit 7 of S-Register	GE571 HE571
R	FXPTO	1001	CB9	Fixed Point Overflow (Test at T2 Del. same cycle)	RX111
R	ALU7	1010	CB10	ALU output PSN7 = 1 (Test at T2 Del. same cycle)	DR751
R	IZT	1011	CB11	Integrating Zero Test	DR751
R	CBY			(REINT or ROSCAR in control) chaining boundary in buffer (output ops)	GE571 HE571
R	IDQ	1100	CB12	Invalid Decimal digit on ALU entry Q	DR751
R	ASCII	1101	CB13	ASCII Stat	DR751
R	Minus	1110	CB14	Q Bus PSN4-7 = 11 or 13	DR751
R	FNB	1111	CB15	Special Control for function branch (see CD Field)	RX011

CB Field

I-O State, CD Field = 0 or 2

Edge Char.	Micro-Order	Bits	Dec. Order	Function - Branch Condition	ALD	
R	0	0000	CB0	No Output (ROAR PSN1 is Reset to 0)	DR571	
R	1	0001	CB1	Set PSN1 of ROAR to 1	DR571	
R	YCD	0010	CB2	Direct Carry	RX111	
R	ADR-I	0011	CB3	Address In Latch Gated by Y2, Y3	FL001	
R	ALU ≠ 0	0100	CB4	ALU output non zero (Test at T2 Del. same cycle)	RX111	
		0101	CB5	Not Used (CPU only)		
R	Halt	0110	CB6	Manual Stop Latch	FL001	
R	Y4	0111	CB7	(Not REINT or ROSCAR in control) Gen. Stat Y4 = 1	DR571	
R	YC113			(REINT or ROSCAR in control) Channel Gen. Stat Y3 = 1	GE571	HE571
R	Y6	1000	CB8	(Not REINT or ROSCAR in control) Gen. Stat Y6 = 1	DR751	
R	SI (7)			(REINT or ROSCAR in control) LS bit of S register	GE571	HE571
R	Load	1001	CB9	Load button (console)	FL001	
R	ALU7	1010	CB10	ALU output bit 7 (Test at T2 Del. same cycle)	DR751	
R	SVC-1	1011	CB11	(Not REINT or ROSCAR in control) Service In AND NOT Command Out or Service Out)	FL001	
R	CBY			(REINT or ROSCAR in control) chaining boundary in buffer (output ops)	GE571	HE571
R	SO∩CO	1100	CB12	Service Out or Command Out	FL001	
		1101	CB13	Not Used (CPU only)		
		1110	CB14	Not Used (CPU only)		
R	FNB	1111	CB15	Special Control for Function Branch	RX011	

CB Field

CD = 1

Edge Char.	Micro-Order	Bits	Dec. Order	Function - Branch Condition	ALD MPX	ALD SC1	ALD SC2
R	ADR-0	0000	CB0	Set Address Out	FB001	GG511	HG511
R	CMD-0	0001	CB1	Set Command Out	FB001	GG523	HG523
R	SVC-0	0010	CB2	Set Service Out	FB011	GG524	HG524
R	Sel	0011	CB3	Set Select Logic to stop interface sequence	FBQ31	GG511	
R	ISO	0100	CB4	Inhibit Select Out	FB031	GG512	HG512
R	1→IR	0101	CB5	Set Interrupt Request	FB021	GF503	HF503
R	0→IR	0110	CB6	Reset Interrupt Request	FB021	GF503	HF503
R	CL-CH	0111	CB7	Clear channel	FK051	GE571	HE571
R	¬OP-O	1000	CB8	Reset Operational Out	FB021	GG513	HG513
R		1001	CB9	Unused	-	-	-
R	ICC	1010	CB10	Set Interface Control Check; Force Error Stat Y12 and force Log-Out	KC081	GG543	HG543
R	REINT	1011	CB11	Reinterpret conditions as shown in CB, CC, CL and CM fields. Remains effective until Restore (Rest) is called		GE571	HE571
R	Dump	1100	CB12	Turn on Dump Control when stat Y10 = 1	FJ021	-	-
R	HIO	1101	CB13	Halt device on interface	FB021	GG531	HG531
		1110	CB14	Unused	-	-	-
R	FNB	1111	CB15	Special control for Function Branch	RX011	RX011	RX011

The B Condition is set to zero whenever CB = 0–14 and CD = 1 or 3 (EXCEPT UNDUMP)

CB Field

CD = 3, CPU and MPX and SC

Edge Char.	Micro-Order	Bits	Dec. Order	Function - Special Controls	ALD CPU/MPX	SC1	SC2
R	STAN	0000	CB0	SC Channel Analyse Status, Force 4-way branch on result		GE591	HE591
R	DC-IN	0001	CB1	Timing Line In for Direct Data Accept	JA114		
R	Log	0010	CB2	Start Log-Out	KC081		
R	SSM	0011	CB3	Set External Mask with ALU bit 7 with bit 1 for Channel 1 and bit 2 for Channel 2	KM131	GE591	HE591
R	SWEA	0100	CB4	Set ASCII, Wait and Enable with ALU bits 4, 5 and 6	KH171		
R	STPC1	0101	CB5	Stop T Clock	KC081		
R	Manual	0110	CB6	Define Normal Stop Loop	KH161		
R	0 SL0	0111	CB7	Reset Select Out gated by Y2, Y3		GG511	HG511
R	Edit	1000	CB8	If Q bus = 001000XY set next ROS address bit 1 and 0 to XY	RX112		
				If Q bus = 001000XY set next ROS address bit 1 and 0 to 11	RX112		
R	SMSC	1001	CB9	Set 1401 Emulator Selector Channel Latch (Y2, Y3 gated)	RX003		
R	DAT	1010	CB10	Enable/Disable Dec. Address Translator (1401 Emulator)	RX003		
R		1011	CB11	Unused	-		
R		1100	CB12	Unused	-		
R	SUP-O	1101	CB13	Set Suppress Out gated by Y2, Y3		GG512	HG512
R	Undump	1110	CB14	Restore ROAR for next cycle	RX011		
R	FNB	1111	CB15	Function Branch	RX011		

The B condition is set to zero whenever CB = 0 -14 and CD = 1 or 3 (EXCEPT UNDUMP)

CC Field

ROS Bits 8-11

CPU State

Sense Latches EB201
Decoder DR111
Decoder Check DS011

Edge Char.	Micro-Order	Bits	Dec. Order	Function - Condition ANDed with CPU/I-O States to form ROS next Address bit 0	ALD
R	0	0000	CC0	No output (ROAR PSN0 is reset to zero	
R	1	0001	CC1	Set PSN0 of ROAR to 1	DR752
R	YCD	0010	CC2	Direct Carry	DR752
R	L4 ≠	0011	CC3	LSAR PSN4, 5, 6 or 7 non zero (Test at T2 Del. same cycle)	DR752
R	ALU ≠ 0	0100	CC4	ALU output non zero (Test at T2 Del. same cycle)	RX113
R	Y1	0101	CC5	MPX Store Address Stat Y1 = 1	DR752
R	Y3	0110	CC6	Condition register PSN 1 = 1	DR752
R	Y5	0111	CC7	General-Purpose stat Y5 = 1	DR752
R	YCH1			(Reinterpret or ROSCAR in control) General-Purpose Channel Stat YCH1	GE571 HE571
R	Y7	1000	CC8	General-Purpose Stat Y7 = 1	DR752
R	CH = WR			(Reinterpret or ROSCAR in control) Channel Command Write Channel Read/Write Stat = 1	GE571 HE571
R	ALU6	1001	CC9	ALU output PSN6 = 1 (Test at T2 Del. same cycle)	RX111
R	ALU0	1010	CC10	ALU output PSN0 = 1 (Test at T2 Del. same cycle)	RX111
R	CDA			(Reinterpret or ROSCAR in control) Data Chaining Indicator	GE571 HE571
R	Q0 ≠	1011	CC11	Q bus PSN0-3 non zero (Test at T2 Del. same cycle)	DR752
R	PRII	1100	CC12	Program Interrupt	DR752
R	YCI	1101	CC13	Indirect Carry	RX111
R	SAT	1110	CC14	Invalid Storage (Memory) Address (IMA) or Protected Storage (Memory) violation (PMA, YM)	DR752
R	QPTY	1111	CC15	Test Q bus for bad parity (Test at T2 Del. same cycle)	DR752

CC Field — I-O State

Edge Char.	Micro-Order	Bits	Dec. Order	Function - Condition ANDed with CPU/I-O States to form ROS Next Address Bit 0	ALD MPX	ALD SC1	ALD SC2
R	0	0000	CC0	No output (ROAR PSN0 is reset to zero			
R	1	0001	CC1	Set PSN0 of ROAR to 1	DR752	DR752	DR752
R	YCD	0010	CC2	Direct Carry	DR752	DR752	DR752
R	IR	0011	CC3	Test Interrupt Request Latch (gated by Y2, Y3)	FL011	GE571	GE571
R	ALU ≠ 0	0100	CC4	ALU output non zero (Test at T2 Del. same cycle)	RX113	RX113	RX113
R	DU/IF	0101	CC5	Unit Unobtainable or I-F Free (Gated by Y2, Y3)	FL011	GE571	HE571
R	MSC	0110	CC6	Is 1401 Emulator Latch = 1			
R	Y5	0111	CC7	General-Purpose Stat Y5 = 1 (Not REINT or ROSCAR in control)	DR572		
R	YCH1			Selector Channel General-Purpose Stat Y1 (REINT called or ROSCAR in control)		GE571	HE571
R	Y7	1000	CC8	General-Purpose Stat Y7 = 1 (Not REINT or ROSCAR in control)	DR752		
R	CH = WR			Selector Channel Read-Write Stat; Write = 1 (REINT called or ROSCAR in control)		GE571	HE571
R	HLD ≠ 1	1001	CC9	Set Direct Control Accept Register on absence of Hold line	FL011		
R	CDA	1010	CC10	Selector Channel Data Chaining Indicator (CDA Flag and cnt. = 0 and not Chaining Boundary)		GE571	HE571
R	STA-I	1011	CC11	Status In gated by Y2, Y3	FL011		
		1100	CC12	Not Used (CPU State only)			
R	BU I	1101	CC13	Selector Channel Buffer empty (gated by Y2, Y3)		GE531	HG531
R	SAT	1110	CC14	Invalid Storage Address (IMA) or Protected Storage Violation (PMA, YM)	DR752		
R	OP-I	1111	CC15	(MX) Operational In and no (Status In or Address In) gated by Y2, Y3	FL011		

CD Field — ROS Bits 12, 13

Sense Latches	EB211
Ctrl Latch & Dec.	DR131
Decoder Check	DS011

ROS Address Control, CB Field Interpretation

Conditions	Source of ROS Address Bits 11	10	9	8	7	6	5	4	3	2	1	0
CD = XX CB = 15	D0	D1	A0	A1	A2	A3	0	Q0	Q1	Q2	Q3	
CD = 0 and CB = 0-14	X0	X1	X2	X3	X4	X5	A0	A1	A2	A3	B Cond	C Cond
CD = 2 and CB = 0-14	X0	X1	A0	A1	A2	A3	X6	X7	X8	X9	B Cond	C Cond
CD = 1 and CB = 0-14	X0	X1	X2	X3	X4	X5	A0	A1	A2	A3	0	C Cond
CD = 3 and CB = 0-13	X0	X1	X2	X3	X4	X5	A0	A1	A2	A3	0	C Cond
CD = 3 and CB = 14	UNDUMP (RESTORE INTERRUPTED CPU ROS ADDRESS INTO ROAR)											

X Bits are unchanged from last ROS address
D Bits are from ROS CD Field
A Bits are from ROS CA Field
Q Bits are from Q bus

CE Field — ROS Bits 14-17 — Emit Field — Sense Latches EB211, Control Latch DR171

Edge Char.	Micro-Order	Bits	Dec. Order	Function - Provide 4 bit Binary Patterns for Control	ALD
E	0000	0000	CE0		
E	0001	0001	CE1	1. Data source for Local Storage address formation controlled by CH Field	RLXXX
E	0010	0010	CE2		
E	0011	0011	CE3		
E	0100	0100	CE4	2. Data source for set or reset of stats controlled by CN Field, SC only:- bit 2 sets but cannot reset Program Check	RYXXX
E	0101	0101	CE5		
E	0110	0110	CE6		
E	0111	0111	CE7		
E	1000	1000	CE8	3. Data source for setting the ALU Function Register controlled by CN = 15	RPXXX
E	1001	1001	CE9		
E	1010	1010	CE10		
E	1011	1011	CE11	4. Data source for ALU P or Q inputs controlled by CP or CQ field	RPXXX RQXXX
E	1100	1100	CE12		
E	1101	1101	CE13		
E	1110	1110	CE14		
E	1111	1111	CE15		

CF Field — ROS Bit 18 — Sense Latch EB221, F Field Entry RX001

Edge Char.	Micro-Order	Bits	Dec. Order	Function - ROS Word Address Parity Bit (Current)	ALD
		0	-	Odd Parity compared with ROAR parity (ROS address CK) ≠	RX001
		1	-	Even Parity compared with ROAR parity (ROS address CK) ≠	RX001

≠ Checks if correct word has been read out

CG Field — ROS Bits 19-20 — Sense Latches EB221, G-Field Entry KP021

Edge Char.	Micro-Order	Bits	Dec. Order	Function - ALU Function Control	ALD
A	?	00	CG0	Indirect Function if Y8 = 0 and not REINT or ROSCAR in control Function Register F set to emit value see fields CN, CE	KP021 KP023
A	Ω			Direct 'OR' function if Y8 = 1 or REINT called or ROSCAR in control. F register = 0000	
A	.	01	CG1	Direct logic function 'AND' P and Q F register = 0101	KP023/32
A	-	10	CG2	Direct arithmetic function 'MINS', P-Q F register = 0011	KP023/32
A	+	11	CG3	Direct arithmetic function 'PLUS', P + Q F register = 1111	KP023/32

CH Field — ROS Bits 21-25 — CPU, MX, SC

Sense Latches EB231
Decoder DS021
Decoder Check DS081

Left Edge Char.	Micro-Order Left	Micro-Order Right	Right Edge Char.	Bits	Dec. Order	Function - Local Storage Address Control	ALD
L	BE→L			00000	CH0	Load LSAR from source BE and use	DS0XX
L				00001	CH1	Not used	-
L	H→L	L→H		00010	CH2	Use H-Register	DS0XX
L				00011	CH3	Not used	-
L	AE→L	L→J		00100	CH4	Load LSAR from source AE, use, load J-register	DS0XX
L	BE→L	L→J		00101	CH5	Load LSAR from source BE, use, load J-register	DS0XX
L	QE→L	L→J		00110	CH6	Load LSAR from source QE, use, load J-register	DS0XX
L	J→L	L→J		00111	CH7	Use J-register	DS0XX
						Decrement = 1	
L	BE→L	INT	C	01000	CH8	Selector Channel only. Allow other channel to break in	HB506
L				01001	CH9	Not used	-
L	H→L	L-1→H		01010	CH10	Use H-register and decrement by 1	DS0XX, CC001
L				01011	CH11	Not used	-
L	AE→L	L-1→J		01100	CH12	Load LSAR as for AE, use, decrement by 1, load J	DS0XX, CC001
L	BE→L	L-1→J		01101	CH13	Load LSAR as for BE, use, decrement by 1, load J	DS0XX, CC001
L	QE→L	L-1→J		01110	CH14	Load LSAR as for QE, use, decrement by 1, load J	DS0XX, CC001
L	J→L	L-1→J		01111	CH15	Use J-register and decrement by 1	DS0XX, CC001
						Increment = 1	
L	BE→L	REST	C	10000	CH16	Selector Channel only. If ROSCAR in control, restore control to ROAR. If ROAR in control reset the REINTERPRET control.	GG512
L				10001	CH17	Not used	-
L	H→L	L+1→H		10010	CH18	Use H-register and increment by 1	DS0XX, CC001
L				10011	CH19	Not used	-
L	AE→L	L+1→J		10100	CH20	Load LSAR as for AE, use, increment by 1, load J	DS0XX, CC001
L	BE→L	L+1→J		10101	CH21	Load LSAR as for BE, use, increment by 1, load J	DS0XX, CC001
L	QE→L	L+1→J		10110	CH22	Load LSAR as for QE, use, increment by 1, load J	DS0XX, CC001
L	J→L	L+1→J		10111	CH23	Use J-register, and increment by 1	DS0XX, CC001
						Increment = 0	
L	JE→L	L→J		11000	CH24	Load LSAR from source JE, use, load J-register	DS0XX
L	JE→L			11001	CH25	Load LSAR from source JE, and use	DS0XX
L	AE→L			11010	CH26	Load LSAR from source AE, and use	DS0XX
L				11011	CH27	Not used	-
L	AE→L	L→H		11100	CH28	Load LSAR from source AE, use, Load H-register	DS0XX
L	BE→L	L→H		11101	CH29	Load LSAR from source BE, use, Load H-register	DS0XX
L	QE→L	L→H		11110	CH30	Load LSAR from source QE, use, Load H-register	DS0XX
						Increment = 2	
L	J→L	L-2→J		11111	CH31	Use J-register, and decrement by 2	CC001, DS071

LSAR Load Definitions — E = Emit field, J = J-register, Q = Q bus

LSAR PSN SOURCE	0	1	2	3	4	5	6	7	COMMENT
QE	E0	E1	E2	Q0	Q1	Q2	Q3	E3	Q bus load for FNB
AE	0	1	0	0	E0	E1	E2	E3	
BE	0	0	0	0	E0	E1	E2	E3	CPU state
BE	0	0	1	0	E0	E1	E2	E3	I-O state MPX or SCI
BE	0	0	1	1	E0	E1	E2	E3	I-O state SC2
JE	J0	J1	J2	J3	J4	J5	E2	E3	J-register Modification

Notes

Address XX01XXXX is invalid
10XXXXXX and 11XXXXXX are equivalent
On 'Increment' a carry is not propagated from bit position 4 to bit position 3

CJ Field ROS Bits 26-28 Y10 = 0 JX = 0 (ROS Bit 54)

Sense Latches EB241
Decoder DR275
Decoder Check DR282

Edge Char.	Micro-Order	Bits	Dec. Order	Function - R Bus Input Control	ALD	
D	Z	0000	CJ0	Zeros with good parity	-	
D	A	0001	CJ1	A register. Storage address	DR281	
D	B	0010	CJ2	R register. Bytes B0, B1 with good parity to RX	DR281	
D	C	0011	CJ3	C register. Bytes CX, C0, C1	DR281	
D	D	0100	CJ4	Storage data register D, bytes D0, D1 with good parity to RX	DR281	
D	HJ	0101	CJ5	Local store address registers, H to R0, J to R1 good parity to RX	DR281	
D	LStor	0110	CJ6	Local store output (see CL field)	KM001	
D	CIB	0111	CJ7	(MPX) Channel Input Bus Y2, Y3 = 00	DR275	DR278
				(SC1) Status to R0 and checks to R1, Y2, Y3 = 01	GE512	
				(SC2) Status to R0 and checks to R1, Y2, Y3 = 10	HE512	
				Timer value to R0, R1 (clock reset to zero when sampled)		

CJ Field Y10 = 0 JX = 1 (ROS Bit 54) Reinterpret

Edge Char.	Micro-Order	Bits	Dec. Order	Function - R Bus Input Control	ALD
D	CST	0000	CJ0	Channel Flag register to R0, Chaining Boundary Flags to R1 (0, 4). Count Control to R1 (5, 7)	GE511
D	S	0001	CJ1	Channel S register (byte address)	GE511
D	T	0010	CJ2	Channel T register (byte count)	GE511
D	W01	0011	CJ3	Buffer bytes 0, 1 to R0, R1 (RD/BK latch = 0) Buffer bytes 0, 1 to R1, R0 (RD/BK latch = 1)	GE511
D	W2	0100	CJ4	Buffer byte W2 to R0, Channel Store Protect Key to R1 (4, 7) bits 0 to 3 MUST = 0	GE511
D	W34	0101	CJ5	Buffer bytes 3, 4 to R0, R1	GE511
D		0110	CJ6	Not used	-
D		0111	CJ7	Not used	-

CJ Field Y10 = 1 JX = 0 (Manual state)

Edge Char.	Micro-Order	Bits	Dec. Order	Function - R Bus Input Control	ALD
D		000	CJ0	No operation	-
D	BAS	001	CJ1	Manual binary address switches	DR282
D	LAS	010	CJ2	a. LOAD Unit Address switches 1. Unit Address 8 bits - R1 2. Channel Select bits R0 positions 5, 6, 7 b. Storage Select rotary switch This switch encoded to give a three-bit field inserted into R0 positions 1, 2, 3 defined as follows: 000 = STOR PROTECT 0 001 = IC - Instruction Counter 100 = MS - Main Storage 011 = FP - Floating Point Registers 010 = GP - General Purpose Registers 101 = PSW - Program Status Word	DR282
D	BDS	011	CJ3	Manual binary data switches	DR281
D	CIT	100	CJ4	Channel input interface tags selected by Y2, Y3	GE512
		101	CJ5	Not used	
D	LStor	110	CJ6	Local store output (see CL field)	KM001
		111	CJ7	Not used	-

CK Field | ROS Bit 29 | Sense Latch EB251 | F-Register KP001

Edge Char.	Micro-Order	Bits	Dec. Order	Function - Skew Control	ALD
A		0	CK0	No SKEW	AQ001
A	*	1	CK1	SKEW (See Note)	AQ001
				Note. Causes the ALU operation to incorporate a 4 bit shift on the Q bus. Q bus positions 0-3 are shifted into skew buffer. Q bus positions 4-7 are shifted into ALU input positions 0-3. Old contents of skew buffer are used for ALU input positions 4-7. When 'Load F' occurs, CK is loaded into F4.	KP001

NOTE: The asterisk calling for a Skew operation can appear either in an arithmetic statement (edge char. A) or in an emit statement (edge char. E) for a Load F micro-order (see CN field)

CL Field | ROS Bits 30-32 | Control Latch and Dec. DR32X

Edge Char.	Micro-Order	Bits	Dec. Order	Function - R Bus Output Control	ALD
D	Z	000	CL0	No destination, parity ignored	(KH025)
D	LStor	001	CL1	Local store input (Call Write limited to 100 microseconds continuous)	(KM001)
D	S	010	CL2	Selector Channel S Register	(HE521)
D	HJ	011	CL3	R0 to H, R1 to J (not REINT called or ROSCAR in control)	(RJ0XX)
D	W34			Selector Channel buffers 3 and 4 (if REINT called or ROSCAR	(HE521)
D				in control)	
D	A	100	CL4	To Register A	(RAXXX)
D	B	101	CL5	To Register B	(RBXXX)
D	C	110	CL6	To Register C	(RCXXX)
D	D	111	CL7	To Register D	(RDXXX)

If CM and CL both call for a transfer, the ALU bit transfer over-rides the byte that it concerns in the A, B, C & D registers only, but the other byte or bytes of the R bus transfer take place normally. The CL transfer to D over-rides the main storage input to D if both occur in 'READ' cycle 2. If IMA or PMA and YM occurs M-D takes priority.

The cases are:				
	ALU→D R→D M→D	ALU→D R→D (One byte over-written by ALU) M Lost	R→D Valid with R→D M→D M Lost	ALU→D M→D Not Permitted

CM Field | ROS Bits 33-36 | CPU, MX and SC | Sense Latch EB261, EB271 | Control Latch DR36X, KU081 | and Decoder

Edge Char.	Micro-Order	Bits	Dec. Order	Function - ALU Output Control	ALD		
A	Z	0000	CM0	Result has no destination	-		
A	AX	0001	CM1	OR of ALU0-ALU6 to AX6; ALU7 to AX7; EX is lost	RA002		
A	A0	0010	CM2	ALU output to A0; EX to AX	RA071		
A	A1	0011	CM3	ALU to A1; EX is lost	RA171		
A	Data	0100	CM4	ALU to Direct Data Out; EX is lost	JA005		
A	CFL			(REINT or ROSCAR in control) ALU to Channel Flags; EX is lost	GE541		HE541
		0100	CM5	Not used (unless REINT called or ROSCAR in control)	-		
A	W4			(REINT or ROSCAR in control) ALU to buffer W4; EX is lost	GE541		HE541
A	B0	0110	CM6	ALU to B0; EX is lost		RB071	
				(REINT or ROSCAR in control) ALU to T0; EX is lost	GE541		HE541
A	B1	0111	CM7	ALU to B1; EX is lost		RB171	
A	T1			(REINT or ROSCAR in control) ALU to T1; EX is lost	GE541		HE541
A	SP	1000	CM8	ALU to Storage Protect keys EX is lost		KU041	
A	CSP			(REINT or ROSCAR in control) ALU to Storage Protect Keys; EX is lost	GE541		HE541
A	CX	1001	CM9	OR of ALU0-ALU6 to CX6; ALU7 to CX7; EX is lost		RC001	
A	SI			(REINT or ROSCAR in control) ALU to SI; EX is lost	GE541		HE541
A	C0	1010	CM10	ALU to C0; EX to CX		RC001	
A	S0			(REINT or ROSCAR in control) ALU to S0; EX to SX	GE541		HE541
A	C1	1011	CM11	ALU to C1; EX is lost		RC171	
A	SX			(REINT or ROSCAR in control) or of ALU0-ALU6, to SX6, ALU7 to SX7, EX is lost	GE541		HE541
A	Y	1100	CM12	ALU to YA and YB stats; EX is lost		RY001	
		1101	CM13	Not used	-		
A	D0	1110	CM14	ALU to D0; EX is lost		RD051	
A	D1	1111	CM15	ALU to D1; EX is lost		RD151	

CN Field | ROS Bits 37-40 | CPU, MX and SC | Latches and Decoder DR401 | ALU Function Dec. KP0X1 | ALU Control Register KP02X

Edge Char.	Micro-Order	Bits	Dec. Order	Function - Control Set/Reset Stats and ALU Free from CE Field Values	ALD
E		0000	CN0	No operation	RY001
E	→YA	0001	CN1	Set stats Y0-3 with Emit field	DR401
E	YA.¬	0010	CN2	Stats Y0-3 ANDed with 'NOT' emit (Reset stats with Emit ones)	DR401
E	YA Ω	0011	CN3	Stats Y0-3 ORed with Emit field (Set stats with Emit ones)	DR401
E	→YB	0100	CN4	Set stats Y4-7 with Emit field	DR401
E	YB.¬	0101	CN5	Stats Y4-7 ANDed with NOT Emit (Reset stats with Emit ones)	DR401
E	YB Ω	0110	CN6	Stats Y4-7 ORed with Emit field (Set stats with Emit ones)	DR401
E	YD.¬	0111	CN7	Stats Y8-11 ANDed with NOT Emit (Reset stats with Emit ones)	DR401
E	YD Ω	1000	CN8	Stats Y8-11 ORed with Emit field (Set stats with Emit ones)	DR402
E	YE.¬	1001	CN9	Stats Y12-15 ANDed with NOT Emit (Reset stats with Emit ones)	DR402
E		1010	CN10	Forces ROS decoder check for diagnostic use (no FOSSL statement)	DR363
E	YCH Ω	1011	CN11	(SC only) YCH1, YCH3 ORed with Emit (Set stats with Emit ones) CE2, Prog. Check, CE4 unused	DR402
E	YCH.¬	1100	CN12	(SC only) YCH1, YCH3 ANDed with NOT Emit (Reset stats with Emit ones) CE2, 4 unused	DR402
E		1101	CN13	Not used	-
E		1110	CN14	Not used	-
E	0000, OR	1111	CN15	ALU indirect function P OR Q ≠ F register = 0000	DR402 KP01X
E		1111	CN15	ALU indirect function ALU decoder check ≠ F register = 0001	KP001
E	0010, DSQ	1111	CN15	ALU indirect function P MINUS Q (Dec) F register = 0010	KP001
E	0011, SUQ	1111	CN15	ALU indirect function P MINUS Q (Bin) F register = 0011	KP001
E	0100, P	1111	CN15	ALU indirect function PASS P ONLY ≠ F register = 0100	KP001
E	0101, AND	1111	CN15	ALU indirect function P AND Q ≠ F register = 0101	KP001
E	0110, DSP	1111	CN15	ALU indirect function Q MINUS P (Dec) F register = 0110	KP001
E	0111, SUP	1111	CN15	ALU indirect function Q MINUS P (Bin) F register = 0111	KP001
E	1000, PNQ	1111	CN15	ALU indirect function P AND (NOT Q) ≠ F register = 1000	KP001
E	1001, Q	1111	CN15	ALU indirect function PASS Q ONLY ≠ F register = 1001	KP001
E	1010, XOR	1111	CN15	ALU indirect function Exclusive OR of PO ≠ F register = 1010	KP001
E	1011, QNP	1111	CN15	ALU indirect function PASS (NOT P) and Q ≠ F register = 1011	KP001
E	1100, RSH	1111	CN15	ALU indirect function RIGHT SHIFT Q (Half Q) F register = 1100	KP001
E	1101, LSH	1111	CN15	ALU indirect function LEFT SHIFT Q (2 Q) F register = 1101	KP001
E	1110, DAD	1111	CN15	ALU indirect function P PLUS Q (Dec) F register = 1110	KP001
E	1111, ADD	1111	CN15	ALU indirect function P PLUS Q (Bin) F register = 1111	KP001
E	*	1111	CN15	ALU indirect function SKEW is applied to that function which has an asterisk to the right of it	AQ001

≠ These functions are logic functions and do not alter the YC1 or YCD carry latches
Direct ALU functions affect the direct YCD carry latch
Indirect ALU functions affect the indirect YCI carry latch
'Load F' and 'Use Indirect' may be given in the same cycle in which case the new function can be executed

CP Field ROS Bits 41–43 PX = 0 (ROS Bit 55) CPU and MX

Sense Latch EB281
Decoder DR431

Edge Char.	Micro-Order	Bits	Dec. Order	Function – ALU Input to P Bus	ALD
A	Z	000	CP0	Zeros, good parity to P and EX ≠	RP001
A	AX	001	CP1	AX to P, zeros to EX	RP001
A	A0	010	CP2	A0 to P, AX to EX	RP001
A	A1	011	CP3	A1 to P, zeros to EX unless CQ = 4 ≠	RP001
A	B0	100	CP4	B0 to P, zeros to EX unless CQ = 4 ≠	RP001
A	B1	101	CP5	B1 to P, zeros to EX unless CQ = 4 ≠	RP001
A	E0	110	CP6	Emit field and 4 zeros (Emit 0000) to P, zeros to EX unless CQ = 4 ≠	RP001
A	0E	111	CP7	4 zeros and Emit field (Emit 0000) to P, zeros to EX unless CQ = 4 ≠	RP001

≠ ALU extension is reset to zero unless CQ = 0100 (C0)

CP Field PX = 1 (ROS Bit 55) SC only

Edge Char.	Micro-Order	Bits	Dec. Order	Function – ALU Input to P Bus	ALD SC1	ALD SC2
A		000	CP0	Not used	-	-
A		001	CP1	Not used	-	-
A	S0	010	CP2	S0 to P, SX to EX	GE561	HE561
A	S1	011	CP3	S1 to P, zeros to EX	GE561	HE561
A	T0	100	CP4	T0 to P, zeros to EX	GE561	HE561
A	T1	101	CP5	T1 to P, zeros to EX	GE561	HE561
A	CSB	110	CP6	Channel Status byte to P, zeros to EX	GE561	HE561
A	W0	111	CP7	Channel Buffer W0 byte to P, zeros to EX	GE561	HE561

CQ Field ROS Bits 44–47

Sense Latches EB291
Decoder DR471
Decoder Check DS015

Edge Char.	Micro-Order	Bits	Dec. Order	Function – ALU Input Control to Q Bus	ALD
A	Z	0000	CQ0	Zeros, good parity to Q and EX unless CP = 2	RQ0XX
A	B0	0001	CQ1	B0 to Q, zeros to EX, unless CP = 2	RQ0XX
A	B1	0010	CQ2	B1 to Q, zeros to EX, unless CP = 2	RQ0XX
A	CX	0011	CQ3	CX to Q, zeros in Q0, 5, CX6, 7 ORed to Q6; zeros to EX	RQ0XX
A	C0	0100	CQ4	C0 to Q, CX to EX	RQ0XX
A	C1	0101	CQ5	C1 to Q, zeros to EX unless CP = 2	RQ0XX
A	D0	0100	CQ6	D0 to Q, zeros to EX unless CP = 2	RQ0XX
A	D1	0111	CQ7	D1 to Q, zeros to EX unless CP = 2	RQ0XX
A	E0	1000	CQ8	Emit field and 4 zeros (Emit 0000) to Q zeros to EX unless CP = 2	RQ0XX
A	0E	1001	CQ9	4 zeros and emit field (Emit 0000) to Q zeros to EX unless CP = 2	RQ0XX
A	Y	1010	CQ10	YA and YB stats to Q, zeros to EX unless CP = 2	RQ0XX
A	Data	1011	CQ11	Direct Data to Q, zeros to EX unless CP = 2	RQ0XX
A	CHI	1100	CQ12	Channel Interrupts to Q, zeros to EX unless CP = 2	RQ0XX
A	EXI	1101	CQ13	External Interrupts to Q, zeros to EX	RQ0XX
A	SP	1110	CQ14	CPU Storage Protect key to Q0, 3 MX Channel key to Q4, 7, zeros to EX unless CP = 2 (REINT or ROSCAR in control) Storage Protect bus to Q4, 7 and channel key unless CP = 2 To Storage Protect bus unless CP = 2	RQ0XX
A		1111	CQ15	Not used	

CR Field — ROS Bits 48-50

Sense Latches EB301
Decoder DR501
Decoder Check DR501

Edge Char.	Micro-Order	Bits	Dec. Order	Function - ALU Miscellaneous Controls	ALD
S		000	CR0	No operation	DR501
S	READ	001	CR1	Call Main Storage READ	KM101
S	WRITE	010	CR2	Call Main Storage WRITE	KM101
C	TRAP	011	CR3	TRAP on IMA or (PMA, YM) ≠	KM001
C	IOS	100	CR4	Set I-O State (effective in cycle defined)	DR501
C	CPU	101	CR5	Reset I-O State	DR501
C	0→SK	110	CR6	Reset skew buffer (effective after current use of Skew) Y10 = 0	AQ001
C	0→ERR			Modify by Y10 = 1 to give reset errors	KH141
C	ADCMP	111	CR7	Set Address Compare	KH142

≠ TRAP during READ phase : ROS Address 10100
TRAP during WRITE phase : ROS Address 10101

CS Field — ROS Bit 51 — Parity Bit of Current ROS Word to Give an Odd Number of Bits for ROS Word

Sense Latch EB301
S-Field Entry DR601

Edge Char.	Micro-Order	Bits	Dec. Order	Function - ROS Word Parity Bit	ALD
Not applicable		0	-	Odd parity	DR601
Not applicable		1	-	Even parity	DR601

CT Field — ROS Bits 52, 53

Sense Latches EB311
T-Field Entry AM311

Edge Char.	Micro-Order	Bits	Dec. Order	Function - Carry Control	ALD
		00	CT0	No operation	-
C	M	01	CT1	(1401 Emulator only) extend ROS address	RX003
C	0	10	CT2	Reset Carry (Direct or Indirect according to ALU function specified)	AM311
C	1	11	CT3	Set Carry (Direct or Indirect according to ALU function specified)	AM311

Extension Field — ROS Bits 54, 55

Edge Char.	Micro-Order	Bits	Dec. Order	Function - Extend Fields J and P	ALD Sense Latch	ALD Decoder
D		54	-	JX Bit Extend J Field (Ref. J Field)	EB311	DR275
A		55	-	PX Bit Extend P Field (Ref. P Field)	EB311	DR431

PRACTICE PROBLEMS

The ROS Word and Its Representation in CAS

1. Fill in the missing binary, decimal order, or micro-order mnemonic representations for the following field values:

Field	*Binary*	*Decimal*	*Micro-order*
CH	0101		
CH		CH25	
CH			QE → L; L → J
CL		CL1	
CM		CM0	
CM		CM9	
CP	010		
CB			FNB
CC	1101		
CE	0100		
CG		CG1	
CN	1101		
CN	1111		

2. Consider the microinstruction of the box at C4 on sample CLD page QD001. (See Fig. 7.26.) What is its
(a) serial number?
(b) leg identifier?
(c) leg selector?
(d) location in ROS?
(e) Give two source boxes prior to the instruction at C4.
(f) Give destination boxes for the instruction at C4.

3. What is the relation between the number of *'s in the leg selector of a box, and the number of branches from that box?

4. Analyze the microinstruction in box C3 of QD001 (Fig. 7.26). In particular, what does the catalog give as the meaning of:
(a) (2) 0:C
(b) (3) A:A1 + 0E → C1
(c) (6) S:READ

7.11 MICROPROGRAMMING THE MODEL 40

Learning Model 40 microprogramming can be simplified by learning how to manipulate the different paths and functional units one at a time. We will begin with moving the data from:

1. Register to register

2. Register to local store
3. Register to main store

The latter two classes of data movement involve the generation of the storage address and the specification of a command for read or write.

Next, we will learn how to manipulate the arithmetic and logic unit functions and the specification of ALU inputs and output destination. The staticizer function and setting will then be brought into focus before a detailed analysis of ROS addressing and conditional branching is presented. Once these steps are completed, a review of the main Model 40 microroutines is presented. Finally, a trace of the instruction fetch and execution cycles for a given RR instruction will be given.

7.11.1 Register-to-Register Transfer

The micro-orders controlling this transfer are located in line 5 of the microinstruction box with the edge character D. See Fig. 7.14. The general format is:

$$\text{D: } C(CJ) \rightarrow C(Cl)^{\dagger}$$

where C(CJ) is a specific mnemonic in the CJ field. (It is the contents of the CJ control field as decoded from the micro-order specification charts.) The CJ refers to the R-register input, and CL refers to the destination of the R-register output.

Examples

1. D: A → B. In the normal CPU mode, the contents of register A, including the A_x extension bits, are transferred to the R register early in the cycle. Late in the same cycle, R0 and R1 are transferred to the B_0, B_1 bytes.

2. D: B → D. The contents of $B(B_0, B_1)$ are transferred via the R register, as in the previous example, to the D register.

3. Z → D. Micro-order Z in the CPU mode implies no operation, that is, no input to R is specified in the cycle. Since the R register is reset early in the cycle prior to the R register set pulse, and since no data are gated prior to its outgating pulse, then zeros are transmitted to the destination register D. The net result, therefore, is to transmit zeros (with good parity) to the D register.

4. Let the contents of register B be (1274) Hex. What will the contents of register A be after executing the following microstatement?

$$\text{D: B} \rightarrow \text{A}$$

Answer: the extension bits will have zeros; A_0 and A_1 bytes will have (1274) Hex. Since R is reset, including its extension bits first, only R0 and R1 bytes will receive the corresponding information from register B. The whole R

†This notation of micro-orders was first proposed by Dr. Leo Hellerman of IBM.

register is then transmitted to the A register, including the zeros found in the R register extension bits (R_x).

Exercise

Let the contents of A = 35235 Hex
B = 7689
C = 29ABC

After executing the following statements in three consecutive microinstructions:

1. (5)* D: C → A
2. (5) D: B → C
3. (5) D: A → B

What is the content of A_0?
What is the content of A_1?
What is the content of B_1?
What is the content of C_0?

7.11.2 Data Movements between Register and Local Store

The movement from register to local storage and vice versa is accomplished, at least partially, like register-to-register movement, except for the addressing of specific words in local storage. To learn how to address local store, we need to examine the local store memory map shown in Fig. 7.7, the emit field setting, and how to load the Q bus of the ALU data path.

Local storage is addressed by an 8-bit local storage address register, LSAR. This yields 256 unique addresses of local storage; however, since only 144 words are implemented physically, some 112 addresses are invalid. Addresses of the format XX01XXXX are invalid, while addresses of the formats 10XXXXXX are considered equivalent to those with 11XXXXXX formats. The 8-bit LSAR yields addresses between 0 and FF, hexadecimal range.

In the local storage map (Fig. 7.7), a number of registers or local storage words are unassigned and can be used by the microprogrammer to store intermediate results. The last 32 words located between E0 and FF Hex addresses are assigned to the 16 fullword, general purpose registers. Each register occupies two consecutive storage locations. The floating-point registers are stored between C0 and CF, four words to a register.

Location 43 is designated to store the first two bytes of the current System/360 instruction being executed. Locations 44 through 47 are used to store PSW information. Location 47 is used very frequently since it holds the

*The (5) indicates that this type of micro-order with a D edge character appears on line 5 of the microinstruction box.

address of the current System/360 instruction being executed. The remaining words in local storage are the unassigned words, the work area, and the assigned words for internal convenience.

Source Code	0	1	2	3	4	5	6	7
H	H0	H1	H2	H3	H4	H5	H6	H7
J	J0	J1	J2	J3	J4	J5	J6	J7
LA (=AE)	0	1	0	0	E0	E1	E2	E3
LB (=BE) (When in CPU State)	0	0	0	0	E0	E1	E2	E3
LB (=BE) (When in I/O State Mpx or SC1)	0	0	1	0	E0	E1	E2	E3
LB (=BE) (When in I/O State, SC2)	0	0	1	1	E0	E1	E2	E3
LQ (=QE)	E0	E1	E2	Q0	Q1	Q2	Q3	E3
LJ (=JE)	J0	J1	J2	J3	J4	J5	E2	E3

Figure 7.15 LASR bits.

Six special local-store address formats are used in the Model 40 (Fig. 7.15). The first two types of LSAR loading are shown in the CH field in the micro-order catalog. In these types, the LSAR is loaded from the J or H register. We will discuss these two types first.

The setting of the LSAR is controlled by the source control signals developed by the CH field. The general format is:

(4) L: C(CH)

This statement is entered on line 4 of the microinstruction box. The four high-order bits of the LSAR (0–3) are controlled independently from its low-order four bits LSAR (4–7). When incrementing through the incrementor circuits, only bits 4–7 are modified. A possible carry-out of bit 4 is not propagated to bit 3.

The incremented address is gated into the H and J registers to hold the local storage address. In addition, the H and J registers can be loaded from or set into the R bus.

It was stated earlier that the local storage communicates with the CPU functions through the R register; hence, it can be treated as a register-to-register transfer controlled by the CJ and CL fields according to the following general formats:

(5) D: C(CJ) $\rightarrow$ C(CL)

Therefore, the total local store transfer would appear on two lines of the microinstruction box as follows:

line (4) L: C(CH)
line (5) D: C(CJ) → C(CL)

Examples of this are:

(4) L: H → L, L → H
(5) D: LSTOR → D

The first micro-order, with L edge character, results in loading bits (0–3) of the H register and bits (4–7) also from the H register into the LSAR (0–7), and then transferring the contents of the LSAR without modifying it to the H register. The second line of code implies transferring the content of the addressed word in local storage to register D via the R register. (*Note:* only R_0 and R_1 bytes are transmitted since D is only 16 bits wide.) This local storage address format is shown at the top of Fig. 7.15.

An example of the second format, the J format, is:

(4) [L: J → L, L + 1 → J]
(5) [D: D → LSTOR]

The third format, namely the LAE format, is also straightforward. The LAE represents the source of the two halves of the LSAR byte. The high-order half L(0–3) will be set to 0100 when the LAE format is encountered by the microprogrammer assembler. The low-order half L(4–7) will be set from the emit field specified in the immediate ROS word. The emit field is entered on line 2 of the microinstruction box following the format E: (CE).

Example

(2) [E: 0110]
(4) [L: AE → L, L → H]
(5) [D: LSTOR → A]

The LSAR register is set with bits

$$\underbrace{0100}_{\text{A}}\ \underbrace{0110}_{\text{emit bits}}$$

or location 46 Hex. The contents of LSAR are transferred to the H register, and the contents of the local storage address 46 are transferred to the A register, all in the same machine cycle.

Example

(2) [E: 0101]
(4) [L: AE → L, L − 1 → H] implies 45 is accessed,
(5) [D: LSTOR → A] 44 goes to the H register

The LBE format is interpreted according to the state of the machine. If the machine is in the CPU mode, the high-order half of the LSAR will be set to 0000, and the low-order bits will be set from the emit field. If the machine is in the I/O state, multiplex, or selector channel 1, then the only change would be in setting the high-order bits of LSAR to 0010 to access the area in local storage assigned to these channels. If I/O selector channel 2 mode is being executed, then the high-order bits will be assigned 0011 to access the unit control word information between locations 30 and 35 in local storage.

The LQE format is perhaps the most general. It receives four bits from the Q bus. Hence, we must first review the setting of the Q bus of the ALU data path. The Q bus receives information from the B, C, and D registers. However, if one does not specify the source of information to the Q or P bus, the default option of setting the P or Q bus to zero will be executed.

The Q bus input is specified as part of the arithmetic statement with edge character A which appears on line 3 of the instructor format. The general format of the arithmetic statement is:

$$\text{(3) A: } C(CP) \cdot C(CG), \quad C(CQ) \rightarrow C(CM)$$

where C(CP), C(CQ) specify the ALU P- and Q-bus input sources, C(CG) specifies the ALU operation, and C(CM) specifies the ALU output destination.

Example

Microinstruction		Comment
E: 1110	←	Set emit field to 1110 (to load the high-order four bits of the Q bus).
A: $Z \cdot E0 \rightarrow Z$	←	AND contents of P (which is zero) with the Q bus (which has been set to 11100000). The answer is zero; however, Q bus remains 11100000.
L: $QE \rightarrow L$; $L + 1 \rightarrow J$		
D: $LSTOR \rightarrow D$		

The third line of code, then, would place 11111100 or FC (Hex) into the LSAR, and FD (Hex) into the J register.

$$\begin{Bmatrix} E_0 & E_1 & E_2 & Q_0 & Q_1 & Q_2 & Q_3 & E_3 \\ 1 & 1 & 1 & 1 & 1 & 1 & 0 & 0 \end{Bmatrix}$$

Therefore, the result of executing the whole microinstruction is to set the contents of the high-order two bytes of the general purpose register 14 into the D register, and set the address of the two low-order bytes of the same register into register J.

If the arithmetic statement was of the form A: $Z \; \Omega \; E0 \rightarrow Z$, then the result of the *local* store fetch would be the same since the LSAR is set from the Q bus and is not affected by the ALU result of the specified operation.

A final format which we will examine here is the LJE format. It is similar to the previously discussed J format, except that the low-order bits L (6, 7) will be set from the two low-order bits of the emit field rather than the J field.

Example

$$\begin{matrix}(2)\\(4)\\(5)\end{matrix}\begin{bmatrix}\text{E: } 0101 & \\ \text{L: } \text{JE} \rightarrow \text{L}; & \text{L} \rightarrow \text{J} \\ \text{D: } \text{LSTOR} \rightarrow \text{A} & \end{bmatrix}$$

Assume that the contents of the J register just prior to executing the microinstruction = 30 Hex (00110000).

The JE format would then yield

$$\overbrace{001100}^{\text{J}}\ \overbrace{01}^{\text{E}} = 31$$

and the contents of the J register now will be set to 31. The contents of the local store address 31 is transferred to the A register.

PRACTICE PROBLEMS

1. What happens in each of the following microinstruction statements?
(a) (5) D: A → C; (b) (5) D: Z → B; (c) (5) D: D → A.

2. Let the initial contents of J = 32 Hex. If the Model 40 executes the following microinstruction:

$$\begin{matrix}(2)\\(4)\\(5)\end{matrix}\begin{bmatrix}\text{E: } 0011 & \\ \text{L: } \text{JE} \rightarrow \text{L}; & \text{L} \rightarrow \text{J} \\ \text{D: } \text{D} \rightarrow \text{LSTOR} & \end{bmatrix}$$

what will the content of J be? What else is accomplished here?

3. Assume that register A is available for intermediate storage. Write a sequence of microinstructions to exchange the contents of registers B and D.

4. Assume that general purpose register 4 in local storage is available for temporary storage. Write the sequence of microinstructions to exchange the contents of hardware registers A and C.

5. Describe the modification taking place after each microinstruction in the following sequence. (Assume that LS(43) = ABCDE Hex, J = 11, and H = A1.)

$$\text{---}\begin{bmatrix}\text{E: } 0011 & \\ \text{L: } \text{AE} \rightarrow \text{L}; & \text{L} \rightarrow \text{H} \\ \text{D: } \text{LSTOR} \rightarrow \text{A} & \end{bmatrix}\text{---}\begin{bmatrix}\text{L: } \text{H} \rightarrow \text{L}; & \text{L} + 1 \rightarrow \text{H} \\ \text{D: } \text{A} \rightarrow \text{LSTOR} & \end{bmatrix}\text{---}$$

$$\begin{bmatrix}\text{L: } \text{H} \rightarrow \text{L}; & \text{L} \rightarrow \text{H} \\ \text{D: } \text{LSTOR} \rightarrow \text{Z} & \end{bmatrix}\text{---}$$

6. Write the sequence of microinstructions necessary to store the contents of registers A,B,C, and D into general purpose registers 10, 11, 12, and 13, respectively.

7. Analyze the following microinstructions:

(a)
$$\begin{bmatrix} \text{E: } 1000 \\ \text{L: } AE \rightarrow L;\ L \rightarrow H \\ \text{D: } LSTOR \rightarrow A \end{bmatrix}$$

(b)
$$\begin{bmatrix} \text{E: } 1001 \\ \text{L: } BE \rightarrow L;\ L \rightarrow J \\ \text{D: } LSTOR \rightarrow Z \end{bmatrix}$$

(Assume that the system is in the CPU state.)

(c) Assume d = 41 Hex.
$$\begin{bmatrix} \text{E: } 0111 \\ \text{L: } JL \rightarrow L;\ L \rightarrow J \\ \text{D: } D \rightarrow LSTOR \end{bmatrix}$$

(d)
$$\begin{bmatrix} \text{E: } 1011 \\ \text{A: } Z \cdot E0 \rightarrow Z \\ \text{L: } QE \rightarrow L;\ L \rightarrow J \\ \text{D: } LSTOR \rightarrow A \end{bmatrix}$$

7.11.3 Main Storage Data Transfer

The main storage unit of the IBM Model 40 is a 2.5-microsecond, random-access, magnetic core unit. It operates on a split cycle basis. This means that there may be a gap of undefined period between the read and write cycles. It must be recognized, however, that the read cycle is still a destructive one and that, for each storage cycle, a read cycle must always precede a write cycle. This allows for reading any number of different storage words, one after another, and for writing them back later in any sequence.

Earlier in this chapter, we introduced the five different memory capacities available on the Model 40, and the capacity of the associated bump storage which is accessible only by the microprogram and is used to hold the unit control for the multiplex channel.

Main storage cycle time is the time required to read or write half a word (two bytes) from any selected location to the D register, over the main storage, 16-bit (plus 2 parity-bit) bus. The read cycle takes 1.4 microseconds, while the write cycle requires 1.1 microseconds, for a total of 2.5 microseconds main storage cycle time. The read-write sequence is specified by the microprogrammer.

The general purpose hardware register A is assigned the special function of main storage address registers when the system is operating in the CPU mode. If the system is placed in the selector channel mode, then the S register assumes this main storage address register function. Both registers (A and S) are equipped with extension bits to permit addressing the 128k and 256k memories. These extension bits are duplicated into the C and R registers to facilitate address arithmetic and temporary storage of main storage addresses.

In summary, then, the Model 40 main storage is addressed by placing the main storage address into the A register, whose contents are decoded as a binary number into the MS address circuits. The main storage read or write options which can be operated on a split cycle are specified by the microprogrammer. The D register is assigned the main storage data register, to hold or to receive the two bytes of information transferred between the CPU and main storage.

Since the D register may receive information from three different sources, and since it is possible, intentionally or otherwise, to specify all three data transfers to the D register in the same machine cycle, a priority is set up as follows. The highest priority is given to the ALU byte, through the R register bus, followed by the main storage, to D register transfer. The microprogrammer, besides having to recognize this dual function of the general purpose register A, would also have to make a specific storage read or write request. This is due to the split cycle operation of the MS.

The mechanics of microprogramming main storage requests is simple and is very similar to the register-to-local storage data movement. The first step is to place the address of the desired halfword of main storage in the A register. This transfer was discussed earlier in this chapter in the section dealing with register-to-register data movement. The remaining step is to indicate, on line 6 and edge character S of the appropriate microinstruction box, the specific choice of *read* or *write* cycle.

The only remaining problem which requires special attention is the timing and sequencing. The *read* micro-order causes the contents of the main storage halfword (whose address is in the A register) to be read out to the D register. By the end of the CPU cycle following the one in which the read command is specified, the contents of the selected halfword of main storage would have already been set into the D register. Hence, at this point, we can split the read-write process to be completed later, or we can go on by specifying the write micro-order in the third cycle from the read cycle. Note that, once a read or write micro-order is specified, the contents of the A register (or S register in the selector channel machine) must remain unchanged. If during the third cycle a write micro-order is specified without modifying the contents of the D register, this will result in reading two bytes and then regenerating the source back into MS. However, if prior to activating the write cycle, following a read cycle, the contents of register D are modified, then the main storage location selected by the contents of register A will be modified to reflect the new contents of register D. This is illustrated in the sequence of microinstructions shown in Fig. 7.16.

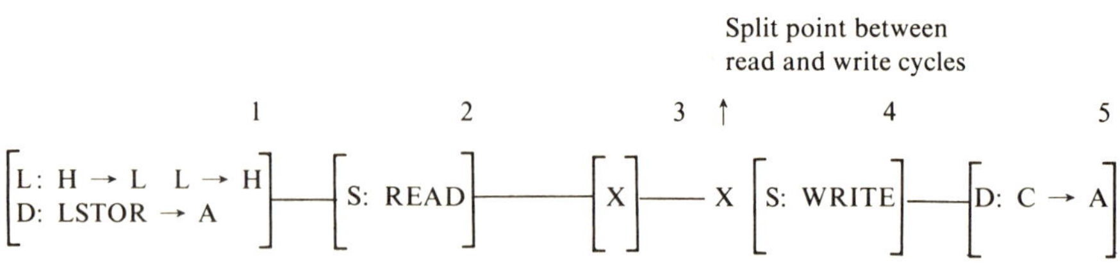

Figure 7.16 Microinstruction sequence for MS read-write split cycle.

The number at the top right corner of each microinstruction box is used to identify the box only in this example. Box 1, the main storage address register, was transferred to register A from the local storage. Box 2 designates the beginning of a main storage read cycle. Box 3 is the second half of the main storage read cycle. In this cycle, all other functional units and data transfers not affecting main storage, A and D registers, are permissible. At the end of main storage, cycle 2 designated by box 3, register D would have received the contents of the selected word in main storage which has been destroyed in the read process. It is permissible at this point (after the end of read cycle 2) to break in with any timing gap between the read and write commands. However, if we choose to continue with the write half of the main storage cycle, we can indicate it in microinstruction box 4.

The two read and write micro-orders are specified in the ALU miscellaneous controls field CR. This field occupies bits 48–50 of the ROS word. They are the only two micro-orders with the edge characters S in the catalog.

Example 1

Write a sequence of microinstructions to transfer two bytes of data from local-storage word 42 into the main storage two bytes whose address is contained in the upper half of general purpose register 8.

The high-order half of general purpose register 8 is stored in local-storage word F0. Three local store addressing options are opened to us. To get this address into the A register, we can use either the LBE or the LQE format. The simplest and most direct is to use the E0 micro-order specified in the (CQ-8 micro-order) catalog. This special micro-order would assemble the emit field followed by 4 zeros on the Q bus (EEEE0000 → Q-bus). Hence, the first microinstruction in the sequence:

$$\begin{matrix}(2)\\(3)\\(4)\\(5)\end{matrix}\begin{bmatrix}\text{E: } 1111\\ \text{A: } Z \cdot E0 \rightarrow Z\\ \text{L: } QE \rightarrow L;\quad L \rightarrow J\\ \text{D: } LSTOR \rightarrow A\end{bmatrix}$$

would place the contents of F0 in register A, as a first step prior to any main storage cycle request.

Another way to transfer local storage F0 contents to register A is to use the more general QE format. (If the address is not F0, the microprogrammer would have to assemble the proper sequence of bits on the Q bus so that the QE format would place in the local-storage address register the address of the proper word in local storage.) A third way is the JE format. This is especially useful if the J register bits J(0–5) happen to be 111100; then the microinstruction:

$$\begin{matrix}(2)\\(4)\\(5)\end{matrix}\begin{bmatrix}\text{E: } 0000\\ \text{L: } E \rightarrow L;\quad L \rightarrow J\\ \text{D: } LSTOR \rightarrow A\end{bmatrix}$$

would place the address F0 in local store address register L and transfer its contents to hardware register A. The choice between these three formats depends on the problem. However, the QE format is the most general.

The contents of LS(42) can be transferred to register D at the appropriate cycle by the simple AE local store address formats as shown in the following microinstruction:

```
(2) E: 0010
(4) L: AE → L
(5) D: LSTOR → D
```

To complete this data transfer, it is still necessary to start with a read micro-order. Hence, the following sequence of five microinstructions will prove to be the minimum time required to finish the intended transfer.

```
1
(2) E: 1111
(3) A: Z·E0 → Z
(4) L: QE → L
(5) D: LSTOR → A

2
(2) E: 0000
(4) L; AE → L;  L → J
(6) S: READ

3
(4) L: J → L;  L → J
(5) D: LSTOR → D

4
(4) L: J → L;  L → J
(5) D → LSTOR
(6) S: WRITE

5
(4) L: H → L;  L → J
(5) D: A → LSTOR
```

Example 2

Assuming that local store word LS(47) contains the current instruction counter IC, trace the following sequence of microinstructions and explain what each microinstruction box does:

```
(2) E: 0111
(4) L: AE → L;  L → H
(5) D: LSTOR → A

(2) E: 0010
(3) A: A1 + 0E → C1
(6) S: READ

(2) E: 0011
(3) A: A0 + Z → C0
(4) L: AE → L;  L → J
(5) D: LSTOR → B
    R: CPU
                     PRI
          0000*

          00000
(2) E: YB ·⌉ 0111   0: C
(3) A: E0 · D0* → B0
(4) L: J → L;  L + 1 → J
(5) D: D → LSTOR
(6) S: WRITE
(7) R: FNB

    → 00001
```

Example 3

A block of N halfwords in main storage are to be transferred to local storage. Given that the address of the first halfword is in LS(45), what is the minimum number of microinstructions needed to complete the transfer of all N halfwords?

The answer to this problem is $4*N + 1$ since it takes a minimum of four machine cycles to transfer one halfword and regenerate (rewrite) the same byte in its original place in MS; hence, to carry this process for N halfwords would require $4*N$ machine cycles. Add to this one initial cycle to get the address of the first halfword in main storage into the A register.

This concludes our discussion of movement of data between registers, registers and local storage, and registers and main storage. The primary purpose for the data transfer is to bring it to the proper microspec function at the proper time to operate on it. In the Model 40, this means to have it in the proper hardware register, to indicate how it is to be transferred on the ALU P- and Q-bus, and to specify what operation the ALU will perform on this data. Our next discussion will therefore deal with the ALU operations, both direct and indirect.

PRACTICE PROBLEMS

1. Let the J register contain the address of the LS location that contains the MS address into which the contents of GPR 6 (general purpose register) is to be stored. Write the necessary sequence of microinstructions to accomplish this data movement.
2. Write the sequence of microinstructions to store the contents of the LS(43) instruction buffer into the MS halfword whose address is given in register C.
3. Write the sequence of microinstructions to exchange the contents of the A and C registers.
4. Write the sequence of microinstructions to exchange the contents of the C register with LS(6F).
5. Write a sequence of microroutines that will transfer the contents of all general-purpose and floating-point registers from local storage into the block of main storage starting with location 00000 Hex.

7.11.4 ALU Operations

An earlier discussion of the ALU in this chapter dealt primarily with the hardware aspect and the types of operations which can be performed in this section of the computer. It consists, as we have seen, of a number of hardware functions designed to perform binary, decimal, arithmetic, logic operations (AND, OR, EXCLUSIVE OR, invert), shifting, rotation, etc.

The basic inputs to the ALU unit are:

1. The P and Q registers, each of one-byte width.
2. The emit field as the source of emit data.

3. The CN field along with the emit field to load the ALU function register.
4. The ALU, YCD, and YCI carry stats.
5. A number of control signals which will activate the proper data path in the process of executing a direct or indirect operation based on the control register bits and the carry stat conditions.

The ALU is completely controlled by the microprogram. Inputs, operations, and gating of output are all determined by specific fields in the ROS word. The four-bit control register (Fig. 7.17) can be set up by the micro-

	ALU Control Bits	Logical Operation		Connect Box Function (F)	Right Shift Function (G)	YC Stat Conditions	YCI or YCD Stat Conditions
		Abb	Meaning				
*P	0 0000	OR	P + Q	P or Q	Zero Output	No Effect on Bit 7	Not Affected
#	1 0001	AND	P . Q	P and Q	Zero Output	No Effect on Bit 7	Not Affected
	2 0010	DSQ	P - Q (Decimal)	Equivalent P Q	Inverse of Q	YC Off Gives Carry to Bit 7	Set if No Carry Bit 0
*P	3 0011	SUQ	P - Q (Binary)	Equivalent P Q	Inverse of Q	YC Off Gives Carry to Bit 7	Set if No Carry Bit 0
	4 0100	P	Pass P	Pass P	Zero Output	No Effect on Bit 7	Not Affected
*P	5 0101	AND	P . Q	P and Q	Zero Output	No Effect on Bit 7	Not Affected
P	6 0110	DSP	Q - P (Decimal)	Equivalent P Q	Q	YC Off Gives Carry to Bit 7	Set if No Carry Bit 0
	7 0111	SUP	Q - P (Binary)	Equivalent P Q	Q	YC Off Gives Carry to Bit 7	Set if No Carry Bit 0
	8 1000	PNQ	P . $\bar{Q}$	P . $\bar{Q}$	Zero Output	No Effect on Bit 7	Not Affected
P	9 1001	Q	Pass P	Pass Q	Zero Output	No Effect on Bit 7	Not Affected
P	10 1010	XOR	Exclusive OR (P Q)	P Q Exclu- sive OR	Zero Output	No Effect on Bit 7	Not Affected
	11 1011	QNP	$\bar{P}$. Q	$\bar{P}$. Q	Zero Output	No Effect on Bit 7	Not Affected
P	12 1100	RSH*	1 Bit Rt Shift of Q	Zero Output	Q Shifted Two Positions Rt	YC On Gives Rt Shift Output Bit 1	Set if There is Carry Bit 0
	13 1101	LSH	1 Bit Left Shift of Q	Zero Output	Q	YC On Gives Carry to Bit 7	Set if There is Carry Bit 0
	14 1110	DAD	P + Q (Decimal)	P Q Exclu- sive OR	Q	YC On Gives Carry to Bit 7	Set if There is Carry Bit 0
*P	15 1111	ADD	P + Q (Binary)	P Q Exclu- sive OR	Q	YC On Gives Carry to Bit 7	Set if There is Carry Bit 0

* Direct Function
\# Gives Function Check

Figure 7.17 ALU control signals and functions.

programmer to specify one of sixteen possible operations, eight logical (AND, OR, etc.), six arithmetic, and two shifts (left and right). These bits are further grouped as twelve indirect and four direct operations, making it possible to use the ALU for two different operations in two different cycles without clearing the function and the carry conditions associated with either function. Figure 7.17 also shows the ALU control bits, the operation, and the effect of these operations on the different hardware units in the ALU section.

Three fields in the TROS control word direct the inputs to the P and Q buses and direct the ALU output back to the data flow. The CP field occupies ROS bits (41–43), and the CQ field in ROS bits (44–47) gates the sources to the P and Q buses. The CM field in ROS bits (33–36) controls the gating of the ALU output to the specified destination.

Three other fields, the CG-ROS bits (19–20), CE-ROS (14–17), and the CN in ROS bits (37–40) are used for control functions. The CG field can specify the ALU arithmetic or logic operations directly, or it may specify an indirect function if CG = 0, and if inhibit dump stat Y8 is on (Y8 = 1). The CE and CN fields are used to set the control register from the function register set previously by an indirect operation. If CN = 1111, then both the control register and function register are set from the emit field in order to specify which indirect function will be performed.

The CK-field ROS bit (29) specifies the skew control. If CK = 1, the eight-bit Q bus will be skewed. Otherwise, the contents of the P and Q buses are fed to the ALU circuitry unaltered, and the direct or indirect carry latches are used for direct or indirect arithmetic operations to combine with the P and Q data according to the control register.

The CR field is ROS bits (48–50) containing all miscellaneous ALU controls. When CR = 110, the skew buffer is reset.

The only other field that affects the ALU operation in some way is the CT field in ROS bits (52–53). This field is used to set or reset the direct or indirect carry stat according to the ALU function specified.

To illustrate the actual microprogramming process for the ALU, we will start with the basic general formula:

$$\text{(3) A: C(CP)C(CG)C(CQ)} \rightarrow \text{C(CM)}$$

This statement is entered on line 3 of the microinstruction box with edge character A. Examples are:

(1) A: Al + Bl → Bl
(2) A: Al − D0 → C0
(3) A: Al Ω D0 → B1
(4) A: B1 ? C0 → C1

The first example is a simple one. The Al byte of the A register is added to the B1 byte of the B register. The result is stored in the B1 byte. The carry bit is stored in state YCI or YCD, depending on the direct or indirect nature of the operation.

In the second example, the A1 byte is placed on the P bus; the 1's complement of D is set with an appropriate carry signal added to perform the 2's complement addition. If a carry bit results, it is carried into the extended carry box, and the result is transmitted to the CX, C0 bits.

In the indirect operation, (with CN = 15), an ALU indirect function Q − P permits subtracting the contents of the P bus from the contents of the Q bus in both the binary and decimal arithmetic mode.

The third example is equivalent to bit-by-bit ORing of the contents of the P and Q buses.

The last example with the ? symbol is used to specify an indirect operation. It says that the operation to be performed is specified in the function register by the general format (2) E: CN15. Another example of this last case is:

$$\left[\begin{array}{l} \text{E: 0111, SUB} \\ \text{A: A1 ? D1} \rightarrow \text{C0} \end{array}\right]$$

ALU Exercise 1

Assume the contents of registers:

B = 1234 Hex
C = 05678 Hex

Compute the contents of register D after executing the following microinstructions:

$$\left[\begin{array}{c} \text{0:C} \\ \text{A: B1 + C1} \rightarrow \text{D1} \end{array}\right] \text{———} \left[\text{A: B0 + C0} \rightarrow \text{D0}\right]$$

The micro-order (0:C) simply resets the carry stat. The first microinstruction adds 34 (B1) to 78 (C1) and stores the answer AC (Hex) in D1. No carry results from this operation.

The second microinstruction adds the contents of the B0 and C0 bytes by gating B0 to the P bus, C0 to the Q bus, and register C extension bits (00) to the extension carry stats. The addition of the two bytes, B0 and C0, results in 68 Hex, which is then transferred to byte D0. Hence, the final answer stored in the D register is 68 AC Hex.

Had the contents of register B been 12FF Hex, then the results at the first microinstruction would have been 77 Hex and a carry. The carry would have been transmitted to the YCD stat, and then to the YC stat. This carry (YC) would have then combined in the second microinstruction with B0 and C0 to produce 69. And the final answer in register D would have been 6977 Hex.

ALU Exercise 2

Assume that the contents of the A1 register is 12 Hex, and B1 = 34 Hex. Execute the microinstruction:

$$\left[\begin{array}{c} \text{0:C} \\ \text{A: A1} - \text{B1} \rightarrow \text{C1} \end{array}\right]$$

and indicate what the contents of C1 and the YCD stats will be after the execution is completed.

Referring back to the first exercise, we note that, since the 0:C micro-order resets the carry state, the subtract signal, when EXCLUSIVE ORed with the carry signal, will yield a 1 (hot one) which is added to the 1's complement of

B1 and the true value of A1 as follows:

```
0001  0010
1100  1011
0000  0001
----  ----
1101  1110   or   DE
```

Note that we do not have a carry-out from this operation, but we do have a carry signal stored in the YCD stat.

ALU Exercise 3

Assume that the contents of B0 = 36 Hex, and C1 = 71 Hex. Execute the microinstruction:

$$\begin{bmatrix} \text{E: } 0111, \text{SUB} \\ \text{A: B0 ? C1}* \rightarrow \text{Y} \end{bmatrix}$$

What will the ALU output be? What ALU output destination is specified?

7.11.5 ALU Right- and Left-Shift Operations

The desired shifting is obtained in the ALU by setting the indirect function register to left shift, skew, skew and left shift, pass Q, pass Q and skew, etc. An eight-bit shift is obtained by moving the whole byte as a register-to-register transfer. Figure 7.4 shows the hardware involved in the shift operations. The contents of the Q bus are passed, either skewed or unmodified, to the right-shift input box. They are then shifted two bit positions to the right. The output of the connect box is zero. The output of the right-shift box is gated into the carry box, which effectively gives a one place left shift. Hence, the Q data shifted two places right have effectively been shifted only one place to the right. The YCI, YCD, and YC stats allow bits to shift from byte to byte in a multiple-byte shift.

In a left-shift operation, the contents of the Q bus are gated directly into the carry box, with the connect box output being zero. Since the carry box effectively performs a one place left shift, the operation is completed in the carry box.

The microprogram control is designated by the emit field and the CN field. For CN = 15 and with the emit field (CE) = 12, the system is set to perform a right-shift Q operation. With CN = 15 and CE = 13, the system is set to perform a left-shift Q operation.

The format for a left shift is:

(2) E: 1101, LSH

and for a right shift:

(2) E: 1100, RSH

Consider, as an example:

$$\begin{bmatrix} \text{E: 1100, RSH;} & \text{1:C} \\ \text{A: Z} \cdot \text{B0} \rightarrow \text{D0} & \end{bmatrix}$$

The RSH micro-order sets the 1100 into the control register. The Z · B0 → D0 is a direct logic function of ANDing zeros on the Q bus with the contents of B register byte B0. The result, which is zero in this case, will be transmitted to D0.

Assume that the following two microinstructions are encountered:

$$\begin{bmatrix} \text{E: 1100, RSH;} & \text{1:C} \\ \text{A: Z ? B0} \rightarrow \text{D0} & \end{bmatrix} \vdash \begin{bmatrix} \\ \text{A: Z ? B1} \rightarrow \text{D1} \end{bmatrix}$$

If the content of B is 9ABC Hex, then the first instruction is:

$$\underbrace{1}_{\text{YCI}} \; \underbrace{1001}_{9} \; \underbrace{1010}_{\text{A}} \quad \text{Shifted right 1}$$

$$\Rightarrow 1100/1101/0 \rightarrow \text{YCI}$$

The second instruction is:

$$\underbrace{0}_{\text{YCI}} \; \underbrace{1011}_{\text{B}} \; \underbrace{1100}_{\text{C}} \quad \text{Shifted right 1}$$

$$\Rightarrow \underbrace{0101}_{5} \; \underbrace{1110}_{\text{E}} \; 0 \rightarrow \text{YCI}$$

The new content of register D is (CD5E Hex).

PRACTICE PROBLEMS

1. Let the registers have the following contents:

Register	*Contents (Hex)*
A	12345
B	6789
C	2468A
D	BCDE

What will the contents of these registers be after each step of the sequence of micro-orders

(a) D: A → B
(b) D: C → A
(c) D: D → Z
(d) D: Z → B
(e) D: D → C

2. Assuming that the C register is available for temporary storage, write a sequence of micro-orders to exchange the contents of the B and D registers.

3. Suppose that only the B and D registers are available for temporary storage. Why can't we write a sequence of micro-orders for exchanging the contents of the A and C registers? Under what special circumstances can we write such a sequence?

4. Describe the modifications taking place after each step of the following sequence of instructions (assuming the following initial contents of storage and registers):

con LS (48) = 12345 (Hex)
con H = C9
con J = 03

(a) [E: 1000
L: AE → L; L → H
D: LSTOR → A]

(b) [L: H → L; L + 1 → H
D: A → LSTOR]

(c) [L: J → L; L → J
D: LSTOR → Z]

(d) [L: J → L; L − 2 → J
D: A → LSTOR]

5. Write a sequence of two microinstructions to store con A in LS (05).

6. Write two microinstructions to read the current System Mask to the B register and the Instruction Buffer to the C register.

7. Write two microinstructions to read and restore LS (6F).

8. Write four microinstructions to read and restore the high halfword of fixed-point register 15 to the C register and the low halfword to the D register.

9. Let the address of a halfword in MS be contained in LS (40). Write a sequence of microinstructions for storing the low halfword of fixed-point register 14 into this MS location. (This takes five microinstructions to the end of the write cycle.)

10. A certain data process is to operate on 10 halfwords of data in MS. What is a lower bound on the number of instructions in the microprogram for this process?

11. What are the hexadecimal addresses of each of the following in LS?
(a) High halfword of fixed point register 10
(b) Low halfword of fixed point register 2
(c) Selector Channel 1 dump area for D register
(d) PSW system mask

12. What will be the final J register value if the initial value and increment are as indicated?

Initial Value	*Increment*	*Final Value in Hexadecimal*
F2	1	
F2	1	
E0	0	
E0	−1	
11	−1	
11	−2	

GENERAL PRACTICE PROBLEMS

ALU Function Specifications

1. What indirect operations are implied in the following micro-orders?
(a) E: 0011, SUQ (b) E: 0100, P (c) E: 0110, DSP
(d) E: 0101, AND (e) E: 1110, DAD

2. What is the difference between the two following microinstructions?

$$\begin{bmatrix}\text{E: 1001, Q*} \\ \text{A: B0} + \text{Z} \rightarrow \text{B1}\end{bmatrix} \text{ and } \begin{bmatrix}\text{E: 1001, Q*} \\ \text{A: B0} + \text{Z*} \rightarrow \text{B1}\end{bmatrix}$$

3. Let C1 = 89 (Hex). After execution of the microinstruction

$$\begin{bmatrix}\text{E: 1101, LSH; 0:C} \\ \text{A: Z ? C1} \rightarrow \text{C1}\end{bmatrix}$$

what will the new contents of YCI and C1 be?

4. Let con A0 = 357 Hex and con C0 = 4E0 Hex. After execution of

$$\begin{bmatrix}\text{1: C} \\ \text{A: A0} + \text{C0} \rightarrow \text{A0}\end{bmatrix}$$

(a) con A0 =
(b) con C0 =
(c) con YCD =

5. Let con A1 = 12 Hex and con B1 = 34 Hex. After execution of

$$\begin{bmatrix}\text{0: C} \\ \text{A: A1} - \text{B1} \rightarrow \text{C1}\end{bmatrix}$$

(a) con A1 =
(b) con B1 =
(c) con C1 =
(d) con YCD =

6. Let con B = 9ABC Hex. After execution of the sequence

$$\begin{bmatrix}\text{E: 1100; RSH; 1:C} \\ \text{A: Z ? B0} \rightarrow \text{D0}\end{bmatrix} \text{ and } \begin{bmatrix}\text{E: 1100; RSH; 1:C} \\ \text{A: Z ? B1} \rightarrow \text{D1}\end{bmatrix}$$

(a) con B =
(b) con D =
(c) con YCI =

7. Let con A0 = 123 Hex, con C1 = 45 Hex, and con SKEW = 6 Hex. After execution of

$$\begin{bmatrix}\text{E: 1001, Q*} \\ \text{A: A0 ? C1*} \rightarrow \text{C0}\end{bmatrix}$$

(a) con A0 =
(b) con C0 =
(c) con C1 =
(d) con SK =

8. Write the sequence of microinstructions to add the contents of general purpose registers 1 and 2 and to store the results in general purpose register 1.

9. Use the skew buffer to shift the contents of the D register four places to the right. Discard the bits shifted out of the low-order positions, and shift 1's into the high-order positions if D is negative; otherwise shift 0's into these vacated high-order positions.

10. Write a microprogram to add the contents of the low-order halfword of general register 3 to a halfword in MS. Assume the MS operand address is contained in LS(40). Store the result in the high halfword of general register 3. Do not destroy any operand. If the addition results in an overflow (carry out setting the YCD stat), set con LS(00) = 1; otherwise, set LS(00) to 0.

11. Let con B1 = AE Hex and con C0 = 12 Hex. After execution of

```
[E: 1001;  0:C      ]
[A: B1 − C0 → C1    ]
```

(a) con B1 =
(b) con C1 =
(c) con YCD =

12. Let con D1 = 56 Hex. After execution of

```
[E: 1101;  LSH;  0:C]
[A: Z ? D1 → D1     ]
```

(a) con D1 =
(b) con YCI =

What happens if these orders are executed a second time, based on the results of the first execution?

13. Let con C0 = 189 Hex. After execution of

```
[E: 0111;  SUP;  0:C]
[A: E0 ? C0 → C0    ]
```

(a) con C0 =
(b) con YCI =

14. Let con C0 = 246 Hex. After execution of

```
[E: 1101;  RSH;  1:C]
[A: 0E ? C0 → C0    ]
```

(a) con C0 =
(b) con YCI =

15. Let con A = 2468A Hex and con C = 3579B Hex. After execution of

```
[E: 0100;  P        ]
[A: A0 ? C0 → C0    ]
```

(a) con C =

16. Let con B1 = AE Hex, con C0 = 12 Hex, and con SK = 3 Hex. After execution of

```
[E: 1001;  0:C      ]
[A: B1 − C0* → C1   ]
```

(a) con B1 =
(b) con C1 =
(c) con YCD =
(d) con SK =

17. Let con B1 = 86 and con SK = 4. After execution of

$$\begin{bmatrix} \text{E: 0000; OR*} \\ \text{A: Z ? B1* } \rightarrow \text{ C0} \end{bmatrix}$$

(a) con B1 =
(b) con C0 =
(c) con SK =

18. Let con A = 13579 Hex and con C = 2468A Hex. After execution of

$$\begin{bmatrix} \text{E: 1000; 1:C} \\ \text{A: AX + CX } \rightarrow \text{ D1} \end{bmatrix}$$

(a) con D1 =
(b) con YCD =

19. Let con C0 = 135. After execution of

$$\begin{bmatrix} \text{E: 1001; Q} \\ \text{A: Z ? C0 } \rightarrow \text{ CX} \end{bmatrix}$$

(a) con C0 =
(b) con YCD =
(c) con YCI =

20. Write a microprogram to shift-right the contents of a halfword in MS. Assume the MS operand address is contained in LS(4F) and the result is to be returned to this MS address. If the operand halfword is positive, shift a 0 into the high-order position; if the halfword is negative, shift in a 1. (We assume the 2's complement representation of negative numbers.) (*Note:* One way to branch on the sign of the operand is to add a 1 to its high-order bit and branch on the resulting value of the YCD. It is also possible to use the νCC = ALU0 micro-order for branching on the sign bit.)

21. Use the skew buffer to shift the contents of B right four bit positions. Discard bits shifted out of the low-order positions and shift 0's into the high-order positions. (This takes three microinstructions.)

7.12 ROS ADDRESSING AND SEQUENCING

Depending upon the machine features, the capacity of the transformer read-only storage, TROS, used in the Model 40 varies up to 8192 ROS words. Only the 4096-word basic ROS model will be discussed here.

TROS on the basic Model 40 contains 16 modules of 256 addressable words in each module. The 256 words are stored on 128 plastic tapes, 2 words to a tape. Each word contains 60 ROS bits, of which only 56 are used. The physical and electrical characteristics of the TROS memory were discussed in Chapter 5. Our interest at this point is how to assign and specify the addresses for the microinstructions that will be stored in this device.

The 4096 TROS words are addressed by a 12-bit ROS address register (ROSAR) shown in Fig. 7.18 (see also Fig. 7.19). It is partitioned into the subfields that are gated into the ROAR from sources specified by the microprogrammer and the hardware status of the system. (1) The leftmost four bits

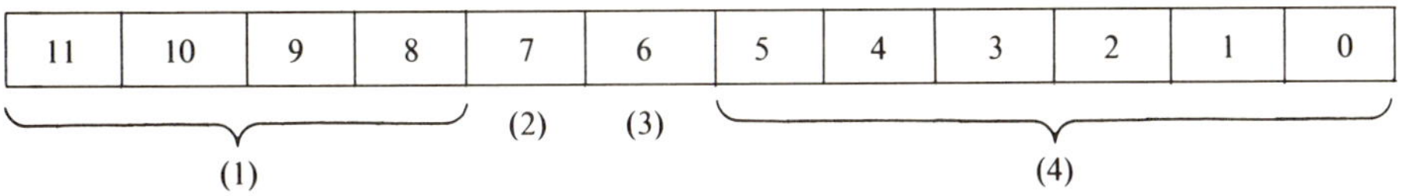

Figure 7.18 The ROSAR.

of the ROSAR (11–8) select one of the 16 modules, module (0–15). (2) Bit 7 of the ROSAR is decoded to select between the upper 128 or lower 128 words in the module. (3) Since two ROS words are stored in each tape, designated by word A and word B, ROSAR bit 6 is decoded to select between them. (4) The remaining 6 ROSAR bits are used to select between the upper or lower 64 TROS tapes (or words) specified by bit 7.

The reason for this partitioning of the TROS module into upper 128 and lower 128 words is to reduce the number of word drivers needed by sharing each driver between the upper and lower corresponding word. The upper and lower 128 words are both divided into 8 blocks with 16 words of storage each, as shown in Fig. 7.19.

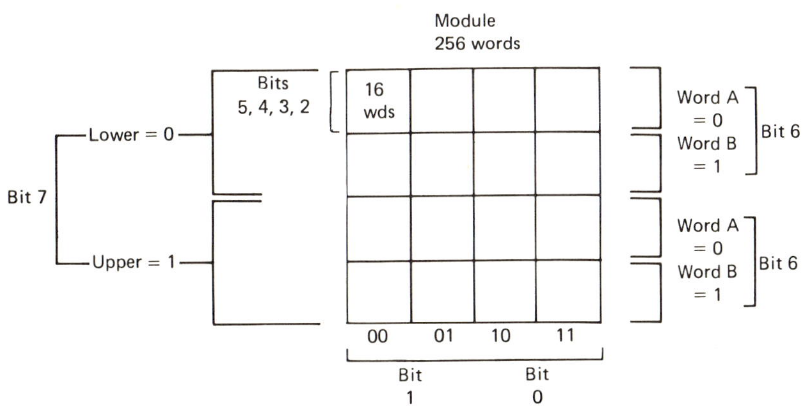

Figure 7.19 ROS addressing.

Figure 7.20 shows the ROS address control formats used in assigning and computing the ROS addresses.

Basically, the twelve-bit ROS address is generated in one of four different ways, depending on the setting of the CA, CB, CC, and CD fields. Certain ROS addresses are forced by hardware. The CB and CC fields determine ROAR (1–0). The CA field is four bits wide, and the CD field is only two bits wide. Yet these two fields (CA and CD), with the combined width of six bits, must determine the ROSAR (11–2). This limitation leads to the loss of flexibility and to the rigid addressing scheme of ROS sequencing. We recall that each ROS word generates the address of its successor ROS word. The four basic and allowable sequencing formats are as follows:

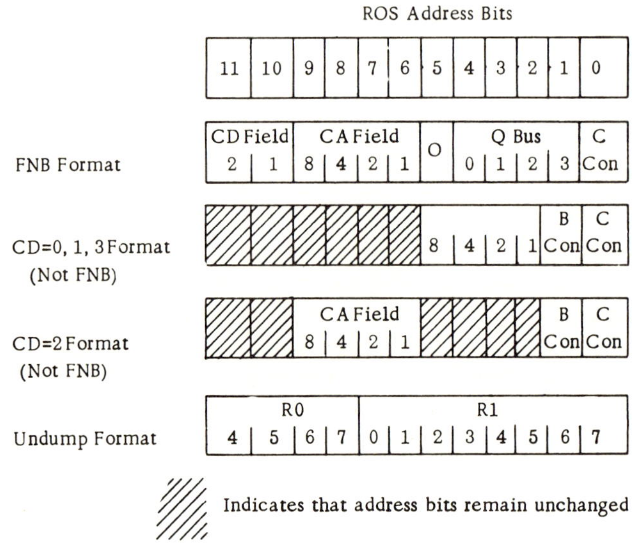

Figure 7.20 Basic ROS address bit generation.

1. *Functional Branch Format, FNB.* This format is often referred to as the 16-way branch format, since the 4 Q-register bits can be set to select any one of 16 unique ROS addresses. More than that, the FNB format can be used to branch to any address in ROS since all ROS bits can be changed by means of this format. When combining the 4 branch bits from the Q register and the C condition branch test, the functional branch format yields the maximum of a 32-way branch set.

The ROS address bits for the next cycle are set as follows:

(a) The CD field is set into ROSAR (10–11).

(b) The CA field is set into ROSAR (6–9).

(c) A zero is set in bit 5.

(d) The four bits Q(0–3) are set into ROSAR (1–4).

(e) The condition specified in the CC field of the ROS control word is tested. The ROSAR bit zero is set to 1 if the test is successful; otherwise, it is set to 0.

The functional branch format can be conveniently used to modify the ROS address within each quadrant of the ROS memory as shown in Fig. 7.21 as long as ROSAR (5) = 0.

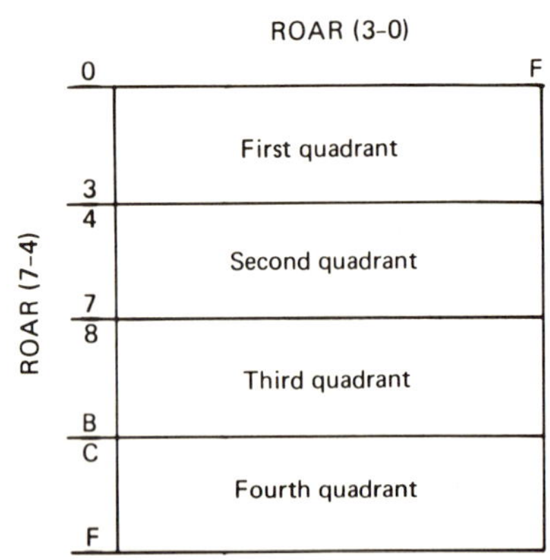

Figure 7.21 ROS memory quadrants.

Example

Assume that the system is executing the microinstructions stored in ROS address 51C Hex. Compute the CA, CB, CC, and CD that will be needed to branch to ROS address 53E Hex. Expressing these addresses in binary first:

```
51C Hex = 0101     0001     1100
53E Hex = 0101     0011     1110
          bits 11-6  bit 5  bits 5-0
```

Note that the two addresses have identical leftmost six bits. Bit 5 of the address to transfer to is 1; hence, the FNB format is not applicable since it transfers 0 to that location. We must therefore look for another format which can make use of the six leftmost bits and permit something other than zero to be stored in ROSAR(5). The best format to use here is the *high CA* with CD = 0, 1, 3 format, discussed next.

2. *CD = 0, 1, 3 Format*. The ROS address is set according to the conditions specified by the CD and CB fields:

(a) In this format, the high-order six bits are unchanged. ROSAR (11–6) are not affected and will retain the ROSAR (11–6) of the old microinstructions.

(b) The CA field is set into ROSAR (5–2).

(c) For CD = 0, the condition specified in the CB field is tested. A 1-bit will be set into ROSAR (1) if the specified condition is present; otherwise, a zero is set in the ROSAR (1) bit.

(d) For CD = 1, 3, ROSAR (1) is set to 0. The condition specified in the CC field is tested. If present, ROSAR (0) is set to 1; otherwise, it is set to 0.

This format is most suited for sequential microinstruction and branches within short routines. Hence, in the previous example, we can specify CA = 1111, CB = 0001, CC = 0000, and CD = 00.

3. *CD = 2 Format*. If CD = 2 and again, as in format 2, CB does not specify FNB, the ROSAR bits are set as follows:

(a) ROSAR (11–10) and ROSAR (5–2) are retained.

(b) The CA field is set into ROSAR (9–6).

(c) The condition specified in the CB field is tested. If the test is satisfied, a 1-bit will be set into ROSAR (1); otherwise, a 0-bit is set in ROSAR (1).

(d) Similarly, the condition specified in field CC is tested. If the test is satisfied, ROSAR (0) bit is set to 1; otherwise, it is set to 0.

This format differs from the second one only in the positioning of the CA field, allowing the microprogrammer to branch within a larger area. It allows him to get to three other modules in this group (the leftmost four bits specify the module).

4. *Undumped Format*. In the last undump cycle (hardware cycle), ROAR is set from the R register as shown in Fig. 7.21. During this undump cycle,

the old ROSAR which has been transmitted to the dump area in local storage is brought back via the R register to resume the operation.

Other special formats with CD = 3 can be specified by setting the CB field to control special data flow functions. Some of these functions are hardware branch conditions.

Examples

1. Specify the CA, CB, CC, and CD fields that will effect a branch from 91D Hex to FDF Hex:

91D = 1001 0001 1101
FDF = 1111 1101 1111

We can use the FNB format with CD = 11, CA = 1111, and Q-bus = 1111 to get the information on the Q bus. The following microinstruction can be used:

$$\begin{bmatrix} \text{E: } 1111 \\ \text{A: } Z \cdot E0 \rightarrow Z \end{bmatrix}$$

2. If a predecessor microinstruction is stored in ROS location 35A, suggest a format which can be used to permit one or more successor microinstructions to be stored in the same module.

Perhaps the second format is the easiest to use. In this format, we retain the leftmost 6 bits of the ROSAR and change the rightmost 6. This method leaves 37 ROS words with addresses 340 to 37F available to be assigned.

3. Assume the machine is in the CPU state and the present TROS address is 281 Hex. Also assume that the CA, CB, CC, and CD fields contain CA = 1111, CB = 0000, CC = 1100, and CD = 10. What is the address of the next ROS word?

No functional branch and CD = 2
Present address = 001010000001
Bits retained: 00XXXX0000XX
CA field --1111------
B condition
C condition depends on PRI
∴ next TROS address 00111100000X

This is a two-way branch to 3C0 or 3C1, depending on PRI.

4. Assume that the machine is in the CPU state during the present cycle which transfers (00011010) to the Q bus. Assume also that CA = 0000, CB = 1111, CC = 00, CD = 10, and CB = 15 = FNB. What is the address of the successor microinstruction?

The new ROS address constructed according to the FNB format (Fig. 7.21) will be 100000000010 = 802 Hex. With this instruction, a 16-way branch is programmed, depending on the Q register's high-order four bits which can assume any value between 0 and 15.

Figure 7.22 gives a flow-chart representation of the next ROS address generation.

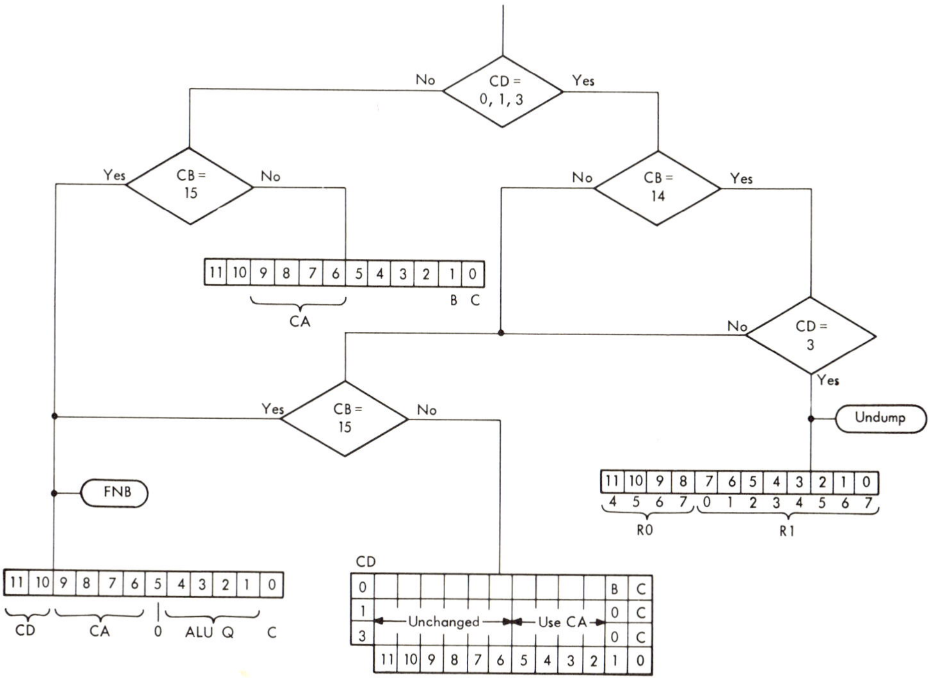

Figure 7.22 Determining the next TROS address.

PRACTICE PROBLEMS

1. Write a microinstruction to branch on the contents of the YCD stat.
2. Discuss the key selector in the microinstruction box which contains the following micro-orders:

   (7) [R: 0 Y7]

   How many successor micro-orders can be expected from the ROS addressing? What will their key identifiers be?
3. Is the address sequence 385 → 37E allowable?
4. Is the address sequence 39C → 385 allowable?
5. Specify the CA, CB, CC, CD, and the Q-bus bits Q (0–3) to transfer control from the ROS control word with hexadecimal address 123 to its successor microinstruction located at 456 hexadecimal.

7.13 MACHINE STATICIZERS

Machine staticizers are status triggers (commonly called stats) which may be set by the microprogram or by the hardware to indicate various system conditions that can be tested to establish the proper microinstruction sequencing. Certain stats may indicate specific hardware conditions.

YA: Set or reset from the emit field as controlled by the CN field, ALU output bits 0–3, or the console switches.	Y0: Storage stat, allows checking for ISA or PSA Y1: MPX stat, also called UBA (use bump address) stat, provides access to MPX storage Y2: General purpose, used to store condition code during CPU microprogram operations Y3: General purpose, used to store condition code during CPU microprogram operations
YB: Set or reset from the emit field as controlled by the CN field, ALU output bits 4–7, or the console switches.	Y4: General purpose, used for B-condition tests Y5: General purpose, used for C-condition tests Y6: General purpose, used for B-condition tests Y7: General purpose, used for C-condition tests
YD: Set or reset from the emit field as controlled by the CN field or the console switches when the system is in manual operation. YD stats are not parity checked.	Y8: ID (inhibit dump stat), prevents MPX break-in, is reset during undump Y9: MI (maskable interrupt), set when a MPX interrupt occurs Y10: Manual stat, allows manual loading and display of registers and storage Y11: Does not exist
YE: Not set by ROS fields; can be set from logic circuits or the console switches. The reset is from the emit field of the ROS word. YE stats are not parity checked.	Y12: Error stat; when set, an error causes a machine hardstop Y13: Does not exist Y14: IZT (integrated zero test) or IDQ (invalid decimal digit on the Q bus); this stat provides two B-condition tests. Y14 cannot be set or reset from the console Y15: Load stat, set when the load button is pressed, provides a B-condition test

YCD	ALU carry latch for direct function.
YCI	ALU carry latch for indirect function.
DISABLE	If on, disables log-out after detection of machine check.
WAIT	Stores bit 14 of current PSW.
CPU	If on, CPU is operating in CPU state. If off, CPU is in I/O state.
PMA	(Protected Memory Address) set when a protected MS location is addressed.
IMA	(Invalid Memory Address) set when invalid storage location is addressed.

Figure 7.23 Stat function and control.

The sixteen general purpose stats in the Model 40 are called Y stats and are numbered Y0–Y15. They are grouped into four groups of four stats each and are referred to as YA(Y0–Y3), YB(Y4–Y7), YD(Y8–Y11), and YE(Y12–Y15). Some of the Y stats perform specific functions and are sometimes referred to by the name of the function they perform. Figure 7.23 shows all the stats, their grouping, and the special functions assigned to them individually. Note that there is no stat group called YC stats; this is done to avoid confusion with the ALU carry stats which are called YC, YCI, and YCD. Other stats are special purpose devices and are always identified with the function performed. The Model 40 detailed data flow, Fig. 7.2 shows how data are set into or transmitted out of these stats. The first eight stats can be set directly as a group by the ALU one-byte output bus. These same eight stats (Y0–Y7) are combined to form a one-byte input to the ALU Q-bus, thus forming an auxiliary one byte of storage for the ALU data path. Figure 7.24 also shows that

CN Field	*4 Bits*	*Function*
	N0, N1, N2, N3	Control setting and resetting general and special purpose stats, ALU function register F, from CE field.

Value	*CPU, MX, and SCA*	*SCB*	*Comment*
0	No Op	No Op	
1	YA = CE	YA = CE	
2	YA/CE	YA/CE	
3	YA V CE	YA V CE	
4	YB = CE	YB = CE	= means set stats with emit value
5	YB/CE	YB/CE	
6	YB V CE	YB V CE	V means OR emit value with stat value
7	YD/CE	YD/CE	
8	YD V CE	YD V CE	/ means AND NOT emit value with stat value
9	YE/CE	YE/CE	
10	no TACT statement		Forces ROS decoder checks (diagnostic use)
11		YCH V CE	(YCH V EMIT) → YCH
12		YCH/CE	$(\text{YCH} \cdot \overline{\text{EMIT}}) \rightarrow \text{YCH}$
13	SPARE		
14	SPARE		
15	F = CE		Emit field into F register

Figure 7.24 Stat control. SCA, SCB refer to selector channels 1 and 2.

each of the four stat groups, YA, YB, YD, and YE, can be set by a separate half-byte bus from the emit field directly, or by ORing or ANDing the contents of the selected group with the emit field. Figure 7.24 also gives the decoding chart for the four-bit CN control field whose function is primarily to control the setting and resetting of the general and special purpose stats. Since YA and YB can combine to communicate with other facilities in the ALU data path, they are the only stats that are parity checked.

7.13.1 YA Stats

1. *Y0 or Ym (Memory or Storage Stat).* The Y0 stat is set and reset by microprogram control from the CE field through the CN control field. This stat is ANDed with PMA (protected memory address) to force the MAT (memory address test) condition. The stat can also be set and reset from the ALU output with micro-order CM = 12. This stat is also hardware set by the diagnostic select switch to set CPU check or MS validate and by the depression of the start key.

2. *Y1 or MXS (Multiplex Storage) Stat.* This is sometimes referred to as the UBA stat. It controls accessing to MPX-UCW (unit control word) storage and the protect keys from the storage protect local storage unit for the multiplex channel. This stat is set through CN field control of the CE field, by a manual switch on the operator-CE console panel, and from ALU output with CM = 12.

3. *Y2 and Y3 Stats.* These are sometimes referred to as condition code stats. They are used as part of the PSW to hold the condition bits. These stats are also used as selection stat control for the following:

Y2	Y3	
0	0	Multiplex channel
0	1	Selector channel 1
1	0	Selector channel 2
1	1	Clock timer

The first and last conditions (Y2, Y3 = 00, 11) are used in the decoding of CJ = 7. The second and third settings (Y2, Y3 = 01, 10) are used in conjunction with the selector channel circuitry to define the channel selected for channel operations with ROAR in control.

These stats are set with the CN control and ALU output. The Y3 stat is also set by hardware upon depression of the start key when the diagnostic select switch on the console is set to CPU Check.

The Y2 stat is also hardware set upon the depression of the start key when the diagnostic select switch is set to MS validate.

7.13.2 YB Stats (Y4, Y5, Y6, Y7)

These are true general purpose stats. They are set through CN control of the CE field and from the ALU output with CM = 12. It is illegal to microprogram a change from the ALU and CE field simultaneously. These stats can also be set through the console data switches. The Y4 and Y6 stats are used for B-condition tests, while Y5 and Y7 are used for C-condition tests.

7.13.3 YD Stats (Y8, Y9, Y10, Y11)

1. *Y8 or the ID (Inhibit Dump Stat).* This stat, when on, is used to inhibit further dumps by MPX or selector channel A after a dump operation, and it is also used by the microprogram to prevent dumps from occurring for a small number of cycles when special operations are being performed. It also reinterprets CG-field ALU function control 00 to give a logical OR instead of an indirect function. The maximum allowable cycles that a first-level dump can be inhibited (including main store holdoff) is eight.

2. *Y9 or MI Stat (Maskable Interrupt Stat).* This mask is set to generate PRI (Program Interrupt) if the system mask allows an I/O program interrupt.

3. *Y10 or MAN Stat (Manual Control Stat).* This stat reinterprets the CJ field to provide a means of sampling switches and to provide access to channel control and interface tags for log-out. It also reinterprets the RSK control of the CR field to provide a way to reset the error latches, including the ERR of the YE field. The effect of this reinterpretation is to reset all error conditions at the beginning of the cycle at T1. This stat is set and reset with the CN control of the CE field.

If Y10 is turned off during a cycle when the control CR = 6 is given, the control RSK is effective immediately. The control REE will also be effective. If Y10 is turned on during a cycle in which the control CR = 6 is given, the skew buffer is not reset. The control REE becomes effective next cycle. If Y10 is on when the write command is given, the main storage address is inhibited from setting address stop.

4. Y11 is not available for use.

7.13.4 YE Stats

1. *Y12 or ERR Stat (Error Stat).* When this stat is on, a machine check causes a hard stop condition. This stat is turned on with any error condition sensed by hardware or by a microprogramming sense of a hardware error which would force a log-out. When the system is in error-disable mode, this stat will *not* be set by the error condition.

When the command *LOG (start log-out) is given, the ERR stat will be set, providing "ENABLE" = 1. When the command *ICC (set interface control check) is given, the ERR stat will be set, providing "ENABLE" = 1. The setting of the ERR stat is controlled by hardware—the resetting is by CN-field control.

Whenever the ERR stat is set, the sensing of an error in the following cycles forces the CPU and channel clock to stop at the end of the cycle in which the error occurs. Whenever the ERR stat is set, all dumps are inhibited, starting with the cycle in which the ERR stat is set. Depression of the system reset key resets the ERR stat.

2. Y13 does not exist.

3. *Y14 or IZT and IDQ Stat (Integrated Zero Test and Invalid Decimal Digit on Q Bus).* This stat consists of two latches which are turned on by hardware as follows.

IZT—This latch is turned on by an ALU output nonzero when performing an indirect ALU function. The condition IZT is available as an early condition for test. The IZT latch is set only when the IZT condition is 1 and does not get reset on IZT = 0.

Testing the IZT condition resets the IZT latch at T3 of the cycle tested. The IZT latch can also be reset through CN control of the CE field.

IDQ—This latch is turned on if either or both decimal digits on the Q ALU input are invalid. Hardware senses for any bit combinations higher than 1001, which is considered an invalid decimal digit on decimal ALU operations.

The condition IDQ is available as an early condition for test. The IDQ latch is set only when the IDQ condition is available and does not get reset on IZT = 0. Testing the IDQ condition resets the IDQ latch at T3 of the cycle tested. The IDQ latch can also be reset through CN control of the CE field.

The YA, YB, YD, and YE stats can be manually changed from the console.

4. *Y15 or Load Stat.* The depression of the IPL (load) button on the operator panel forces the load stat on. The load stat is reset by the CN control of the CE field. The load stat forces a hard stop if a machine error is sensed.

7.13.5 Other Stats (Not Referred to as Y Stats)

These stats are latched up by certain particular machine conditions and are indicated on the console. These stats are referred to by the condition they perform.

1. *YCD and YCI.* These are two separate ALU carry latches for the direct and indirect functions of ALU.

2. *Enable Stat.* When this stat is on, a machine error causes either a log-out or hard stop. The overall check is ignored when enable is off. Enable is set

by the micro-order SWEA (which also sets WAITE and ASCII) and ALU output when bit five is on. It is reset by SWEA and ALU (5) = 0.

SET: SWEA · (ALU bit 5 = 1)
RESET: SWEA · (ALU bit 5 = 0)

3. *Wait Latch.* The stat indicates that the system is in the waite loop by using SWEA (CB = 4, CD = 3) in conjunction with the ALU output bit 6:

SET: SWEA · (ALU output bit 6 = 1)
RESET: SWEA · (ALU output bit 6 = 0)

4. *ASCII.* It is used to control the system to operate on data in the ASCII mode. ASCII is set by using SWEA (CB = 4, CD = 3) in conjunction with ALU output bit 4:

SET: SWEA · (ALU output bit 4 = 1)
RESET: SWEA · (ALU output bit 4 = 0)

5. *CPU-I/O Stat.* Interpretation of this stat controls whether the CPU data paths act as a CPU or as part of the channel facilities. The stat also provides for different interpretations of the CB and CC fields of the ROS word. When the CPU stat is off, the system is operating in the I/O mode.

6. *PMA and IMA (Protected Memory Address and Invalid Memory Address) Stats.* The first stat, PMA, is set when a protected main storage location is addressed. The second stat, IMA, is set when an invalid main storage location is set. The setting of IMA may be tested by either of the following micro-orders:

(7) R: SAT
(6) C: TRAP

If the IMA stat is set, or if both the Y0 and the PMA stats are set, then the trap test forces a microprogram branch to ROS address 404 Hex or 405 Hex.

7. *External Mask.* This is set to reflect the state of bit 7 of the PSW.

7.14 MICROINSTRUCTION TRACE

Figure 7.25 shows a sample CLD page, which happens to be the first half-word instruction fetch. It is offered as an example of a CLD and as a tool to trace the beginning of an instruction fetch routine. The following explanation is offered on the CLD level to reinforce the reader's understanding and to tie the CLD to the actual data flow.

Entries to this I-fetch routine can be made from many possible sources. This is shown by the number of input net lines at the left side of the page. Entries to this routine are made at the end of each execution routine; another entry is made by the *execute* instruction whose objective is actually an I-fetch

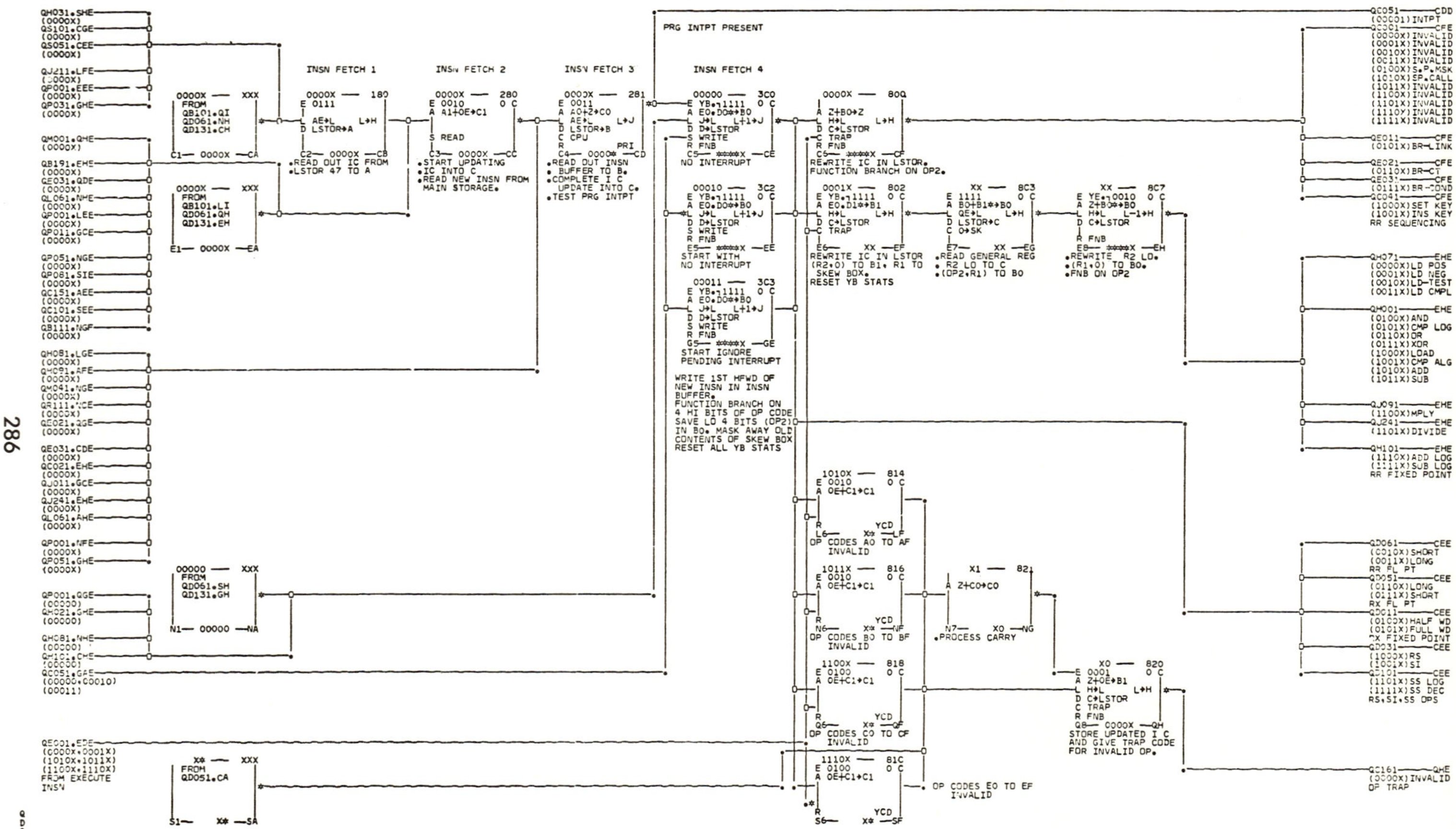

Figure 7.25 A sample CLD page. First halfword instruction fetch.

operation of a single instruction outside the normal instruction sequence. A third entry is made from the program segment that loads a new PSW into the data flow. In all cases, the updated instruction count (IC) is held in local storage and points to the main storage halfword address from which the next machine language instruction or part of an instruction is to be fetched.

The I-fetch operation is then defined as the microroutine and the associated hardware that will:

1. Read the machine language instruction from main storage in halfword chunks.
2. Do all operations common to a given class of machine instruction after decoding it.
3. Update the IC before issuing the final functional branch to the routine designed to execute the fetched instruction.

The instruction fetch routine also checks on the validity of the instruction code as well as the main storage address in the IC register. If either is found invalid, a program check interrupt is executed. If during an I-fetch the PRI is on to signify the event that a program interrupt is requested, the microprogram will branch to the appropriate interrupt routine before issuing a main storage read call.

Depending on the instruction format (RR, RX, RS, SI, or SS), one to three halfwords have to be read from main storage to the CPU in order to finish the I-fetch operations and prior to branching to the execution routines. Therefore, the I-fetch segments were designed with an exit point attached to the tail end of each of the three routines:

FIRST HALFWORD	I-fetch
SECOND HALFWORD	I-fetch
THIRD HALFWORD	I-fetch

An instruction halfword is sometimes called a *syllable*; hence, Fig. 7.26 is the first-syllable I-fetch and first-level functional branch.

Again, there are many exits from this first-syllable I-fetch routine as shown in Fig. 7.26. The top exit, marked by the QC051-CDD net, is a branch to the interrupt-handling routine. It is shown as the output of box C4 (ROS address 281 in the top right corner of the box). This exit is a result of a C-condition branch test on the status of the PRI stat. If the PRI is on, the next block to be executed is on page QC051-CDD net. However, if the PRI is off, the next ROS word to be executed is the one in box C5.

Another type of exit shown in Fig. 7.25 is the case of an invalid operation trap. A third type of exit is to the two and three halfword routines. A fourth type of exit, in the case that the first halfword is found to be an RR type, is made directly to that instruction execution routine.

Before tracing the microinstructions, we shall review a few basic facts about the microinstruction box.

1. Each box is identified by its two character coordinates shown at the lower left corner of each box. For example, the coordinate of the top rightmost box (Fig. 7.2) is C1; the coordinate of the lower rightmost microinstruction box in the same figure is Q8.

2. The ROS absolute address assigned to the box is shown in the upper rightmost corner of the box. For example, the addresses of the two boxes with coordinates C5 and EG are $(3C0)_{16}$ and $(802)_{16}$, respectively.

3. Leg selectors and leg identifiers determine the number of branches from an instruction and identify the successor microinstruction to be executed by the binary bit identifier of the branch leg.

4. Note that blocks E5 and G5 can be addressed only from the incoming connection.

We shall trace this sequence by starting with block C2. In this box, the IC value which resides in local storage word 47 (E:0111, L:AE → L,..., L → H) is read out to register A. We are now set to read the halfword from MS. Box C2 connects only to C3. Box C3 issues the read command to main storage. It also updates the address IC by two.

Box C4 completes the IC update and reads the contents of the instruction buffer register (LS-43) to register B. Note that the two-way branch in box C4 is shown as a four-way branch. The leg selector in C4 is 0*. Therefore, the only two possible successor microinstructions are the two with leg identifiers 00 (box C5) or 01, identified by the branch page QC051 on another CLD page. If no interrupt was found pending by microinstruction C4, then the logical successor is microinstruction C5.

In C5, the new instruction (halfword) fetched from main storage is written into the instruction buffer, all YB stats are reset save LS (4) contents in B0, C5 masks away the old contents of skew and then executes a four-way functional branch on the four main storage bits of operation code. This branch determines not only the type of instruction format but also the general class of System 360 instruction in a given format.

To replace the asterisks with a four-bit number, the Q register contents must be known. In box C5, the arithmetic statement specifies D0 to Q. To determine what value is currently in D0, we must go back to block C3 to find what was transferred to D0. Assume that D0 contains 10101111. Then the branch leg selector becomes 1010X, and the next block to be executed after C5 is L6.

If we assume that D0 was 00011010 (which is the Op Code for the RR fixed-point add instruction), then the next microinstruction to be executed following the C5 microinstruction will be E6.

Boxes E6, E7, and E8 will be executed as shown in Fig. 7.27. The final step in box E8 is to execute a functional branch on the second half of the operation code to determine the exact instruction to be executed and to use that code to address the proper microinstruction execution routine.

Figures 7.26, 7.27, and 7.28 are additional CLD sample pages which can be used for further traces and further practice.

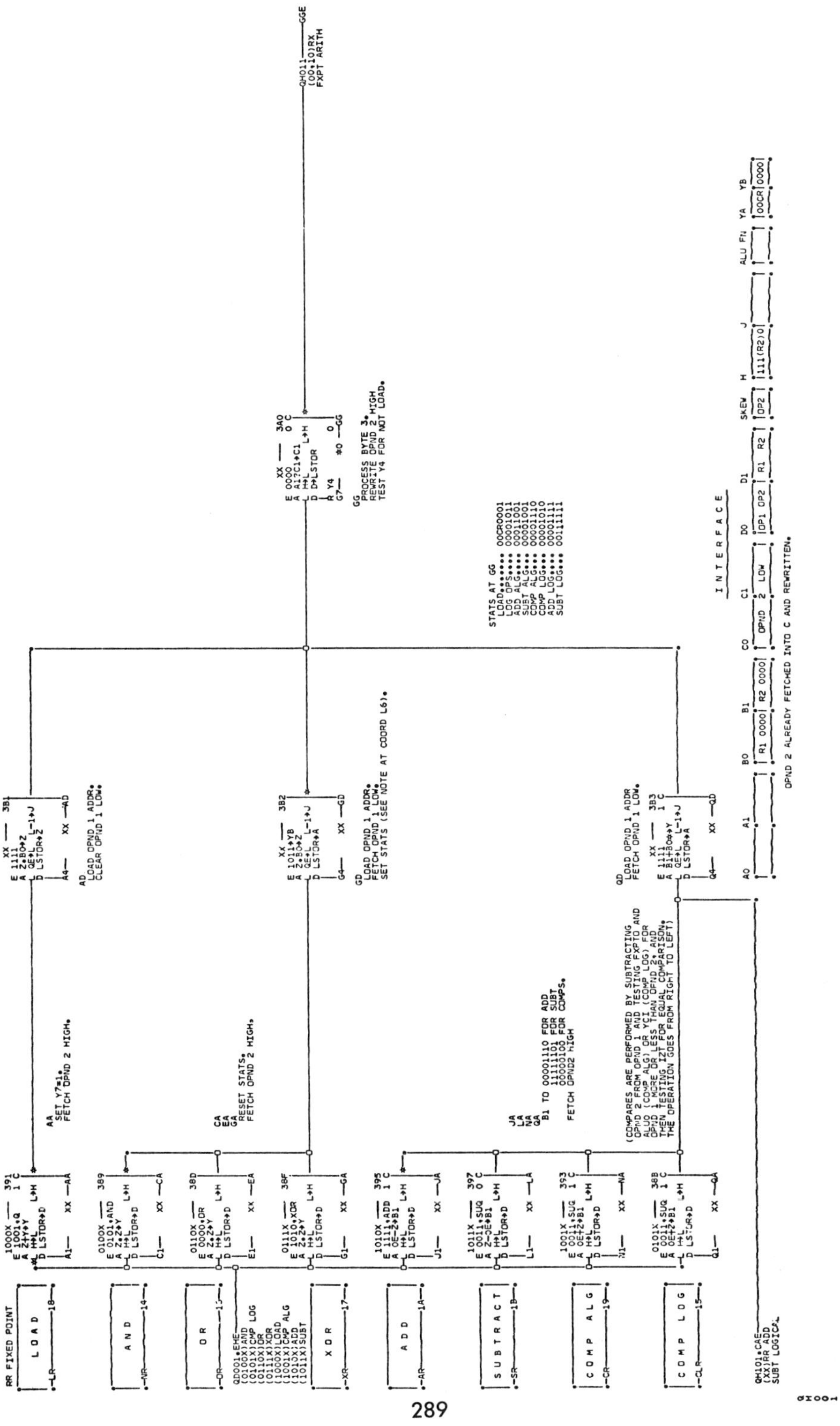

RR FIXED POINT
LOAD
AND
OR
XOR
ADD
SUBTRACT
COMP ALG
COMP LOG
STATS AT GG
INTERFACE
OPND 2 ALREADY FETCHED INTO C AND REWRITTEN.

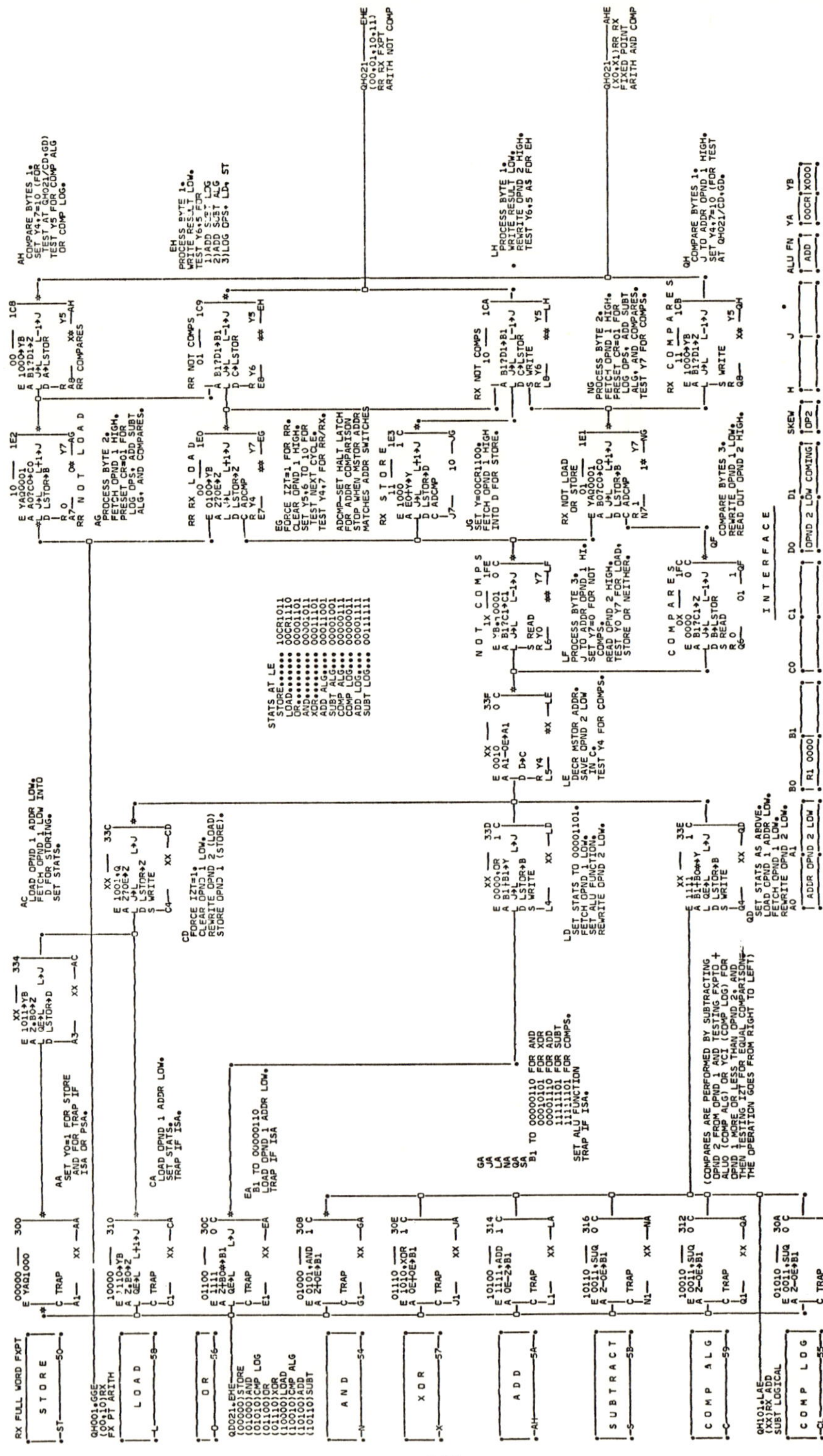

Figure 7.27 A sample CLD page.

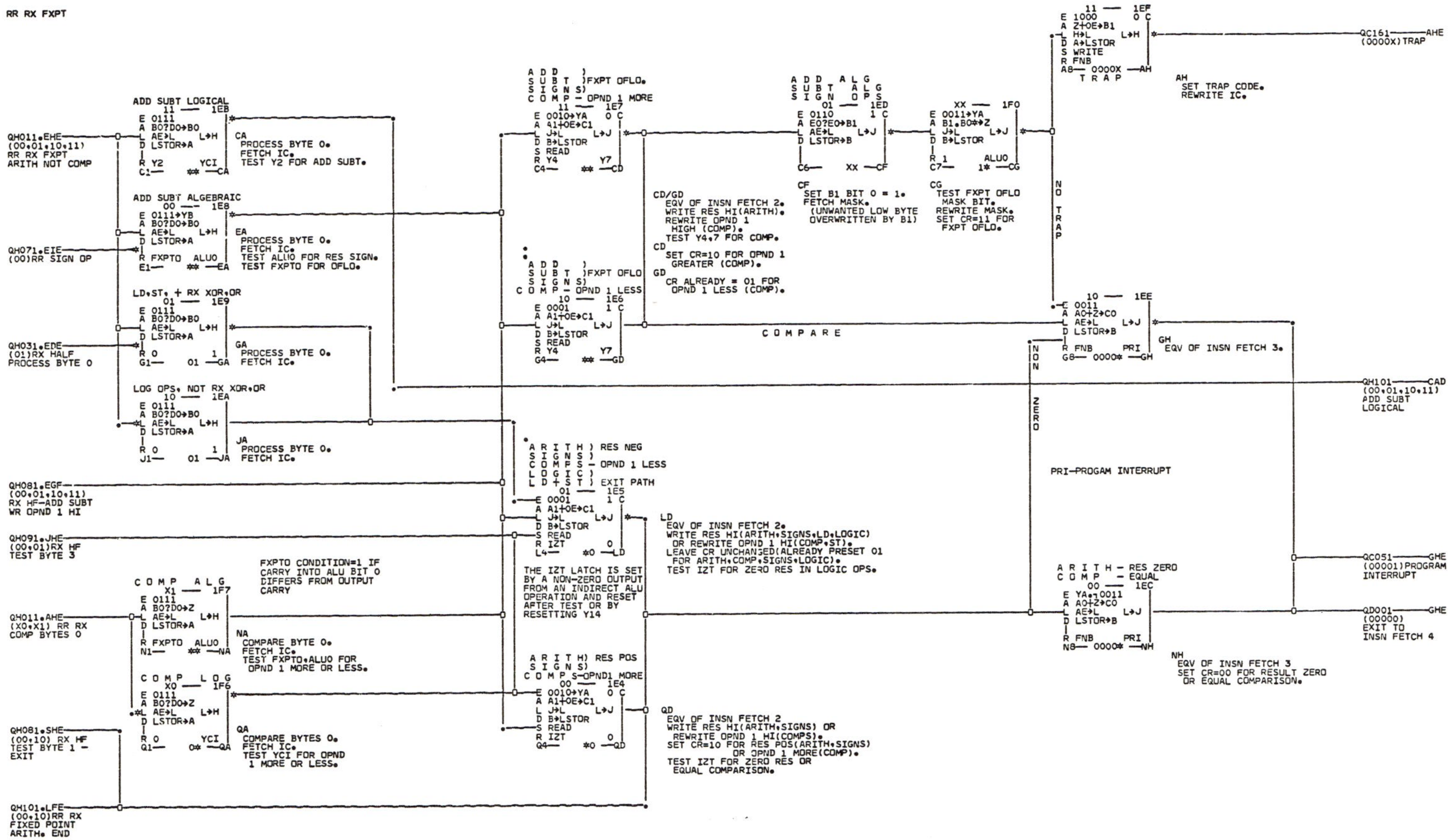

Figure 7.28 A sample CLD page.

GENERAL PROBLEMS*

In the following problems, the entrance to your microprogram will be from box LF on CLD page QD001 (see Fig. 7.26). This box represents the instruction of ROS address 814, which has been modified to

```
     1010X  814
  [ A: Z:Z → Z ]
  [ R: FNB     ]
 LG   0000X   LF
```

This box has destination denoted by the right-edge line note

XY0001—LFE
(0000X)

Assume the complete F module (module 15) of ROS is available for your microprogram. Your program is to be written on CLD page XY0001, where XY are your initials. Analysis of QD001 shows the following interface conditions to hold upon entrance to your program.

con A = IC (instruction counter value)
con C = IC + 2
con B0 (0–3) = 2nd Hex digit of op code
con B0 (4–7) = 0
con D = 1st hfwd of instruction
LS (47) is read out, but not restored
YB stats are reset

In the following, "GPR" denotes general purpose register or fixed point register.

1. Instruction:

MS(IC) MS(IC + 2)

A	A	GPR1	X	GPR2	D2	

The halfword second operand is examined. If it is zero, the instruction counter is set to the address contained in the first operand. If it is not zero, normal sequencing takes place. *Note:* AA is a "transfer on zero" instruction.

2. Instruction:

MS(IC) MS(IC + 2)

A	B	GPR1	X	GPR2	D2	

*These practice problems were designed by Dr. Leo Hellerman as teaching aids. They were used by him while teaching the Model 40 microprogramming course at the IBM Systems Research Institute. They were used in this chapter with his permission for which I am grateful.

Let the second operand address be denoted by ADDR 2 = D2 + con GPR2. Then this instruction computes

$$\text{RESULT} = \text{MS(ADDR2)} + \text{MS(ADDR2} + 2) - \text{MS(ADDR} + 4)$$

and stores result in GPR1. *Note:* It would be interesting to compare the execution time for this special-purpose instruction with the time for execution of the equivalent OS program.

3. Instruction:

MS(IC)				MS(IC + 2)		
A	C	GPR1	GPR3	GPR2	D2	

The sign of operand 2 is examined. If it is positive, operand 2 is added to the contents of GPR1. If it is negative, operand 2 is added to the contents of GPR3. *Note:* GPR1 is positive accumulator; GPR3 is negative accumulator.

4. Instruction:

MS(IC)			MS(IC + 2)		
A	D	╳	GPR2	D2	

The single byte operand 2 is examined, a hexadecimal digit at a time. If a digit is found to be 0 or 8 it is replaced by F (15 decimal); otherwise it is replaced by 0.

PRACTICE PROBLEMS

1. True or false:
   (a) The source of A-register data can be specified from the R bus and the ALU output in the same cycle.
   (b) The source of D-register data can be specified from the R bus, ALU, and MS data output in the same cycle.
   (c) If statements (a) and (b) are true, there is no coincidence circuitry to control the priority of the incoming data; rather, it is up to the microprogrammer in both cases to ensure such priority of the incoming data.
   (d) If both P and Q specify an extension, Q blocks P.
   (e) The ROAR register is reset every machine cycle and is set by hardware logic or microprogram.
   (f) Information may be gated directly from:

MS to C	D1 to Q
D to MS	BX to AX
D0 to P	

   (g) ROS fields CA, CB, and CD are used to determine the address of the next ROS word.

(h) The read-only back-up address register ROBAR is just a diagnostic tool for the customer engineer; it is used to record the TROS address used in the cycle in which the machine is stopped.

REFERENCES

1. *IBM Field Engineering Manual of Instruction—System/360, Model 40.* Form 223-2843-0.
2. *IBM Field Engineering Manual of Instruction—System/360, Model 40, Comprehension Introduction.* Form 2-23-2840-0.
3. Husson, S. S., "Microprogramming Manual for IBM System/360 Model 50," IBM Technical Report.
4. *IBM Field Engineering Maintenance Manual.* Form 223-2841-0.
5. *IBM Field Engineering Manual of Instruction—System/360, The 2040 Processing Unit.* Form 223-2842-0.
6. *IBM Field Engineering Manual of Instruction—System/360, Model 40. Theory of Operations.* Form 223-2844-0.

MICROPROGRAMMING THE IBM SYSTEM/360, MODEL 50

8

8.1 INTRODUCTION

The Model 50 is a medium to large scale member of the IBM System/360 product line. It has a wide range of computing and data processing capabilities. The Model 50 can be used for communications and control as well as for commercial and scientific purposes. The system can operate on fixed or variable field data with equal facility. Arithmetic operations may be performed with decimal numbers, short or long floating point numbers, or fullword or half-word binary numbers.

The Model 50 processing unit operates with a .5-microsecond internal cycle. The basic Model 50 employs a 2.0-microsecond core storage unit with a capacity of 65,536 bytes. This core storage capacity is expandable to 512 kilobytes. The CPU has sixteen fullword general purpose registers and four double word floating-point registers, as well as other work and channel registers. These registers are all available in the 64 fullword scratch pad memory which operates at a speed of 5-microsecond.

The Model 50 data flow, instruction fetch and execution, and channel operations are controlled by a balanced capacitor read-only storage unit with a .5-microsecond cycle time. Each ninety-bit word is a microinstruction and controls the action of the system for that .5-microsecond. The major distinguishing features of the Model 50 are:

1. Both arithmetic and storage operations are carried out using a full thirty-two bit word.

2. The data flow is organized into a number of parallel data paths that can operate concurrently. This parallelism was a factor in producing an internal performance which is equal to about six times that of the IBM 709 on scientific problems and seven times that of the IBM 7070 on commercial problems.

3. The flexible and efficient control of the input/output operations is made possible by the use of versatile and the system-integrated multiplexor and selector channels. The channel operations in the Model 50 are overlapped with the CPU operation.

8.2 SYSTEM ORGANIZATION AND DATA FLOW

Figure 8.1, a data flow diagram of the IBM Model 50 processor, shows the functional units, registers, latches, counters, adder, mover, shifter, in-gate lines and positions, and out-gate lines and positions. It also shows the interface between the CPU and main storage, the CPU and local storage, and the CPU and both the selector and multiplexor channels.

A few general observations will help the reader to understand this data flow diagram.

1. There are three data-bus lines that are different in width, and through which data are channeled from one functional unit to another. These are the 32-bit adder-out bus, the 24-bit instruction-address bus, and the 8-bit mover-out bus.

2. Registers having the same shading are gated at the same time. For example, all or specific bits on the adder-out bus can be gated into the H and M registers simultaneously if the gates to the H and M registers are opened at a given time during the machine cycle.

3. Data are gated into the top of every functional unit and out from the bottom.

4. The horizontal lines and the numbers associated with them for an in-gate or out-gate show where the data are going and what section of the register can be gated out.

5. There are many blocks of different widths marked registers, and fewer blocks marked latches. A register is a temporary storage device consisting of one or more storage elements (triggers) which may hold either control information or data. Information is gated in or out by a specific micro-order and is altered only by another micro-order in some later microinstruction. The setting and resetting of a register is under the microprogrammer control.

A latch is also a temporary storage device consisting of a set of flip-flops which are reset at the same time during every machine cycle. Information can be stored in a latch for only one cycle and must be moved to a permanent

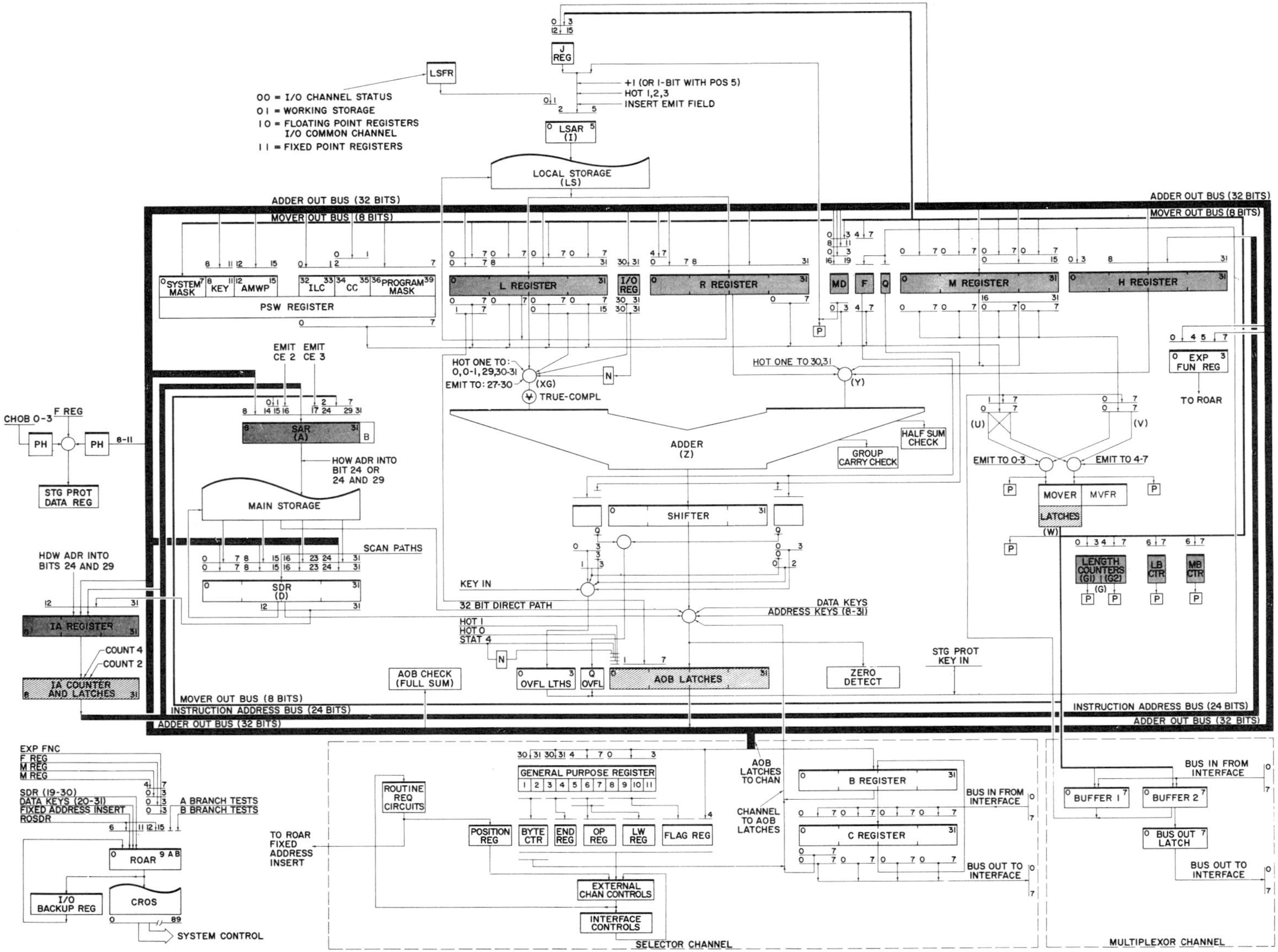

Figure 8.1 IBM System/360, Model 50 data flow.

register if it is to be preserved for later use. The setting and resetting of a latch is under the hardware control.*

6. Stats are storage elements (indicators, toggles) that are set automatically or by micro-orders to reflect the status of the machine. There are eight general purpose stats and four other stats designed for specific functions.

The Model 50 CPU data flow consists primarily of two parallel paths which may be activated simultaneously. One is a 32-bit-wide adder-shifter path fed by several 32-bit registers. The other is an 8-bit-wide logical mover path which may have two simultaneous byte inputs, selected from various CPU registers. The mover manipulates one-byte blocks in half-byte increments.

To simplify the discussion and understanding of Model 50 data flow, it is advisable to break it up into six basic data paths and examine each one separately. These basic paths are: ADDER, SHIFTER, MOVER, INSTRUCTION ADDRESS, LOCAL STORAGE, and MAIN STORAGE.

8.3 ADDER DATA PATH

The adder, Fig. 8.2, is capable of performing both binary and decimal arithmetic. Decimal arithmetic is performed by doing a binary add and generating a decimal correction factor in the L register in the same cycle. Another cycle is needed to subtract the correction factor from the results of the preceding cycle. The adder is made up of:

32 individual adder units
4 parity-checking circuits (one for each byte)
4 parity-generating circuits (one for each byte)
carry look-ahead circuitry

When performing arithmetic functions, data are gated to the right-adder input (Y) from the 32-bit register H, M, R, or the two low-order bytes of M. The left-adder input (XG) contains a true/complement gate and is fed by a 32-bit L register.

In a single 500-ns cycle, two 32-bit numbers are gated into the XG- and Y-adder inputs; they pass through the adder and continue on to set the adder output latches (AOB). At the end of the CPU cycle, the adder output is in the AOB ready to be gated out and into an operating register. In one microinstruction, the microprogrammer will specify the origin of data to be gated into XG and Y and the destination of data coming out of the AOB latches.

In the System/360, subtraction is achieved by the 2's complement; the adder

*The distinction between a register and a latch is not consistent with all product lines, or even with all models of the IBM System/360 product line.

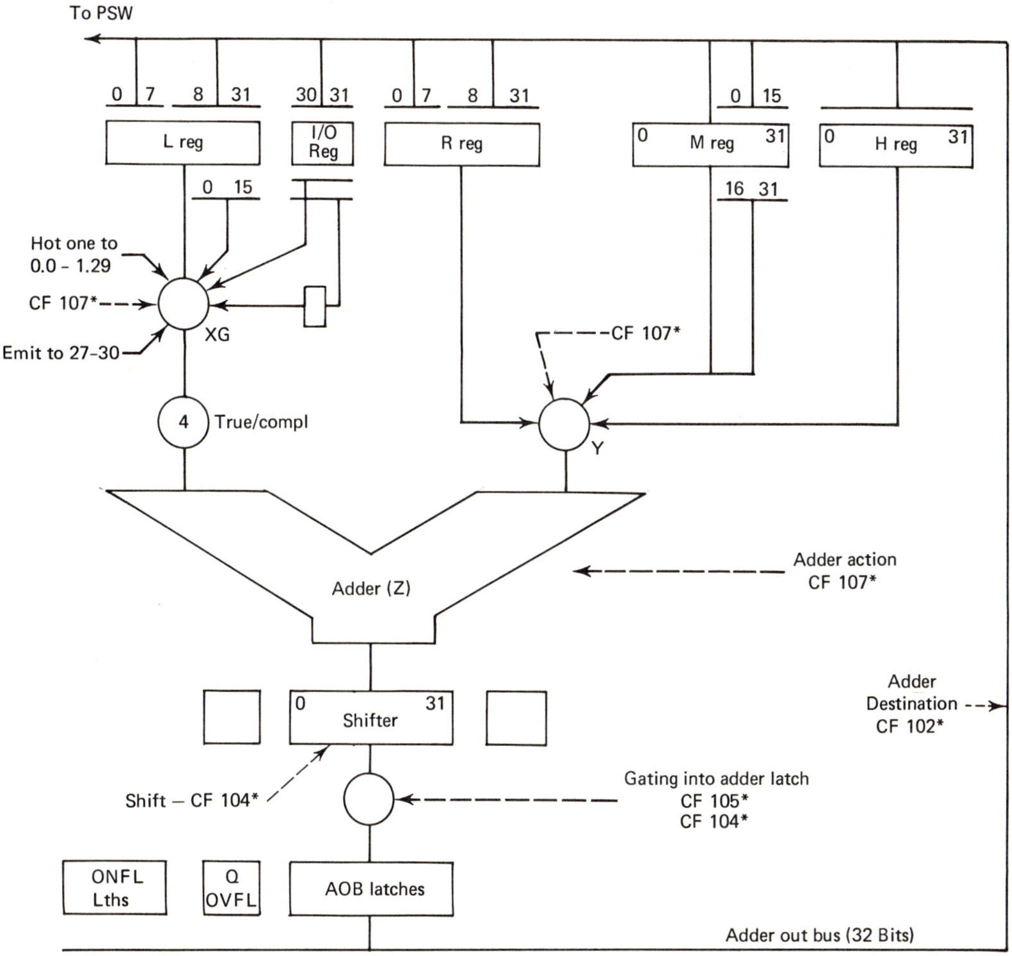

Figure 8.2 Adder data path.*

is built to accommodate this operation. The true/complement gate on the XG input must be set by the microprogrammer for each adder operation. A (Hot 1) or a carry into the low-order adder position is available for this purpose.

When a complement gate is set, bits gated into XG will be inverted (i.e., ones become zeros and zeros become ones), thus forming the 1's complement of the original XG input. The microprogrammer will continue the process of forming the 2's complement by inserting a carry (Hot 1) into the Y-adder input.

*The CF-XXX is the page number in the micro-order catalog which has the micro-orders pertaining to this data path. The catalog is in the appendix. It is divided into two sections: CF 101 to CF 113 deal with the CPU mode, and CF 201 to CF212 deal with the I/O mode.

Example

To subtract 3 from 7, we add the 2's complement of the binary representation of the number 3 to the binary representation of 7.

Binary representation of 7	1 1 1 √
Binary representation of 3	0 1 1
1's complement of 3	1 0 0
2's complement of 3 (100 + 001)	1 0 1 √
Hence 7 − 3 = (111 + 101) =	Carry → (1) 1 0 0
= 4	

(The carry bit is ignored in this case.)

Since multiplication and division are accomplished by successive additions and subtractions, the adder data path is used for all arithmetic operations.

Notice the in-gates for the L and M registers in Fig. 8.1. Separate controls are assigned to the high-order bytes of these registers to permit floating-point numbers to be handled. In floating-point notation, the high-order byte contains a binary exponent, and the remaining 24 bits contain a fraction expressed in hexadecimal form.

8.4 SHIFTER DATA PATH

The shifter (Fig. 8.3) enables the adder output to be shifted to the left or right either one or four places. A group of four overflow latches are used to

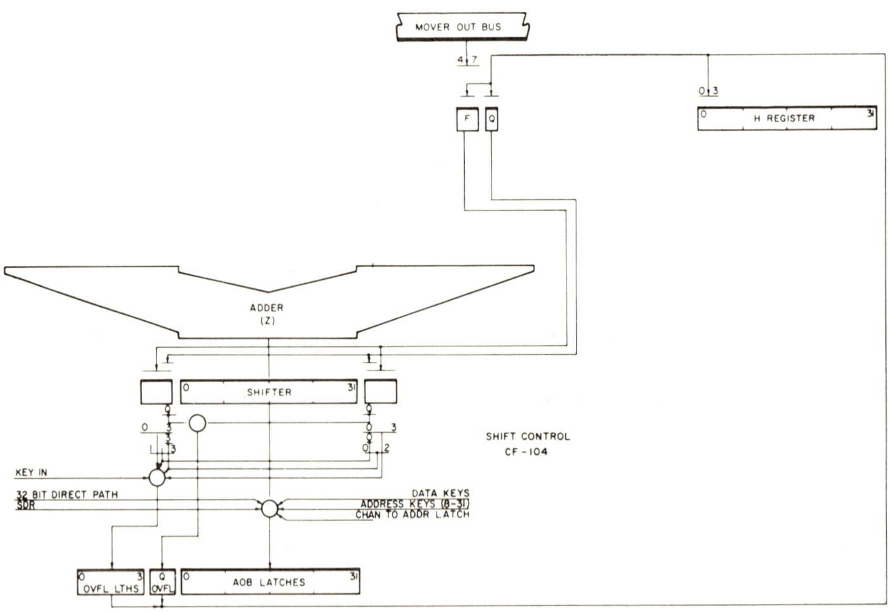

Figure 8.3 Shifter data path.[1]

store the bit or bits spilled out from the left or right end of the adder output. The direction of shifting and the number of shift positions are indicated by specific micro-orders. Bits entering the overflow latches may be stored into the F register, and a bit entering the Q latch will be stored into a 1-bit Q register.

The overflow bits may be entered into AOB latch positions that were left vacant by a shift. The shifted and patched-up data are gated into the AOB latches which hold this information for gating into some operating register at the beginning of the next CPU cycle.

8.4.1 One-Place Shift

Depending on the specific micro-order used, the adder outputs will enter the shifter logic and will be skewed to the left or right one place. In a shift-left-one command, the high-order adder output bit (position 0) is shifted left in either true or complement form into the low-order overflow latch (position 3). Similarly in a shift-right-one command, the low-order adder output bit (position 31) is shifted left into the high-order overflow latch (position 0). The remaining three overflow latch positions have bits skewed into them from the corresponding bits of the F register. Either the Q-register bit or the overflow bit shifted into the F register may be entered into the vacated position of the shifted adder-output latches (Fig. 8.4).

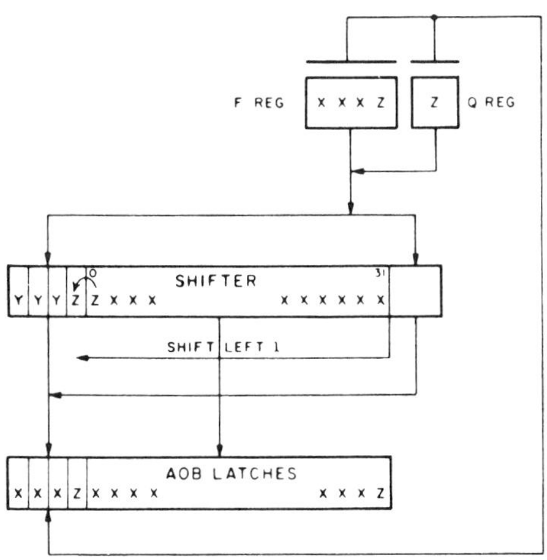

Figure 8.4 Shifted adder-output latches.

For a shift-left-one, the only bit that may be lost is the high-order position bit of the F register. Similarly, for a shift-right-one, the only bit that may be lost is the low-order position bit of the F register.

8.4.2 Four-Place Shift

For a four-place shift, the adder outputs enter the shifter logic and are skewed to the left or right four places, depending on the specific micro-order used. The four high-order or four low-order adder-output positions are gated into the overflow latches and may be immediately spilled into the F register.

Depending on the specific micro-order, the four vacant bits in the shifter may have:

1. The contents of the F register entered into the shifter output.
2. The contents of a 4-bit emit field entered.
3. Four zeros inserted.

A four-place shift for a floating-point Op Code does not shift positions 0–7. Only bits 8–31 are shifted, and the overflow bits are lost.

Because the micro-orders involving the adder are many, they have been divided here into groups: adder-XG inputs, adder-Y inputs, adder functions, gating into adder latches and shifting, and adder latch destination. See Table 8.1,[3] which gives a listing of microactions pertinent to the adder data path.

Table 8.1 ADDER MICRO-ORDERS

Adder-XG Inputs

1. L (0–31) to XG.
2. One to bit zero of XG.
3. Emit to XG (27–30).
4. L (16–31) to XG (0–15).
5. L (0–29) to XG (0–29), emit 1's to XG (30–31) and to Y.
6. Emit to XG position 29 (Hot 4).
7. Emit 1's to XG (0–1).
8. I/O register to XG (30–31).
9. I/O register complemented to XG (30–31).

Adder-Y Inputs

1. R (0–31) to Y.
2. M (0–31) to Y.
3. M (16–31) to Y (16–31).
4. H (0–31) to Y.

Adder Functions

1. True (XG input).
2. Complement (XG input).
3. Binary add, no carry saved.
4. Binary add, no carry saved, Hot 1 in if F = 0.
5. Binary add, save carry zero.
6. Binary add, save carry zero, EOR carry one.
7. Binary add, save carry one, block carry eight.

Adder Functions (continued)

8. Binary add, save carry eight, pass carry eight.
9. Decimal halve, low-order correct.
10. Decimal add and correct, save carry zero in stat 1.
11. Decimal double and correct, save carry in stat 1.
12. Decimal halve, high-order correct.
13. Decimal add and correct, save carry per byte stats in stat 1.

Gating into Adder Latches

1. Shift zero.
2. Shift right one, spill to F, and enter from Q.
3. Shift zero, insert inverted sign.
4. Shift zero, insert plus sign.
5. Shift zero, insert minus sign.
6. Shift zero, insert normal sign.
7. Shift zero and gate LAR to H directly.
8. Shift left one, spill complement to F, and enter from Q.
9. Shift left one, spill to F, and enter from Q.
10. Shift left one, spill to F, and enter from F.
11. Shift left one, spill to Q, no enter.
12. Shift left one, no spill, and enter from Q.
13. Shift right one, spill to F, and enter from F.
14. Shift right one, spill to F, no enter.
15. Shift right one, spill to Q, no enter.

Table 8.1 ADDER MICRO-ORDERS (*Continued*)

Gating into Adder Latches (*continued*)

16. Shift right one, spill to Q, and enter from Q.
17. Shift left one, spill to Q, and enter from F while shifting F.
18. Shift left four, spill to F, and no enter.
19. Floating-point shift left four, no enter.
20. Floating-point shift left four, and enter from F.
21. Shift right four, spill to F, and no enter.
22. Shift right four, spill to F, and enter from F.
23. Floating-point right shift four, and no enter.
24. Floating-point right shift four, and enter 00001.
25. Shift right four, spill to H (0–3), and no enter.
26. Shift right four, no spill, and enter from F.
27. Floating-point shift left four, no spill, and enter from emit field.
28. Shift right one, spill to Q, and enter from F.
29. Data keys to adder latches, bits 28–31 to F.
30. Channel (MIB) to adder latches.
31. MDR to adder latches (read holdoff).
32. Address keys to adder latches.

Adder-Latch Output Gating

1. Adder latches to H.
2. Adder latches (29–31) to I/O register.
3. Adder latches to I/O register and R.
4. Adder latches (12–15) to J.
5. Adder latches to L.
6. Adder latches (8–31) to L (8–31).
7. Adder latches to L and M.
8. Adder latches to M.
9. Adder latches to MDR.
10. Adder latches to MDR and R.
11. Adder latches to R.
12. Adder latches (0–7) to R (0–7).
13. Adder latches (8–31) to R (8–31).
14. Adder latches to SAR and L.
15. Adder latches to SAR and R.
16. Adder latches (0–7) to L (0–7).
17. Adder latches (8–31) to SAR.
18. Adder latches (8–31) to IAR.
19. Adder latches to R and SAR without invalid address trap.
20. Adder latches to R and SAR with word address special check.
21. Adder latches to R and SAR with double-word special check.

Adder-Latch Output Gating (*continued*)

22. Adder latches (0–31) to L, (0–15) to M right, (0–3) to MD.
23. Adder latches (8–11) to MD.
24. I-fetch gate. Adder latches (0–31) to L and M, (12–15) to J, (16–19) to MD, etc.

The Micro-Orders Involving the L Register

1. L (0–31) to the adder-XG input.
2. L (16–31) to adder-XG input (0–15).
3. Adder output latches to L (0–31).
4. Adder output latches to SAR and L.
5. Adder output latches to L and M.
6. Adder output latches (0–7) to L (0–7).
7. Adder output latches (8–31) to L (8–31).
8. L to U per L-byte counter.
9. L to U per M-byte counter.
10. L (0–7) to U.
11. L (8–15) to U.
12. L (16–23) to U.
13. L (24–31) to U.
14. W to L (0–7).
15. W to L (8–15).
16. W to L (16–23).
17. W to L (24–31).
18. LS to L, write back from R.
19. LS to R, write back from L.
20. L to LS.
21. LS to L, regenerate from L.
22. I-fetch gate: adder output latches (T) to L and M, T (12–15) to J, T (16–19) to MD, etc.
23. Adder output latches (T) to L, T (0–15) to M right, T (0–3) to MD.

The Micro-Orders Involving the R Register

1. R (0–31) to adder-Y input.
2. Adder output latches to R.
3. Adder output latches (0–7) to R (0–7).
4. Adder output latches (8–31) to R (8–31).
5. Adder output latches to SAR and R.
6. Adder output latches to R and SAR, without invalid address trap.
7. Adder output latches to R and SAR, with word-address special check.
8. Adder output latches to R and SAR, with double-word special check.
9. Adder output latches to I/O register and R.
10. R (24–31) to mover-U input.
11. R to local storage (LS).
12. LS to L, write back from R.
13. LS to R, regenerate from R.

Table 8.1 ADDER MICRO-ORDERS (*Continued*)

The Micro-Orders Involving the R Register (continued)

14. LS to R, write back from L.
15. Set R sign stat if R (0) equals zero.
16. Turn on the A-branch bit if R position 31 is one.

The Micro-Orders Involving the Q Register

1. Shift right one, spill to F, enter from Q.
2. Shift right one, spill to Q, no enter.
3. Shift right one, spill to Q, enter from Q.
4. Shift right one, spill to Q, enter from F.
5. Shift left one, spill to F, enter from Q.
6. Shift left one, spill complement to F, enter from Q.
7. Shift left one, spill to Q, no enter.
8. Shift left one, no spill, enter from Q.
9. Shift left one, spill to Q, enter from F shifting F.

The Micro-Orders Involving the F Register

1. Shift left one, spill to F, enter from Q.
2. Shift left one, spill complement to F, enter from Q.
3. Shift left one, spill to F, enter from F.
4. Shift right one, spill to F, no enter.
5. Shift left one, spill to Q, enter from F shifting F.
6. Shift left four, spill to F, no enter.
7. Shift left four, spill to F, enter from F.
8. Shift right four, spill to F, no enter.
9. Shift right four, spill to F, enter from F.
10. Shift right four, no spill, enter from F.
11. Shift right one, spill to Q, enter from F.
12. Shift right one, spill to F, enter from Q.
13. Floating-point left shift four, enter from F.
14. Data keys to latch, bits 28–31 to F.
15. W (4–7) to F.

The Micro-Orders Involving the F Register (continued)

16. W to MD counter and F.
17. Read tag memory to F.
18. Write F to tag memory.

The Micro-Orders Involving the M Register

1. M (0–31) to Y.
2. M (16–31) to Y (16–31).
3. Adder output latches to M.
4. Adder output latches to L and M.
5. Adder output latches (T) to M, T (8–11) to storage protect tag.
6. M to V per M-byte counter.
7. M to V per L-byte counter.
8. W to M per M-byte counter.
9. Set the A-branch bit if stat 0 is on and M (0–1) equals 01 (for I-fetch function branch).
10. If ROSDR (16–18) equals 000 and ROSDR (12–15) equals 0110, insert M (0–3) into ROAR (4–7). (This is a function branch.)
11. If ROSDR (16–18) equals 000 and ROSDR (12–15) equals 1000, insert M (4–7) into ROAR (4–7). (This is a function branch.)

Positions 16–31 and 40–63 of the program status word are not included in the PSW register.

The Micro-Orders Involving the PSW Register

1. PSW (32–39) to U.
2. W (2–7) to PSW (34–39).
3. W to PSW (0–7).
4. Adder latches (12–15) to PSW (12–15).
5. Adder latches (0–31) to M; also (8–11) to storage protect tag PSW (8–11).
6. Set the A-branch bit if PSW (15) is 1 (monitor bit).

The actual mnemonics, their ROS field assignments, and binary codes will be presented later. A number of microprogram examples will be given and discussed to help the reader familiarize himself with the practical aspect of microprogramming the IBM System/360, Model 50.

To perform any operation through the adder, we gate the two operands to the XG and Y inputs of the adder, specify the operation, and specify where the result is to be stored. Assume that the two operands reside in registers L and R. Assume also that we want to *add* the contents of these two registers and store the results in M. Can this be done? This question can be answered by consulting one of two sources of information. The first is the data flow diagram in Fig. 8.1 to see if the existing gating and the in-gate and out-gate line positions would permit this operation. The second source is to search the appropriate fields in the ROS decoding charts in the appendix to see if the micro-orders necessary for this gating do exist. For example, suppose we want to transmit the low-order byte of the R register and set it in the low-order byte of the L register. Can this be done? Yes, because we can transmit the contents of the R register to the right input of the adder Y, without specifying any input to XG. Hence, the AOB latches will receive the contents of the R register intact. If we look at the in-gate positions over the L register, we will find a line drop from the AOB bus to the in-gate line position (0–7). This is a sure indication that such a micro-order does exist. On the other hand, if the same problem states that the low-order byte of the R register is to be set into the low-order byte of the H register instead of the L register, it would not be possible, at least not over the adder data path, and not without affecting the whole contents of the H register.

PROBLEMS

1. Consult Fig. 8.1 and Table 8.1 in answering the following questions:
   (a) Given the two registers L and M, is it possible to form a new word consisting of the right halfword of the L [L(16–31)] followed by the right half of the M [M(16–31)], in one cycle?
   (b) Is there one micro-order that will effectively do a rotate shift-left-four positive?
   (c) What micro-order would you use to test for the sign bit of register R?
   (d) List all the micro-orders dealing with the input and output gating to and from the H register.
2. True or false. Information may be gated directly from the:
   (a) H register to the R register.
   (b) L to RY.
   (c) M to RY.
   (d) M (16–31) to XG (0–15).
   (e) AOB (0–7) to the L register (0–7).
   (f) AOB (16–31) to M (16–31).
   (g) R (8–31) to XG (8–31).
   (h) LS to the M register.
3. True or false. The output of the adder can be gated to the:
   (a) Condition-code register (PSW bits 34 and 35).
   (b) J register.
   (c) MD register.
   (d) U input of the mover.
   (e) SDR (storage data register).
   (f) Local store function register.
   (g) Mover function register.

8.5 MOVER DATA PATH

The eight-bit mover data path (Fig. 8.5) is used primarily for the execution of variable-field-length instructions. Two byte sources may be selected simultaneously for a logical operation by the mover.[1]

The left-mover input (U) may be a byte selected from the L register under the control of one of two byte counters LB and MB, a byte formed by the contents of the two 4-bit registers MD and F, or the rightmost byte of the R register. The right-mover input (V) is a byte selected from the M register under the control of either LB or MB.

A three-bit mover function register is provided to control the mover action. The mover function is set by gating the rightmost three bits of the emit field. The specific micro-order used is E(13) → WFN.

Table 8.2 MOVER MICRO-ORDERS

Left Side Inputs

1. R (24–31) to U.
2. L to U per L-byte counter.
3. L to U per M-byte counter.
4. MD and F to U.
5. Interrupt register to U, reset the interrupt register.
6. PSW (32–39) to U.
7. Direct data in lines to U.
8. L (0–7) to U.
9. L (8–15) to U.
10. L (16–23) to U.
11. L (24–31) to U.
12. Multiplex buffer in bus to U.

Right Side Inputs

1. M to V per M-byte counter.
2. M to V per L-byte counter.
3. Multiplexor buffer to V.
4. SDR parity to emit (scan).

Mover Actions

1. U (0–3) to W (0–3).
2. V (0–3) to W (0–3).
3. Emit to W (0–3).
4. Gate to W (0–3) per mover function register (WFN).
5. U (4–7) to W (4–7).
6. V (4–7) to W (4–7).
7. Emit to W (4–7).
8. Gate to W (4–7) per mover function register (WFN).

Actions 4 and 8 specify, among other things, all the logical operations.

Mover Output Gating

1. W to M per M-byte counter.
2. W (6–7) to M-byte counter.
3. W (6–7) to L-byte counter.
4. W (2–7) to PSW (34–39).
5. W to PSW (0–7).
6. W (0–3) to J register.
7. W to channel control.
8. W and emit to SAR (bump address select).
9. W (0–3) to length counter 1.
10. W (4–7) to length counter 2.
11. W to length counters.
12. Modify emit field per ASCH, insert this field into U or V (for zone or sign), gate W to M per M-byte counter.
13. W (0–3) to MD counter.
14. W (4–7) to F register.
15. W to MD counter and F register.
16. W to L (0–7).
17. W to L (8–15).
18. W to L (16–23).
19. W to L (24–31).
20. W and emit to SAR (bump address).
21. W and emit to SAR (bump address, split read/write).

The Micro-Orders Involving the Mover Function Register

1. Gate to W (0–3) per mover function register (WFN).
2. Gate to W (4–7) per mover function register (WFN).
3. Emit (1–3) to mover function register (WFN) per mode trigger.

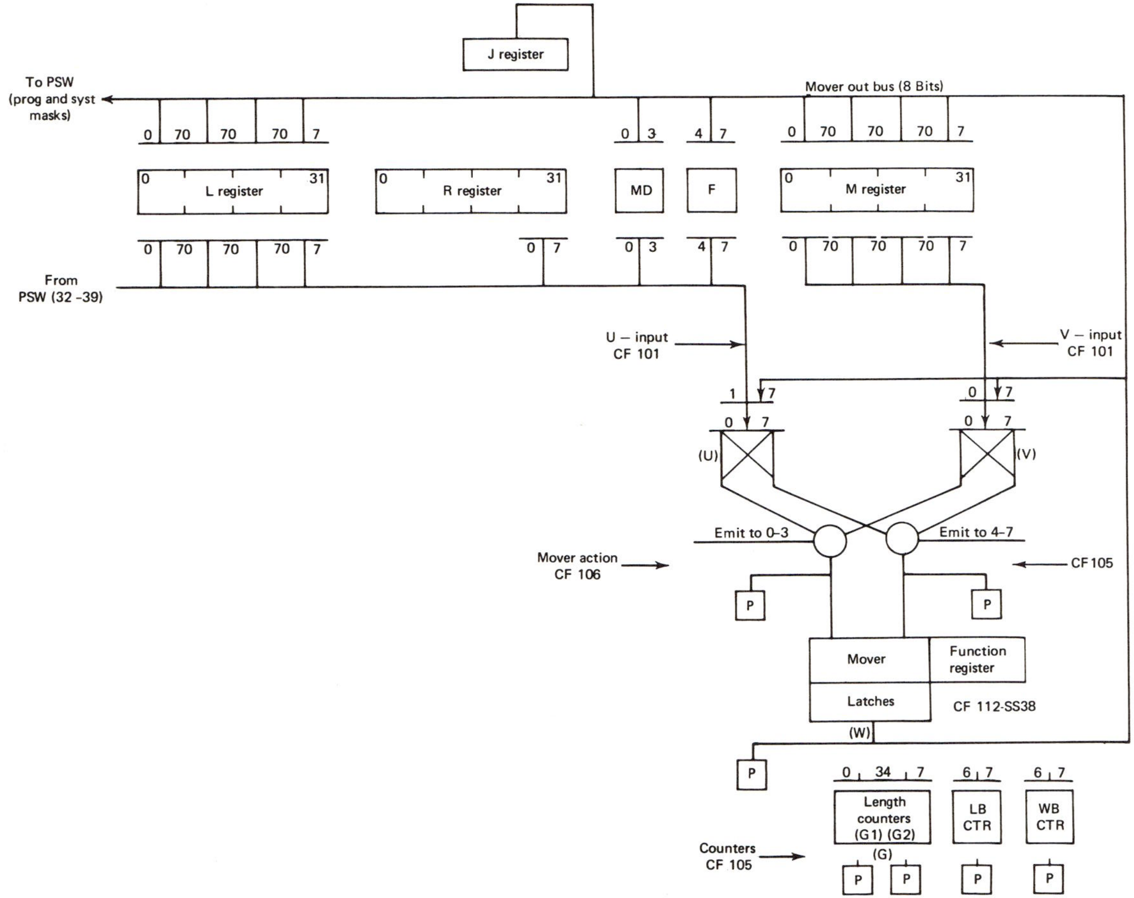

Figure 8.5 Mover data path.

The timing here is identical to the adder path timing. The U and V gates are opened at set-register pulse time (000) of the processing unit cycle, and the bytes enter the mover to be acted upon. At latch pulse time (375), the new byte is set into the mover output latches (W), where it is held until it is gated into one or more registers at the register set pulse of the following cycle. If no data are gated into the mover, then the contents of the mover latches by the end of that cycle will be zeros. Mover micro-orders are listed in Table 8.2.[2]

Another mover input is the ROS emit field. The four-bit emit field may be inserted into either the right or left digit of the mover output.

With two 1-byte inputs gated into the U and V, the mover can perform one of many operations such as:

1. Logically AND, OR, or EXCLUSIVE OR the two inputs together to form a new byte.
2. Select any two half-bytes available at inputs U and V to form a new byte. The move-numeric and move-zone instructions are executed in this manner.
3. The digits of bytes gated to the left mover input U (0–7) may be cross-gated. This will produce a new byte U (4–7, 0–3). Note that cross-gating is not available at the right mover input V.
4. Select any half-byte from either input, and merge it with the half-byte specified in the emit field to form a new byte.

Additional mover functions are valid digit, sign detection, and zero detection.

The mover output latch byte (W) may be gated into a byte of the M register under the control of the MB counter. The mover output also sets the length counters G1 and G2, the MD counter, the F register, the J register, and the byte counters LB and MB.

8.6 INSTRUCTION ADDRESS DATA PATH

The basic instruction address bus is 24 bits wide (Fig. 8.6). Thus it can contain the address of any one of a maximum of 512K-byte main storage available on the Model 50.

The length of an instruction format can be one, two, or three halfwords; the main storage data bus is four bytes wide. Hence, in one memory cycle, one can fetch four bytes. The four bytes may contain two halfword RR instructions; one RR instruction followed by a halfword belonging to the next instruction; a halfword belonging to a previous instruction followed by an RR instruction; a fullword RX, RS, or SI instruction; or the first two halves of a three-halfword SS instruction. Therefore, the address counter latches are updated by either two or four. The instruction address is gated out of the instruction address register (IAR) and into the instruction address counter where it is updated. At the beginning of the next processing unit instruction

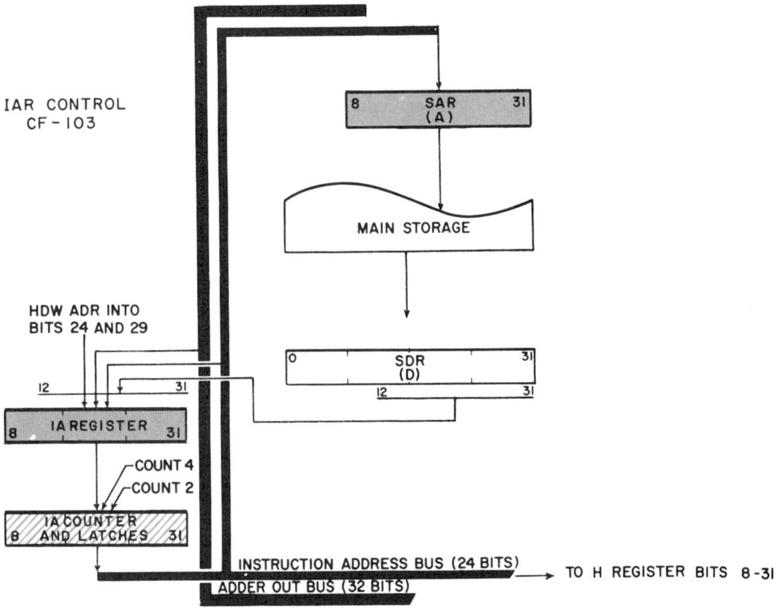

Figure 8.6 Instruction address register data path.

cycle, the updated address is gated from the IAR counter latches to both the storage address registers (SAR) and back to the instruction address registers.[1]

When the processing unit executes an instruction out of sequence, such as when a branch or an interruption occurs, then the instruction address register is set from the adder-out bus. Also, when an interrupt occurs, a path from the IA counter to the H register allows the existing instruction address to be stored for later use. A data path from bits 12–31 of the SDR to the IAR is available for checking system operation with fault location tests.

8.7 MAIN STORAGE IN-OUT DATA PATH

The Model 50 has a maximum 512K-byte main storage that operates on a 2.0 μs cycle (Fig. 8.7). This is four times the CPU cycle time (500 ns). The storage is addressed from a 24-bit storage address register. Data are transferred to and from storage via the storage data register (SDR).[1]

Each time storage is cycled, it both reads out to the storage data register and writes back into the same address from SDR. The only decision, then, that must be made between a storage read or a storage write cycle is whether to change the data in SDR between the read half or the write half of the storage cycle.

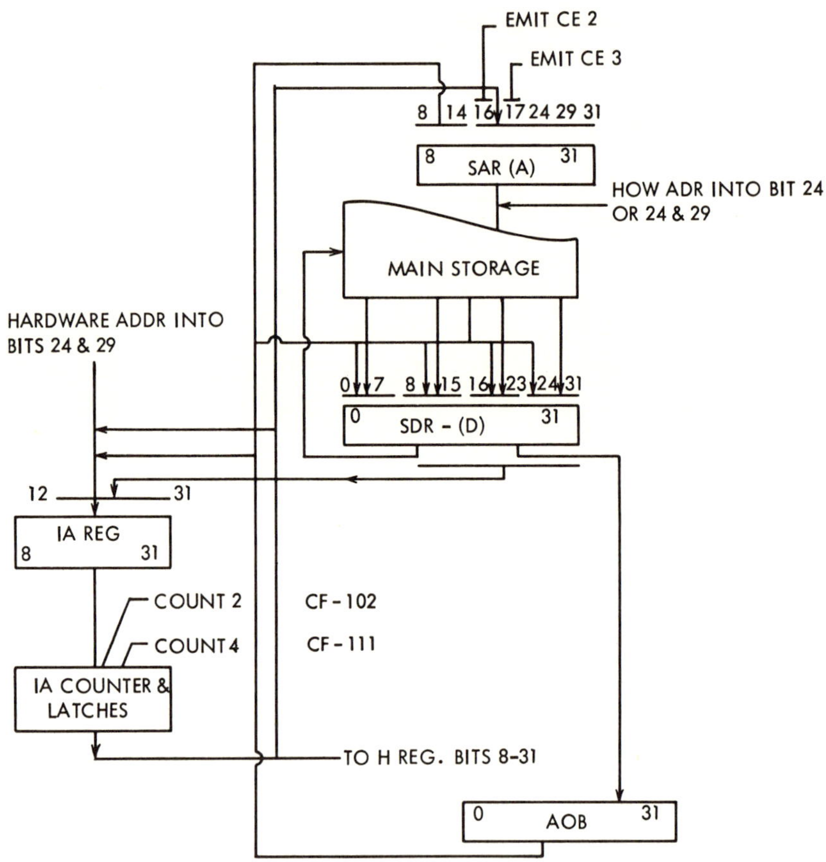

Figure 8.7 Main storage in-out data path. *The micro-orders involved with the storage address register are:*

1. Adder output latches to SAR and R.
2. Adder output latches to R and SAR, without invalid address trap.
3. Adder output latches to R and SAR, with word-address special check.
4. Adder output latches to R and SAR, with double-word special check.
5. Adder output latches (8–31) to SAR.
6. Adder output latches to SAR and L.
7. Increment IAR by four, and gate into IAR and SAR.
8. Gate W (0–7), and emit field to SAR (bump address).

To initiate storage request, an address is set into the storage address register (SAR) at the CPU-register set-pulse time. Nearly two CPU cycles later (1 μs), the data addressed from storage are set into the SDR, where they can be gated into AOB latches before the third CPU cycle.

Data are moved into storage by gating the CPU data in the AOB latches into SDR at set-register pulse time of the second CPU cycle. Note that data from the AOB latches are gated into the SDR by four separate in-gates which are specified by the B3 counter.

Addresses for main storage are normally set into the SAR from either the instruction address bus or the adder-out bus, depending on whether we are fetching the next instruction or operands. In addition, the mover-out bus and the emit field supply bits to the SAR when a multiplex channel-bump storage word is required.

We noted earlier that a storage cycle is initiated by a micro-order which gates an address to the storage address register. The following 2-μs main storage cycle can be divided into the read half (1.0 μs), followed by a write half (also 1.0 μs; see Fig. 8.8). R1 and R2 are the read cycles, followed by

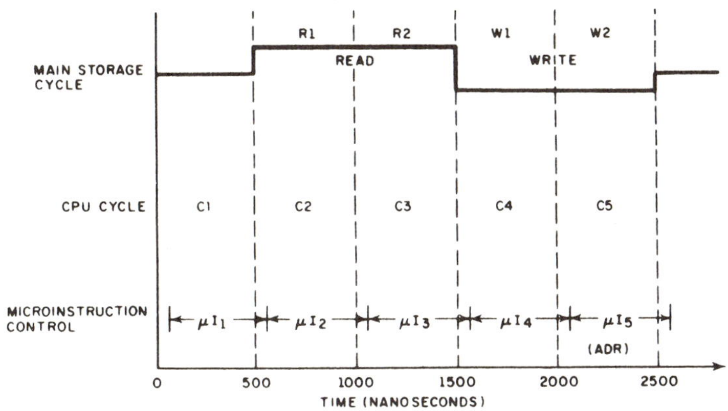

Figure 8.8 Model 50 CPU and main storage cycles.

W1 and W2, the write (or regeneration) cycles. W2 will be followed by R1 if another storage has been initiated during the last half of the regeneration, or by W2 until a memory request is made.

If a storage request is made while the memory clock is not in the W2 cycle, a storage holdoff occurs automatically and persists until the clock reaches the W2 cycle. At that time, this new main-storage service request is dispatched. Also, if a transfer of the SDR content is requested during the R1 cycle, a storage read holdoff is made for one cycle. Data from the adder output latches can be gated into the SDR in an R2 cycle. The SDR cannot be referenced after the R2 cycle. Note that data from the AOB latches are gated into SDR by four separate in-gates. This effectively allows data to be stored on a byte basis (per byte stat) even though main storage handles data by word.

8.8 LOCAL STORAGE DATA PATH

The Model 50 has a local storage of 64 words. Each word is made up of 32 data bits plus 4 parity bits and is addressed by a 6-bit address register,

LSAR. The local storage completes a conventional read/write cycle in 0.5 μs cycle time—the same as a basic CPU cycle.

The local storage is divided into four quadrants. The first quadrant is used for channel controls. The second contains the 16 working storage registers that are used as a scratch pad. The third quadrant contains the four 64-bit floating-point registers that are part of the System/360 architecture, and it contains some spares and channel buffer registers. The fourth quadrant contains the 16 general purpose registers which are also part of the System/360 architecture (see Table 8.3).[3]

Figure 8.9 shows the basic path data take when local storage supplies a word to the processing unit. Local storage has essentially two buffer registers, the L and R registers.

When requested, local storage starts its cycle halfway through a processing unit cycle, so that the read portion is completed by the register set pulse. The

Table 8.3 LOCAL STORAGE ASSIGNMENTS

0	Channel 1—command adder	32	Floating-point register 0—high order
1	Channel 1—data adder	33	Floating-point register 0—low order
2	Channel 1—unit adder count	34	Floating-point register 2—high order
3	Channel 1—data buffer A	35	Floating-point register 2—low order
4	Channel 2—command adder	36	Floating-point register 4—high order
5	Channel 2—data adder	37	Floating-point register 4—low order
6	Channel 2—unit adder count	38	Floating-point register 6—high order
7	Channel 2—data buffer A	39	Floating-point register 6—low order
8	Channel 3—command adder	40	Spare
9	Channel 3—data adder	41	Spare
10	Channel 3—unit adder count	42	Spare
11	Channel 3—data buffer A	43	Spare
12	MPX channel—command adder	44	R register break-in buffer
13	MPX channel—data adder	45	MPX channel L-register buffer
14	MPX channel—count	46	MPX channel interrupt buffer
15	MPX channel—unit adder	47	MPX channel working storage
16	Working storage 0	48	General register 0
17	Working storage 1	49	General register 1
18	Working storage 2	50	General register 2
19	Working storage 3	51	General register 3
20	Working storage 4	52	General register 4
21	Working storage 5	53	General register 5
22	Working storage 6	54	General register 6
23	PSW backup (bits 0–15)	55	General register 7
24	Working storage 8	56	General register 8
25	Working storage 9	57	General register 9
26	Working storage 10	58	General register 10
27	Working storage 11	59	General register 11
28	Working storage 12	60	General register 12
29	Working storage 13	61	General register 13
30	Instruction buffer 14	62	General register 14
31	Working storage 15	63	General register 15

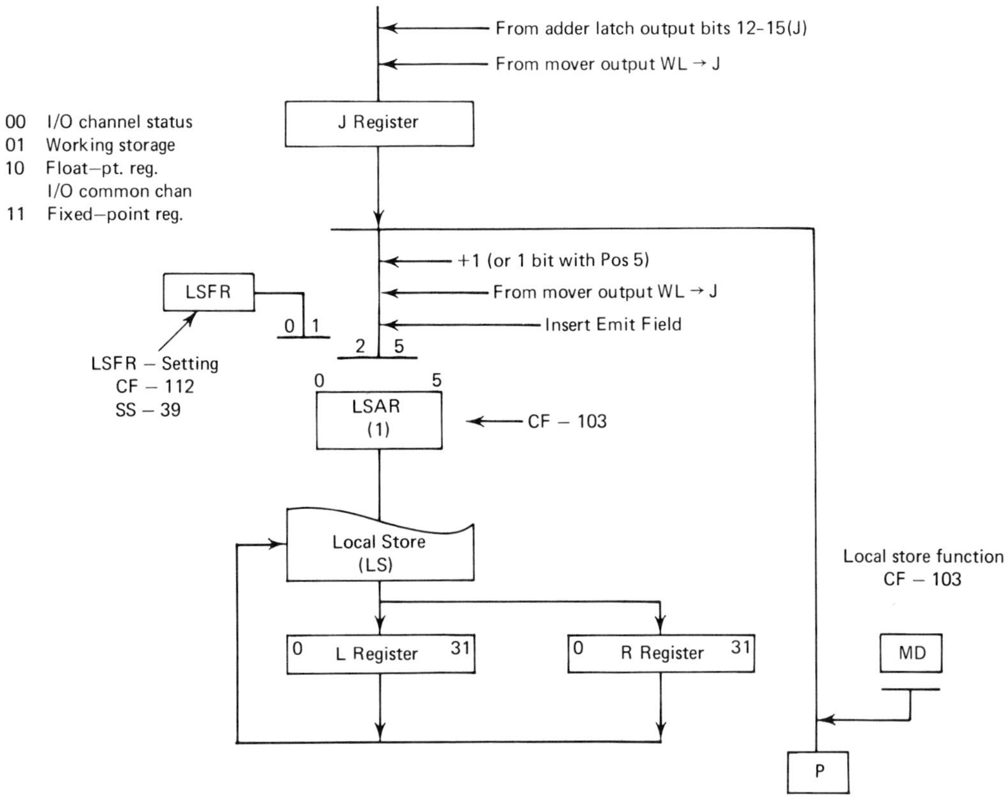

Figure 8.9 Local storage data path.

data word gates into either the L or R register, where it is now available to the processing unit.

The data word to be written may be obtained from either the L or R register, or from latches on the local-store sense amplifiers. The selection of the source is controlled by the microprogram. This makes it possible to read out one word during the first half of the cycle and to store a different word during the second half of the same cycle.

Addresses for the local storage address register (LSAR) are 6 bits wide. The two high-order bits determine which quadrant of the local store is to be used. (See Table 8.4.) The four remaining bits specify which of the 16 words within the quadrant are selected.

The two high-order bits are set into the local storage function register (LSFR) from the emit field before they gate into LSAR. The four remaining bits come from either the emit field, the mover output bus (0–3), the adder output bus (12–16) via the J register, or the MD register. Notice the provision for ORing one bit with a unit bit of the LSAR. This provides an easy way to automatically address the second word of a floating-point register.

Table 8.4 LOCAL STORAGE QUADRANT SELECTION

LSAR (0, 1)	*Quadrant Selected*
0 0	I/O channel status
0 1	Working storage
1 0	Floating-point registers plus I/O common channels
1 1	Fixed-point registers

8.9 COUNTERS

Four registers are used as general counters. These are the two 4-bit length counters G1 and G2, and the L- and M-byte counters. These four counters can be set directly from the mover output (Table 8.5).

The G1 and G2 counters may be used as two separate counters and can be incremented or decremented simultaneously or separately; or they can be coupled together to form a one 8-bit counter, G. When G is set, its sign is reset to zero. When the counter counts below zero, the sign bit is turned on.

The 2-bit M-byte (MB) counter and the L-byte (LB) counter are used to

Table 8.5 COUNTER MICRO-ORDERS

The Micro-Orders Involving the Length Counters

1. Decrement length counter 1.
2. Decrement length counter 2.
3. Decrement length counters coupled.
4. Decrement both length counters.
5. Decrement length counters, hot carry.
6. W to length counters.
7. W (0–3) to length counter 1.
8. W (4–7) to length counter 2.

The Micro-Orders Involving the L- and the M-Byte Address Counters

1. L register to mover U input per L-byte counter.
2. L register to mover U input per M-byte counter.
3. M register to mover V input per L-byte counter.
4. M register to mover V input per M-byte counter.
5. Mover output bus W (6–7) to L-byte counter.
6. Mover output bus W (6–7) to M-byte counter.
7. W to M per M-byte counter.
8. Select L-byte counter.

The Micro-Orders Involving the L- and the M-Byte Address Counters (continued)

9. Select M-byte counter.
10. Set selected counter(s) to zero.
11. Set selected counter(s) to three.
12. Count up.
13. Count down.

The Micro-Orders Involving the MD Counter

1. Adder output latches (0–31) to L, (0–15) to M right, (0–3) to MD.
2. Adder output latches (8–11) to MD.
3. MD and F to U.
4. W (0–3) to MD counter.
5. W to MD counter and F register.
6. I-fetch gate, adder output latches (0–31) to L and M, (12–15) to J, (16–19) to MD.
7. Set the A-branch bit if MD is even.
8. Set the A-branch bit if MD is the floating-point register.
9. Local-storage function register and MD to the local-storage address register.
10. Local-storage function register and MD plus 1 to the local-storage address register.

specify the bytes being moved from and into the L and M registers. The 4-bit MD register could also be used as a counter.

8.10 STATS

Stats (or indicators) are 1-bit storage elements set to reflect the status of the machine. Eight general purpose stats are available to the microprogrammer as an aid in remembering conditions from one CPU cycle to the next.

Seventeen micro-orders can affect the setting of one or more general purpose stats (Table 8.6). These stats are interrogated by other microinstructions which will result in two-way or multiway branches.

The *refetch stat* indicates whether to go to the main storage for the next I-fetch or use the leftover halfword stored in the top of the *Op buffer* located in local store 011110 (working register 14). If this leftover halfword can be used, 1.5 μs can be saved. This is the difference between the 2.0-μs main storage

Table 8.6 STAT CONTROL MICRO-ORDERS

The Micro-Orders Involving the General Purpose Stats

1. Emit to stats 0–3.
2. Stats 0–3 on per emit.
3. Stats 0–3 off per emit.
4. Stats 0–3 on, reset byte stats.
5. Set stats 0–3 and the L sign stat on.
6. Set stat 0 if the index field is 0, stat 1 if the B field is 0, and the one-syllable OP stat if T (16–17) is 0.
7. Set stat 1 if T (0–3) is 0, and set the one-syllable OP stat if T (16–17) is 0.
8. Set stat 3 if T (0–31) is 0.
9. Set byte stats per emit and stat 3 to the value in T (30).
10. Decimal add and correct, save carry 0 in S1.
11. Decimal add and correct, save carry per byte stats in S1.
12. Decimal double correct, carry S1.
13. Emit to stats 4–7.
14. Stats 4–7 on per emit.
15. Stats 4–7 off per emit.
16. Set stats 4–7 per floating-point rules, set exponent difference register.
17. Set stats 4–7 per console manual controls.

In addition to the micro-orders listed above, micro-orders exist to turn on the A-branch bit or to turn on the B-branch bit if a specified general purpose stat is on.

The Micro-Orders Involving the One-Syllable OP Stat

1. Set stat 0 if the index field is 0, set stat 1 if the B field is 0, and set the one-syllable OP stat.
2. Set stat 1 if the B field is 0, and set the one-syllable OP stat.

The Micro-Orders Involving the Byte-Store Stats

1. Set stats 0–3 on, reset the byte-store stats.
2. Emit to the byte stats.
3. Set byte stats from emit field, and set stat 3 to T (30).
4. Latch to MDR, write per CPU byte stats.
5. Turn on the byte stat specified by the M-byte counter.
6. Turn on the A-branch bit if the AOB bytes specified by the byte stats are 0.
7. Turn on the B-branch bit if the AOB bytes specified by the byte store stats are 0.
8. Decimal add and correct, save the carry from the byte specified by the byte store stats in stat 1.

Micro-Orders Involving the Carry Stat

1. Binary add, save carry zero.
2. Binary add, save carry zero, EOR carry 1.
3. Binary add, carry 8, pass 8.

cycle time and the 0.5 μs required to fetch the same information from the Op buffer in local storage.

The *one-syllable OP stat* is used to speed up CPU operations by eliminating main storage cycles not needed to fetch new instructions.

Each time an instruction is moved from the adder output latches to local storage, the first two bits of the second halfword (positions 16 and 17) are checked for zeros. If they contain zeros, the one-syllable OP stat is set to tell the CPU that the right halfword in the local storage OP buffer *may* be an entire instruction of an RR format. If this right halfword is on a halfword boundary, the next I-fetch does not require a main storage access.

The *carry stat* is used to help the microprogrammer remember some carry indications from one CPU cycle to the next. Four *byte-store stats* (BS) determine which of the SDR in-gates are to be used in a given store cycle, and when a given SDR byte is to be reset.

8.11 BASIC SYSTEM TIMING[1]

A four MHz (megahertz) crystal-controlled oscillator (Fig. 8.10) with a frequency divider provides a basic source of timed pulses to be used in the entire CPU. A clock cycle is 500 ns in duration. The actual clock for the processing unit consists of only four basic pulses (see Fig. 8.11):

1. The register set pulse.
2. The late-register set pulse.
3. The set error-register pulse.
4. The latch pulse.

These pulses are described in the following paragraphs:

1. *Register Set Pulse.* Figure 8.11 shows that the register set pulse comes up at the beginning of the 500-ns CPU cycle and stays up for 100 ns. During register set-pulse time, data will be gated into the selected registers such as the L, M, R, or H registers. The data may be coming from the adder output latches or from local storage. Read-only storage data register out-gates are also conditioned at this time.

2. *Late-Register Set Pulse.* Because data processed in one processing unit cycle must be gated into a register at the beginning of the next cycle, a means is provided to set control information into a register some time after the beginning of a processing unit cycle. The late-register set pulse is available at about 125 time of every 500-ns processing unit cycle. Control information set into a register at late-register set-pulse time will be available into a register at late-register set-pulse time of the following cycle. At zero time of the following cycle, then, data in a register are not in the process of being changed, but are still available to be gated elsewhere at register set-pulse time.

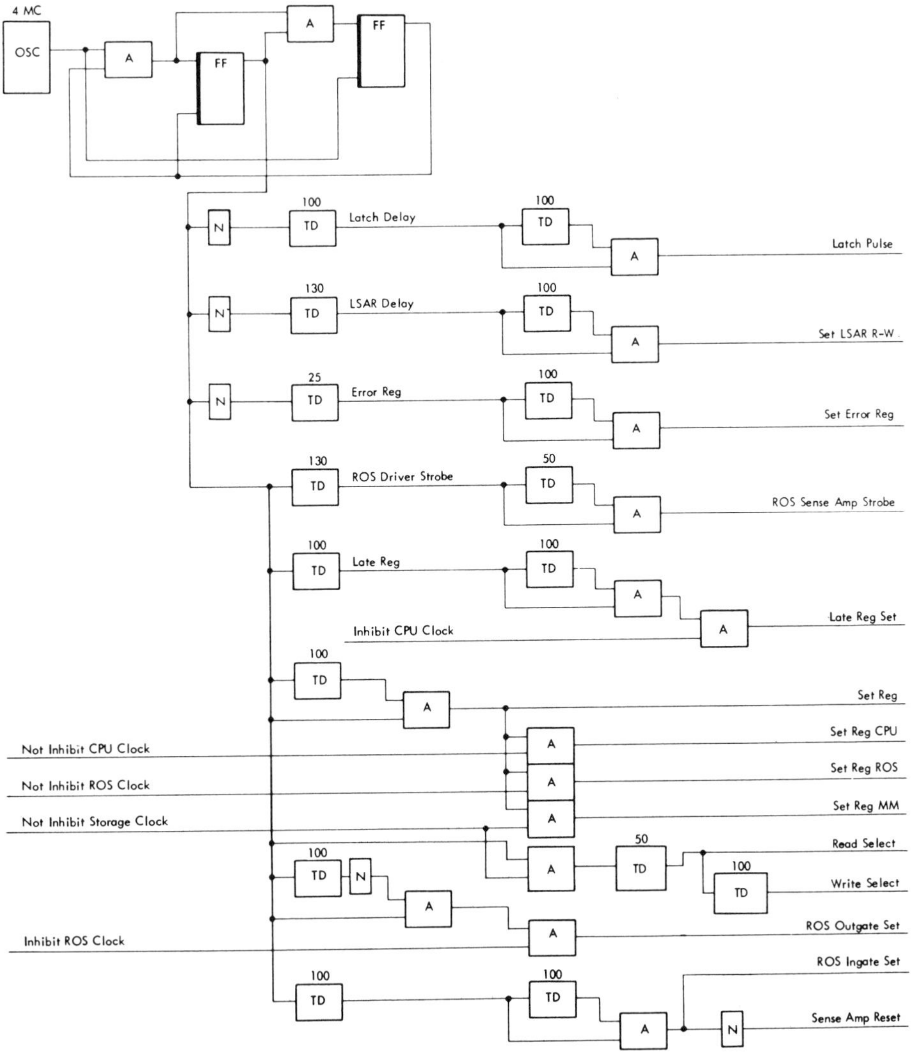

Figure 8.10 Model 50 clock.

3. *Set Error-Register Pulse.* The set error-register pulse is available to sample all possible error conditions as early as possible in the processing unit cycle, while still allowing the machine to stop on an error at the end of the same cycle. Errors as late as the adder half-sum are sampled and will stop the machine at the end of the cycle.

4. *Latch Pulse.* The latch pulse is available at about 375 time to set adder output into the adder output latches (T). Any data placed in the adder output latches are available until about 375 time of the following cycle. The register

set pulse at zero time of the following cycle can be used to gate data from the adder output latches into a working register.[2]

8.11.1 Local Storage Clocking

Local storage cycles occur at the same 500-ns rate as CPU. However, the local storage cycle is skewed from the CPU cycle by about 250 ns (Fig. 8.11). The reason for the skew is to allow local storage to:

1. Have a word read out and available to CPU at the beginning of the next cycle.
2. Be able to store a word just generated by the previous CPU cycle.

Figure 8.11 shows the basic local storage clocking relative to the CPU clock. Note that the start memory pulse comes halfway through the CPU cycle

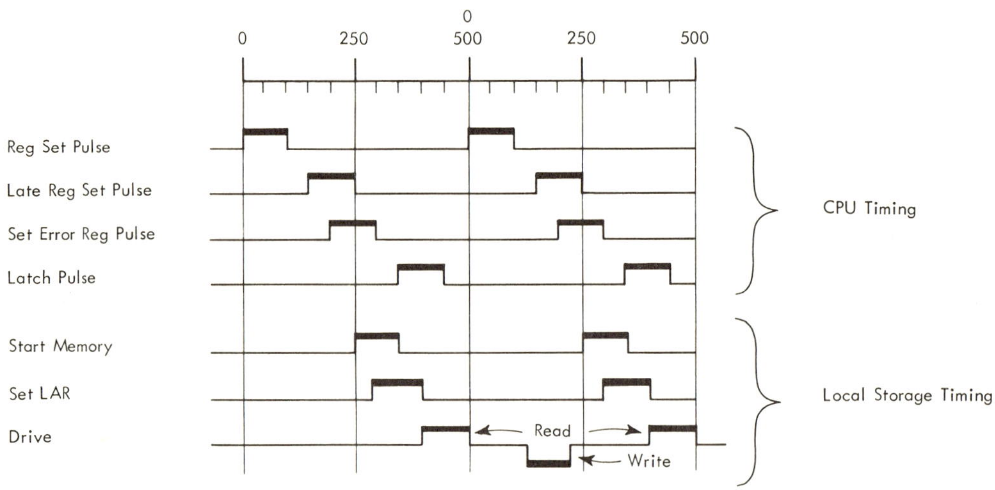

Figure 8.11 Relationship of local storage to CPU cycles.

(at 250 time). Next, the set LAR pulse gates the local storage address into the local storage address register (LSAR). Read drivers pulse the array, and at about 420 time, the sense amplifiers are strobed to set the sense amplifier latches which hold the data until they can be gated into either the L or R register at 500 time (set-register pulse time).

The data in either the L or the R register may be written back into local storage (depending on the micro-order). Write drive lines for local storage come up at 120 time.

8.11.2 Main Storage Timing

The basic timing consideration for main storage and its relationship to CPU timing was presented in the Main Storage Data Path section of this chapter.[1] To summarize, we must remember that:

1. The main storage cycle is 2.0 μs—equivalent to four CPU cycles.
2. A four-step ring (R1, R2, W1, and W2), called the main storage clock, keeps CPU posted on the status of the main storage cycle.
3. To initiate storage request, an address is set into storage address register SAR at the CPU register set-pulse time. Nearly two CPU cycles later, the data addressed from storage are set in the storage data register (SDR) where they can be gated to the AOB latches before the third CPU cycle.
4. Data are moved into storage by gating the CPU data in the AOB latches into SDR at set-register pulse time of the third CPU cycle or before.
5. Figure 8.12 shows the main storage–CPU relationship. At the start of the main storage cycle, the main storage clock is set to R1, and the set-SAR

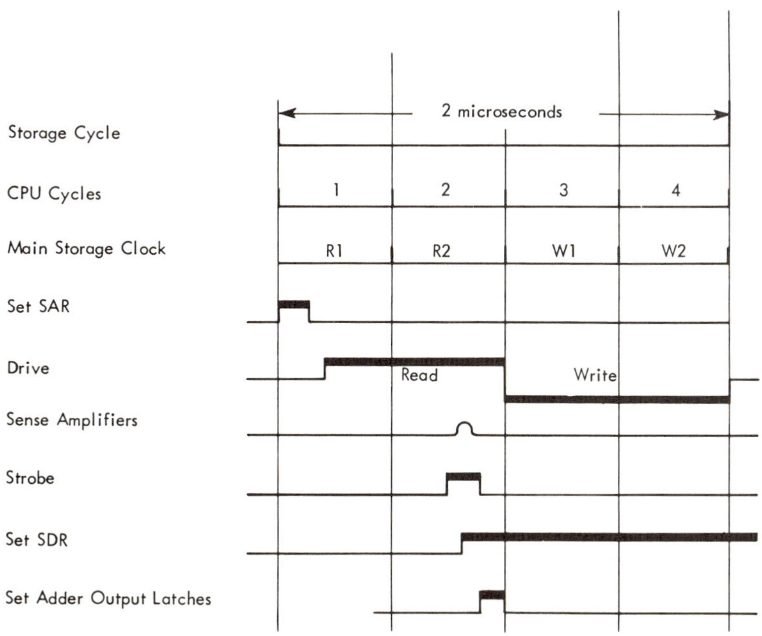

Figure 8.12 Main-storage—CPU relationship.

pulse comes up at set-register pulse time. Drivers pulse the array in the read direction through R2 time. The sense amplifiers produce their outputs slightly more than halfway through R2 time.

6. If data are being read out of storage, the sense amplifier outputs are strobed into the SDR beginning about 300 ns into R2 time. By 385 ns of R2, all bits from the storage location are in the SDR, and the adder output latches can be set to SDR, thus providing the CPU with a word from storage at set-register pulse time W3.

7. If data are being written into storage, the sense amplifier outputs are not strobed into the SDR. Instead, the contents of the adder output latches are gated into the SDR at set-register pulse time R2 or W1.

8. The W2 cycle will be followed by R1 if another storage has been initiated during the last half of the regeneration (W2), or it will be followed by another W2 until a memory request is made.

9. If a storage request is made while the memory clock is not in the W2 cycle, a storage holdoff occurs automatically until the clock reaches the W2 cycle. At that time, this new main storage request will be serviced.

If a transfer of the SDR content is requested during the R1 cycle, a storage-read holdoff is made for one cycle. Data from the adder output latches AOB can be gated into the SDR in an R1 cycle. The SDR cannot be referenced after the R2 cycle.

8.11.3 Read-Only Storage Cycle

An ROS word controls the action of the processor for one machine cycle (Fig. 8.13). Hence, the read-only storage must begin cycling during the previous CPU cycle. At set-register pulse time of this CPU cycle, most of the next read-only storage address is set into the read-only storage address register, ROAR. At about 350 time, a timing pulse, called *driver strobe*, ANDs with the ROS addressing logic to produce *array drive*, which pulses a specific ROS drive line to produce 180 ROS bits (two ROS words).

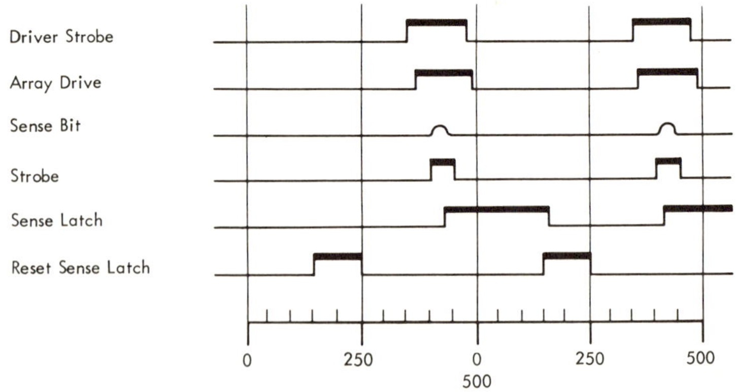

Figure 8.13 Read-only storage timing.

The specific word chosen is defined by a strobe pulse, which is a clock pulse gated with a B-branch bit (the twelfth bit of the ROS address). This strobe pulse selects and gates the outputs of the appropriate 90 sense amplifiers into the sense latches. The reason for delaying the final selection of the ROS word until so late in the cycle (about 400 time) is to allow the result of this previous CPU cycle to determine what the next cycle will do. For the current CPU cycle, then, the ROS word is available in the ROS sense latches where it is gated at set-register pulse time into the read-only storage data register, ROSDR. Once in ROSDR, the ROS word can be decoded for use through the current CPU cycle.

Some ROSDR positions have a second latch which is set according to the content of the first latch at late-register set pulse time. This is done because certain control information is needed to carry an operation through the next set-register pulse time.

8.12 CROS CONTROL WORD

Capacitor read-only storage (CROS) was used to contain the IBM System/360, Model 50 microprogram. CROS has a storage capacity of 2816 words of 90 bits each. Each 90-bit word controls the gates and the sequencing of the CROS memory itself for one machine cycle. The gating for each functional unit (such as the left or right input to the adder or the mover's left or right input to the shifter) is controlled by the bit combination in an encoded control field. The mnemonics that designate each bit combination in each sub-segment in the control word are called micro-orders. The set of micro-orders that designates the action of the machine for one machine cycle is called a microinstruction. The set of microinstructions that defines a sequence of operations to perform a given mathematical or logical function is called a microprogram.

A 90-bit microprogram word is read out of ROS and set into the read-only storage data register every machine cycle. The address of the next microprogram word in CROS is composed partly of the ROS addressing field in ROSDR and by the branch logic activated by the test and branch microcommands.

Note that machine cycles are in either CPU mode or I/O mode. Both modes are under CROS control. The I/O-mode cycles are machine cycles which use the processing unit circuits to transfer data to or from an I/O channel. The CPU-mode cycles are cycles in which the I/O channels do not use the processing unit circuits and data paths.

The CROS-word decode logic combines specific bits from the ROSDR fields with the CPU-mode and I/O-mode triggers to activate control lines. The I/O mode differs from the CPU mode with regard to a number of fields. Figures 8.14 and 8.15 show the difference in the microprogram word fields for the

Field	ROS Bits	Function of Field
	0	Parity of bits 0-30
LU	1-3	Left input to mover
MV	4-5	Right input to mover
ZP	6-11	Bits 0-5 of next ROS address
ZF	12-15	Source of bits 6-9 of the next ROS address
ZN	16-18	ROS addressing mode
TR	19-23	Destination of adder latch contents
	24	Spare
WS	25-27	Source of local storage address
SF	28-30	Local storage function
	31	Parity of bits 32-55
IV	23-24	Invalid digit test and instruction address register control
AL	35-39	Shift control and gating into adder latch
WM	40-43	Mover destination
UP	44-45	Byte counter function
MD	46	MD counter control
LB	47	LB counter control
MB	48	MB counter control
DG	49-51	Length counter and carry insertion control
UL	52-53	Mover function -- left digit
UR	54-55	Mover function -- right digit
	56	Parity of bits 57-89
CE	57-60	Emit field (used as data)
LX	61-63	Left input to adder
TC	64	True/complement control of left adder input
RY	65-67	Right input to adder
AD	68-71	Adder function
AB	72-77	Condition branch test A (furnishes bit 10 of next ROS address)
BB	78-82	Condition branch test B (furnishes bit 11 of next ROS address)
	83	Spare
SS	84-89	Stat setting and miscellaneous control

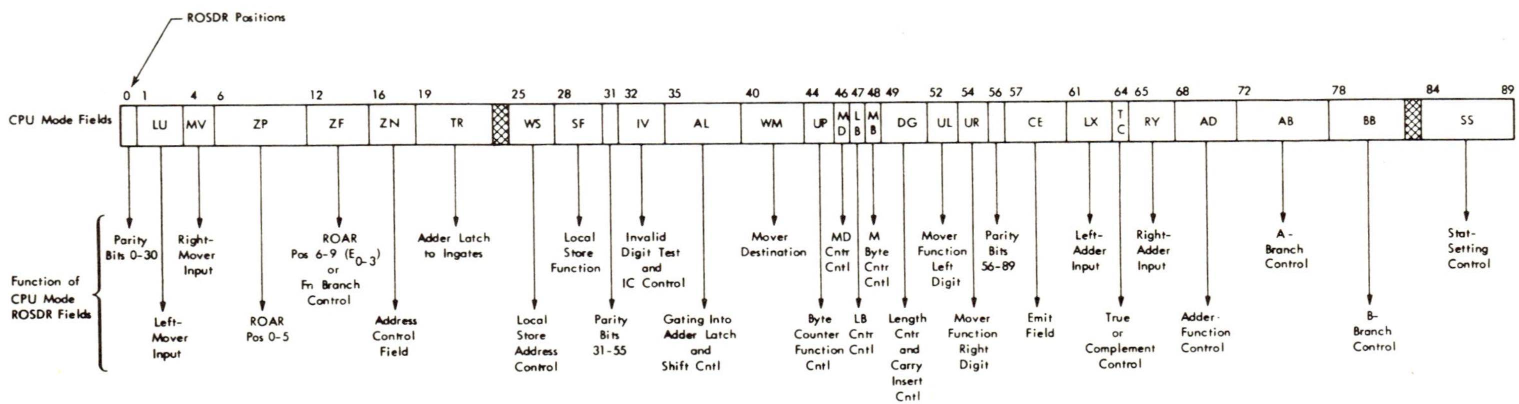

Figure 8.14 Microinstruction format—CPU mode.

Field	ROS Bits	Function of Field
	0	Parity of bits 1-30
LU	1-3	Left input to mover
MV	4-5	Right input to mover
ZP	6-11	Bits 0-5 of next ROS address
ZF	12-15	Source of bits 6-9 of next ROS address
ZN	16-18	ROS addressing mode
TR	19-23	Destination of adder latch contents
	24	Spare
CS	25	Local storage address sector
SA	26-27	Local storage address
SF	28-30	Local storage function
	31	Parity of bits 32-55
CT	32-34	Timing signals to channel
AL	35-39	Shift control and gating into adder latch
WL	40-42	Mover destination
HC	43	Carry insertion control
MS	44-46	Multiplexor channel stat setting

Field	ROS Bits	Function of Field
CG	47-48	Control signals to channel
MG	49-51	Multiplexor channel gate control
UL	52-53	Mover function -- left digit
UR	54-55	Mover function -- right digit
	56	Parity of bits 57-89
CE	57-60	Emit field (used as data)
LX	61-63	Left input to adder
TC	64	True/complement control of left adder input
RY	65-67	Right input to adder
CL	68-70	Selector channel adder latch tests
	71	Spare
AB	72-77	Condition branch test A (furnishes bit 10 of next ROS address)
BB	78-82	Condition branch test B (furnishes bit 11 of next ROS address)
	83	Spare
SS	84-89	Stat setting and miscellaneous control

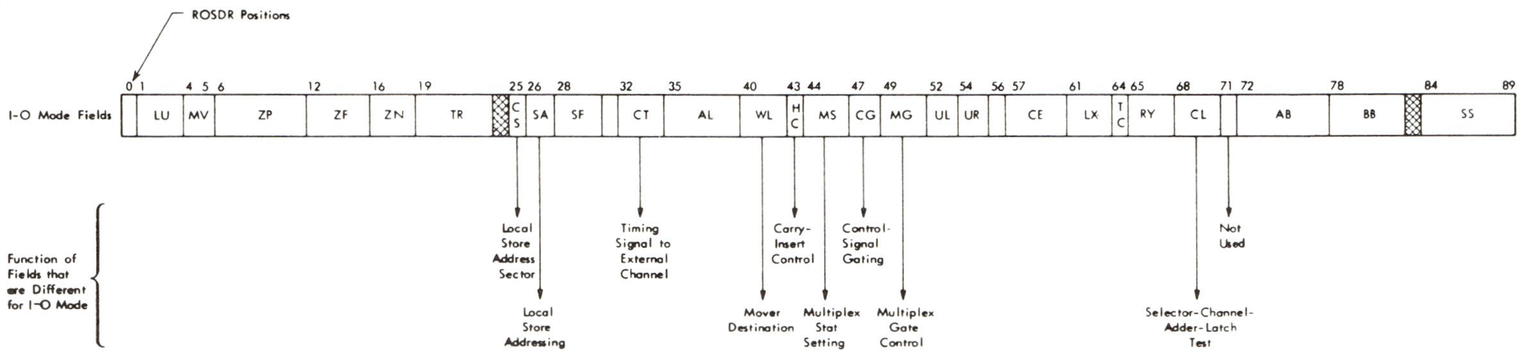

Figure 8.15 Microinstruction format—I/O mode.

CPU and I/O modes. Some fields are completely redefined. In other fields, certain code points are defined differently.

Note that most fields in the I/O mode are identical to those in the CPU mode. The exceptions are those that deal with the timing signal to multiplexor and selector channel controls.

There are 88 one-bit out-gate registers in ROSDR (CROS-word positions 24 and 83 are spare positions for future expansion). There are also 20 in-gate register positions: ROSDR positions 19–23, 28–30, and 40–51. The in-gate registers allow a CROS word to control some gates during the register set time of the next cycle. For example, one CROS word may contain the following microinstruction: Add L and R registers, and store the answer in the L register. The CROS word adds the registers in one machine cycle and stores the sum from the adder-out bus at register set time of the next cycle. The in-gate registers are reset from the out-gate registers at 130-ns time of every machine cycle.

The out-gate registers are set in four groups:

1. *Parity Bits*. Positions 0, 31, and 56 are parity bits for positions 0–30, 31–55, and 56–89, respectively.

ROSDR (0–30) has a special circuit that forces correct parity on any cycle that inserts a fixed address into ROAR. This circuit is necessary because the ROSDR (6–18) is prevented from storing the corresponding ROS latches when the insert fixed-address time is active.

2. *ROS Address Field*. Positions 6–18 control the address of the next CROS word. If the next microinstruction is not a conditional branch instruction, the present ROS word could contain the complete address of the next word. However, it is very desirable to have the capability of branching to one of many microinstructions based on the setting of machine status latches.

The twelve bits of the next microinstruction address are specified by a number of fields. The ZP field specifies the high-order six bits of the next instruction. The ZF in most cases specifies the next four bits (called low-order base) of the next address.[3] The final two bits of the address are the A and B bits (called branch control segment). They are specified by the branch control field, either unconditionally or as functions of the stats of the machine. (See Fig. 8.16.) Although fields A and B are generally specified independently, it is possible to force one bit as a function of the value of the other bit. It is also possible to create a 64-way branch by gating data from the data flow into ROSAR (6–9). This is called functional branch (FNB).

On special conditions, such as detecting an invalid address, the address of a fixed microroutine is inserted into ROAR; an insert-fixed add line prevents the setting of ROSDR out-gate positions.

To illustrate the B-bit branch, assume that a microinstruction (ROS word) is to be followed by one of two microinstructions depending on the sign of the content of the adder output latches (AOB). If the AOB bit 0 is zero, set the

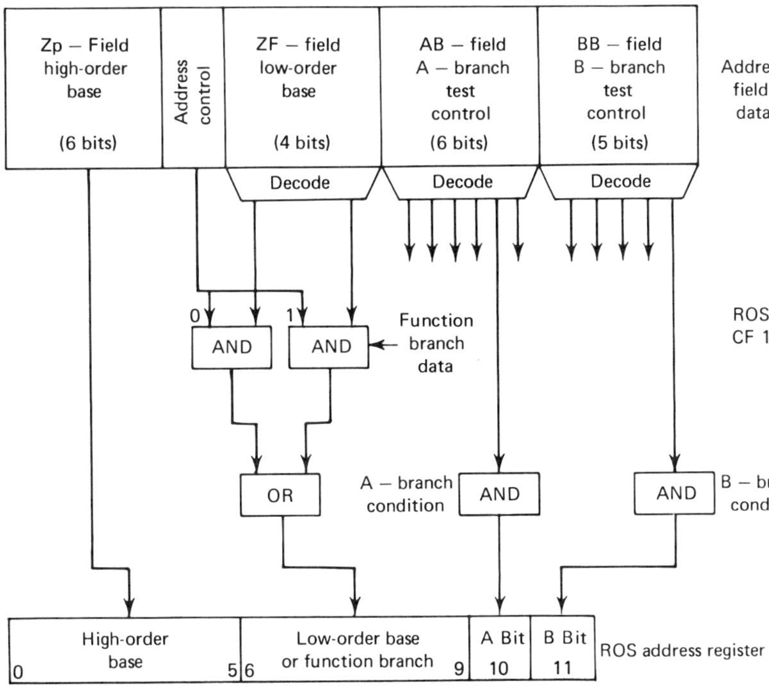

Figure 8.16 Formation of the address of the next ROS word.[3]

condition code (bits 34 and 35 of the program status word) to 00; otherwise, set the condition code bits to 11 (see Fig. 8.17). Another example of function branching is found in the instruction fetch and decode microprogram. All System/360 instruction formats have an 8-bit Op Code field and, thus, a maxi-

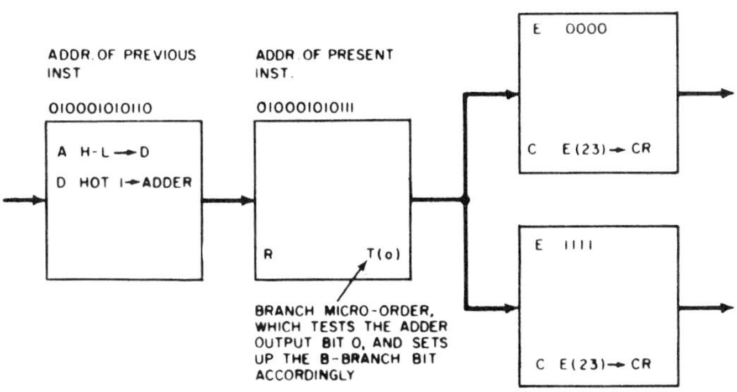

Figure 8.17 Example of B-bit branch control.

mum of 256 possible instruction codes. The System/360 instruction set contains 141 unique Op Codes recognizable by the microprogram; the remaining ones are invalid codes.

The 8-bit System/360 operation code is initially loaded into bits 0–7 of the M register. A function branch is taken on M bits 0–3 to determine the class of instruction, i.e., floating-point long, etc. A second function branch is taken on the remaining Op Code bits 4–7 to address the first microinstruction of a routine that will execute a specific instruction within an instruction class such as VFL logical AND.

3. *Emit Field.* The 4-bit (positions 57–61) field of the microprogram word is called the emit field. It is the only ROS-word segment that is not decoded. In the previous example, we used the micro-order E(23) → CR, which orders the last two bits of the emit field to be transmitted and stored in condition code bits 34 and 35 of the PSW.

The emit field may be used in many ways, for example:

(a) To determine the address of a register in a sector of local store.

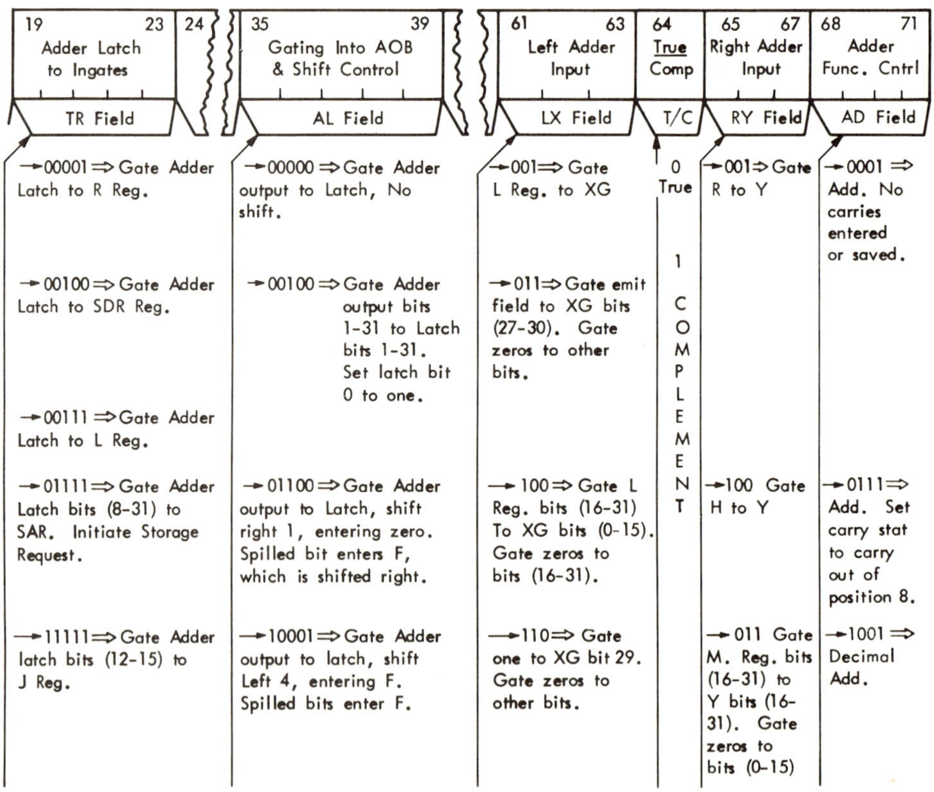

Figure 8.18 Control fields and micro-orders for the Model 50 adder data path.

(b) To enter a constant into the adder path or into the mover path.

(c) To set stats on or off, according to the 1 bits present in the emit field.

4. *Data Flow Control Fields.* This segment of the ROS word contains the static control fields, each controlling a particular area of the data flow such as gating to the left or right input of the adder, destination of the adder output latches, and mover inputs.

The control fields can be grouped according to their functions. The following list represents this logical grouping of the control fields in the ROS control word.

(a) Figure 8.18 shows the primary control fields and some sample micro-orders for the Model 50 adder data path.

(b) Figure 8.19 shows the fields assigned to control the actions of the mover data path.

(c) Figure 8.20 shows the fields that control local store addressing.

(d) *Counters.* The DG control field (positions 49–51) is called the length counter and carry-insert control. This field controls two 4-bit counters. They may be counted down individually as two 4-bit fields, or as an extended 8-bit counter. Associated with each counter is a sign bit which can be used to effect a conditional branch by setting the A or B bit of the ROS address field.

There are three 1-bit control fields: the MD (position 46), the LB (posi-

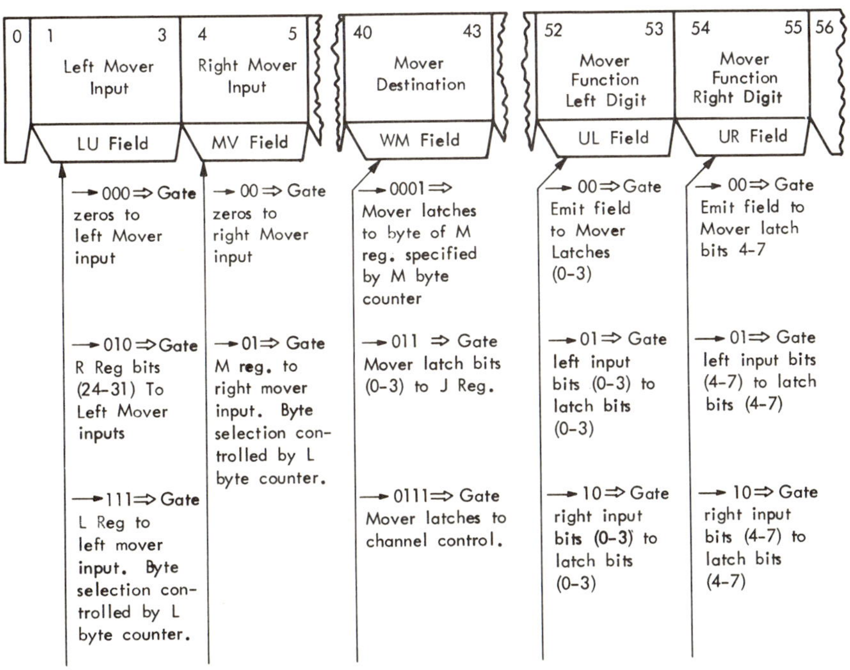

Figure 8.19 The segment of the ROS word that controls the mover data path and micro-orders.

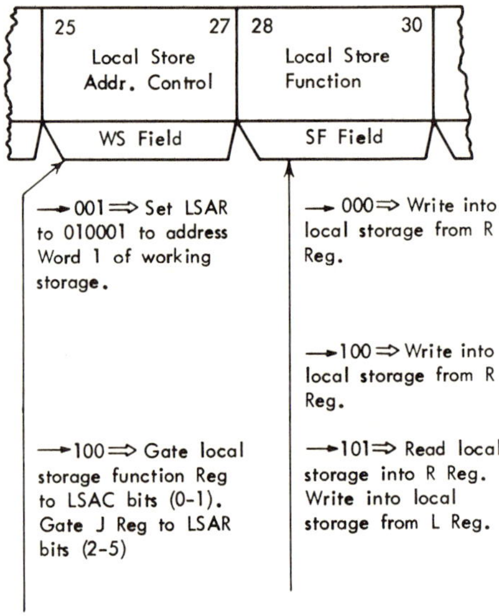

Figure 8.20 Fields that control local-store addressing.

tion 47), and the MB (position 48). They are called the MD counter control, the LB counter control, and the M-byte counter control. These byte counters are vital to the mover's input and output gating. The MD is used to control the byte counters.

(e) *Miscellaneous Controls Field.* This is the SS field which occupies positions 84–89. There are 64 micro-orders specified by this field. A large number of the micro-orders in this category deal with stat settings. However, this field is also a catchall for all micro-orders that do not fit anywhere else.

(f) Finally, there is the invalid-digit test and IC control field which occupies positions 32–34. Called the IV field, it is used for both invalid-digit detection in the mover for decimal operations, and for instruction-counter control.

8.13 MICROPROGRAMMING STEPS

The microprogrammer writes the microprogram control routines by specifying the micro-orders for each microinstruction. These micro-orders are similar to the mnemonics in an assembly language and must be translated into binary bits. Also, an absolute address in CROS must be assigned to each microinstruction. The IBM Control Automation System (CAS) will accept the micro-orders, produce the 90-bit pattern (zeros and ones) for each control word, and assign the address for each ROS word. The input to CAS is a listing of the microinstructions in the form of a logical flow chart prepared on the logic sketch sheet, IBM Form 620-8262-0 (see Fig. 6.6).

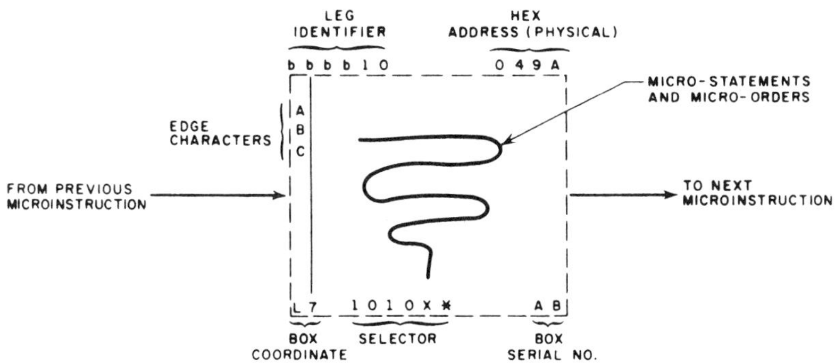

Figure 8.21

Figure 8.21 shows the microinstruction box. It has a fixed format with which the microprogrammer must comply. The print area within the box is 13 characters wide; the maximum number of lines in the box is 18, including the top and bottom lines.

The top line of the box contains:

1. The *leg identifier* set (up to six characters), right-justified in character positions 1–6 of the line.
2. The *absolute address.* This address is assigned by the CAS program. It is printed in hexadecimal and is right-justified in the last four character positions of the line. The address gives the physical location of this particular microinstruction in ROS.

The bottom line contains:

1. The two-character *box coordinate* at the left end of the line. It defines the relative position of the microinstruction printed on the logic sketch sheet.
2. The *leg selector* (up to six characters), right-justified in print positions 5–10 on the line. The left four characters specify the functional branch bits of the ROS address of the succeeding ROS word. The last two bits specify the conditional branch bits A and B. The first four positions for the functional branch may be left blank. An X indicates a "don't care" condition, leaving it up to the addressing routine in the CAS system to fill these bit positions. An asterisk (*) implies a conditional branch test, and the bits so indicated will be assigned a zero or one, based on the machine condition tested. As an example, if one box contains a conditional branch micro-order which will set bit A and has no micro-order for setting the B bit, then the leg selector field will have bbbb*X (b stands for blank). The corresponding branch bits in the leg-identifier fields of the successor candidates must be filled out by the microprogrammer (see Fig. 8.22).
3. The box serial number at the right end of the line is permanently assigned by the CAS system as a symbolic name for that box. Although the box can be moved to another position on the logic sheet with new box coordinates, its box serial number remains fixed.

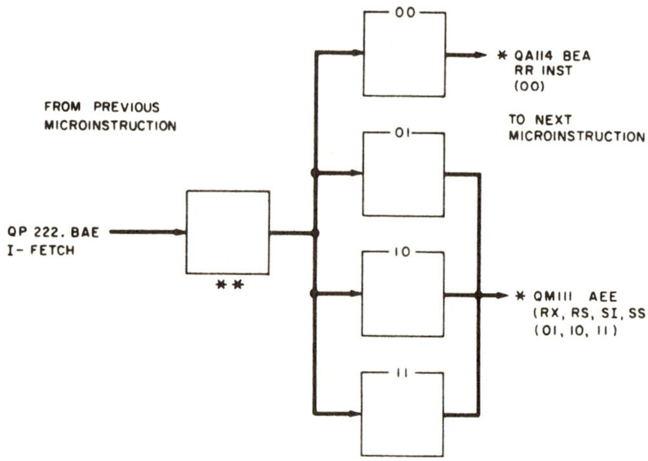

Figure 8.22 Microprogram net identification.

Figure 8.22 also shows the net flow which is from left to right, as indicated by the arrows on the lines connecting the boxes. Each net has associated with it a unique number called a net number. This net number consists of a five-character sheet number (net number QP222) on which the source (the previous microinstruction) resides, the two-character source-block serial number, and the one-character output pin, making an eight-character total.

The microprogrammer may draw horizontal or vertical lines to join any number of box input or output points and lines intersecting points. Lines may be drawn through a box area, but this prevents the use of that box.

When it is necessary to specify some or all predecessor or successor boxes appearing on another CAS logic sheet, an edge-like identification is given.

Assume that in Fig. 8.22, the five boxes and the nets connecting these boxes form a complete CAS logic sheet. The left edge has a two-line note which identifies the source of this net. The first line here identifies the sheet number (QP222) followed by a dot (.), followed by a three-character source of net on the preceding sheet, e.g., QP222.BAE. The second line (I-fetch) is the name of the function or the subroutine.

The two output lines on the right side of Fig. 8.22 are called right-edge lines. Note the asterisk on each of these lines. The asterisk here means that the lines are a source to another sheet. The sheet number and the net entry point are given on that line. The second and succeeding lines of the edge line contain a list of the leg identifiers for all the following instructions associated with this net, e.g., 01, 10, 11.

We have thus far defined all the box and net qualifiers in Fig. 8.22, except for the left-edge characters and the location of the micro-orders within this box.

Figures 8.23* and 8.24* show the templates used in microprogramming the

*These figures were prepared by R. B. Stuart of IBM and are used with his permission.

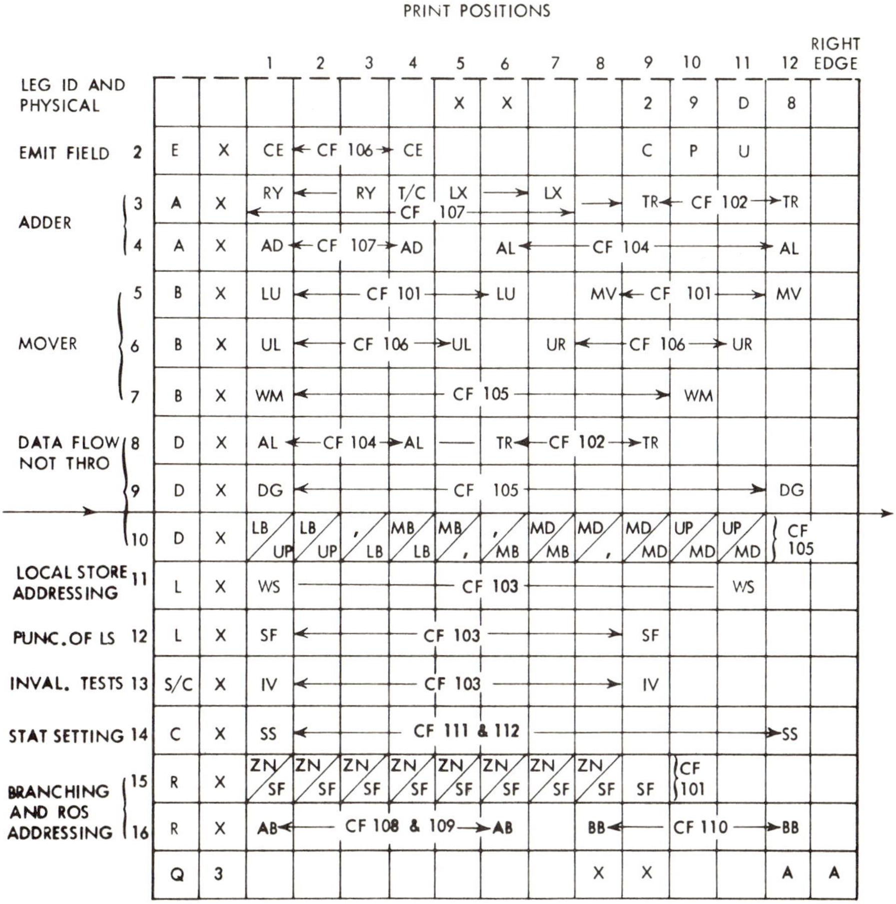

Figure 8.23 CPU-mode print box template.

Model 50. The first, Fig. 8.23, is viewed as a template for the CPU mode, and Fig. 8.24 is for the I/O mode. We will examine Fig. 8.23 in some detail. The microprogrammer must follow this fixed format when transcribing a microprogram on the logic sheets.

Note that there are 18 lines of print, each line consisting of 14 print positions. The top line, containing the leg identifier and the physical address, does not appear on this template; nor does the bottom line, which contains the box coordinate, leg selector, and box serial number. Hence, the microprogrammer can transcribe his micro-orders on 16 lines. Each line has a left-edge character which defines the class of micro-order that can be printed on that line. These left-edge characters are:

E emit field lines
A arithmetic function—or the adder

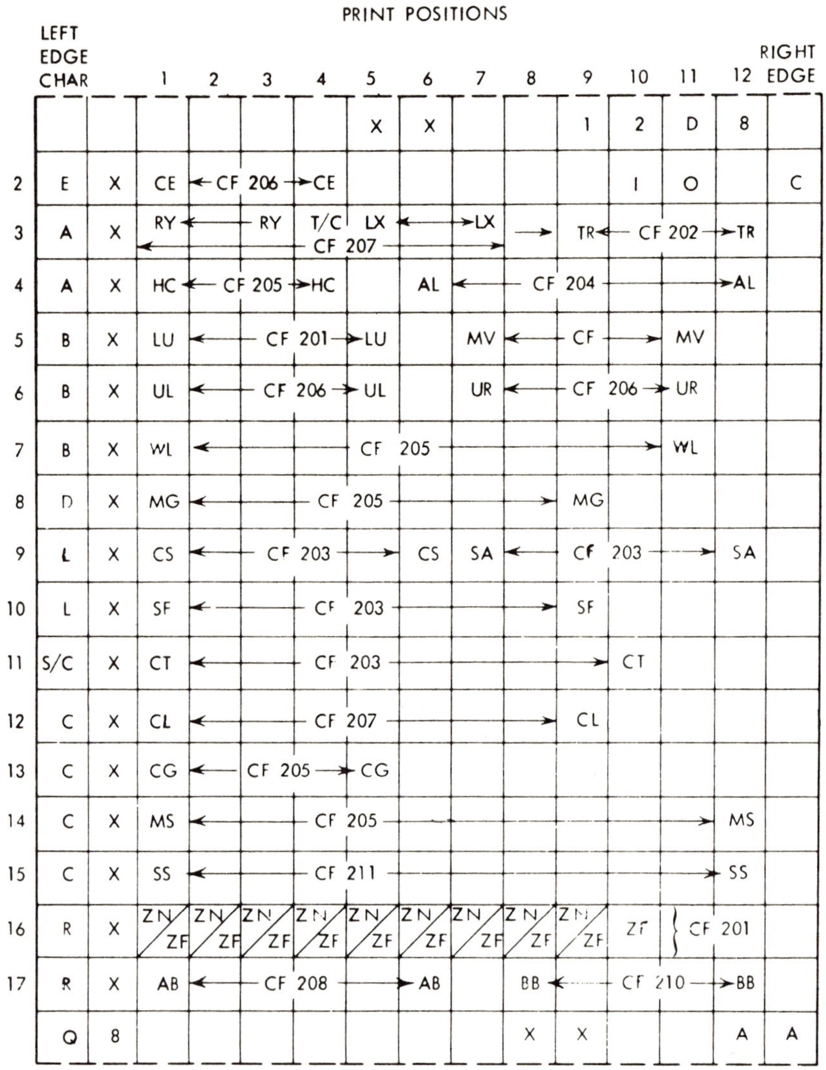

Figure 8.24 I/O-mode number 1 box template.

B mover function
D data flow not through adder or mover
L local-store addressing and data flow
S storage control or addressing
C miscellaneous controls
R ROS addressing

The left-edge characters specify the order in which the microinstruction types appear in the microinstruction box. For example, we cannot have a

D-type microinstruction on print line 2; nor can we have the emit field on print line 6. The rightmost edge of the line (print position 13) must be left blank.

Print line 2 contains a four-bit configuration of the emit field which occupies print positions 1–4, plus the CPU or I/O mode designation. The next two print lines in the CPU mode are reserved for the inputs to the adder function and for the shift control and gating into the adder. Print lines 5, 6, and 7 specify all controls to the mover.

The next three print lines deal with micro-orders that control all data flow not going through the adder or mover. Print lines 11 and 12 deal with local-store addressing and local-storage functions, respectively. These are followed by two lines of miscellaneous controls and stat-setting controls. The last two print lines, with the R-edge characters, deal with ROS addressing and branching.

The arrows in line 2, position 8, and in line 8, position 5, are desirable but are not required since they are not checked by the CAS interpreter.

Note that the AL field with left-edge character A, line 4, the UR field, the MV field, and the BB field must all be right-justified when transcribing the micro-orders.

8.14 SELECTOR CHANNEL

All I/O operations of the selector channel are initiated and controlled by instructions, commands, and control words. Although I/O operations proceed concurrently with the processing of data, the CPU controls the channel by the following set of I/O instructions (Fig. 8.25):

1. START I/O—used to initiate the execution of a command or a chain of commands.

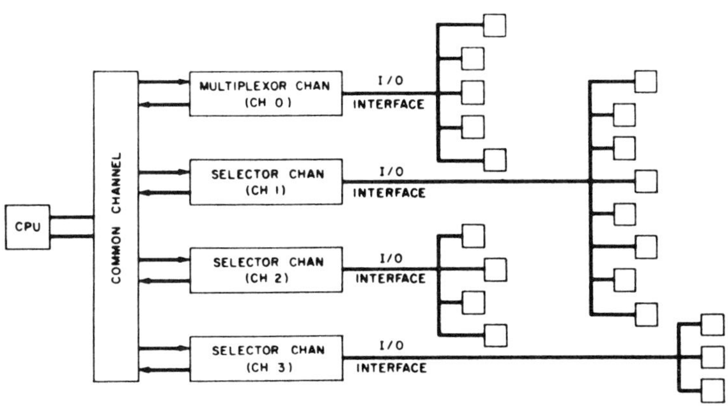

Figure 8.25 Channel relationship to I/O interface and common channel.[4]

2. TEST I/O—used to test for and clear I/O interruptions.
3. HALT I/O—used to terminate the execution of the current command.
4. TEST CHANNEL—used to test the state of the addressed channel.

When the execution of the I/O instruction is completed, the CPU is released to proceed with the program, and the channel operation proceeds independently with the execution of the command or chain of I/O commands. The selector channel is capable of executing the following six commands: write, read, read backward, control, sense, and transfer in channel.

The selector channel activity is controlled by special circuitry in each channel, as well as by microprogram routines. There are twelve microprogram routines associated with the selector channel operations. These control the flow of data and control words to and from main storage and local storage, and they control the modifications to a control storage word resulting from channel activity. The microprogram has the ability to perform conditional branches based upon conditions in the selector channel. Branching allows a general microprogram routine to be used for several different purposes.

Switching the microprogram control between the CPU and the channels (I/O mode) is handled by control circuits in the common channel. The process of switching from CPU to channel control is called *break-in*. The process of switching from channel to CPU is called *break-out*. When a channel requires a service, it sends a request signal to the common channel. Such a request may be honored during a nonmemory cycle. Such a cycle is termed a priority cycle and is overlapped with CPU processing. During the priority cycle, the common channel determines which of several possibly competing channel requests will be honored, and it initiates the break-in process. The break-in cycle that follows is not overlapped with CPU processing. During this cycle, the following actions occur:

1. The starting address of the channel routine is sent to the read-only storage.
2. The read-only storage that was generated (but not used) in the priority cycle will go to the backup register.
3. The I/O-mode trigger is turned on, controlling the decoding of the microinstructions.
4. The contents of the R register are stored in a backup location in local store.

In the next cycle, the control reverts to the read-only storage, which commences execution of the channel routine.

The last microinstruction of all channel routines is the same. It controls the break-out process, which consists of:

1. Gating the contents of the backup register to the read-only storage address register so that CPU processing will resume in the next cycle.
2. Reloading the R register from local storage.
3. Turning off the I/O-mode trigger.

If another channel request is recognized by the common channel during the priority cycle, item 3 and the next break-in cycle are omitted. If another channel request is recognized before the break-out cycle, both break-out and break-in cycles are omitted.

The selector channel has three data buffers: A, B, and C (Fig. 8.26). The four-byte C register is used for byte assembly and disassembly. On output operations, the C register receives data from the B register and transmits the data to the I/O interface. For input operations, this data flow is reversed.

The B register is connected to the C register and to the CPU adder latch. In this way, the channel has access to the various registers and local storage in the CPU.

Each selector channel has a unique location in local storage which it uses as a data buffer. This location is called register A. For output operations, the data word fetched from main storage is placed in register A. Input data are placed in register A prior to storing in main storage.

Each selector channel uses three other words of local storage to hold parts of the channel control words, the channel address word, and the unit address. One word holds the command address with the protection key; the second word holds the data address and the protection key; the third contains the address of the device in operation, as well as the byte count.

Other registers in each selector channel are (Fig. 8.1):

1. *Byte Counter.* This two-bit register controls the selection and gating of bytes 0–3 between the C register and the I/O interface.

2. *End Register.* This two-bit register is set with the value that the byte counter should contain after the last byte of the data has been transmitted to or from the I/O interface.

3. *General Purpose Register.* This eleven-bit register receives the initial byte count, the end-register count, the last work indications, operation codes, and flags the CPU facilities. This information is sent to the proper register from the general purpose register.

4. *Operation Register.* This three-bit register reflects the operation state of the channel: input (read or serve), output (write or control), or read backward.

5. *Flag Register.* The five flag conditions from the channel control word are stored here.

6. *Position Register.* This determines the type of microprogram routine being used to control the CPU facilities.

7. *Request Register.* This contains the priority of the routine currently being requested by the channel.

8. *Channel Status Register.* The error conditions detected in this channel, and which form part of the channel status word, are held here. This information is subsequently gated to the CPU.

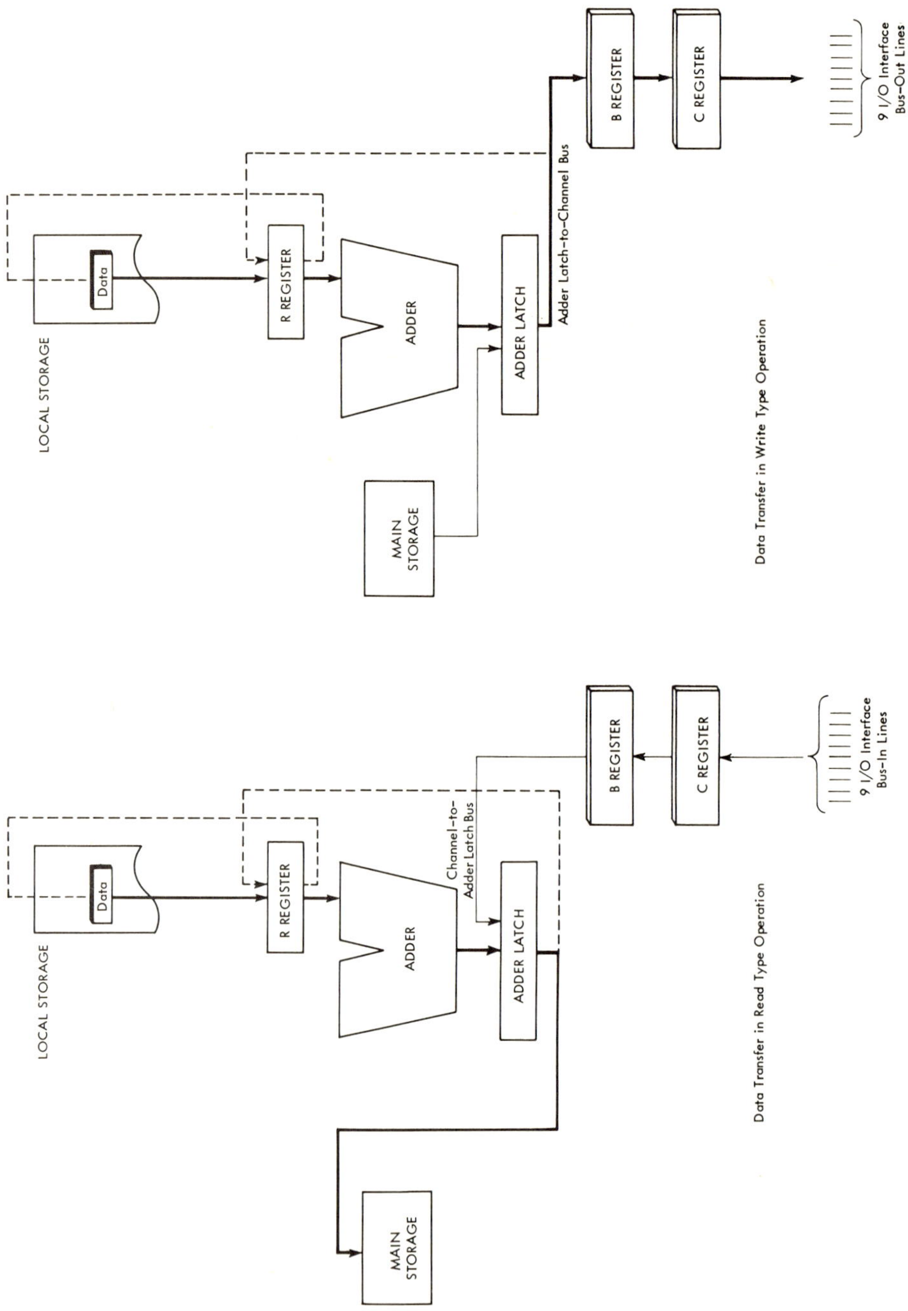

Figure 8.26 Data flow to and from channel.

9. *Last-Word Latches.* Three latches are set by the microprogram to indicate the termination of a data transfer.

8.14.1 Channel Output Operation (Write)

During an output operation (Fig. 8.26), the data are transmitted from main storage to the I/O device using the A, B, and C data buffers. The data are transmitted four bytes at a time within the Model 50.[5]

Registers A, B, and C are set up initially with the first three words, and the data address and count are adjusted accordingly. The data bytes are disassembled in the C register and are transmitted one byte at a time over the I/O interface. After the transmission of the last byte from the C register, the word in the B register is set in the C register. The channel then requests the microprogram routine, which transfers the word in A (local store) to B. Since A is now empty, the channel requests the routine, which loads A from main storage. This routine also updates the data address and count.

At the end of the operation, the last-word latches provide for the correct handling of the last three words. The end register is used to cause the transmission to terminate with the correct byte of the last word.

8.14.2 Channel Input Operation (Read)

During the input operation (Fig. 8.26), data are transmitted from the I/O device and are assembled one byte at a time in the C register under the control of the byte count. After the word is assembled, it is transmitted to the B register, freeing the C register. The channel then requests the microprogram routine to transfer the contents of B to local storage A. The data are then transferred to main storage from local storage by another microprogram routine, which also updates the data address and count.

8.15 MULTIPLEXOR CHANNEL

The multiplexor channel is physically an integral part of the CPU; it uses the CPU clock and power supplies, and it shares the CPU data paths. These data paths are controlled by microprogramming. An area of local storage is set aside for multiplexor channel use.

Two 8-bit plus-parity registers, buffer register 1 and buffer register 2, are provided for the multiplexor channel use. Both registers can be set from either the interface-bus-in or the mover output.

The path of data for an input operation is from interface bus-in to buffer register 2, to the input of the mover, through the mover and into the proper byte of the CPU L register, and then to the SDR and main storage. The path for an output operation is from main storage to the L register, where the proper byte is gated through the mover to buffer register 2, and then to the interface-bus-out. Every byte of data transmitted or received requires a main-storage cycle.

Signals on the interface lines of the I/O units are usually initiated under control of the microprogram and are terminated in response to signals from the I/O units. One exception to this is the select-out signal, which is initiated to poll the interface. The interface polling sequence is controlled by clock timing and the polling latch.

A seven-bit buffer register is provided to hold the ROS address of the next microprogram routine desired. The output of this register and the outputs of the three-bit priority register are sent to priority controls in the common channel to request the next routine. The output of the routine-request buffer is also used to determine what routine is being executed in preparation for the next routine. In this case, it is combined with other conditions in the routine generator. These other conditions include the interface in tags, the emit field of the microinstructions, several status latches, and operation code lines from the CPU.

To permit conditional branching in the channel microprogram, a set of four I/O stats is provided. These I/O stats can be set, reset, and tested by the microprogrammer.

Seven words in local storage are provided for use by the multiplexor channel. They are:

1. The interruption buffer which is loaded with the unit address and the interruption conditions sent from an I/O device.

2. Two words set aside as backup storage for the L and R registers of the CPU.

3. Four words containing the unit control word (UCW) for the device currently using the interface.

8.16 MULTIPLEXOR CHANNEL INPUT/OUTPUT OPERATIONS

When the CPU decodes an I/O instruction, a signal is sent to the multiplexor channel. After the channel has polled the interface and has found no device needing service, a request for a microprogram is made. When this request is honored by the common channels, the ROS is switched to I/O mode and assumes control of the CPU and the multiplexor channel. The I/O command is routed through the data paths to the interface-bus-out, and the proper microroutine address is sent, all under microprogram control.

Toward the end of this routine, the next routine is requested as a result of decisions made during the routine and as a result of interface responses. This continues through several routines which perform such functions as the comparison of unit addresses, the analysis of status, and the preparation for the data-handling routines.

The ROS remains in I/O mode until the unit status has been analyzed. If during that time the operation is terminated, the status is loaded directly into the CSW, the proper condition code is set, and the CPU is signalled that the operation is completed. If the operation is not terminated, the channel goes back to polling, and after the unit status is analyzed, the CPU is signalled that the channel has accepted the instruction and is performing it. The CPU then goes on to its next instruction.

8.17 INSTRUCTION FETCH

The Model 50 is controlled entirely by stored programming. A sequence of operations is stored or loaded into core storage before the system processes data. The processing unit then proceeds, step by step, through the stored program routine, analyzing each instruction, and performing the functions called for before proceeding to the next instruction.

The length of the System/360 instruction can be one, two, or three halfwords. Length is related to the number of storage addresses necessary for performing the operation. An instruction consisting of one halfword (RR format) cannot cause any references to main storage. An instruction that occupies a fullword (RX, RS, SI formats) provides one storage address, and a three-halfword instruction of the SS format type provides two or more storage accesses.

Normally, the operation of the processing unit is controlled by instructions taken in sequence. An instruction is fetched from a location specified by the current instruction address in the instruction address register (IAR). The instruction address is then increased by the number of bytes in the instruction to address the next instruction in sequence. The instruction is then executed, and the same steps are repeated using the new address in the IAR.

$$\text{MS} \quad (\text{IAR}) \rightarrow \text{IR}$$
$$\text{IAR} + 2/4 \rightarrow \text{IAR}$$

Conceptually, all halfwords of an instruction are fetched from storage after the preceding operation is completed and before execution of the current operation. However, the physical storage word size and overlap of instruction execution with storage access may cause the actual instruction fetching to be different. Thus, it is possible to modify an instruction in storage by the immediately preceding instruction.

Instruction fetching, then, is defined as that operation and the associated hardware of the CPU responsible for retrieving machine instructions from the storage units, and performing operations common to many instructions. More specifically, the instruction counter (IC) obtains instructions from storage, determines the format (RR, RX, RS, SI, and SS), and loads the appropriate registers with operands and data fields. At this point, control is transferred to the execution phase, where specific machine instructions are decoded and executed. At the completion of the previous instruction, the contents of the IAR are gated to the storage address register (SAR) and a storage fetch is initiated. At the same time, a check is made to see whether any interruption requests from the channels or external signals are pending. If a request is pending, the interruption is handled first; otherwise, the instruction fetch is continued. That is, the instruction fetch routine reads the instruction from storage, decodes the instruction, and branches to the proper execute routine. Figure 8.27 shows the first-level I-fetch logic flow.

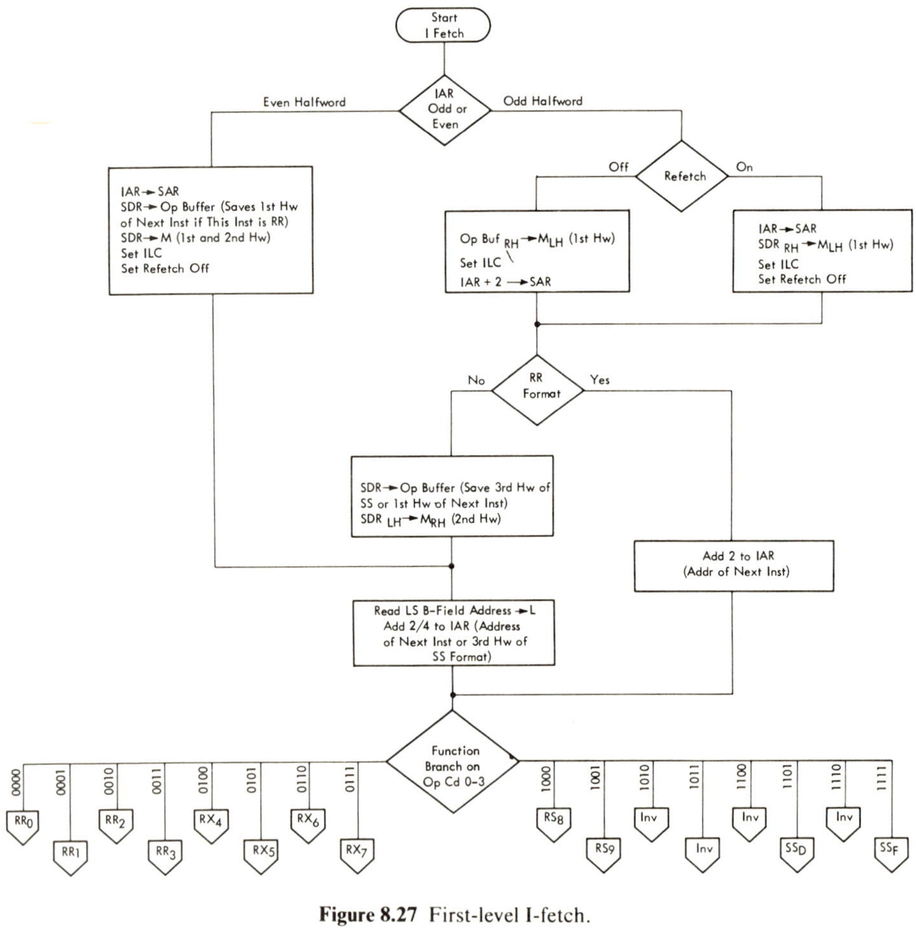

Figure 8.27 First-level I-fetch.

If the last halfword of the last instruction was an even halfword, the accompanying odd halfword has normally been buffered in a local storage location (operation-buffer register; word 14 of the working storage sector of the local storage). Three cases of instruction fetch may be distinguished (Fig. 8.28):

1. A one-halfword instruction (RR type). This halfword is placed in the left half of the M register. The previously initiated storage request is ignored. The instruction address register is incremented by two.

$$\begin{aligned} &\mathrm{IR}(16\text{–}31) \rightarrow \mathrm{M}(0\text{–}15) \\ &\mathrm{IAR} + 2 \rightarrow \mathrm{IAR} \end{aligned}$$

2. The first halfword of a two- or three-halfword instruction. This halfword is placed in the left half of the M register. The left half of the word fetched from storage is placed in the right half of the M register. The word from storage is also placed in the instruction buffer in local storage. The instruction address register is incremented by four.

$$\begin{aligned} &\mathrm{IR}(16\text{–}31) \rightarrow \mathrm{M}(0\text{–}15) \\ &\mathrm{MS(IAR)} \rightarrow \mathrm{IR}; \qquad \mathrm{MS(IAR)}(0\text{–}15) \rightarrow \mathrm{M}(16\text{–}31) \\ &\mathrm{IAR} + 4 \rightarrow \mathrm{IAR} \end{aligned}$$

3. The refetch stat is on, indicating that the instruction buffer may not reflect the current state of the main storage location from which the next instruction is to come. The word is refetched from main storage, placed in M, and tested to see whether the right halfword is a one-halfword instruction. If it is, the sequence continues as in case 1. If it is not, the sequence continues as in case 2.

$$\begin{aligned} &\mathrm{MS(IAR)} \rightarrow \mathrm{M, IR} \\ &\mathrm{IAR} + 2/4 \rightarrow \mathrm{IAR} \end{aligned}$$

A fourth case occurs when the previous instruction ended with an odd halfword.

4. The word fetched from storage is placed in the M register. The word is also placed in the instruction buffer in local storage. The IAR is incremented by two if the instruction in the M register is a one-halfword instruction; otherwise, it is incremented by four.

$$\begin{aligned} &\mathrm{MS(IAR)} \rightarrow \mathrm{M, IR} \\ &\mathrm{IAR} + 2/4 \rightarrow \mathrm{IAR} \end{aligned}$$

At this point, M contains the instruction (or the first two halfwords, if it is of the SS type). The instruction length code of the program status word has been set according to bits 0–1 of M. Bits 12–15 of M have been gated to the J register, and bits 16–19 to the MD counter. The general register specified by MD (the base-address register for the RX, RS, SI, or SS formats) is read from

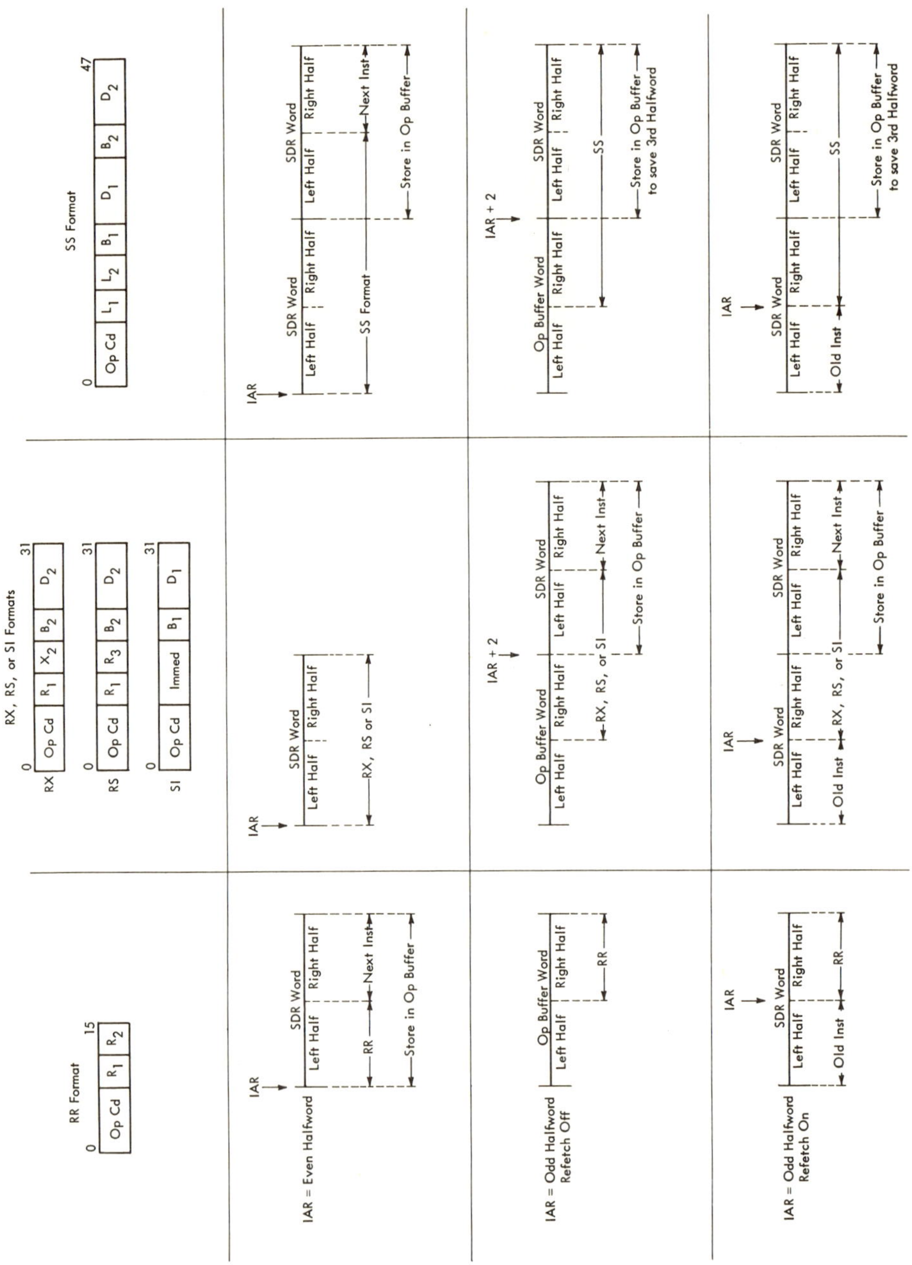

Figure 8.28 Instruction formats and source of instruction halfwords.

local storage to the L register. Simultaneously, bits 8–11 of M (corresponding to the R1 field of certain instruction formats) are gated into MD.

$$
\begin{aligned}
&M(0, 1) \rightarrow PSW(32, 33)\\
&M(12\text{–}15) \rightarrow J\\
&M(16\text{–}19) \rightarrow MD\\
&GPR(MD) \rightarrow L\\
&M(8\text{–}11) \rightarrow MD
\end{aligned}
$$

A function branch is made in the microprogram using the first four bits of the operation code from M. This separates the instructions according to format classes.

$$M(0\text{–}3) \rightarrow ROSAR\,(6\text{–}9)$$

Simultaneously, a four-way condition branch is made depending on whether bits 12–15 and/or bits 16–19 of the M register are zero. This branch is not meaningful in all cases, but it distinguishes the cases of $B = 0$ in the RS, SI, and SS formats, and $X = 0$ in the RX format.

The action up to this point is referred to as the first-level instruction-fetch sequence. The second-level instruction-fetch sequences are different for each format class, but, in general, they include the following sections:

1. The base address and index address computations are completed. An initial data-storage cycle may be initiated. The local-storage function register (LSFR) is set to address the proper segment of local storage. Various stats and counters are set to required initial values. In the case of three-halfword instructions, the third halfword is fetched from storage or from the instruction buffer in local storage, and the instruction address register is incremented by two.

2. The second-level instruction-fetch sequence is terminated by a function branch using the second four bits of the operation code. This branches the microprogram into the individual instruction sequences.

Second-level I-fetch completes the decoding of the instruction, stores the IAR address if the instruction format could result in a branch instruction, and branches to the proper execute microprogram.

Figures 8.29 and 8.30 show the second-level I-fetch flow paths for all valid and invalid Op Codes (invalid Op Codes are the unassigned configurations of the eight-bit Op-Code field).

8.17.1 RR Formats

The RR0 format (0X codes) stores the contents of the general purpose register specified by the R2 field in the R register. The contents of IAR are stored in the H register to save the next instruction address if this instruction results in a branch operation.

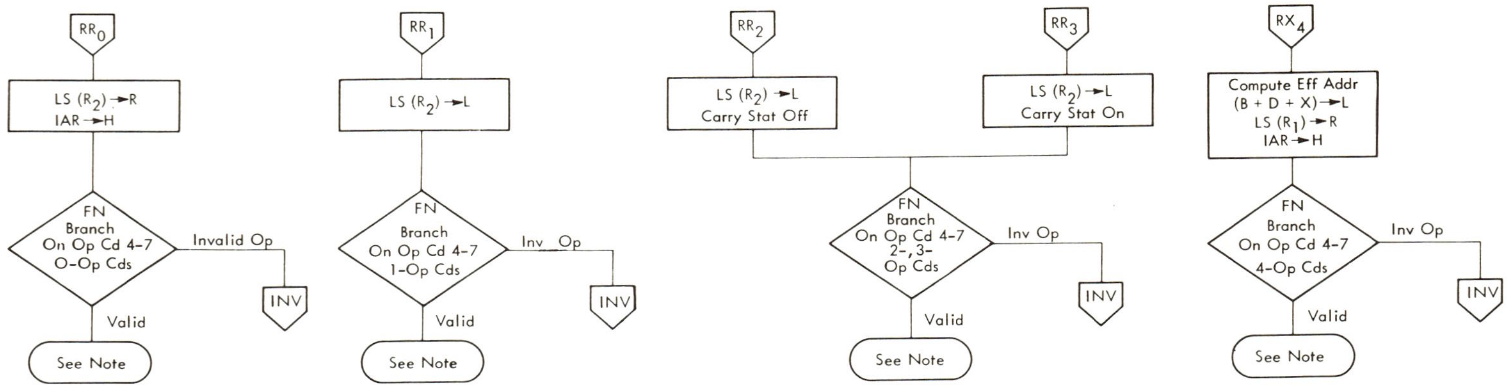

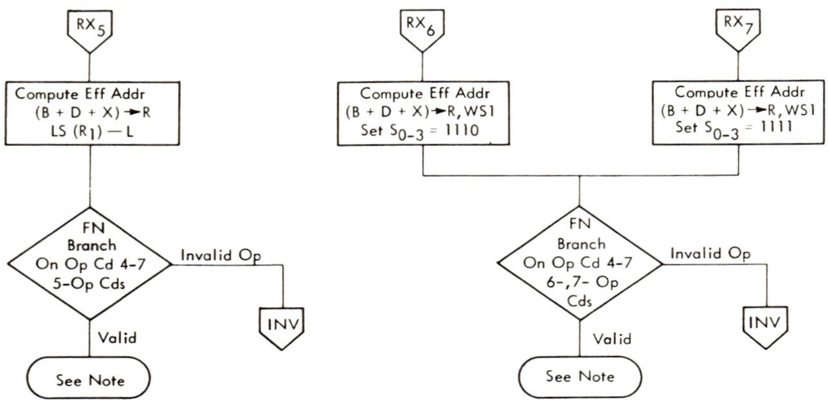

Note: The function branch decodes Op-Code bits 4-7 and branches to the proper routine for the Op-Code.
Some combinations of the eight bit Op-Code are invalid. The invalid codes are:

Valid	Invalid
04-0A	00-03, 0B-0F
10-1F	None
20-24, 28-2F	25-27
30-34, 38-3F	35-37
40-4C, 4E-4F	4D
50, 54-5F	51-53
60, 68-6F	61-67
70, 78-7F	71-77

Figure 8.29 Second-level I-fetch RR and RX formats.

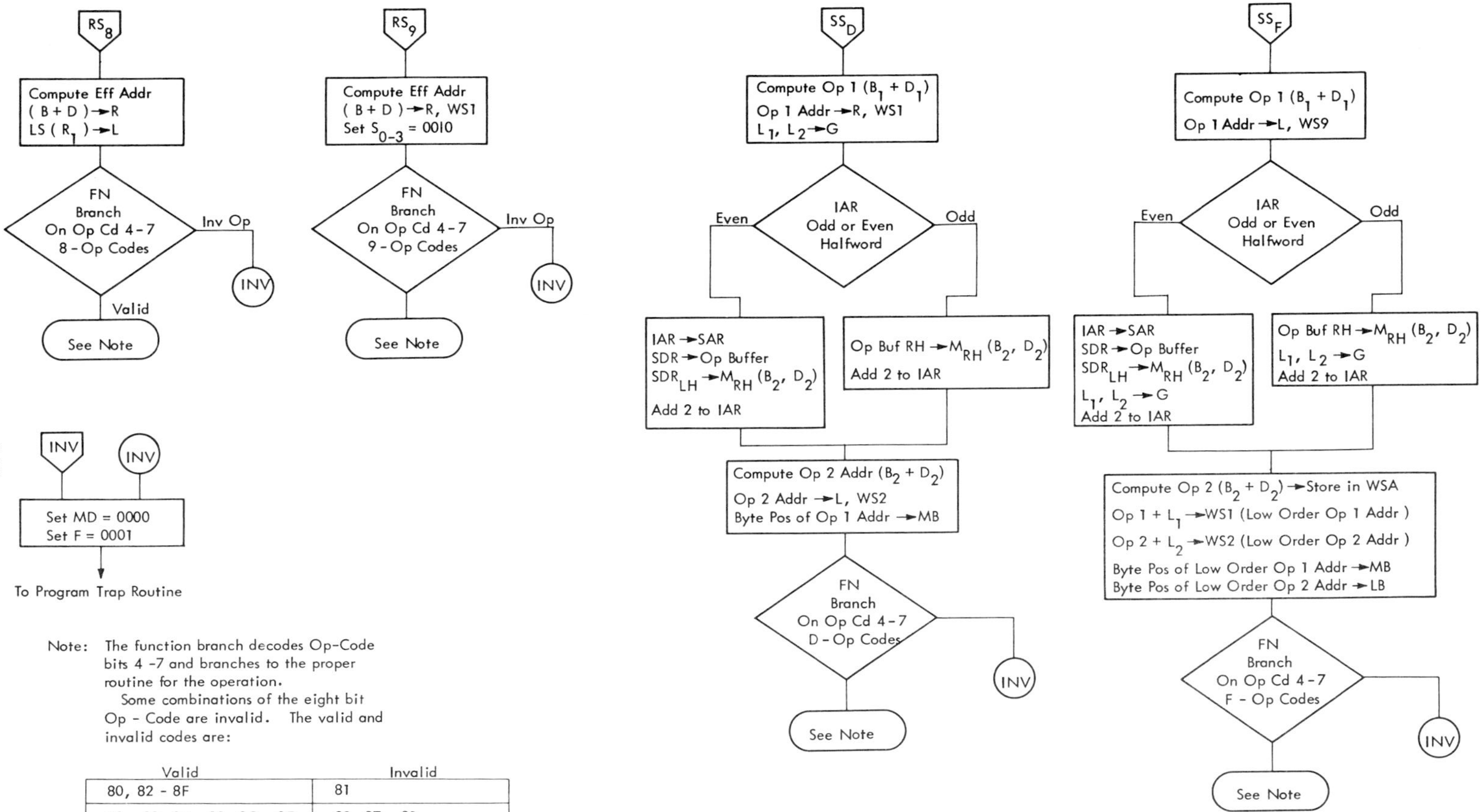

Note: The function branch decodes Op-Code bits 4 -7 and branches to the proper routine for the operation.
Some combinations of the eight bit Op - Code are invalid. The valid and invalid codes are:

Valid	Invalid
80, 82 - 8F	81
90 - 92, 94 - 98, 9C - 9F	93, 97 - 9B
D1 - D7, DC - DF	D0, D8 - DB
F1 - F3, F8 - FD	F0, F4 - F7, FE - FF

Figure 8.30 Second-level I-fetch RS, ST, and SS formats.

RR formats 1–3 (1X–3X codes) store the contents of the general purpose register specified by the R2 field in the L register.

RR formats 2–3 (2X–3X codes) set the carry stat as a signal to the floating-point routine indicating whether to operate in floating-point-long or floating-point-short mode.

8.17.2 RX Formats

The RX4 format (4X codes) computes the effective address by adding the contents of the general purpose registers specified by the B and X fields to the displacement field, and it stores the result in the L register. The general purpose register specified by the R1 field is stored in the R register. The contents of IAR are stored in the H register to save the next instruction address if this instruction results in a branch operation.

The RX5 format (5X codes) computes the effective address by adding the contents of the general purpose registers specified by the B and X fields to the contents of the displacement field, and it stores the result in the R register. The contents of the general purpose register specified by the R1 field are stored in the L register.

RX formats 6–7 (6X–7X codes) compute the effective address by adding the contents of the general purpose registers specified by the B and X fields to the contents of the displacement field, and it stores the result in the R register and WS1 of local storage. General purpose stat 3 is set as a signal to the floating-point routine indicating whether to operate in floating-point-long or floating-point-short mode.

8.17.3 RS/SI Formats

The RS8 formats (8X codes) compute the effective address by adding the contents of the general purpose registers specified by the B field to the contents of the displacement field, and they store the result in the R register. The contents of the general purpose register specified by the R1 field are stored in the L register.

The RS9 format (9X codes) computes the effective address by adding the contents of the general purpose register specified by the B field to the contents of the displacement field, and it stores the result in the R register and in WS1 of local storage.

8.17.4 SS Formats

The SSD format (DX codes) computes the effective operand-1 address (B1 + D1) and stores the result in the R register and in WS1 of local storage. The L1 and L2 fields are stored in the G register. The third halfword is read into the M register (positions 16–31), and the effective operand-2 address is computed and stored in the L register and in WS2 of local storage. The byte position of the operand-1 address is stored in the MB counter to be used in locating the first data-byte position in the operand-1 word.

The SSF format (FX codes) computes the effective operand-1 address (B1 + D1) and stores the result in the L register and in WS9 of local storage. The L1 and L2 fields are stored in the G register. The third halfword is read into the right half of the M register, and the operand-2 effective address is computed and stored in WS1. The byte position of the low-order address is stored in the MB counter. The low-order address of the operand-2 field is computed and stored in WS2. The byte position of the low-order address is stored in the LB counter.

Figures 8.31 through 8.34 trace the microinstructives for the first- and second-level fetch cycles for the RR/RX fixed-point add/subtract instructions A, AR, S, and SR.

Figure 8.35 gives the logic flow for the execution of the fixed-point add/subtract microroutine. After becoming familiar with the Model 50, the reader should attempt to produce the microprograms for the add instruction (1A).

8.18 EXAMPLES OF MODEL-50 MICROPROGRAMMING

1. What is the content of register M after executing the following microinstructions?

$$\left[\text{A: } -4 \rightarrow \text{M}\right]$$

The minus symbol is a micro-order to gate the complement of XG to the left adder input (CF 107–TC0). The 4 micro-order is to gate a 1 to XG-bit 29 (Hot 4) and to gate 0's to all other bits (CF 107–LX6). Hence, the microinstruction, when executed, will place 1's in all the register-M bit positions, except for bit 29 of register M, which will have a 0, or FFFFFB Hex.

2.

$$\left[\begin{array}{l}\text{E: } 0010 \\ \text{C: } \text{E}(23) \rightarrow \text{CR}\end{array}\right]$$

This microinstruction will transfer bits (2, 3) of the emit field (10) to the condition code bits (34, 35) of the PSW.

3.

$$\left[\begin{array}{l}\text{E: } 0001 \\ \text{L: WS, E} \rightarrow \text{LSA} \\ \text{L: LS} \rightarrow \text{R} \rightarrow \text{LS}\end{array}\right]$$

The first line in the microinstruction box indicates that the ROS word emit field is set to 0001. The second line is one micro-order, which sets LSAR bits (0, 1) to 01, or the second quadrant in the local storage; also, LSAR bits 2–5, forming the rest of the local-storage address, are set from the emit field. In this case, the LSAR will be set to 010001 which references the second word in the work space area of the local storage.

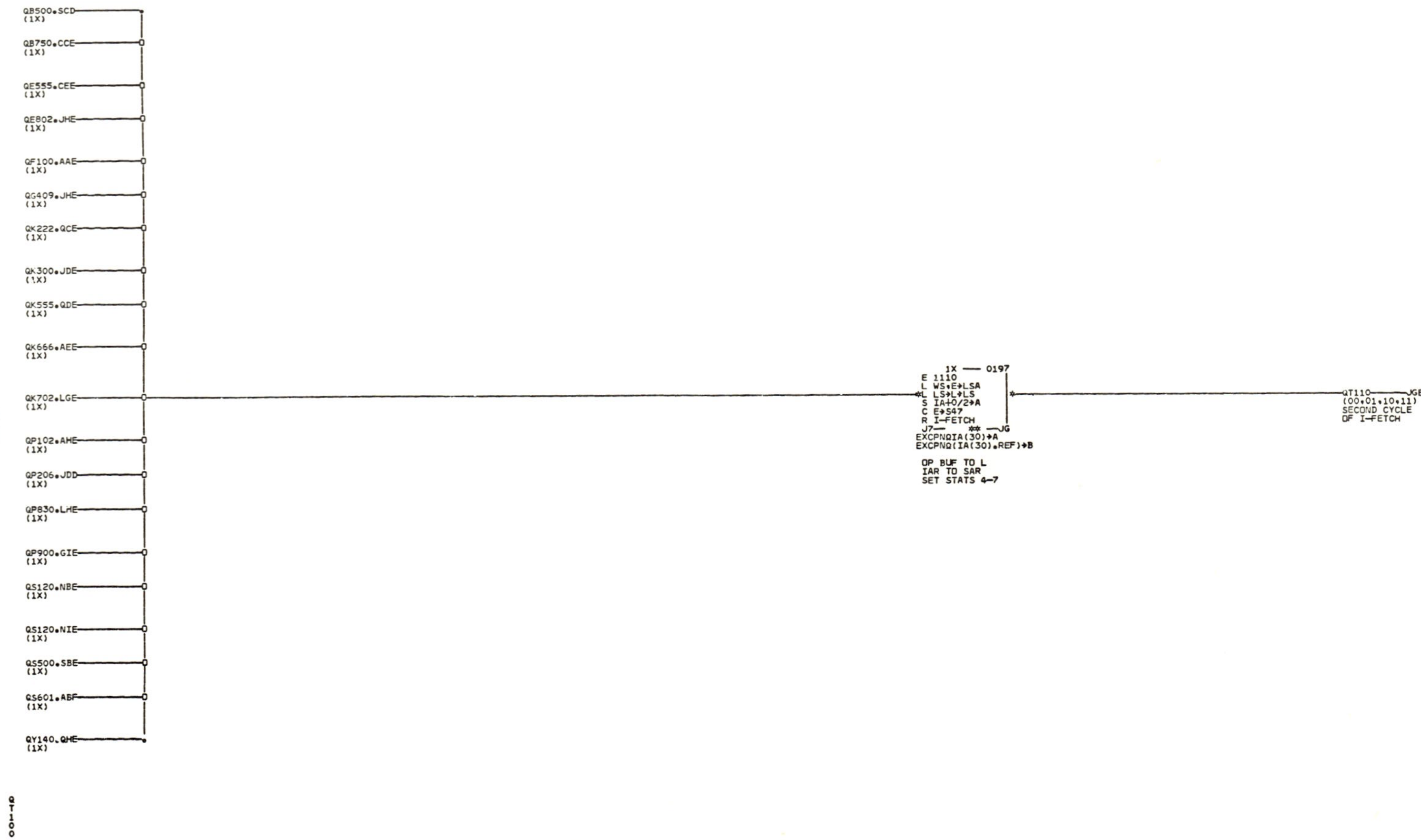

Figure 8.31 Model 50 instruction fetch.

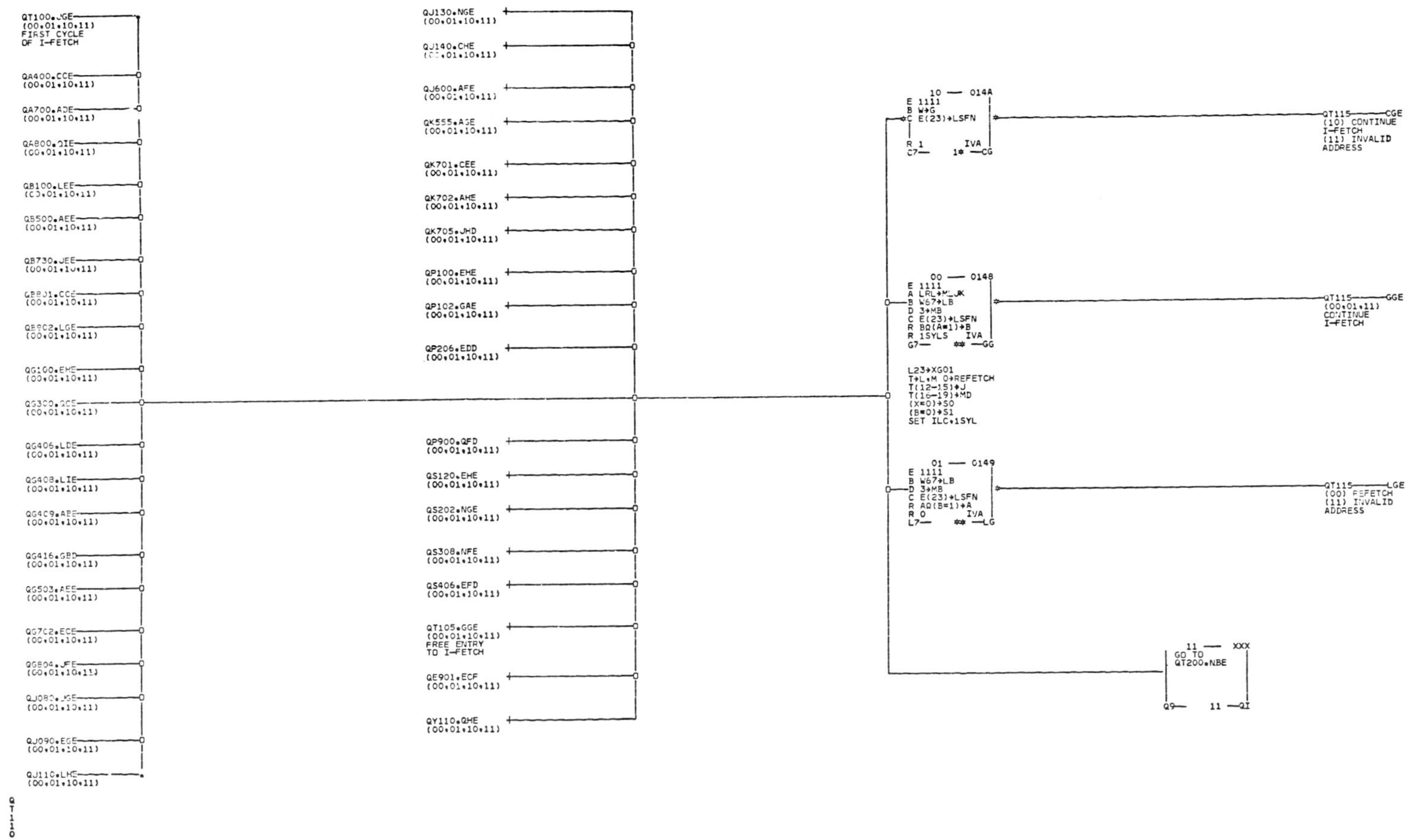

QT100.JGE
(00,01,10,11)
FIRST CYCLE
OF I-FETCH
QA400.CCE
(00,01,10,11)
QA700.ADE
(00,01,10,11)
QA800.DIE
(00,01,10,11)
QB100.LEE
(00,01,10,11)
QB500.AEE
(00,01,10,11)
QB730.JEE
(00,01,10,11)
QB801.CCE
(00,01,10,11)
QB802.LGE
(00,01,10,11)
QG100.EHE
(00,01,10,11)
QG300.GCE
(00,01,10,11)
QG406.LDE
(00,01,10,11)
QG408.LIE
(00,01,10,11)
QG409.ABE
(00,01,10,11)
QG416.GBD
(00,01,10,11)
QG503.AEE
(00,01,10,11)
QG702.ECE
(00,01,10,11)
QG804.JFE
(00,01,10,11)
QJ080.JGE
(00,01,10,11)
QJ090.EGE
(00,01,10,11)
QJ110.LHE
(00,01,10,11)
QJ130.NGE
(00,01,10,11)
QJ140.CHE
(00,01,10,11)
QJ600.AFE
(00,01,10,11)
QK555.AJE
(00,01,10,11)
QK701.CEE
(00,01,10,11)
QK702.AHE
(00,01,10,11)
QK705.JHD
(00,01,10,11)
QP100.EHE
(00,01,10,11)
QP102.GAE
(00,01,10,11)
QP206.EDD
(00,01,10,11)
QP900.QFD
(00,01,10,11)
QS120.EHE
(00,01,10,11)
QS202.NGE
(00,01,10,11)
QS308.NFE
(00,01,10,11)
QS406.EFD
(00,01,10,11)
QT105.GGE
(00,01,10,11)
FREE ENTRY
TO I-FETCH
QE901.ECF
(00,01,10,11)
QY110.QHE
(00,01,10,11)
10 — 014A
E 1111
B W→G
C E(23)→LSFN
R 1 IVA
C7— 1* —CG
QT115—CGE
(10) CONTINUE
I-FETCH
(11) INVALID
ADDRESS
00 — 0148
E 1111
A LRL→M_JK
B W67→LB
D 3→MB
C E(23)→LSFN
R BQ(A=1)→B
R 1SYLS IVA
G7— ** —GG
QT115—GGE
(00,01,11)
CONTINUE
I-FETCH
L23→XG01
T→L,M 0→REFETCH
T(12–15)→J
T(16–19)→MD
(X=0)→S0
(B=0)→S1
SET ILC,1SYL
01 — 0149
E 1111
B W67→LB
D 3→MB
C E(23)→LSFN
R AQ(B=1)→A
R 0 IVA
L7— ** —LG
QT115—LGE
(00) REFETCH
(11) INVALID
ADDRESS
11 — XXX
GO TO
QT200.NBE
Q9— 11 —QI
QT110

Figure 8.32

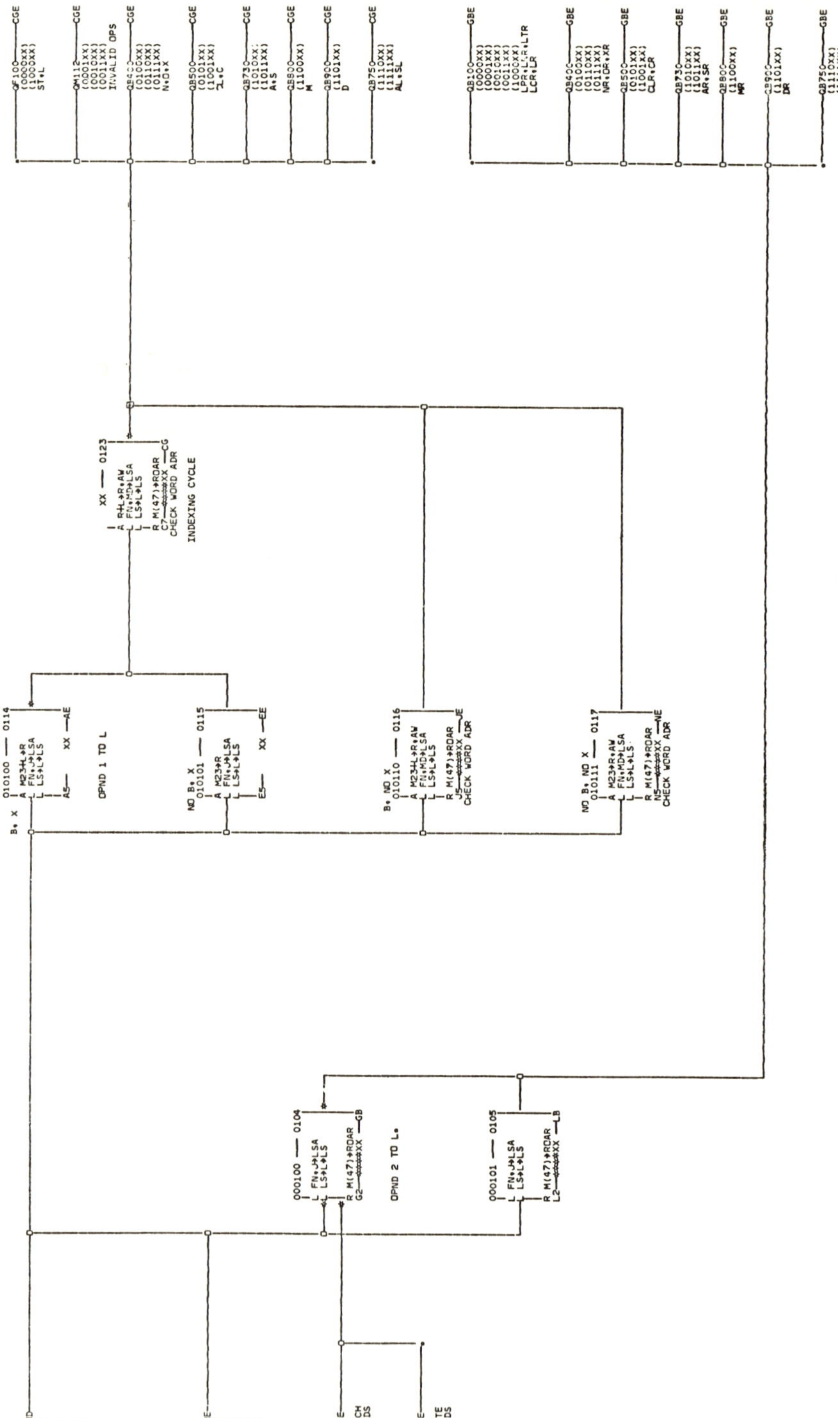

Figure 8.33

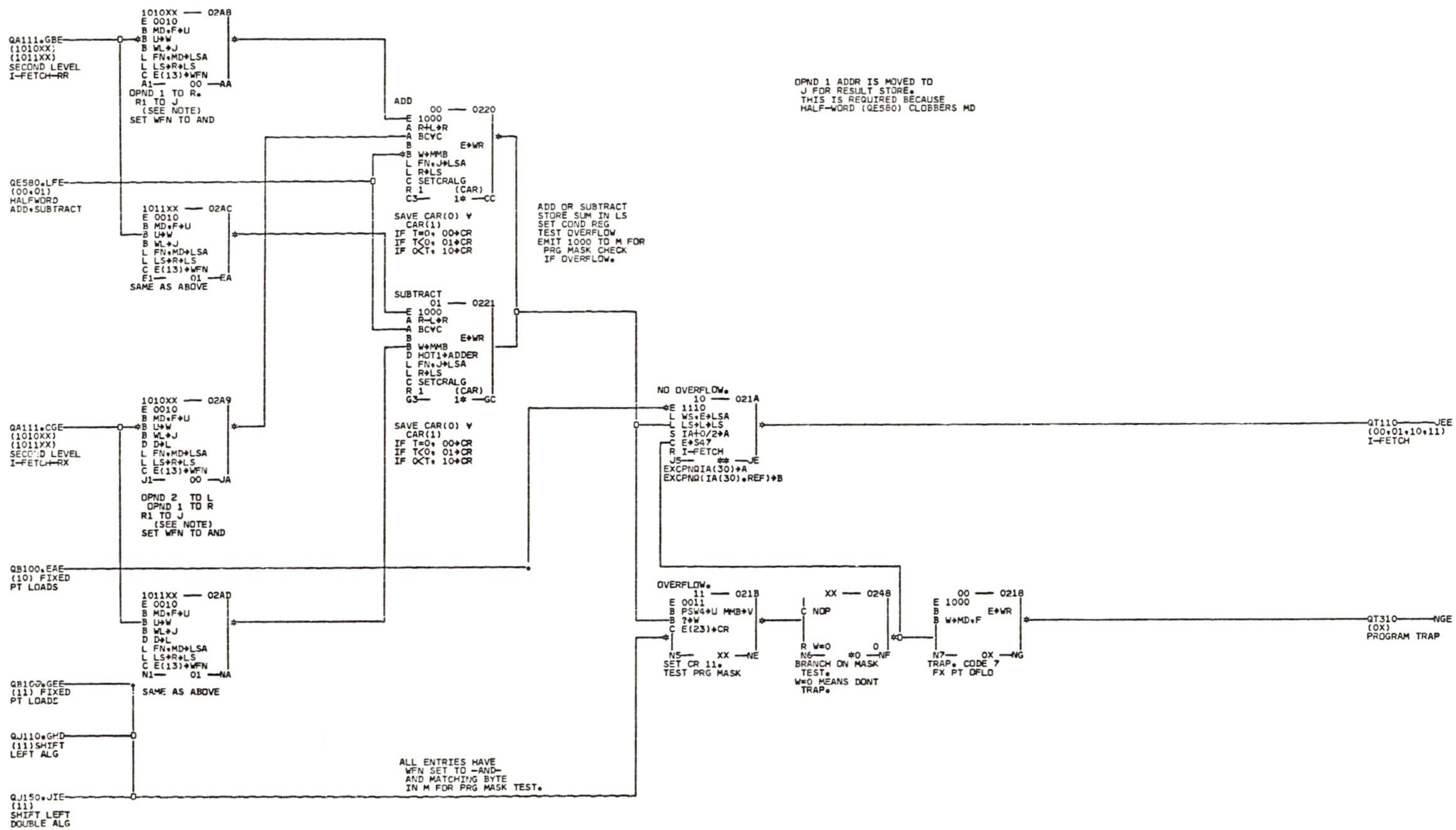
QA111.GBE
(1010XX)
(1011XX)
SECOND LEVEL
I-FETCH—RR
1010XX — 02A8
E 0010
B MD+F→U
B U→W
B WL→J
L FN+MD→LSA
L LS→R+LS
C E(13)→WFN
A1— 00 —AA
OPND 1 TO R.
R1 TO J
(SEE NOTE)
SET WFN TO AND
QE580.LFE
(00,01)
HALFWORD
ADD,SUBTRACT
1011XX — 02AC
E 0010
B MD+F→U
B U→W
B WL→J
L FN+MD→LSA
L LS→R+LS
C E(13)→WFN
E1— 01 —EA
SAME AS ABOVE
QA111.CGE
(1010XX)
(1011XX)
SECOND LEVEL
I-FETCH—RX
1010XX — 02A9
E 0010
B MD+F→U
B U→W
B WL→J
D D→L
L FN+MD→LSA
L LS→R+LS
C E(13)→WFN
J1— 00 —JA
OPND 2 TO L
OPND 1 TO R
R1 TO J
(SEE NOTE)
SET WFN TO AND
QB100.EAE
(10) FIXED
PT LOADS
1011XX — 02AD
E 0010
B MD+F→U
B U→W
B WL→J
D D→L
L FN+MD→LSA
L LS→R+LS
C E(13)→WFN
N1— 01 —NA
SAME AS ABOVE
QB100.GEE
(11) FIXED
PT LOADS
QJ110.GHD
(11)SHIFT
LEFT ALG
QJ150.JIE
(11)
SHIFT LEFT
DOUBLE ALG
ADD
00 — 0220
E 1000
A R+L→R
A BC∀C
B E→WR
B W→MMB
L FN,J→LSA
L R→LS
C SETCRALG
R 1 (CAR)
C3— 1* —CC
SAVE CAR(0) ∀
CAR(1)
IF T=0, 00→CR
IF T<0, 01→CR
IF 0<T, 10→CR
SUBTRACT
01 — 0221
E 1000
A R−L→R
A BC∀C
B E→WR
B W→MMB
D HOT1→ADDER
L FN,J→LSA
L R→LS
C SETCRALG
R 1 (CAR)
G3— 1* —GC
SAVE CAR(0) ∀
CAR(1)
IF T=0, 00→CR
IF T<0, 01→CR
IF 0<T, 10→CR
ADD OR SUBTRACT
STORE SUM IN LS
SET COND REG
TEST OVERFLOW
EMIT 1000 TO M FOR
PRG MASK CHECK
IF OVERFLOW.
OPND 1 ADDR IS MOVED TO
J FOR RESULT STORE.
THIS IS REQUIRED BECAUSE
HALF-WORD (QE580) CLOBBERS MD
NO OVERFLOW.
10 — 021A
E 1110
L WS,E→LSA
L LS+L→LS
S IA+0/2→A
C E→S47
R I-FETCH
J5— ** —JE
EXCPNΩIA(30)→A
EXCPNΩ(IA(30),REF)→B
QT110——JEE
(00,01,10,11)
I-FETCH
OVERFLOW.
11 — 021B
E 0011
B PSW4→U MMB→V
B ?→W
C E(23)→CR
N5— XX —NE
SET CR 11.
TEST PRG MASK
XX — 0248
C NOP
R W=0
N6— ≠0 0 —NF
BRANCH ON MASK
TEST.
W=0 MEANS DONT
TRAP.
00 — 0218
E 1000
B E→WR
B W→MD,F
N7— 0X —NG
TRAP. CODE 7
FX PT OFLO
QT310——NGE
(0X)
PROGRAM TRAP
ALL ENTRIES HAVE
WFN SET TO —AND—
AND MATCHING BYTE
IN M FOR PRG MASK TEST.

Figure 8.34

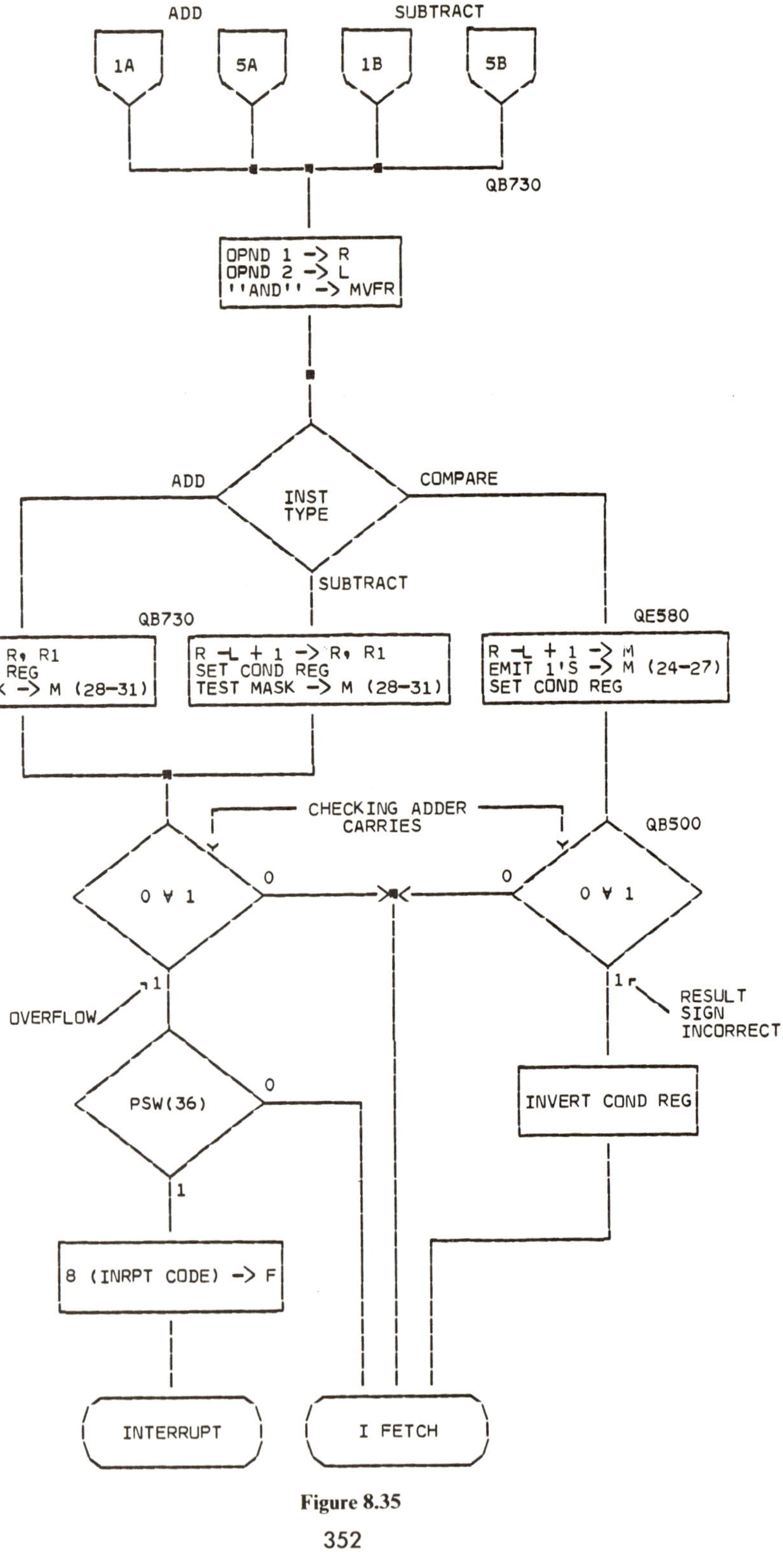
ADD
SUBTRACT
1A
5A
1B
5B
QB730
OPND 1 -> R
OPND 2 -> L
''AND'' -> MVFR
INST TYPE
ADD
COMPARE
SUBTRACT
QB730
R + L -> R, R1
SET COND REG
TEST MASK -> M (28-31)
R ¬L + 1 -> R, R1
SET COND REG
TEST MASK -> M (28-31)
QE580
R ¬L + 1 -> M
EMIT 1'S -> M (24-27)
SET COND REG
CHECKING ADDER CARRIES
QB500
0 ∀ 1
0
0
0 ∀ 1
1
OVERFLOW
1
RESULT SIGN INCORRECT
PSW(36)
0
1
INVERT COND REG
8 (INRPT CODE) -> F
INTERRUPT
I FETCH

Figure 8.35

The third line is also a single micro-order which, when executed, transfers the contents of the addressed local-storage word, places it into the R register, and regenerates from the R register.

4.

$$\begin{bmatrix} \text{E: } 0100 \\ \text{A: } R + L \rightarrow R, \text{AW} \\ \text{C: } E \rightarrow S03 \end{bmatrix}$$

Line 1: Sets the emit field to 0100.

Line 2: Gates the contents of the R register to the right input of the adder, and the uninverted (true) bits of the L register to the left input of the adder. Performs the binary addition, places the result back into the R register and SAR register, and initiates the storage cycle with a command to trap if the address is not on the word boundary.

Line 3: Turns on stats 0–3 per emit field.

5.

$$\begin{bmatrix} \text{E: } 0011 \\ \text{L: FN, J} \rightarrow \text{LSA} \\ \text{L: LS} \rightarrow \text{R,} \quad \text{L} \rightarrow \text{LS} \\ \text{C: E(23)} \rightarrow \text{LSFN} \end{bmatrix}$$

Line 1: Sets the emit field to 0011.

Line 2: Assembles the local-storage address bits by transferring the two bits in the local-storage function register into the two leftmost (high-order) bit positions of the local-storage address register. Note that the two bits in the local-storage function register would have been set early in the cycle by the micro-order in line four. The remaining four bits of the local-storage address register are brought from register J as specified by the second line.

Line 3: Reads (destructively) the contents of the addressed local-storage word into register R and regenerates the local-storage word from register L.

If we assume the initial contents of the registers to be:

Register J = 1111
Register R = 12345678
Register L = 13579ACE
G.P. Register 16 = ABCDEF13

then, after this microinstruction has been executed:

Register R = ABCDEF13
Register L = 13579ACE (remains the same)
G.P. Register 16 = 13579ACE

6.

$$\begin{bmatrix} \text{E: } 0010 \\ \text{B: LLB} \rightarrow \text{U;} \quad \text{MMB} \rightarrow \text{V} \\ \text{B: ?} \rightarrow \text{WL;} \quad \text{?} \rightarrow \text{WR} \\ \text{B: W} \rightarrow \text{MMB} \\ \text{D: } 0 \rightarrow \text{LB, MB} \\ \text{C: E(13)} \rightarrow \text{WFN} \end{bmatrix}$$

This microinstruction deals with the mover. It deals with setting the mover function register, resetting the LB and MB controls, and specifying the sources of data for the left and right inputs of the mover. It performs an indirect mover function and then specifies the W latches' (mover output) destinations.

Line 1: Sets the emit field to 0010.

Line 2: Gates the L register to the left mover input U. The byte selection is controlled by the L-byte counter. Gates M register to the right mover input V. Byte selection is controlled by the M-byte counter.

Line 3: Gates to the mover output latches the result of the operation specified indirectly by the mover function register setting.

Now assume that the mover function register is set to 100. This will result in straight gating of U bits (0–3) to W latch bits (0–3), and V bits (4–7) to W latch bits (4–7), i.e., forms a new byte by concatenating the high-order half byte of U with the low-order half byte of V.

Line 4: Specifies the mover output destination in specific; gates the mover latches to the byte of the M register specified by the M-byte counter.

Line 5: Sets both the LB and MB byte counters to zero.

Line 6: Transmits the three leftmost bits of the emit field (010) to the mover function register. This transfer is not effective until the beginning of the next cycle.

7.
[L: FN, MD → LSA
L: LS → L → LS
R: M(47) → ROAR]

The first and second lines deal with the assembly of the local-storage address from the two-bit local-storage function register and four-bit general purpose MD register (or counter). Then they deal with reading the contents of the selected local-storage word into the L register and regenerating the source information into local storage.

The novel feature in this microinstruction is the micro-order in line 3, which gates M-register bits (4–7) to the next ROS address bits (6–9). This micro-order is very useful in instruction fetch and decode cycles.

8.
[E: 0111
A: R − L → L, M
B: E → WL
B: WL → J
D: Hot 1 → ADDER
R: T = 0
x*]

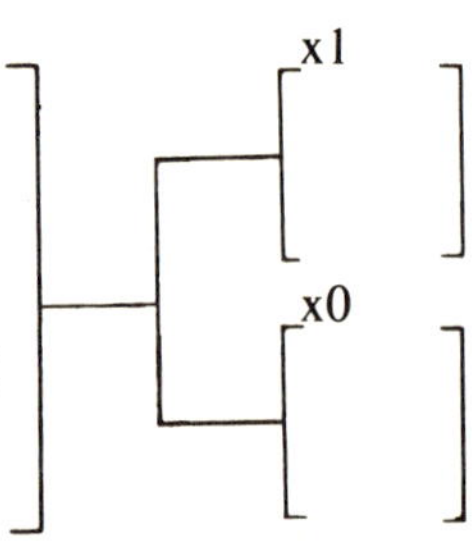

This example illustrates the subtraction of the contents of L from the contents of R by the 2's complement of L added to the true contents of R. The

2's complement operation is shown on two lines in the microinstruction box. The minus sign in line 2 implies taking the 1's complement of L, and line 5 forces a binary 1 (Hot 1) to the low-order bit position of the adder.

Lines 3 and 4 gate the emit field to the left half of the W latches (note that zeros will be gated to the right half of the W latches). Then they gate the output of the left half of the W latches to the four-bit register J.

Line 6 specifies a B-bit condition test. If the adder output bus is zero, it sets the B bit of the ROAR to one; otherwise, it sets the B bit to zero. Note that this micro-order tests the output of the adder operation specified in the previous cycle. This point will be illustrated in the next exercise.

9. Assume that the R register contains the address of the first operand in main storage, and that the L register contains the address of the second operand in main storage.

Write the microroutine that will fetch these two operands, subtract the second operand from the first, and then store the result back in main storage where the first operand was. Also, set the condition lock bits of the PSW (34, 35) to zero if the result is zero, and to one if the result is not zero.

There are many ways to solve this problem. The solution given in Fig. 8.36 is not the most efficient; however, it illustrates a few points worth examining.

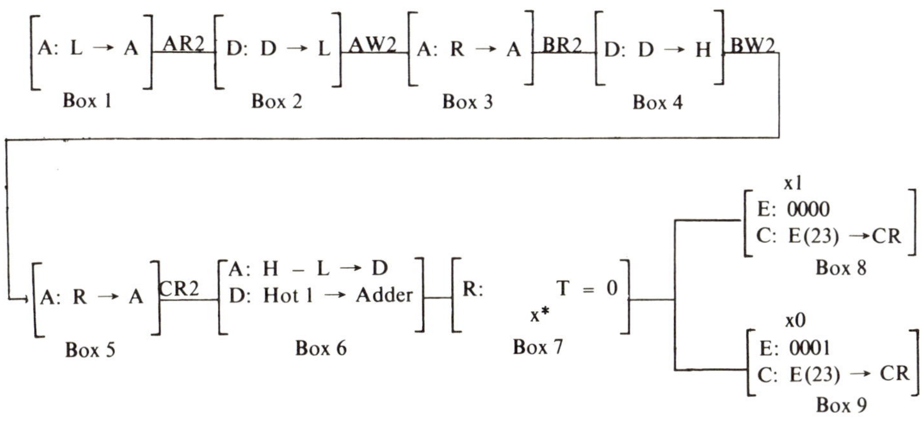

Figure 8.36

Microinstruction 1 gates the address of the second operand to the SAR register and initiates a request cycle. It forms the R1 cycle of the main storage ring (R1R2W1W2). Microinstruction 2 gates the output of the storage data register D to the L register. In the main-storage timing section of this chapter, it was indicated that this operation is not permissible until the beginning of the W1 cycle. Where then is the R2 cycle? The answer is: since that microinstruction would be null in this problem and to avoid wasting the ROS work, a hardware wait cycle to fill the R2 cycle required by the main-storage timing is initiated at the AR2, BR2, and CR2 points. Similarly, to fill the requirement for W2 cycle during the write (regenerate) main-storage cycle, a wait cycle is forced at the AW2 and BW2 points of the microroutine. Therefore, to bring each of the two operands from main storage, two microinstructions were used to specify this action. At the execution time, however, two wait cycles will be forced to fulfill the four main-storage cycles (R1R2W1W2) required.

The second point illustrated in Fig. 8.36 is the data condition branch test specified in microinstruction box 7. It tests the results of the adder operation specified in microinstruction box 6 and sets up a leg selector to indicate that two successor microinstructions must be incorporated in the microroutine to carry the different actions specified for each condition. Note that this branch test affects the final setting of the low-order bit of the ROSAR. This is called the B bit; hence, the micro-order is a B condition test one.

Another method of effecting this branch operation, and one that will save an ROS storage word, is to preserve the state of the adder output in a special stat 3 by adding the micro-order C: (T = 0 → S3) (if the adder latch is zero, turn on stat 3; otherwise, turn off stat 3) to the microinstruction box. We can assume that more than 50 percent of the times such a test is performed the answer is no (i.e., adder output is zero) and, hence, stat 3 is off. In such a case, the condition code bits in the PSW can be set to one in the cycle in which the branch test per stat 3 is made. This is shown in the following sequence:

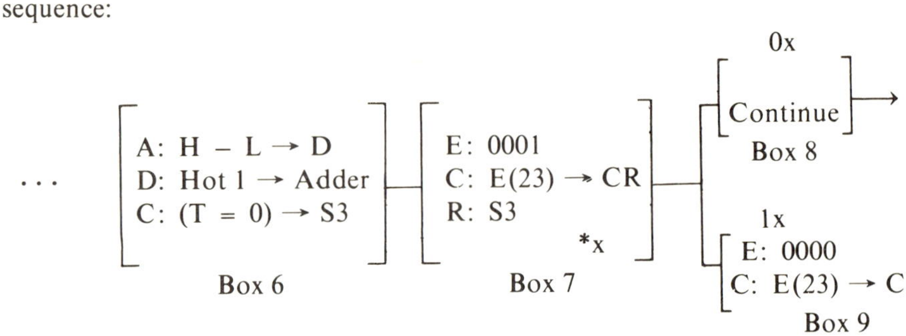

10. Trace the action indicated in the following sequence of microinstructions:

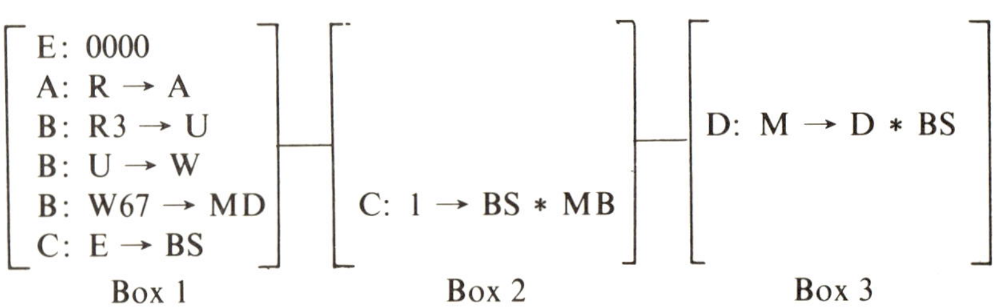

Microinstruction box 1 specifies the number of parallel operations. The new type of micro-order not present in the previous example is shown on the last line with edge character C—namely, E → BS, which gates the emit field to the CPU byte stats.

Box 2 turns on the byte stat indicated by the MB counter. Box 3 gates the contents of the M register to the adder output latches and then gates the adder latch byte to the SDR register under the control of the CPU byte stats.

11.

```
A: 0 → R
L: RN, MD → LSA
L: R → LS
S: IA + 0/2 → A
S: SMIF
R: I-fetch
                **
```

To see how this action is carried out, we must review every micro-order, line by line. The first three lines are simple. Register R is zeroed, and the selected local-storage word is also zeroed. Line 3 has the micro-order IA + 0/2 → A which gates IAR to SAR, incremented by 2 if the refetch stat is off, and not incremented if the refetch stat is on. It initiates storage request and inhibits invalid address trap. Instead, the invalid address stat is set.

If the refetch stat is off and bit 30 is 1 (indicating that the next instruction is off the word boundary), then the SMIF micro-order in line 5 suppresses the initiation of the main storage cycle by the IA + 0/2 → A micro-order.

The last micro-order, I-fetch, is a four-way branch that will set the A and B bits of the ROSAR as follows:

0 → A, 0 → B;	IAR bit 30 is 1 and the refetch stat is off.
0 → A, 1 → B;	IAR bit 30 is 1 and the refetch stat is on.
1 → A, 0 → B;	IAR bit 30 is zero.
1 → A, 1 → B;	Exception trigger is on.

PROBLEMS

1. Given the following contents:

LB Counter = 00
MB Counter = 10
MFR Register = 000

(a) Trace the action in the following microinstruction:

```
E: 0010
B: LLB → U;  MMB → V
B: ? → WL     ? → WR
B: W → MMB
D: 0 → LB, MB
C: E(13) → WFN
```

(b) The effective mover function action for this cycle is the cross-gating of U(4–7) to W(0–3), and U(0–3) to W(4–7). True or false?

2. Trace the microinstructions:

(a)
```
D: D → L
L: FN, MD → LSA
L: LS → R → LS
              XX
```

(b)
```
A: R → R
A:                Q → SR1 → Q
D: LB − 1
R:                    LB = 0
                        x*
```

(c)
```
E: 0001
C: S47.7E
R: S3    S5
      **
```

3. Assume that A and B are two positive integers that occupy registers R and L, respectively. Write the microroutine that will execute $2 * (A - B)$ and store the result in the general purpose register 8.

4. Assume that the two positive integers A and B are located in the first and the second words of the second quadrant in local storage. Write the microroutine that will retrieve these two operands, perform the function $(A - B)^2$, and then store the result in the main-storage word whose address is specified by the H register.

5. The System/360, RR fixed-point add instruction has just been fetched and decoded. Register MD contains the R1 field, and the J register contains the R2 field of the instruction. Write the execution routine that will retrieve the contents of the general purpose registers specified by the MD and J registers, respectively. Add the results, set the condition coding bits of the PSW to 00 if the result is zero, to 01 if the result is negative, and to 10 if the result is positive. Then store the sum into the general purpose register specified by the MD register.

6. Trace the first- and second-level I-fetch microroutines shown in Figs. 8.31–8.34 for the fixed-point RR Op Code LNR $(11)_{16}$.

7. The two System/360, RS instructions, store-multiple and load-multiple, with operation codes 90 and 98, share the same instruction-fetch sequence with one exception: stat 3 is set to zero if the load-multiple instruction is decoded; it is set on if the store-multiple instruction is decoded. Assume that the starting address in main storage is specified by register R and that the starting address of the general purpose register is given in the MD register. Write the common microroutine to execute the particular instruction depending on the setting of stat 3. At the end of the execution, branch to the instruction-fetch routine.

REFERENCES

1. *System/360 Model 50—Comprehensive Introduction, Field Engineering Manual, Form Y23-2821-0.*

2. *System/360 Model 50—Functional Units, Field Engineering Manual, Form Y22-2822-1.*

3. Schnabel, D. L., "The Design of Processor Controls Using a Read Only Storage," *IBM Technical Report TR00.1318* (August 19, 1965).

4. *System/360 Model 50—Multiplexor Channel, Field Engineering Manual, Form Y22-2827-1.*

5. *System/360 Model 50—Selector Channel, Field Engineering Manual, Form Y22-2826-1.*

APPENDIX 8A: CPU STATUS AFTER SECOND-LEVEL I-FETCH*

8A.1 Op Codes 0X–3X

		0X	*1X*	*2X, 3X*
Registers				
R		R2 contents	?	0
L		?	R2 contents	Ho contents of FP register (R2)
M	(0–7)	Op Code	Op Code	Op Code
	(8–11)	R1 address	R1 address	R1 address
	(12–15)	R2 address	R2 address	R2 address
	(16–19)	0	0	0
H	(8–31)	IAR	IAR	?
J		R2 address	R2 address	R2 address
F		?	?	0
Counters				
G1		0	0	0
G2		0	0	0
MD		R1 address	R1 address	R1 address
LB		1	1	1
MB		0	3	3
Stats				
0		1 if R2 address = 0	1 if R2 address = 0	1 if R2 address = 0
1		0	?	?
2		0	?	?
3		0	?	?
4		1	1	1
5		1	1	1
6		1	1	1
7		0	0	0
Carry stat		?	?	Off for 2X; on for 3X
Byte stats		1111	1111	1111
Other				
LSFR		11 (fixed-point)	11 (fixed-point)	10 (floating-point)
Main Storage				
Cycle		W2	W2	W2
Working Storage (for all Op Codes)				
7		Ho PSW (0–15)		
14		Last 32-bit word fetched during an I-fetch (excluding a refetch)		

*This material was prepared by R. B. Stuart of IBM and is used with his permission.

8A.2 Op Codes 4X–7X

		4X	5X	6X, 7X
Registers				
R		Operand 1	Operand-2 address	Ho contents of FP register (R1)
L		Operand-2 address	Operand 1	B2 or X2 contents
M	(0–7)	Op Code	Op Code	0
	(8–11)	R1 address	R1 address	0
	(12–15)	X2 address	X2 address	0
	(16–19)	0	0	0
	(20–31)	D2	D2	0
H	(8–31)	IAR	?	?
J		X2 address	X2 address	X2 address
Counters				
G1		0	0	0
G2		0	0	0
MD		R1 address	R1 address	R1 address
LB		3	1	1
MB		0	3	3
Stats				
0		0	1 if X2 address = 0	1
1		0	1 if B2 address = 0	1
2		0	?	1
3		0	?	0 for 6X; 1 for 7X
4		1	1	1
5		1	1	1
6		1	1	1
7		0	0	0
Byte stats		1111	1111	1111
Other				
LSFR		11 (fixed-point)	11 (fixed-point)	10 (floating-point)
Main Storage				
SAR		?	Operand-2 address (interrupt if not a fullword address)	Operand-2 address (interrupt if not a doubleword address on 6X, or a fullword address on 7X)
Cycle		W2	R1	R2
Working Storage (for all Op Codes)				
7		Ho PSW (0–15)		
14		Last 32-bit word fetched during an I-fetch (excluding a refetch)		

8A.3 Op Codes 8X–9X

	8X	*9X*
Registers		
R	(Operand-2 address)	
L	(Contents of the R1 general register)	
M	(Op code/R1R3/base and decrement)	
H	?	?
J	(R3 address)	
F	0	?
Counters		
G1	0	0
G2	0	0
MD	(R1 address)	
LBC	1	1
MBC	3	3
Stats		
0	On if R3 = 0	0
1	On if B2 = 0	0
2	?	1
3	?	0
4	1	1
5	1	1
6	1	1
7	0	0
Carry	?	?
Byte stats	1111	1111
R sign	?	?
L sign	?	?
Other		
LSFN Register	11	11
WFN Register	?	?
Cond Register	?	?
Main Storage		
SAR	?	?
Cycle	W2	W2
Working Store		
1	?	Operand-2 address
7	(Bits 0–15 of the PSW for a backup area)	
E	(The last word fetched during this I-fetch)	

8A.4 Op Codes DX–FX

	DX	*FX*
Registers		
R	B1–D1 address	Rightmost operand-2 byte address
L	B2–D2 address	Rightmost operand-1 byte address
M	B2–D2 address	Rightmost operand-1 byte address
H	?	Rightmost operand-1 byte address
J	L2 field	L2 field
F	---------------(Second digit of Op Code)---------------	
Counters		
G1	L field	Li field
G2		L2 field
MD	B2 address	L1 field
LBC	0	(Byte address of rightmost operand-2 word)
MBC	(Byte address of leftmost operand-1)	(Byte address of rightmost operand-1 word)
Stats		
GP stats 0–7	0000 0001	0000 0000
Carry	?	?
Byte stats	0000	0000
R sign	?	Off
L sign	?	?
Other		
LSFN register	11	11
WFN register	?	000 (cross from U)
Cond register	?	?
Main Storage		
SAR	?	(Rightmost operand-2 address)
Cycle	(Completed)	R2
Working Store		
1	(Leftmost operand-1 byte address)	(Rightmost operand-1 byte address)
2	(Leftmost operand-2 byte address)	(Rightmost operand-2 byte address)
7	-----------(Bits 0–15 of the PSW for a backup area)-----------	
9	?	(B1–D1 address)
A	?	(B2–D2 address)
B	?	(Op Code/L1 L2/0 ?/L1 L2)
E	-----------(The last word fetched during this I-fetch)-----------	

APPENDIX 8B: MODEL 50 FIELD CONTROL SPECIFICATIONS (MICRO-ORDERS)

LU FIELD ROSDR 1-3 MOVER INPUT - LEFT SIDE

EDGE CHAR	MNEMONIC	BITS	DEC ORDER	FUNCTION	ALD PAGE
		000	LU0	GATE ZEROS TO LEFT MOVER INPUT U.	DS001
B	MD.F→U	001	LU1	GATE MD REG TO LEFT MOVER INPUT U BITS 0-3. GATE F REG TO MOVER INPUT U BITS 4-7.	DS001
B	R3→U	010	LU2	GATE R REG BITS 24-31 TO LEFT MOVER INPUT U.	DS001
B	DCI→U	011	LU3	GATE DIRECT CONTROL DATA IN LINES TO LEFT MOVER INPUT U.	DS001
B	XTR→U	100	LU4	GATE EXTERNAL INTERRUPT REG TO LEFT MOVER INPUT U. RESET EXTERNAL INTERRUPT REG.	DS001
B	PSW4→U	101	LU5	GATE PSW BITS 32-39 (INSTRUCTION LENGTH,CONDITION REG.PROGRAM MASK) TO LEFT MOVER INPUT U.	DS001
B	LMB→U	110	LU6	GATE L REG TO LEFT MOVER INPUT U. BYTE SELECTION CONTROLLED BY M BYTE COUNTER.	DS001
B	LLB→U	111	LU7	GATE L REG TO LEFT MOVER INPUT U. BYTE SELECTION CONTROLLED BY L BYTE COUNTER.	DS001

MV FIELD ROSDR 4-5 MOVER INPUT - RIGHT SIDE

EDGE CHAR	MNEMONIC	BITS	DEC ORDER	FUNCTION	ALD PAGE
		00	MV0	GATE ZEROS TO RIGHT MOVER INPUT V.	DS021
B	MLB→V	01	MV1	GATE M REG TO RIGHT MOVER INPUT V. BYTE SELECTION CONTROLLED BY L BYTE COUNTER.	DS021
B	MMB→V	10	MV2	GATE M REG TO RIGHT MOVER INPUT V. BYTE SELECTION CONTROLLED BY M BYTE COUNTER.	DS021
		11	MV3	UNDEFINED. NOTE 1	

ZP FIELD ROSDR 6-11 ROS ADDRESS BITS 0-5

THIS FIELD IS GATED DIRECTLY TO FORM BITS 0-5 OF THE NEXT ROS ADDRESS.

NOTE 1 1410-7010 USAGE OF UNDEFINED ORDERS

EDGE CHAR	MNEMONIC	BITS	DEC ORDER	FUNCTION	ALD PAGE
B	R3→V	11	MV3	GATE R REG BIT 24-31 TO RIGHT MOVER INPUT V.	BV001

CF101 01

ZF FIELD   ROSDR 12-15   FUNCTION BRANCH CONTROL OR ROS ADDRESS BITS 6-9

IF THE ZN FIELD IS NON-ZERO, THIS FIELD IS GATED DIRECTLY TO FORM BITS 6-9 OF THE NEXT ROS ADDRESS.
IF THE ZN FIELD IS ZERO, THIS FIELD IS DECODED AS SHOWN BELOW.

EDGE CHAR	MNEMONIC	BITS	DEC ORDER	FUNCTION	ALD PAGE
		0000	ZF0	UNDEFINED	
R	D→ROAR.SCAN	0010	ZF2	GATE STORAGE DATA REG BITS AS FOLLOWS: BITS 19-30 TO ROS ADDRESS	KK002
		0100	ZF4	UNDEFINED	
R	M(03)→ROAR	0110	ZF6	GATE M REG BITS 0-3 TO NEXT ROS ADDRESS BITS 6-9.	KK002
R	M(47)→ROAR	1000	ZF8	GATE M REG BITS 4-7 TO NEXT ROS ADDRESS BITS 6-9.	KK002
R	F→ROAR	1010	ZF10	GATE F REG TO NEXT ROS ADDRESS BITS 6-9.	KK002
R	ED→ROAR	1100	ZF12	GATE EXPONENT DIFFERENCE REG TO NEXT ROS ADDRESS BITS 6-9.	KK002
R	RETURN→ROAR	1110	ZF14	GATE ROS ADDRESS BUFFER REG TO NEXT ROS ADDRESS. THIS REGISTER IS SET ON I/O BREAK-IN.	KK002

ZN FIELD   ROSDR 16-18   ROS ADDRESS CONTROL   (NULL VALUE: ZN4)

EDGE CHAR	MNEMONIC	BITS	DEC ORDER	FUNCTION	ALD PAGE
		000	ZN0	DECODE ZF FIELD	
S	SMIF	001	ZN1	SUPPRESS INITIATION OF MAIN STORAGE CYCLE BY DECIMAL ORDER IV7 IN THIS MICROINSTRUCTION IF REFETCH STAT IS OFF AND IAR BIT 30 IS 1. (NEXT INSTRUCTION IS OFF WORD BOUNDARY.)	KK003
R	AΩ(B=0)→A	010	ZN2	FORCE A BIT TO 1 IF B BIT IS 0.	KK003
R	AΩ(B=1)→A	011	ZN3	FORCE A BIT TO 1 IF B BIT IS 1.	KK003
		100	ZN4	NORMAL ROS ADDRESSING.	
R	FNTRAP	101	ZN5	FORCE ROS ADDRESS TO INVALID OP TRAP VALUE IF THIS MICROINSTRUCTION WAS REACHED BY A FUNCTION BRANCH.	KK003
R	BΩ(A=0)→B	110	ZN6	FORCE B BIT TO 1 IF A BIT IS 0.	KK003
R	BΩ(A=1)→B	111	ZN7	FORCE B BIT TO 1 IF A BIT IS 1.	KK003

11 MAR 65 255099
3 AUG 65 255449
11 JAN 66 255458
10 MAY 66 255793

CONTROL FIELD SPECIFICATION	
CPU MODE	
DATE 10 MAY 66	MACH. 2050
	FRAME 01
	P.N. 5365003
IBM CORP. SDD	PAGE 1

CF101

```
TEMPLATE FOR 'A' LINE STATEMENT:  RRR±LLL→TTTT

TEMPLATE FOR 'D' LINE STATEMENT:  AAAA→TTTT

   WHERE: R = RY FIELD     L = LX FIELD     T = TR FIELD
          I = TC FIELD     A = AL FIELD
```

TR FIELD ROSDR 19–23 ADDER LATCH DESTINATION

EDGE CHAR	MNEMONIC	BITS	DEC ORDER	FUNCTION	ALD PAGE
A	Z			DUMMY, USED IN 'A' LINE STATEMENT WHEN THERE IS ALSO A 'D' LINE STATEMENT.	
		00000	TR0	NO GATING FROM ADDER LATCH TO REGISTERS.	
A.D	T	00000	TR0	NO GATING FROM ADDER LATCH TO REGISTERS. USED WHEN OTHER TERMS OF STATEMENT ARE PRESENT.	
A.D	R	00001	TR1	GATE ADDER LATCH TO R REG.	DR121
A.D	R0	00010	TR2	GATE ADDER LATCH BITS 0–7 TO R REG BITS 0–7.	DR121
A.D	M	00011	TR3	GATE ADDER LATCH TO M REG.	DR121
A.D	D	00100	TR4	GATE ADDER LATCH TO SDR.	DR111
A.D	L0	00101	TR5	GATE ADDER LATCH BITS 0–7 TO L REG BITS 0–7.	DR111
A.D	R.A	00110	TR6	GATE ADDER LATCH TO R REG AND SAR. INITIATE STORAGE REQUEST.	DR111
A.D	L	00111	TR7	GATE ADDER LATCH TO L REG.	DR111
S	HA→A	01000	TR8	GATE HARDWARE-GENERATED ADDRESS TO SAR. INITIATE STORAGE REQUEST. DECODE EMIT FIELD TO DETERMINE STORAGE FUNCTION.	DR121

EMIT VALUE	FUNCTION
X000	WRITE IN MAIN STORAGE ADDRESS 80.
X001	READ FROM MAIN STORAGE ADDRESS 80.
X010	OR FROM MAIN STORAGE ADDRESS 80.
X011	UNDEFINED
X100	WRITE IN MAIN STORAGE ADDRESS 84. SET IAR TO 84.
X101	READ FROM MAIN STORAGE ADDRESS 84. SET IAR TO 84.
X110	UNDEFINED
X111	UNDEFINED

EDGE CHAR	MNEMONIC	BITS	DEC ORDER	FUNCTION	ALD PAGE
A.D	R.AN	01001	TR9	GATE ADDER LATCH TO R REG AND SAR. INITIATE STORAGE REQUEST. SUPPRESS INVALID ADDRESS TRAP. SET INVALID ADDRESS STAT INSTEAD.	DR121
A.D	R.AW	01010	TR10	GATE ADDER LATCH TO R REG AND SAR. INITIATE STORAGE REQUEST. TRAP IF ADDRESS NOT ON WORD BOUNDRY.	DR121
A.D	R.AD	01011	TR11	GATE ADDER LATCH TO R REG AND SAR. INITIATE STORAGE REQUEST TRAP IF ADDRESS NOT ON DOUBLE WORD BOUNDARY.	DR121
D*	D→IAR	01100	TR12	GATE SDR TO IAR. INTERLOCK WITH STORAGE TIMING RING.	DR121
D*	SCAN→D	01101	TR13	GATE SCAN BUS TO SDR IN FOLLOWING ARRANGEMENT: PARITY OF BYTE 0 TO BIT 0. BITS 0–7 TO BITS 1–8. PARITY OF BYTE 1 TO BIT 9. BITS 8–15 TO BITS 10–17. PARITY OF BYTE 2 TO BIT 18. BITS 16–23 TO BITS 19–26. PARITY OF BYTE 3 TO BIT 27. BITS 24–27 TO BITS 28–31. GOOD PARITY IS INSERTED IN SDR.	DR121
A.D	R13	01110	TR14	GATE ADDER LATCH BITS 8–31 TO R REG BITS 8–31.	DR121
A.D	A	01111	TR15	GATE ADDER LATCH BITS 8–31 TO SAR. INITIATE STORAGE REQUEST.	DR131
A.D	L.A	10000	TR16	GATE ADDER LATCH TO L REG AND SAR. INITIATE STORAGE REQUEST.	DR131
A.D	R.D	10001	TR17	GATE ADDER LATCH TO R REG AND SDR .	DR121
		10010	TR18	UNDEFINED	
A.D	R.IO	10011	TR19	GATE ADDER LATCH TO R REG. GATE ADDER LATCH BITS 30–31 TO I/O REG.	DR131

* NOT PART OF A STATEMENT

CFC 102 – 2050 CONTROL FIELD SPECIFICATION – CPU MODE (TR)

CF102 01

TR FIELD (CONTINUED)

EDGE CHAR	MNEMONIC	BITS	DEC ORDER	FUNCTION	ALD PAGE
A,D	H	10100	TR20	GATE ADDER LATCH TO H REG.	DR141
A,D	IA	10101	TR21	GATE ADDER LATCH BITS 8-31 TO IAR.	DR141
A,D	FOLD	10110	TR22	GATE SCAN BUS BITS 28-31 TO SDR BITS 24-27. GATE ZEROS TO SDR BITS 0-23 AND 28-31. GOOD PARITY IS INSERTED IN SDR.	DR141
		10111	TR23	UNDEFINED	
A,D	L,M	11000	TR24	GATE ADDER LATCH TO L REG AND M REG.	DR111
A,D	MLJK	11001	TR25	GATE ADDER LATCH TO L REG AND M REG. GATE LATCH BITS 12-15 TO J REG. GATE LATCH BITS 16-19 TO MD COUNTER. TURN OFF REFETCH STAT. IF LATCH BITS 12-15 ALL ZERO, TURN STAT 0. OTHERWISE TURN OFF STAT 0. IF LATCH BITS 16-19 ALL ZERO, TURN ON STAT 1. OTHERWISE TURN OFF STAT 1. IF LATCH BITS 16-17 ALL ZERO, TURN ON ONE-SYLLABLE STAT. OTHERWISE TURN OFF ONE-SYLLABLE STAT. IF LATCH BITS 0-1 EQUAL 00, SET ILC TO 01. IF LATCH BITS 0-1 EQUAL 01 OR 10, SET ILC TO 10. IF LATCH BITS 0-1 EQUAL 11, SET ILC TO 11.	DR111
A,D	MHL	11010	TR26	GATE ADDER LATCH TO L REG. GATE ADDER LATCH BITS 0-15 TO M REG BITS 16-31. GATE ADDER LATCH BITS 0-3 TO MD COUNTER.	DR111
A,D	MD	11011	TR27	GATE ADDER LATCH BITS 8-11 TO MD COUNTER.	DR111
A,D	M,SP	11100	TR28	GATE ADDER LATCH TO M REG. GATE ADDER LATCH BITS 8-11 TO STORAGE PROTECTION KEY (PSW BITS 8-11).	DR111
A,D	D*BS	11101	TR29	GATE ADDER LATCH TO SDR UNDER CONTROL OF CPU BYTE STATS.	DR111
A,D	L13	11110	TR30	GATE ADDER LATCH BITS 8-31 TO L REG BITS 8-31.	DR111
A,D	J	11111	TR31	GATE ADDER LATCH BITS 12-15 TO J REG.	DR111

ZR FIELD ROSDR 24

0	SELECT BASIC ROS DATA	XG401
1	SELECT EXTENDED ROS DATA	KK302

11 MAR 65 255099
3 AUG 65 255449
11 JAN 66 255458
9 MAY 66 255793

CONTROL FIELD SPECIFICATION CPU MODE	
DATE 9 MAY 66	MACH. 2050
	FRAME 01
	P.N. 5365004
IBM CORP. SDD	PAGE 1

CF102

WS FIELD ROSDR 25–27 LOCAL STORAGE ADDRESSING (NULL VALUE: WS4)

(SEE WT FIELD PAGE CFC 113

EDGE CHAR	MNEMONIC	BITS	DEC ORDER	FUNCTION	ALD PAGE
		000	WS0	UNDEFINED	
L	WS1→LSA	001	WS1	SET LSAR TO 010001 TO ADDRESS WORD 1 OF WORKING STORAGE	KLOXX
L	WS2→LSA	010	WS2	SET LSAR TO 010010 TO ADDRESS WORD 2 OF WORKING STORAGE.	KLOXX
L	WS.E→LSA	011	WS3	SET LSAR BITS 0–1 TO 01. SET LSAR BITS 2–5 TO EMIT FIELD VALUE TO ADDRESS WORKING STORAGE PER THE EMIT FIELD.	KLOXX
L	FN.J→LSA	100	WS4	GATE LOCAL STORAGE FUNCTION REG TO LSAR BITS 0–1. GATE J REG TO LSAR BITS 2–5.	KLOXX
L	FN.JΩ1→LSA	101	WS5	GATE LOCAL STORAGE FUNCTION REG TO LSAR BITS 0–1. GATE J REG TO LSAR BITS 2–5. OR A ONE INTO LSAR BIT 5.	KL031
L	FN.MD→LSA	110	WS6	GATE LOCAL STORAGE FUNCTION REG TO LSAR BITS 0–1. GATE MD REG TO LSAR BITS 2–5.	KLOXX
L	FN.MDΩ1→LSA	111	WS7	GATE LOCAL STORAGE FUNCTION REG TO LSAR BITS 0–1. GATE MD REG TO LSAR BITS 2–5. OR A ONE INTO LSAR BIT 5.	KL031

CF103 01

SF FIELD ROSDR 28–30 LOCAL STORAGE FUNCTION (NULL VALUE: SF7)

EDGE CHAR	MNEMONIC	BITS	DEC ORDER	FUNCTION	ALD PAGE
L	R→LS	000	SF0	WRITE INTO LOCAL STORAGE FROM R REG.	KLOXX
L	LS→L,R→LS	001	SF1	READ LOCAL STORAGE INTO L REG. WRITE INTO LOCAL STORAGE FROM R REG.	KLOXX
L	LS→R→LS	010	SF2	READ LOCAL STORAGE INTO R REG. REGENERATE FROM R REG.	KLOXX
		011	SF3	UNDEFINED	
L	L→LS	100	SF4	WRITE INTO LOCAL STORAGE FROM L REG.	KLOXX
L	LS→R,L→LS	101	SF5	READ LOCAL STORAGE INTO R REG. WRITE INTO LOCAL STORAGE FROM L REG.	KLOXX
L	LS→L→LS	110	SF6	READ LOCAL STORAGE INTO L REG. REGENERATE FROM L REG.	KLOXX
		111	SF7	NO OPERATION.	

IV FIELD ROSDR 32–34 INVALID DIGIT TEST AND INSTRUCTION ADDRESS REG CONTROL

EDGE CHAR	MNEMONIC	BITS	DEC ORDER	FUNCTION	ALD PAGE
		000	IV0	NO OPERATION	
C	WL→IVD	001	IV1	TRAP IF VALUE OF MOVER OUTPUT BITS 0–3 IS GREATER THAN 9.	KP011
C	WR→IVD	010	IV2	TRAP IF VALUE OF MOVER OUTPUT BITS 4–7 IS GREATER THAN 9.	KP011
C	W→IVD	011	IV3	TRAP IF VALUE OF EITHER MOVER OUTPUT BITS 0–3 OR MOVER OUTPUT BITS 4–7 IS GREATER THAN 9.	KP011
S	IA+4→A,IA	100	IV4	INCREMENT IAR BY 4. GATE RESULT TO IAR AND SAR. INITIATE STORAGE REQUEST. INHIBIT INVALID ADDRESS TRAP. SET INVALID ADDRESS STAT INSTEAD.	KP012
C	IA+2/4	101	IV5	IF INSTRUCTION LENGTH CODE VALUE IS 0 OR 1. INCREMENT IAR BY 2. IF ILC VALUE IS 2 OR 3. INCREMENT IAR BY 4. GATE RESULT BACK TO IAR.	KP012
C	IA+2	110	IV6	INCREMENT IAR BY 2. GATE RESULT BACK TO IAR.	KP012
S	IA+0/2→A	111	IV7	GATE IAR TO SAR. INCREMENTED BY 2 IF REFETCH STAT IS OFF. NOT INCREMENTED IF REFETCH STAT IS ON. INITIATE STORAGE REQUEST. INHIBIT INVALID ADDRESS TRAP. SET INVALID ADDRESS STAT INSTEAD. IAR IS NOT ALTERED.	KP012

11 MAR 65 255099
3 AUG 65 255449
11 DEC 65 255458
29 MAR 66 256490

TEMPLATE FOR 'D' LINE STATEMENT: AAAA→TTTT

WHERE: A = AL FIELD T = TR FIELD

AL FIELD ROSDR 35-39 SHIFT CONTROL AND GATING INTO ADDER LATCH

EDGE CHAR	MNEMONIC	BITS	DEC ORDER	FUNCTION	ALD PAGE
		00000	AL0	GATE ADDER OUTPUT TO LATCH, NO SHIFT.	
A	Q→SR1→F	00001	AL1	GATE ADDER OUTPUT TO LATCH, SHIFTED RIGHT 1, ENTERING Q. SPILL BIT ENTERS F, WHICH IS SHIFTED RIGHT.	DR071
A	LO,¬S4→	00010	AL2	GATE ADDER OUTPUT BITS 8-31 TO LATCH BITS 8-31. GATE L REG BITS 1-7 TO LATCH BITS 1-7. SET LATCH BIT 0 TO THE COMPLEMENT OF STAT 4.	DR061
A	+SGN→	00011	AL3	GATE ADDER OUTPUT BITS 1-31 TO LATCH BITS 1-31. SET LATCH BIT 0 TO ZERO.	DR061
A	-SGN→	00100	AL4	GATE ADDER OUTPUT BITS 1-31 TO LATCH BITS 1-31. SET LATCH BIT 0 TO ONE.	DR061
A	LO,S4→	00101	AL5	GATE ADDER OUTPUT BITS 8-31 TO LATCH BITS 8-31. GATE L REG BITS 1-7 TO LATCH BITS 1-7. GATE STAT 4 TO LATCH BIT 0.	DR061
D*	IA→H	00110	AL6	GATE ADDER OUTPUT TO LATCH. NO SHIFT. GATE IAR TO H REG BITS 8-31. H REG BITS 0-7 REMAIN UNCHANGED.	DR081
A	Q→SL→-F	00111	AL7	GATE ADDER OUTPUT TO LATCH, SHIFTED LEFT 1, ENTERING Q. COMPLEMENT OF SPILL BIT ENTERS F, WHICH IS SHIFTED LEFT.	DR081
A	Q→SL1→F	01000	AL8	GATE ADDER OUTPUT TO LATCH, SHIFTED LEFT 1, ENTERING Q. SPILL BIT ENTERS F, WHICH IS SHIFTED LEFT.	DR081
A	F→SL1→F	01001	AL9	GATE ADDER OUTPUT TO LATCH, SHIFTED LEFT 1, ENTERING BIT 0 OF F. SPILL BIT ENTERS F, WHICH IS SHIFTED LEFT.	DR081
A	SL1→Q	01010	AL10	GATE ADDER OUTPUT TO LATCH. SHIFTED LEFT 1. ENTERING ZERO. SPILL BIT ENTERS Q.	DR071
A	Q→SL1	01011	AL11	GATE ADDER OUTPUT TO LATCH, SHIFTED LEFT 1, ENTERING Q. SPILL BIT IS DISCARDED.	DR081
A	SR1→F	01100	AL12	GATE ADDER OUTPUT TO LATCH. SHIFTED RIGHT 1. ENTERING ZERO. SPILL BIT ENTERS F, WHICH IS SHIFTED RIGHT.	DR071
A	SR1→Q	01101	AL13	GATE ADDER OUTPUT TO LATCH. SHIFTED RIGHT 1. ENTERING ZERO. SPILL BIT ENTERS Q.	DR071
A	Q→SR1→Q	01110	AL14	GATE ADDER OUTPUT TO LATCH. SHIFTED RIGHT 1. ENTERING Q. SPILL BIT ENTERS Q.	DR071
A	F→SL1→Q	01111	AL15	GATE ADDER OUTPUT TO LATCH. SHIFTED LEFT 1. ENTERING BIT 0 OF F. SPILL BIT ENTERS Q. F IS SHIFTED LEFT 1. ENTERING ZERO.	DR051
A	SL4→F	10000	AL16	GATE ADDER OUTPUT TO LATCH. SHIFTED LEFT 4. ENTERING ZEROS. SPILL BITS ENTER F.	DR0XX
A	F→SL4→F	10001	AL17	GATE ADDER OUTPUT TO LATCH. SHIFTED LEFT 4. ENTERING F. SPILL BITS ENTER F.	DR081
A	FPSL4	10010	AL18	GATE ADDER OUTPUT BITS 0-7 TO LATCH 0-7. GATE ADDER OUTPUT BITS 8-31 TO LATCH BITS 8-31. SHIFTED LEFT 4. ENTERING ZEROS. SPILL BITS FROM ADDER OUTPUT 8-11 ARE DISCARDED.	DR081
A	F→FPSL4	10011	AL19	GATE ADDER OUTPUT BITS 0-7 TO LATCH BITS 0-7. GATE ADDER OUTPUT BITS 8-31 TO LATCH BITS 8-31. SHIFTED LEFT 4. ENTERING F. SPILL BITS FROM ADDER OUTPUT 8-11 ARE DISCARDED.	DR081

* NOT PART OF A STATEMENT.

CFC 104 - 2050 CONTROL FIELD SPECIFICATION - CPU MODE (AL)

CF104 01

AL FIELD (CONTINUED)

EDGE CHAR	MNEMONIC	BITS	DEC ORDER	FUNCTION	ALD PAGE
A	SR4→F	10100	AL20	GATE ADDER OUTPUT TO LATCH, SHIFTED RIGHT 4, ENTERING ZEROS. SPILL BITS ENTER F.	DROXX
A	F→SR4→F	10101	AL21	GATE ADDER OUTPUT TO LATCH, SHIFTED RIGHT 4, ENTERING F. SPILL BITS ENTER F.	DROXX
A	FPSR4→F	10110	AL22	GATE ADDER OUTPUT BITS 8-31 TO LATCH BITS 8-31, SHIFTED RIGHT 4, ENTERING ZEROS. GATE ZEROS TO LATCH BITS 0-7. SPILL BITS ENTER F.	DROXX
A	1→FPSR4→F	10111	AL23	GATE ADDER OUTPUT BITS 0-7 TO LATCH BITS 0-7. GATE ADDER OUTPUT BITS 8-31 TO LATCH BITS 8-31, SHIFTED RIGHT 4, ENTERING 0001. SPILL BITS ENTER F. A HOT CARRY IS FORCED INTO POSITION 7 OF THE ADDER. (THIS ORDER CAN BE USED ONLY WITH AN AD FIELD VALUE OF AD6.)	DR081
A	SR4→H	11000	AL24	GATE ADDER OUTPUT TO LATCH, SHIFTED RIGHT 4, ENTERING ZEROS. SPILL BITS ENTER H REG BITS 0-3. GATE LATCH BITS 4-7 TO R REG BITS 0-3. (THIS ORDER CAN BE USED ONLY WITH A TR FIELD VALUE OF TR1, AND LATCH BITS 4-7 MUST HAVE AN EVEN NUMBER OF ONES.)	DR052
A	F→SR4	11001	AL25	GATE ADDER OUTPUT TO LATCH, SHIFTED RIGHT 4, ENTERING F. SPILL BITS ARE DISCARDED. F IS NOT ALTERED.	DR052
A	E→FPSL4	11010	AL26	GATE ADDER OUTPUT BITS 0-7 TO LATCH BITS 0-7. GATE ADDER OUTPUT BITS 8-31 TO LATCH BITS 8-31, SHIFTED LEFT 4, ENTERING THE EMIT FIELD. SPILL BITS FROM ADDER OUTPUT 8-11 ARE DISCARDED.	DR041
A	F→SR1→Q	11011	AL27	GATE ADDER OUTPUT TO LATCH, SHIFTED RIGHT 1, ENTERING BIT 3 OF F. SPILL BIT ENTERS Q. F IS NOT CHANGED.	DR051
D	DKEY	11100	AL28	GATE DATA KEYS TO LATCH. GATE BITS 28-31 OF DATA KEYS TO F.	DR062
D	CH	11101	AL29	GATE BUS FROM SELECTOR CHANNELS TO LATCH.	DR061
D	D	11110	AL30	GATE STORAGE DATA REG TO LATCH. INTERLOCK WITH STORAGE TIMING RING TO CAUSE POSSIBLE STORAGE HOLDOFF.	DT041
D	AKEY	11111	AL31	GATE ADDRESS KEYS TO LATCH BITS 8-31. GATE ZEROS TO LATCH BITS 0-7.	DR062

11 MAR 255099
3 AUG 65 255449

CONTROL FIELD SPECIFICATION		
CPU MODE		
DATE 16 AUG 65	MACH.	2050
	FRAME	01
	P.N.	5365006
IBM CORP. SDD	PAGE	1

CF104

WM FIELD ROSDR 40–43 MOVER OUTPUT DESTINATION

EDGE CHAR	MNEMONIC	BITS	DEC ORDER	FUNCTION	ALD PAGE
		0000	WM0	NO OPERATION	DS201
B	W→MMB	0001	WM1	GATE MOVER LATCHES TO BYTE OF M REG SPECIFIED BY M BYTE COUNTER. (SEE NOTE 1)	DS201
B	W67→MB	0010	WM2	GATE MOVER LATCH BITS 6–7 TO M BYTE COUNTER.	DS201
B	W67→LB	0011	WM3	GATE MOVER LATCH BITS 6–7 TO L BYTE COUNTER.	DS201
B	W27→PSW4	0100	WM4	GATE MOVER LATCH BITS 2–7 TO PSW REGISTER BITS 34–39 (CONDITION CODE AND PROGRAM MASK).	DS201
B	W→PSW0	0101	WM5	GATE MOVER LATCH TO PSW REGISTER BITS 0–7 (SYSTEM MASK).	DS201
B	WL→J	0110	WM6	GATE MOVER LATCH BITS 0–3 TO J REG.	DS201
B	W→CHCTL	0111	WM7	GATE MOVER LATCH TO CHANNEL CONTROL.	DS201
B	W,E→A(BUMP)	1000	WM8	GATE MOVER LATCH BITS 0–1 TO SAR BITS 14–15. GATE MOVER LATCH BITS 2–7 TO SAR BITS 24–29. GATE EMIT FIELD BITS 2–3 TO SAR BITS 16–17. INITIATE BUMP STORAGE REQUEST.	DS201
B	WL→G1	1001	WM9	GATE MOVER LATCH BITS 0–3 TO LENGTH COUNTER G1. RESET G1 SIGN.	DS201
B	WR→G2	1010	WM10	GATE MOVER LATCH BITS 4–7 TO LENGTH COUNTER G2. RESET G2 SIGN.	DS201
B	W→G	1011	WM11	GATE MOVER LATCH TO COMBINED LENGTH COUNTER G. RESET G1,G2 SIGNS.	DS201
B	W→MMB(E?)	1100	WM12	GATE MOVER LATCH TO BYTE OF M REG SPECIFIED BY M BYTE COUNTER. IN CONJUNCTION WITH ORDERS ULO AND/OR URO AND VALUE OF 1 FOR BIT 12 OF PSW (ASCII BIT), THE EMIT FIELD IS MODIFIED IN THE MOVER AS FOLLOWS, WHERE 0,1,2,3 ARE EMIT FIELD BITS AND 0',1',2',3' ARE VALUES USED IN MOVER: 0' = ¬0 Ω ¬2 1' = 2 Ω ¬1 2' = ¬2 3' = 3	DS201
B	WL→MB	1101	WM13	GATE MOVER LATCH BITS 0–3 TO MD COUNTER.	DS201
B	WR→F	1110	WM14	GATE MOVER LATCH BITS 4–7 TO F REG.	DS201
B	W→MD,F	1111	WM15	GATE MOVER LATCH BITS 0–3 TO MD COUNTER. GATE MOVER LATCH BITS 4–7 TO F REG.	DS201

NOTE 1 7070–7074 EMULATOR USAGE OF ORDER

EDGE CHAR	MNEMONIC	BITS	DEC ORDER	FUNCTION	ALD PAGE
B	W→M3	0001	WM1	GATE MOVER LATCHES TO 3RD BYTE OF M REG. THIS IS HARDWARE FORCE WHICH OVERIDES THE MB COUNTER WHEN T IS ALSO GATED TO M REG.	DS201

```
TEMPLATE FOR 'D' LINE STATEMENT:  UPLB.MB.MD
                             OR:  LB.MB.MDUP

   WHERE:  UP = UP FIELD      MB = MB FIELD
           LB = LB FIELD      MD = MD FIELD
           ANY COMBINATION OF LB.MB AND MD
           MAY BE PRESENT.
```

UP FIELD ROSDR 44–45 LB.MB.MD COUNTER FUNCTION CONTROL (NULL VALUE: UP2)

EDGE CHAR	MNEMONIC	BITS	DEC ORDER	FUNCTION	ALD PAGE
D	0→	00	UP0	SET SELECTED COUNTERS TO ZERO.	DS311
D	3→	01	UP1	SET SELECTED COUNTERS TO THREE.	DS311

CF105 01

EDGE CHAR	MNEMONIC	BITS	DEC ORDER	FUNCTION	ALD PAGE
D	-1	10	UP2	DECREMENT SELECTED COUNTERS BY ONE. IF UNDERFLOW, SET TO ALL ONES.	DS311
D	+1	11	UP3	INCREMENT SELECTED COUNTERS BY ONE. IF OVERFLOW, SET TO ALL ZEROS.	DS311

MD FIELD ROSDR 46 MD COUNTER CONTROL

EDGE CHAR	MNEMONIC	BITS	DEC ORDER	FUNCTION	ALD PAGE
		0	MD0	NO OPERATION	
D	MD	1	MD1	SELECT MD COUNTER FOR OPERATION SPECIFIED BY UP FIELD.	DS311

LB FIELD ROSDR 47 LB COUNTER CONTROL

EDGE CHAR	MNEMONIC	BITS	DEC ORDER	FUNCTION	ALD PAGE
		0	LB0	NO OPERATION	
D	LB	1	LB1	SELECT LB COUNTER FOR OPERATION SPECIFIED BY UP FIELD.	DS311

MB FIELD ROSDR 48 MB COUNTER CONTROL

EDGE CHAR	MNEMONIC	BITS	DEC ORDER	FUNCTION	ALD PAGE
		0	MB0	NO OPERATION	
D	MB	1	MB1	SELECT MB COUNTER FOR OPERATION SPECIFIED BY UP FIELD.	DS311

DG FIELD ROSDR 49-51 LENGTH COUNTER AND CARRY INSERTION CONTROL

EDGE CHAR	MNEMONIC	BITS	DEC ORDER	FUNCTION	ALD PAGE
		000	DG0	NO OPERATION	
D	CSTAT→ADDER	001	DG1	INSERT CARRY STAT INTO ADDER.	DS301
D	HOT1→ADDER	010	DG2	INSERT HOT CARRY INTO ADDER.	DS301
D	G1-1	011	DG3	DECREMENT LENGTH COUNTER G1 BY ONE. IF UNDERFLOW, SET G1 SIGN TO MINUS AND COUNTER TO ALL ONES.	DS302
D	HOT1.G-1	100	DG4	DECREMENT COUPLED LENGTH COUNTERS BY ONE. IF UNDERFLOW OCCURS IN BITS 4-7, SET G2 SIGN TO MINUS. IF ALL 8 BITS UNDERFLOW, SET G1 SIGN TO MINUS. INSERT HOT CARRY INTO ADDER.	DS301
D	G2-1	101	DG5	DECREMENT LENGTH COUNTER G2 BY ONE. IF UNDERFLOW, SET G2 SIGN TO MINUS AND COUNTER TO ALL ONES.	DS302
D	G-1	110	DG6	DECREMENT COUPLED LENGTH COUNTERS BY ONE. IF UNDERFLOW OCCURS IN BITS 4-7, SET G2 SIGN TO MINUS. IF ALL 8 BITS UNDERFLOW, SET G1 SIGN TO MUNUS.	DS302
D	G1.2-1	111	DG7	DECREMENT EACH LENGTH COUNTER BY ONE. IF A COUNTER UNDERFLOWS, SET ITS SIGN TO MINUS AND ITS VALUE TO ALL ONES.	DS302

11 MAR 255099
3 AUG 65 255449
14 DEC 65 255458
29 MAR 66 256490

CONTROL FIELD SPECIFICATION	
CPU MODE	
DATE 29 MAR 66	MACH. 2050
	FRAME 01
	P.N. 5365007
IBM CORP. SDD	PAGE 1

CF105

UL FIELD ROSDR 52-53 MOVER ACTION - LATCH BITS 0-3 (NULL VALUE: UL1)

EDGE CHAR	MNEMONIC	BITS	DEC ORDER	FUNCTION	ALD PAGE
B	E→WL	00	UL0	GATE EMIT FIELD TO MOVER LATCH BITS 0-3. (NOTE: EMIT FIELD IS RECODED IF ORDER WM12 IS USED AND ASCII BIT IS ON. SEE WM12.)	DS101
B	UL→WL	01	UL1	GATE LEFT INPUT (U) BITS 0-3 TO LATCH BITS 0-3.	DS101
B	VL→WL	10	UL2	GATE RIGHT INPUT (V) BITS 0-3 TO LATCH BITS 0-3.	DS101
B	?→WL	11	UL3	GATE TO LATCH BITS 0-3 ACCORDING TO CONTENTS OF MOVER FUNCTION REGISTER AS FOLLOWS:	DS101

FUNCTION REGISTER	ACTION
000	CROSS. GATE U BITS 4-7 TO LATCH BITS 0-3.
001	OR. GATE OR OF U BITS 0-3 AND V BITS 0-3 TO LATCH BITS 0-3.
010	AND. GATE AND OF U BITS 0-3 AND V BITS 0-3 TO LATCH BITS 0-3.
011	EXCLUSIVE OR. GATE EXCLUSIVE OR OF U BITS 0-3 AND V BITS 0-3 TO LATCH BITS 0-3.
100	STRAIGHT. GATE U BITS 0-3 TO LATCH BITS 0-3.
101	U LEFT.V RIGHT. GATE U BITS 0-3 TO LATCH BITS 0-3.
110	V LEFT.U RIGHT. GATE V BITS 0-3 TO LATCH BITS 0-3.
111	UNDEFINED

UR FIELD ROSDR 54-55 MOVER ACTION - LATCH BITS 4-7 (NULL VALUE: UR1)

EDGE CHAR	MNEMONIC	BITS	DEC ORDER	FUNCTION	ALD PAGE
B	E→WR	00	UR0	GATE EMIT FIELD TO MOVER LATCH BITS 4-7. (NOTE: EMIT FIELD IS RECODED IF ORDER WM12 IS USED AND ASCII BIT IS ON. SEE WM12.) NOTE 2	DS101
B	UR→WR	01	UR1	GATE LEFT INPUT (U) BITS 4-7 TO LATCH BITS 4-7. NOTE 2	DS101
B	VR→WR	10	UR2	GATE RIGHT INPUT (V) BITS 4-7 TO LATCH BITS 4-7. NOTE 2	DS101
B	?→WR	11	UR3	GATE TO LATCH BITS 4-7 ACCORDING TO CONTENTS OF MOVER FUNCTION REGISTER AS FOLLOWS:	DS101

FUNCTION REGISTER	ACTION
000	CROSS. GATE U BITS 0-3 TO LATCH BITS 4-7.
001	OR. GATE OR OF U BITS 4-7 AND V BITS 4-7 TO LATCH BITS 4-7.
010	AND. GATE AND OF U BITS 4-7 AND V BITS 4-7 TO LATCH BITS 4-7.
011	EXCLUSIVE OR. GATE EXCLUSIVE OR OF U BITS 4-7 AND V BITS 4-7 TO LATCH BITS 4-7.
100	STRAIGHT. GATE U BITS 4-7 TO LATCH BITS 4-7.
101	U LEFT.V RIGHT. GATE V BITS 4-7 TO LATCH BITS 4-7.
110	V LEFT.U RIGHT. GATE U BITS 4-7 TO LATCH BITS 4-7.
111	UNDEFINED

NOTE 1: WHEN EQUAL DECIMAL ORDERS ARE USED IN UL AND UR FIELDS. THE FOLLOWING COMBINATION MNEMONICS ARE USED:

UL1 AND UR1: U→W

UL2 AND UR2: V→W

UL3 AND UR3: ?→W

CF106 01

NOTE 2 1410-7010 EMULATOR FEATURE EXTENDS UR FIELD FUNCTION IF THE UX FIELD (ROSDR 83) IS ON (1).

EDGE CHAR	MNEMONIC	BITS	DEC ORDER	FUNCTION	ALD PAGE
B	+→WR	00	UR0,UX1	ADD THE MOVER (U) INPUT BITS 4-7 TO THE MOVER (V) INPUT BITS 4-7. INPUT VALUES GREATER THAN 9 ARE REDUCED BY 8. THE SETTING OF STAT 4 IS ENTERED AS A CARRY IN. STAT 4 IS SET TO 1 IF A CARRY RESULTS.	XE201
B	-→WR	01	UR1,UX1	COMPLIMENT BITS 4-7 OF THE MOVER (U) INPUT AND ADD RESULT TO THE MOVER (V) INPUT BITS 4-7. INPUT VALUES > 9 ARE REDUCED BY 8. THE SETTING OF STAT 4 IS ENTERED AS A CARRY IN. STAT 4 IS SET TO 1 IF A CARRY RESULTS.	XE201
B	UD→WR	10	UR2,UX1	A 1 IS ORED TO THE R SIGN STAT IF THE VALUE OF THE MOVER (U) INPUT BITS 4-7 IS > 9. ALSO GATE MOVER (U) INPUT BITS 4-7 TO LATCH BITS 4-7.	DS101

CE FIELD ROSDR 57-60

THIS FIELD IS USED AS DATA BY MICRO ORDERS.

11 MAR 65 255099
3 AUG 65 255449
11 JAN 66 255458

```
TEMPLATE FOR 'A' LINE STATEMENT: RRR±LLL→TTTT

   WHERE: R = RY FIELD        L = LX FIELD
          ± = TC FIELD        T = TR FIELD
```

LX FIELD   ROSDR 61-63   LEFT INPUT TO ADDER (XG)

EDGE CHAR	MNEMONIC	BITS	DEC ORDER	FUNCTION	ALD PAGE
		000	LX0	NO OPERATION	DR011
A	1	000	LX0	NO OPERATION – USED WITH TC0 (COMPLEMENT) FOR –1.	DR011
A	L	001	LX1	GATE L REG TO XG.	DR011
A	SGN	010	LX2	GATE ONE TO XG BIT 0. GATE ZEROS TO OTHER BITS OF XG.	DR011
A	E	011	LX3	GATE EMIT FIELD TO XG BITS 27-30. GATE ZEROS TO OTHER BITS.	DR011
A	LRL	100	LX4	GATE L REG BITS 16-31 TO XG BITS 0-15. GATE ZEROS TO BITS 16-31.	DR011
A	LWA	101	LX5	GATE L REG TO XG. OR ONES INTO XG BITS 30-31 AND RIGHT INPUT (Y) BITS 30-31. THIS MICROINSTRUCTION MUST CONTAIN ORDER TC0 AND EITHER RY1 OR RY4. (USED TO COMPARE WORD ADDRESSES)	DR011
A	4	110	LX6	GATE ONE TO XG BIT 29 (HOT 4). GATE ZEROS TO OTHER BITS.	DR011
A	64C	111	LX7	GATE ONES TO XG BITS 0-1. GATE ZEROS TO OTHER BITS.	DR011

TC FIELD   ROSDR 64   TRUE/COMPLEMENT GATING TO LEFT ADDER INPUT   (NULL VALUE: TC1)

EDGE CHAR	MNEMONIC	BITS	DEC ORDER	FUNCTION	ALD PAGE
A	–	0	TC0	GATE COMPLEMENT OF XG TO LEFT ADDER INPUT.	DR021
A	+	1	TC1	GATE XG (TRUE) TO LEFT ADDER INPUT.	DR021

RY FIELD   ROSDR 65-67   RIGHT INPUT TO ADDER (Y)

EDGE CHAR	MNEMONIC	BITS	DEC ORDER	FUNCTION	ALD PAGE
		000	RY0	NO OPERATION	DR021
A	R	001	RY1	GATE R REGISTER TO Y. (NOTE 1)	DR021
A	M	010	RY2	GATE M REGISTER TO Y.	DR021
A	M23	011	RY3	GATE M REG BITS 16-31 TO Y BITS 16-31. GATE ZEROS TO BITS 0-15.	DR021
A	H	100	RY4	GATE H REGISTER TO Y. (SEE NOTE BELOW)	DR021
B	SEMT	101	RY5	OR THE FOUR SDR PARITY BITS INTO THE FOUR EMIT FIELD BITS.	DS021
		110	RY6	UNDEFINED	
		111	RY7	UNDEFINED. (NOTE 2)	

NOTE 1 IF ORDER LX5 IS USED. ONES ARE ORED INTO Y BITS 30-31. SEE LX5.

NOTE 2 1410-7010 USAGE OF UNDEFINED ORDERS.

EDGE CHAR	MNEMONIC	BITS	DEC ORDER	FUNCTION	ALD PAGE
A	ROS	111	RY7	GATE ROSDR BITS 72-82 AND 92-97 TO Y BITS 16-30 AND PARITY A AND B BRANCHING IS INHIBITED	DR021
A	ROS=NNNN	111	RY7	NNNN IS A 4 DGT HEX VALUE GATED FROM ROSDR TO THE Y INPUT. UNITS DIGIT MUST BE EVEN.	XE001

CFC 107 – 2050 CONTROL FIELD SPECIFICATION – CPU MODE (LX.TC.RY.AD)

CF107 01

AD FIELD ROSDR 68-71 ADDER FUNCTION (NULL VALUE: AD1)

EDGE CHAR	MNEMONIC	BITS	DEC ORDER	FUNCTION	ALD PAGE
		0000	AD0	UNDEFINED	
		0001	AD1	ADD. NO CARRIES ENTERED OR SAVED	
A	BCF0	0010	AD2	ADD. NO CARRY SAVED. IF F REG EQUALS ZERO, INSERT CARRY INTO POSITION 31.	DR031
		0011	AD3	UNDEFINED	
A	BC0	0100	AD4	ADD. SET CARRY STAT TO CARRY OUT OF POSITION 0.	DR032
A	BC∨C	0101	AD5	ADD. SET CARRY STAT TO EXCLUSIVE OR OF CARRIES OUT OF POSITIONS 0 AND 1.	DR032
A	BC1B	0110	AD6	ADD. SET CARRY STAT TO CARRY OUT OF POSITION 1. BLOCK CARRY FROM POSITION 8 TO POSITION 7.	DR032
A	BC	0110	AD6	AS ABOVE. USED WITH ORDER AL23.	DR032
A	BC8	0111	AD7	ADD. SET CARRY STAT TO CARRY OUT OF POSITION 8.	DR032
A	DHL	1000	AD8	DECIMAL HALVE (LOW ORDER). BIT 2 OF EACH DIGIT OF THE SUM IS TESTED. IF THE BIT IS ONE, THE NEXT DIGIT POSITION TO THE RIGHT IN THE L REG IS SET TO 0110. IF THE BIT IS ZERO, THE DIGIT IN L REG IS SET TO 0000. THE LEFTMOST DIGIT IN THE L REG IS SET IN THE SAME WAY FROM THE AUXILIARY TRIGGER.	DR031
A	DC0	1001	AD9	DECIMAL ADD. SET STAT 1 TO CARRY OUT OF POSITION 0. INSERT PREVIOUS VALUE OF STAT 1 AS CARRY INTO POSITION 31. TEST CARRY OUT OF EACH DIGIT POSITION. IF CARRY, SET CORRESPONDING DIGIT POSITION IN L REG TO 0000. IF NO CARRY, SET DIGIT IN L REG TO 0110.	DR032
A	DDC0	1010	AD10	DECIMAL DOUBLE. SET STAT 1 TO CARRY OUT OF POSITION 0. INSERT PREVIOUS VALUE OF STAT 1 AS CARRY INTO POSITION 31. TEST EACH DIGIT OF SUM. IF 5 OR GREATER, SET CORRESPONDING DIGIT POSITION IN L REG TO 0110. IF LESS THAN 5, SET DIGIT IN L REG TO 0000.	DR032
A	DHH	1011	AD11	DECIMAL HALVE (HIGH ORDER). BIT 2 OF EACH DIGIT OF THE SUM IS TESTED. IF THE BIT IS ONE, THE NEXT DIGIT POSITION TO THE RIGHT IN THE L REG IS SET TO 0110. IF THE BIT IS ZERO, THE DIGIT IN L REG IS SET TO 0000. THE LEFTMOST DIGIT IN L REG IS SET TO 0000. THE AUXILIARY TRIGGER IS SET TO BIT 2 OF THE RIGHTMOST SUM DIGIT (SUM BIT 30).	DR032
A	DCBS	1100	AD12	DECIMAL ADD. SET STAT 1 TO CARRY OUT OF LEFTMOST BYTE POSITION FOR WHICH A BYTE STAT IS ON. INSERT PREVIOUS VALUE OF STAT 1 AS CARRY INTO POSITION 1. TEST CARRY OUT OF EACH DIGIT POSITION. IF CARRY, SET CORRESPONDING DIGIT POSITION IN L REG TO 0000. IF NO CARRY, SET DIGIT IN L REG TO 0110.	DR032
		1101	AD13	UNDEFINED	
		1110	AD14	UNDEFINED	
		1111	AD15	UNDEFINED	

NOTE: IN ALL CASES ADDITION IS BINARY. THE TERM 'DECIMAL ADD' USED ABOVE REFERS TO THE MANNER OF GENERATING THE CORRECTION FACTOR IN THE L REG.

11 MAR 255099
3 AUG 65 255449
11 JAN 66 255458

CONTROL FIELD SPECIFICATION		
CPU MODE		
DATE 11 JAN 66	MACH.	2050
	FRAME	01
	P.N.	5365009
IBM CORP. SDD	PAGE	1

CF107

AB FIELD ROSDR 72–77 CONDITION BRANCH TEST A

EDGE CHAR	MNEMONIC	BITS	DEC ORDER	FUNCTION	ALD PAGE
R	0	000000	AB0	SET A BIT TO 0.	
R	1	000001	AB1	SET A BIT TO 1.	KK201
R	S0	000010	AB2	SET A BIT TO VALUE OF STAT 0.	KK201
R	S1	000011	AB3	SET A BIT TO VALUE OF STAT 1.	KK201
R	S2	000100	AB4	SET A BIT TO VALUE OF STAT 2.	KK201
R	S3	000101	AB5	SET A BIT TO VALUE OF STAT 3.	KK201
R	S4	000110	AB6	SET A BIT TO VALUE OF STAT 4.	KK201
R	S5	000111	AB7	SET A BIT TO VALUE OF STAT 5.	KK201
R	S6	001000	AB8	SET A BIT TO VALUE OF STAT 6.	KK201
R	S7	001001	AB9	SET A BIT TO VALUE OF STAT 7.	KK201
R	CSTAT	001010	AB10	SET A BIT TO VALUE OF CARRY STAT.	KK211
		001011	AB11	UNDEFINED. NOTE 4. CF109 (SEE NOTE 5)	
R	1SYLS	001100	AB12	SET A BIT TO VALUE OF ONE SYLLABLE OP IN BUFFER STAT.	KK211
R	LSGNS	001101	AB13	SET A BIT TO VALUE OF L SIGN STAT.	KK211
R	⊻SGNS	001110	AB14	SET A BIT TO EXCLUSIVE OR OF L SIGN STAT AND R SIGN STAT.	KK211
		001111	AB15	UNDEFINED. NOTE 4. CF109 (SEE NOTE 5)	
R	CRMD	010000	AB16	PERFORM AND OF MD REG (MASK) AND DECODED CONDITION REG. IF RESULT IS NON–ZERO, SET A BIT TO ONE. IF RESULT IS ZERO, SET BIT A TO ZERO.	KK251
R	W=0	010001	AB17	IF MOVER LATCHES ARE ZERO, SET A BIT TO ONE. IF NON–ZERO, SET A BIT TO ZERO.	KK211
R	WL=0	010010	AB18	IF MOVER LATCH BITS 0–3 ARE ZERO, SET A BIT TO ONE. IF NON–ZERO, SET A BIT TO ZERO.	KK211
R	WR=0	010011	AB19	IF MOVER LATCH BITS 4–7 ARE ZERO, SET A BIT TO ONE. IF NON–ZERO, SET A BIT TO ZERO.	KK211
R	MD=FP	010100	AB20	IF BOTH BIT 0 AND BIT 3 OF MD REG ARE ZERO, SET A BIT TO ONE. OTHERWISE SET A BIT TO ZERO. (VALID FLOATING POINT REG)	KK221
R	MB=3	010101	AB21	IF MB COUNTER EQUALS 3, SET A BIT TO ONE. OTHERWISE SET A BIT TO ZERO.	KK221
R	MD3=0	010110	AB22	SET A BIT TO COMPLEMENT OF MD REG BIT 3.	KK221
R	G1=0	010111	AB23	IF LENGTH COUNTER G1 IS ZERO, SET A BIT TO ONE. OTHERWISE SET A BIT TO ZERO.	KK221
R	G1<0	011000	AB24	SET A BIT TO VALUE OF LENGTH COUNTER G1 SIGN.	KK221
R	G<4	011001	AB25	IF COUPLED LENGTH COUNTERS HAVE VALUE LESS THAN 4, SET A BIT TO ONE. OTHERWISE SET A BIT TO ZERO. SIGN OF G1 IS INCLUDED IN VALUE.	KK221

CF108 01

AB FIELD (CONTINUED)

EDGE CHAR	MNEMONIC	BITS	DEC ORDER	FUNCTION	ALD PAGE
R	G1MBZ	011010	AB26	IF EITHER LENGTH COUNTER G1 OR MB COUNTER IS ZERO, SET A BIT TO ONE. OTHERWISE SET A BIT TO ZERO.	KK221
R	IOS0	011011	AB27	SET A BIT TO VALUE OF MULTIPLEXOR CHANNEL STAT 0.	KK191
R	IOS2	011100	AB28	SET A BIT TO VALUE OF MULTIPLEXOR CHANNEL STAT 2.	KK191
R	R(31)	011101	AB29	SET A BIT TO VALUE OF R REG BIT 31.	KK221
R	F(2)	011110	AB30	SET A BIT TO VALUE OF F REG BIT 2.	KK231
R	L(0)	011111	AB31	SET A BIT TO VALUE OF L REG BIT 0.	KK231
R	F=0	100000	AB32	IF F REG IS ZERO, SET A BIT TO ONE. OTHERWISE SET A BIT TO ZERO.	KK231
R	UNORM	100001	AB33	IF STAT 0 IS OFF AND ADDER OUTPUT BITS 8-11 ARE ZERO, SET A BIT TO ONE. OTHERWISE SET A BIT TO ZERO.	KK231
R	TZ*BS	100010	AB34	IF ALL BYTES OF ADDER OUTPUT SPECIFIED BY BYTE STATS ARE ZERO, SET A BIT TO ONE. OTHERWISE SET A BIT TO ZERO.	KK231
R	EDITPAT	100011	AB35	SET A BIT TO VALUE OF EDIT STAT 1. SET B BIT TO VALUE OF EDIT STAT 2. (SEE NOTE 1)	KK231
R	PROB	100100	AB36	SET A BIT TO VALUE OF PSW BIT 15.	KK231
R	TIMUP	100101	AB37	SET A BIT TO VALUE OF TIMER UPDATE SIGNAL. RESET TIMER UPDATE SIGNAL.	KK231
		100110	AB38	UNDEFINED, NOTE 4, CF109	
R	GZ/MB3	100111	AB39	IF EITHER THE VALUE OF THE COUPLED LENGTH COUNTER G IS ZERO OR THE VALUE OF THE MB COUNTER IS 3 SET A BIT TO ONE. OTHERWISE SET A BIT TO ZERO.	KK231
		101000	AB40	UNDEFINED (SEE NOTE 5)	
R	LOG	101001	AB41	SET A BIT TO VALUE OF LOG/SCAN STAT.	KK291
R	STC=0	101010	AB42	IF SCAN TEST COUNTER IS ZERO, SET A BIT TO ONE. OTHERWISE SET A BIT TO ZERO.	KK291
R	G2<=LB	101011	AB43	IF LENGTH COUNTER G2 IS LESS THAN OR EQUAL TO LB COUNTER, SET A BIT TO ONE. OTHERWISE SET A BIT TO ZERO. SIGN OF LENGTH COUNTER IS IGNORED.	KK251
		101100	AB44	UNDEFINED	
R	D(7)	101101	AB45	SET A BIT TO VALUE OF SDR BIT 7.	KK221
R	SCPS	101110	AB46	SET A BIT TO VALUE OF SCAN PASS STAT.	KK291
R	SCFS	101111	AB47	SET A BIT TO VALUE OF SCAN FAIL STAT.	KK291
R	STORV	110000	AB48	SET A BIT TO VALUE OF I/O STORAGE VIOLATION TRIGGER.	KK291
R	W(67)→AB	110001	AB49	SET A BIT TO VALUE OF MOVER LATCH BIT 6. SET B BIT TO VALUE OF MOVER LATCH BIT 7. (SEE NOTE 1) (SEE NOTE 5)	KK291

11 MAR 255099
3 AUG 65 255449
11 JAN 66 255458
29 MAR 66 256490

CONTROL FIELD SPECIFICATION	
CPU MODE	
DATE 29 MAR 66	MACH. 2050
	FRAME 01
	P.N. 5365010
IBM CORP. SDD	PAGE 1

CF108

AB FIELD (CONTINUED)

EDGE CHAR	MNEMONIC	BITS	DEC ORDER	FUNCTION	ALD PAGE
R	Z23≠0	110010	AB50	IF ADDER OUTPUT BITS 16–31 ARE NON–ZERO SET A BIT TO ONE. OTHERWISE SET A BIT TO ZERO.	KK211
R	CCW2OK	110011	AB51	IF SDR BITS 16–31 AND 5–7 ARE ALL ZERO SET A BIT TO ONE. OTHERWISE SET A BIT TO ZERO. (COUNT AND FLAG TEST)	KK291
R	MXBIO	110100	AB52	SET A BIT TO VALUE OF MULTIPLEXOR INPUT BUFFER BIT 0.	KK191
R	IBFULL	110101	AB53	SET A BIT TO VALUE OF INPUT BUFFER FULL STAT.	KK191
R	CANG	110110	AB54	IF ADDER OUTPUT BITS 29–31 ARE NON–ZERO OR INVALID ADDRESS TRIGGER IS ON, SET A BIT TO ONE. OTHERWISE SET A BIT TO ZERO.	KK291
R	CHLOG	110111	AB55	SET A BIT TO VALUE OF CHANNEL ERROR LOG REQUEST LINE.	KK191
R	I–FETCH	111000	AB56	FOUR–WAY BRANCH: (SEE NOTE 1) 0→A,0→B IAR BIT 30 IS ONE AND REFETCH STAT OFF 0→A,1→B IAR BIT 30 IS ONE AND REFETCH STAT ON 1→A,0→B IAR BIT 30 IS ZERO 1→A,1→B EXCEPTION TRIGGER ON	KK151
R	IA(30)	111001	AB57	SET A BIT TO VALUE OF IAR BIT 30.	KK211
R	EXT,CHIRPT	111010	AB58	IF EITHER A TIMER UPDATE REQUEST HAS OCCURRED, OR AN EXTERNAL INTERRUPT IS REQUESTED WITH MASK BIT ON, SET A BIT TO ONE. OTHERWISE SET A BIT TO ZERO. IF A CHANNEL INTERRUPT IS REQUESTED SET B BIT TO ONE. OTHERWISE SET B BIT TO ZERO. (SEE NOTE 1)	KK151
R	DCHOLD	111011	AB59	SET A BIT TO VALUE OF DIRECT CONTROL HOLD LINE.	KK191
R	PSS	111100	AB60	SET A BIT TO VALUE OF PROGRESSIVE SCAN STAT. IF PSS IS ON WHEN TESTED RESET IT, UNLESS SUPERVISORY STAT IS ON AND SUPERVISORY ENABLE STORAGE STAT IS OFF.	KK191
R	IOS4	111101	AB61	SET A BIT TO VALUE OF MULTIPLEXOR CHANNEL STAT 4.	KK191
		111110	AB62	UNDEFINED, NOTE 4	
R	RX,SO	111111	AB63	IF STAT 0 IS ON AND M REG BITS 0–1 HAVE VALUE 01 SET A BIT TO ONE. OTHERWISE SET A BIT TO ZERO. (RX FORMAT AND INDEXING REQUIRED)	KK251

NOTE 1 IN BRANCHES WHERE BOTH A AND B BITS ARE SET, THE B BIT VALUE IS THE OR OF VALUES SPECIFIED BY AB AND BB FIELDS.

NOTE 2 A BIT AND B BIT VALUES MAY BE SUBSEQUENTLY ALTERED BY ZN FIELD ORDERS.

NOTE 3 ALL AB FIELD BRANCHES TEST CONDITIONS EXISTING AT THE END OF THE PREVIOUS CYCLE.

NOTE 4 1410–7010 EMULATER USAGE OF UNDEFINED ORDERS.

EDGE CHAR	MNEMONIC	BITS	DEC ORDER	FUNCTION	ALD PAGE
R	W(0)	001011	AB11	SET A BIT TO VALUE OF MOVER LATCH BIT 0.	XE201
R	W(1)	001111	AB15	SET A BIT TO VALUE OF MOVER LATCH BIT 1.	XE201
R	W23 11	100110	AB38	SET A BIT TO ONE IF MOVER LATCH BITS 2–3 ARE SET TO ONES (11).	XE201
R	R(4)	111110	AB62	SET A BIT TO VALUE OF R REG BIT 4	XE201

CFC 109 – 2050 CONTROL FIELD SPECIFICATION – CPU MODE (AB)

AB FIELD (CONTINUED)

NOTE 5 7070-7074 EMULATOR USAGE OF ORDERS

EDGE CHAR	MNEMONIC	BITS	DEC ORDER	FUNCTION	ALD PAGE
R	RD7=9	001011	AB11	SET A BIT TO 1 IF DIGIT 7 OF R REG = 1001	XF711
R	EMSTAT	001111	AB15	SET A BIT TO VALUE OF EMULATOR STAT	XF111
R	L(22)	101000	AB40	SET A BIT TO VALUE OF L REG. BIT 22	XF111
R	W(6)→A,1→B	110001	AB49	SET A BIT TO VALUE OF MOVER BIT 6. SET B BIT TO 1 (BB FIELD BITS 78-82 OF ROS WORD SET TO VALUE 00001)	KK291
R	W6→A,W7Ω52→B	110001	AB49	SET A BIT TO VALUE OF MOVER BIT 6. SET B BIT TO THE ''OR'' OF MOVER BIT 7 AND STAT 2. (BB FIELD BITS 78-82 OF ROS WORD SET TO VALUE 00100)	KK291

11 MAR 65 255099
3 AUG 65 255449
11 JAN 66 255458
9 MAR 66 256490
9 MAY 66 255793

CONTROL FIELD SPECIFICATION CPU MODE	
DATE 10 MAY 66	MACH. 2050
	FRAME 01
	P.N. 5365011
IBM CORP. SDD	PAGE 1

CF109

BB FIELD ROSDR 78–82 B CONDITION TEST

EDGE CHAR	MNEMONIC	BITS	DEC ORDER	FUNCTION	ALD PAGE
R	0	00000	BB0	SET B BIT TO 0.	
R	1	00001	BB1	SET B BIT TO 1.	KK141
R	S0	00010	BB2	SET B BIT TO VALUE OF STAT 0.	KK141
R	S1	00011	BB3	SET B BIT TO VALUE OF STAT 1.	KK141
R	S2	00100	BB4	SET B BIT TO VALUE OF STAT 2.	KK141
R	S3	00101	BB5	SET B BIT TO VALUE OF STAT 3.	KK141
R	S4	00110	BB6	SET B BIT TO VALUE OF STAT 4.	KK141
R	S5	00111	BB7	SET B BIT TO VALUE OF STAT 5.	KK141
R	S6	01000	BB8	SET B BIT TO VALUE OF STAT 6.	KK141
R	S7	01001	BB9	SET B BIT TO VALUE OF STAT 7.	KK141
R	RSGNS	01010	BB10	SET B BIT TO VALUE OF R SIGN STAT.	KK151
R	HSCH	01011	BB11	IF HIGH SPEED CHANNEL IS OPERATING OR 256 SUBCHANNEL OPTION IS INSTALLED, SET B BIT TO ONE. OTHERWISE SET B BIT TO ZERO.	KK141
R	EXC	01100	BB12	SET B BIT TO VALUE OF EXCEPTION TRIGGER.	KK151
R	WR=0	01101	BB13	IF MOVER LATCH BITS 4–7 ARE ZERO, SET B BIT TO ONE. OTHERWISE SET B BIT TO ZERO.	KK151
		01110	BB14	UNDEFINED	
R	T13=0	01111	BB15	IF ADDER OUTPUT BUS BITS 8–13 ARE ZERO, SET B BIT TO ONE. OTHERWISE SET B BIT TO ZERO.	KK151
R	T(0)	10000	BB16	SET B BIT TO VALUE OF ADDER OUTPUT BUS BIT 0.	KK151
R	T=0	10001	BB17	IF ADDER OUTPUT BUS IS ZERO, SET B BIT TO ONE. OTHERWISE SET B BIT TO ZERO.	KK151
R	TZ*BS	10010	BB18	IF THE BYTES OF THE ADDER OUTPUT BUS SPECIFIED BY THE BYTE STATS ARE ALL ZERO, SET B BIT TO ONE. OTHERWISE SET B BIT TO ZERO.	KK151
R	W=1	10011	BB19	IF THE MOVER LATCHES CONTAIN 0000 0001 SET B BIT TO ONE. OTHERWISE SET B BIT TO ZERO.	KK151
R	LB=0	10100	BB20	IF LB COUNTER IS ZERO SET B BIT TO ONE. OTHERWISE SET B BIT TO ZERO.	KK161
R	LB=3	10101	BB21	IF LB COUNTER EQUALS 3 SET B BIT TO ONE. OTHERWISE SET B BIT TO ZERO.	KK161
R	MD=0	10110	BB22	IF MD REG IS ZERO SET B BIT TO ONE. OTHERWISE SET B BIT TO ZERO.	KK161
R	G2=0	10111	BB23	IF LENGTH COUNTER G2 IS ZERO SET B BIT TO ONE. OTHERWISE SET B BIT TO ZERO.	KK161
R	G2<0	11000	BB24	SET B BIT TO VALUE OF LENGTH COUNTER G2 SIGN.	KK161
R	G2LBZ	11001	BB25	IF EITHER LENGTH COUNTER G2 OR LB COUNTER IS ZERO, SET B BIT TO ONE. OTHERWISE SET B BIT TO ZERO.	KK161

CFC 110 – 2050 CONTROL FIELD SPECIFICATION – CPU MODE (BB)

CF110 01

BB FIELD (CONTINUED)

EDGE CHAR	MNEMONIC	BITS	DEC ORDER	FUNCTION	ALD PAGE
R	IOS1	11010	BB26	SET B BIT TO VALUE OF MULTIPLEXOR CHANNEL STAT 1.	KK151
R	MD/JI	11011	BB27	IF EITHER BIT 1 OR BIT 3 OF EITHER MD REG OR J REG IS ONE, SET B BIT TO ONE (ILLEGAL FP REG). OTHERWISE SET B BIT TO ZERO.	KK161
R	IVA	11100	BB28	SET B BIT TO VALUE OF INVALID ADDRESS STAT.	KK161
R	IOS3	11101	BB29	SET B BIT TO VALUE OF MULTIPLEXOR CHANNEL STAT 3.	KK161
R	(CAR)	11110	BB30	SET B BIT TO VALUE OF CARRY LATCH AS SET THIS CYCLE.	KK171
R	(Z00)	11111	BB31	SET B BIT TO VALUE OF ADDER SUM BIT 0 (BEFORE SHIFT) THIS CYCLE.	KK171

NOTE WITH THE EXCEPTION OF ORDERS BB30 AND BB31, ALL TESTS ARE MADE ON THE STATUS OF THE CPU AT THE END OF THE PREVIOUS CYCLE.

UX FIELD ROSDR 83 1410-7010 FEATURE ONLY

0	UX0	NO OPERATION	XG401
1	UX1	EXTEND THE UR FIELD FUNCTION	XG401

SEE CF106 NOTE 2.

11 MAR 255099
3 AUG 65 255449
11 JAN 66 255458

CONTROL SIGNAL SPECIFICATION
CPU MODE
DATE 11 JAN 66 MACH. 2050
FRAME 01
P.N. 5365012
IBM CORP. SDD PAGE 1

CF110

SS FIELD ROSDR 84–89 STAT SETTING AND MISCELLANEOUS CONTROL

EDGE CHAR	MNEMONIC	BITS	DEC ORDER	FUNCTION	ALD PAGE
		000000	SS0	NO OPERATION	
		000001	SS1	UNDEFINED (SEE NOTE 2, CF112-2)	
		000010	SS2	UNDEFINED, NOTE 1 CF112 (SEE NOTE 2, CF112-2)	
C	D→CR*BS	000011	SS3	SET CONDITION REG FOR TEST AND SET INSTRUCTION. SET COND REG BIT 0 TO 0. SET BIT 1 OF COND REG TO ONE IF THE FIRST BIT IS ON IN ANY SDR BYTE SELECTED BY THE CPU BYTE STATS. WHEN USED WITH LARGE CAPACITY STORAGE, THIS ORDER INTERLOCKS THE SAME AS AN SDR TO LATCH TRANSFER.	RP251
C	E→SCANCTL	000100	SS4	CONTROL SCAN STATS PER EMIT FIELD AS FOLLOWS:	KS601

EMIT	FUNCTION
0000	IF SDR IS ALL ONES, STEP THE BINARY TRIGGER.
0001	INCREMENT SCAN TEST COUNTER BY 1, THEN IF BINARY TGR. IS OFF, TURN ON THE PASS TGR.. IF BINARY TGR. IS ON, TURN ON THE FAIL TRG.
0010	UNDEFINED
0011	TURN OFF IGNORE I/O ERROR TRIGGER.
0100	TURN ON IGNORE I/O ERROR TRIGGER.
0101	TURN OFF PASS IAR AND FAIL IAR.
0110	TURN ON INVERT SAR BIT 16 TRIGGER.
0111	TURN OFF INVERT SAR BIT 16 TRIGGER AND RESET THE BINARY TGR.
1000	SET SUPERVISORY STAT TO VALUE OF SDR BIT 6. SET PROGRESSIVE SCAN STAT TO VALUE OF SDR BIT 5. SET SEQUENCE COUNTER TO VALUE OF SDR BITS 0–2. SET SUPERVISORY, ENABLE STORAGE STAT TO VALUE OF SDR 4.
1001	SET LCS PARITY CONTROL TGR..
1010	SET LCS BYTE PARITY MODE.
1011	RESET LCS PARITY CONTROL TGR. AND LCS BYTE PARITY MODE.
1100	TURN OFF LOG TGR.
1101–1111	UNDEFINED

CF111 01

SS FIELD (CONTINUED)

EDGE CHAR	MNEMONIC	BITS	DEC ORDER	FUNCTION	ALD PAGE
C	L,RSGNS	000101	SS5	IF LEFT MOVER INPUT U BITS 4-7 HAS VALUE LESS THAN 1010, FORCE INVALID DATA ROS TRAP. IF VALUE EQUALS 1011 OR 1101 (MINUS SIGN) TURN ON L SIGN STAT AND INVERT R SIGN STAT. OTHERWISE TURN OFF L SIGN STAT.	KS601
C	IVD/RSGNS	000110	SS6	IF LEFT MOVER INPUT U BITS 4-7 HAS VALUE LESS THAN 1010, FORCE INVALID DATA ROS TRAP. IF VALUE EQUALS 1011 OR 1101 (MINUS SIGN) INVERT R SIGN STAT.	KS601
C	EDITSGN	000111	SS7	IF MOVER LATCH BITS 4-7 HAS VALUE GREATER THAN 1001 (VALID SIGN) TURN ON R SIGN STAT. OTHERWISE TURN OFF R SIGN STAT. IF MOVER LATCH BITS 4-7 EQUAL TO 1010, 1100, 1110 OR 1111 (PLUS SIGN) TURN OFF L SIGN STAT.	KS601
C	E→S03	001000	SS8	GATE EMIT FIELD TO STATS 0-3.	KS611
C	S03ΩE,1→LSGN	001001	SS9	TURN ON STATS 0-3 PER EMIT FIELD. TURN ON L SIGN STAT.	KS611
C	S03ΩE	001010	SS10	TURN ON STATS 0-3 PER EMIT FIELD.	KS611
C	S03ΩE,0→BS	001011	SS11	TURN ON STATS 0-3 PER EMIT FIELD. TURN OFF CPU BYTE STATS.	KS611
C	X0,B0,1SYL	001100	SS12	IF ADDER LATCH BITS 12-15 ARE ZERO, SET STAT 0 TO ONE. (X=0) OTHERWISE SET STAT 0 TO ZERO. IF ADDER LATCH BITS 16-19 ARE ZERO, SET STAT 1 TO ONE. (B=0) OTHERWISE SET STAT 1 TO ZERO. IF ADDER LATCH BITS 16-17 ARE ZERO, SET ONE-SYLLABLE-OP-IN-BUFFER STAT TO ONE. OTHERWISE SET THIS STAT TO ZERO.	KS611
C	FPZERO	001101	SS13	IF ADDER LATCH BITS 8-31 ARE ZERO, F REG IS ZERO, AND STAT 3 IS ON, TURN ON STAT 0. OTHERWISE TURN OFF STAT 0.	KS611

11 MAR 65 255099
3 AUG 65 255449
11 JAN 66 255458
5 APR 66 256490
9 MAY 66 255793

CONTROL FIELD SPECIFICATION	
CPU MODE	
DATE 12 MAY 66	MACH. 2050
	FRAME 01
	P.N. 5365013
IBM CORP. SDD	PAGE 1

CF11

SS FIELD ROSDR 84–89 STAT SETTING AND MISCELLANEOUS CONTROL

EDGE CHAR	MNEMONIC	BITS	DEC ORDER	FUNCTION	ALD PAGE
C	FPZERO.E→FN	001110	SS14	IF ADDER LATCH BITS 8–31 ARE ZERO, F REG IS ZERO, AND STAT 3 IS ON, TURN ON STAT 0. OTHERWISE TURN OFF STAT 0. GATE EMIT FIELD BITS 2–3 TO LOCAL STORAGE FUNCTION REG. (EFFECTIVE FOR LOCAL STORAGE ADDRESSING THIS CYCLE.)	KS611
C	BO.1SYL	001111	SS15	IF ADDER LATCH BITS 0–3 ARE ZERO, TURN ON STAT 1. (B=0) OTHERWISE TURN OFF STAT 1. IF ADDER LATCH BITS 16–17 ARE ZERO, TURN ON ONE-SYLLABLE-OP-IN-BUFFER STAT. OTHERWISE TURN THIS STAT OFF.	KS611
C	S03.¬E	010000	SS16	TURN OFF STATS 0–3 PER EMIT FIELD.	KS621
C	(T=0)→S3	010001	SS17	IF ADDER LATCH IS ZERO TURN ON STAT 3. OTHERWISE TURN OFF STAT 3.	KS621
C	E→BS.T30→S3	010010	SS18	GATE EMIT FIELD TO CPU BYTE STATS. SET STAT 3 TO VALUE OF ADDER LATCH BIT 30.	KS621
C	E→BS	010011	SS19	GATE EMIT FIELD TO CPU BYTE STATS.	KS621
C	1→BS*MB	010100	SS20	TURN ON BYTE STAT INDICATED BY VALUE OF MB COUNTER	KS621
C	DIRCTL*E	010101	SS21	DIRECT CONTROL PER EMIT AS FOLLOWS:	KS631

EMIT BIT	FUNCTION IF BIT IS ON
0	GATE H REG BITS 24–31 TO DIRECT CONTROL DATA OUT REG.
1	INITIATE PULSE ON WRITE OUT LINE.
2	GATE H REG BITS 8–15 TO DIRECT CONTROL TIMING SIGNAL BUS OUT LINES IN PULSE FORM.
3	INITIATE PULSE ON READ OUT LINE. IF AT THE END OF THIS PULSE THE HOLD IN LINE IS NOT ON, LATCH THE DIRECT CONTROL BUS IN.

EDGE CHAR	MNEMONIC	BITS	DEC ORDER	FUNCTION	ALD PAGE
		010110	SS22	UNDEFINED	
C	MANUAL→STOP	010111	SS23	SET STOP TRIGGER TO VALUE OF MANUAL TRIGGER.	KS721
C	E→S47	011000	SS24	GATE EMIT FIELD TO STATS 4–7.	KS621
C	S47ΩE	011001	SS25	TURN ON STATS 4–7 PER EMIT FIELD.	KS621
C	S47.¬E	011010	SS26	TURN OFF STATS 4–7 PER EMIT FIELD.	KS621
C	S47.ED*FP	011011	SS27	SET STATS 4–7 AND EXPONENT DIFFERENCE REG FOR FLOATING POINT AS FOLLOWS.	KS621

STAT 4 TURNED ON IF:
STAT 0 OR STAT 1 IS ON AND RIGHT ADDER INPUT BIT 0 IS ONE AND THERE IS A CARRY OUT OF POSITION 1.
OR STAT 0 OR STAT 1 IS ON. THERE IS A CARRY OUT OF POSITION 1. AND EITHER LEFT ADDER INPUT BIT 0 IS ONE OR STAT 1 IS ON BUT NOT BOTH (ADD TYPE. RESULT MINUS)
OR BOTH STAT 0 AND STAT 1 ARE OFF AND LEFT ADDER INPUT BIT 0 IS NOT EQUAL TO RIGHT ADDER INPUT BIT 0 (MULTIPLY OR DIVIDE. SIGNS UNLIKE).

STAT 5 TURNED ON IF LEFT ADDER INPUT BIT 0. RIGHT ADDER INPUT BIT 0 AND STAT 1 CONTAIN AN EVEN NUMBER OF ONES. (TRUE ADD REQUIRED)

STAT 6 TURNED ON IF VALUE OF EXPONENT DIFFERENCE REG IS LESS THAN 16 (DEC) IN ABSOLUTE VALUE.

STAT 7 TURNED ON IF VALUE OF EXPONENT DIFFERENCE REG IS ZERO.

ABSENCE OF TURN ON CONDITION CAUSES STAT TO BE TURNED OFF.

EXPONENT DEFFERENCE REG SET AS FOLLOWS:
BIT 0 SET TO ONE IF CARRY FROM ADDER POS 1 AND SUM BITS 1–4 NON ZERO. OR IF NO CARRY FROM POS 1 AND SUM BITS 1–4 EQUAL 1111.
BITS 1–3 SET EQUAL TO ADDER SUM BITS 5–7.

SS FIELD (CONTINUED)

EDGE CHAR	MNEMONIC	BITS	DEC ORDER	FUNCTION	ALD PAGE
C	OPPANEL→S47	011100	SS28	SET STATS 4-7 PER MAINTENANCE CONSOLE SWITCHES AS FOLLOWS: STATS — CONDITION 0000 NONE OF THOSE DESCRIBED BELOW. 0001 INSTRUCTION STEP, WAIT BIT NOT ON, START. 0010 SET INSTRUCTION COUNTER, MANUAL TRIGGER ON. 0011 REPEAT INSTRUCTION, MANUAL TRIGGER ON. 0100 IAR COMPARE (SYNC OR STOP), MANUAL TRIGGER AND WAIT BIT BOTH OFF. 0110 ENTER CHANNEL, MANUAL TRIGGER ON. 1XXY DISPLAY OR STORE WITH MANUAL TRIGGER ON AS FOLLOWS: XX - 00 MAIN STORAGE 01 PROTECTION TAG STORAGE 10 LOCAL STORAGE 11 MULTIPLEXOR BUMP STORAGE Y - 0 DISPLAY 1 STORE	KS621
C	CAR.(T≠0)→CR	011101	SS29	SET LEFT BIT OF CONDITION REG (PSW BIT 34) TO VALUE OF CARRY OUT OF ADDER POSITION 0. SET RIGHT BIT (PSW 35) TO ONE IF ADDER LATCH IS NON-ZERO, TO ZERO IF LATCH IS ZERO.	KS621
C	KEY→F	011110	SS30	GATE FROM TAG STORAGE DATA LINES TO F REG. IF SAR ADDRESSES LCS, INITIATE SIGNAL TO STORAGE INDICATING READ STORAGE KEY OPERATION.	KS641
C	F→KEY	011111	SS31	GATE F REG TO TAG STORAGE DATA LINES. INITIATE WRITE STORAGE KEY OPERATION. (NOTE: THIS ORDER IS GIVEN ON W2 CYCLE OF STORAGE AND CAUSES HOLDOFF UNTIL NEXT W2 CYCLE.)	KS641
C	1→LSGNS	100000	SS32	TURN ON L SIGN STAT.	KS641
C	0→LSGNS	100001	SS33	TURN OFF L SIGN STAT.	KS641
C	1→RSGNS	100010	SS34	TURN ON R SIGN STAT.	KS641
C	0→RSGNS	100011	SS35	TURN OFF R SIGN STAT.	KS641
C	L(0)→LSGNS	100100	SS36	SET L SIGN STAT TO VALUE OF L REG BIT 0.	KS641
C	R(0)→RSGNS	100101	SS37	SET R SIGN STAT TO VALUE OF R REG BIT 0.	KS641
C	E(13)→WFN	100110	SS38	GATE EMIT FIELD BITS 1-3 TO MOVER FUNCTION REG.	KS641

CF112 01

SS FIELD (CONTINUED)

EDGE CHAR	MNEMONIC	BITS	DEC ORDER	FUNCTION	ALD PAGE
C	E(23)→LSFN	100111	SS39	GATE EMIT FIELD BITS 2-3 TO LOCAL STORAGE FUNCTION REG. (EFFECTIVE FOR LOCAL STORAGE ADDRESSING THIS CYCLE.)	KS641
C	E(23)→CR	101000	SS40	GATE EMIT FIELD BITS 2-3 TO CONDITION REG (PSW BITS 34-35).	KS651
C	SETCRALG	101001	SS41	SET THE CONDITION REG ACCORDING TO ALGEBRAIC CONDITIONS AS FOLLOWS: COND CODE — CONDITION 00 ADDER LATCH ZERO. 01 ADDER LATCH BIT 0 IS ONE. (MINUS) 10 ADDER LATCH BIT 0 IS ZERO, BITS 1-31 NON-ZERO. (PLUS)	KS651
	NOTE 1			1410-7010 USAGE OF UNDEFINED ORDERS	
C	RSGNVW23 01	000001	SS2	INVERT THE R SIGN STAT IF MOVER LATCH BITS 2-3 ARE SET TO 01.	XE201
C	SETCRLOG	101010	SS42	SET THE CONDITION REG ACCORDING TO LOGICAL CONDITIONS AS FOLLOWS: COND CODE — CONDITION 00 ADDER LATCH BYTES SELECTED BY BYTE STATS ARE ZERO. 01 ADDER LATCH BYTES SELECTED BY BYTE STATS ARE NOT ALL ZERO AND THERE IS NO CARRY OUT OF ADDER POSITION 0. 10 ADDER LATCH BYTES SELECTED BY BYTE STATS ARE NOT ALL ZERO AND THERE IS A CARRY OUT OF ADDER POSITION 0.	KS651
C	¬S4,S4→CR	101011	SS43	SET LEFT BIT OF CONDITION REG (PSW BIT 34) TO THE COMPLEMENT OF STAT 4. SET THE RIGHT BIT (PSW BIT 35) EQUAL TO STAT 4.	KS651
C	S4,¬S4→CR	101100	SS44	SET LEFT BIT OF CONDITION REG (PSW BIT 34) EQUAL TO STAT 4. SET THE RIGHT BIT (PSW 35) TO THE COMPLEMENT OF STAT 4.	KS651
C	1→REFETCH	101101	SS45	TURN ON THE REFETCH TRIGGER.	KS651
C	SYNC→OPPANEL	101110	SS46	SEND ADDRESS COMPARE SYNC/STOP PULSE TO CONSOLE.	KS651
C C	SCAN*E.10	101111	SS47	TURN ON FLT OP REG POS. 1. IF THIS MICROINSTRUCTION CONTAINS THE FOLD MICRO-ORDER, TURN ON FLT OP REG POS. 0, OTHERWISE TURN IT OFF. GATE EMIT FIELD TO FLT OP REG POS 2-5.	KS521
	1→SUPOUT	110000	SS48	TURN ON MULTIPLEXOR CHANNEL SUPPRESS OUT LINE.	KS661
C	MPXSELRESET	110001	SS49	SELECTIVE RESET OF MULTIPLEXOR CHANNEL.	KS661
C	E(0)→IBFULL	110010	SS50	SET INTERRUPT BUFFER FULL STAT TO VALUE OF EMIT FIELD BIT 0.	KS661
C		110011	SS51	UNDEFINED	

11 MAR 65 255099
3 AUG 65 255449
11 JAN 66 255458
30 MAR 66 256490
10 MAY 66 255793

CONTROL FIELD SPECIFICATION
CPU MODE
DATE 9 MAY 66 MACH. 2050
FRAME 01
P.N. 5365014
IBM CORP. SDD PAGE 1

CF112

SS FIELD (CONTINUED)

EDGE CHAR	MNEMONIC	BITS	DEC ORDER	FUNCTION	ALD PAGE
C	E→CH	110100	SS52	GATE THE EMIT FIELD TO THE COMMON CHANNEL.	KS661
		110101	SS53	UNDEFINED (SEE NOTE 2)	
C	1→TIMERIRPT	110110	SS54	TURN ON THE TIMER BIT IN THE EXTERNAL INTERRUPT REG.	KS661
C	T→PSW,IPL→T	110111	SS55	GATE IPL UNIT ADDRESS BITS 0-7 TO L REG BITS 0-7. GATE CHANNEL ADDRESS TO L REG BITS 21-23. ORDER TR7 IS USED WITH THIS ORDER TO PROVIDE REGISTER PULSE FOR L. BUT T→L GATING IS INHIBITED.	KS661
C	T→PSW	111000	SS56	GATE ADDER LATCH BITS 12-15 TO PSW BITS 12-15 (MODE BITS)	KS671
C	SCAN*E,00	111001	SS57	TURN OFF FLT OP REG POS 1. IF THIS MICRO-INSTRUCTION CONTAINS THE FOLD MICRO-ORDER, TURN ON FLT OP REG POS 0. OTHERWISE TURN IT OFF. GATE EMIT FIELD TO FLT OP REG POS 2-5.	KS521
C	1→IOMODE	111010	SS58	TURN ON I/O MODE STAT.	KS671
C	0→IOMODE	111011	SS59	TURN OFF I/O MODE STAT.	KS671
C	1→SELOUT	111100	SS60	TURN ON MULTIPLEXOR CHANNEL SELECT OUT LINE.	KS671
C	1→ADROUT	111101	SS61	TURN ON MULTIPLEXOR CHANNEL ADDRESS OUT LINE.	KS671
C	1→COMOUT	111110	SS62	TURN ON MULTIPLEXOR CHANNEL COMMAND OUT LINE.	KS671
C	1→SERVOUT	111111	SS63	TURN ON MULTIPLEXOR CHANNEL SERVICE OUT LINE.	KS671

NOTE 2 7070 / 7074 EMULATOR USAGE OF ORDER

EDGE CHAR	MNEMONIC	BITS	DEC ORDER	FUNCTION	ALD PAGE
C	E(3)→EMSTAT	000001	SS1	SET EMULATOR STAT = TO BIT 3 OF EMIT FIELD.	XF111
A,C	C*M→COP	000010	SS2	M REG BITS 23-91 (9 BITS) TO 70 OP CODE CONVERTER. EMIT FIELD TO BYTE STATS. ADDER LATCH BIT 30 TO STAT 3	XF111
C	U=OF,ALPHA	110101	SS53	IF U=OF, SET STAT 0=1. IF U BITS 2,3 ≠ 1 ; SET STAT 4 = 1.	XF111

* MAY BE USED TO FILL TEMPLATE

TEMPLATE FOR 'A' LINE STATEMENT SSS→TTT
WHERE S=SS FIELD ; T = TR FIELD

LU FIELD ROSDR 1-3 MOVER INPUT - LEFT SIDE

EDGE CHAR	MNEMONIC	BITS	DEC ORDER	FUNCTION	ALD PAGE
		000	LU0	GATE ZEROS TO LEFT MOVER INPUT U.	DS001
B	MD,F→U	001	LU1	GATE MD REG TO LEFT MOVER INPUT U BITS 0-3. GATE F REG TO MOVER INPUT U BITS 4-7.	DS001
B	R3→U	010	LU2	GATE R REG BITS 24-31 TO LEFT MOVER INPUT U.	DS001
B	BIB→U	011	LU3	GATE MULTIPLEX CHANNEL BUFFER IN BUS TO LEFT MOVER INPUT U.	DS001
B	L0→U	100	LU4	GATE L REG BITS 0-7 TO LEFT MOVER INPUT U.	DS001
B	L1→U	101	LU5	GATE L REG BITS 8-15 TO LEFT MOVER INPUT U.	DS001
B	L2→U	110	LU6	GATE L REG BITS 16-23 TO LEFT MOVER INPUT U.	DS001
B	L3→U	111	LU7	GATE L REG BITS 24-31 TO LEFT MOVER INPUT U.	DS001

MV FIELD ROSDR 4-5 MOVER INPUT - RIGHT SIDE

EDGE CHAR	MNEMONIC	BITS	DEC ORDER	FUNCTION	ALD PAGE
		00	MV0	GATE ZEROS TO RIGHT MOVER INPUT V.	DS021
		01	MV1	UNDEFINED	
B	BIB→V	10	MV2	GATE MULTIPLEX CHANNEL BUFFER IN BUS TO RIGHT MOVER INPUT V.	DS021
		11	MV3	UNDEFINED. NOTE 1	

ZP FIELD ROSDR 6-11 ROS ADDRESS BITS 0-5

THIS FIELD IS GATED DIRECTLY TO FORM BITS 0-5 OF THE NEXT ROS ADDRESS.

NOTE 1 1410-7010 USAGE OF UNDEFINED ORDERS

EDGE CHAR	MNEMONIC	BITS	DEC ORDER	FUNCTION	ALD PAGE
B	R3→V	11	MV3	GATE R REG BITS 24-31 TO MOVER RIGHT	BV001

CI CFC 201 - 2050 CONTROL FIELD SPECIFICATION - I/O MODE (LU.MV.ZP.ZF.ZN)

CF201 01

ZF FIELD ROSDR 12-15 FUNCTION BRQNCH CONTROL OR ROS ADDRESS BITS 6-9

IF THE ZN FIELD IS NON-ZERO, THIS FIELD IS GATED DIRECTLY TO FORM BITS 6-9 OF THE NEXT ROS ADDRESS.
IF THE ZN FIELD IS ZERO, THIS FIELD IS DECODED AS SHOWN BELOW.

EDGE CHAR	MNEMONIC	BITS	DEC ORDER	FUNCTION	ALD PAGE
		0000	ZF0	UNDEFINED	
R	D→ROAR.SCAN	0010	ZF2	GATE STORAGE DATA REG BITS AS FOLLOWS: BITS 19-30 TO ROS ADDRESS	KK002
		0100	ZF4	UNDEFINED	
R	M(03)→ROAR	0110	ZF6	GATE M REG BITS 0-3 TO NEXT ROS ADDRESS BITS 6-9.	KK002
R	M(47)→ROAR	1000	ZF8	GATE M REG BITS 4-7 TO NEXT ROS ADDRESS BITS 6-9.	KK002
R	F→ROAR	1010	ZF10	GATE F REG TO NEXT ROS ADDRESS BITS 6-9.	KK002
R	ED→ROAR	1100	ZF12	GATE EXPONENT DIFFERENCE REG TO NEXT ROS ADDRESS BITS 6-9.	KK002
R	RETURN→ROAR	1110	ZF14	GATE ROS ADDRESS BUFFER REG TO NEXT ROS ADDRESS. THIS REGISTER IS SET ON I/O BREAK-IN.	KK002

ZN FIELD ROSDR 16-18 ROS ADDRESS CONTROL (NULL VALUE: ZN4)

EDGE CHAR	MNEMONIC	BITS	DEC ORDER	FUNCTION	ALD PAGE
		000	ZN0	DECODE ZF FIELD	
S	SMIF	001	ZN1	SUPPRESS INITIATION OF MAIN STORAGE CYCLE BY DECIMAL ORDER IV7 IN THIS MICROINSTRUCTION IF REFETCH STAT IS OFF AND IAR BIT 30 IS 1. (NEXT INSTRUCTION IS OFF WORD BOUNDARY.)	KK003
R	AΩ(B=0)→A	010	ZN2	FORCE A BIT TO 1 IF B BIT IS 0.	KK003
R	AΩ(B=1)→A	011	ZN3	FORCE A BIT TO 1 IF B BIT IS 1.	KK003
		100	ZN4	NORMAL ROS ADDRESSING.	
R	FNTRAP	101	ZN5	FORCE ROS ADDRESS TO INVALID OP TRAP VALUE IF THIS MICROINSTRUCTION WAS REACHED BY A FUNCTUON BRANCH.	KK003
R	BΩ(A=0)→B	110	ZN6	FORCE B BIT TO 1 IF A BIT IS 0.	KK003
R	BΩ(A=1)→B	111	ZN7	FORCE B BIT TO 1 IF A BIT IS 1.	KK003

3 AUG 65 255449
11 DEC 65 255458
9 MAY 66 255793

CONTROL FIELD SPECIFICATION I/O MODE	
DATE 9 MAY 66	MACH. 2050
	FRAME 01
	P.N. 5365018
IBM CORP. SDD	PAGE 1

CF201

```
TEMPLATE FOR 'A' LINE STATEMENT:  RRR±LLL→TTTT

TEMPLATE FOR 'D' LINE STATEMENT:  AAAA→TTTT

   WHERE: R = RY FIELD    L = LX FIELD    T = TR FIELD
          ± = TC FIELD    A = AL FIELD
```

TR FIELD ROSDR 19–23 ADDER LATCH DESTINATION

EDGE CHAR	MNEMONIC	BITS	DEC ORDER	FUNCTION	ALD PAGE
A	Z			DUMMY, USED IN 'A' LINE STATEMENT WHEN THERE IS ALSO A 'D' LINE STATEMENT.	
		00000	TR0	NO GATING FROM ADDER LATCH TO REGISTERS.	
A,D	T	00000	TR0	NO GATING FROM ADDER LATCH TO REGISTERS. USED WHEN OTHER TERMS OF STATEMENT ARE PRESENT.	
A,D	R	00001	TR1	GATE ADDER LATCH TO R REG.	DR121
A,D	R0	00010	TR2	GATE ADDER LATCH BITS 0–7 TO R REG BITS 0–7.	DR121
A,D	M	00011	TR3	GATE ADDER LATCH TO M REG.	DR121
A,D	D	00100	TR4	GATE ADDER LATCH TO SDR.	DR111
A,D	L0	00101	TR5	GATE ADDER LATCH BITS 0–7 TO L REG BITS 0–7.	DR111
A,D	R,A	00110	TR6	GATE ADDER LATCH TO R REG AND SAR. INITIATE STORAGE REQUEST.	DR111
A,D	L	00111	TR7	GATE ADDER LATCH TO L REG.	DR111
S	HA→A	01000	TR8	GATE HARDWARE GENERATED ADDRESS TO SAR. INITIATE STORAGE REQUEST. DECODE EMIT FIELD TO DETERMINE STORAGE FUNCTION.	DR121

EMIT VALUE	FUNCTION
X000	WRITE IN MAIN STORAGE ADDRESS 80.
X001	READ FROM MAIN STORAGE ADDRESS 80.
X010	OR FROM MAIN STORAGE ADDRESS 80.
X011	UNDEFINED
X100	WRITE IN MAIN STORAGE ADDRESS 84. SET IAR TO 84.
X101	READ FROM MAIN STORAGE ADDRESS 84. SET IAR TO 84.
X110	UNDEFINED
X111	UNDEFINED

EDGE CHAR	MNEMONIC	BITS	DEC ORDER	FUNCTION	ALD PAGE
A,D	R,AN	01001	TR9	GATE ADDER LATCH TO R REG AND SAR. INITIATE STORAGE REQUEST. SUPPRESS INVALID ADDRESS TRAP. SET INVALID ADDRESS STAT INSTEAD.	DR121
A,D	R,AW	01010	TR10	GATE ADDER LATCH TO R REG AND SAR. INITIATE STORAGE REQUEST. TRAP IF ADDRESS NOT ON WORD BOUNDARY.	DR121
A,D	R,AD	01011	TR11	GATE ADDER LATCH TO R REG AND SAR. INITIATE STORAGE REQUEST. TRAP IF ADDRESS NOT ON DOUBLE WORD BOUNDARY.	DR121
D*	D→IAR	01100	TR12	GATE SDR TO IAR. INTERLOCK WITH STORAGE TIMING RING.	DR121
D*	SCAN→D	01101	TR13	GATE SCAN BUS TO SDR IN FOLLOWING ARRANGEMENT: PARITY OF BYTE 0 TO BIT 0. BITS 0–7 TO BITS 1–8. PARITY OF BYTE 1 TO BIT 9. BITS 8–15 TO BITS 10–17. PARITY OF BYTE 2 TO BIT 18. BITS 16–23 TO BITS 19–26. PARITY OF BYTE 3 TO BIT 27. BITS 24–27 TO BITS 28–31. GOOD PARITY IS INSERTED IN SDR.	DR121
A,D	R13	01110	TR14	GATE ADDER LATCH BITS 8–31 TO R REG BITS 8–31.	DR121
A,D	A	01111	TR15	GATE ADDER LATCH BITS 8–31 TO SAR. INITIATE STORAGE REQUEST.	DR131
A,D	L,A	10000	TR16	GATE ADDER LATCH TO L REG AND SAR. INITIATE STORAGE REQUEST.	DR131
A,D	R,D	10001	TR17	GATE ADDER LATCH TO R REG AND SDR.	DR121
		10010	TR18	UNDEFINED	
A,D	R,IO	10011	TR19	GATE ADDER LATCH TO R REG. GATE ADDER LATCH BITS 30–31 TO I/O REG.	DR131

* NOT PART OF A STATEMENT

CFC 202 – 2050 CONTROL FIELD SPECIFICATION – I/O MODE (TR)

CF202 01

TR FIELD (CONTINUED)

EDGE CHAR	MNEMONIC	BITS	DEC ORDER	FUNCTION	ALD PAGE
A,D	H	10100	TR20	GATE ADDER LATCH TO H REG.	DR141
A,D	IA	10101	TR21	GATE ADDER LATCH BITS 8-31 TO IAR	DR141
A,D	FOLD	10110	TR22	GATE SCAN BUS BITS 28-31 TO SDR BITS 28-31. GATE ZEROS TO SDR BITS 0-27. GOOD PARITY IS INSERTED IN SDR.	DR141
		10111	TR23	UNDEFINED	
A,D	L,M	11000	TR24	GATE ADDER LATCH TO L REG AND M REG.	DR111
A,D	MLJK	11001	TR25	GATE ADDER LATCH TO L REG AND M REG. GATE LATCH BITS 12-15 TO J REG. GATE LATCH BITS 16-19 TO MD COUNTER. TURN OFF REFETCH STAT. IF LATCH BITS 12-15 ALL ZERO, TURN ON STAT 0. OTHERWISE TURN OFF STAT 0. IF LATCH BITS 16-19 ALL ZERO, TURN ON STAT 1. OTHERWISE TURN OFF STAT 1. IF LATCH BITS 16-17 ALL ZERO, TURN ON ONE-SYLLABLE STAT. OTHERWISE TURN OFF ONE-SYLLABLE STAT. IF LATCH BITS 0-1 EQUAL 00, SET ILC TO 01. IF LATCH BITS 0-1 EQUAL 01 OR 10,SET ILC TO 10. IF LATCH BITS 0-1 EQUAL 11, SET ILC TO 11.	DR111
A,D	MHL	11010	TR26	GATE ADDER LATCH TO L REG. GATE ADDER LATCH BITS 0-15 TO M REG BITS 16-31. GATE ADDER LATCH BITS 0-3 TO MD COUNTER.	DR111
A,D	MD	11011	TR27	GATE ADDER LATCH BITS 8-11 TO MD COUNTER.	DR111
A,D	D*BI	11100	TR28	GATE ADDER LATCH TO SDR UNDER CONTROL OF I/O BYTE STATS.	DR121
A,D	D*BS	11101	TR29	GATE ADDER LATCH TO SDR UNDER CONTROL OF CPU BYTE STATS.	DR111
A,D	L13	11110	TR30	GATE ADDER LATCH BITS 8-31 TO L REG BITS 8-31.	DR111
A,D	IO	11111	TR31	GATE ADDER LATCH BITS 29-31 TO I/O REG.	DR131

ZR FIELD ROSDR 24

USED TO EXTEND ROS ADDRESSING

0 ADDRESSES BASIC ROS XG401

1 ADDRESSES EXTENDED ROS KK302

11 MAR 255099
3 AUG 65 255449
11 JAN 66 255458

CONTROL FIELD SPECIFICATION
I/O MODE
DATE 18 FEB 66 MACH. 2050
FRAME 01
P.N. 5365019
IBM CORP. SDD PAGE 1

CF202

NOTE: CS FIELD AND SA FIELD MNEMONICS ARE WRITTEN TOGETHER AS A STATEMENT.

CS FIELD ROSDR 25 LOCAL STORAGE ADDRESS SECTOR

EDGE CHAR	MNEMONIC	BITS	DEC ORDER	FUNCTION	ALD PAGE
L	2.3.	0	CS0	SET LSAR BITS 0-3 TO VALUE 1011. (TO ADDRESS REGISTERS 44-47)	KL001
L	0.CH.	1	CS1	SET LSAR BITS 0-1 TO VALUE 00. (I/O REGISTERS) SET LSAR BITS 2-3 ACCORDING TO THE CHANNEL CURRENTLY IN CONTROL: 11 MULTIPLEXOR CHANNEL 00 SELECTOR CHANNEL 1 01 SELECTOR CHANNEL 2 10 SELECTOR CHANNEL 3	KLOXX

SA FIELD ROSDR 26-27 LOCAL STORAGE ADDRESS

EDGE CHAR	MNEMONIC	BITS	DEC ORDER	FUNCTION	ALD PAGE
L	0→LSA	00	SA0	SET LSAR BITS 4-5 TO VALUE 00.	KLOXX
L	1→LSA	01	SA1	SET LSAR BITS 4-5 TO VALUE 01.	KLOXX
L	2→LSA	10	SA2	SET LSAR BITS 4-5 TO VALUE 10.	KLOXX
L	3→LSA	11	SA3	SET LSAR BITS 4-5 TO VALUE 11.	KLOXX

CFC 203 - 2050 CONTROL FIELD SPECIFICATION - I/O MODE (CS.SA.SF.CT)

CF203 01

SF FIELD ROSDR 28–30 LOCAL STORAGE FUNCTION (NULL VALUE: SF7)

EDGE CHAR	MNEMONIC	BITS	DEC ORDER	FUNCTION	ALD PAGE
L	R→LS	000	SF0	WRITE INTO LOCAL STORAGE FROM R REG.	KLOXX
L	LS→L,R→LS	001	SF1	READ LOCAL STORAGE INTO L REG. WRITE INTO LOCAL STORAGE FROM R REG.	KLOXX
L	LS→R→LS	010	SF2	READ LOCAL STORAGE INTO R REG. REGENERATE FROM R REG.	KLOXX
		011	SF3	UNDEFINED	
L	L→LS	100	SF4	WRITE INTO LOCAL STORAGE FROM L REG.	KLOXX
L	LS→R,L→LS	101	SF5	READ LOCAL STORAGE INTO R REG. WRITE INTO LOCAL STORAGE FROM L REG.	KLOXX
L	LS→L→LS	110	SF6	READ LOCAL STORAGE INTO L REG. REGENERATE FROM L REG.	KLOXX
		111	SF7	NO OPERATION	

CT FIELD ROSDR 32–34 TIMING SIGNALS TO EXTERNAL CHANNEL

EDGE CHAR	MNEMONIC	BITS	DEC ORDER	FUNCTION	ALD PAGE
		000	CT0	NO OPERATION	
C	FIRSTCYCLE	001	CT1	SEND FIRST CYCLE CHECK SIGNAL TO CHANNEL. IF THIS IS NOT THE FIRST CYCLE OF A CHANNEL ROUTINE, A CHANNEL CONTROL CHECK AND LOG-OUT WILL RESULT.	KE531
C	DTC1	010	CT2	SENT INGATE TIMING PULSE TO CHANNEL.	KE441
C	DTC2	011	CT3	SEND OUTGATE TIMING PULSE TO CHANNEL.	KE441
S	IA 4→A.IAR	100	CT4	INCREMENT IAR BY 4. GATE RESULT TO IAR AND SAR. INITIATE STORAGE REQUEST. INHIBIT INVALID ADDRESS TRAP. SET INVALID ADDRESS STAT INSTEAD.	KP012
		101	CT5	UNDEFINED	
		110	CT6	UNDEFINED	
		111	CT7	UNDEFINED	

11 MAR 65 255099
3 AUG 65 255449
9 MAY 66 255793

CONTROL FIELD SPECIFICATION
I/O MODE
DATE 10 MAY 66 MACH. 2050
FRAME 01
P.N. 5365020
IBM CORP. SDD PAGE 1

CF203

TEMPLATE FOR 'D' LINE STATEMENT: AAAA→TTTT
WHERE: A = AL FIELD T = TR FIELD

AL FIELD ROSDR 35-39 SHIFT CONTROL AND GATING INTO ADDER LATCH

EDGE CHAR	MNEMONIC	BITS	DEC ORDER	FUNCTION	ALD PAGE
		00000	AL0	GATE ADDER OUTPUT TO LATCH, NO SHIFT.	
A	Q→SR1→F	00001	AL1	GATE ADDER OUTPUT TO LATCH, SHIFTED RIGHT 1, ENTERING Q. SPILL BIT ENTERS F, WHICH IS SHIFTED RIGHT.	DR071
A	L0,¬S4→	00010	AL2	GATE ADDER OUTPUT BITS 8-31 TO LATCH BITS 8-31. GATE L REG BITS 1-7 TO LATCH BITS 1-7. SET LATCH BIT 0 TO THE COMPLEMENT OF STAT 4.	DR061
A	+SGN→	00011	AL3	GATE ADDER OUTPUT BITS 1-31 TO LATCH BITS 1-31. SET LATCH BIT 0 TO ZERO.	DR061
A	-SGN→	00100	AL4	GATE ADDER OUTPUT BITS 1-31 TO LATCH BITS 1-31. SET LATCH BIT 0 TO ONE.	DR061
A	L0,S4→	00101	AL5	GATE ADDER OUTPUT BITS 8-31 TO LATCH BITS 8-31. GATE L REG BITS 1-7 TO LATCH BITS 1-7. GATE STAT 4 TO LATCH BIT 0.	DR061
D*	IA→H	00110	AL6	GATE ADDER OUTPUT TO LATCH. NO SHIFT. GATE IAR TO H REG BITS 8-31. H REG BITS 0-7 REMAIN UNCHANGED.	DR081
A	Q→SL→-F	00111	AL7	GATE ADDER OUTPUT TO LATCH, SHIFTED LEFT 1, ENTERING Q. COMPLEMENT OF SPILL BIT ENTERS F, WHICH IS SHIFTED LEFT.	DR081
A	Q→SL1→F	01000	AL8	GATE ADDER OUTPUT TO LATCH, SHIFTED LEFT 1, ENTERING Q. SPILL BIT ENTERS F, WHICH IS SHIFTED LEFT.	DR081
A	F→SL1→F	01001	AL9	GATE ADDER OUTPUT TO LATCH, SHIFTED LEFT 1, ENTERING BIT 0 OF F. SPILL BIT ENTERS F, WHICH IS SHIFTED LEFT.	DR081
A	SL1→Q	01010	AL10	GATE ADDER OUTPUT TO LATCH, SHIFTED LEFT 1, ENTERING ZERO. SPILL BIT ENTERS Q.	DR071
A	Q→SL1	01011	AL11	GATE ADDER OUTPUT TO LATCH, SHIFTED LEFT 1, ENTERING Q. SPILL BIT IS DISCARDED.	DR081
A	SR1→F	01100	AL12	GATE ADDER OUTPUT TO LATCH. SHIFTED RIGHT 1. ENTERING ZERO. SPILL BIT ENTERS F. WHICH IS SHIFTED RIGHT.	DR071
A	SR1→Q	01101	AL13	GATE ADDER OUTPUT TO LATCH. SHIFTED RIGHT 1. ENTERING ZERO. SPILL BIT ENTERS Q.	DR071
A	Q→SR1→Q	01110	AL14	GATE ADDER OUTPUT TO LATCH. SHIFTED RIGHT 1. ENTERING Q. SPILL BIT ENTERS Q.	DR071
A	F→SL1→Q	01111	AL15	GATE ADDER OUTPUT TO LATCH. SHIFTED LEFT 1. ENTERING BIT 0 OF F. SPILL BIT ENTERS Q. F IS SHIFTED LEFT 1. ENTERING ZERO.	DR051
A	SL4→F	10000	AL16	GATE ADDER OUTPUT TO LATCH. SHIFTED LEFT 4. ENTERING ZEROS. SPILL BITS ENTER F.	DR0XX
A	F→SL4→F	10001	AL17	GATE ADDER OUTPUT TO LATCH. SHIFTED LEFT 4. ENTERING F. SPILL BITS ENTER F.	DR081
A	FPSL4	10010	AL18	GATE ADDER OUTPUT BITS 0-7 TO LATCH 0-7. GATE ADDER OUTPUT BITS 8-31 TO LATCH BITS 8-31. SHIFTED LEFT 4. ENTERING ZEROS. SPILL BITS FROM ADDER OUTPUT 8-11 ARE DISCARDED.	DR081
A	F→FPSL4	10011	AL19	GATE ADDER OUTPUT BITS 0-7 TO LATCH BITS 0-7. GATE ADDER OUTPUT BITS 8-31 TO LATCH BITS 8-31. SHIFTED LEFT 4. ENTERING F. SPILL BITS FROM ADDER OUTPUT 8-11 ARE DISCARDED.	DR081

* NOT PART OF A STATEMENT.

CFC 204 - 2050 CONTROL FIELD SPECIFICATION - I/O MODE (AL)

CF204 01

AL FIELD (CONTINUED)

EDGE CHAR	MNEMONIC	BITS	DEC ORDER	FUNCTION	ALD PAGE
A	SR4→F	10100	AL20	GATE ADDER OUTPUT TO LATCH, SHIFTED RIGHT 4, ENTERING ZEROS. SPILL BITS ENTER F.	DROXX
A	F→SR4→F	10101	AL21	GATE ADDER OUTPUT TO LATCH, SHIFTED RIGHT 4, ENTERING F. SPILL BITS ENTER F.	DROXX
A	FPSR4→F	10110	AL22	GATE ADDER OUTPUT BITS 8-31 TO LATCH BITS 8-31, SHIFTED RIGHT 4, ENTERING ZEROS. GATE ZEROS TO LATCH BITS 0-7. SPILL BITS ENTER F.	DROXX
A	1→FPSR4→F	10111	AL23	GATE ADDER OUTPUT BITS 0-7 TO LATCH BITS 0-7. GATE ADDER OUTPUT BITS 8-31 TO LATCH BITS 8-31, SHIFTED RIGHT 4, ENTERING 0001. SPILL BITS ENTER F. A HOT CARRY IS FORCED INTO POSITION 7 OF THE ADDER. (THIS ORDER CAN BE USED ONLY WITH AN AD FIELD VALUE OF AD6.)	DR081
A	SR4→H	11000	AL24	GATE ADDER OUTPUT TO LATCH, SHIFTED RIGHT 4, ENTERING ZEROS. SPILL BITS ENTER H REG BITS 0-3. GATE LATCH BITS 4-7 TO R REG BITS 0-3. (THIS ORDER CAN BE USED ONLY WITH A TR FIELD VALUE OF TR1, AND LATCH BITS 4-7 MUST HAVE AN EVEN NUMBER OF ONES.)	DR052
A	F→SR4	11001	AL25	GATE ADDER OUTPUT TO LATCH, SHIFTED RIGHT 4, ENTERING F. SPILL BITS ARE DISCARDED. F IS NOT ALTERED.	DR052
A	E→FPSL4	11010	AL26	GATE ADDER OUTPUT BITS 0-7 TO LATCH BITS 0-7. GATE ADDER OUTPUT BITS 8-31 TO LATCH BITS 8-31, SHIFTED LEFT 4, ENTERING THE EMIT FIELD. SPILL BITS FROM ADDER OUTPUT 8-11 ARE DISCARDED.	DR041
A	F→SR1→Q	11011	AL27	GATE ADDER OUTPUT TO LATCH, SHIFTED RIGHT 1, ENTERING BIT 3 OF F. SPILL BIT ENTERS Q. F IS NOT CHANGED.	DR051
D	DKEY	11100	AL28	GATE DATA KEYS TO LATCH. GATE BITS 28-31 OF DATA KEYS TO F.	DR062
D	CH	11101	AL29	GATE BUS FROM SELECTOR CHANNELS TO LATCH.	DR061
D	D	11110	AL30	GATE STORAGE DATA REG TO LATCH. INTERLOCK WITH STORAGE TIMING RING TO CAUSE POSSIBLE STORAGE HOLDOFF.	DT041
D	AKEY	11111	AL31	GATE ADDRESS KEYS TO LATCH BITS 8-31. GATE ZEROS TO LATCH BITS 0-7.	DR062

11 MAR 65 255099
3 AUG 65 255449

CONTROL FIELD SPECIFICATION I/O MODE	
DATE 16 AUG 65	MACH. 2050
	FRAME 01
	P.N. 5365021
IBM CORP. SDD	PAGE 1

CF204

WL FIELD ROSDR 40-42 MOVER OUTPUT DESTINATION

EDGE CHAR	MNEMONIC	BITS	DEC ORDER	FUNCTION	ALD PAGE
		000	WL0	NO OPERATION	
B	W→L0	001	WL1	GATE MOVER LATCH TO L REG BITS 0-7.	DS211
B	W→L1	010	WL2	GATE MOVER LATCH TO L REG BITS 8-15.	DS211
B	W→L2	011	WL3	GATE MOVER LATCH TO L REG BITS 16-23.	DS211
B	W→L3	100	WL4	GATE MOVER LATCH TO L REG BITS 24-31.	DS211
B	W,E→A(BMP)	101	WL5	GATE MOVER LATCH BITS 0-1 TO SAR BITS 14-15. GATE MOVER LATCH BITS 2-7 TO SAR BITS 24-29. GATE EMIT FIELD BITS 2-3 TO SAR BITS 16-17. SET REMAINING SAR BITS TO ZERO. INITIATE BUMP STORAGE REQUEST.	DS211
B	W,E→A(BMP)S	110	WL6	GATE MOVER LATCH BITS 0-1 TO SAR BITS 14-15. GATE MOVER LATCH BITS 2-7 TO SAR BITS 24-29. GATE EMIT FIELD BITS 2-3 TO SAR BITS 16-17. SET REMAINING SAR BITS TO ZERO. INITIATE BUMP STORAGE REQUEST FOR SPLIT READ-WRITE CYCLE.	DS211
		111	WL7	UNDEFINED	

HC FIELD ROSDR 43 CARRY INSERTION CONTROL

EDGE CHAR	MNEMONIC	BITS	DEC ORDER	FUNCTION	ALD PAGE
		0	HC0	NO OPERATION	
A	HOT1	1	HC1	INSERT HOT CARRY INTO ADDER.	DS301

MS FIELD ROSDR 44-46 MULTIPLEXOR CHANNEL STAT CONTROL

EDGE CHAR	MNEMONIC	BITS	DEC ORDER	FUNCTION	ALD PAGE
		000	MS0	NO OPERATION	
C	BIB(03)→IOS	001	MS1	GATE BUFFER IN BUS BITS 0-3 TO MULTIPLEXOR CHANNEL STATS.	FA081
C	BIB(47)→IOS	010	MS2	GATE BUFFER IN BUS BITS 4-7 TO MULTIPLEXOR CHANNEL STATS.	FA081
C	BIB03→IOS*E	011	MS3	GATE BUFFER IN BUS BITS 0-3 TO THOSE MULTIPLEXOR CHANNEL STATS CORRESPONDING TO ONE BITS IN THE EMIT FIELD. LEAVE UNCHANGED THOSE STATS CORRESPONDING TO ZERO BITS.	FA081
C	BIB47→IOS*E	100	MS4	GATE BUFFER IN BUS BITS 4-7 TO THOSE MULTIPLEXOR CHANNEL STATS CORRESPONDING TO ONF BITS IN THE EMIT FIELD. LEAVE UNCHANGED THOSE STATS CORRESPONDING TO ZERO BITS.	FA081
C	IOSΩE	101	MS5	TURN ON THOSE MULTIPLEXOR CHANNEL STATS CORRESPONDING TO ONE BITS IN THE EMIT FIELD.	FA081
C	IOS.¬E	110	MS6	TURN OFF THOSE MULTIPLEXOR CHANNEL STATS CORRESPONDING TO ONE BITS IN THE EMIT FIELD.	FA081
C	BIB4.ERR→IOS	111	MS7	GATE MULTIPLEX BUS IN BIT 4 TO MULTIPLEXOR CHANNEL STAT 4. (CHANNEL END) SET MULTIPLEXOR CHANNEL STAT 1 TO MULTIPLEX BUS IN (BIT 0 OR BIT 2 OR BIT 3 OR BIT 6 OR BIT 7 OR (BIT 1 AND NOT BIT 5)). BIT 0 = ATTN BIT 6 = UNIT CHK BIT 2 = CH END BIT 7 = UNIT EXCPT BIT 3 = BUSY BIT 1 AND NOT BIT 5 = STATUS MODIFIER AND NOT DEVICE END THIS COMBINATION CORRESPONDS TO A STATUS ERROR.	FA081

CFC 205 - 2050 CONTROL FIELD SPECIFICATION - I/O MODE (WL. HC. MS. CG. MG)

CF205 01

CG FIELD ROSDR 47-48 CONTROL SIGNALS TO CHANNEL

EDGE CHAR	MNEMONIC	BITS	DEC ORDER	FUNCTION	ALD PAGE
		00	CG0	NO OPERATION	
C	CH→BI	01	CG1	SELECTOR CHANNEL BYTE GATE CONTROL. CAUSES CHANNEL TO SET BGC REGISTER FROM BYTE COUNTER AND END REGISTER. DECODED OUTPUT OF BGC REG IS GATED TO BYTE STATS BY ORDER TR28. (WHEN READ BKWD, OUTPUT IS CROSSED BEFORE GATING TO BYTE STATS.)	KE381
C	1→PRI	10	CG2	SEND END OF CHANNEL ROUTINE SIGNAL TO CHANNEL PRIORITY CIRCUITS. (USED IN NEXT TO LAST CYCLE OF ROUTINE.)	KE321
C	1→LCY	11	CG3	SEND LAST CYCLE SIGNAL TO CHANNEL PRIORITY CIRCUITS. (USED IN LAST CYCLE OF ROUTINE.)	KE321

MG FIELD ROSDR 49-51 MULTIPLEXOR CHANNEL GATE CONTROL (NULL VALUE: MG3)

EDGE CHAR	MNEMONIC	BITS	DEC ORDER	FUNCTION	ALD PAGE
D	BFR2→BIB	000	MG0	GATE MULTIPLEXOR CHANNEL BUFFER 2 TO BUFFER IN BUS.	FA071
C	CHPOSTEST	001	MG1	INITIATE A SELECTOR CHANNEL POSITION TEST.	FA071
D	BFR2→BUSO	010	MG2	GATE MULTIPLEXOR CHANNEL BUFFER 1 TO BUFFER IN BUS. GATE MULTIPLEXOR CHANNEL BUFFER 2 TO BUS OUT (TO INTERFACE).	FA071
D	BFR1→BIB	011	MG3	GATE MULTIPLEXOR CHANNEL BUFFER 1 TO BUFFER IN BUS.	FA071
D	BOB→BFR1	100	MG4	GATE MULTIPLEXOR CHANNEL BUFFER 1 TO BUFFER IN BUS. THEN GATE BUFFER OUT BUS TO MULTIPLEXOR CHANNEL BUFFER 1.	FA071
D	BOB→BFR2	101	MG5	GATE MULTIPLEXOR CHANNEL BUFFER 1 TO BUFFER IN BUS. THEN GATE BUFFER OUT BUS TO MULTIPLEXOR CHANNEL BUFFER 2.	FA071
D	BUSI→BFR1	110	MG6	GATE MULTIPLEXOR CHANNEL BUFFER 1 TO BUFFER IN BUS. THEN GATE BUS IN (FROM INTERFACE) TO MULTIPLEXOR CHANNEL BUFFER 1.	FA071
D	BUSI→BFR2	111	MG7	GATE MULTIPLEXOR CHANNEL BUFFER 1 TO BUFFER IN BUS. THEN GATE BUS IN (FROM INTERFACE) TO MULTIPLEXOR CHANNEL BUFFER 2.	FA071

11 MAR 65 255099
3 AUG 65 255449

CONTROL FIELD SPECIFICATION	
I/O MODE	
DATE 16 AUG 65	MACH. 2050
	FRAME 01
	P.N. 5365022
IBM CORP. SDD	PAGE 1

CF205

UL FIELD ROSDR 52-53 MOVER ACTION - LATCH BITS 0-3 (NULL VALUE: UL1)

EDGE CHAR	MNEMONIC	BITS	DEC ORDER	FUNCTION	ALD PAGE
B	E→WL	00	UL0	GATE EMIT FIELD TO MOVER LATCH BITS 0-3. (NOTE: EMIT FIELD IS RECODED IF ORDER WM12 IS USED AND ASCII BIT IS ON. SEE WM12.)	DS101
B	UL→WL	01	UL1	GATE LEFT INPUT (U) BITS 0-3 TO LATCH BITS 0-3.	DS101
B	VL→WL	10	UL2	GATE RIGHT INPUT (V) BITS 0-3 TO LATCH BITS 0-3.	DS101
B	?→WL	11	UL3	GATE TO LATCH BITS 0-3 ACCORDING TO CONTENTS OF MOVER FUNCTION REGISTER AS FOLLOWS:	DS101

FUNCTION REGISTER	ACTION
000	CROSS. GATE U BITS 4-7 TO LATCH BITS 0-3.
001	STRAIGHT. GATE U BITS 0-3 TO LATCH BITS 0-3.
010	AND. GATE AND OF U BITS 0-3 AND V BITS 0-3 TO LATCH BITS 0-3.
011	V LEFT.U RIGHT. GATE V BITS 0-3 TO LATCH BITS 0-3.
100	OR. GATE OR OF U BITS 0-3 AND V BITS 0-3 TO LATCH BITS 0-3.
101	U LEFT.V RIGHT. GATE U BITS 0-3 TO LATCH BITS 0-3.
110	EXCLUSIVE OR. GATE EXCLUSIVE OR OF U BITS 0-3 AND V BITS 0-3 TO LATCH BITS 0-3.
111	UNDEFINED

UR FIELD ROSDR 54-55 MOVER ACTION - LATCH BITS 4-7 (NULL VALUE: UR1)

EDGE CHAR	MNEMONIC	BITS	DEC ORDER	FUNCTION	ALD PAGE
B	E→WR	00	UR0	GATE EMIT FIELD TO MOVER LATCH BITS 4-7. (NOTE: EMIT FIELD IS RECODED IF ORDER WM12 IS USED AND ASCII BIT IS ON. SEE WM12.)	DS101
B	UR→WR	01	UR1	GATE LEFT INPUT (U) BITS 4-7 TO LATCH BITS 4-7.	DS101
B	VR→WR	10	UR2	GATE RIGHT INPUT (V) BITS 4-7 TO LATCH BITS 4-7.	DS101
B	?→WR	11	UR3	GATE TO LATCH BITS 4-7 ACCORDING TO CONTENTS OF MOVER FUNCTION REGISTER AS FOLLOWS:	DS101

FUNCTION REGISTER	ACTION
000	CROSS. GATE U BITS 0-3 TO LATCH BITS 4-7.
001	STRAIGHT. GATE U BITS 4-7 TO LATCH BITS 4-7.
010	AND. GATE AND OF U BITS 4-7 AND V BITS 4-7 TO LATCH BITS 4-7.
011	V LEFT.U RIGHT. GATE U BITS 4-7 TO LATCH BITS 4-7.
100	OR. GATE OR OF U BITS 4-7 AND V BITS 4-7 TO LATCH BITS 4-7.
101	U LEFT.V RIGHT. GATE V BITS 4-7 TO LATCH BITS 4-7.
110	EXCLUSIVE OR. GATE EXCLUSIVE OR OF U BITS 4-7 AND V BITS 4-7 TO LATCH BITS 4-7.
111	UNDEFINED

NOTE:- WHEN EQUAL DECIMAL ORDERS ARE USED IN UL AND UR FIELDS. THE FOLLOWING COMBINATION MNEMONICS ARE USED:

UL1 AND UR1: U→W

UL2 AND UR2: V→W

UL3 AND UR3: ?→W

CE FIELD ROSDR 57-60 EMIT FIELD

THIS FIELD IS USED AS DATA BY MICRO ORDERS.

CFC 206 - 2050 CONTROL FIELD SPECIFICATION - I/O MODE (UL. UR. CE)

```
TEMPLATE FOR 'A' LINE STATEMENT: RRR±LLL→TTTT

   WHERE: R = RY FIELD        L = LX FIELD
          ± = TC FIELD        T = TR FIELD
```

LX FIELD ROSDR 61–63 LEFT INPUT TO ADDER (XG)

EDGE CHAR	MNEMONIC	BITS	DEC ORDER	FUNCTION	ALD PAGE
		000	LX0	NO OPERATION	DR011
A	1	000	LX0	NO OPERATION – USED WITH TC0 (COMPLEMENT) FOR –1.	DR011
A	L	001	LX1	GATE L REG TO XG.	DR011
A	SGN	010	LX2	GATE ONE TO XG BIT 0. GATE ZEROS TO OTHER BITS OF XG.	DR011
A	E	011	LX3	GATE EMIT FIELD TO XG BITS 27–30. GATE ZEROS TO OTHER BITS.	DR011
A	LRL	100	LX4	GATE L REG BITS 16–31 TO XG BITS 0–15. GATE ZEROS TO BITS 16–31.	DR011
A	LWA	101	LX5	GATE L REG TO XG. OR ONES INTO XG BITS 30–31 AND RIGHT INPUT (Y) BITS 30–31. THIS MICROINSTRUCTION MUST CONTAIN ORDER TC0 AND EITHER RY1 OR RY4. (USED TO COMPARE WORD ADDRESSES)	DR011
A	IOC	110	LX6	GATE ONES COMPLEMENT OF I/O REG TO XG BITS 30–31. GATE ZEROS TO BITS 0–29.	DR011
A	IO	111	LX7	GATE I/O REG TO XG BITS 30–31. GATE ZEROS TO XG BITS 0–29.	DR011

TC FIELD ROSDR 64 TRUE/COMPLEMENT GATING TO LEFT ADDER INPUT (NULL VALUE: TC1)

EDGE CHAR	MNEMONIC	BITS	DEC ORDER	FUNCTION	ALD PAGE
A	-	0	TC0	GATE COMPLEMENT OF XG TO LEFT ADDER INPUT.	DR021
A	+	1	TC1	GATE XG (TRUE) TO LEFT ADDER INPUT.	DR021

RY FIELD ROSDR 65–67 RIGHT INPUT TO ADDER (Y)

EDGE CHAR	MNEMONIC	BITS	DEC ORDER	FUNCTION	ALD PAGE
		000	RY0	NO OPERATION	DR021
A	R	001	RY1	GATE R REGISTER TO Y. (SEE NOTE BELOW)	DR021
A	M	010	RY2	GATE M REGISTER TO Y.	DR021
A	M23	011	RY3	GATE M REG BITS 16–31 TO Y BITS 16–31. GATE ZEROS TO BITS 0–15.	DR021
A	H	100	RY4	GATE H REGISTER TO Y. (SEE NOTE BELOW)	DR021
B	SEMT	101	RY5	OR THE FOUR SDR PARITY BITS INTO THE FOUR EMIT FIELD BITS.	DS021
		110	RY6	UNDEFINED	
		111	RY7	UNDEFINED	

NOTE: IF ORDER LX5 IS USED. ONES ARE ORED INTO Y BITS 30–31. SEE LX5.

CF207 01

CL FIELD ROSDR 68–70 SELECTOR CHANNEL ADDER LATCH TESTS

EDGE CHAR	MNEMONIC	BITS	DEC ORDER	FUNCTION	ALD PAGE
		000	CL0	NO OPERATION	
		001	CL1	UNDEFINED	
		010	CL2	UNDEFINED, NOTE 1	
C	CCW2TEST	011	CL3	SEND PROGRAM CHECK SIGNAL TO SELECTOR CHANNEL IF SDR BITS 5–7 ARE NOT ZERO OR IF SDR BITS 16–31 ARE ZERO. (COUNT AND FLAG TEST)	DR031
C	CATEST	100	CL4	SEND PROGRAM CHECK SIGNAL TO SELECTOR CHANNEL IF ADDER OUTPUT BITS 29–31 ARE NOT ZERO. (ADDRESS TEST)	DR031
C	UATEST	101	CL5	SET BIT 8 OF ADDER-LATCH-TO-CHANNEL BUS TO ONE IF ADDER OUTPUT BITS 0–7 ARE ALL ZERO. OTHERWISE SET BIT 8 TO ZERO. ALL OTHER BITS OF BUS ARE SET TO ZERO. (UNIT ADDRESS TEST)	DR031
C	LSWDTEST	110	CL6	GATE ADDER LATCH BITS 15–31 TO ADDER-LATCH-TO-CHANNEL BUS BITS 15–31. SEND LAST 1 WORD SIGNAL TO SELECTOR CHANNEL IF ADDER LATCH BITS 15–31 HAVE VALUE GREATER THAN ZERO AND LESS THAN 5. SEND LAST 2 WORDS SIGNAL IF THESE BITS HAVE VALUE GREATER THAN 4 AND LESS THAN 9. SEND LAST 3 WORDS SIGNAL IF THESE BITS HAVE VALUE GREATER THAN 8 AND LESS THAN 13. (LAST WORDS TEST)	DR031
		111	CL7	UNDEFINED	

NOTE 1 1410–7010 USAGE OF UNDEFINED ORDERS

C	CCW2EMUTEST	010	CL2	GATE CCW BIT 38 FROM SDR BIT 6 TO CHANNEL TRANSLATOR CONTROL.	

11 MAR 65 225099
3 AUG 65 255449
11 JAN 66 255458

CONTROL FIELD SPECIFICATION I/O MODE	
DATE 11 JAN 66	MACH. 2050
	FRAME 01
	P.N. 5365024
IBM CORP. SDD	PAGE 1

CF207

AB FIELD   ROSDR 72-77   CONDITION BRANCH TEST A

EDGE CHAR	MNEMONIC	BITS	DEC ORDER	FUNCTION	ALD PAGE
R	0	000000	AB0	SET A BIT TO 0.	
R	1	000001	AB1	SET A BIT TO 1.	KK201
R	S0	000010	AB2	SET A BIT TO VALUE OF STAT 0.	KK201
R	S1	000011	AB3	SET A BIT TO VALUE OF STAT 1.	KK201
R	S2	000100	AB4	SET A BIT TO VALUE OF STAT 2.	KK201
R	S3	000101	AB5	SET A BIT TO VALUE OF STAT 3.	KK201
R	S4	000110	AB6	SET A BIT TO VALUE OF STAT 4.	KK201
R	S5	000111	AB7	SET A BIT TO VALUE OF STAT 5.	KK201
R	S6	001000	AB8	SET A BIT TO VALUE OF STAT 6.	KK201
R	S7	001001	AB9	SET A BIT TO VALUE OF STAT 7.	KK201
R	CSTAT	001010	AB10	SET A BIT TO VALUE OF CARRY STAT.	KK211
		001011	AB11	UNDEFINED	
R	1SYLS	001100	AB12	SET A BIT TO VALUE OF ONE SYLLABLE OP IN BUFFER STAT.	KK211
R	LSGNS	001101	AB13	SET A BIT TO VALUE OF L SIGN STAT.	KK211
R	∀SGNS	001110	AB14	SET A BIT TO EXCLUSIVE OR OF L SIGN STAT AND R SIGN STAT.	KK211
		001111	AB15	UNDEFINED	
R	CRMD	010000	AB16	PERFORM AND OF MD REG (MASK) AND DECODED CONDITION REG. IF RESULT IS NON-ZERO, SET A BIT TO ONE. IF RESULT IS ZERO, SET BIT A TO ZERO.	KK251
R	W=0	010001	AB17	IF MOVER LATCHES ARE ZERO, SET A BIT TO ONE. IF NON-ZERO, SET A BIT TO ZERO.	KK211
R	WL=0	010010	AB18	IF MOVER LATCH BITS 0-3 ARE ZERO, SET A BIT TO ONE. IF NON-ZERO, SET A BIT TO ZERO.	KK211
R	WR=0	010011	AB19	IF MOVER LATCH BITS 4-7 ARE ZERO, SET A BIT TO ONE. IF NON-ZERO, SET A BIT TO ZERO.	KK211
R	MD=FP	010100	AB20	IF BOTH BIT 0 AND BIT 3 OF MD REG ARE ZERO, SET A BIT TO ONE. OTHERWISE SET A BIT TO ZERO. (VALID FLOATING POINT REG)	KK221
R	MB=3	010101	AB21	IF MB COUNTER EQUALS 3, SET A BIT TO ONE. OTHERWISE SET A BIT TO ZERO.	KK221
R	MD3=0	010110	AB22	SET A BIT TO COMPLEMENT OF MD REG BIT 3.	KK221
R	G1=0	010111	AB23	IF LENGTH COUNTER G1 IS ZERO, SET A BIT TO ONE. OTHERWISE SET A BIT TO ZERO.	KK221
R	G1<0	011000	AB24	SET A BIT TO VALUE OF LENGTH COUNTER G1 SIGN.	KK221
R	G<4	011001	AB25	IF COUPLED LENGTH COUNTERS HAVE VALUE LESS THAN 4, SET A BIT TO ONE. OTHERWISE SET A BIT TO ZERO. SIGN OF G1 IS INCLUDED IN VALUE.	KK221

CFC 208 – 2050 CONTROL FIELD SPECIFICATION – I/O MODE (AB)

CF208 01

AB FIELD (CONTINUED)

EDGE CHAR	MNEMONIC	BITS	DEC ORDER	FUNCTION	ALD PAGE
R	G1MBZ	011010	AB26	IF EITHER LENGTH COUNTER G1 OR MB COUNTER IS ZERO, SET A BIT TO ONE. OTHERWISE SET A BIT TO ZERO.	KK221
R	IOS0	011011	AB27	SET A BIT TO VALUE OF MULTIPLEXOR CHANNEL STAT 0.	KK191
R	IOS2	011100	AB28	SET A BIT TO VALUE OF MULTIPLEXOR CHANNEL STAT 2.	KK191
R	R(31)	011101	AB29	SET A BIT TO VALUE OF R REG BIT 31.	KK221
R	F(2)	011110	AB30	SET A BIT TO VALUE OF F REG BIT 2.	KK231
R	L(0)	011111	AB31	SET A BIT TO VALUE OF L REG BIT 0.	KK231
R	F=0	100000	AB32	IF F REG IS ZERO, SET A BIT TO ONE. OTHERWISE SET A BIT TO ZERO.	KK231
R	UNORM	100001	AB33	IF STAT 0 IS OFF AND ADDER OUTPUT BITS 8-11 ARE ZERO, SET A BIT TO ONE. OTHERWISE SET A BIT TO ZERO.	KK231
R	TZ*BS	100010	AB34	IF ALL BYTES OF ADDER OUTPUT SPECIFIED BY BYTE STATS ARE ZERO, SET A BIT TO ONE. OTHERWISE SET A BIT TO ZERO.	KK231
R	EDITPAT	100011	AB35	SET A BIT TO VALUE OF EDIT STAT 1. SET B BIT TO VALUE OF EDIT STAT 2. (SEE NOTE 1)	KK231
R	PROB	100100	AB36	SET A BIT TO VALUE OF PSW BIT 15.	KK231
R	TIMUP	100101	AB37	SET A BIT TO VALUE OF TIMER UPDATE SIGNAL. RESET TIMER UPDATE SIGNAL.	KK231
		100110	AB38	UNDEFINED	
R	GZ/MB3	100111	AB39	IF EITHER THE VALUE OF THE COUPLED LENGTH COUNTER G IS ZERO OR THE VALUE OF THE MB COUNTER IS 3 SET A BIT TO ONE. OTHERWISE SET A BIT TO ZERO.	KK231
		101000	AB40	UNDEFINED	
R	LOG	101001	AB41	SET A BIT TO VALUE OF LOG/SCAN STAT.	KK291
R	STC=0	101010	AB42	IF SCAN TEST COUNTER IS ZERO, SET A BIT TO ONE. OTHERWISE SET A BIT TO ZERO.	KK291
R	G2<=LB	101011	AB43	IF LENGTH COUNTER G2 IS LESS THAN OR EQUAL TO LB COUNTER, SET A BIT TO ONE. OTHERWISE SET A BIT TO ZERO. SIGN OF LENGTH COUNTER IS IGNORED.	KK251
		101100	AB44	UNDEFINED	
R	D(7)	101101	AB45	SET A BIT TO VALUE OF SDR BIT 7.	KK221
R	SCPS	101110	AB46	SET A BIT TO VALUE OF SCAN PASS STAT.	KK291
R	SCFS	101111	AB47	SET A BIT TO VALUE OF SCAN FAIL STAT.	KK291
R	STORV	110000	AB48	SET A BIT TO VALUE OF I/O STORAGE VIOLATION TRIGGER.	KK291
R	W(67)→AB	110001	AB49	SET A BIT TO VALUE OF MOVER LATCH BIT 6. SET B BIT TO VALUE OF MOVER LATCH BIT 7. (SEE NOTE 1)	KK291

11 MAR 255099
3 AUG 65 255449

CONTROL FIELD SPECIFICATION I/O MODE	
DATE 16 AUG 65	MACH. 2050
	FRAME 01
	P.N. 5365025
IBM CORP. SDD	PAGE 1

CF208

AB FIELD (CONTINUED)

EDGE CHAR	MNEMONIC	BITS	DEC ORDER	FUNCTION	ALD PAGE
R	Z23≠0	110010	AB50	IF ADDER OUTPUT BITS 16-31 ARE NON-ZERO SET A BIT TO ONE. OTHERWISE SET A BIT TO ZERO.	KK211
R	CCW2OK	110011	AB51	IF SDR BITS 16-31 AND 5-7 ARE ALL ZERO SET A BIT TO ONE. OTHERWISE SET A BIT TO ZERO. (COUNT AND FLAG TEST)	KK291
R	MXBIO	110100	AB52	SET A BIT TO VALUE OF MULTIPLEXOR INPUT BUFFER BIT 0.	KK191
R	IBFULL	110101	AB53	SET A BIT TO VALUE OF INPUT BUFFER FULL STAT.	KK191
R	CANG	110110	AB54	IF ADDER OUTPUT BITS 29-31 ARE NON-ZERO OR INVALID ADDRESS TRIGGER IS ON, SET A BIT TO ONE. OTHERWISE SET A BIT TO ZERO.	KK291
R	CHLOG	110111	AB55	SET A BIT TO VALUE OF CHANNEL ERROR LOG REQUEST LINE.	KK191
R	I-FETCH	111000	AB56	FOUR-WAY BRANCH: (SEE NOTE 1) 0→A,0→B IAR BIT 30 IS ONE AND REFETCH STAT OFF 0→A,1→B IAR BIT 30 IS ONE AND REFETCH STAT ON 1→A,0→B IAR BIT 30 IS ZERO 1→A,1→B EXCEPTION TRIGGER ON	KK151
R	IA(30)	111001	AB57	SET A BIT TO VALUE OF IAR BIT 30.	KK211
R	EXT.CHIRPT	111010	AB58	IF EITHER A TIMER UPDATE REQUEST HAS OCCURRED, OR AN EXTERNAL INTERRUPT IS REQUESTED WITH MASK BIT ON, SET A BIT TO ONE. OTHERWISE SET A BIT TO ZERO. IF A CHANNEL INTERRUPT IS REQUESTED SET B BIT TO ONE. OTHERWISE SET B BIT TO ZERO. (SEE NOTE 1)	KK151
R	DCHOLD	111011	AB59	SET A BIT TO VALUE OF DIRECT CONTROL HOLD LINE.	KK191
R	PSS	111100	AB60	SET A BIT TO VALUE OF PROGRESSIVE SCAN STAT AND RESET PSS.	KK191
R	IOS4	111101	AB61	SET A BIT TO VALUE OF MULTIPLEXOR CHANNEL STAT 4.	KK191
		111110	AB62	UNDEFINED. NOTE 4	
R	RX.SO	111111	AB63	IF STAT 0 IS ON AND M REG BITS 0-1 HAVE VALUE 01 SET A BIT TO ONE. OTHERWISE SET A BIT TO ZERO. (RX FORMAT AND INDEXING REQUIRED)	KK251

NOTE 1 IN BRANCHES WHERE BOTH A AND B BITS ARE SET, THE B BIT VALUE IS THE OR OF VALUES SPECIFIED BY AB AND BB FIELDS.

NOTE 2 A BIT AND B BIT VALUES MAY BE SUBSEQUENTLY ALTERED BY ZN FIELD ORDERS.

NOTE 3 ALL AB FIELD BRANCHES TEST CONDITIONS EXISTING AT THE END OF THE PREVIOUS CYCLE.

NOTE 4 1410-7010 EMULATOR USAGE OF UNDEFINED ORDERS

EDGE CHAR	MNEMONIC	BITS	DEC ORDER	FUNCTION	ALD PAGE
R	AR4	111110	AB62	SET A BIT TO VALUE OF R REGISTER BIT 4	XE201

BB FIELD ROSDR 78-82 B CONDITION TEST

EDGE CHAR	MENMONIC	BITS	DEC ORDER	FUNCTION	ALD PAGE
R	0	00000	BB0	SET B BIT TO 0.	
R	1	00001	BB1	SET B BIT TO 1.	KK141
R	S0	00010	BB2	SET B BIT TO VALUE OF STAT 0.	KK141
R	S1	00011	BB3	SET B BIT TO VALUE OF STAT 1.	KK141
R	S2	00100	BB4	SET B BIT TO VALUE OF STAT 2.	KK141
R	S3	00101	BB5	SET B BIT TO VALUE OF STAT 3.	KK141
R	S4	00110	BB6	SET B BIT TO VALUE OF STAT 4.	KK141
R	S5	00111	BB7	SET B BIT TO VALUE OF STAT 5.	KK141
R	S6	01000	BB8	SET B BIT TO VALUE OF STAT 6.	KK141
R	S7	01001	BB9	SET B BIT TO VALUE OF STAT 7.	KK141
R	RSGNS	01010	BB10	SET B BIT TO VALUE OF R SIGN STAT.	KK151
R	HSCH	01011	BB11	IF HIGH SPEED CHANNEL IS OPERATING OR 256 SUBCHANNEL OPTION IS INSTALLED, SET B BIT TO ONE. OTHERWISE SET B BIT TO ZERO.	KK141
R	EXC	01100	BB12	SET B BIT TO VALUE OF EXCEPTION TRIGGER.	KK151
R	WR=0	01101	BB13	IF MOVER LATCH BITS 4-7 ARE ZERO, SET B BIT TO ONE. OTHERWISE SET B BIT TO ZERO.	KK151
		01110	BB14	UNDEFINED	
R	T13=0	01111	BB15	IF ADDER OUTPUT BUS BITS 8-13 ARE ZERO, SETS B BIT TO ONE. OTHERWISE SET B BIT TO ZERO.	KK151
R	T(0)	10000	BB16	SET B BIT TO VALUE OF ADDER OUTPUT BUS BIT 0.	KK151
R	T=0	10001	BB17	IF ADDER OUTPUT BUS IS ZERO, SET B BIT TO ONE. OTHERWISE SET B BIT TO ZERO.	KK151
R	TZ*BS	10010	BB18	IF THE BYTES OF THE ADDER OUTPUT BUS SPECIFIED BY THE BYTE STATS ARE ALL ZERO, SET B BIT TO ONE. OTHERWISE SET B BIT TO ZERO.	KK151
R	W=1	10011	BB19	IF THE MOVER LATCHES CONTAIN 0000 0001 SET B BIT TO ONE. OTHERWISE SET B BIT TO ZERO.	KK151
R	LB=0	10100	BB20	IF LB COUNTER IS ZERO SET B BIT TO ONE. OTHERWISE SET B BIT TO ZERO.	KK161
R	LB=3	10101	BB21	IF LB COUNTER EQUALS 3 SET B BIT TO ONE. OTHERWISE SET B BIT TO ZERO.	KK161
R	MD=0	10110	BB22	IF MD REG IS ZERO SET B BIT TO ONE. OTHERWISE SET B BIT TO ZERO.	KK161
R	G2=0	10111	BB23	IF LENGTH COUNTER G2 IS ZERO SET B BIT TO ONE. OTHERWISE SET B BIT TO ZERO.	KK161
R	G2<0	11000	BB24	SET B BIT TO VALUE OF LENGTH COUNTER G2 SIGN.	KK161
R	G2LBZ	11001	BB25	IF EITHER LENGTH COUNTER G2 OR LB COUNTER IS ZERO, SET B BIT TO ONE. OTHERWISE SET B BIT TO ZERO.	KK161

CFC 210 - 2050 CONTROL FIELD SPECIFICATION - I/O MODE (BB)

CF210 01

BB FIELD (CONTINUED)

EDGE CHAR	MNEMONIC	BITS	DEC ORDER	FUNCTION	ALD PAGE
R	IOS1	11010	BB26	SET B BIT TO VALUE OF MULTIPLEXOR CHANNEL STAT 1.	KK151
R	MD/JI	11011	BB27	IF EITHER BIT 1 OR BIT 3 OF EITHER MD REG OR J REG IS ONE SET B BIT TO ONE (ILLEGAL FP REG). OTHERWISE SET B BIT TO ZERO.	KK161
R	IVA	11100	BB28	SET B BIT TO VALUE OF INVALID ADDRESS STAT.	KK161
R	IOS3	11101	BB29	SET B BIT TO VALUE OF MULTIPLEXOR CHANNEL STAT 3.	KK161
R	(CAR)	11110	BB30	SET B BIT TO VALUE OF CARRY LATCH AS SET THIS CYCLE.	KK171
R	(Z00)	11111	BB31	SET B BIT TO VALUE OF ADDER SUM BIT 0 (BEFORE SHIFT) THIS CYCLE.	KK171

NOTE WITH THE EXCEPTION OF ORDERS BB30 AND BB31, ALL TESTS ARE MADE ON THE STATUS OF THE CPU AT THE END OF THE PREVIOUS CYCLE.

11 MAR 65 255099
3 AUG 65 255449

CONTROL SIGNAL SPECIFICATION I/O MODE	
DATE 17 FEB 66	MACH. 2050
	FRAME 01
	P.N. 5365027
IBM CORP. SDD	PAGE 1

CF210

SS FIELD ROSDR 84–89 STAT SETTING AND MISCELLANEOUS CONTROL

EDGE CHAR	MNEMONIC	BITS	DEC ORDER	FUNCTION	ALD PAGE
		000000	SS0	NO OPERATION	
		000001	SS1	UNDEFINED	
		000010	SS2	UNDEFINED, NOTE 1 CF212	
C	D→CR*BS	000011	SS3	SET CONDITION REG FOR TEST AND SET INSTRUCTION. SET COND REG BIT 0 TO 0. SET BIT 1 OF COND REG TO ONE IF THE FIRST BIT IS ON IN ANY SDR BYTE SELECTED BY THE CPU BYTE STATS. WHEN USED WITH LARGE CAPACITY STORAGE, THIS ORDER INTERLOCKS THE SAME AS AN SDR TO LATCH TRANSFER.	RP251
C	E→SCANCTL	000100	SS4	CONTROL SCAN STATS PER EMIT FIELD AS FOLLOWS:	KS601
				EMIT — FUNCTION	
				0000 IF SCAN TEST STAT IS OFF AND SDR IS ALL ONES SET SCAN TEST STAT ON. IF SCAN TEST STAT IS ON AND SDR IS ZERO LEAVE SCAN TEST STAT ON. OTHERWISE SET SCAN TEST STAT OFF.	
				0001 IF SCAN TEST STAT IS OFF TURN ON SCAN PASS STAT. IF SCAN TEST STAT IS ON TURN ON SCAN FAIL STAT. INCREMENT SCAN TEST COUNTER BY 1.	
				0010 TURN OFF SCAN TEST STAT.	
				0011 TURN OFF IGNORE I/O ERROR TRIGGER.	
				0100 TURN ON IGNORE I/O ERROR TRIGGER.	
				0101 TURN OFF SCAN PASS STAT AND SCAN FAIL STAT.	
				0110 TURN ON INVERT SAR BIT 13 TRIGGER.	
				0111 TURN OFF INVERT SAR BIT 13 TRIGGER.	
				1000 SET SUPERVISORY SCAN TRIGGER TO VALUE OF SDR BIT 6. SET PROGRESSIVE SCAN STAT TO VALUE OF SDR BIT 5. SET CLOCK ADVANCE COUNTER TO VALUE OF SDR BITS 0–2.	
				1001–1011 UNDEFINED	
				1100 TURN OFF LOG/SCAN STAT.	
				1101–1111 UNDEFINED	
C	L.RSGNS	000101	SS5	IF LEFT MOVER INPUT U BITS 4–7 HAS VALUE LESS THAN 1010, FORCE INVALID DATA ROS TRAP. IF VALUE EQUALS 1011 OR 1101 (MINUS SIGN) TURN ON L SIGN STAT AND INVERT R SIGN STAT. OTHERWISE TURN OFF L SIGN STAT.	KS601
C	IVD/RSGNS	000110	SS6	IF LEFT MOVER INPUT U BITS 4–7 HAS VALUE LESS THAN 1010, FORCE INVALID DATA ROS TRAP. IF VALUE EQUALS 1011 OR 1101 (MINUS SIGN) INVERT R SIGN STAT.	
C	EDITSGN	000111	SS7	IF MOVER LATCH BITS 4–7 HAS VALUE GREATER THAN 1001 (VALID SIGN) TURN ON R SIGN STAT. OTHERWISE TURN OFF R SIGN STAT. IF MOVER LATCH BITS 4–7 EQUAL TO 1010, 1100, 1110 OR 1111 (PLUS SIGN) TURN OFF L SIGN STAT.	KS601
C	E→S03	001000	SS8	GATE EMIT FIELD TO STATS 0–3	KS611
C	S03ΩE.1→LSGN	001001	SS9	TURN ON STATS 0–3 PER EMIT FIELD. TURN ON L SIGN STAT.	KS611
C	S03ΩE	001010	SS10	TURN ON STATS 0–3 PER EMIT FIELD.	KS611
C	S03ΩE.0→BS	001011	SS11	TURN ON STATS 0–3 PER EMIT FIELD. TURN OFF CPU BYTE STATS.	KS611
C	X0.B0.1SYL	001100	SS12	IF ADDER LATCH BITS 12–15 ARE ZERO, SET STAT 0 TO ONE. (X=0) OTHERWISE SET STAT 0 TO ZERO. IF ADDER LATCH BITS 16–19 ARE ZERO, SET STAT 1 TO ONE. (B=0) OTHERWISE SET STAT 1 TO ZERO. IF ADDER LATCH BITS 16–17 ARE ZERO, SET ONE-SYLLABLE-OP-IN-BUFFER STAT TO ONE. OTHERWISE SET THIS STAT TO ZERO.	KS611
C	FPZERO	001101	SS13	IF ADDER LATCH BITS 8–31 ARE ZERO, F REG IS ZERO, AND STAT 3 IS ON, TURN ON STAT 0. OTHERWISE TURN OFF STAT 0.	KS611

CF211 01

SS FIELD (CONTINUED)

EDGE CHAR	MNEMONIC	BITS	DEC ORDER	FUNCTION	ALD PAGE
C	FPZERO,E→FN	001110	SS14	IF ADDER LATCH BITS 8–31 ARE ZERO, F REG IS ZERO, AND STAT 3 IS ON, TURN ON STAT 0. OTHERWISE TURN OFF STAT 0. GATE EMIT FIELD BITS 2–3 TO LOCAL STORAGE FUNCTION REG. (EFFECTIVE FOR LOCAL STORAGE ADDRESSING THIS CYCLE.)	KS611
C	B0,1SYL	001111	SS15	IF ADDER LATCH BITS 0–3 ARE ZERO, TURN ON STAT 1. (B=0) OTHERWISE TURN OFF STAT 1. IF ADDER LATCH BITS 16–17 ARE ZERO, TURN ON ONE-SYLLABLE-OP-IN-BUFFER STAT. OTHERWISE TURN THIS STAT OFF.	KS611
C	S03,¬E	010000	SS16	TURN OFF STATS 0–3 PER EMIT FIELD.	KS621
C	(T=0)→S3	010001	SS17	IF ADDER LATCH IS ZERO TURN ON STAT 3. OTHERWISE TURN OFF STAT 3.	KS621
C	E→BS,T30→S3	010010	SS18	GATE EMIT FIELD TO CPU BYTE STATS. SET STAT 3 TO VALUE OF ADDER LATCH BIT 30.	KS621
C	E→BS	010011	SS19	GATE EMIT FIELD TO CPU BYTE STATS.	KS621
C	1→BS*MB	010100	SS20	TURN ON BYTE STAT INDICATED BY VALUE OF MB COUNTER.	KS621
C	DIRCTL*E	010101	SS21	DIRECT CONTROL PER EMIT AS FOLLOWS:	KS631

EMIT BIT	FUNCTION IF BIT IS ON
0	GATE H REG BITS 24–31 TO DIRECT CONTROL DATA OUT REG.
1	INITIATE PULSE ON WRITE OUT LINE.
2	GATE H REG BITS 8–15 TO DIRECT CONTROL TIMING SIGNAL BUS OUT LINES IN PULSE FORM.
3	INITIATE PULSE ON READ OUT LINE. IF AT THE END OF THIS PULSE THE HOLD IN LINE IS NOT ON, LATCH THE DIRECT CONTROL BUS IN.

EDGE CHAR	MNEMONIC	BITS	DEC ORDER	FUNCTION	ALD PAGE
		010110	SS22	UNDEFINED	
C	MANUAL→STOP	010111	SS23	SET STOP TRIGGER TO VALUE OF MANUAL TRIGGER.	KS721
C	E→S47	011000	SS24	GATE EMIT FIELD TO STATS 4–7.	KS621
C	S47ΩE	011001	SS25	TURN ON STATS 4–7 PER EMIT FIELD.	KS621
C	S47,¬E	011010	SS26	TURN OFF STATS 4–7 PER EMIT FIELD.	KS621
C	S47,ED*FP	011011	SS27	SET STATS 4–7 AND EXPONENT DIFFERENCE REG FOR FLOATING POINT AS FOLLOWS:	KS621

STAT 4 TURNED ON IF:
STAT 0 OR STAT 1 IS ON AND RIGHT ADDER INPUT BIT 0 IS ONE AND THERE IS A CARRY OUT OF POSITION 1.
OR STAT 0 OR STAT 1 IS ON, THERE IS A CARRY OUT OF POSITION 1, AND EITHER LEFT ADDER INPUT BIT 0 IS ONE OR STAT 1 IS ON BUT NOT BOTH (ADD TYPE, RESULT MINUS)
OR BOTH STAT 0 AND STAT 1 ARE OFF AND LEFT ADDER INPUT BIT 0 IS NOT EQUAL TO RIGHT ADDER INPUT BIT 0 (MULTIPLY OR DIVIDE, SIGNS UNLIKE)

STAT 5 TURNED ON IF LEFT ADDER INPUT BIT 0, RIGHT ADDER INPUT BIT 0 AND STAT 1 CONTAIN AN EVEN NUMBER OF ONES. (TRUE ADD REQUIRED)

STAT 6 TURNED ON IF VALUE OF EXPONENT DIFFERENCE REG IS LESS THAN 16 (DEC) IN ABSOLUTE VALUE.

STAT 7 TURNED ON IF VALUE OF EXPONENT DIFFERENCE REG IS ZERO.

ABSENCE OF TURN ON CONDITION CAUSES STAT TO BE TURNED OFF.

EXPONENT DIFFERENCE REG SET AS FOLLOWS:
BIT 0 SET TO ONE IF CARRY FROM ADDER POS 1 AND SUM BITS 1–4 NON-ZERO,OR IF NO CARRY FROM POS 1 AND SUM BITS 1–4 EQUAL 1111.
BITS 1–3 SET EQUAL TO ADDER SUM BITS 5–7.

11 MAR 65 255099
3 AUG 65 255449
11 JAN 66 255458

CONTROL FIELD SPECIFICATION I/O MODE	
DATE 11 JAN 66	MACH. 2050
	FRAME 01
	P.N. 5365028
IBM CORP. SDD	PAGE 1

CF211

SS FIELD (CONTINUED)

EDGE CHAR	MNEMONIC	BITS	DEC ORDER	FUNCTION	ALD PAGE
C	OPPANEL→S47	011100	SS28	SET STATS 4-7 PER MAINTENANCE CONSOLE SWITCHES AS FOLLOWS: STATS — CONDITION 0000 NONE OF THOSE DESCRIBED BELOW. 0001 INSTRUCTION STEP. WAIT BIT NOT ON. START. 0010 SET INSTRUCTION COUNTER. MANUAL TRIGGER ON. 0011 REPEAT INSTRUCTION. MANUAL TRIGGER ON. 0100 IAR COMPARE (SYNC OR STOP). MANUAL TRIGGER AND WAIT BIT BOTH OFF. 0110 ENTER CHANNEL. MANUAL TRIGGER ON. 1XXY DISPLAY OR STORE WITH MANUAL TRIGGER ON AS FOLLOWS: XX - 00 MAIN STORAGE 01 PROTECTION TAG STORAGE 10 LOCAL STORAGE 11 MULTIPLEXOR BUMP STORAGE Y - 0 DISPLAY 1 STORE	KS621
C	CAR.(T≠0)→CR	011101	SS29	SET LEFT BIT OF CONDITION REG (PSW BIT 34) TO VALUE OF CARRY OUT OF ADDER POSITION 0. SET RIGHT BIT (PSW 35) TO ONE IF ADDER LATCH IS NON-ZERO, TO ZERO IF LATCH IS ZERO.	KS621
C	KEY→F	011110	SS30	GATE FROM TAG STORAGE DATA LINES TO F REG. IF SAR ADDRESSES LCS, INITIATE SIGNAL TO STORAGE INDICATING READ STORAGE KEY OPERATION.	KS641
C	F→KEY	011111	SS31	GATE F REG TO TAG STORAGE DATA LINES. INITIATE WRITE STORAGE KEY OPERATION. (NOTE: THIS ORDER IS GIVEN ON W2 CYCLE OF STORAGE AND CAUSES HOLDOFF UNTIL NEXT W2 CYCLE.)	KS641
C	1→LSGNS	100000	S 2	TURN ON L SIGN STAT.	KS641
C	0→LSGNS	100001	SS33	TURN OFF L SIGN STAT.	KS641
C	1→RSGNS	100010	SS34	TURN ON R SIGN STAT.	KS641
C	0→RSGNS	100011	SS35	TURN OFF R SIGN STAT.	KS641
C	L(0)→LSGNS	100100	SS36	SET L SIGN STAT TO VALUE OF L REG BIT 0.	KS641
C	R(0)→RSGNS	100101	SS37	SET R SIGN STAT TO VALUE OF R REG BIT 0.	KS641
C	E(13)→WFN	100110	SS38	GATE EMIT FIELD BITS 1-3 TO MOVER FUNCTION REG.	KS641
C	E(23)→LSFN	100111	SS39	GATE EMIT FIELD BITS 2-3 TO LOCAL STORAGE FUNCTION REG. (EFFECTIVE FOR LOCAL STORAGE ADDRESSING THIS CYCLE.)	KS641
C	E(23)→CR	101000	SS40	GATE EMIT FIELD BITS 2-3 TO CONDITION REG (PSW BITS 34-35).	KS651
C	SETCRALG	101001	SS41	SET THE CONDITION REG ACCORDING TO ALGEBRAIC CONDITIONS AS FOLLOWS: COND CODE — CONDITION 00 ADDER LATCH ZERO 01 ADDER LATCH BIT 0 IS ONE. (MINUS) 10 ADDER LATCH BIT 0 IS ZERO. BITS 1-31 NON-ZERO. (PLUS)	KS651

NOTE 1 — 1410-7010 EMULATOR USAGE OF UNDEFINED ORDERS

EDGE CHAR	MNEMONIC	BITS	DEC ORDER	FUNCTION	ALD PAGE
C	SPLIT R/W	000001	SS01	SPLIT MEMORY CYCLE CONTROL FOR 1410 WORD MARK PRESERVATION. STORES BITS 1-7 OF EACH BYTE INTO MEMORY WHILE REGENERATING BIT 0	XE401

CF212 01

SS FIELD (CONTINUED)

EDGE CHAR	MNEMONIC	BITS	DEC ORDER	FUNCTION	ALD PAGE
C	SETCRLOG	101010	SS42	SET THE CONDITION REG ACCORDING TO LOGICAL CONDITIONS AS FOLLOWS: COND CODE — CONDITION 00 ADDER LATCH BYTES SELECTED BY BYTE STATS ARE ZERO. 01 ADDER LATCH BYTES SELECTED BY BYTE STATS ARE NOT ALL ZERO AND THERE IS NO CARRY OUT OF ADDER POSITION 0. 10 ADDER LATCH BYTES SELECTED BY BYTE STATS ARE NOT ALL ZERO AND THERE IS A CARRY OUT OF ADDER POSITION 0.	KS651
C	¬S4,S4→CR	101011	SS43	SET LEFT BIT OF CONDITION REG (PSW BIT 34) TO THE COMPLEMENT OF STAT 4. SET THE RIGHT BIT (PSW BIT 35) EQUAL TO STAT 4.	KS651
C	S4,¬S4→CR	101100	SS44	SET LEFT BIT OF CONDITION REG (PSW BIT 34) EQUAL TO STAT 4. SET THE RIGHT BIT (PSW 35) TO THE COMPLEMENT OF STAT 4.	KS651
C	1→REFETCH	101101	SS45	TURN ON THE REFETCH TRIGGER.	KS651
C	SYNC→OPPANEL	101110	SS46	SEND ADDRESS COMPARE SYNC/STOP PULSE TO CONSOLE.	KS651
C	SCAN*E,10	101111	SS47	TURN ON SCAN CONTROL TRIGGER 1. IF THIS MICROINSTRUCTION CONTAINS TR22 ORDER, TURN ON SCAN CONTROL TRIGGER 0. OTHERWISE TURN IT OFF. GATE EMIT FIELD TO SCAN CONTROL TRIGGERS 2-5.	KS521
C	1→SUPOUT	110000	SS48	TURN ON MULTIPLEXOR CHANNEL SUPPRESS OUT LINE.	KS661
C	MPXSELRESET	110001	SS49	SELECTIVE RESET OF MULTIPLEXOR CHANNEL.	KS661
C	E(0)→IBFULL	110010	SS50	SET INTERRUPT BUFFER FULL STAT TO VALUE OF EMIT FIELD BIT 0.	KS661
		110011	SS51	UNDEFINED	
C	E→CH	110100	SS52	GATE THE EMIT FIELD TO THE COMMON CHANNEL.	KS661
		110101	SS53	UNDEFINED	
C	1→TIMERIRPT	110110	SS54	TURN ON THE TIMER BIT IN THE EXTERNAL INTERRUPT REG.	KS661
C	T→PSW,IPL→T	110111	SS55	GATE IPL UNIT ADDRESS BITS 0-7 TO L REG BITS 0-7. GATE CHANNEL ADDRESS TO L REG BITS 21-23. ORDER TR7 IS USED WITH THIS ORDER TO PROVIDE REGISTER PULSE FOR L. BUT T→L GATING IS INHIBITED.	KS661
C	T→PSW	111000	SS56	GATE ADDER LATCH BITS 12-15 TO PSW BITS 12-15 (MODE BITS).	KS671
C	SCAN*E,00	111001	SS57	TURN OFF SCAN CONTROL TRIGGER 1. IF THIS MICROINSTRUCTION CONTAINS TR22 ORDER, TURN ON SCAN CONTROL TRIGGER 0. OTHERWISE TURN IT OFF. GATE EMIT FIELD TO SCAN CONTROL TRIGGERS 2-5.	KS521
C	1→IOMODE	111010	SS58	TURN ON I/O MODE STAT.	KS671
C	0→IOMODE	111011	SS59	TURN OFF I/O MODE STAT.	KS671
C	1→SELOUT	111100	SS60	TURN ON MULTIPLEXOR CHANNEL SELECT OUT LINE.	KS671
C	1→ADROUT	111101	SS61	TURN ON MULTIPLEXOR CHANNEL ADDRESS OUT LINE.	KS671
C	1→COMOUT	111110	SS62	TURN ON MULTIPLEXOR CHANNEL COMMAND OUT LINE.	KS671
C	1→SERVOUT	111111	SS63	TURN ON MULTIPLEXOR CHANNEL SERVICE OUT LINE.	KS671

11 MAR 65 255099
3 AUG 65 255449
11 JAN 66 255458

CONTROL FIELD SPECIFICATION
I/O MODE
DATE 11 JAN 66 MACH. 2050
FRAME 01
P.N. 5365029
IBM CORP. SDD PAGE 1

CF212

MICROPROGRAMMING THE RCA SPECTRA 70/ MODEL 45

9

9.1 INTRODUCTION

The RCA Spectra 70/Model 45 is a general purpose, medium-scale processor capable of accepting a wide variety of industrially-accepted codes and programming language. This multilingual system can be used to solve a wide variety of business, scientific, communication, and real-time application problems. Architecturally, it is compatible with the other processors in the Spectra 70 product line, which includes the 70/15, 70/25, 70/35, 70/45, 70/55, and 70/60. However, its internal organization and data flow, its timing, its microprogramming and microprogramming aids are considerably different from all other processors.

The RCA Spectra 70 is architecturally similar to the IBM System/360. Both systems apply a common code structure based on the Extended Binary Code Decimal Interchange Code (EBCDIC) of eight bits plus parity. The standard unit of storage in both systems is called a byte. Both families offer the facilities for using the American Standard Code for Information Interchange (ASCII). Fixed length data of 8, 16, 32, or 64 bits may be processed. Variable-length data of up to 256 characters (bytes) can be processed by both systems.

Both systems have similar and complex instruction formats 2, 4, or 6 bytes wide, signaling register to register, register to storage, and storage to storage information interchange respectively. Each instruction begins with the operation code byte, followed first by one or two general purpose registers, then by memory addresses as required.

Both systems have a base register which can be indexed by any of the general purpose registers.

The system interrupts, their classification into five classes (program interrupts, external interrupts, supervisor call interrupts, and input/output interrupts), and the priorities of handling them are also similar.

The microprogrammed processors in both product lines apply similar flowchart microprogramming language and a rigidly-formatted instruction box. Both systems apply an encoded control ROS word organization with a built-in sequencing and decision-logic specification.

Other architectural features such as the sixteen 32-bit general purpose registers (GPR's), the four 64-bit floating-point registers (FPR's), the memory-protection features, and other similarities are clearly documented and easily discernible by reading the principles-of-operation manuals for both systems.*

These architectural similarities mask considerable differences in their respective design implementations, circuit technology, and ROS implementation. They benefit the reader since architectural similarities permit the reader to focus immediately on their respective hardware and microprogramming structures.

Some of the more noticeable dissimilarities in the design and implementations of the two systems are:

1. The Spectra 70/45 is a single bus system. Its data flow, its arithmetic and logic box (ALB), and its three memories all communicate with each other over one 18-bit-wide data bus (DB). This must be contrasted with the System/360 Models 40 or 50, with their multidata-path structures which communicate over two or more data buses of different widths.

Another difference in the data flow of the two systems is the abundance of general purpose hardware registers with extensive gating in the System/360 versus the minimum hardware registers in the Spectra 70/45. This is not to suggest that the performance of the 70/45 suffers from the apparent lack of hardware registers and bussing, or that the complexity of data flow of the System/360 by itself makes it in any way superior to the Spectra 70, for the two systems are judged by their performance throughput, turn-around, response time, and other parameters external to the respective systems data-flow structure.

2. The System/360 architecture provides only *one* set of sixteen 32-bit-wide GPR's plus *four* double-word (64-bit) FPR's, all in the high-speed local store. This single set of GPR's services all the demands of the system. When an

*IBM Systems/360 Principles of Operations Manual, Form #A22–6821–4, File #S360–01, and RCA Spectra 70, Systems Information Manual, Doc. 70–00–601, Dec. 1964.

interrupt occurs, the GPR's are stored away and loaded with the new data in the process of shifting from one mode of operation to the next. This implies an expenditure of time necessary to STORE MULTIPLE and LOAD MULTIPLE. This time was saved in the Spectra 70/45 by providing *four* complete sets of GPR's, each set assigned to service one of the four system states.* Addressing of the scratch pad memory (SPM) is in many cases program-state-dependent, allowing the same read-only storage (ROS) algorithm to operate with different physical registers, depending on the operating state. The Spectra 70/45 also stores these four sets of sixteen 32-bit GPR's plus one single set of four double-word FPR's, all in a high-speed destructive read-storage unit—the SPM or fast memory (FM). The SPM is also used as a substitute for the number of hardware registers present in System/360. This use of SPM for intermediate storage and for the supply of data to the ALB is justified on the basis of timing, as we will see later. However, if the two operands are in SPM, two machine cycles are needed to perform any operation through the ALB which involves both operands.

3. The third difference between the two system implementations is not as obvious as the first two. This deals with the microprogramming of the two systems. The System/360 is micro-order-oriented. The micro-orders are primarily organized around the control gate that permits information to flow from one storage and functional station to the next. As such, the System/360 is not highly specified. That is to say, almost every new conceivable microinstruction can be structured from the available micro-orders, provided that the timing and concurrency of information-gating is permitted. This flexibility by itself does not imply or ensure efficient microprogrammed control. This is to be contrasted with the Spectra 70/45 microinstruction structure [RCA's term for microinstruction is elementary operation (EO)]. The RCA ROS words are highly structured and EO-oriented. Its ROS [cr read-only memory (ROM)] word is divided into fields. Each field when decoded specifies an action or specifies a toggle setting but not the state of unique gates. These fields may have a number of different interpretations, depending on the setting of another field. That is to say, the decoding of the Spectra 70/45 fields are, in general, interdependent. Therefore one can conclude that the Spectra 70/45 is highly specified, and its microprogramming, because of its sophisticated ROS word structure, is difficult to master.

The highly specified structure of the Spectra 70 system does not in any way imply an inferior design. It merely points out that the system designers chose to optimize the data processing function by a highly organized, interdependent EO structure. The intention here in contrasting the Spectra 70 with other equipment is once again to underscore the flexibility of microprogramming. The designers of the IBM System/360 chose a general, simple micro-

*The four operating states provided in Model 45 are: the program state, the executive state, the interrupt state, and the machine state. The System/360 has only two states: the program state and the supervisor state.

order structure, and the Spectra 70 designers chose a more sophisticated interdependent EO ROS structure. The appropriateness of each design must be viewed in the context of the total system design and the constraints and trade-offs that were imposed on the respective system designers.

4. There are a number of other dissimilarities between the two systems that will become apparent after one gains more insight into the microprogramming of the two systems. These differences, although they will not be discussed here, will include: the concurrency of operations; the high-speed local storage mapping and addressing; the microinstruction organization and ROM addressing and branching; the extent of hardware facilities, toggles, and triggers to share the control with the microroutines in the spectra 70/45 and others.

9.2 70/45 HARDWARE (FIG. 9.1)

The 70/45 has three memories: the main memory (MM), the SPM or fast memory (FM) (called local storage in the IBM System/360), and the ROM. The first two are used for data and the last for microinstruction storage. The 70/45 has a number of hardware registers, plus an eight-bit-wide arithmetic and logic box (ALB). All these functional units are interconnected by a single data bus two bytes wide.

9.2.1 The Main Memory

The main memory (MM) is expandable to 262,144 bytes. It has a 1.44 microsecond cycle time and a 16-bit word storage structure representing two bytes; a parity bit is stored with each byte in the MM. The DB is 16 bits wide and normally transmits either half of the data register (DR) directly to the utility register (UR), intermediate register (IR), main memory address register (MMAR), or the main memory register (MMR), thus providing a complete interchange flexibility among the ALB and the two data memories. In addition, a second data bus (DBA) is provided for transmittal of the two most significant bits of 18-bit memory address to MMAR. This second bus is driven primarily by the DR and B register (BR). This model 70/45 has a memory-protection feature to prevent destruction of information by programming or input devices. This feature consists of a set of registers that are constantly scanned during instruction-addressing to ascertain that the address falls within the allowable limits. If a violation is detected, an interrupt condition is set, and the appropriate action is taken.

The following describes a typical operation requiring data to be read from MM for transfer to the UR using an address accessed from the SPM. The address accessed from the SPM is transferred to the MMAR via the DR and is incremented during its regeneration into the SPM during the first microinstruc-

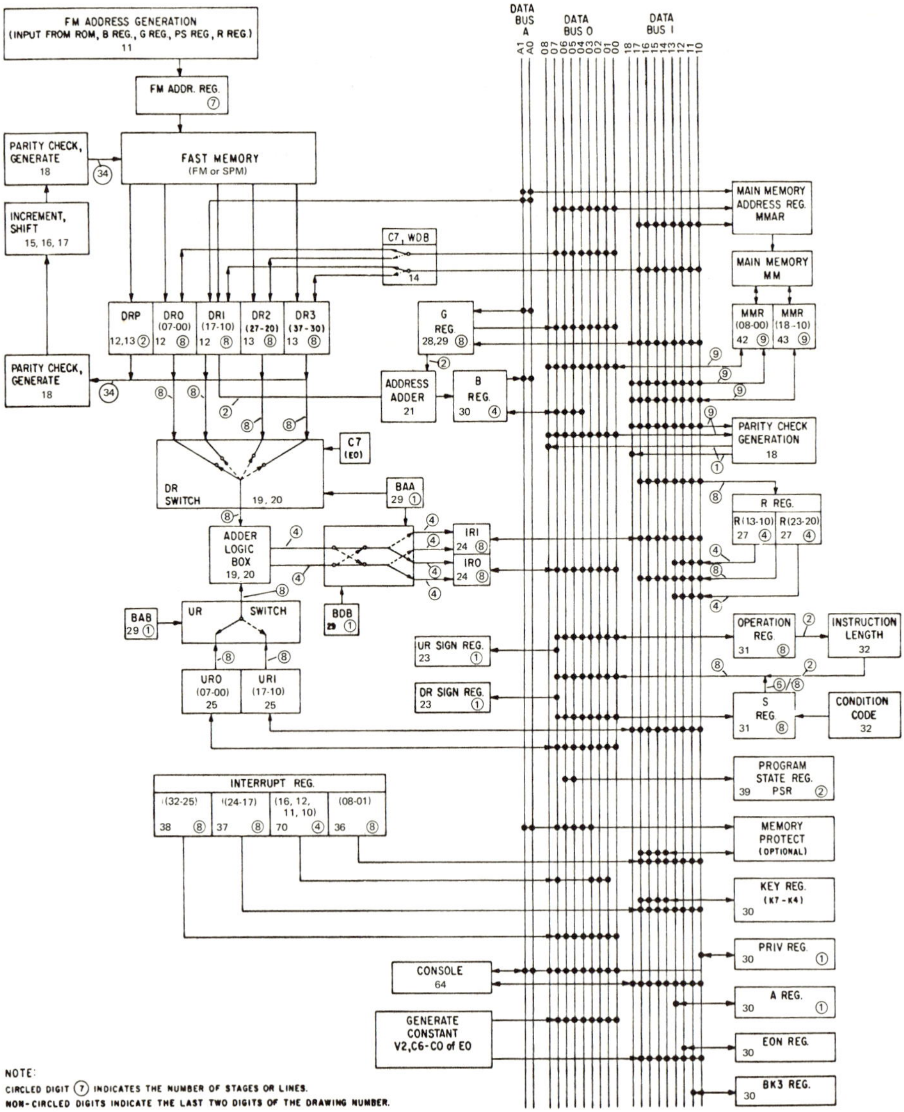

Figure 9.1 Spectra 70/45 block diagram.

tion. The read cycle takes three machine cycles. As far as a main memory read (MMR) is concerned, no useful work is accomplished during the second microinstruction. This second EO cycle can be used to perform other functions not related to the MM data path. If no other function is needed during this second EO cycle, a memory address register W (MARW) micro-order can be specified which causes the machine to wait between the first and third EO time, thus saving one word of ROS. At the beginning of the third microinstruction, the contents of the addressed location in MM are transmitted to the MMR. The

MMR is specified as the source and the UR as the destination during the third microinstruction. This example will be expanded and the regenerate cycle will be discussed in a later section on the 70/45 timing.

9.2.2 The Scratch Pad Memory (SPM)

The 70/45 utilizes a high-speed SPM as an integral part of the data flow. It is used as a substitute for one hundred and twenty-eight 32-bit-wide flip-flop registers. It provides for the storage of four sets of sixteen 4-byte-wide GPR's plus one set of four 8-byte-wide FPR's, input and output control registers, and other utility functions such as storing the source- and destination-memory-address registers. Each of the four system states is also provided with a unique set of registers in SPM as well as their instruction counter, interrupt mask register, and interrupt status register. An SPM map is shown in Table 9.1.

The memory is operated in a linear select fashion, with a fullword access at 120 nanoseconds and an overall cycle time of 480 nanoseconds. This timing permits three SPM cycles to be executed for each MM cycle. This will be discussed in a later section devoted to the systems' timing.

When the system is placed in the emulation mode (EM), the SPM contains several tables (such as a decimal-to-binary-address table), and a mapping of hardware registers of the emulated machines required by this mode.

The scratch pad memory address register (SPMAR) is driven directly from the applicable registers: base address register (BR), general purpose register (GR), program status register (PSR), R register (RR), status register (SR), and the ROM. See Fig. 9.1. This causes a word from the SPM to be read out into the DR and passed through the incrementor and shifter during the same machine cycle. In addition, the DR with its new contents can act as the source register for a transfer to any other register on the data bus, or to the ALB to be operated upon. Note that the DR is 32 bits wide and that two switches (the DR switch and the WDB (word driver bit) toggle switch) are built into the data flow to allow for the selection of one of the four bytes, to be gated into or out of the DR register.

The shift incrementor network in the SPM regeneration path has the ability to shift a 32-bit word one place, either right or left, and to increment or decrement the word by one or two bits. If no shift and increment or decrement microcontrol are specified by the microinstruction, the word read from the SPM will be regenerated unchanged.

9.2.3 The Read-only Memory

The third memory in the 70/45 processor is the read-only memory (ROM). It is part of the control system and stores the sequences of microinstructions used to control the transfer of information, the arithmetic operations, and the machine status switching.

Table 9.1 FAST MEMORY LAYOUT AND REGISTER ASSIGNMENT

Hexadecimal	0	1	2	3	4	5	6	7	8	9	A	B	C	D	E	F
0	PROCESSOR UTILITY						AAD A ADDRESS REGISTER	BAD B ADDRESS REGISTER	GENERAL PURPOSE REGISTER NO. 8 P4	GENERAL PURPOSE REGISTER NO. 9 P4	GENERAL PURPOSE REGISTER NO. 10 P4	GENERAL PURPOSE REGISTER NO. 11 P4	INTERRUPT MASK REGISTER P4	INTERRUPT STATUS REGISTER P4	PROGRAM COUNTER P4	GENERAL PURPOSE REGISTER NO. 15 (WEIGHT) P4
	NO. 1	NO. 2	NO. 3	NO. 4	NO. 5	NO. 6										
1		I/O CHANNEL REGISTERS – MULTIPLEXOR							PROCESSOR UTILITY		I/O CHANNEL REGISTERS – SELECTOR NO. 1					
	PROCESSOR UTILITY NO. 7	UTILITY NO. 1	CHANNEL ADDRESS REGISTER	CHANNEL COMMAND REGISTER II	CHANNEL COMMAND REGISTER I	STATUS REGISTER	UTILITY NO. 2	UTILITY NO. 3	NO. 8	NO. 9	CHANNEL ADDRESS REGISTER	CHANNEL COMMAND REGISTER II	CHANNEL COMMAND REGISTER I	ASSEMBLY STATUS REGISTER	UTILITY NO. 2	UTILITY NO. 3
2	INTERRUPT MASK REGISTER P1	INTERRUPT STATUS REGISTER P1	PROGRAM COUNTER P1	INTERRUPT FLAG REGISTER	INTERRUPT MASK REGISTER P2	INTERRUPT STATUS REGISTER P2	PROGRAM COUNTER P2	GENERAL PURPOSE REGISTER NO. 7 P3	INTERRUPT MASK REGISTER P3	INTERRUPT STATUS REGISTER P3	PROGRAM COUNTER P3	GENERAL PURPOSE REGISTER NO. 11 P3	GENERAL PURPOSE REGISTER NO. 12 P3	GENERAL PURPOSE REGISTER NO. 13 P3	GENERAL PURPOSE REGISTER NO. 14 P3	GENERAL PURPOSE REGISTER NO. 15 (WEIGHT) P3
3	PROCESSOR UTILITY		I/O CHANNEL REGISTERS – SELECTOR NO. 2						PROCESSOR UTILITY		I/O CHANNEL REGISTERS – SELECTOR NO. 3					
	NO. 10	NO. 11	CHANNEL ADDRESS REGISTER	CHANNEL COMMAND REGISTER II	CHANNEL COMMAND REGISTER I	ASSEMBLY STATUS REGISTER	UTILITY NO. 2	UTILITY NO. 3	NO. 12	NO. 13	CHANNEL ADDRESS REGISTER	CHANNEL COMMAND REGISTER II	CHANNEL COMMAND REGISTER I	ASSEMBLY STATUS REGISTER	UTILITY NO. 2	UTILITY NO. 3
4	GENERAL PURPOSE REGISTER NO. 0 P2	GENERAL PURPOSE REGISTER NO. 1 P2	GENERAL PURPOSE REGISTER NO. 2 P2	GENERAL PURPOSE REGISTER NO. 3 P2	GENERAL PURPOSE REGISTER NO. 4 P2	GENERAL PURPOSE REGISTER NO. 5 P2	GENERAL PURPOSE REGISTER NO. 6 P2	GENERAL PURPOSE REGISTER NO. 7 P2	GENERAL PURPOSE REGISTER NO. 8 P2	GENERAL PURPOSE REGISTER NO. 9 P2	GENERAL PURPOSE REGISTER NO. 10 P2	GENERAL PURPOSE REGISTER NO. 11 P2	GENERAL PURPOSE REGISTER NO. 12 P2	GENERAL PURPOSE REGISTER NO. 13 P2	GENERAL PURPOSE REGISTER NO. 14 P2	GENERAL PURPOSE REGISTER NO. 15 P2
5	PROCESSOR UTILITY		I/O CHANNEL REGISTERS – SELECTOR NO. 4						PROCESSOR UTILITY		I/O CHANNEL REGISTERS – SELECTOR NO. 5					
	NO. 14	NO. 15	CHANNEL ADDRESS REGISTER	CHANNEL COMMAND REGISTER II	CHANNEL COMMAND REGISTER I	ASSEMBLY STATUS REGISTER	UTILITY NO. 2	UTILITY NO. 3	NO. 16	NO. 17	CHANNEL ADDRESS REGISTER	CHANNEL COMMAND REGISTER II	CHANNEL COMMAND REGISTER I	ASSEMBLY STATUS REGISTER	UTILITY NO. 2	UTILITY NO. 3
6	GENERAL PURPOSE REGISTER NO. 0 P1	GENERAL PURPOSE REGISTER NO. 1 P1	GENERAL PURPOSE REGISTER NO. 2 P1	GENERAL PURPOSE REGISTER NO. 3 P1	GENERAL PURPOSE REGISTER NO. 4 P1	GENERAL PURPOSE REGISTER NO. 5 P1	GENERAL PURPOSE REGISTER NO. 6 P1	GENERAL PURPOSE REGISTER NO. 7 P1	GENERAL PURPOSE REGISTER NO. 8 P1	GENERAL PURPOSE REGISTER NO. 9 P1	GENERAL PURPOSE REGISTER NO. 10 P1	GENERAL PURPOSE REGISTER NO. 11 P1	GENERAL PURPOSE REGISTER NO. 12 P1	GENERAL PURPOSE REGISTER NO. 13 P1	GENERAL PURPOSE REGISTER NO. 14 P1	GENERAL PURPOSE REGISTER NO. 15 P1
7	FLOATING-POINT REGISTER NO. 0		FLOATING-POINT REGISTER NO. 2		FLOATING-POINT REGISTER NO. 4		FLOATING-POINT REGISTER NO. 6		PROCESSOR UTILITY		I/O CHANNEL REGISTERS – SELECTOR NO. 6					
									NO. 18	NO. 19	CHANNEL ADDRESS REGISTER	CHANNEL COMMAND REGISTER II	CHANNEL COMMAND REGISTER I	ASSEMBLY STATUS REGISTER	UTILITY NO. 2	UTILITY NO. 3

* Word Address is in Hexadecimal; e.g., 2A = Program Counter P3.

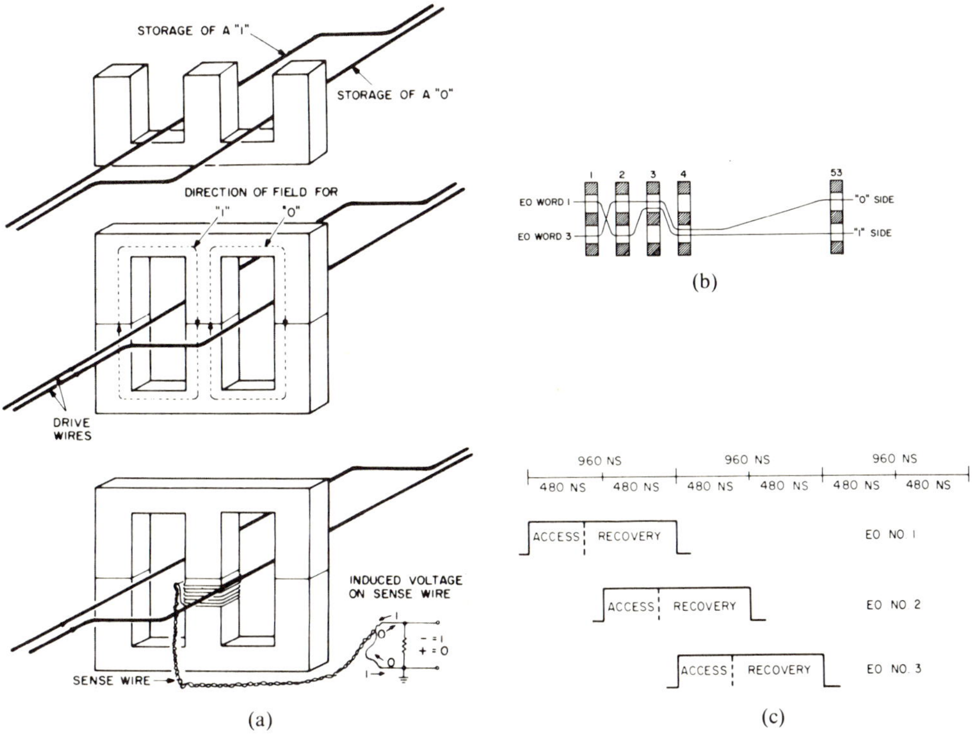

(a) (b) (c)

Figure 9.2 E-core ROM technology. (a) E-core wiring; (b) wiring of the elementary operation (EO) words; (c) ROM cycle time.

Figure 9.2 illustrates the basic ROM technology. It uses an E core transformer-type ROM. The bit pattern stored in a ROM word (EO) is determined by the way each word-drive is threaded through each E core that corresponds to one bit in the ROM word. This transformer read only storage (TROS) technology provides bipolar signal outputs. A drive-line threaded through one leg of the E core would produce an output signal of a polarity opposite to that produced if the same word-drive-line is threaded through the other leg of the E core. This difference is sensed by a differential sense amplifier whose output represents the bit stored. The decision as to which polarity of the output signal represents a 1 or a 0 is arbitrary.

The ROM is organized into units of 2048 words of 56 bits each. One 2048-word-bank is used for the 70/45 standard architectural implementation and up to two additional ROM banks may be supplied for the emulator's microcode.

The ROM bank is organized into two separate memory stacks (each memory stack containing 1024 56-bit words) which generally operate on an alternating basis. This organization (Fig. 2.3) effectively halves the access time, by overlapping the access time of one with the recover time of the other half. Because of conditional branches, this microprogramming restriction is sometimes dispensed with, and sequential addresses are accessed from the same bank. This would imply an automatic 480-macrosecond time-delay penalty.

The 12-bit ROM address register (ROMAR) is used to address ROM. The input to the ROMAR comes from the address generator, the regenerate address (REGAD), or multiplexor regenerator address (MREGAD) registers. (See Fig. 9.3.)

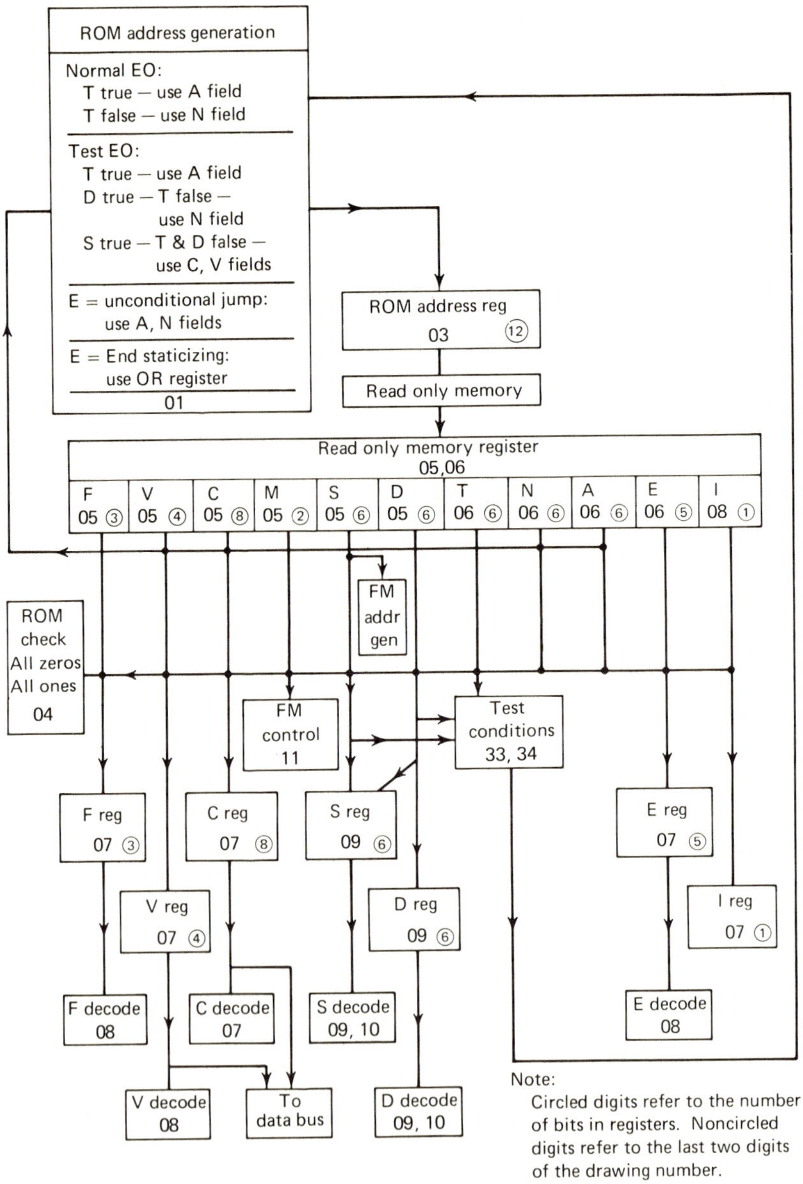

Figure 9.3 Spectra 70/45 UI format and control.

The address generator uses fields of the current microinstruction to generate the address of the next microinstruction. Where only one field is used, it is placed in the least significant section of the 12-bit address (the 6 most significant bits remaining from the current address). The contents of the operation register (OR) may also be used for addressing. The REGAD register normally holds the ROM address that has also been sent to the ROMAR.

The ROM word (56 bits) is organized into eleven fields: F, V, C, M, S, D, T, N, A, E, and I. Some of these fields are logically grouped: F, V, and C belonging to one logical group; M, S, and D belonging to another logical group; and T, N, and A belonging to a third one. The remaining two fields, E and I, are independent functions.

All fields of the ROM word are available in every microinstruction and are read into the ROM DR. In some cases, a field may be multipurpose and have different functions at different times. The alphabetical letter used to indicate the field is an abbreviation, and the full word will give an indication of the function:

F = Function (of microinstruction)
V = Variation (of function)
C = Counter-control (also used as a constant and for addressing)
M = Memory control (controls SPM operation)
S = Source (for DB transfers; also used to specify conditions in special test micro-orders and to address SPM)
D = Destination (for DB transfers; also used to specify conditions in special test microinstructions)
T = Test (specifies test conditions in all microinstructions)
N = Normal (address of next microinstruction, depending on the test results or E field)
A = Alternate (address of next microinstruction, depending on the test results or E field)
E = Exception (specifies checking for validity of addresses or data; also used to specify unconditional jumps)
I = Inhibit (inhibits an I/O servicing break prior to executing the next microinstruction).

These functional field groups are designed to operate as follows:

FVC	MSD	TNA	E	I

1. *The Function Control Fields (F, V, C):* The F, V, and C fields are the most complex to specify because they define a variety of control functions which include the ALB data path, the counters, shift and increment control, and toggle settings. Some of the functions specified by F, V, and C fields can be performed concurrently. In general the F defines the basic function to be performed, the V field further defines the variations and/or further specifications of that function, and the C is the counter control field.

The F field, for example, can specify either shift or increment. If it is a shift function, the V field specifies right or left shift and the disposition of the bits shifted out of or into the SPM location specified. If the F field specifies an increment function, the V will be used to specify whether the counter is incremented or decremented and the value by which the counter is modified. Another example of the FVC control is to specify the ALB operation. The F field indicates a set, perform, or set and perform operation, whereas the V field specifies one of eight different arithmetic, logic, decimal binary, or transfer through ALB operations.

(field bit positions)

F = $(F)_8$ = F (2–0)—Function
V = $(V)_{16}$ = V (3–0)—Subfunction (Variation)
C = $(C)_{16} \cdot (C)_{16}$ = C(7–4) for the switch- and toggle-setting control the data flow into and out of ALB, also used to introduce a literal onto DB; concatenated with C (3–0) for the counter initialization and decrement control.

(field radix)
(concatenation)

The reader must note the bit significance of the C (7–4) field.

C(7) controls the ingating and outgoing of the leftmost or rightmost two bytes—to and from the DR register. It also sets the WDB toggle which controls the gating of the two bytes of data on the DB (0, 1) into the right or left half of the four-byte register DR, on transfers to the DR during a perform EO.

The toggles BAA, BAB, BDB are specified by the C(6), C(5), and C(4), respectively. They are set to the state of C(0) if specified.

The RR and GR counters are decremented as specified by C(3) and C(2) and are decremented by 0, −1, −2, −3 which is specified by C(1) · C(0).

2. *C(7), MSD Fields (data transfer) F ≠ 0*

The M, S, and D fields control the transfer of information between the functional units. The M designates the operation of the SPM, and the S and D fields designate the source and destination registers for DB transfers. That is, by using the S and D fields, it is possible to transfer from one register to any other register along the DB. Note that any SPM operation must use the DR so that, if a read from the SPM is specified, DR automatically becomes the source register for any transfer on the DB bus. Similarly, in an SPM write operation, any of the hardware registers can be designated as a source register. The contents of the specified hardware register are transmitted to one or the other half of the DR; the remaining half of the DR is accessed from the specified SPM location, and the full word is written into the SPM.

M = $(M)_4$ = M(1–0)—FM or SPM control
S = $(S)_4 \cdot (S)_{16}$ = S(5–4) · S(3–0)—FM address or source register
D = $(D)_4 \cdot (D)_{16}$ = D(5–4) · D(3–0)—specifies source or destination register

The S and D fields can have a number of different interpretations, depending on the M and F fields of the associated microinstruction (EO).

3. *TNA Fields—Branch Control*

Every EO specifies the address for the next EO to be executed.

The T is a 6-bit field which denotes the specific test conditions to be tested. This width of field permits any one of 63 different combinations of hardware states to be tested. If the test is true (T-true), the A field is the low-order 6-bit portion of the address of the next EO within the block of 64. If the test is a false/no test (T-false), the N field is the low-order 6-bit portion of the address of the next EO. Address is determined prior to execution of the EO.

A: If T-true: A(5–0) → ROMAR(5–0) · ROMAR(11–6) remains unchanged

N: If F $\neq$ 5 and T-false/F = 5. T-false · D-true; N(5–0) → ROMAR(5–0) · ROMAR(11–6) remains unchanged

$T = (T)_4 \cdot (T)_{16} = T(5\text{–}4) \cdot T(3\text{–}0)$—test condition code
$N = (N)_8 \cdot (N)_8 = N(5\text{–}3) \cdot N(2\text{–}0)$—next EO address in ROM if test false
$A = (A)_8 \cdot (A)_8 = A(5\text{–}3) \cdot A(2\text{–}0)$—next EO address if test is true

4. *E Field*

$E = (E)_2 \cdot (E)_{16} = E(4) \cdot E(3\text{–}0)$—exception code

5. *I Field*

$I = (I)_2 = I(0)$—inhibit I/O servicing.

A check for correct operations on all the bits of the ROM is made at the appropriate intervals. The check in effect makes use of two special microinstructions, one consisting of a pattern of all *zeroes* and another consisting of all *ones*. These two patterns cause an error to be indicated when not in check mode. In the 56-bit version, there are three parity bits which are checked on each read cycle.

Certain of the fields (F, V, C, S, D, and I) make use of a second level of registers. It is the output of these operation decoders that generally controls the execution of the appropriate operations in the processor.

When an SPM operation (M(0) = 1) is specified, the S field is used for SPM addressing. If the operation is a write, a source is required. The source in this instance is indicated by the D field; the D field bits are transferred to the S decoding logic for the operation. If the source is any of the other registers on the DB, 16 bits from the DB are placed in half of the DR as directed by either C7 of the EO or the WDB (the other half of the DR containing the contents of the SPM word addressed), and the entire 32 bits are then placed in the SPM. In the SPM read cycles, the D field is interpreted normally, and the S field is automatically set to indicate the DR as the source. These operations are defined in the S and D field descriptions.

9.2.4 The Arithmetic and Logic Box (ALB)

The arithmetic and logic portion of the processor operates on eight bits in parallel. Any one of the four bytes in the DR register can be selected and transmitted through the 4/1 DR switch to form one byte input to the ALB. The other inputs to the ALB are taken from the UR, which is two bytes wide, and the selection of the byte to be gated to the ALB is specified in the microinstruction field which sets the byte address B (BAB) switch.

The output of the ALB goes to the 16-bit intermediate register (IR). As indicated on the block diagram (Fig. 9.1), the 8-bit output of the ALB is divided into two 4-bit halves which can be switched as dictated by the byte divider bit trigger (BDB). The 8 bits are then further switched to the appropriate half of the IR under the control of the byte address A(BAA) trigger.

One operand of the ALB function is the byte of the UR selected by the BAB toggle; BAB = 0 selects UR0 while BAB = 1 selects UR1. The other operand is the byte of the DR selected by C7 and the BAA toggle.

If C7 = 0 and BAA = 0, select DR 0
If C7 = 0 and BAA = 1, select DR 1
If C7 = 1 and BAA = 0, select DR 2
If C7 = 1 and BAA = 1, select DR 3

The ALB output is set into the byte of the IR selected by the BAA toggle: BAA = 0 selects IR0, BAA = 1 selects IR1. The transfer paths from ALB to IR are further defined by the contents of the V field of the EO and BDB toggle, so that when $V_2V_1V_0 = 000$, the full 8-bit result from the ALB is transferred to IR0 or IR1, ignoring BDB.

The remaining variations of $V_2V_1V_0$ bits select only four bits of the ALB determined by the BDB toggle [BDB = 0 selects ALB (7–4), BDB = 1 selects ALB (3–0)].

Example: If V(2–0) = 100 and BDB = 1, this would transfer ALB (3–0) to IR (3–0), leaving IR(7–4) unchanged.

The 1-bit DR sign register (DRSR) is set under specific conditions during transfers into the IR. Similarly, the 1-bit UR sign register (URSR) is set under specific conditions during transfers into the UR. The DRSR and URSR are set if the sign is negative. The ALB is designed to perform one of eight operations specified by the microinstruction (the V field). These functions are: binary add, binary subtract, decimal add, decimal subtract, transfer utility register UR, transfer the complement of UR, logical OR, and *exclusive* OR. When the ALB is set to any one of these eight functions, it remains in that mode until it is reset by the microprogram. This modification is specified in the F field. The F field would either set, perform, or set and perform. In order to perform an operation on successive data types, the perform microinstruction will continue operations on the data operands using the *carried-over* setting of the control-carry and result-zero flops. The control flops will be reinitialized only by another set, or set and perform, microinstruction.

9.2.5 Other Hardware Registers

A number of other registers are shown on the simplified 70/45 data flow in Fig. 9.1. These are all connected to the DB bus for various purposes. They include the following:

General Purpose Register (GR) is an 8-bit decrementing counter used as a general purpose counter, and is also used to address SPM.

Status Register (SR) also has properties peculiar to the Spectra 70 (and should not be used in emulators) in that it contains:

1. the condition code representing the results of previous operations (positive, negative, zero, and overflow) the equivalent of the previous result indicators in earlier processors, and
2. four interrupt masks.

The 8-bit SR is fed from DB0. SR provides the lower order bits on the DB(5–0). The two high-order bits of the 8-bit byte may be provided by the instruction length hardware or the S register (Fig. 9.1). The instruction length information is derived from the two high-order bits of the OR. Two bits of S contain the condition code.

Operation Register (OR) is an 8-bit register normally holding the operation code, with the special property that the individual bits of the register can be tested by EO (except for OR7, which cannot be tested by itself). The OR is also used in the staticizing routine as a portion of the 64-way branch facility.

Base Address Register (BR) is a 4-bit BR used for address-indexing and also during shift operations to supply and receive the end data.

R Register (RR) is an 8-bit general purpose, decrementing counter that may also be split into two independent 4-bit counters.

The BR or RR may also be used in addressing the SPM but, when so used, are more restricted than the GR.

Interrupt Register (IR) is a 32-bit register that records the many internal and external conditions capable of causing a program interrupt; when an interrupt occurs, its cause or reason is automatically set into the register.

Other miscellaneous registers in the system include:

Arithmetic Mode Flip-Flop (AR) is a 1-bit register indicating whether the arithmetic operations are being performed in the ASCII or EBCDIC Code.

Program Status (PSR) is a 2-bit register indicating the current program state of the processor; it is also used in SPM addressing, i.e., FPC → MAR.

Key Register (KR) is a 4-bit register holding the memory protection key and is used with the optional memory protect feature.

Privilege Register (PR) is a 1-bit register used to indicate specific operations that cannot be executed when set. The KR, AR, and PR are operated simultaneously as far as transfers to and from the DB are concerned.

Emulator On Register (EON) is a 1-bit register used to indicate that the operation of the 70/45 is in the emulator mode. (This is used in machines provided with the emulator option.)

Bank 3 Register (BK3) is a 1-bit register used to indicate the correct ROM bank to be addressed. (This is used in machines provided with the emulator option.)

9.2.6 Wired-in Control Functions

The following is a set of wired-in control functions designed to assist the microprogram control specified in the EO:

EDR Flip-Flop. Regardless of the F or V fields of an EO, the EDR flip-flop is reset whenever (BR = 0000), (S field = X1000X) and (I/O INST); namely, BR = 0000, and BR is being used to address the SPM. Resetting of EDR causes the DR operand to appear as all zeros to the ALB. The condition continues until EDR is set by a set function or set and perform EO.

RZ Flip-Flop. The RZ flip-flop is set by the set function, or set and perform EO, and V fields $\neq 0$. It is reset by a perform or set and perform EO and a nonzero result at the output of the ALB.

Sign Flip-Flops. When the destination code calls for setting the sign flip-flops (DRS or URS), the sign box is set if the sign is negative. If E = 6/7, the sign flip-flops are set according the 4-bit decimal sign at the input to the ALB (the DR switch or UR switch output); otherwise, the sign flip-flops are set from DB07.

WDB Toggle (Word Driver Bit). The WDB toggle is set according to C7 only in a set function, perform, or set and perform EO.

Transfer Controls From DB to DR. The WDB toggle controls the transfer from the DB to the DR only if FMCY = 1 in a perform EO. The transfer is controlled by C0 and C1 in a merge EO. Otherwise, the transfer is controlled by C7.

Transfer Controls From DR to DB and ALB. The transfer from DR to the DB is always controlled by C7. The transfer from the DR to the ALB is always controlled by EDR, C7, and BAA.

Final EO True Test Condition. The last EO of an instruction should contain E = 00010 (end instruction) and must contain a test with the true branch leading to staticizing (instruction fetch microroutine). The unconditional jump (E = 00001) should not be used to go to staticizing because various starting signals must be generated.

E-Code Requirements. The E codes for end staticizing (E = 00011) and end instruction (E = 00010) require a true test condition T. The A field must be either 000000 or 000001 to specify a standard even or odd entrance address.

I/O Servicing. Not more than four consecutive EO's (1.92 micros) must be permitted without I/O servicing.

Triggering GR or RR. It is not permissible to transfer into or out of the GR or RR and trigger that counter in the same EO.

BAA, BAB, BDB Triggering. It is not permissible to trigger a BAA, BAB, or BDB toggle in one EO and test that toggle with a test and branch function that drops through in the immediately following EO.

Interrupt Register Bit #21. Bit #21 of the interrupt register (supervisor call instruction interrupt in Spectra 70 operations) will be set if the E field equals 00000011 and the OR equals 0000 1010.

Interrupt Register Bit #32. Bit #32 of the interrupt register (debug mode option of the program control instruction in Spectra 70 operations) will be set if the E field equals 0000 0010 and the OR equals 1000 0010 and RR10 equals one.

Testing of a Destination Register. A register that is used as a destination for a DB transfer may not be tested in the same EO.

9.2.7 Data Transfer

All the above logic and storage facilities are interconnected by the two-byte-wide data bus (DB) (Fig. 9.1). A second two-bits-wide bus called DBA is provided to transmit the two most significant bits of the main memory address into the MMAR (the remaining 16 bits of the address are transmitted over the DB bus). The DB bus can be viewed as two separate byte buses called DB0 and DB1. Two bytes can be gated on the DB from a number of different sources and transmitted as a single 16-bit field or as two different bytes to two different destinations. It is imperative that the reader study carefully the extent of the gating of the contents of the different hardware facilities on this bus and the gating of the information on the bus into the different permissible facilities. This gating, although extensive enough for the functions performed by the Spectra 45, does present some limitations of which the microprogrammer must be aware. Note, for example, that the OR is fed from DB0; hence, UR1 cannot be transmitted to OR. If a byte is to be transmitted from ALB to OR, the BAA bit control switch must be set to transmit the ALB output byte to OR through UR0. The R register, when used as one eight-bit decrementing counter, communicates with other functions via the DB1 bus. The R register can also be split into two independent four-bit counters, each of which is loaded or unloaded by the four least-significant bits of DB1 [DB (10–3)]. Note also that the G register can be loaded from the DB1 and unloaded to the DB0; therefore a transfer of UR1 to GR presents no problem whereas a direct UR0 to GR transfer is not possible. The reader is urged at this point to examine Fig. 9.1 carefully in order to master the capabilities, limitations, and alternate paths through which the data can be transmitted. He must be familiar with the function of every box in Fig. 9.1, its ingating and outgating capabilities, and the switches that affect this gating.

9.3 70/45 OPERATIONAL TIMING

The 70/45 basic EO time interval (central processing unit (CPU) cycle) is 480 nanoseconds. This is split into four equal time periods (TP1, TP2, TP3, and TP4) of 120 nanoseconds each. One exception to this is in the case of a "SP2" XXX (set and perform twice) or set followed by "PR2" XXX (perform twice) where the following timing sequence occurs: TP1, TP2, TP3, TP4, TP3, TP4. The 'SP2' and 'PR2' are used to process two-byte operands in 1.5 EO. The 70/45 has one clock pulse generator to provide and synchronize the timing of the different register and memory transfers. These timing pulses also synchronize the microprogrammed and the hardware control functions in the system.

Figure 9.4(a,b,c,d) shows the timing relationships between the ALB, SPM, MM, ROM, and other toggle set and reset timing controls. These figures are also helpful in tracing the concurrent action of all the functions involved in any given operation.

9.3.1 SPM—Timing

The SPM has a 480-nanosecond memory cycle, and an access time of 120 nanoseconds. The SPM read command would transfer the SPM word (four bytes) into the DR in the first 120 nanoseconds. The regenerate cycle is not started until 180 nanoseconds later, allowing time for incrementing or shifting the data before writing it back into memory. The actual regeneration time is 180 nanoseconds. Hence the SPM is capable of reading and regenerating (or writing) one word each CPU cycle.

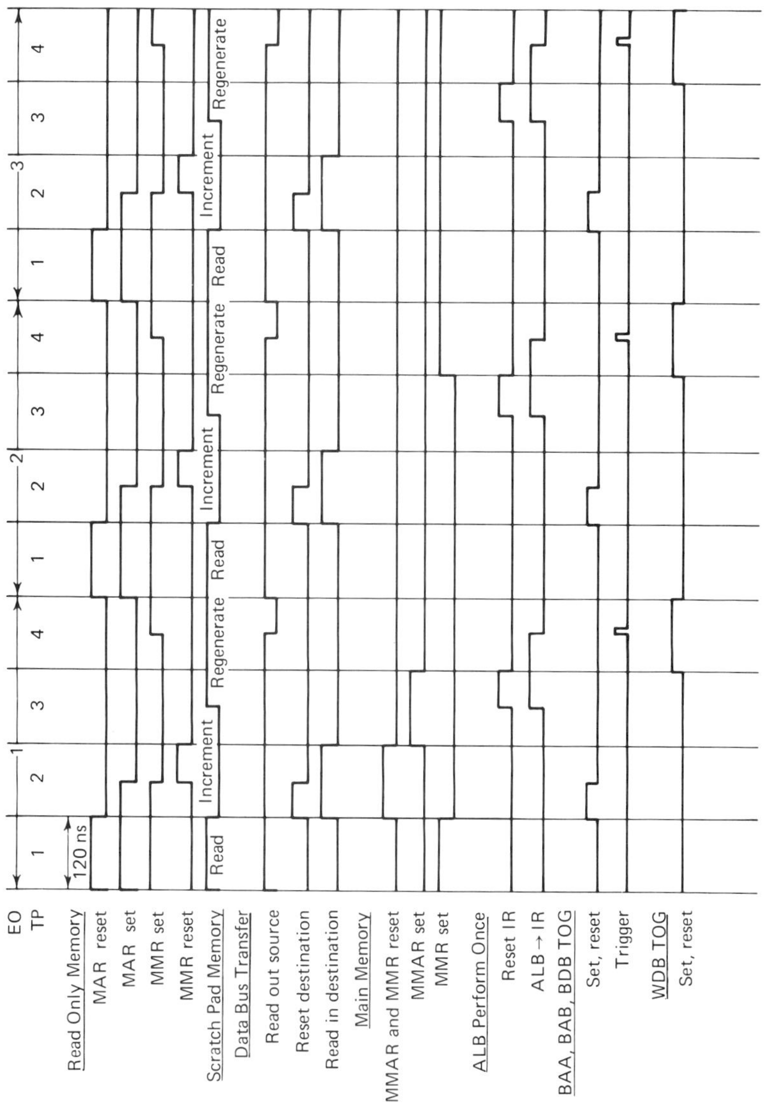

Figure 9.4(a) Spectra 70/45 BPU operational timing diagram.

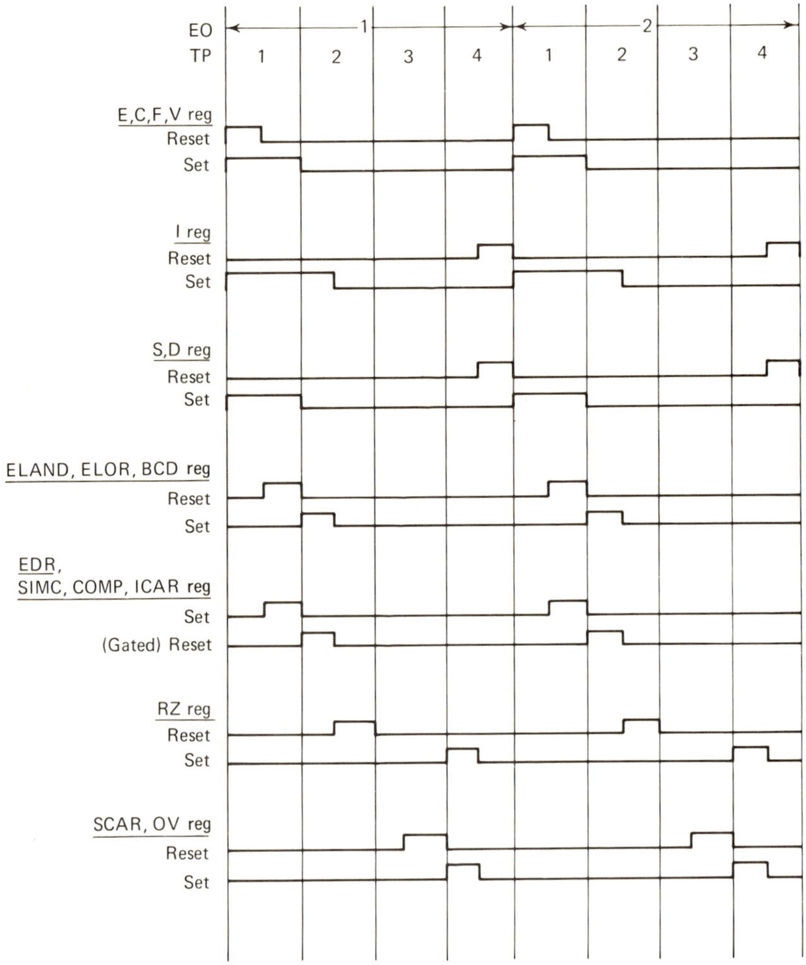

Figure 9.4(b) Spectra 70/45 BPU operational timing diagram.

9.3.2 Main Memory Timing

The main memory (MM) cycle time is 1.44 microseconds. It takes three CPU cycles to read and regenerate the information in MM. In Section 9.2 we considered the data flow and the hardware involved in a typical operation requiring data to be read from MM for transfer to the UR, using an address accessed from SPM. The timing chart in Fig. 9.4(a) shows that the MM address is read out during the first timing point (first 120 nanoseconds) of the first EO cycle. The MMAR and MMR are both reset during the second timing period of the first EO cycle. During the third timing period of the first EO cycle, the MM address read-out from SPM is transferred to MMAR. The

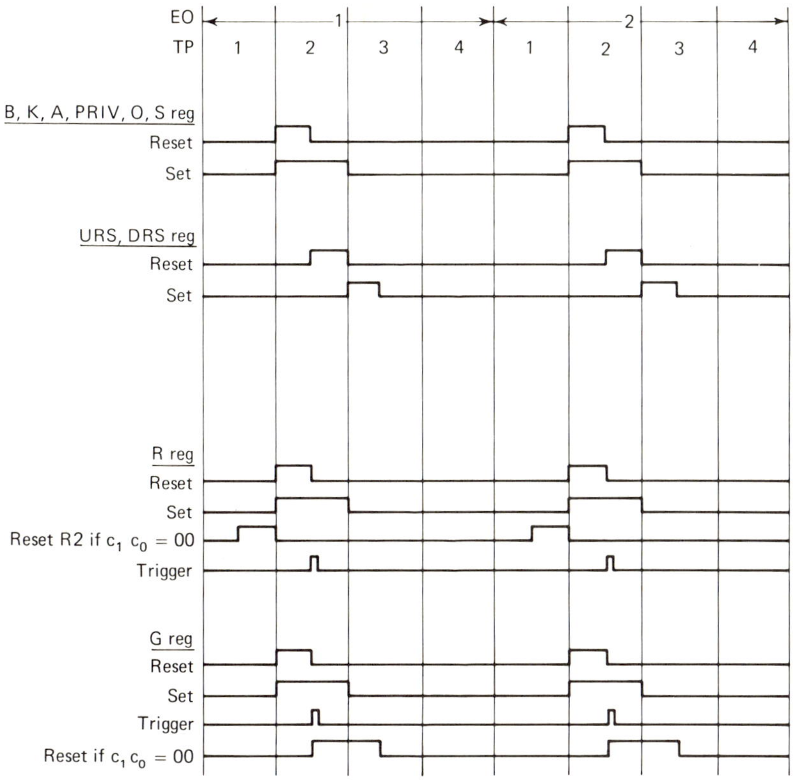

Figure 9.4(c) Spectra 70/45 BPU operational timing diagram.

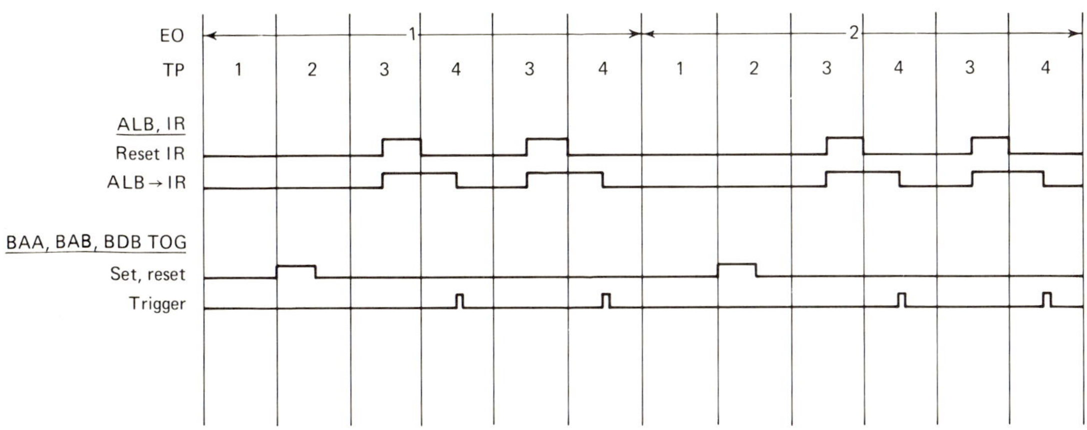

Figure 9.4(d) Spectra 70/45 BPU operational timing diagram.

SPM word is incremented and replaced in the SPM during the first EO cycle. As far as this operation is concerned, no useful work is accomplished during the second EO until its last time period, when the actual MM read-out takes place. If no other useful work is planned for the second EO time, a wait cycle is used. This automatically causes a wait of 480 nanoseconds. The MMR is specified as the source and the UR as the destination during the third EO cycle. As indicated in Fig. 9.4(c), transfers to UR via the DB are accomplished during the second time period of the EO cycle.

Enough time remains in the third EO cycle for an operation through the ALB to the IR with one byte. If a perform-twice EO is specified, time periods three and four are repeated for a total of six time periods rather than the normal four time periods of the EO (i.e., the second execution of the function requires one-half the usual EO time interval since no DB transfer is made). There are two transfers of a byte from ALB to the IR. This byte transfer is controlled by the BAA and BAB toggles and occurs during the fourth timing period of the first EO. The arithmetic function is subsequently performed on the second byte in the DR and UR and the result transferred to the upper half of the IR.

9.3.3 Read-only Memory

The basic ROM cycle time is 960 nanoseconds. The 70/45 uses two banks of ROM that generally operate on an alternating basis to effectively halve the ROM access time to 480 nanoseconds. Therefore, it is possible to access one EO every machine cycle (480 nanoseconds). In case of a jump or branch, it is necessary to access sequential addresses from the same bank, thereby incurring an automatic 480 nanosecond delay as a time penalty. The ROM timing shown in Fig. 9.4(a) is valid only for addressing alternate stacks (even-odd or odd-even). The contents of ROM addressed in the first 120 nanoseconds are available at the end of the EO cycle.

One of the functions that can be specified by the F field is test EO, which permits a test of three conditions for a four-way branch. All other EO's however, permit a test with a two-way branch. In either case, the tests that are specified by the EO are completed before the second basic time period, and at this time period, the ROMAR is set for the next EO. The conditions affecting the test and jump are, therefore, those that have been resolved in previous operations and are not a result of the current EO.

9.3.4 Simultaneity

All operations that are completely independent with regard to data paths and ROM field-bit functions can be carried on simultaneously. Although SPM operations such as increment and shift are normally independent of DB transfers between hardware registers (other than DR), addressing the memory involves the S field and, therefore, concurrent execution of the two functions with one EO is ruled out by the M field, which indicates the SPM operation.

Data bus transfers and ALB operations can be executed in the same EO and can be independent of, or dependent on, each other. The ALB operation takes place after the DB transfer. If the transfer involves either the DR or the UR, the result is involved in an operation through the ALB. If the transfer does not involve either register, the transfer and the ALB operation are independent of each other.

If an ALB operation is specified in one EO together with a bus transfer from the IR, the data transferred from the IR is the data present from the previous operation, and the data in the IR at the end of the operation is the result of the current ALB operation. If an ALB operation is specified on one EO together with a bus transfer *to* the IR, the bus transfer to the IR takes place first and then the IR receives the output of the ALB.

The increment/shift function provides a facility for shifting or incrementing a value as it is written in SPM. The word being shifted or incremented need not have previously been in SPM, but may be operated upon as it is written.

The merge EO function will occur before the incrementing. It provides the only way to modify a single byte of the DR and leave the remaining three undisturbed. Normally, either the upper or the lower half of the DR is loaded from the DB. The merge EO provides for clearing and setting either the DR0 or the DR1 byte from the corresponding bus without altering any other bytes within the DR. Figure 9.5 illustrates the execution time of an EO which

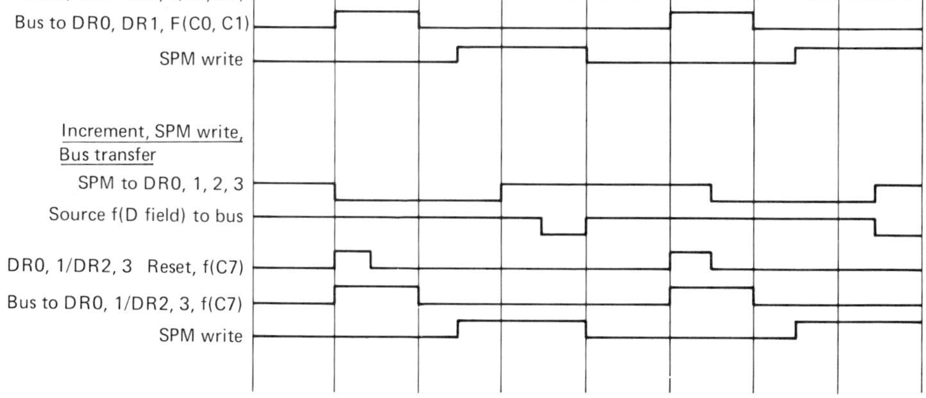

Figure 9.5 Spectra 70/45 EO execution timing.

includes the merge/increment operation with SPM write and DB transfer to the DR. During the first timing period (first 120 nanoseconds), the addressed SPM word is accessed into the DR. Also during the same time, data is put in the DB from the register specified by the destination field. During the second time period in the EO cycle, part of the DR is reset, then is set by the bus. Data in the DR is then transferred to the same location in the SPM from which the original word was read. On the way it is incremented as required by the EO.

9.4 MICROPROGRAMMING THE SPECTRA 70/45

Microprogramming the Spectra 70/45 involves the transfer of information over the DB, the transformation of information through the ALB, and finally the conditional transfer of control. To effect these operations, the EO word is set to specify the combinations of functions that are permitted to happen in one EO time period.

The process of microprogramming this system requires intimate knowledge of the system's data flow, its hardware facilities and functional capability, its data busing system, its ingating and outgating, its timing, and finally the hardware control signals which have higher priority over the microprogrammed control in certain machine states. The microprogrammer must also have a clear understanding of the ROM stack and the ROM word organization. He must master the functions of each micro-operation implied by any unique combination of binary bits in each field in the ROM word. There are complex interrelationships and interdependencies between each group of ROM fields that affect their decoding, even their function. An example here will be the function of the S and D fields. These two fields normally indicate the source and destination registers during an EO where the SPM is not specified or when field M contains 00. However, when M = 01, which implies an SPM read cycle, the contents of the S field are interpreted as the address of an SPM word from which data is to be transmitted, and the D field specifies the destination facility. A third possible interpretation of these two S and D fields is found in the case of the SPM write operation, when M = 11. Here, the DR is automatically selected as the destination register, the S specifies the SPM address (the destination of data transferred in this cycle), and the D field specifies the source.

This interdependency between the functions performed and the decoding of the fields in the ROM word is perhaps the most difficult aspect of mastering the Spectra 70/45 microprogramming. It requires a thorough analysis of the field decoding charts. We will defer this task to the latter part of this section.

Another basic requirement for developing the microprogramming skill for this system is the familiarity with the mechanics of writing a Spectra 70/45 microprogram—the different phases of preparing, assembling, simulating, and documenting the microroutines. This requires a familiarity with the RCA 70/45 Microprogramming Design Aid System known as the RCA-MIDAS system. Like any other computer-aided design program, the MIDAS requires a fixed input format and specific control card sequence to select the desired facilities, functions, and other options available.

We will begin the analysis of these requirements with the MIDAS system.

9.4.1 MIDAS

MIDAS is a microprogramming design aid system developed by RCA programmers to assist in assembling, debugging, and simulating the microroutines and, finally, to prepare the necessary documentations and data files needed for the manufacture and maintenance of the control section of the 70/45 system. It includes an assembler to permit microprogramming in an easy meaningful symbolic notation, a simulator to test and debug the microroutines and the internal operations of the system, and a number of facilities to update, to delete, or to add new EO's in the microroutines. The MIDAS system is designed to be open-ended with the intent of adding other functions and facilities to it as the need arises.

MIDAS Assembler–Input/Output Forms. The assembler accepts the sequence of microinstructions, punched on standard 80-column punch cards in a fixed format, converts this input into an internal form used as an output to the microprogrammers, as an input to the ROM word and stack manufacturing, as an input to the simulator program, and as an input to the logic flow-chart output routine.

The microprogrammer prepares the input to the MIDAS assembler on a special form (Fig. 9.6) where each line represents an image of the EO record to be punched on one punched card. The form is divided into fields, most of which correspond to the ROM word fields. It also contains a label field in columns C1–6. This field contains a symbolic name for the EO documented on the line. The rules governing this label are:

1. The label can be 1 to 6 alphanumeric characters wide, starting with an alphabetic one and excluding the two characters $ and / (these two characters have been assigned a certain meaning in the MIDAS job control program).
2. The label must be punched left justified, i.e., starting with column 1.
3. It must not contain any blanks.
4. No relative addressing is permitted; that is, if the address of the first EO in a microroutine is assigned a label, for example TALLY, we cannot refer to the next EO as TALLY + 1.

Coding sheets for this source language input will have two sets of headings. The purpose of this is to allow the programmer to write each field in either mnemonic or binary representation.

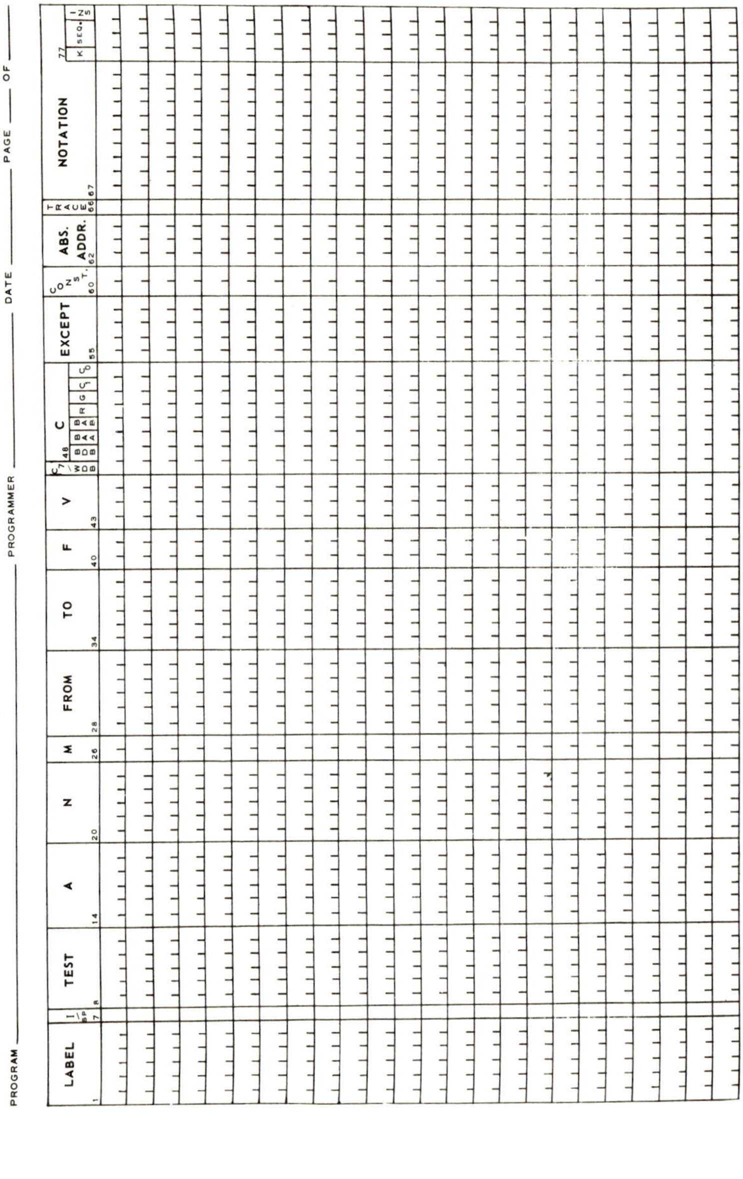

Figure 9.6 Microcoding assembler form.

Notes: 1. It should be noted that when coding mnemonically, FROM and TO columns are unique in that FROM does not necessarily mean the S field and TO does not necessarily mean the D field. The programmer should only consider these fields as FROM and TO, and the assembler will make the determination whether to place the bits in the S or D fields.

2. Descriptions of various fields do not apply when a special form of EO called a TEST EO is used. A complete description of a TEST EO may be found in Appendix K.

The field on the microprogramming form with the heading of "absolute address" permits the microprogrammer to specify an absolute ROM address for any number of EO's in the microroutine. It can be used in conjunction with a label or by itself. Another field entitled "CONST" is used to specify two hexadecimal characters to be put on either the DB0 or DB1 byte bus as specified by the entry in the FV fields. The "notation" or "remark" field is for the microprogrammer and has no effect on the MIDAS assembler. The remaining fields, T, N, A, M, S, D, F, V, I, and E, are used to specify the functions to be performed in that EO.

This input is punched by trained keypunch operators on standard 80-column keypunch cards.

The assembler first checks the syntax of each EO Over twenty-seven validity checks are performed by the MIDAS system. These checks include incompatible combinations of micro-orders within an EO, invalid microcodes, violation of basic timing rules, and checks for certain illegal sequences of EO's.

The normal output of the assembler is illustrated in Fig. 9.7. It contains the assembly listing, the ROM absolute address assigned to each EO, the symbolic micro-orders, and the encoded binary bits in the control word. Another optional output is the microflow-chart printout illustrated in Fig. 9.8. This provides the microprogrammer with a visual aid to check the logical flow of the

```
 DATE 06/20/69       501 READ ONLY MEMORY  (ROM)

LABEL  TEST  A      N       MFROM MOVETO F    V  C FIELD  EXC    ABS T    COMMENT          ROMAR FVC MS D T N A E I


 TITLE 501 READ ONLY MEMORY  (ROM)                                        00000000
AFT     BRTESTTRESETSHR1    FFR1  UR     SP1RADDS SS             4600     00000100         4600  12B111A102B0107000
AFLAG I      ASHRD                FFR2   CONDB0 R              804620     00000200         4620  3D0031C00002146011
AFTI         AFT            FFR1  IR     SLLZB2 S                4775     00000300         4775  428011A0E000046010
AFUT9               AFUT91  FUT9  UR     SP1RADDS SS             4657     00000400         4657  12B1118100006046010
AFUT9I              AFUT9   DR    FUT9   SLLZB2 S                4656     00000500         4656  428031808005746010
AFUT91              AFUT92  IR    FUT9   SRLZB2 S                4660     00000600         4660  4A80318OA006146010
AFUT92              AFUT93  FUT9  G      SRLZB2 R                4661     00000700         4661  4A0011818006246010
AFUT93I             AFUT94  GA    MAR    CONDB0             SHADE404662   00000800         4662  394000F1C006363041
AFUT94I             AFUT95        FUT9   CONDB0 R              804663     00000900         4663  3D0031800006446011
AFUT95IMR17E1AFUT98AFUT96   MMR   GUR    MGEDB1                  4664     00001000         4664  300200C13346567001
AFUT96 OR0EQ0AFUT97AFUT99   DR    IR     SP1BADDS RR             4665     00001100         4665  12B000B0E017066000
AFUT97              AFB0I   BIR   FUT9   SETSAMES                4666     00001200         4666  788031812003547010
AFUT98              P2CALL        F1     CONDB1                444667     00001300         4667  3A4430800000045010
AFUT99              AFB0I   BIR   FUT9   CONDB1 S              034670     00001400         4670  3A8331812003547010
AIORT  OR4EQ0AIORT1AEREX          IRDS   CONDB0                804671     00001500         4671  3D000000D052572000
```

Figure 9.7 Sample MIDAS assembler output listing.

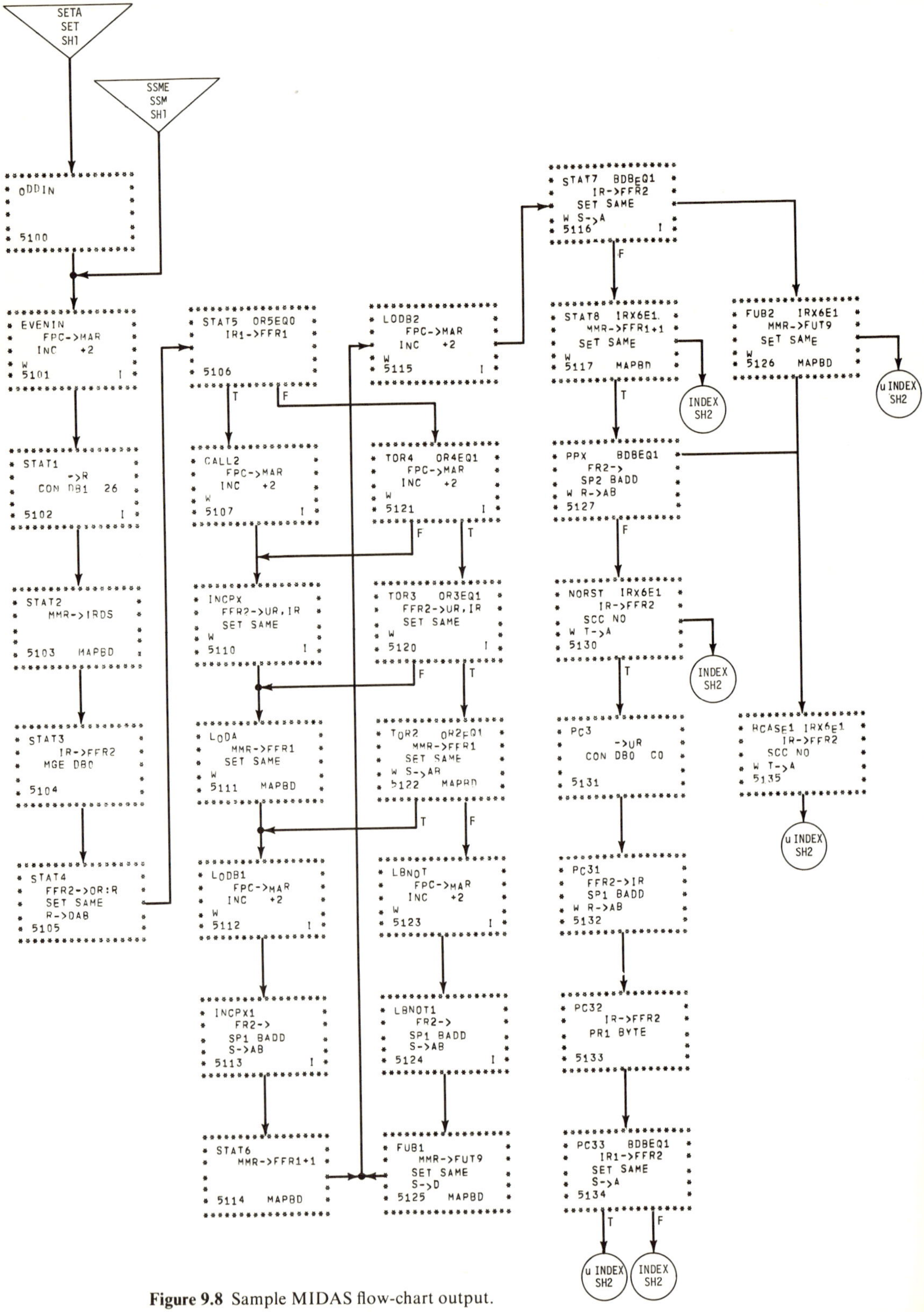

Figure 9.8 Sample MIDAS flow-chart output.

microroutines. The flow-chart program output is similar to the IBM control automation systems output. It consists of EO boxes which have a fixed format with fixed printout positions for each micro-order. For example, the register-to-register transfers are always printed on line 2.

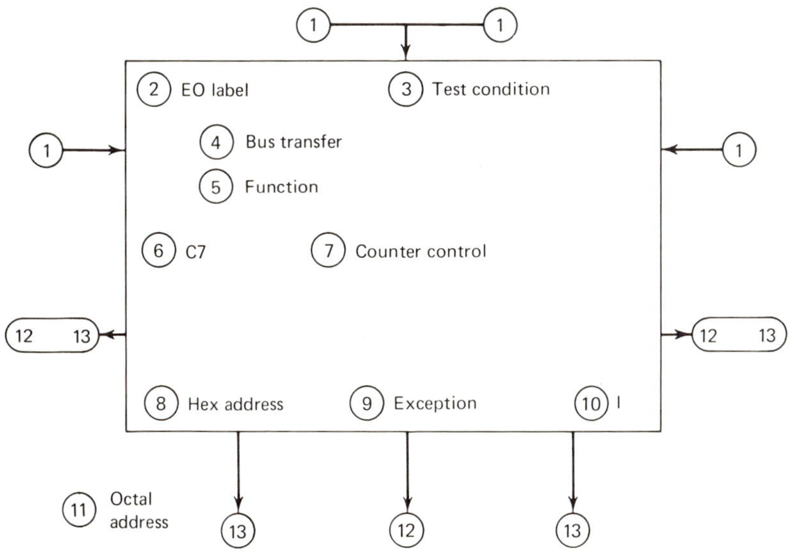

Figure 9.9 EO flow-chart format.

Figure 9.9 shows the format of this microinstruction box. The fields therein have the following meaning:

1. *Entry from Previous Microinstruction (EO).* One or more entries may be specified for an EO. All entries from previous EO's are shown on the EO flow-charts. The entry may be shown as a direct line connecting the EO with the previous EO, or the mnemonic EO label of the previous EO may be shown in a circle or triangle. A circle is used to designate an entry or exit point with the same algorithm. A triangle is used for entry or exit to another algorithm. An EO may be duplicated in different portions of the same algorithm in order to clarify the logical EO flow. An EO may also be shared and will, therefore, appear in different algorithms. In order to avoid confusion in the flow, only the paths that apply to that algorithm will be shown. (Thus, an erroneous exit or entry from or to the algorithm will not be cross-referenced.)

2. *EO Label.* The EO label identifies the EO and the EO flow-charts. It is not represented in the EO word in ROM. The first letter (or group of letters) of the label identifies the algorithm in which the EO is used.

3. *Test Condition.* The test condition mnemonic specifies the test to be performed during execution of the EO. This test condition is represented in the

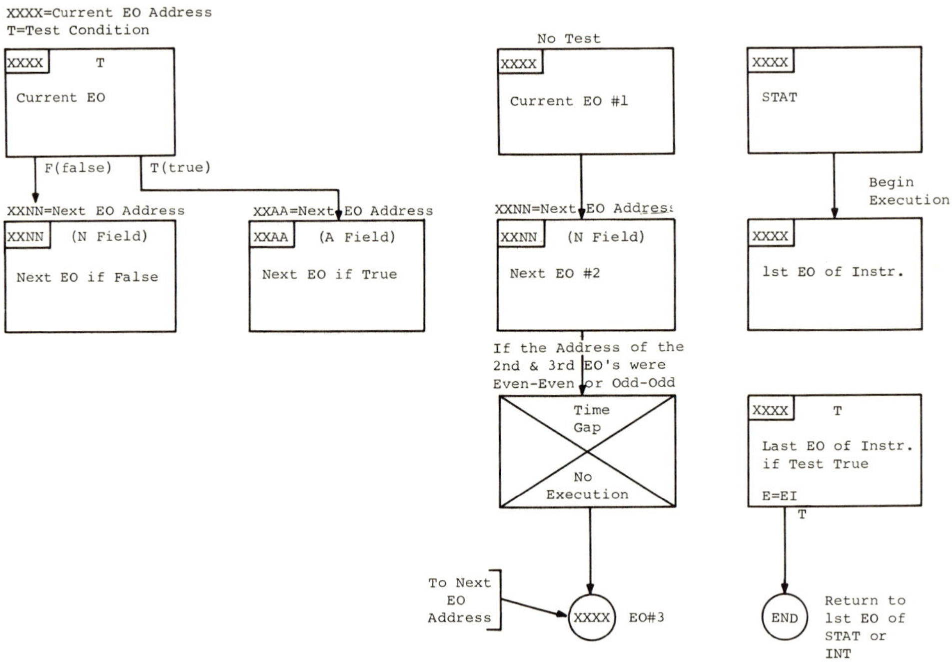

Figure 9.10 TNA fields (branch control).

T field of the EO word in the ROM (Fig. 9.3). Figure 9.10 illustrates the EO flow when a test and branch condition is specified.

4. *Bus Transfer*. The bus transfer mnemonic specifies the data movement to be executed by the EO. The form of the mnemonic is SOURCE DESTINATION. The bus transfer mnemonics are represented in the EO word in the ROM in the M, S, and D fields (Fig. 9.11).

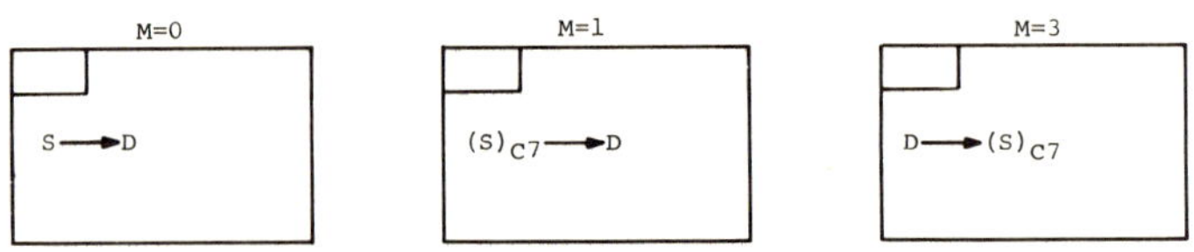

Figure 9.11 C7, MSD fields (data transfer); $F \neq 5$.

If the source is a scratch pad FM location, i.e., the first letter of mnemonic is F, the M field is $(01)_2$.

If the destination is a scratch pad location, the M field is $(11)_2$.

If neither the source nor the destination is a scratch pad location, the M field is $(00)_2$.

If a scratch pad location is specified in a bus transfer, the code for the scratch pad location is in the S field regardless of whether the scratch pad location is a source or a destination.

5. *Function*. The function mnemonic which appears on the EO flow-chart designates the functions to be performed during the EO. These mnemonics are represented in the EO word in the ROM in the F and V fields, and in some cases, in all or a portion of the C field (Fig. 9.12).

6. *C7*. The C7 mnemonic is represented by the C7 bit of the EO word in the ROM. The EO flow-chart mnemonics for C7 are either space or W. If a W is shown in the EO flow-chart, the C7 bit is 1, and DR2 and DR3 are specified. If the EO flow-chart shows a space (blank) in the C7 position, the C7 bit is 0 and DR0 and DR1 are specified.

7. *Counter Control*. The counter control mnemonic specifies the setting, resetting, or triggering of the various toggles and R and G counters. The counter control mnemonics are represented in the C field of the EO word in the ROM.

8. *Hex Address*. This mnemonic in the EO flow-chart is the ROM address of the EO. The form of the address is $X_1X_2X_3X_4$, where

$(X_1)_4$	represents bits	2^{11}-2^{10}	of address
$(X_2)_{16}$	represents bits	2^9-2^6	of address
$(X_3)_4$	represents bits	2^5-2^4	of address
$(X_4)_{16}$	represents bits	2^3-2^0	of address

This form of the address is used in the 70/45 assembler and simulator and is included in the EO flow-charts to facilitate cross-referencing with the assembler listing.

9. *Exceptions*. The exception mnemonic in the EO flow-chart is represented in the ROM in the E field of the EO. If no exception appears in the EO flow-chart, the E field is 00000 if a test condition is specified; if no test condition is specified, the E field is 00001.

10. *I (Inhibit)*. The EO flow-chart shows either a space (blank) or 1 in the inhibit position. If the EO flow-chart has a blank in this position, the I field of the EO in the ROM is 0, and I/O servicing is inhibited between this EO and the next EO in the flow-chart.

11. *Octal Address*. This mnemonic in the EO flow-chart is the ROM address of the EO. The form of the address is $Y_1Y_2Y_3Y_4$, where

$(Y_1)_8$	represents bits	2^{11}-2^9	of address
$(Y_2)_8$	represents bits	2^8-2^6	of address
$(Y_3)_8$	represents bits	2^5-2^3	of address
$(Y_4)_8$	represents bits	2^2-2^0	of address

12. *Exit to Next EO if No Test Condition is Specified*. If no test condition is specified, there is only one exit to the next EO. This is shown on the EO flow-chart by a line connecting this EO box directly to the next EO box or to a circle or triangle containing the EO label of the next EO and flow-chart sheet number where the next EO is shown. This line is in one of the positions

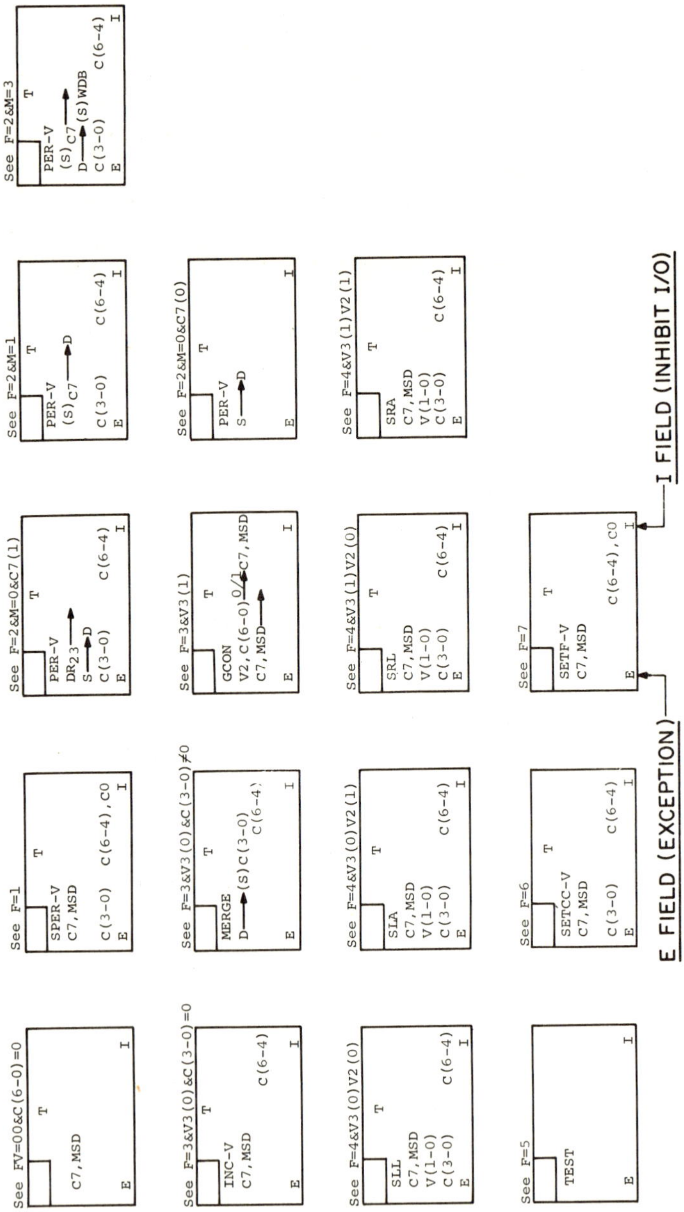

Figure 9.12 FVC field (function).

labeled 12 in Fig. 9.9. The address of the next EO is represented in the EO word in the ROM in the A and N fields as follows:

a. If no test condition is specified and no exception is specified in the EO flow-chart, the A field contains the six most-significant bits of the ROM address of the next EO to be executed. The six least-significant bits of the next EO address are in the N field. (The EO in the ROM has E field coded 00001.)
b. If no test condition is specified and an exception is specified in the EO flow-chart, both the A and the N fields contain the six least-significant bits of the ROM address of the next EO to be executed. (The A field is the same as the N field.) The six most-significant bits of the next EO address are the same as those of the current EO.

13. *Exit to Next EO if a Test Condition is Specified.* If a test condition is specified, there are two possible exits to the next EO. These are shown by 2 separate lines in any of the positions labeled 13 in Fig. 9.9. Each line connects this EO box directly to the next, or to a circle or triangle containing the EO label of the next EO and the flow-chart sheet number where the next EO is shown. A "T" alongside the line indicates the exit if the test is true; an "F" alongside the line indicates the exit if the test is false (Fig. 9.8). The address of the next EO is represented in the EO word in the ROM in the A and N fields as follows:

a. If a test condition is specified in the EO flow-chart and the exception is not ENDIN nor EES, the A field contains the six least-significant bits of the ROM address of the next EO to be executed if the test condition is true, and the N field contains the six least-significant bits of the ROM address of the next EO to be executed if the test condition is false. The six most-significant bits of the ROM address of the next EO to be executed (test condition true or false) are the same as those of the current EO address.
b. If a test condition is specified in the EO flow-charts and the exception is ENDIN and the test condition is true, the A field contains the six least-significant bits of the ROM address of the next EO to be executed. The six most-significant bits of the ROM address of the next EO, if the test is true, are 10100. If the test is false, the N field contains the six least-significant bits of the ROM address of the next EO to be executed; the six most-significant bits of the next EO address are the same as those of the current EO address.
c. If a test condition is specified and the exception is EES and the test is true, the ROM address of the next EO to be executed is obtained as follows:

   2^0 of address is 2^0 of A field,
   2^5-2^1 of address are a result of "logical OR" of 2^5-2^1 bits
      of A field with 2^4-2^0 bits of OR, respectively.

2^8-2^6 of address are 2^7-2^5 of OR, respectively,
2^{11}-2^9 of address are 100, respectively.

If the test is false, the N field contains the six least-significant bits of the ROM address of the next EO to be executed; the six most-significant bits of the next EO address are the same as those of the current EO address.

MIDAS Simulator. Another main MIDAS function is provided by the *Simulator* program, which is used as another debugging and performance evaluation aid. It contains the following optional features:

1. *Trace option* is used to trace the sequences of microinstructions as they are executed. The trace may consist of a printout of the contents of specified registers as the microroutines are being executed. Tracing can be performed for the complete microroutine or any part of it between any specified addresses.
2. *Memory Snapshots.* The simulator provides a snapshot of the MM and the SPM points specified by the microprogrammer.
3. *Termination Options.* A simulator problem can be terminated for many reasons such as time and page limits. Termination may also be specified by other exceptional conditions specified for the simulator.
4. *Checking Options.* This option provides a very useful tool in the simulator for checking the contents of the MM and the SPM against predicted results on termination of the problem.

9.5 MICROPROGRAMMING AND EO FIELD DECODING

This section has three objectives:

1. To present the ROM field decoding charts and the conditions which affect and alter the decoding of that field.
2. To relate the fields or any grouping of fields to the hardware facilities shown in Fig. 9.1 that are affected by them.
3. To provide practical illustrations and help the reader develop experience in microprogramming this system.

9.5.1 Data Movement Control Fields (M,S,D)

This group of control fields controls the transfer of data between:

1. register to register
2. register and FM (SPM)
3. register and main storage.

1. The first type is simple if one is familiar with the data-flow diagram (Fig. 9.1), the data transfer capabilities and limitations of the system, and also

the setting of the control switches which affect the byte selecting and gating pattern.

The micro-orders found in these fields are selected to identify which of the three data transfer types is requested in a particular EO. Note the following cases: If the micro-order which is written in the S or the D field starts with an F, it indicates that the FM is involved in a read or write operation. If this micro-order starting with an F is in the S field, it further implies an FM-to-register read operation, and the M field is set to 01. Here the FM is read out to the DR, or two bytes of the addressed FM word are read to the right or left half of the DR as specified by the setting of C7 (bit 7 of the C field), operated upon (if any action is specified in the F and V fields), and then regenerated (with or without shifting or decrementing) into the FM. In this FM cycle the D field specifies the destination register and the S field specifies the address of the source data in the FM. The actual transfer of data here is from the FM to the DR to the destination register specified by the D field. All this is accomplished in one machine cycle.

2. If the micro-order specified in the D field starts with an F, it implies an FM write cycle, and the M field is set to 11. In this case the DR is automatically selected as the destination register in anticipation of a possible DB transfer of information into the DR from a source register specified by the D field. That is, the contents of the register specified by the D field are transmitted to the DR (right or left half, depending on C7), and then the contents of DR are regenerated into the FM location specified by the S field.

3. If neither the S nor the D micro-orders start with an F, this indicates that no FM operation is specified for the associated EO cycle. The M field is therefore set to 00. The S field indicates the source; the D field indicates the destination register in a register-to-register operation. All other EO fields are interpreted normally.

Note that M = 10 is an invalid code, and the results are unpredictable.

We will first illustrate these three cases before introducing the main memory data transfers.

α1: UR(S); IR(D)
α2: IR(S); UR(D)
α3: UR(S); OR : R(D)
α4: FFR1(S); UR, IR(D); s(C7)
α5: BIR(S); FPC(D); s(C7)

The above notation is to be interpreted as follows:

1. αi is the symbolic address of the associated EO.
2. β(S); β(D); s(C)

That is, the micro-order β in the source field S specifies the source data to be transferred to the destination specified by the micro-order β appearing in the D field. The s(C) implies the setting or resetting of the bits in the C field to effect the setting of data-flow selection switches. In general, one can represent the whole EO format to correspond to the EO input form as follows:

$$\alpha i;\ \beta(I);\ \beta(T);\ \beta(N);\ \beta(A);\ \beta(M);\ \beta(S);\ \beta(D);\ \beta(F);\ \beta(V);$$
$$s(C7, B, A, D, C3, C2, C1, C0);\ \text{Absl};\ \beta(E); \text{CONST}$$

Note the Absl field here implies the absolute address specified by the microprogrammer. The CONST is a two hexadecimal digit constant that can be specified by the microprogrammer to be loaded on the DB.

The first EO with symbolic address $\alpha 1$ specifies that the contents of the 16-bit UR are to be transferred to the 16-bit IR register. This is an example of a register-to-register transfer. Other examples of this type of transfer are found in $\alpha 2$ and $\alpha 3$. $\alpha 3$-EO says that the two bytes in the source register UR are to be transmitted to two registers: the UR0 byte goes to the OR and simultaneously the UR1 byte goes to the R register.

EO-$\alpha 4$ represents an FM request. Here the contents of the fast floating-point register 1 (FFR1) in the FM are to be transferred to the DR, then depending on the setting of the C7 switch, either the DR0,1 or DR2,3 contents are transmitted to the destination register(s). In this specific example, DR2,3 is to be transmitted to the two specified destination registers UR, IR. Note that the character (:) in the micro-order OR:R (in $\alpha 3$) specifies a concatenation operation, whereas the comma (,) as in UR, IR (in $\alpha 4$) specifies a parallel transfer.

The fifth example ($\alpha 5$) specifies that two bits from register B, namely, B(1–0), and the full IR register are transmitted over bus A(1–0) and full data bus (forming a total of 18 bits) to the destination register specified by the D field. The fast program counter (FPC) micro-order specifies a program counter register in the FM. This program counter is state-dependent.

An important side issue is the use of the DR to hold the 18-bit memory address field. Figure 9.1 shows that the two-bit A data bus is always gated to DR1 bits (0, 1). This means that the remaining 16 bits transferred on DB0 and DB1 must always be set in the DR2 and DR3 bytes so as to leave DR1 available for the A bus. A general rule which covers this case is to always set bit C7 on, whenever the microprogrammer is arranging for an MM address transfer.

Appendix A gives the FM address codes for the CPU normal EO cycle ($\overline{I/O}$). The leftmost column specifies the encoded S field mnemonic as it appears on the assemble output. Note that the fast memory address register (FMAR) is seven bits wide, FMAR (6–0). It is divided into two address fields, and FMAR (6–4) is used to specify one of four states. For example, there are four independent sets of 16 GPR's; each set will be accessible to the microroutine depending on the state in which the system is operating. The first state's GPR's are stored in the SPM addresses 60 to 6F; the second state's in 40 to 4F; the third in 20 to 2F; and the final state in 00 to 0F. The program status register (PSR) was defined as a 2-bit register indicating the current program state of the processor. These two bits then are used to define the setting of the high order bits of the FMAR.

The floating-point registers are stored in locations 70 to 7F. Only one set of the floating-point registers is used to service all the states.

The remaining four low-order bits of the fast memory address register, FMAR (3–0), are specified by the microprogrammer and are transmitted to FMAR (3–0) from the B register [B (3–0) → FMAR (3–0)], from the left half of the R register, from the right half of the R register, or from the G register. Note that the left half of the R register is designated as R1, and the right half is designated as R2.

Examples

Assume the R register is set to "26":

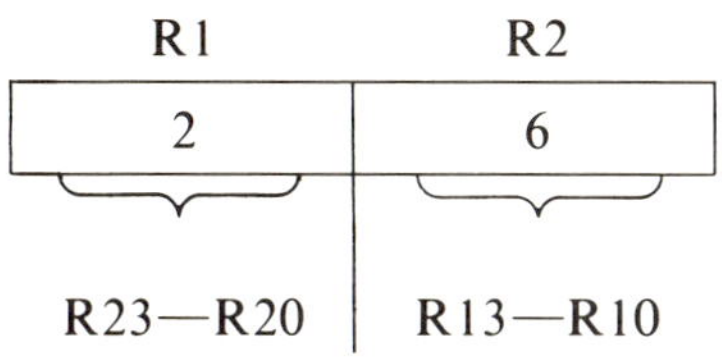

Then,

1. FFR1 means 72
2. FFR1 + 1 means 73 (72 + 1)
3. FFR2 means 76
4. FFR2 + 1 means 77
5. FR1 means 62, 42, 22, or 02, depending on which state the system is in.

Another type of FM address code is the FB form. The only difference between this and the FR1 or FR2 is that the four low-order FMAR bits come from the B register rather than the R1 or R2. Note that the FB is also program-state-dependent.

The most general format is the FG. It is state-independent and permits the microprogrammer to modify any register in FM. It must be used carefully. The FMAR is totally specified by transmitting the seven leftmost bits of the G register to the FMAR. For example, assume G = "D0" hex (11000000). Then

60

the seven leftmost bits when transmitted to FMAR will address location 60. Hence the EO (FG → UR) will transmit the contents of the FM word 60 to UR.

Note that, in the FG and FG + 1 formats, the G(0) = 1 case overrides the C7 setting.

For example, let G = "E2" hex. Then, which FM registers are addressable by the FG format? The correct answer is 71. The following example incorporates the setting of the G register and its use to address the FM hex.

$\alpha 1$: → G(D); CON(E), DB1(V); (C7 = 1); "E2" (CONST)
$\alpha 2$: FG(S); UR (D)

The first EO ($\alpha 1$) loads the DB1 with the specified constant "E2" hex and gates the contents of DB1 into the G register.

The second EO ($\alpha 2$) transfers G (7–1) to FMAR (6–0) to address location 71. It is important to recognize that the constant E2 could not have gotten to the G register if we loaded it on the DB0 bus since the G register only receives data from the DB1 bus.

Appendix A also specifies the mnemonics for other UR's in the FM. The mnemonic specifies the type of UR, and the specific program-state further defines the exact FM location desired.

Appendices B and C tabulate hardware, source and destination register codes respectively. The structure of these tables is self-explanatory. It consists of the source or destination mnemonic followed by the hexadecimal SD fields code, followed by the source or destination hardware registers. The columns labeled transfer bus show the transfer between the designated register over the data bus A and the data bus bits.

Main Memory Data Transfer. The third and final type of data transfer is the main memory to register transfer. The main memory has a 1.44 microsecond memory cycle and a 16-bit word representing two bytes. The MM cycle is equivalent to three EO cycles. The MM request is made by placing the 18-bit address field in the MMAR. This function is specified by placing the MAR micro-order in the D field. The read cycle takes three machine cycles before the particular MM word is destructively read out and the same or different word is regenerated in that memory location. As far as an MM read cycle is concerned, no useful work is accomplished during the second microinstruction; however, the second EO in this three-cycle MM sequence can be used to specify any action not involving the MMAR or main memory read (MMR). For example,

$\alpha 1$: I; X(S); MAR(D)
$\alpha 2$: I; anything
$\alpha 3$: MMR(S); X(D)

The question is, what if no other micro-operation is to be specified in $\alpha 2$? Will we waste an EO to satisfy the MM timing requirement? No; to alleviate this loss of an ROM wire, a read-with-wait micro-order was provided by the designer. This is the memory address read-with-wait (MARW) micro-order. An example of how it is used is the following:

$\alpha 1$: I; X(S); MARW(D)
$\alpha 2$: MMR(S); X(D)

Note that this sequence of read-with-wait takes two ROM words; however, it still requires three EO cycle times to execute. The MARW says transfer the 18-bit MM address from the source register X into the MMAR; however, wait one EO time before performing the next EO ($\alpha 2$).

The last MM operation is that of write. This write operation is specified by requesting a memory cycle in $\alpha 1$ (by transferring an address into MMAR), and transferring the 16-bit word to be written into selected memory location in the second EO time ($\alpha 2$). The following action takes place. The address in MMAR is decoded and the corresponding X, Y drive lines are activated and destructively read-out the contents of that MM location. Early in the second EO cycle time, the data → MMR is sensed and the MM sense lines are inhibited from setting the MMR; rather the contents of DB0 and DB1 are allowed to set MMR instead. During the third EO time the regeneration cycle from MMR into main storage is completed. For example:

$\alpha 1$: I; X(S); MAR
$\alpha 2$: I; Y(S); MMR
$\alpha 3$: Anything not affecting the memory registers.

9.5.2 I Field

The I field consists of 1 bit. It is staticized in the I bit register and is used to inhibit a break in the EO sequence with a request for servicing. The following EO will be given over to the start of the I/O servicing routine if the new EO does not contain another I inhibit micro-order. This means with the setting of I in each EO, the I/O servicing routine is postponed one more EO cycle time. The maximum time for the execution of EO's without a break for I/O servicing is 1.92 microseconds. This allows for four EO's if they are the normal four-time period type. If any of the EO's are of the perform twice (PR2) or 6 basic machine time periods, then less than four EO's are permitted.

The I/O servicing required the MM and all its interface registers; hence I/O servicing cannot be permitted to occur during a memory cycle. This fact explains why the I field was specified in the $\alpha 1$ and $\alpha 2$ EO's in the last memory example. In the MARW example, the I in the $\alpha 1$—EO controls the wait cycle as well. Such servicing also requires the FM-DR data path since most channel information is stored in FM.

In the normal mode, the next normal EO address is preserved in REGAD and MREGAD—ROMAR (11–0) → REGAD (11–0), MREGAD (11–0). When MULTIPLEX (MUX) servicing breaks in, REGAD is undisturbed. During MUX servicing, the next MUX EO address is preserved in MREGAD —ROMAR (11–0) → MREGAD (11–0). I bit is used during MUX servicing to inhibit SELECTOR (SEL) servicing. When SELECTOR (SEL) servicing breaks in, REGAD and MREGAD are undisturbed.

When returning from SEL service to MUX service, MREGAD (11–0) → ROMAR (11–0), and the BPU continues MUX service. When returning from

SEL/MUX service to normal mode, REGAD (11–0) → ROMAR (11–0), and the BPU continues in the normal mode.

9.5.3 Function Control Fields: F, V, C

The F, V, and C fields are the most complex because they define a variety of functions that can be performed, and because separate functions can be performed in different parts of the processor. In general terms, F defines the basic functions to be performed, V further defines the variations within the broad general function, and C is a counter control field.

One way to present these fields is to look at the eight major classes of functions encoded in the 3-bit F field, and within each class, to examine the variations in action specified by the V and the C field settings.

The F field is encoded to signal the following action:

1. F = 0	No action
2. F = 1	Set and Perform (SPER)
3. F = 2	Perform (PER)
4. F = 3 · V3(0)	Increment/Merge
5. F = 3 · V3(1)	GCON
6. F = 4	SLL/SLA/SRL/SRA
7. F = 5	Test
8. F = 6	SETCC (set condition code)
9. F = 7	SET FUNCTION

In the remainder of this section we will take these major classes of actions and analyze the variations within each one based on the V and C field settings.

1. *F = 000—No Operation.* This EO performs the usual test, data movement, and exceptional checks but does not use or modify the ALB or the shift/increment network. The V and C fields (except C7) are ignored, and the other fields have their usual interpretation.

2. *F = 001—Set and Perform Function.* This EO sets the ALB function control flip-flops to perform functions specified by the V field, sets the toggles, and triggers the counters as specified in the C field. It then executes (performs) the function once or twice and transfers the result into IR, as specified by the V field. After each execution of the ALB function, the toggles are modified as specified in the V field. The other fields have their usual meaning.

The toggles BAA, BAB, and BDB specified by C6, C5, and C4 are set to the state of C0; the toggle WDB is set to the state of C7. The toggles can be set before the ALB function is executed, and complemented after each execution of the function. This complement operation is designed to provide an automatic control when processing two-byte register data through a one-byte ALB (by a set and perform twice microcommand).

The V(3) bit specifies the number of times to execute the ALB function and complement the toggles.

V3 = 0 Execute function once, and complement toggles once.
V3 = 1 Execute function twice, and complement toggles twice.

The V2, V1, V0 bits specify the function to be performed. (See also Set Function F = 111 for definition of these functions.)

V = X000 Function does not change
V = X001 BCD add absolute; set ICAR if IR03(1)
V = X010 Binary address add
V = X011 Binary address subtract
V = X100 Transfer UR into IR
V = X101 Transfer two's complement of UR into IR
V = X110 BCD add absolute
V = X111 BCD subtract absolute

Appendix E (with FV bit patterns between 0010000–0011111) shows the 16 micro-orders variations, eight for SP1 and eight for SP2.

3. *F = 010—Perform Function (PER).* This EO executes the function previously set into the ALB and transfers the result into the IR. The definition of the function set will be reviewed in F = 111 (Set Function). In this operation the V(3) bit specifies the number of times this function is performed. The V (2–0) defines transfer from the ALB into IR. The first execution (perform) of the ALB function is controlled by the initial setting of the three toggles BAA, BAB, BDB, and by the C7 bit of the EO. After each execution, the three toggles and the WDB are modified as specified in the C field; that is, they are not automatically complemented as in set and perform where F = 001. Each execution is controlled by the new states of the three toggle switches.

The timing of the various functions is such that an operand can be read from the SPM or any other source register into the DR or UR, and the 8-bit result of the ALB function using the operand can be set into the IR, all in a single EO time interval. If a second execution of the ALB function is specified by V(3) = 1, no additional DB transfer is made, but the operands already in the DR and UR are used. The second execution of the function therefore requires one-half the usual EO time interval (TP2 and TP3). Hence a 16-bit result of the set-and-perform-twice operation can be set into the IR in a total of 1.5 EO-time or 720 nanoseconds (TP1, TP2, TP3, TP4, TP3, TP4).

One operand of the ALB function is the byte of the UR selected by the BAB toggle:

BAB = 0 selects UR0
BAB = 1 selects UR1

The other operand is the byte of the DR selected by C7 and the BAA toggle:

C7 = 0 BAA = 0 select D0
C7 = 0 BAA = 1 select D1

C7 = 1 BAA = 0 select D2
C7 = 1 BAA = 1 select D3

The result is set into the byte of the IR selected by the BAA toggle:

BAA = 0 select IR0
BAA = 1 select IR1

The transfer path from ALB to IR is further defined by the V field of the EO and the BDB toggle. This transfer is shown in Appendix E.

Note the following rules which govern the toggle setting:

a. In a set, or set and perform, the only allowable mnemonics are S for set, R for reset, or blank. All the toggles must be set, reset, or unchanged. No arbitrary mix is allowed.
b. In the set function any toggle set in that EO is still set at the end of the EO.
c. In a set and perform, any toggle that has an entry (set or reset) will be complemented at the end of the EO during the TP4 timing point.
d. Also, during any perform EO, all toggles that have an entry will be complemented during TP4.

Appendix F gives a list of all C field columns and their valid mnemonics.

Before reviewing the remaining F action implied by F = 3 through F = 6, we will review the F = 111 set function operation.

4. *F = 111—Set Function (SETF).* This EO sets the ALB function control flip-flops to perform the function specified by the V field, and sets the toggles specified by the C field to the state of C0, but does not execute (perform) this function. The other fields have their usual meaning.

The V field designates the setting of the ALB function control flip-flops as follows:

F, V = 70	The function does not change.
F, V = 71	*BCD (decimal) add absolute*, set ICAR if IR03 = 1. The EDR, BCD, RZ flip-flops are set. The six decimal fill method described in Chapter 8 (used with the IBM System 360 Model 50) is used.
F, V = 72	*Binary address add* The EDR and RZ flip-flops are set, and all others are reset. This prepares the ALB and the address adder to perform a 10-bit binary addition of the selected portions of the DR and UR.
F, V = 73	*Binary address subtract* EDR, COMP, ICAR and RZ flip-flops are set, and all others are reset. This operation is done by the two's complement arithmetic.
F, V = 74	*Transfer UR into IR* The RZ flip-flop is set, and all others are reset. This micro-order prepares the ALB and the address adder to perform a

10-bit transfer of the selected UR register operand. The transfer is performed by binary addition of the UR operand to all zero DR operands.

F, V = 75 *Transfer two's complement of UR into IR*
Similar to F, V = 74, except that the COMP and ICAR triggers are set to permit the two's complement of the UR to be added to all zero DR operands.

F, V = 76 *BCD add absolute*
EDR, BCD, and RZ flip-flops are set, and all others are reset. The ALB is prepared to perform an 8-bit decimal addition of the selected DR and UR operands.

F, V = 77 *BCD subtract absolute*
The EDR, COMP, BCD, ICAR, and RZ flip-flops are set, and all others are reset. This prepares the ALB to perform an 8-bit decimal subtraction of the selected UR operand from the selected DR operands.

F, V = 78 *The function does not change.*

F, V = 79 *Logical OR*
The EDR, ELOR, SIMC, COMP, ICAR, and RZ flip-flops are set, and all others are reset. This prepares the ALB to perform an 8-bit logical OR of the selected UR and DR operands, and prepares the address adder to perform a 2-bit binary subtract of the GR operand from the DB operands.

F,V = 7A *Logical AND*
The EDR, ELAND, SIMC, ICAR, and RZ flip-flops are set, and all others are reset. This function prepares the ALB to perform an 8-bit logical AND of the selected UR and DR operands, and prepares the address adder to perform a 2-bit binary addition of the GR and DR operands with a forced carry from the ALB.

F,V = 7B *Exclusive OR*
The EDR, SIMC, COMP, ICAR, and RZ flip-flops are set, and all others are reset. This prepares the ALB to perform an 8-bit exclusive OR of the selected UR and DR operands, and prepares the address adder to perform a 2-bit binary subtraction of the GR operand from the DR operands.

F,V = 7C *Binary Subtract with ICAR set*
The EDR, COMP, and RZ flip-flops are set, and all others are reset. This prepares the ALB and the address adder to perform a 10-bit binary subtraction of the UR from the DR without the usual initial carry. This is equivalent to adding the one's complement of UR to the DR.

F,V = 7D, 7E, or 7F are not assigned.

The following three examples will serve to review and establish familiarity with the data movements and functional control field specifications. After the problem is stated, the microprogram prepared on the MIDAS input form is given and explained. Note that these examples contain micro-orders in fields such as the TNA and E, which are not yet covered. Their inclusion will serve as an introduction to the remaining ROM fields to be discussed immediately after the examples.

Example 1: Branch on NOT ZERO (Fig. 9.13)

PROGRAM $\overline{\text{ZERO}}$ BRANCH (Branch on 1) PROGRAMMER DATE

If (FFRI) − 1 ≠ 0; FFRZ → FPC
Else FPC = Next Instruction

LABEL 1	I/SP 7	TEST 8	A 14	N 20	M 26	FROM S 28	TO D 34	F 40	V 43	C7 / C 48 (WDB, BDB, BAA, BAB, R, G, C1, C0)	EXCEPT E 55	CONST. 60
NZB			NZB1				UR	CON	DB1			01
NZB1			NZB2			FFR1		SP2	BSUB	S S S		
NZB2		RZ=1	000000	NZB3							ENDIN	
NZB3			NZB4			FFR2						
NZB4		TRUE	000000			DR	FPC				ENDIN	

Figure 9.13 Example 1: $\overline{\text{ZERO}}$ branch microroutine.

If the contents of FFR1 minus 1 are not equal to zero, store the contents of FFR2 in FPC; otherwise FPC = the next instruction.

The five EO's needed to implement this operation are given the labels NZB, NZB1, ..., NZB4.

NZB loads the specified constant (01 hex) on to the DB1 bus, which is in turn gated into the UR register. (Remember that the UR is first cleared; hence, at the end of this operation UR = 0001 hex.) This EO specifies NZB1 as its successor.

NZB1 The F, V fields specify a set and perform binary subtract twice. The C7, BAB, and BAA are set. The contents of FFR1 (specified by S field) are automatically transmitted into DR2,3, and in this and the following EO cycle, the binary subtraction is performed. The RZ (result = zero) trigger is set to reflect the output. This EO specifies NZB2 as its successor.

NZB2 This EO tests if RZ = 1; if the result is true, we terminate this operation and branch to instruction fetch routine (hardware address generation); otherwise, we branch to NZB3.

NZB3 This EO brings the contents of FFR2 and places it into the DR register. The next address specified is NZB4.

NZB4 Stores the contents of the DR register into the FPC and terminates the routine by transferring control to the instruction fetch routine (staticizer routine).

Example 2: Byte move (Fig. 9.14)

PROGRAM Byte Move Microroutine PROGRAMMER DATE

LABEL 1	I/SP 7	TEST 8	A 14	N 20	M 26	FROM S 28	TO D 34	F 40	V 43	C7/WDB	C 48 (BDB, BAA, BAR, R, G, C1, C0)	EXCEPT E 55	CONST. 60
MOVE	I		MOVE1			FFR1	MARW	INC	+1	S			
MOVE1			MOVE2			MMR	UR	SP1	TR		S G−1		
MOVE2	I		MOVE3			FFR2	MAR	INC	+1	S			
MOVE3	I		MOVE4			IR	MMR1	NOP					
MOVE4		G=0	0000000	MOVE				NOP				ENDIN	

Figure 9.14 Byte move microroutine.

Move number of characters specified in G (assuming G is set to some count) from memory as addressed by FFR1 to memory as addressed by FFR2. This is a memory-to-memory transfer with byte control.

The first EO labeled MOVE transfers the contents of the SPM register FFR1 and sets the leftmost two bytes of FFR1 into DR2,3(C7 = 1), increments this address by +1 and regenerates it into the FFR1 location, transfers the 18-bit address to MMAR, and issues a memory request with a wait cycle. The B in the C field would transfer the zero bit of MAR to BAB trigger to control the UR switch. The I in the I field indicates an I/O inhibit cycle. In this case, since MARW microcommand is used, the I would affect two EO cycles. The function of the third cycle is documented in the EO labeled MOVE1. The MMR contains the two-byte word brought from MM. The MMR contents are transferred to the UR. The ALB function triggers are set and perform one ALB function "Transfer" (UR to IR, 8 bits). The byte transferred is determined by the zero bit of the MAR register (even byte if zero, odd byte if one). Note that, since we incremented the MM address stored in FFR1 by 1, the next time we go through this loop we will select the second UR byte. The G counter is decremented by 1 and an unconditional branch is made to EO labeled MOVE2.

The MOVE2 EO transfers the contents of the SPM word FFR2 (containing the MM move to address) to the DR2,3 and on to the MAR register. Thus, a second MM request has been initiated, and the I inhibit trigger is set. The contents of the FFR2 are incremented by 1 and regenerated into FFR2. An unconditional branch is made to the second EO cycle in the MM write cycle labeled MOVE3.

In MOVE3, the contents of IR1 are stored in MMR and regenerated in the byte position in the memory specified by the MAR zero bit. Again, an unconditional branch is made to MOVE4.

In MOVE4, a test G = 0 is made. If the test failed, a branch to MOVE is taken; otherwise, a branch to the instruction fetch microroutine is taken.

Exercise: Write a move character-by-character microroutine taking advantage of a two-byte write capability for the two cases.

1. In the properly aligned case
2. In the nonaligned case.

Again, assume in both cases that the G counter is set to the number of bytes to be moved, and that the FFR1 and FFR2 contain the initial addresses of the two main memory blocks.

Example 3: Tally routine (Fig. 9.15)

PROGRAM Tally Microroutine PROGRAMMER DATE

LABEL (1)	I/SP (7)	TEST (8)	A (14)	N (20)	M (26)	FROM S (28)	TO D (34)	F (40)	V (43)	C7/WDB	C (48) BDB	BAA	BAB	R	G	C1	C0	EXCEPT E (55)	CONST. (60)
TALLY	I		TALLY1			FFR1	MAR	INC	+0	S									
TALLY1	I		TALLY2				UR	CON	DB1	S									01
TALLY2			TALLY3			MMR	DR	SP2	BSUB	S		S	S						
TALLY3		RZ≠1	TALLY6	TALLY4				NOP											
TALLY4	I			TALLY5		FFR2		INC	+0	S									
TALLY5				TALLY6		DR	FPC	NOP		S									
TALLY6	I			TALLY7		FFRJ	MAR	NOP		S									
TALLY7	I			TALLY8		IR	MMR	NOP											
TALLY8		TRUE	000000															ENDIN	

Figure 9.15 Tally microroutine.

The following tally routine subtracts a 1 from memory as addressed by R1 and branches to address in R2 when R1 reaches zero. Let us analyze the following microroutine.

The first EO labeled TALLY transfers the contents of FFR1 in main storage into DR2,3, then on to MAR. The address is regenerated into FFR1 unchanged. An MM cycle request is initiated, and the I/O inhibit trigger is set. An unconditional branch to TALLY1 is made by the end of this EO.

TALLY1—EO zeros the UR register, loads DB1 with the specified constant "01" hex, and then gates the data bus into UR. At this point, UR is equal to "0001" hex. Since this is only the second EO cycle of the MMR cycle, the I/O inhibit trigger is set. An unconditional transfer to TALLY2 is made.

TALLY2 specifies that the contents of the MMR is to be transmitted to DR2,3. The BAB and BAA are set to control the ALB data transfer. The ALB function triggers are conditioned to set and perform twice the ALB function "Binary Subtract." The setting of BAB and BAA and their automatic complement by the end of the first perform cycle insures that the subtraction is carried out systematically (the two low-order bytes in the first perform cycle, then the two high-order bytes). Note that this subtraction is carried out by adding the two's complement of UR to the true DR2,3. At the end of this EO an unconditional branch to TALLY3 is executed.

In TALLY3, if the result of the binary subtraction is not zero—that is, $RZ \neq 1$, test is false, branch to TALLY6; if $RZ = 1$ is true, branch to TALLY4.

In TALLY4, the contents of FFR2 in SPM are transferred to DR2,3, then regenerated into FFR2 without any modifications. The I/O inhibit trigger is set. A nonconditional branch to TALLY5 is then executed.

TALLY5 simply stores this address (in DR2,3) into the program counter register (FPC) in SPM, and transfers control to EO TALLY6.

In TALLY6 we find that the contents of FFR1 are again transferred to MAR via the DR2,3; the I/O inhibit trigger is set; DR2,3 is regenerated into FFR1 without any modifications, and finally an unconditional branch to TALLY7 is executed.

TALLY7 forms the second EO cycle of the MM write cycle. The contents of IR, that is, the result of the binary subtract operation, are set into the MMR to be generated (written) into the specified MM location. The I inhibit trigger is set, and an unconditional branch is made to TALLY8.

TALLY8 forces the system to transfer unconditionally to the beginning of the I fetch routine.

We have thus far discussed the function control field action implied in F = 0, 1, 2, and 7 (no action, set and perform, perform, and set function). We will continue this section with the other actions implied in the remaining groups of F classes of functions.

5. *F = 011, V3 = 0—Merge and Increment.* This EO inserts the byte from DB0 into DR0 and/or the byte from DB1 into DR1, or increments or decrements the contents of the DR and stores the result in the SPM. The particular operation is specified by the V field.

The RR and GR counters are not affected by the C field, but they may be specified as destination registers.

The WDB toggle is not changed. Toggles BDB, BAA, and BAB as specified in the C field are set to the value of the DB(10).

For the merge operation (V = 0X000), the DR need not be specified as the destination register. The result of a merge is left in the DR. The timing is such that a word is read out from SPM, or one or two bytes may be merged into that word, and the resulting word is stored in the SPM.

For an increment operation (V = 0X01 or 0X10), the word may be loaded entirely from the SPM, or partly from the MM and partly from any source register as specified by the M, S, and D fields before incrementing, or the contents of the DR may be held over from previous cycles. The contents of the DR are incremented as specified by the V field and stored in the SPM.

The V and C fields have the following interpretations:

V = 0000 merge as specified by C1, C0, increment by 0
V = 0001 merge as specified by C1, C0, increment by 1
V = 0010 merge as specified by C1, C0, increment by 2
V = 0011 merge as specified by C1, C0, increment not defined
V = 0100 merge as specified by C1, C0, increment by 0
V = 0101 merge as specified by C1, C0, increment by 1
V = 0110 merge as specified by C1, C0, increment by 2
V = 0111 merge as specified by C1, C0, increment not defined

C7 = 0 read from DB into DR0 and DR1 if RIDR and C1, C0 = 00
1 read from DB into DR2 and DR3 if RIDR and Cl, C0 = 00
(in either case, the state of WDB is not changed)

C6 = 0 do not change DBD
1 set BDB accordingly to DB10

C5 = 0 do not change BAA
1 set BAA according to DB10

C4 = 0 do not change BAB
1 set BAB according to DB10

C3, C2 ignored

C1, C0 = 00 set DR from bus according to C7 if RIDR
01 clear and set DR0 from DB0
10 clear and set DR1 from DB1
11 clear and set DR0 from DB0 and DR1 from DB1

Appendix E gives a listing of the merge/increment micro-orders.

Example 1

Assuming FR1 (in the SPM) contains the value "ABCD/EF12" hex, what will be the contents of DR after the following EO has been executed:

α1: FR1(s); MERGE(F); DB01; C7 = 1

The answer is that after this EO is executed, the FR1 and DR will have EF12EF12.

Example 2

If register UR contains "1234" hex, what will be the value of FR1 after executing the following EO:

α2: UR(s); FR1(D); MERGE(F); DB0(V)

The answer is FR1 = 1200 0000. How do you justify the answer?

6. *F = 011, V3 = 1 Generate Constant.* This EO transmits the byte consisting of V2, C6, C5, C4, C3, C2, C1, C0 onto DB1 and/or DB0 as specified by V1 and V0. Counters and toggles are not changed unless they are specified as destination registers. If the source register specified in the S field and the constant are transmitted on the same bus, the logical "OR" will occur on the same bus.

C7, E, I, and the remaining EO fields have the usual meaning. The V field controls the transmission of the constant on the DB. V2 is the most significant bit of the constant.

V =	0XXX	Do not transmit constant to DB (see M/INC.)
	1X000	Do not transmit constant to either DB
	1X01	Transmit constant to DB0 only
	1X10	Transmit constant DB1 only
	1X11	Transmit constant to both DB0 and DB1

Table 9.2 shows the three basic micro-orders which fall in this area of F functions. These mnemonics are also shown in Appendix E.

Table 9.2. GENERATE CONSTANT

	Bit Patterns		F = 3 · V3(1) (G CON)
	F Field	*V Field*	
Mnemonic	*2 1 0*	*3 2 1 0*	*Function and Remarks*
CON DB0 (YY)	0 1 1	1 X 0 1	Generate constant $(YY)_{16}$ on bus 0 $(YY)_{16} = (V_2C_6C_5C_4C_3C_2C_1C_0)_2$
CON DB1 (YY)	0 1 1	1 X 1 0	Generate constant $(YY)_{16}$ on bus 1 $(YY)_{16} = (V_2C_6C_5C_4C_3C_2C_1C_0)_2$
CON DB01 (YY)	0 1 1	1 X 1 0	Generate constant $(YY)_{16}$ on bus 0 and bus 1 $(YY)_{16} = (V_2C_6C_5C_4C_3C_2C_1C_0)_2$

A number of microprogramming examples have already used these micro-orders; hence we will continue with shift control.

7. *F = 100 Shift.* This field shifts the contents of the DR one place left or right into the SPM. The shift is specified in the V field; all other fields have the same meaning. The DR is loaded as usual from the SPM or any other source register, partially or entirely before shifting it. If no SPM cycle is specified, no shifting occurs.

The V field designates the shift to be performed as follows:

			Lost Bit Stored in	*Vacated Bit Filled from*
V =	0000	Left Logical	—	BR2
	0001	Left Logical	BR0	BR2
	0010	Left Logical	BR2	0

		Lost Bit Stored in	*Vacated Bit Filled from*
0011	Left Logical	BR2	BR0
0100	Left Algebraic*	—	BR2
0101	Left Algebraic	BR0	BR0
0110	Left Algebraic	BR2	IR07 ≠ URS†
0111	Left Algebraic	BR2	IR07 = URS†
1000	Right Logical	—	BR2
1001	Right Logical	BR0	BR2
1010	Right Logical	BR2	0
1011	Right Logical	BR2	BR0
1100	Right Algebraic	—	BR2
1101	Right Algebraic	BR0	BR2
1110	Right Algebraic	BR2	DR07†
1111	Right Algebraic	BR2	DR07 ≠ OV†

*A left algebraic shift will set the overflow (OV) if the two most-significant bits of the DR are not equal (DR07 ≠ DR06); this EO does not reset OV. Resetting OV can be done with a set function or a set and perform function.

†*Note:* A 1 is supplied if the boolean expression is true; a 0 is supplied if the boolean expression is false.

Appendix E gives the 16 micro-orders designed to perform the shift control.

8. *F = 101 Test and Branch.* This has a very limited usage. It was designed to handle a few special cases. The EO makes three successive tests and selects the next EO according to the first test condition that is true. The three test conditions are specified in the T, D, and S fields, and are executed in that order. If the condition specified by the T field is true, the next EO is selected from the ROM location specified by the A field. If the T condition is false but the D condition is true, the next ROM address is selected from the N field. If both the T and D conditions are false, but the S field test condition is true, the next ROM address comes from the C and V fields. In this case, the C provides the upper 8 bits whereas the V provides the lower 4 bits of the 12-bit ROM address field. If all three conditions are false, the next EO ROM address is obtained by inserting a binary 1 in the least-significant bit position of the current ROM address without changing any other bit.

These conditions and the types of tests performed are summarized in Appendix E.

9. *F = 110 Set Condition Code.* This EO may or may not set the condition code represented by the two bits SREG3 and SREG4. The two bits may be set directly from the EO, or as a complex function of the contents of the OR and several other machine indications. The variations are specified in the V field.

V3 bit specifies whether or not the condition code will be set as follows:

V3 = 0 Set condition code
V3 = 1 Do not set condition code

The V2, V1, and V0 bits define the way in which the condition code is to be set if V3 = 0.

V = 00XX Set condition code as a function of OR, IRO7, UR07, URS, SCAR, OV, RZ, STR
V = 0100 Set condition code to 00
V = 0101 Set condition code to 01
V = 0110 Set condition code to 10
V = 0111 Set condition code to 11
V = 1XXX Condition code is not changed.

Appendix E shows the classes of the machine operations and the conditions in each class under which the condition code is set.

9.5.3 TNA-Branch Control Fields

It was stated earlier that the TNA are test and branch fields, which permit the microprogrammer to specify one of 63 different tests representing hardware or flip-flop conditions that can be tested. The tests are encoded in the 6-bit T field.

The N and the A fields are also 6 bits wide. If the test condition is false or if no test condition is specified, the next EO is taken from the address designated in N. If the test condition is true, the next EO is taken from the specified address in A. In both cases, these fields provide only the low-order 6 bits of the 12-bit ROMAR. This means branching is constrained to a block of 64 ROM words.

Another branch option can be specified in the E field. The specific micro-order therein will cause the A and the N fields to provide a 12-bit jump address. This gives the flexibility to jump anywhere in ROM.

The conditional branch is implemented in the system as follows:

A: If T-true: A (5–0) → ROMAR (5–0). ROMAR (11–6)
ROMAR (11–6) remains unchanged.
N: If F $\neq$ 5 and T-false/F = 5. T-false. D-true
N (5–0) → ROMAR (5–0). ROMAR (11–6)
ROMAR (11–6) remains unchanged.

Note that when M = 0/1, the T field cannot select a register or an indicator to be tested if the register or indicator is selected by the D field.

Appendix G gives the specific tests and the conditions associated with each. The format is self-explanatory.

In the previous exercises we have seen a number of test and branch conditions and conditional test options. The reader should familiarize himself with the 63 test and branch conditions encoded in the T field and specified in Appendix G.

9.5.4 Exception Check Control—E Field

The E field for the most part is concerned with various exception conditions in the Spectra 70. There are many data and programming exceptions, conditions that may be considered errors or simply indicators to the program and may result in an interrupt. They include overflow, underflow, and data not being in the correct format or not being in the correct memory location. Other conditions tested include the correct decimal code. The E field is also a catchall for all other miscellaneous functions which do not fit in any other EO field, such as the conditional jump.

We have seen how the ENDIN micro-order which is specified in the E field forces an unconditional jump to the beginning of the instruction fetch routine. The ENDIN micro-order is used at the end of the microroutine and will automatically jump back to the appropriate (hardware-generated) address in the ROM to fetch the next instruction. Another unconditional jump ENDST can be used at the end of the instruction fetch to cause a 256-way branch based on the Op code in OR. Appendix H gives a listing of all the exception memories available.

Appendix I summarizes the FM (fast memory) field settings. Appendix J provides a further explanation of, and the mnemonics for, the T, A, N, E field variations. The last, Appendix K, gives a summary of the test and branch EO's discussed earlier in the chapter.

APPENDIX 9A: SCRATCH PAD LOCATION MNEMONICS

Mnemonic	*S Field Bit Pattern 543210*	*Fast Memory Address*						
		6	*5*	*4*	*3*	*2*	*1*	*0*
FB	010000	(PS2)	(PS1)	0	(B3)	(B2)	(B1)	(B0)
FB + 1	010001	(PS2)	(PS1)	0	(B3)	(B2)	(B1)	1
FR1	010010	(PS2)	(PS1)	0	(R13)	(R12)	(R11)	(R10)
FR1 + 1	010011	(PS2)	(PS1)	0	(R13)	(R12)	(R11)	1
FR2	010100	(PS2)	(PS1)	0	(R23)	(R22)	(R21)	(R20)
FR2 + 1	010101	(PS2)	(PS1)	0	(R23)	(R22)	(R21)	1
FG	010110	(G7)	(G6)	(G5)	(G4)	(G3)	(G2)	(G1)
Note: For FG – C7 = 1 if (G0) = 1; C7 = C7 if (G0) = 0								
FG + 1	010111	(G7)	(G6)	(G5)	(G4)	(G3)	(G2)	1
Note: For FG + 1 – C7 = 1 if (G0) = 1; C7 = C7 if (G0) = 0								
FUT9	011000	(PS2)	(PS1)	1	1	0	0	1
FFR1	011010	1	1	1	(R13)	(R12)	(R11)	(R10)
FFR1 + 1	011011	1	1	1	(R13)	(R12)	(R11)	1
FFR2	011100	1	1	1	(R23)	(R22)	(R21)	(R20)
FFR2 + 1	011101	1	1	1	(R23)	(R22)	(R21)	1
FR	011110	0	0	0	0	0	(R23)	(R22)
Note: For FR – C7 = 1 if (R21) = 1; C7 = C7 if (R21) = 0								
FR + 1	011111	0	0	0	0	0	(R23)	1
Note: For FR + 1 – (C7) = 1 if (R21) = 1; (C7) = ROM C7 if (R21) = 0								
F1	001000	0	0	0	0	0	0	0
F2	001001	0	0	0	0	0	0	1
F3	001010	0	0	0	0	0	1	0
F4	001011	0	0	0	0	0	1	1
F5	001100	0	0	0	0	1	0	0
F6	001101	0	0	0	0	1	0	1
F7	001110	0	0	0	0	1	1	0
F8	001111	0	0	0	0	1	1	1
FUT10	000000	(PS2)	$\overline{(PS2)}$	1	0	0	0	0
FUT11	000001	(PS2)	$\overline{(PS2)}$	1	0	0	0	1
FUT8	000010	(PS2)	(PS1)	1	1	0	0	0
FIFR	000011	0	1	0	0	0	1	1
FIM	000100	0	K	0	$\overline{(PS2)}$	$\overline{(PS1)}$	0	0
Note: For FIM – K = 1 if (PS2) or (PS1) = 1								
FIS	000101	0	K	0	$\overline{(PS2)}$	$\overline{(PS1)}$	0	1
Note: For FIS – K = 1 if (PS2) or (PS1) = 1								
FPC	000110	0	K	0	$\overline{(PS2)}$	$\overline{(PS1)}$	1	0
Note: For FPC – K = 1 if (PS2) or (PS1) = 1								
FIW	000111	0	Y	0	1	1	1	1
Note: For FIW – Y = 1 if not PS4 Interrupt								

APPENDIX 9B: SOURCE MNEMONICS

Mnemonic	*S or D Field Bit Pattern 543210*	*Source Register*	*Transfer Bus*
B	010001	B (3–0)	Bus 0 (7–4)
BIR	010010	2 bits from B and Full IR	
		B (1–0)	Bus A (1–0)
		IR0 (7–0)	Bus 0 (7–0)
		IR1 (7–0)	Bus 1 (7–0)
B:KAP	011110	B (3–0)	Bus 0 (3–0)
		K (7–4)	Bus 1 (7–4)
		A	Bus 13
		EON	Bus 12
		BK3	Bus 11
		PRIV	Bus 10
CFF1	011001	Control flip-flops set 1	
		BCD	Bus 07
		ICAR	Bus 06
		RZ	Bus 05
		SCAR	Bus 04
		MARA (1–0)	Bus 1 (1–0)
CFF2	011010	Control flip-flops set 2	
		COMP	Bus 07
		SIMC	Bus 06
		ELOR	Bus 05
		OV	Bus 04
		MREGAD (11–8)	Bus 0 (3–0)
		MREGAD (7–0)	Bus 1 (7–0)
CFF3	011011	Control flip-flops set 3	
		BAA	Bus 07
		BAB	Bus 06
		BDB	Bus 05
		UDB	Bus 04
		US	Bus 03
		DS	Bus 02
		EDR	Bus 01
		ELAND	Bus 00
		FMMAR (6–0)	Bus 1 (6–0)
DR	001011	If C7 = 0, DR01	
		DR0 (7–0)	Bus 0 (7–0)
		DR1 (7–0)	Bus 1 (7–0)
		If C7 = 1, DR23	
		DR2 (7–0)	Bus 0 (7–0)
		DR3 (7–0)	Bus 1 (7–0)
GA	001110	G (7–0)	Bus 0 (7–0)

APPENDIX 9B: SOURCE MNEMONICS (Continued)

Mnemonic	*S or D Field Bit Pattern 543210*	*Source Register*	*Transfer Bus*
GB	001111	G (7–0)	Bus 1 (7–0)
INT1	010100	Interrupt register, lower half	
		INT (32–25)	Bus 0 (7–0)
		INT (24–17)	Bus 1 (7–0)
INT2	010101	Interrupt register, upper half	
		INT (16–09)	Bus 0 (7–0)
		INT (08–01)	Bus 1 (7–0)
IR	001010	Full IR	
		IR0 (7–0)	Bus 0 (7–0)
		IR1 (7–0)	Bus 1 (7–0)
IR1	010110	IR1 (7–0)	Bus 1 (7–0)
MACHE	011111	Machine error and IR	
		IR0 (7–0)	Bus 0 (7–0)
		MRPE	Bus 01
		DRPE	Bus 00
		IR1 (7–0)	Bus 1 (7–0)
MAR	011000	MAR0 (7–0)	Bus 0 (7–0)
		MAR1 (7–0)	Bus 1 (7–0)
MMR	001100	MMR	
		MR0 (8–0)	Bus 0 (8–0)
		MR1 (8–0)	Bus 1 (8–0)
MPMR	100000	Memory protect register	
		MPMR (3–0)	Bus 1 (7–4)
OR:R	011100	OR (7–0)	Bus 0 (7–0)
		R1 (3–0)	Bus 1 (7–4)
		R2 (3–0)	Bus 1 (3–0)
PS	001101	PS (2–1)	Bus 0 (6–5)
R	001001	Full R register	
		R1 (3–0)	Bus 1 (7–4)
		R2 (3–0)	Bus 1 (3–0)
R1	000111	R1 (3–0)	Bus 1 (3–0)
R2	001000	R2 (3–0)	Bus 1 (3–0)
S	010011	S (7–0)	Bus 0 (7–0)
S:G	011101	S and G registers	
		S (7–0)	Bus 0 (7–0)
		G (7–0)	Bus 1 (7–0)
UR	010000	Full UR	
		UR0 (7–0)	Bus 0 (7–0)
		UR1 (7–0)	Bus 1 (7–0)

APPENDIX 9C: DESTINATION MNEMONICS

Mnemonic	*D Field Bit Pattern 543210*	*Destination Register*	*Transfer Bus or Data*
BUR	010010	Full B reg. and lower 12 bits of UR	
		B (3–0)	Bus 0 (7–4)
		UR0 (3–0)	Bus 0 (3–0)
		UR1 (7–0)	Bus 1 (7–0)
		UR0 (7–4)	0000
DR	001000	If C7 = 0, DR01	
		DR0 (7–0)	Bus 0 (7–0)
		DR1 (7–0)	Bus 1 (7–0)
		If C7 = 1, DR23	
		DR2 (7–0)	Bus 0 (7–0)
		DR3 (7–0)	Bus 1 (7–0)
G	011000	G (7–0)	Bus 1 (7–0)
GUR	010011	Full UR, lower 2 bits of G	
		UR0 (7–0)	Bus 0 (7–0)
		UR1 (7–0)	Bus 1 (7–0)
		G (1–0)	Bus A (1–0)
		G (7–2)	000000
IR	001110	Full IR register	
		IR0 (7–0)	Bus 0 (7–0)
		IR1 (7–0)	Bus 1 (7–0)
IRDS	001101	IR register and DR sign reg.	
		IR0 (7–0)	Bus 0 (7–0)
		IR1 (7–0)	Bus 1 (7–0)
		If (E Field ≠ 06/07), DS	Bus 07
		If (E Field = 06/07) · [DRX (3–0) = 3/5/B/D], DS	1
IR, DR	001100	16 bits to both DR and IR	
		If C7 = 0: DR0 (7–0) and IR0 (7–0)	Bus 0 (7–0)
		DR1 (7–0) and IR1 (7–0)	Bus 1 (7–0)
		If C7 = 1: DR2 (7–0) and IR0 (7–0)	Bus 0 (7–0)
		DR3 (7–0) and IR1 (7–0)	Bus 1 (7–0)
KAP	011011	K (7–4)	Bus 1 (7–4)
		A	Bus 13
		EON	Bus 12
		BK3, BK3A	Bus 11
		PRIV	Bus 10

APPENDIX 9C: DESTINATION MNEMONICS (Continued)

Mnemonic	*D Field Bit Pattern 543210*	*Destination Register*	*Transfer Bus or Data*
MAR	011100	Main memory address reg.; perform next EO immediately	
		MARA (1–0)	Bus A (1–0)
		MAR0 (7–0)	Bus 0 (7–0)
		MAR1 (7–0)	Bus 1 (7–0)
MARW	011101	Main memory address reg.; wait one EO time before performing next EO	
		MARA (1–0)	Bus A (1–0)
		MAR0 (7–0)	Bus 0 (7–0)
		MAR1 (7–0)	Bus 1 (7–0)
MMR	011111	Two byte to MMR	
		MR0 (8–0)	Bus 0 (8–0)
		MR1 (8–0)	Bus 1 (8–0)
MMR1	011110	Single byte to MR0 or MR1;	
		If MAR10 (0): MR0 (8–0)	Bus 1 (8–0)
		If MAR10 (1): MR1 (8–0)	Bus 1 (8–0)
MPMAR	100000	MPMAR (6–5)	Bus A (1–0)
		MPMAR (4–0)	Bus 0 (7–3)
MPMR	100001	MPMR (3–0)	Bus 1 (7–4)
OR:R	001011	OR and R registers	
		OR (7–0)	Bus 0 (7–0)
		R1 (3–0)	Bus 1 (7–4)
		R2 (3–0)	Bus 1 (3–0)
PS	001001	Program state registers;	
		If R20 (1): PS (2–1)	Bus 0 (6–5)
		If R20 (0): PS2	(R22)
		PS1	(R21)
R	001010	Full R register	
		R1 (3–0)	Bus 1 (7–4)
		R2 (3–0)	Bus 1 (3–0)
S	011011	S (7–0)	Bus 0 (7–0)
UR	010000	Full UR register	
		UR0 (7–0)	Bus 0 (7–0)
		UR1 (7–0)	Bus 1 (7–0)
UR, IR	001111	Full UR and IR registers	
		UR0 (7–0) and IR0 (7–0)	Bus 0 (7–0)
		UR1 (7–0) and IR1 (7–0)	Bus 1 (7–0)
URUS	010001	UR register and U sign register	
		UR0 (7–0)	Bus 0 (7–0)
		UR1 (7–0)	Bus 1 (7–0)
		If E Field $\neq$ 06/07: US	Bus 07
		If (E Field = 06/07) · [URX (3–0) = 3/5/B/D]: US	1

APPENDIX 9D: F FIELD MNEMONICS AND REQUIRED BIT SETTING

Mnemonic	*F*	*V*	*Remarks*
NOP	000	0000	No Function
SP1	001	0XXX	Set Function and Perform Once
SP2	001	1XXX	Set Function and Perform Twice
SET	111	XXXX	Set Function
PR1	010	0XXX	Perform Once
PR2	010	1XXX	Perform Twice
INC	011	0XXX	Increment = Set C_1C_0 to 00
CON	011	1XXX	Generate Constant onto Bus (Constant is $V_2C_6C_5C_4C_3C_2C_1C_0$)
MIN	011	0XXX	
MGE	011	0XXX	Merge and Increment
SLL	100	00XX	Shift Left Logical
SLA	100	01XX	Shift Left Algebraic
SRL	100	10XX	Shift Right Logical
SRA	100	11XX	Shift Right Algebraic
TBR	101	XXXX	Test and Branch
SCC	110	XXXX	Set Condition Code
			Note: V field positions shown as X will be set according to the contents of the subfunction field. (Columns 43–46.)

APPENDIX 9E: FV FIELD COMBINATIONS

Mnemonic	*Bit Patterns* *F*	*V*	*Function and Remarks*
CON DB0 (YY)	011	1X01	Generate constant $(YY)_{16}$ on bus 0 $(YY)_{16} = (V_2C_6C_5C_4C_3C_2C_1C_0)_2$
DB1 (YY)	011	1X10	Generate constant $(YY)_{16}$ on bus 1 $(YY)_{16} = (V_2C_6C_5C_4C_3C_2C_1C_0)_2$
DB01 (YY)	011	1X11	Generate constant $(YY)_{16}$ on bus 0 and bus 1 $(YY)_{16} = (V_2C_6C_5C_4C_3C_2C_1C_0)_2$
INC + 0	011	0000	Increment scratch pad contents by 0. $C_1C_0 = 00$
+ 1	011	0001	Increment scratch pad contents by 1. $C_1C_0 = 00$
+ 2	011	0010	Increment scratch pad contents by 2. $C_1C_0 = 00$
− 1	011	0101	Decrement scratch pad contents by 1. $C_1C_0 = 00$
− 2	011	0110	Decrement scratch pad contents by 2. $C_1C_0 = 00$
MGE DB0	011	0000	Merge bus 0 to DR0. $C_0 = 1$
DB1	011	0000	Merge bus 1 to DR1. $C_1 = 1$
DB01	011	0000	Merge bus 0 to DR0 and bus 1 to DR1. $C_1C_0 = 11$
MIN H + 0	011	0000	Merge bus 0 to DR0 and increment scratch pad contents by 0. $C_0 = 1$
H + 1	011	0001	Merge bus 0 to DR0 and increment scratch pad contents by 1. $C_0 = 1$
H + 2	011	0010	Merge bus 0 to DR0 and increment scratch pad contents by 2. $C_0 = 1$
H − 1	011	0101	Merge bus 0 to DR0 and decrement scratch pad contents by 1. $C_0 = 1$
H − 2	011	0110	Merge bus 0 to DR0 and decrement scratch pad contents by 2. $C_0 = 1$
L + 0	011	0000	Merge bus 1 to DR1 and increment scratch pad contents by 0. $C_1 = 1$
L + 1	011	0001	Merge bus 1 to DR1 and increment scratch pad contents by 1. $C_1 = 1$
L + 2	011	0010	Merge bus 1 to DR1 and increment scratch pad contents by 2. $C_1 = 1$
L − 1	011	0101	Merge bus 1 to DR1 and decrement scratch pad contents by 1. $C_1 = 1$
L − 2	011	0110	Merge bus 1 to DR1 and decrement scratch pad contents by 2. $C_1 = 1$
NOP	000	0000	No Function
PR1 BYTE	010	0000	Perform previously set ALB function once on 8 bits through ALB to IR.

APPENDIX 9E: FV FIELD COMBINATIONS (Continued)

Mnemonic	*Bit Patterns* *F*	*V*	*Function and Remarks*
B4	010	0100	Perform previously set ALB function once on 4 bits straight through ALB to IR.
B4NS	010	0011	Perform previously set ALB function once on 4 bits through ALB to left 4 bits of IR. Right 4 bits of IR from BCD resultant sign. Sign is inverse of that in PR1 B4S.
B4S	010	0010	Perform previously set ALB function once on 4 bits through ALB to left 4 bits of IR. Right 4 bits of IR from BCD resultant sign.
PR1 B4X	010	0110	Perform previously set ALB function once on 4 bits through ALB crossover to IR.
B4XZ	010	0111	Perform previously set ALB function once on 4 bits through ALB crossover to IR and 0000 to other 4 bits of IR.
B4Z	010	0101	Perform previously set ALB function once on 4 bits through ALB straight to IR and 0000 to other 4 bits of IR.
ZONE	010	0001	Perform previously set ALB function once on right 4 bits through ALB to IR. Left 4 bits of IR from zone generator.
PR2 BYTE	010	1000	Perform previously set ALB function twice on 8 bits through ALB to IR.
B4	010	1100	Perform previously set ALB function twice on 4 bits straight through ALB to IR.
B4NS	010	1011	Perform previously set ALB function twice on 4 bits through ALB to left 4 bits of IR. Right 4 bits of IR from BCD resultant sign. Sign is inverse of that in PR2B4S.
B4S	010	1010	Perform previously set ALB function twice on 4 bits through ALB to left 4 bits of IR. Right 4 bits of IR from BCD resultant sign.
B4X	010	1110	Perform previously set ALB function twice on 4 bits through ALB crossover to IR.
B4XZ	010	1111	Perform previously set ALB function twice on 4 bits through ALB crossover to IR and 0000 to other 4 bits of IR.
B4Z	010	1101	Perform previously set ALB function twice on 4 bits through ALB straight to IR and 0000 to other 4 bits of IR.

Mnemonic	Bit Patterns F	V	Function and Remarks
ZONE	010	1001	Perform previously set ALB function twice on right 4 bits through ALB to IR. Left 4 bits of IR from zone generator.
SCC NO	110	1000	Condition code does not change.
OR	110	0000	Set condition code as function of OR.
0	110	0100	Set condition code to 0.
1	110	0101	Set condition code to 1.
2	110	0110	Set condition code to 2.
3	110	0111	Set condition code to 3.
SET AND	111	1010	Set ALB function "Logical And"
BADD	111	0010	Set ALB function "Binary Add"
BSUB	111	0011	Set ALB function "Binary Subtract"
DADD	111	0110	Set ALB function "Decimal Add"
DSUB	111	0111	Set ALB function "Decimal Subtract"
EXOR	111	1011	Set ALB function "Exclusive Or"
OR	111	1001	Set ALB function "Logical Or"
SADD	111	0001	Set ALB function "Special Add"
SAME	111	1000	ALB function does not change
SET SSUB	111	1100	Set ALB function "Special Subtract"
TR	111	0100	Set ALB function "Transfer"
TRC	111	0101	Set ALB function "Transfer Complement"
SLA B2B0	100	0101	Shift Left Algebraic. Fill from B2. Shifted bit to B0.
B2L	100	0100	Shift Left Algebraic. Fill from B2. Shifted bit lost.
EQB2	100	0111	Shift Left Algebraic. Fill with 1 if US = IR07. Shifted bit to B2.
NEB2	100	0110	Shift Left Algebraic. Fill with 1 if US ≠ IR07. Shifted bit to B2.
SLL B0B2	100	0011	Shift Left Logical. Fill from B0. Shifted bit to B2.
B2B0	100	0001	Shift Left Logical. Fill from B2. Shifted bit to B0.
B2L	100	0000	Shift Left Logical. Fill from B2. Shifted bit lost.
ZB2	100	0010	Shift Left Logical. Fill with 0. Shifted bit to B2.
SP1 BADD	001	0010	Set and perform once ALB function "Binary Add"
BSUB	001	0011	Set and perform once ALB function "Binary Subtract"

Mnemonic	Bit Patterns F	V	Function and Remarks
DADD	001	0110	Set and perform once ALB function "Decimal Add"
DSUB	001	0111	Set and perform once ALB function "Decimal Subtract"
SADD	001	0001	Set and perform once ALB function "Special Add"
SP1 SAME	001	0000	Perform previously set ALB function once on 8 bits.
TR	001	0100	Set and perform once ALB function "Transfer" (UR to IR, 8 bits).
TRC	001	0101	Set and perform once ALB function "Transfer Complement" ($\overline{\text{UR}}$ to IR, 8 bits).
SP2 BADD	001	1010	Set and perform twice ALB function "Binary Add"
BSUB	001	1011	Set and perform twice ALB function "Binary Subtract"
DADD	001	1110	Set and perform twice ALB function "Decimal Add"
DSUB	001	1111	Set and perform twice ALB function "Decimal Subtract"
SADD	001	1001	Set and perform twice ALB function "Special Add"
SP2 SAME	001	1000	Perform previously set ALB function twice on 8 bits.
TR	001	1100	Set and perform twice ALB function "Transfer" (UR to IR, 8 bits).
TRC	001	1101	Set and perform twice ALB function "Transfer Complement" (UR to IR, 8 bits).
SRA B2B0	100	1101	Shift Right Algebraic. Fill from B2. Shifted bit to B0.
B2L	100	1100	Shift Right Algebraic. Fill from B2. Shifted bit lost.
DRB2	100	1110	Shift Right Algebraic. Fill from DR07. Shifted bit to B2.
SRA OVB2	100	1111	Shift Right Algebraic. Fill with 1 if DR07 $\neq$ OV; 0 if DR07 = OV. Shifted bit to B2.
SRL B0B2	100	1011	Shift Right Logical. Fill from B0. Shifted bit to B2.
B2B0	100	1001	Shift Right Logical. Fill from B2. Shifted bit to B0.
B2L	100	1000	Shift Right Logical. Fill from B2. Shifted bit lost.
ZB2	100	1010	Shift Right Logical. Fill with 0. Shifted bit to B2.

APPENDIX 9F: C FIELD (COLUMNS 47–54)

1. A breakdown of C field columns and their valid mnemonics is as follows:

(a) *WDB or C7**
R = set WDB or C7 to 0
S = set WDB or C7 to 1

(b) *BDB, BAA, BAB*
R = set to 0
S = set to 1
T = trigger
B = set to the value of DB10

The setting of BDB, BAA, and BAB is controlled by the same bit in the C field (C0). Therefore, if one of these three columns has a mnemonic, the other two columns must have either the same mnemonic or the blank. For example, if an R appears in the column for BAA, either an R or blank must appear in the columns for BDB and BAB.

In addition, if an S (set to 1) appears in either of these columns (BDB, BAA, BAB) with an F field setting of set and perform (F = 001), this indicates that the specified counters should be set *and* triggered. The mnemonic B may only be used if the function is increment or merge (F = 011).

(c) *R Column* (C3)
S = Set R2 to all ones [f (0R)]
T − 1 = Decrement R register by −1 [f (0R)]
T − 2 = Decrement R register by −2 [f (0R)]
T − 4 = Decrement R register by −4 [f (0R)]

The programmer should be extremely cautious in triggering the R register since the trigger of R is a function of the 0R register.

A trigger can be generated during a Set Condition Code EO (SCC), Shift EO, Set and Perform EO, or a Perform EO. The C3 bit selects R as the register to be triggered and the C1C0 bits select the value to be decremented. The 0R register determines if R is to be treated as an 8-bit counter or two 4-bit counters R1 and R2. Table 9F.1 shows the various combinations of the 0R register and their effects on the R trigger. It should be noted that the mnemonics R and B are invalid for the R column (C3).

*See notes at the end of this appendix.

Table 9F.1. R TRIGGER CONTROL

C3–0 Mnemonic	*OR7–5 = 110*	*OR7–5 = 111*	*OR7–6 = 11*
TΔ-0	R1(– 1) R2 = F(1111)	R1(– 1) R2 = F(1111)	R2 = F(1111) R1 – unchanged
TΔ-1	R(– 1)	R1 (– 1) R2 (– 1)	R2(– 1) R1 – unchanged
TΔ-2	R(– 2)	R1(– 2) R2(– 2)	R2(– 2) R1 – unchanged
TΔ-4	R (– 4)	R1 (– 4) R2 (– 4)	R2 (– 4) R1 – unchanged

(d) *G Column*

Modification of the G register is a function of C2, C1, C0, and the F field.

C2–0 Mnemonic	*Effect on G register*
RΔΔ	Set G register to a value of 0
T-1	Trigger G register by –1
T-2	Trigger G register by –2
T-4	Trigger G register by –4

2. Table 9F.2 shows the allowable C field combinations associated with the particular function shown. The race condition subroutine checks for proper C field combinations.

Table 9F.2. ALLOWABLE CODING FOR C FIELD

Function	*C7*	*C6 BDB*	*C5 BAA*	*C4 BAB*	*C3 R*	*C2 G*	
NOP	S/R	—	—	—	—	—	
SP*	S/R	T/S/R	T/S/R	T/S/R	—	—	
PR	S/R	T	T	T	T/S	T/R	
MIN/MGE/INC	S/R	B	B	B	—	—	
CON	S/R	—	—	—	—	—	
SHIFT	S/R	T	T	T	T/S	T/R	(Same as PR)
TBR	—	—	—	—	—	—	
SCC	S/R	T	T	T	T/S	T/R	(Same as PR)
SET	S/R	S/R	S/R	S/R	—	R	

*Either S or R in a Set and Perform function causes a trigger after the function is executed.

3. The assembler should set the following bits:

WDB or C7	*BDB*	*BAA*	*BAB*	*R*	*G*	*C1*	*C0*	*Assembler Action*
R								Set C7 = 0
S								Set C7 = 1
	B							Set C6 = 1
	R							Set C6 = 1, C0 = 0
	S							Set C6 = 1, C0 = 1
	T							Set C6 = 1
		B						Set C5 = 1
		R						Set C5 = 1, C0 = 0
		S						Set C5 = 1, C0 = 1
		T						Set C5 = 1
			B					Set C4 = 1
			R					Set C4 = 1, C0 = 0
			S					Set C4 = 1, C0 = 1
			T					Set C4 = 1
				S				Set C3 = 1, C1C0 = 00
				T		—	1	Set C3 = 1, C1C0 = 01
				T		—	2	Set C3 = 1, C1C0 = 10
				T		—	4	Set C3 = 1, C1C0 = 11
					R			Set C2 = 1, C1C0 = 00
					T	—	1	Set C2 = 1, C1C0 = 01
					T	—	2	Set C2 = 1, C1C0 = 10
					T	—	4	Set C2 = 1, C1C0 = 11

4. Since the particular setting of all registers specified in the C field except WDB may be controlled by C1C0, the programmer must use extreme care when specifying the setting of each register. He may only call for a setting in the C field that will cause the required setting of C1C0 to be compatible with all other registers specified in the C field. For example, the mnemonic S when used in BAA column causes C0 to be set to 1. The programmer may not use an R in the G column of the same EO word since this causes C1C0 to be set to 00.

Notes:

1. Programmers must use extreme care in programming for WDB or C7 (Column 47). The following rules apply.

(a) WDB will control the affected half of the DR register *only* when the function (Columns 40–42) is PR1 or PR2 (Perform), and then only when a transfer from the Bus to DR is specified. It should also be noted that this control will be on a *previous* setting of WDB, and *not* the setting of WDB as specified within the EO word to be executed. In all other cases, the desired half of DR is under control of C7.

(b) WDB may be set only when the functions (Columns 40–42) SP1, SP2 (Set function and perform), PR1, PR2 (Perform), or SET (Set function) are specified.

Note that any transfers to or from DR register specified in an EO word that includes one of the above functions, except as outlined in (a) above, will be under control of C7 and not WDB, but WDB will be set to the value specified in Column 47 of the coding sheet. In all other cases, WDB will not be affected when a value is specified in Column 47.

2. The assembler sets C7 to 0 or 1, depending on whether the mnemonic R or S is used by the programmer. If Column 47 is blank, C7 is reset to zero.

APPENDIX 9G: TEST CONDITION MNEMONICS

Mnemonic	*T Field Bit Pattern*	*Test Condition True If*
B0EQ1	010110	B0 (1)
B0EQ0	010110	Complement test* of B0 (1)
BIR = +1	011001	B3 (1) · B1 (0) · B0 (0) · (IR = 00000000/00000001)
BAAEQ1	100010	BAA (1)
BAAEQ0	100010	Complement test of BAA (1)
BABEQ1	100011	BAB (1)
BABEQ0	100011	Complement test of BAB (1)
BDBEQ1	100100	BDB (1)
BDBEQ0	100100	Complement test of BDB (1)
BRTEST	101011	B2 (1)/(R2 = 0000)
DR1 = FS	010000	DR1 = 00100010
DR0 = FS	010001	DR0 = 00100010
DSEQ0	010010	DS (0)
DSEQ1	010010	Complement test of DS (0)
GORTST	001100	(G = XX1XXXXX)/(OR = XX1XXXXX)/(OR = 01000001)
GTEST	010101	[(OR ≠ 111XXXXX) · (G4 = 0)]/[(OR = 111XXXXX) · (GR6 = 0)]
G = 0	011101	(G = 00000000)/(G = 10000000)
G = ONES	011110	G = 11111111
GGR4	011111	GR $\neq (00)_{16}/(01)_{16}/(02)_{16}/(03)_{16}/(04)_{16}/(80)_{16}/(81)_{16}/(82)_{16}/(83)_{16}/(84)_{16}$
IR7EQ1	010111	IR07 (1)
IR7EQ0	010111	Complement test IR07 (1)
IREQ0	011010	IR = $(00)_{16}$
IRNE0	011010	Complement test of IR = $(00)_{16}$
IRXR = 0	011011	{BAA (0) · [IR1 (3–0) = 0]}/{BAA (1) · [IR0 (3–0) = 0]}
†IRX6E1	111100	[BAA (0) · IR06 (1)]/[BAA (1) · IR16 (1)]
MR7EQ1	001000	MR07 = (1)
MR7EQ0	001000	Complement test of MR07 = (1)
†MRX6E0	110000	[BAB (0) · MR06 (0)]/[BAB (1) · MR16 (0)]
†MR06E0	110001	MR06 (0)
†MR16E0	110010	MR16 (0)
†MR0616	111011	MR06 (1) · MR16 (1)
†MR17E1	110100	MR17 (1)
†MRX6X7	110101	[BAB (0) · MR06 (1) · MR07 (1)]/[BAB (1) · MR16 (1) · MR17 (1)]
†MRSCWM	111101	MR06 (1) · MR07 (1) · MR16 (1) · MR17 (1)
†MR14E1	110111	MR14 (1)
†MRX5E1	111000	[MAR10 (0) · MR05 (1)]/[MAR10 (1) · MR15 (1)]
†MRX4X5	111001	[MAR10 (0) · MR05 (0) · MR04 (0)]/[MAR10 (1) · MR15 (0) · MR14 (0)]
†MRENC	111010	MR05 (0) · MR04 (0) · MR15 (0) · MR14 (0)
†MRXESP	110011	[MAR10 (0) · MMR0 = XX000001]/[MAR10 (1) · MMR1 = XX000001]
†MRXESI	110110	[MAR10 (0) · MMR0 = XX000001]/[MAR10 (0) · MMR0 = XX111100]/[MAR10 (1) · MMR1 = XX000001]/[MAR10 (1) · MMR1 = XX111100]
†MRXESM	111110	[MAR10 (0) · MMR0 = XX000001]/[MAR10 (0) · MMR0 = XX001100]/[MAR10 (1) · MMR1 = XX000001]/[MAR10 (1) · MMR1 = XX001100]

APPENDIX 9G: TEST CONDITION MNEMONICS (Continued)

Mnemonic	*T Field Bit Pattern*	*Test Condition True If*
OR0EQ0	000001	OR0 (0)
OR0EQ1	000001	Complement test of OR0 (0)
OR1EQ0	000010	OR1 (0)
OR1EQ1	000010	Complement test of OR1 (0)
OR2EQ0	000011	OR2 (0)
OR2EQ1	000011	Complement test of OR2 (0)
OR3EQ0	000100	OR3 (0)
OR3EQ1	000100	Complement test of OR3 (0)
OR4EQ0	000101	OR4 (0)
OR4EQ1	000101	Complement test of OR4 (0)
OR5EQ0	000110	OR5 (0)
OR5EQ1	000110	Complement test of OR5 (0)
OR6EQ0	000111	OR6 (0)
OR6EQ1	000111	Complement test of OR6 (0)
OVEQ1	101001	OV (1)
OVEQ0	101001	Complement test of OV (1)
RZTEST	001001	(RR = 00000000)/[(OR = 111XXXXX) · (R1 = 0000/R2 = 0000)]
RNE0/1	001010	(RR ≠ 00000000) · (RR ≠ 11111111)
REQ1S	001011	RR = XXXXXXXX
R1EQ1S	001101	R1 = 1111
R2EQ0	001110	R2 = 0000
R2EQ1S	001111	R2 = 1111
R21EQ0	010011	R21 (0)
R21EQ1	010011	Complement test of R21 (0)
R1EQ0	010100	R1 = 0000
R1NE0	010100	Complement test of R1 = 0000
R1EQ1	011100	R1 = 0001
R1NE1	011100	Complement test of R1 = 0001
RZ = 1	100000	RZ (1)
RZ = 0	100000	Complement test of RZ (1)
RZSCAR	100001	RZ (0)/(RR = 11111111)/SCAR (0)
SCAR = 0	100101	SCAR (0)
SCAR = 1	100101	Complement test of SCAR (0)
TRUE	111111	No Test. Result always true
USEQ0	100110	US (0)
USEQ1	100110	Complement test of US (0)
USEQ07	101010	[US (0) · IR07 (0)]/[US (1) · IR07 (1)]
URTEST	011000	{BAB (0) · [UR (03–00) > 9]}/{BAB (1) · [UR (13–10) > 9]}

*When any Complement Test is specified, the assembler places the bit pattern of the A field into the N field of the EO word and the bit pattern of the N field is placed in the A field. For all other cases, the A and N fields are placed in their respective positions in the decoded EO word.

†These tests only valid when OR ≠ 100111XX.

APPENDIX 9H: EXCEPTION MNEMONICS

Mnemonic	*E Field Bit Pattern 43210*	*Action and Remarks*
	00000	No exception. If test is true, next EO address is: 2^5–2^0 = 2^5–2^0 of A field 2^{11}–2^6 remain unchanged. If test is false, next EO address is: 2^5–2^0 = 2^5–2^0 of N field 2^{11}–2^6 remain unchanged.
	00001	Unconditional jump. Address of next EO is: 2^5–2^0 = 2^5–2^0 of N field 2^{11}–2^6 = 2^5–2^0 of A field (See Appendix 9K).
EES	10011	Emulator End Staticizing. If test is false, address of next EO is: 2^5–2^0 = 2^5–2^0 of N field 2^{11}–2^6 remain unchanged. If test is true, address of next EO is: 2^0 = 2^0 of A field 2^5–2^1 = result of logical or of 2^5–2^1 of A field with 2^4–2^0 of OR, respectively. 2^8–2^6 = 2^7–2^5 of OR, respectively. 2^{11}–2^9 = 100, respectively.
EXCXX	YYYYY	XX are two hexadecimal digits which, when converted to binary, equal 000YYYYY. For action resulting from these exceptions, refer to 70/45 Processor EO Flow Chart Manual, EDP form 70-45-101.
ENDIN	00010	End of instruction. Address of next EO is: If test is true: 2^5–2^0 = 2^5–2^0 of A field 2^{11}–2^6 = 101001, respectively If test is false: 2^5–2^0 = 2^5–2^0 of N field 2^{11}–2^6 remain unchanged.
ENDST	00011	End of staticizing. If test is false, address of next EO is: 2^5–2^0 = 2^5–2^0 of N field 2^{11}–2^6 remain unchanged. If test is true, address of next EO is: 2^0 = 2^0 of A field 2^5–2^1 = result of logical or of 2^5–2^1 of A field with 2^4–2^0 of OR, respectively.

APPENDIX 9H: EXCEPTION MNEMONICS (Continued)

Mnemonic	*E Field Bit Pattern 43210*	*Action and Remarks*
		2^8–2^6 = 2^7–2^5 of OR, respectively. 2^{11}–2^9 = 000 respectively.
		If OR = $(0A)_{16}$, set INT21
MAPBD	10001	Mapped Memory boundary. If MR1 = $(8F)_{16}$, set INT24.
NOOP	01110	No-op interrupt. Set INT23.
SHADE	00100	Address shaded memory.
TBK3	10010	Trigger BK3 flip-flop.

APPENDIX 9I: M FIELD SETTINGS

The M field setting controls the operation on Fast Memory and the assembler should accomplish the following:

*1. If *neither* FROM or TO is Fast Memory, set M = 00.
2. If FROM *is* Fast Memory and TO is *not* Fast Memory, set M = 01.
3. If FROM is *not* Fast Memory and TO *is* Fast Memory, set M = 11.
*4. If FROM and TO are both the same Fast Memory location, set M = 01. Definition of M field is as follows:

M = 00—Not a Fast Memory operation
M = 01—Fast Memory read
M = 11—Fast Memory write
M = 10—Not used

APPENDIX 9J: T, A, N, AND E FIELD VARIATIONS

Two tables have been included in this appendix to show the assembler action on various combinations of the test field (col. 8–13), the A field (col. 14–19), the N field (col. 20–25) and the Exception field (col. 55–59).

Table 9J.1 contains the possible variations of T, A, and N if the E field was *not coded.* The column headed by "EXCEPT" shows how the assembler automatically codes the EXCEPT field in the decoded EO. The last column explains how the A and N fields are decoded under these conditions.

*The only time the FROM and TO fields may both be Fast Memory is when they are both the same Fast Memory location. M must contain the proper binary code if EO is coded in binary.

Table 9J.1. ASSEMBLER ACTIONS FOR T, A, N, AND E FIELD VARIATIONS WITH UNCODED EXCEPT FIELD

Test	*A*	*N*	*EXCEPT**	*A and N Fields of EO Word Contain*
Coded	Coded	Coded	00	A = 2^5–2^0 of coded A field N = 2^5–2^0 of coded N field
Coded	Coded	Blank	00	A = 2^5–2^0 of coded A field N = EO address +1 (2^5–2^0)
Coded	Blank	Coded	00	A = EO address +1 (2^5–2^0) N = 2^5–2^0 of coded N field
Coded	Blank	Blank	00	A = EO address +1 (2^5–2^0) N = EO address +1 (2^5–2^0)
Blank	Coded	Coded	01	A = 2^{11}–2^6 of coded A field† N = 2^5–2^0 of coded A field
Blank	Coded	Blank	01	A = 2^{11}–2^6 of coded A field N = 2^5–2^0 of coded A field
Blank	Blank	Coded	01	A = 2^{11}–2^6 of coded N-field N = 2^5–2^0 of coded N field
Blank	Blank	Blank	01	A = 2^{11}–2^6 of EO address +1 N = 2^5–2^0 of EO address +1

*EXCEPT column indicates the action taken by the assembler in decoding the EXCEPT field if the programmer has left it blank.

†N field ignored in this case.

Table 9J.2. ASSEMBLER ACTIONS FOR T, A, N, AND E FIELD VARIATIONS WITH CODED EXCEPT FIELD

Test	*A*	*N*	*EXCEPT*	*A and N Fields of EO Word Contain*
Coded	Coded	Coded	01	A = 2^5–2^0 of coded A field* N = 2^5–2^0 of coded N field
Coded	Coded	Blank	01	A = 2^5–2^0 of coded A field* N = 2^5–2^0 of EO address +1
Coded	Blank	Coded	01	A = 2^5–2^0 of EO address +1* N = 2^5–2^0 of coded N field
Coded	Blank	Blank	01	A = 2^5–2^0 of EO address +1* N = 2^5–2^0 of EO address +1
Blank	Coded	Coded	01	A = 2^{11}–2^6 of coded A field† N = 2^5–2^0 of coded A field
Blank	Coded	Blank	01	A = 2^{11}–2^6 of coded A field N = 2^5–2^0 of coded A field
Blank	Blank	Coded	01	A = 2^{11}–2^6 of coded N field N = 2^5–2^0 of coded N field
Blank	Blank	Blank	01	A = 2^{11}–2^6 of EO address +1 N = 2^5–2^0 of EO address +1
Coded	Coded	Coded	≠01	A = 2^5–2^0 of coded A field N = 2^5–2^0 of coded N field
Coded	Coded	Blank	≠01	A = 2^5–2^0 of coded A field N = 2^5–2^0 of EO address +1
Coded	Blank	Coded	≠01	A = 2^5–2^0 of EO address +1 N = 2^5–2^0 of coded N field

Table 9J.2. ASSEMBLER ACTIONS FOR T, A, N, AND E FIELD VARIATIONS WITH CODED EXCEPT FIELD (Continued)

Test	*A*	*N*	*EXCEPT*	*A and N Fields of EO Word Contain*
Coded	Blank	Blank	≠01	A = 2^5-2^0 of EO address +1 N = 2^5-2^0 of EO address +1
Blank	Coded	Coded	≠01	A = 2^5-2^0 of coded A field N = 2^5-2^0 of EO address +1
Blank	Coded	Blank	≠01	A = 2^5-2^0 of coded A field N = 2^5-2^0 of EO address +1
Blank	Blank	Coded	≠01	A = 2^5-2^0 of coded N field N = 2^5-2^0 of EO address +1
Blank	Blank	Blank	≠01	A = 2^5-2^0 of EO address +1 N = 2^5-2^0 of EO address +1

*These cases are probably coding errors since the A and N fields of the EO word are combined to form one 12-bit address when the exception code is 01 and the combination is neither the address of a label nor of N label.

†N field ignored in this case.

Table 9J.2 contains all possible variations of T, A, N, and E if the programmer *codes* the EXCEPT field himself. The table shows the difference in the way the A and N fields are decoded if the EXCEPT field is 01 or if it is not 01 (≠01), since 01 is the only code which changes the addressing scheme.

Notes:

1. The assembler decodes the A and N fields as shown except when a complement test is used. Then the A and N fields are treated as if they were reversed, i.e., A field coding is used as if it were in the N field and vice versa.
2. All 6-bit addresses are checked to see if they are in block.

APPENDIX 9K: TEST AND BRANCH EO

The test and branch EO is a special EO, the fields in the TBR EO having meanings different from the normal EO.

There may be up to three tests and branches within the same EO and if none of the tests is true, Read Only Memory will automatically "drop through" to the next sequential EO. This EO may be coded either symbolically or "binarily" and the special requirements for both cases are described below.

I. *Symbolically Coded Test and Branch EO*

When the test and branch EO is symbolically coded, the mnemonic TBR is used in the F field. Up to three lines of coding may be used for each test EO and TBR *must appear* in the F field (col. 40–42) of all three lines.

In order for the assembler to distinguish between separate EO's, each EO *must* contain a label in the LABEL field on the first line of coding and the LABEL field *must* be blank on the second and third lines of coding if they are used.

Only five fields besides the LABEL field on the coding sheet may be used. These are the TEST field, the A field, the F field, the EXCEPT field, and the ABS ADDR field. These fields are used to describe the particular test flow desired and are set up as follows:

A. *First Card*

1. The *LABEL* field (col. 1–6) *must* be coded and with a valid label. All characters are valid for a label except the use of a numeric for the first character of a label name. The address of this EO is assigned to this label and then placed in the tag table.
2. The *TEST* field (col. 8–13) must be coded with a valid test mnemonic. The bit pattern for this test is placed in T field of the EO word or with the proper binary bit pattern (see Appendix 9B). Complement tests are not legal for a test EO.
3. The *A* field (col. 14–19) must be coded with a valid label of a current EO. The EO referred to must be located within the same ROM block that the test and branch EO is placed. This field will be the jump address of the first test which is contained in the TEST field if the test is true. This is placed in the A field of the assembled word.
4. The *F* field (col. 40–42) *must* contain the mnemonic TBR. The bit pattern is placed in the F field by the assembler.
5. The *EXCEPT* field (col. 55–59) may contain any valid exception mnemonic or may be coded in binary or hexadecimal according to the rules for this field as explained in the specification, or may be left blank. This field is placed in the E field of the EO word.
6. The *ABS ADDR* field (col. 62–65) may contain a valid *octal* number to specify the desired address of the test and branch EO or may be left blank. If blank, the next sequential address is assumed and assigned to this EO.

All other fields *must* be left blank except the I field (col. 7) which may be coded normally to inhibit I/O servicing.

B. *Second Card**

1. The *LABEL* field *must* be blank if this is to be assembled as part of the same EO.
2. The *TEST* field must contain a valid test pattern for the second test. The assembler will place this bit pattern in the D field. Valid tests for the second line of coding are:

*This card not necessary if only one test is desired. If no second test is desired but a third test is used, this card must be present and only the F field is coded with TBR.

Bit Pattern	*Meaning*
000001	SSDEC (R1/R2 = 0)/R = 0
000010	DR1 = significant start symbol
000011	DR0 = significant start symbol

3. The *A* field must contain a valid label of a current EO. The EO referred to must be located within the same ROM block that the test and branch EO is placed. This field will be the jump address of the second test which is contained in the TEST field of the second card if the test is true. The assembler will place this address in the N field of the EO word.
4. The *F* field *must* contain the mnemonic TBR.

All other fields should be left blank.

C. *Third Card**

1. The *LABEL* field *must* be blank.
2. The *TEST* field must contain a valid test pattern for the third test. The assembler will place this bit pattern in the S field. Valid tests for the third line of coding are as follows:

Bit Pattern	*Meaning*
000100	R1 NE condition code
000101	R GR 0
000110	DR1 NE digit select symbol
000111	DR0 NE digit select symbol

3. The *A* field must contain a valid mnemonic label. This field will be the jump address of the third test which is contained in the TEST field of the third card. Because this is a 12-bit address, it is not necessary that this label be in the same block. The assembler will place this address in the VC fields of the EO word in the format $C_7C_6C_5C_4C_3C_2C_1C_0V_3V_2V_1V_0$ with the low-order bits in the V field. It should be noted, however, that the V field is placed directly in front of the C field on the coding form and in the assembled EO word.

II. *Binary Coded Test and Branch EO*

A test and branch EO may also be coded on a single card by using the binary format; requirements for the binary coding are given below in a field-by-field description.

*This card not necessary if only a one or two test EO is desired. The assembler will automatically fill in an insert if this card is missing.

Note: A mnemonic test EO may consist of one, two, or three lines of coding at any time. Error messages are printed after the assembled EO word to aid the microprogrammer in debugging his coding.

1. The LABEL field (col. 1–6) may contain any valid label name.
2. The I field (col. 7) is coded normally to inhibit I/O servicing if necessary. This may be coded either in binary or mnemonically and will be placed in the I field of the EO word.
3. The TEST field (col. 8–13) will contain a valid binary test pattern as found in Appendix 9A or may be coded mnemonically. This is the first test of the EO and will be placed in the T field of the assembled word.
4. The A field (col. 14–19) will contain a 6-bit binary pattern describing the jump address of the first test if it is true. (The most significant 6 bits of the address are omitted because it is required that this jump be within block and the most significant 6 bits describe the block address.) These 6 bits are placed in the A field of the EO word. If desired, this field may also be coded with a valid label of a current EO within block.
5. The N field (col. 20–25) is similar to the A field except this is the jump address of the second test (col. 34–39). This address is decoded into the N field of the EO word and the same coding rules apply for this as for the A field.
6. The M field (col. 26–27) *must* contain the binary 00. Without this, the FROM and TO fields will not be decoded.
7. The FROM field (col. 28–33) must contain a valid bit pattern for the third test as shown above. This bit pattern is placed in the S field of the EO word.
8. The TO field (col. 34–39) must contain a valid bit pattern for the second test as shown above. This bit pattern is placed in the D field of the EO word.
9. The F field (col. 40–42) *must* contain the bit pattern "101" for a binary test and branch EO. Using the mnemonic coding TBR will invalidate the assembly and cause an error flag to be set because it will be assumed to be a mnemonically coded EO.
10. The VC fields (col. 43–54) will contain the 12-bit binary jump address of the third test coded in the FROM field. This address is coded in the format $C_7C_6C_5C_4C_3C_2C_1C_0V_3V_2V_1V_0$ with the least significant four bits coded in the V field (col. 43–46). This address can only be coded by a valid *binary* 12-bit address which represents the four character octal address of the EO that is being referred to. No mnemonic code may be used in this field.
11. The EXCEPT field (col. 55–59) is coded with any valid mnemonic exception as described in Appendix 9H or in straight binary.
12. The CONST field (col. 60–61) is not used in the test and branch EO and is ignored.
13. The ABS ADDR field (col. 62–65) must be blank or contain a valid four character *octal* number.

The remaining fields are not used.

Field	*ROMAR*	*F*	*VC*	*M*	*S*	*D*	*T*	*N*	*A*	*E*	*I*
Description	*EO adder*	*Function*	*3rd test jump adder*	*Not used*	*3rd test*	*2nd test*	*1st test*	*2nd jump adder*	*1st jump adder*	*EXCEPT field*	*I bit*
Prntr. Char.	XXXX†	5*	XXX	XX	XX	XX	XX	XX†	XX†	XX	X
No. of Bits	12	3	12	2	6	6	6	6	6	5	1

Note: X indicates a printer graphic; * indicates function field always containing a "5" for a test EO; † indicates the field in octal.

Assembly Listing Format for Test and Branch EO

The format of the assembled test and branch EO is as follows:

ROMAR	*F*	*V*	*C*	*M*	*S*	*D*	*T*	*N*	*A*	*E*	*I*
XXXX	X	X	XX	X	XX	XX	XX	XX	XX	XX	X

where X indicates the number of printer graphics associated with the appropriate fields. All fields are hexadecimal except ROMAR, N, and A, which are octal. The test and branch EO decoding is shown on page 486.

PRACTICE PROBLEMS

1. Supply the register sizes and bit numbering systems for the following:

(a) IR (d) DR (g) BR
(b) UR (e) OR (h) MAR
(c) MMR (f) R (i) GR

2. How many bits does an MMR read?

3. How many bits does a scratch pad-read read?

4. How many bits does an ROM-read read?

5. What is the cycle of time of MM? ROM? SPM?

6. List the functions specified in the F field.

7. What is the binary configuration in the F field for a set function?

8. What are the differences between set function and the set and perform function?

9. Determine the operations specified by the F and V fields:

	F	V
a	0 0 1	0 1 0 1
b	1 0 0	0 0 0 1
c	1 1 1	0 1 0 1
d	1 1 0	0 1 1 1
e	0 1 0	1 1 1 0
f	1 0 1	1 0 1 1
g	0 0 1	1 0 0 0
h	0 1 1	0 0 0 0
i	0 1 1	1 1 1 0
j	0 1 0	1 0 0 0
k	1 1 1	1 1 0 0
l	1 0 0	1 1 0 0

10. Match the bit positions of the C field with the appropriate control flip-flops:

BAA	C4
BDB	C7
BAB	C5
WDB	C6

11. Explain the use of WDB.

12. If BAA/BAB/BDB is specified in a set and perform function, will it be triggered and then set to the value of C0?

13. List the bit settings of the M field for the following:
(a) SPM read
(b) SPM write
(c) Register-to-register operation
What is the only bit configuration that is illegal in the M field?

14. Write the binary configuration of the S, D, and M fields to accomplish the following transfers (assume PS1, HSM = 00's):

	S	D	M
(a) IR → UR			
(b) UR → OR:R			
(c) IR → MARW			
(d) MMR → FPC (C7 = 1)			
(e) FPC → MAR (C7 = 1)			
(f) IR → MMR			
(g) → G			
(h) G → IR			

15. Assume in Problem 14 that IR contains (1234_{16}). What will the contents of each of the following be at completion of Problem 14?

IR	____________	FPC	____________
UR	____________	G	____________
OR	____________	DR	____________
R	____________	MAR	____________
MMR	____________		

16. Write an EO program in binary to add UR to DR right (binary add)

$$DR = 11223344_{16}$$
$$UR = 4334_{16}$$

17. What are the settings of the following registers at the completion of Problem 16?

IR ____________
BAA ____________
BAB ____________
UR ____________

18. Write a program in assembler language to multiply the contents of SPM 22 by 4.

19. Assume

$$\begin{aligned} UR &= FFFF_{16} \\ DR &= FFFFFFFF_{16} \\ IR &= 432A_{16} \end{aligned}$$

(a) Write a program that reverses the contents of IR so that IR contains $34A2_{16}$.

(b) Assuming the same operands, write a program that results in the following conditions:

(1) UR $A234_{16}$

(2) IR $A234_{16}$

(3) DR A234 $A234_{16}$

(4) MM at location $234A_{16} = A234_{16}$

20. Assume

$$\begin{aligned} SPM\ 22 &= 0000124E_{16} \\ SPM\ 70 &= 0000126A_{16} \\ SPM\ 71 &= 00001280_{16} \end{aligned}$$

Write a program in assembly language that compares the MM locations specified by SPM 72 and 73.

If (72) = (73), Branch to $124E_{16}$.

If (72) > (73), Branch to $126A_{16}$.

If (72) < (73), Branch to 1280_{16}.

21. Assume

$$\begin{aligned} SPM\ 70 &= 0000126A_{16} \\ SPM\ 71 &= 0000128A_{16} \\ SPM\ 22 &= 00003100_{16} \end{aligned}$$

Write a program that subtracts 1 from the diad specified by the MM address in SPM 70. If the diad at the address specified by SPM 70 = 0, branch to the address specified by SPM 71.

22. Write a program that moves the contents of MM location 1000 through 1002 to MM locations 1003 through 1005. Assume

$$\begin{aligned} SPM\ 70 &= 1000_{16} \\ SPM\ 71 &= 1003_{16} \\ MM\ 1000 &= 010203040506070809_{16} \end{aligned}$$

23. Write a program that transfers N number of bytes from MM 1000_{16} to MM 2000_{16}. Assume

$$\begin{aligned} N &= 63_{10} \\ SPM\ 70 &= 1000_{16} \\ SPM\ 71 &= 2000_{16} \\ SPM\ 74 &= N \end{aligned}$$

24. Code Problem 1 for the assembler and simulator where:

$$\begin{aligned} SPM\ 74 &= 13_{16} \\ MM\ 1000 &= 090807060504030201_{16} \end{aligned}$$

$$1009 = 0A0B0C0D0E0F1011_{16}$$
$$1010 = 1213141516171819_{16}$$
$$2000\text{–}2020 = FF_{16\text{'s}}$$

25. Trace the data processing action in the two microroutines shown in Fig. 9.16.

PROGRAM Microroutine Trace Exercise PROGRAMMER DATE

Exercise-1 ADD: (R1) + R(2) ⇒ R1

LABEL 1	I/SP 7	TEST 8	A 14	N 20	M 26	FROM S 28	TO D 34	F 40	V 43	C7/WDB; C 48 (BDB, BAA, BAB, R, G, C1, C0)	EXCEPT E 55	CONST. 60
RADD			RADD1			FFR1	UR	SET	BADD	S SS		
RADD2			RADD2			FFR2		PR2	BYTE	S TT		
RADD2		TRUE	000000			IR	FFR1	NOP		S	ENDIN	
EXERCI	S	E-2	ASSUME	E SPM	F	2=(000	01234)16					
				SPM	F	1=(000	05544)16					
			D	O W←	U	.V						
ALFA1			ALFA2			F1	UR	SET	AND	S SS		
ALFA2			ALFA3			F2		PR2	BYTE	TT		
ALFA3		TRUE	000000		JR		F1	SCC	NO	S	ENDIN	

Figure 9.16 Trace microroutine.

REFERENCES

1. "Spectra 70/45 Processor EO Flow Charts," RCA Document #70-45-101, March 1966.

2. Benjamin, R. I., "The Spectra 70/45 Emulator for the RCA 301." Communications of the ACM, December 1965. Vol. 8, No. 12, p. 748–752.

3. Vollbrecht, John R., "MIDAS—Microprogramming Design Aid System for RCA Spectra 70/45." Presented at the ACM Workshop on Microprogramming, Oct. 7–8, 1968; Bedford, Mass. (No proceedings were published.)

4. RCA internal document which "provides a detailed description of the microprogrammed algorithms for the 501 emulator in the Spectra 70/45."

5. "70/45 EO Format Description," RCA Internal Document, December 11, 1967.

MICROPROGRAMMING THE HONEYWELL H4200

10

10.1 INTRODUCTION

The Honeywell H4200 system is a medium-to-large-scale business data processing system. Its instruction set, character definition, and data format are completely compatible with all the Honeywell Series 200 processors. The H4200 main storage is four characters wide. The character is eight bits wide. It includes six data bits, an item mark bit, and a word mark bit (end of field mark). The last two bits are used to delimit instructions and data fields.

The Series 200 Basic Instruction Set includes binary, decimal, and logical operations for variable-length character fields. The scientific option provides floating-point operations on 48-bit operands (12-bit exponent and 36-bit mantissa). The system design of the H4201 (the CPU) is based on a four-character (24-data bit) memory fetch and an adder which may process 6-bit, 24-bit, or 36-bit operands. Many binary, logical, and decimal operations are executed with a four-character fetch followed by serial processing (one 6-bit character at a time) through the adder. Indexing and floating-point operations are performed in parallel, requiring 24- and 36-bit adds, respectively.

The general instruction has the format:[1]

F	A	B	V

where F is the operation code, A is the first operand field address, B the second operand field address, and V is the variant or operation code modifier. The A and B address fields can be two, three, or four characters wide, in accordance with the character mode controls set by the program. The two-character mode provides twelve bits of direct addressing. The three-character mode provides a total of 18 bits (15 bits for direct addressing and the remaining three to select indirect or indexed addressing). The last character mode (four characters) yields 19 bits of direct addressing, plus five bits of indirect or indexed addressing. Data fields are variable in length, fetched in four-character blocks. They are fetched and executed generally from right to left, with the rightmost character occupying the highest memory address. This data fetch is terminated by the end of field (word mark). The instructions on the other hand are fetched from left to right (low memory address to higher memory address). Special orders, such as move operations, can operate on data fields in either direction. The counters controlling instruction fetch and/or data fetch are normally specified to terminate with an address one character beyond the last character fetched. This allows an abbreviated instruction format to be issued (F, FA) which will use the counter status of the preceding instruction for one or both of its field addresses. This technique, referred to as chaining, is useful in processors with limited memory capacity. Further details on the instruction set, instruction format, and data formats of the H4200 system will be discussed in Section 10.5.

The H4200 was designed to provide efficient high-speed serial processing of variable-field-length data, and to facilitate efficient processing of scientific operands. Its data flow, its register structure, and its microprogramming design and implementation differ from the IBM System/360 and the Spectra 70/45 systems discussed in previous chapters.

The H4200, in contrast to the IBM/360, is a highly-specified system. It emphasizes the microinstruction function over the more basic, more flexible, and more fundamental micro-order structure of the System/360.

The application of the ROS Control to the H4200 was based not on generalizing the design but, rather, on enhancing the performance of a highly specialized system structure. The ROS word structure and ROS algorithms were specified by the same designers responsible for processor logic design. The ROS algorithms were not viewed as a program to be executed on a basic register structure but rather as an extension of that structure. The interaction between the hardware and ROS algorithms is extensive since the designer could specialize either area to achieve performance objectives. The result is a system design which would appear familiar to a processor logic designer and quite foreign to a "firmware" programmer. In contrast the IBM/360 emphasizes a more basic and flexible ROS structure which has limited and easily specified interaction with the hardware. In the IBM/360 the ROS algorithms may clearly be viewed as a program running on a basic machine (the processor register set). This characteristic facilitated the development of emulators for the IBM/360. The microprogram of the H4200 was modified for application in the multiprocessor H8200 system. The modification involved the addition of algorithms for the direct execution of system monitor functions and interprocessor communications. In spite of the fact that these sequences were different in format and cycle-flow from normal character processing, they were implemented with no revisions required in the ROS word structure. This indicates that, although the H4200 ROS word format was designed to fulfill a specific need (high-speed character processing), it may well be applicable to a more general class of processing problems.

The two main differences in the microprogramming of the IBM System/360 and the Honeywell H4200 are:

1. The sequencing and branching schemes, and
2. The basic subcommand and microinstruction structure.

Since the System/360 approach to both problems has been discussed in Chapter 5 and illustrated on Models 40 and 50, it will be sufficient to discuss how these two problems were approached by Honeywell, the options that were considered, and the final solution that was implemented in the H4200 System.

The first—namely, the branching and sequencing—is influenced by the data flow; the relationship between the central processing unit (CPU), scratch pad memory (SPM), main memory (MM), and the read-only memory (ROM); and finally by the functional requirements for the system. The MM cycle time is 750 nanoseconds for a four-character word. The CPU clock period is 125 nanoseconds (called PDA). The integrated SPM cycle time is 125 nanoseconds. The adder in the CPU adds a character or a word in 250 nanoseconds. This timing made it necessary to design the CPU data flow and the control to keep up with the memory's ability to supply four characters every six PDA's (750 nanoseconds).

It is also evident that the control of a variable-field-length machine is more complex to design and that the cycle patterns for any given instruction execution are highly irregular. This irregularity implies a need for greater flexibility in changing the cycle machine function (state) quickly.

The ROM is a stack of 2048 words by 120 bits with an access time of 135 nanoseconds. The read-out cycle is overlapped by 10 nanoseconds providing an effective cycle time of 125 nanoseconds. It uses a two-core-per-bit linear select memory with nondestructive read-out. The ROM stack is written into by a loader system from magnetic tape before it is installed in the machine. The interface between the ROM and the processor consists of the 11-bit ROM address register (Fig. 10.1—QR register) and a 120-bit ROM data register (Fig. 10.4—K register).

The design of the ROS word format and the sequencing and branching in this machine were based on the flow-charted algorithms rather than the more general structure that the system's data flow would permit. It reflects the programming need to microprogram a set of specified processes.

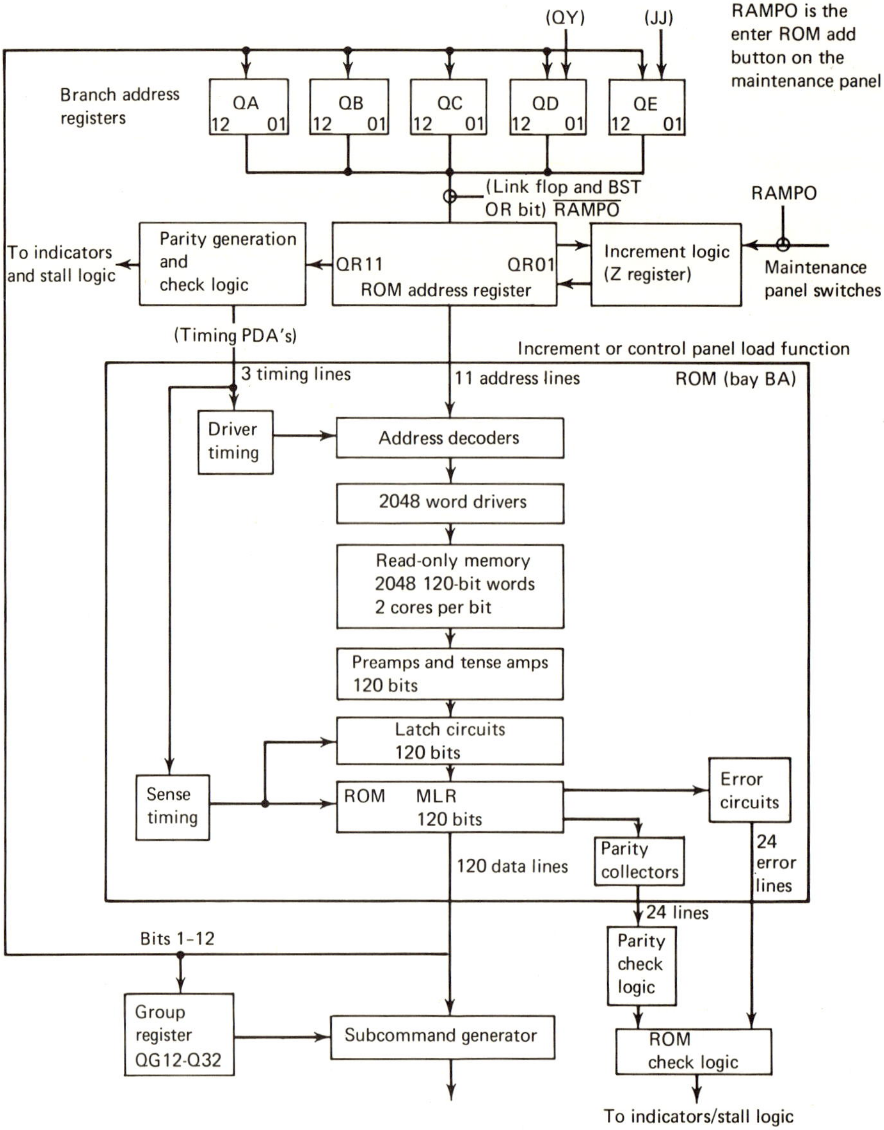

Figure 10.1 ROM block diagram.

10.2 BRANCHING

The analysis of the branching requirements leads to the following design (Figs. 10.1 and 10.4):

An incrementor was provided for incrementing through its sequential ROM addresses. The basic scheme would be to increment the ROM address register (QR) through the sequential steps of the microroutine until a branch order is reached, at which point a new ROM address is loaded into the QR under the microprogram control.

There were over 100 branch hardware indicators that could be set to reflect the status of the system at different times. In many cases it was found that these stats (indicators) had to be set one or more clock periods (PDA's) prior to the branch cycle. Another requirement was to group the test and branch commands so that several of these indicators could be tested simultaneously.

These branching requirements led to designing into the system a set of five full-width (11-bit) branch address registers (QA–QE) which could be loaded by the microprogram from an 11-bit ROM field. In the case of the QD and QE registers, they can also receive information from the Y register (QY) and the JJ register. The JJ register is six bits wide. It contains the instruction operation code. The instruction fetch cycle is terminated by branching to the instruction execution routine, by gating JJ to QE to QR. The absolute address of the beginning of the execution routine is generated by shifting the Op-code (contents of the JJ register) left two bit positions during the transfer from the JJ to the QE register. For example, the absolute address for the beginning of execution of a compare instruction is 0154 (octal) which is formed by adding two low-order zeros to the Op-code 33 (octal). Thus,

$$(33)\text{ octal} = 011 \mathrel{\vdots} 011$$

and a shift left by two positions produces $01 \mathrel{\vdots} 101 \mathrel{\vdots} 100 = (154)$ octal.

The branch path of QY to QD to QR is used to locate the scientific instruction execution routines.

The five branch address registers, QA–QE, have five corresponding test matrix sections A through E. Each section could specify one of 28 tests to be made. Each section could set a stat (called a link flop) associated with each address register, QA–QE, when the specified test was true. These stats could also be set unconditionally when executing a specific ROM cycle. For example, stat A may be set by the subcommand OABPO if the B pointer register contains zero, or by the subcommand ONAWM if a word mark is detected in the M register. These stats A–E associated with the branch registers QA–QE have a fixed priority, and A having the highest and E the lowest. The execution of a branch command would determine the highest priority stat which is on, and transfer the contents of its associated branch register to the QR. The system timing is such that the ROM addressing (set QR) proceeds one step (one machine cycle) beyond the microinstruction cycle where the branch command is specified. The ROM word organizations and branch subcommands which support this design philosophy will be reviewed in a later section. However, to clarify this design and the relative timing of the sequences of events, consider the following example.[2]

A given decimal operation has its last four microinstructions stored in locations 100, 101, 102, and 103. After location 103 it is desirable to perform certain tests on the accuracy and validity of the results. We wish to transfer to an overflow error routine if the overflow stat is on, or go to the I fetch if there was no overflow and if the operation had terminated properly, or branch to fetch new operand if the operation has not terminated and no overflow condition occurred. Assume that the highest priority (i.e., branch register QA, branch test set, and branch—stat A) is assigned to the overflow condition. Let the termination condition be assigned the next highest priority—stat B. If location 104 is the last to be processed, the branch test must be specified in 103. The branch addresses and stats must be set at least one machine cycle prior to branch command execution. This would lead to the following valid sequence of microinstructions:

101 Set branch register A to overflow error routine entry.

102 { Set branch register B to next I fetch entry.
Test overflow into stat A, test terminate into stat B.

(We can set up one branch address and perform five simultaneous tests in each microinstruction.)

103 Branch on test indicators.

104

105 Start of instruction fetch microroutine.

This arrangement implies that the total branch register set up, test condition specification, and branch execution is spread over three machine cycles. This timing proved to be unsatisfactory for processing the floating-point operands stored in the high-speed SPM. This potential performance loss was corrected by providing a high-speed test circuitry for the floating-point indicators. The test provided a test and branch sequence that took two machine cycles rather than three. One bit in the ROM word (bit 51) was used to specify the high-speed test process.

10.3 MICROINSTRUCTION STRUCTURE

The second main difference between this system and the IBM System/360 microprogram organization is in the way the microinstruction is designed.

Once the branching scheme was established, the action commands and the test and branch condition command list was created. This list contained over 300 commands. This high number of commands was due to a large number of allowable variations on the action commands. These variations are permitted to happen in the same sequence without a need for branching. The microprogram with these command variations can respond to the different systems conditions without branching, thus achieving a higher processing speed.

The condition test and indicator set commands were coded into 17 bits of ROM word. Another 12 bits (11 + parity) were reserved for branch address specification. A third field of five bits was assigned to control the SPM addressing. The remaining 86 bits of the 120-bit ROM word were to be used to encode the over 300 action commands which appear in over 2000 patterns.

The first approach taken was to partition the action group into mutually exclusive sets (that never appeared in the same ROM word) and encode these sets into smaller fields, which was feasible. However, it was ruled out as an inflexible solution that would limit the checkout corrections and any future architectural extensions.

The second encoding technique that was finally adopted was to investigate the 4200 instruction set and the ROM word sequencing and group them into classes of similar processing patterns. The basic arithmetic and logic operations were grouped into one class. It was also recognized that there exists a basic set of 65 mutually exclusive commands that appeared in virtually all groups; examples would be register-to-register transfer, shift right, and so on. These 65 basic commands were coded into 34 bits. The remaining control action was organized into 12 groups of nonbasic commands. No group contained more than 50 such actions. If 50 bits were assigned to indicate a specified action within a group, and if a 12-bit field is assigned to indicate the type of group, then the interpretation of the 50-bit command field would be unambiguous and the ROM word organization would have been completed. This final ROM word organization is shown in Fig. 10.2. The following additional comments on the ROM word organization may be helpful.

Bit 1	Branch address parity for the branch field in bits 2–12
Bits 2–12	Branch address field which is transmitted and set into QA–QE as specified
Bit 13	Branch on stored test. When this branch command (OOBST) is present, a branch will occur to the address stored in the branch register associated with the highest priority branch stat (A to E) that is set.
Bits 14–15	Link quadrant selection is a field which selects one of four groups of seven test actions in each test section.

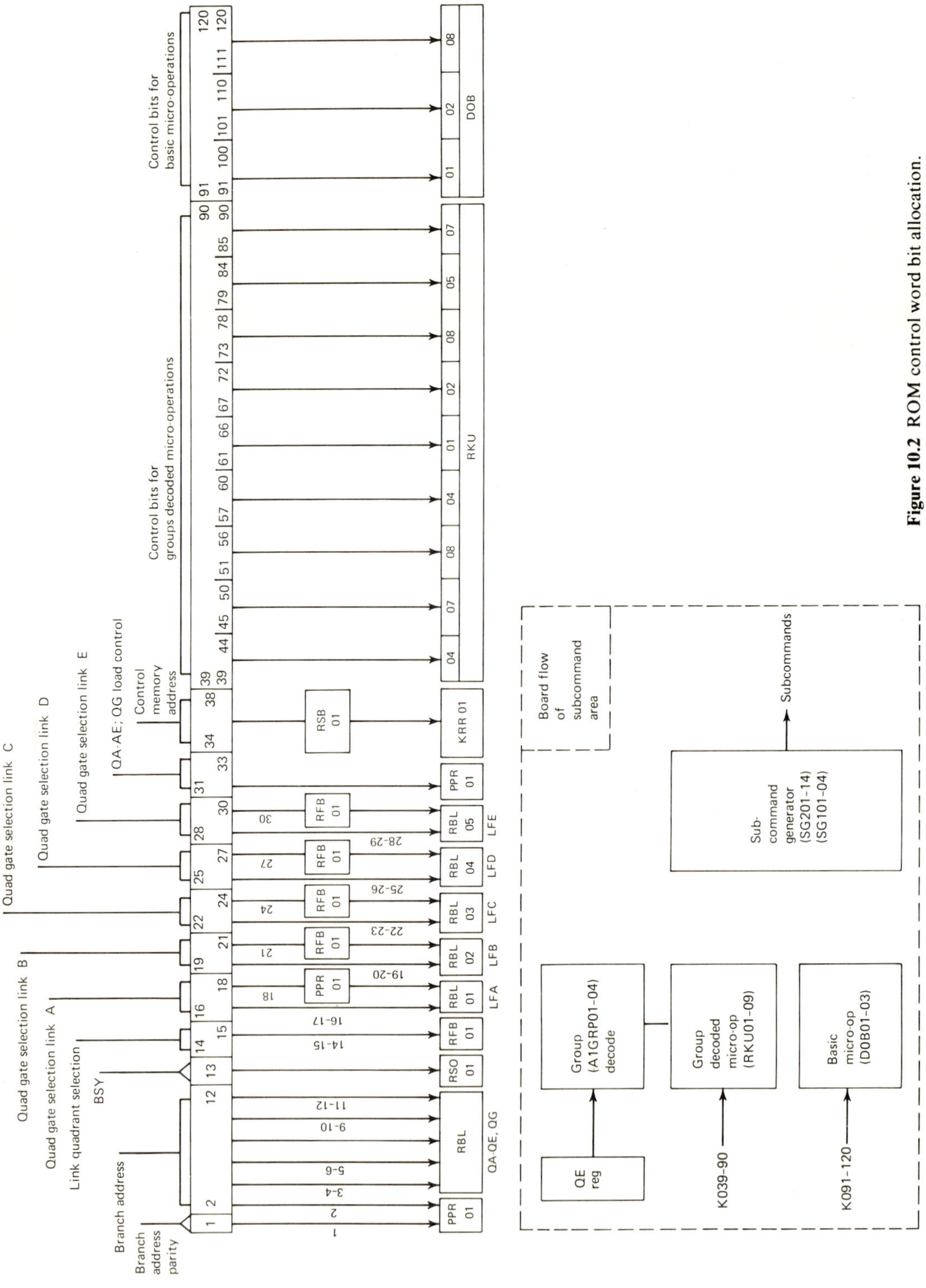

Figure 10.2 ROM control word bit allocation.

Bits 16–18	Quad Gate Selection Link A	These five 3-bit fields select which seven tests in each test section were activated.
Bits 19–21	Quad Gate Selection Link B	
Bits 22–24	Quad Gate Selection Link C	
Bits 25–27	Quad Gate Selection Link D	
Bits 28–30	Quad Gate Selection Link E	
Bits 31–33	QA–QE—load branch addresses in each ROM word.	
Bits 34–38	High-speed SPM address	
Bits 39–90	Control bits for group-decoded micro-operations	
Bits 91–120	Control bits for basic micro-operations	

The last 82 bits (bits 39 through 120) are used directly in the generation of micro-operations.

Bit position 51 is a fast branch subcommand (OOBDT), similar to (OOBST) except that (OOBDT) is dependent on the group interpretation. It was indicated earlier that this branch works with floating-point execution to reduce the test and branch from three cycles to two.

The 12-bit active group identification field is loaded from ROM into a 12-bit group register, at the start of a sequence. During the early cycles in the opera-

Table 10.1. ROM GROUPS

Group	*QG Bit*	*Instruction or operation*	*Special function*
A	01	HLT-CPO-CSM-DLR-LCR-SCR-SIB-SST-SVI-BCC-BCE-BBE-RVI	Control panel function
B	02	CAM-EXT-HRD-IDA-IDB-IXA-LRD-MCL-PCB-PDT-NOP-DCB	Order fetch diagnostic tests
C	03	CSM-DTM	Diagnostic trace control
D	04	EXM-MOS-MCE	
E	05	ADS-COM-DIV-TLU-MPY-MCW (A,S, D,BA,BS,HS,HA,EXT,C,M,ZA,ZS,LCA)	
F	06*	DAA (floating divide)	
G	07*	MAA-TAA-TAL-TLA (floating multiply, floating AC → AC)	
H	08*	ASA-BMS (floating add-sub, floating binary mantissa shift)	
J	09*	BTD-DTB-FBA-FBI-TMA (floating BTD, BTB conversion, floating branch, floating mem → AC)	
K	10*	BIM-TAM-TLM (floating int mult, floating AC → mem)	
L	11	MAT-MIT	
M	12	BCT-LIB-MPM-RNM (B,SI,CI,SW,CW, RNM)	

*QG bits 06–10 are required for floating-point option only.

tion (code execution is sequential to fetch the operands and prepare for execution), the branch address field is not used for its primary purpose and therefore could be used to identify the active group. The contents of the ROM address field are transferred to a sixth register, called a group register, to maintain this group code setting until it is changed by the microroutine when the branch address field is available. Table 10.1 shows the ROM groups, each group containing a similar class operation code. The entries are either Series 200 operation codes or codes for functions common to a number of Series 200 operations. The Series 200 operation code entries are underlined. The brackets contain the Series 200 operation codes that are serviced partially or completely by one of these entries in the group. For further details on the Series 200 operation codes, the reader may refer to Appendix A in this chapter or to a Series 200 programmer's reference manual.

At this point, prior to introducing the basic operations, group decode microinstructions, and actual microprogramming examples, it would be necessary to study the system's data flow and timing, and the basic Honeywell Series 200 architecture.

10.4 H4200 DATA FLOW (Fig. 10.3)

The main features of the H4200 are: the simultaneous memory access by the two independent processing units, the CPU and the Input/Output Controller (IOC); the electronically-alterable ROM; the interleaved memory for memory access overlap; the floating read/write channels, the diagnostic maintenance instructions, and the separate maintenance processor.

Figure 10.3 shows the general organization. Two to four memory modules of 16K or 32K four-character words can be attached to the memory bus controller (MBC). Since the memory modules cycle are independent of each other, the MBC is therefore capable of providing simultaneous service to both the CPU and the IOC. If these two processors are attempting to access the same memory module simultaneously, the MBC grants access to the IOC. If, while one processor is being serviced, the other requests access to the same memory module, the MBC stalls the second request until the initial access is completed.

The H4200 CPU organization is shown in Fig. 10.4. This simplified data flow of the CPU can be partitioned into four main functional sections: (1) the main memory (MM) which includes the MBC; (2) the arithmetic and logic unit (ALB) which includes the five four-character registers (A, B, D, E, N) and a four-character adder; (3) the high-speed scratch pad memory (SPM) (called the control memory) used for address generation; and finally, (4) the control section implemented in a read-only memory.

The *adder* is designed to perform logical, binary, and decimal operations in a serial manner (character for character) and in parallel. Floating-point operands, for example, are processed in parallel. When the system is equipped

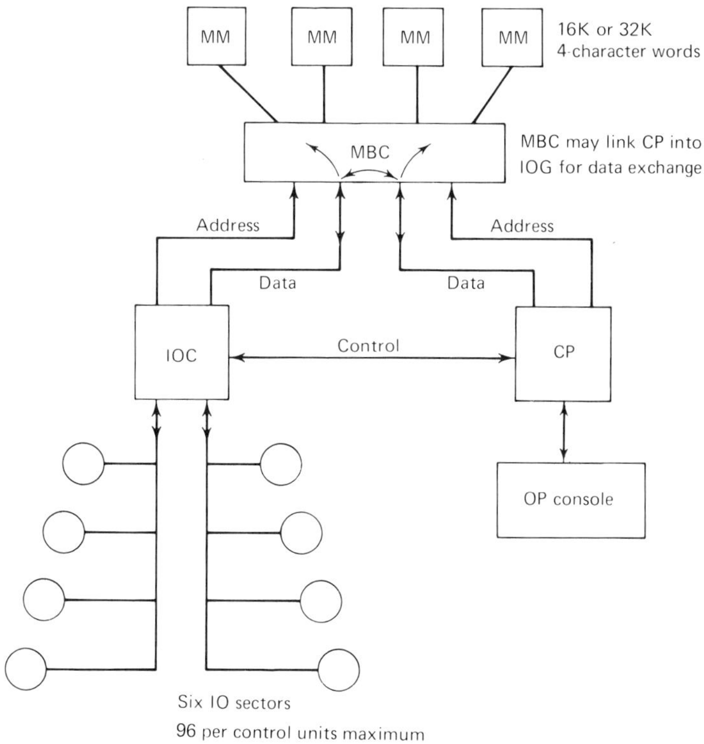

Figure 10.3 General organization.

with the scientific option, the A and B registers, the adder and the adder data bus are all extended to six characters (36 bits). The adder is designed to perform a character or word operation in two basic processor clock cycles (PDA)—250 nanoseconds. The adder receives the two operands from the A and B registers and gates the results back to A or B. The N register is the main memory data register. The E and D registers are backup registers to B and A operand registers respectively. The ALB section also includes several one-character registers including the six-bit Op-code register (JJ), the six-bit Op-code modifier or variant register (VV), and a general purpose, eight-bit character register (QU). These registers are loaded from the N register through the N select gates (NS). This character selection is specified by the two low-order bits of the B pointer (BP). Notice the line from the A and B register output bus to the N register. Note also the gating from VV to QU (data bits) and QU character to A register.

Note the line from N directly into A, bypassing the D register. There is a separate microcommand to specify this direct gating. All four registers (A, B, D, and E) can perform bit or character shift right or left. They can also perform shift round to place the shifted-out character or bit in the vacated character or bit.

The *High Speed SPM* is a 32-word × 19-bit array with a cycle time of 125

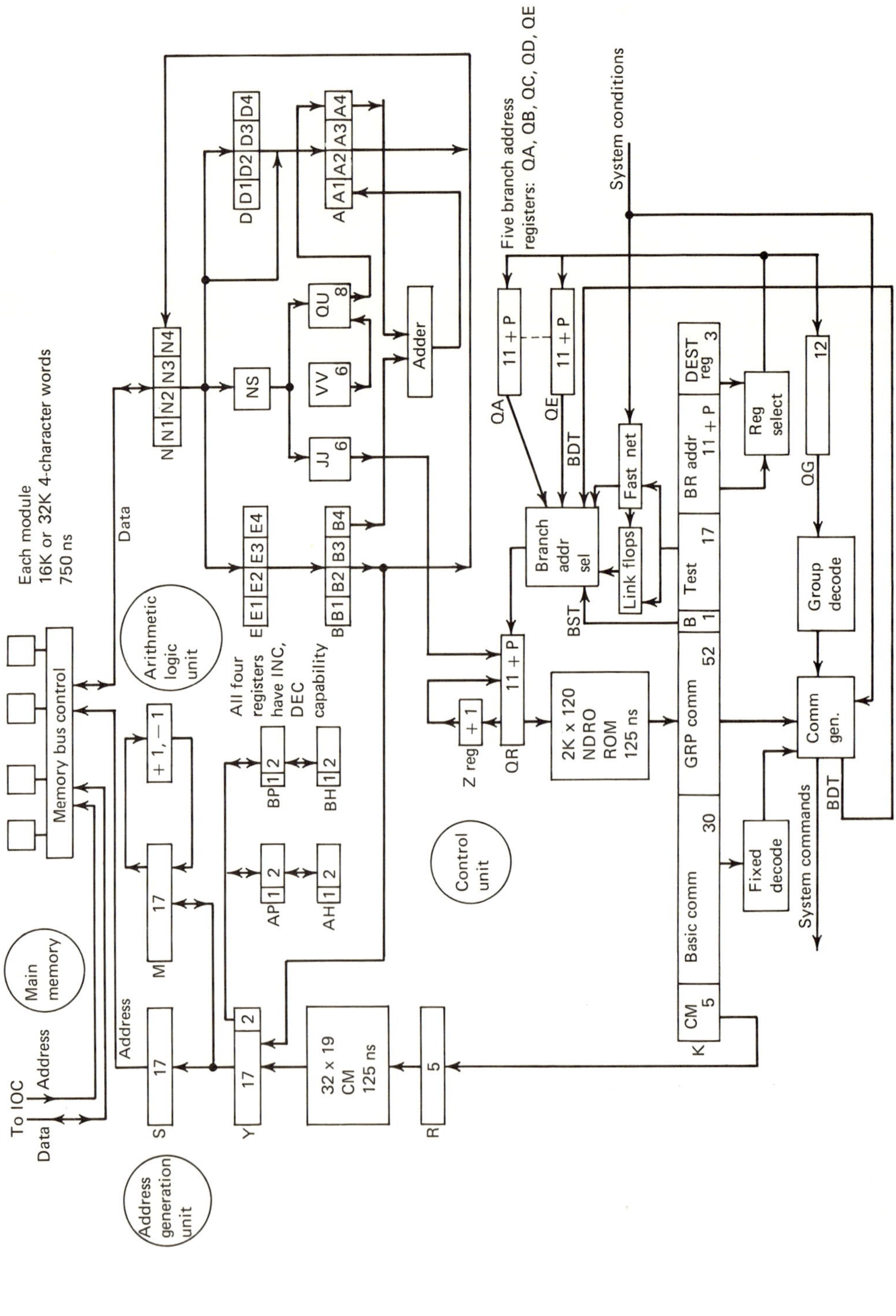

Figure 10.4 Partial block diagram.

nanoseconds. Its main function is that of memory address generation. It also serves as a substitute for a number of hardware registers for intermediate storage. The SPM stores the instruction address and the two operand addresses called the A field counter (AC) and the B field counter (BC). SPM addressing is controlled by a five-bit R register which may be loaded from the CM field in ROM word. SPM output is fed into a 19-bit register designated Y. The high-order 17 bits of the Y register provide the word address for memory and are the input to the MM address registers. The low-order two bits are used to indicate which of the four characters is to be selected. The Y register also gates into the memory address counter register M which is equipped with increment/decrement circuitry. The two-bit character address field in Y is transmitted to the two low-order bits of either the A pointer counter (AP) or to the B pointer counter (BP). The AP and BP counters each have an associated history pointer counter, AH and BH, respectively. All four-character address counters have an additional leftmost bit which can indicate when a word boundary has been crossed.

In reality, the SPM has sixty-four 19-bit words, although only 32 words are located in, and directly addressable by, the central processor. Locations 40 through 77 (octal) are located in the processor, while locations 00 through 37 (octal) are located in the IOC. For this reason only, five bits are required to specify the SPM addressing. Figure 10.5 shows the SPM map for the processor.

The *ROM subsystem* provides the subcommands needed to direct the internal processor operation each machine cycle (each clock period). Earlier in this chapter, we discussed the microcommand structure, ROM word organization, and sequencing and branching methods implemented in the H4200. This section of the data flow should be very simple to understand and tie to the rest of the processor's functions. The ROM storage capacity is 2048 120-bit words. Its access time is 135 nanoseconds, with a 10 nanosecond overlap, thus providing an effective cycle time of 125 nanoseconds. The ROM is implemented from a two-core-per-bit linear select technology with nondestructive read-out. The interface between ROM and the processor consists of the ROM address register (QR) and the ROM data register (K). The QR is 12 bits (11 + parity) wide, fitted with an increment (+1) circuitry for sequential ROM word access. This incrementing logic allows the contents of sequential ROM locations to be made available for subcommand generation every successive clock period. Since we discussed ROM sequencing and branching before, it would be sufficient to review some of the hardware in this section. The ROM sequential operation could be altered by the branch address selection box which would load QR from one of five ROM branch address registers (QA through QE) according to the link flop (stats) settings and the fixed (A to E) highest to lowest priority. The five link flops are set to reflect the test that can be performed on the system indicators. There are 129 test commands allocated to these link flops. The JJ to QE to QR register path (shown in Fig. 10.4 as a direct path from JJ to QR) is used to transmit the operation code (stored in

SPM Word Location	*Name*	*Function*
40	—	Not used
41, 42, 43	ACO	Floating point accumulator 0
44	—	Not used
45, 46, 47	AC1	Floating point accumulator 1
50	—	Not used
51, 52, 53	AC2	Floating point accumulator 2
54	—	Not used
55, 56, 57	AC3	Floating point accumulator 3
60	WL3	Work location 3 } Provide intermediate storage for the microprogram
61	WL4	Work location 4
62	WL5	Work location 5
63	WL6	Work location 6
64	CSC	Consequence counter register used as a buffer for sequence counter for swapping and chaining.
65	WL7	Work location 7
66	EIR	External interrupt register
67	WL2	Work location 2
70	BC	B counter register to store the B operand address
71, 72, 73	—	Not used
74	AC	A counter register to store A operand address
75	WL1	Work location 1
76	ITR	Internal interrupt register
77	SC	Sequence counter—this is the main program counter.

Figure 10.5 The SPM map.

JJ early in the instruction fetch) to be used as the address of the first microinstruction in that particular instruction execution microroutine. There is another branch path from QY to QD to QR to address the scientific instruction execution microroutines in ROM. The BST (branch on stored test) subcommand generated directly from the ROM bit-position-13 is used to select the branch register (QA through QE) associated with the highest priority link flop that is on. The timing diagram in Fig. 10.6 shows a delay of one basic clock cycle (PDA = 125 nanoseconds) between the setting of BST and actual branch.

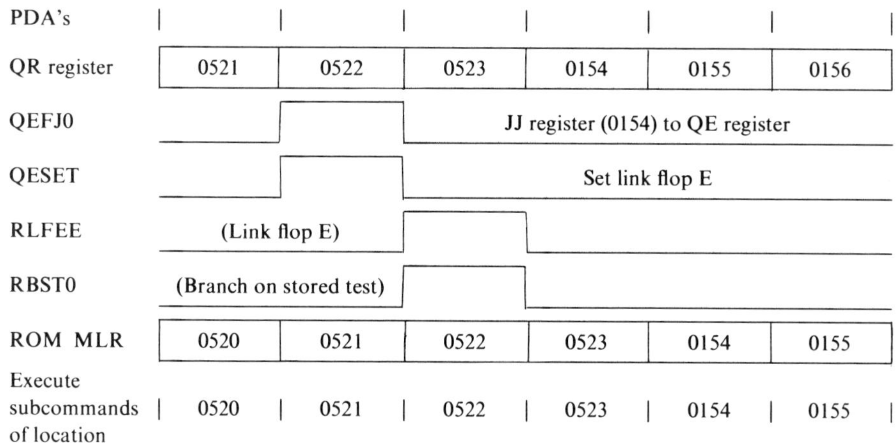

Figure 10.6 ROM branch timing.

Another branch subcommand is the BDT (branch on direct test) shown in the data flow. It is generated by the ROM bit position 51. This fast branch subcommand functions like the BST except that it can be implemented in less time to meet the timing requirements of the scientific instructions. The fast network reduces the test branch span to two words, one for test and BDT command, and one for following the BDT.

10.5 H4200 SYSTEM ARCHITECTURE[1]

This section is intended as a brief review of the pertinent architectural features of the H4200. It will help the reader to comprehend better some of the microprogramming illustrations given later in the chapter. For a complete understanding of the Honeywell 200 Series architecture, the reader is advised to read the Honeywell EPD publication entitled *Programmers' Reference Manual* (Models 200/1200/2200/4200), Hardware Manual (File no. 113.0005.-0000.1-139).

The H4200 is a variable-field-length machine. That is, any number of characters, from one up to a maximum number, are permitted to form a field in memory. The word mark bit which is indicated by a circle around the character is used to terminate the retrieval of an instruction, to terminate the execution of an instruction, and finally to delimit data fields.

In the latter case, the word mark is associated with the leftmost character in the field. In the case of an instruction termination, the word mark is set in the storage position immediately following its rightmost (its last) character. Hence, if the address of the rightmost character of the data field is given, the fetching of data proceeds from right to left until a character with a word mark is encountered.

10.5.1 Instruction Formats

There are six basic machine instruction formats, as shown in Fig. 10.7. They vary from one another in the number of operand address fields, the

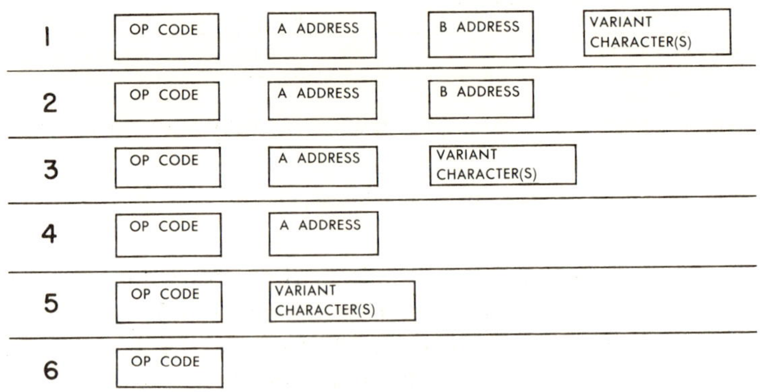

Figure 10.7 Series 200 instruction formats.

number and mode of addressing (this can be two-, three-, or four-character address fields), and the presence or absence of the variant field character(s).

10.5.2 Addressing

An instruction is stored in a field from 1 to N characters wide depending on the format of the instruction and the mode of addressing (two, three, or four characters). The instruction is addressed by specifying the address of its Op-code (leftmost) character. The machine reads the instruction from left to right until it senses a word mark (which is contained in the Op-code of the following instruction). A data field, on the other hand, is processed from its rightmost character (the address of which is specified by the programmer) leftward until a word mark (field mark) is encountered in its leftmost character.

Figure 10.8 shows a 9-character add instruction to add the contents of field A specified by address $(1776)_8$ to the B field specified by address $(1492)_8$.

The data group contains the concept of the *item* which consists of one or more consecutive storage locations where boundaries can be defined by giving their leftmost character positions, and which is operated on from left to right until an item mark is encountered. Alternatively, the boundaries can be defined by their rightmost positions in which case the operations proceed from

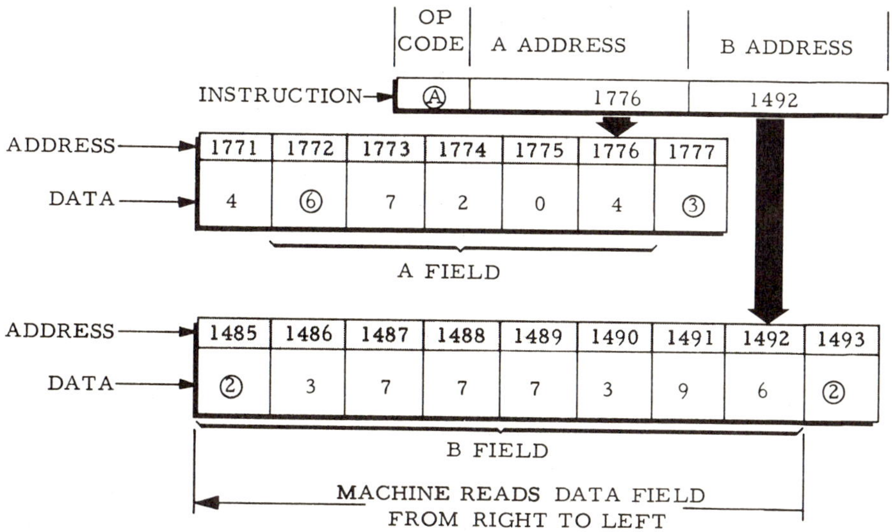

Figure 10.8 Extraction of data fields in typical add instruction.

right to left until an item mark is encountered. An item mark is shown as a dash under the character. Therefore an item can be one or more fields. The final concept in the H4200 data structure is the *record* which is defined as any unit of information to be transferred between the MM and I/O. A record is addressed by specifying its leftmost character location. The record is read from left to right until a record mark is sensed. Figure 10.9 shows the two item formats in the MM.

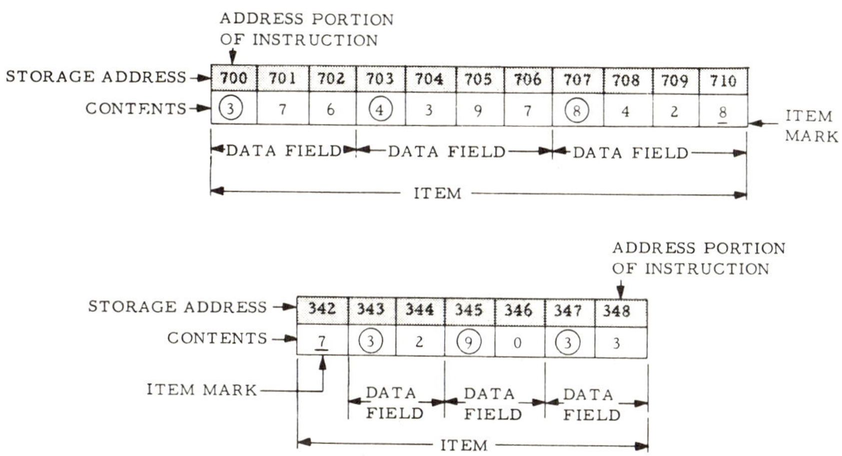

Figure 10.9 Two item formats in main memory.

10.5.3 Address Specifications and Modifications

As stated earlier in this section, an instruction is stored in a field of from 1 to N characters, depending on the instruction format and the addressing mode. This is specified by the programmer (by means of a change addressing mode instruction). The leftmost character (six bits) always contains the Op-code designation. The variant characters or IOC characters when present do occupy the rightmost character of the instruction.

10.5.4 Two-Character Address Mode (2 CM)

Each operand address is specified by the binary contents of two consecutive characters or 12 consecutive binary bits. This structure would permit the addressing of memory locations in the range of $0–4095_{10}$. During the extraction phase, the two characters are placed in the rightmost 12-bit positions of the address registers without disturbing the remaining leftmost positions of that register. The remaining bits are called *bank* bits. They are used to specify the base address. If bank bit 13 is 1, the address in the rightmost 12 bits specifies the address relative to 4096; if bit 14 is 1, then the base address $(8192)_{10}$ will be added to the relative address in the two-character address in order to derive the absolute address.

10.5.5 Three-Character Address Mode (3 CM)

This contains the rightmost fifteen bits to represent an address of an MM location in the range of $0–32{,}767_{10}$. The remaining three leftmost bits are referred to as an *address modifier* and are called the *sector bits*. The sector bits are not modified during the extraction of the instruction. The sector bits play a roll similar to the bank bits in the two-character mode. Each sector bit implies a base address to which the 15-bit address is added to form the absolute address. For example, if the 15 address bits specify an address of 9000_{10} and the rightmost bit of the sector (bit 16) is 1, then the actual operand address will be $(32{,}768_{10} + 9000)$.

10.5.6 Four-Character Address Mode (4 CM)

An operand address written in four-character mode is stored in the rightmost 19 bits of the possible 24 bits in the four characters. The rightmost 19 bits represent an address of an MM location in the range of $0–524{,}288_{10}$. The leftmost five bits are called an *address modifier* and are used to specify whether the address is direct, indirect, or indexed. Figure 10.10 illustrates the extrac-

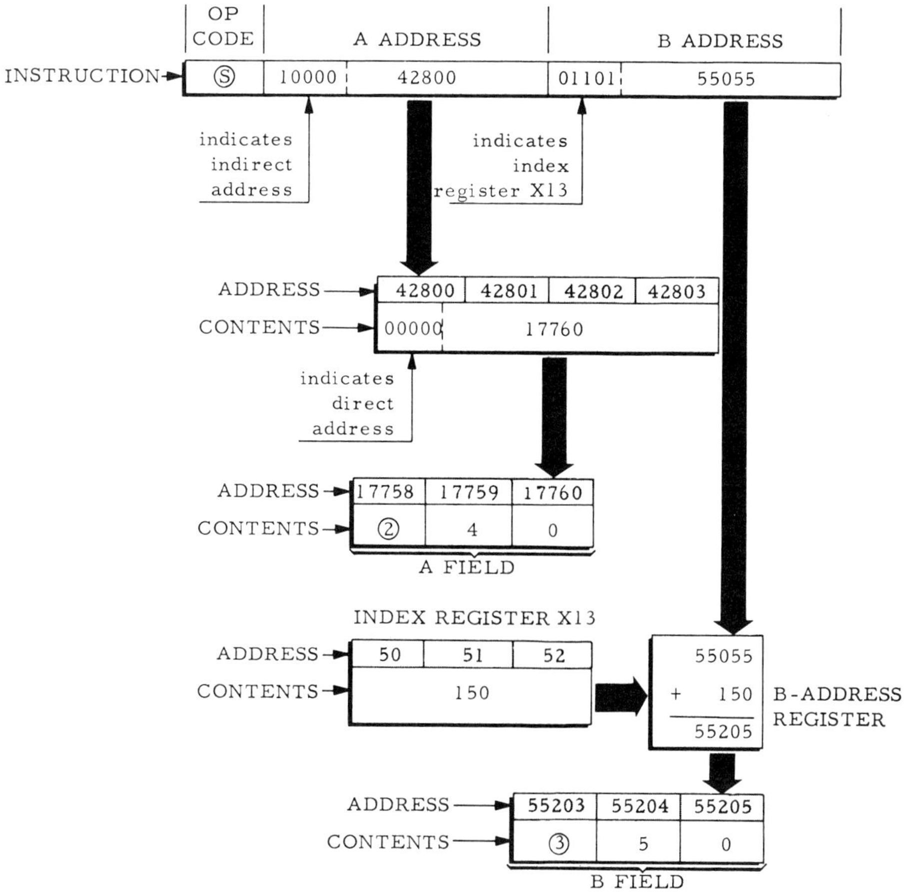

Figure 10.10 Extraction of indirect and indexed four-character addresses.

tion of indirect and indexed four-character addresses. If these 5 leftmost bits are:

00000—they specify a direct address
10000—they specify an indirect address

00001 to 11111 (skipping over 10000) they specify an indexed address. *Indirect addressing* for the three- and four-character modes is similar, in that the address given in the operand address field is the address of a second field in the MM that contains the actual or chained indirect address of the operand. The *indexed addressing* is more difficult. The rightmost four bits of the 5-bit field specify one of the 15 index registers (X1–X15) whose contents are to be added to the address field of the instruction. (*Note:* only the number of active address bits of the index register and the address field are combined.)

Instructions: Appendix A gives a list of the H4200 valid instructions and timing formulas. These instructions are classified into six functional categories: (1) arithmetic, (2) logic, (3) control, (4) interrupt control, (5) editing, and (6) input/output.

The binary subtraction is performed by adding the two's complement of the subtrahend to the minuend (B operand).

10.6 MICROPROGRAMMING THE H4200

The H4200 is a difficult machine to microprogram. This difficulty stems from the complex register organization and the variable-field-data structure. The item mark and word mark bits that are used to delimit instructions and data field keep the control section anticipating the alteration of the process with every new character to be handled. The microprogramming is also complicated by the large number of test and branch commands that can be operated simultaneously. Another difficulty stems from the six-to-one ratio between the main storage access time and the basic cycle time; unless properly microprogrammed, the CPU could become main storage bound, and the H4200 performance could be adversely affected.

The ROM control consists of a basic idle loop, an instruction fetch sequence common to all H4200 operation codes, an execution microroutine for each operation code, and some maintenance sequences.

The idle loop is entered when the processor is placed in the STOP Mode. Control panel or console requests for direct entry, or inquiry operations to SPM and MM, are executed in this loop. When the RUN command is issued, the processor branches from the idle loop to the instruction fetch sequence.

The instruction fetch routine, or the order extraction routine as it is called by Honeywell, consists of fetching the instruction from the main storage, extracting the operation code, and transferring it into the JJ register, transferring the variants or Op-code modifier into the VV, then storing the A field address in the address counter (AC) in the SPM location 74 (octal), and storing the **B** field address into the BC address counter in the SPM location 70 (octal). Finally the Op-code in JJ is shifted two positions left to produce the address of the first microinstruction in the microroutine that will execute the operation code. This instruction fetch is much more complex than this description indicates. The complexity stems from the many internal and external conditions that could affect this process. This extraction cycle is one of the most complex and difficult microroutines to follow; hence, we will postpone its treatment until we have looked at some simpler sequences. Before we do that, we will first look at the microprogramming assembly language and design aids available with the H4200.

Microprogramming the H4200 requires a good familiarity with the machine data flow and timing and with its architecture and design philosophy. It requires familiarity with the ROM stack and ROM word organization, the subcommands, their structure, and their formats. It also requires a thorough knowledge of the ROM automatic design (ROMAD) System, its capabilities and limitations, and its output format. The following sections are designed to deal with some of these topics which have not been discussed.

10.7 ROMAD

This acronym refers to an interim ROMAD System employed as an aid in designing the H4201 microcontrol. It is basically an extension of the Logic Design Automation System (DAII) developed by Honeywell to support the logic design and partitioning. Each ROM word is assigned an absolute address of seven alphanumeric characters that will consist of three leftmost alphabetic characters followed by a four-octal-character address. The three alphabetic characters specify the microroutine to which this specific microinstruction belongs.

Example: EXT 0100 refers to the *ext*raction microroutine which starts at ROM octal location 0100.

A file is created as a series of symbolic locations, each of which has an associated field which may include one branch address (in symbolic form) and any number of micro-orders. For example:

```
MCE0140:    /    )
         /YXNU01
         /YXNU02
         /YXNU03 = SFYY0, UMRQ0, MFYY0, MDEC0, GRP0D,
                   BT051, BT063, YYY04, YYY05, YYY06
```

(This input format and the individual fields and subcommands will be discussed later.)

The ROMAD generates from the microinstruction an object tape which shows the ROM bit pattern of every ROM word stored in the absolute address assigned by the microprogrammer. (The ROMAD does not have any general address assignment capability except in the sequential blocks in a microroutine whose first block address is assigned by the microprogrammer.) The ROMAD replaces all the mnemonic branch assignments by their octal value, determines the word bit pattern, and generates an output tape which is the ROM manufacturing interface tape.

The ROMAD provides several edits: it checks the data flow net, checks the micro-orders, and performs other input validity tests. Another design aid is the pattern file which extracts each word in which a given command or test was present. Parity checking on ROM address unassigned codes or a double assignment of the same address location are all tested and monitored by the ROMAD.

The system consists of the glossary file containing the symbolic address assignment, the symbolic file containing the five-character subcommands, their ROM bit pattern, and also a simple, easy-to-read English text of the function performed. Finally, the system contains the Honeywell Design Automation System tape.

The microprogrammer inputs his microroutine as a series of logic sequences of input records written symbolically. An input record (shown in the example below) would consist of the symbolic location of that microinstruction on one line, the branch address on a second line, and a series of subcommands (micro-orders) in any order he wishes, provided that they are consistent with the machine function and the symbolic file. After the assembly and ROM bit translation, the ROMAD would then translate these microinstructions into an

```
MCE01401:   /   )
         /YXNU01
         /YXNU02
         /YXNU03 = SFYY0  ,UMRQ0  ,MFYY0  ,MDEC0  ,GRP0D  ,
                   BT051  ,BT063  ,YYY04  ,YYY05  ,YYY06  ,
                   YYY07  ,YYY08  ,YYY09  ,YYY10  )

MCE0141:    /   )
         /YXNU01
         /YXNU02
         /YXNU03 = ESRC0  ,EFE40  ,MF1D0  ,GRP0D  ,BT051  ,
                   YYY03  ,YYY04  ,YYY05  ,YYY06  ,YYY07  ,
                   YYY08  ,YYY09  ,YYY10  )

MCE01421:   /   )
         /YXNU01
         /MCE0155
         /YXNU03 = 0ESET  ,YFMM0  ,USMST  ,ESRC0  ,EFE40  ,
                   QEFK0  ,0QAD1  ,1CMW2  ,RFKK0  ,CFYA0  ,
                   GRP0D  ,BT063  ,YYY07  ,YYY08  ,YYY09  ,
                   YYY10  )

MCE01431:   /   )
         /YXNU01
         /YXNU02
         /YXNU03 = 00BST  ,ESRC0  ,EFE40  ,GRP0D  ,BT063  ,
                   YYY03  ,YYY04  )

MCE0144:    /   )
         /YXNU01
         /MCE0410
         /YXNU03 = 1CMBC  ,RFKK0  ,GRP0D  ,QAFK0  ,BT063  ,
                   YYY04  ,YYY05  ,YYY06  ,YYY07  ,YYY08  ,
                   YYY09  ,YYY10  )
```

Figure 10.11 Sample microcommand input to ROMAD system.

easy-to-read English text showing the action flow and the branching modes and nets.

ROMAD provides the microprogrammer with the capability of replacing or deleting any micro-order within the microinstruction or any microinstruction

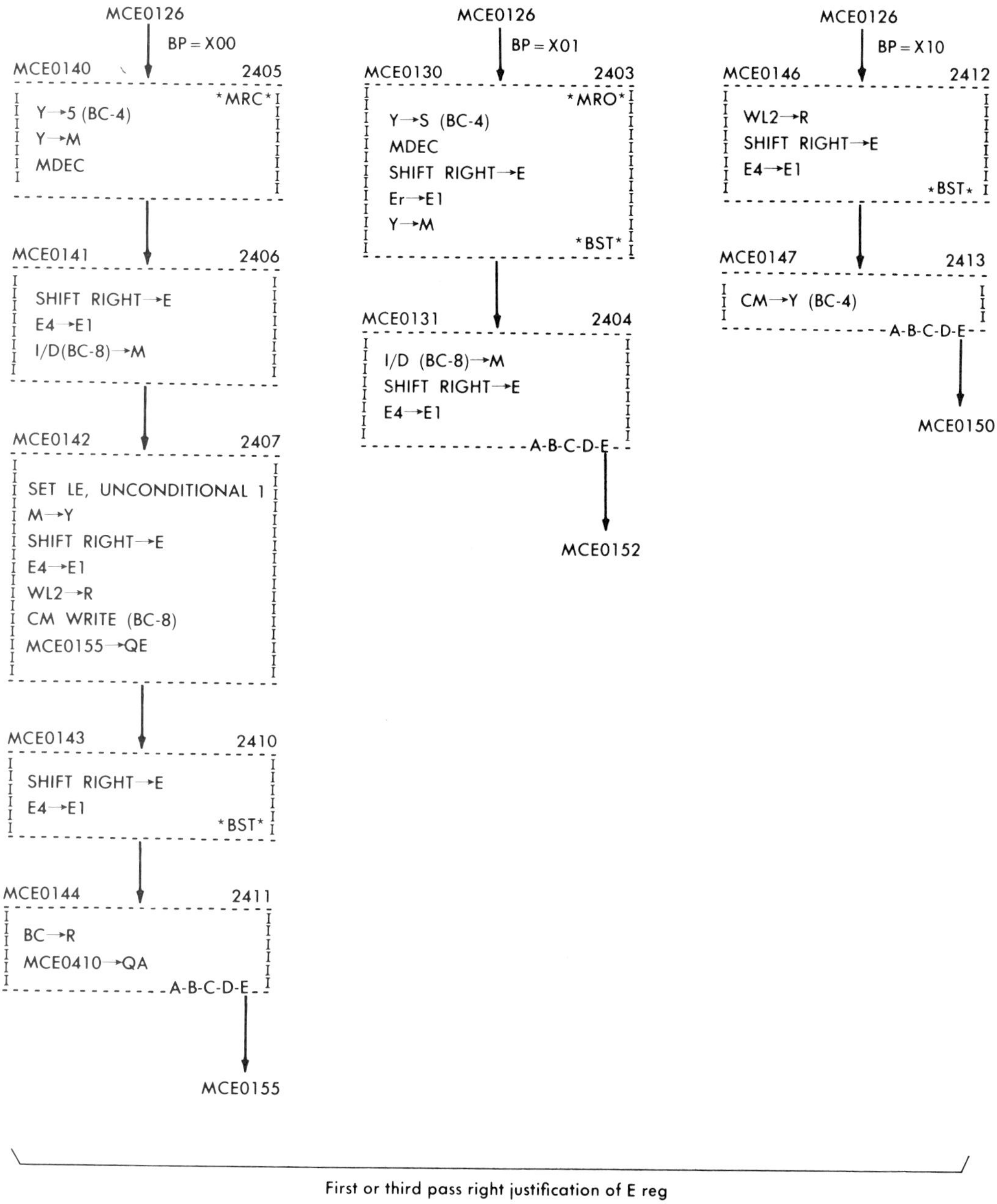

Figure 10-12 Sample ROMAD output.

within a microroutine. Figures 10.11 and 10.12 show samples of the input and output to and from the ROMAD respectively.

We stated earlier that in order to microprogram the H4200, one would have to be familiar with the glossary file (Appendix B), the flow-chart equivalent file (Appendix C), the input format, and finally, the output format.

I. *The Glossary File*

This contains all the microsubcommands needed to perform any logical function, arithmetic function, or data transfers. Figure 10.13 shows a sample page of the glossary file. It contains:

1. The function's symbolic name (subcommand) consisting of 5 alphanumeric characters.
2. The comment (CMT) followed by a description of the action or condition under which this action is executed.
3. The bit assignment field.

The subcommands are grouped into three classes:

(a) *The test commands*. These have a general formal 0λXXX, where the 0 indicates that it is a test command, the λ represents the grouping of such

```
FUNCTION

00ASRB1:SU     /    /CMT (IF EX IS LESS THAN EY BXGEY21              )
        = YXNUEE

00ASRC0:       /    /CMT (A SHIFT RIGHT ONE CHARACTER                )
        = 1113006

00ASRC1:       /    /CMT (IF BP DOES NOT EQUAL X11,SFBP300           )
        = 0611001

00ASRC2:       /    /CMT (IF AP DOES NOT EQUAL X11, SFAP300          )
        = 0913002

00ASRC3:       /    /CMT (IF A IS TAKEN FOR EDIT,PN1,SEATF10         )
        = 0661001

00ASTA0:       /    /CMT (ADDER SET ADVANCE CARRY                    )
        = 0761001

00ASTA1:SU     /    /CMT (IF FP SUBTRACT RQY0130                     )
        = YXNUEE

00ASTA2:       /    /CMT (IF OVERFLOW                                )
        = 0441001

00ASTA3:SU     /    /CMT (IF SNA=1, OTHERWISE RESET ACI,SGNAN30      )
        = 0741001

00ASTA4:SU     /    /CMT (IF SNX=0 ,SGNBN40                          )
        = 0411001
```

Figure 10.13 Sample glossary file bit assignment entry for all tests and commands.

a test command according to the link flop they effect, and the XXX field designates the condition to be tested. For example, the subcommand (0CCH1) implies that there is a test command in which the C link flop will be set if the test CH = 001 is satisfied. Other examples here are 0BAP3 = Set LB, if AP = X11; and 0DB3R = Set LD, if BH = 0XX (the X indicating the "don't care" condition where the corresponding bit can be "0" or "1").

(b) *The action subcommands.* These have the general format of a four alphabetic character followed by one numeric character. The alphabetic characters specify the action to be taken. The numeric character is either zero or nonzero. If it is zero, it specifies an unconditional subcommand; otherwise, it specifies a conditional subcommand. Examples are BFAD0 and AFAD1. The first microcommand says to set the character B4 from the adder unconditionally, whereas the second subcommand (AFAD1) says to set the character A1 from the adder output only if (ZA + ZS). Other examples are ASLC0 and BSLC1. The ASLC0 says to shift the left contents of A one character unconditionally. The BSLC1 says to shift the contents of register B left if the A pointer rightmost two characters are not equal to 0.

(c) *Miscellaneous subcommands.* This last class of subcommands contains the miscellaneous ones that do not fit into the previous two classes, for example, the no-operation subcommands and the dummies. An example here would be the setting of the 5-bit CM field in the ROM word to address the different locations of the SPM. (*Note:* the SPM can be changed every cycle.) The format for this class is 1CMλλ, where 1CM signals this class of action, namely, setting the CM field. The last two characters specify the address of the specific SPM word or the source of this address. For example, the 1CMW1, 1CMW2, . . . , 1CMW7 microsubcommands select the work locations 1 through 7 in the SPM. The ROMAD System would place the correct absolute address into the CM field to interrogate the proper SPM word. Other examples are 1CMLR and 1CMHR; the subcommands would address the two SPM locations for the external high-level interrupts (EIR), and the internal lower-level interrupts (IIR), respectively.

In the glossary file (Fig. 11.13 and Appendix B), these subcommand formats are preceded by two 0's. This is simply a requirement of the ROMAD output format and can be ignored. The subcommand name in the glossary file is followed by the description of the action taken or the condition specified. This description is bracketed and is preceded by CMT—an abbreviation for comment.

Examples

00ASRC0: / /CMT (A shift right one character) unconditional
= 1113006

and

```
OOASRC1:   /   /CMT    (If BP does not equal X11, SFBP300)
  = 0611001
```

Notes:

1. That the basic description of the first command (namely, a shift right one character) is not repeated in the second example. Only the condition is specified in the comment field of the second example. In this specific subcommand (OOASRC1) the condition is: if BP is not equal to X11; SFBP300. The X11 says that the test does not apply to the high-order bits of BP; only the two low-order bits will be tested for 11.

2. The SFBP300 is of little meaning to the microprogrammer; it specifies a hardware function to be activated or tested.

3. The last piece of information given here is the 7-digit ROM bit assignment. It shows the required ROM bit structure needed to generate the subcommand. The 7 bits break down in the following manner:

```
                                        XXX      X       X
Starting ROM bit ------------------------┘      |       |
Number of consecutive ROM bits -----------------┘       |
Octal configuration of ROM bits ------------------------┘
Example: 0923006
ROM bits                                092      093     094
Structure                                1        1       0
(or three consecutive ROM bits starting with location 092 will
contain the code 110 or octal 6).
```

If the terms YXNUHDW or YXNUDMY appear in place of this ROM bit structure, it means that this subcommand is not generated by the ROM bit structure but by hardware in the case of HDW; or that this condition is always taking place (in the case of DMY). Instead of generating it, other subcommands must be generated to inhibit this condition. An example of the first is a subcommand BDT3C, which is generated by the hardware to transfer QC to QR and set B35–B33 to 111. An example of the second is the MDEC0, which is used for clarification since the M register normally decrements.

II. *The Input Records*

The microcommand inputs to the ROMAD are illustrated in Fig. 10.11. Each record consists of 4 lines and represents one microinstruction. The first line gives the symbolic location, consisting of 3 alphabetic characters followed by 4 numeric characters. The 3 alphabetic characters represent the symbolic name for the microroutine. In Fig. 10.11, MCE is a symbolic name given to the microroutine which executes the move-character-and-edit instruction, followed by the 4-digit octal absolute address assigned to this microinstruction. Note that the ROMAD has limited flexibility in assigning the addresses. The microprogrammer would, in fact, have to assign the address of the first microinstruction in each microroutine, and the ROMAD would then be able to assign the consecutive ROM addresses for the consecutive microinstructions.

Review: The absolute address for the beginning of the execution routine of any instruction is generated by shifting the instruction code (OP-CODE) left two bit positions and adding two 0's at the vacated two rightmost bit positions. For example, the compare instruction OP-CODE is 33 octal or 011011 binary. If we shift the binary representation left two positions, and regroup the resulting field, we should have 001 | 101 | 100 or 0154 octal. This symbolic address is followed by three symbolic address fields, most of which appear in the form YXNU01, YXNU02, or YXNU03. These formats, when they appear, imply that this specific address field is not used. The first address field, which appears in line two of the input record, is used to give the absolute address when assigned to the start of the block of microinstructions. The second entry (where the YXNU02 may appear) gives the symbolic address for the "branch to location" if the specified branch conditions are met. The last address position (where YXNU03 may appear), represents the absolute address of the preceding "branch to address." This, however, is infrequently used and generally appears as YXNU03 format.

The last symbolic address field (namely the YXNU03) is followed by an equal sign followed by an unlimited number of five-character fields, separated by commas, five fields to a line. These fields contain the symbolic microsubcommands, which were discussed earlier and which are found in the glossary file, Appendix B.

The microcommand fields which have the format YYYXX are dummy fields. They are specified by the microprogrammer to allow him to add new microcommands at a later date without having to restate the entire microinstruction. This procedure reduced the time required for the file update.

Let us review the MCE0144 microinstruction (Fig. 10.11). It specifies that this microinstruction has a symbolic name MCE0144. Since the third line contains a symbolic branch address MCE0410, it must have a microcommand to specify the condition under which this branch is taken (QAFK0). The fourth line has the following microcommands:

ICMBC	implies CM address field equal to BC
RFKK0	implies ROM CM address field → R register
GRP0D	implies that group D is active; it is for checking purposes only, and is not a command
QAFK0	implies that the branch address is to be placed in QA register
BT063	is a parity bit; each half-ROM word (60 bits) is coded with even parity for hardware checking, and these bits are placed in "don't care" positions.

Finally, YYY04–YYY10 are no-operation symbols placed in this record for future expansion only.

III. *The Flow-Chart Equivalent (Appendix C)*
The final step for the ROMAD in performing the microroutine assembly is to organize and to format the action specified into a flow-chart-like output in which each box contains the actions specified in one microinstruction (example, Fig. 10.14). Each line in the box spells out in a readable form the actions specified by some subcommand. This is obtained by replacing each subcommand by its equivalent file form, Appendix C. Appendix C shows three columns (or four in a few instances) which are the flow-chart equivalent columns arranged alphabetically, the subcommand name (five characters), and one or two ROM-bit field instructions, as discussed earlier.

10.8 MICROPROGRAM TRACE AND EXAMPLES

The following section gives a few simple and direct microroutines. We will use them to illustrate the H4200 microroutine structure and trace the action as specified by the ROM on a cycle-by-cycle basis.

Example 1

Transfer the contents of one Control Memory Location (SPM) to another. The means by which the transfer is accomplished in the 4201 is illustrated in Fig. 10.14. In EXT0102 the A counter (SPM location 74 octal) is addressed (AC → R). In EXT0103 the contents of the A counter are stored in the Y

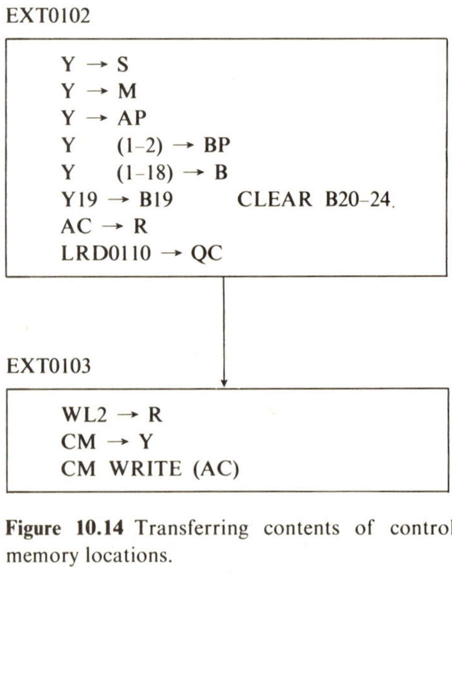

Figure 10.14 Transferring contents of control memory locations.

register (CM → Y), WL2 (SPM location 67 octal) is addressed (WL2 → R), and the command to write into SPM from Y is issued (CM write). It should be noted that other actions pertinent to the extraction microroutine are also performed.

The corresponding input records and subcommands for these two microinstructions are:

```
EXT0102:    /    )
        /YXNU01
        /LRD0110
        /YX    03 = SFYY0,  MFYY0,  APFY0,  UMRQ0, BPFY0,  QCFK0,  BLFY0,
                    BMFY0,  GRPCG,  1CMAC,  RFKK0, YYY06,  YYY07,  YYY08,
                    YYY09,  YYY10)
EXT0103:    /    )
        /YXNU01
        /EDM0110
        /YXNU03 =   ICMW2,  YFCM0,  RFKK0,  CFYA0,  YYY21,  YYY20,  0ASTR,
                    0BLRD,  0CHRD,  0DEDM,  0DFK0,  MINC0,  0QAD2,  GRPCG,
                    YYY03,  YYY04,  YYY05,  YYY06,  YYY07,  YYY08,  YYY09,
                    YYY10)
```

Example 2

This example, given in the form of an exercise, is designed to familiarize the reader with the input and output formats and the subcommand equivalent file in Appendix C. It also provides a good illustration of the setting, the timing, and the execution of a multiway branch in the system.

Note the timing on the branching operation. The branch addresses were set in the first four microinstructions. The link flops were set in the fourth microinstruction, and the branch on test subcommand (BST) was set in the fifth microinstruction. Note the delay of one PDA between the setting of the BST and the actual branching which does not take place until the sixth and last cycle in Fig. 10.15.

In this microroutine and the A counter register located in SPM-74 (which contains the A operand address) is stored in WL2 (SPM-67) if the order is SCR or LCR (Appendix A). It also checks for and exits to the proper microroutine in case external or internal interrupts or diagnostic extraction or stop requests are made.

The reader is asked to prepare the input records (similar to those in Fig. 10.11). Assign the first microinstruction the symbolic address EXT0100.

Example 3

Positioning of a result character for delivery. The technique for positioning a character in an operand register to place it in the proper character position for delivery to memory is relatively simple and is used many times in instruction

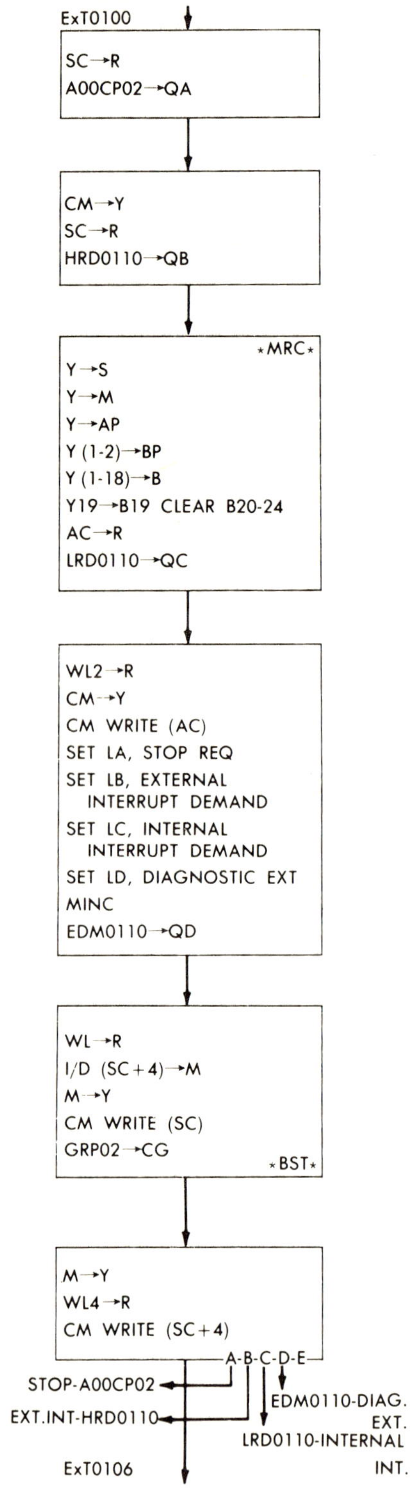

Figure 10.15 Data flow for Example 2.

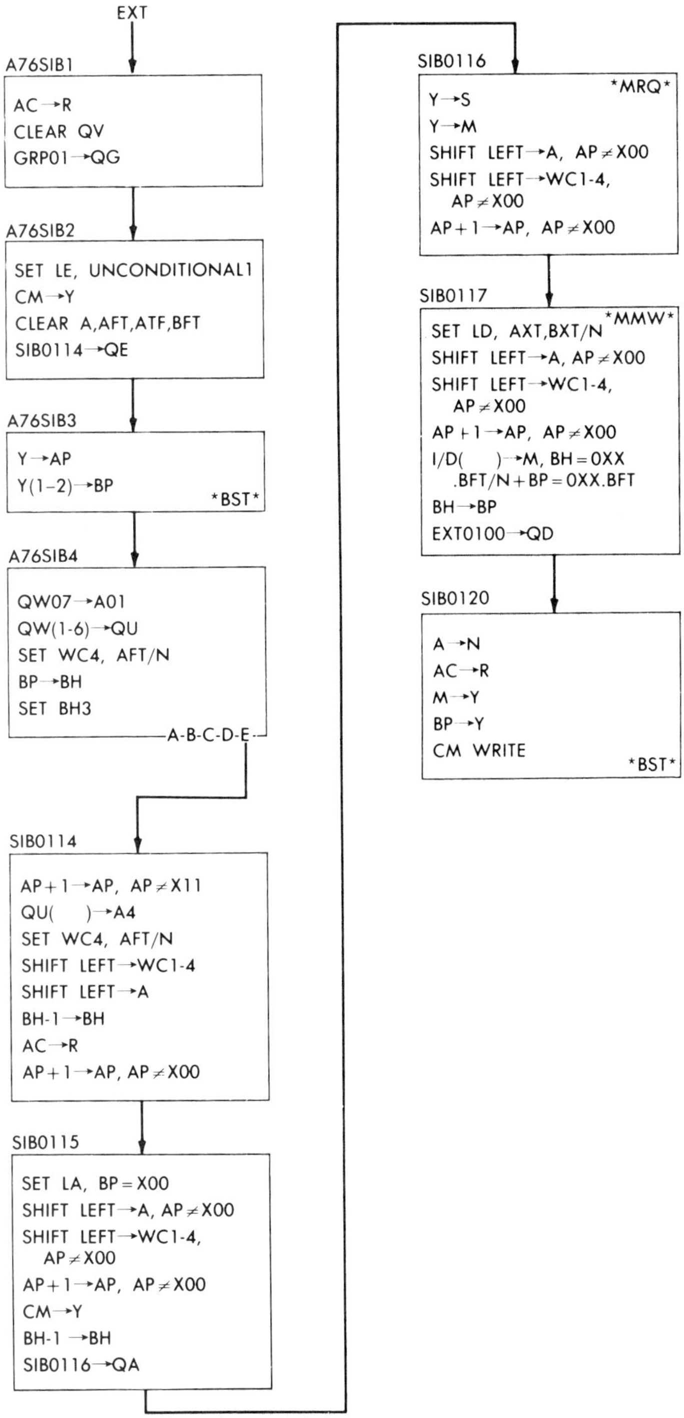

Figure 10.16 Positioning result characters.

flow. The technique described here is for the SIB instruction. In this case the character to be stored in memory is delivered to the A4 character position (BAA01 through BAA06). However the A address of the instruction may specify storing of the character in any of the four character positions of the resulting half-word. A method must be used to shift the character into the position specified by the low-order two bits in the SIB A address bits.

The method used is to load the A and B pointer registers with the low-order two A address Bits. The difference between the pointer setting and character position 11 (A4) is equal to the number of shifts required to position the character. The difference can be found by incrementing the pointer until it equals 11. Each time the pointer is allowed to increment, the character is shifted to the left once. This is accomplished by the subcommands of AP + 1 → AP if AP ≠ X11 and shift left → A if AP ≠ X11. However, the latter of the two does not exist; only the subcommand shift left → A if AP ≠ X00 exists. Since the 00 character is only one character position to the right of the 11-character position, an accurate difference can be derived by unconditionally making the pointer contents one greater before the shifting is allowed to start.

Figure 10.16 shows the action just described. In A76SIB3 the A and B pointers are loaded with the low-order two A address bits. The unconditional increment occurs in SIB0114 (the A register is also shifted because the result is two characters). In cycles SIB0114 through SIB0117 the A pointer is incremented and the A register is shifted as long as the A pointer is not equal to X00. When the A pointer equals X00, shifting and pointer incrementing stops, and the character is transferred to the N register. Trace the action as specified by the data flow shown in Fig. 10.16.

Example 4

Branch if characters equal (BCE). This Op-code was chosen to illustrate the microinstruction flow and the action performed during the executing cycles. This H4200 instruction specifies that the single character specified by the B address is compared to the variant character. If the bit configuration of the two characters is equal, the program branches to the location specified by the A address and the contents of the sequence counter are stored in the B counter. If the bit configuration is not equal, the program continues in sequence. Issuing this instruction when the processor is in Special Mode (SCM) causes the comparison indicators to be set to define how the variant character compares to the B character. The comparison indicators can then be interrogated by subsequent branch on condition test (BCT) instructions. Here is how it is executed:

1. *Data flow (Fig. 10.4)*

The variant character, which is to be subtracted from the B character, is transferred from the VV register to the QU register and then to the A4 position of the A register. It is complemented in the A register and then transferred to B4 via the adder. The B character is extracted from the MM, placed in the QU register with an N Select subcommand, and then transferred to A4 where it is added to the complement of the variant character (ACI is set to provide the true binary difference). The difference is then transferred from the adder to A1 where it is examined by Y Had a One (YH1) for not being zero.

2. *Pointer and history usage*

The A pointer register is used to perform the update of the B counter following the extraction of the B field character that is to be compared to the variant. The B address pointer bits are transferred to the A pointer register, and then the contents of the A pointer register are decremented by one. If bit three of the A pointer register sets because of this decrement, the B field address is moved to the next half-word (B field pointer bits originally 00); if bit three does not set, the B field address is left defining the original half-word.

3. *Cycle flow (Figs. 10.17 and 10.18)*

(a) A55BCE1 through A55BCE4. The contents of the B counter are extracted from control memory and used to address the MM in A55BCE3. The high-order 17 bits of this address are transferred to the M register, and the low-order two bits are transferred to the A and B pointer registers in this same cycle. The low-order two bits of the address are transferred to the B pointer register to control the N Select subcommand that is issued in BCE0116; they are transferred to the A pointer register to control the B counter update that is performed in BCE0117 and BCE0120.

Link E is set in A55BCE2 to allow an entry to BCE0114 from A55BCE4; it is set in A55BCE4 to allow cycles BCE0014 through BCE0016 to be executed twice.

Exiting A55BCE4, the original B address is in the M and B Pointer registers, the next sequential B field pointer bits are in the A point register, link flop E is set, the specified B field character has been requested from main memory, and the variant is in the QU register (see A55BCE4). The next cycle is BCE0114 (Link E, unconditionally).

(b) BCE0114 through BCE0125. The variant character is transferred from the QU register to A4 during the first pass through BCE0115. It is complemented in BCE0116, and the complemented variant is transferred from the adder to the B register during the second pass through BCE0115.

The B field character is received from the MM during the second BCE0115 cycle, transferred to the QU register in BCE0116, and then transferred to A4 in BCE0117. A carry is forced into the low-order bit position of the adder (ACI) in BCE0117 to validate the difference

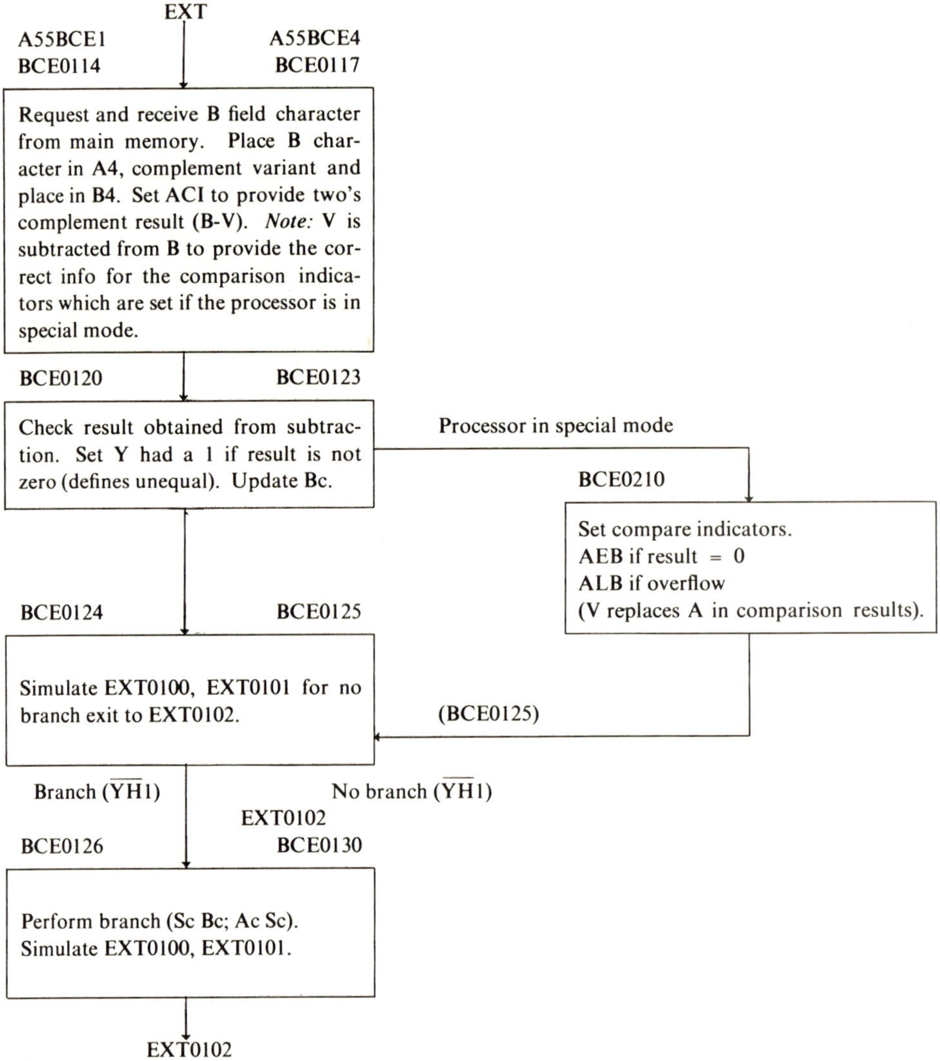

Figure 10.17 Block diagram of BCE execution.

that is transferred from the adder to A1 in BCE0121. YH1 is reset in BCE0121 and is set in BCE0122 if the variant and B field character are not equal. If they are equal, their difference is zero and YH1 remains rest. YH1 is examined for a reset by link flop D in BCE0123. If YH1 is reset, the variant and B field characters are equal, and BCE0126 through BCE0130 are entered to perform the Branch. Link flop E is unconditionally set in BCE0123 to allow

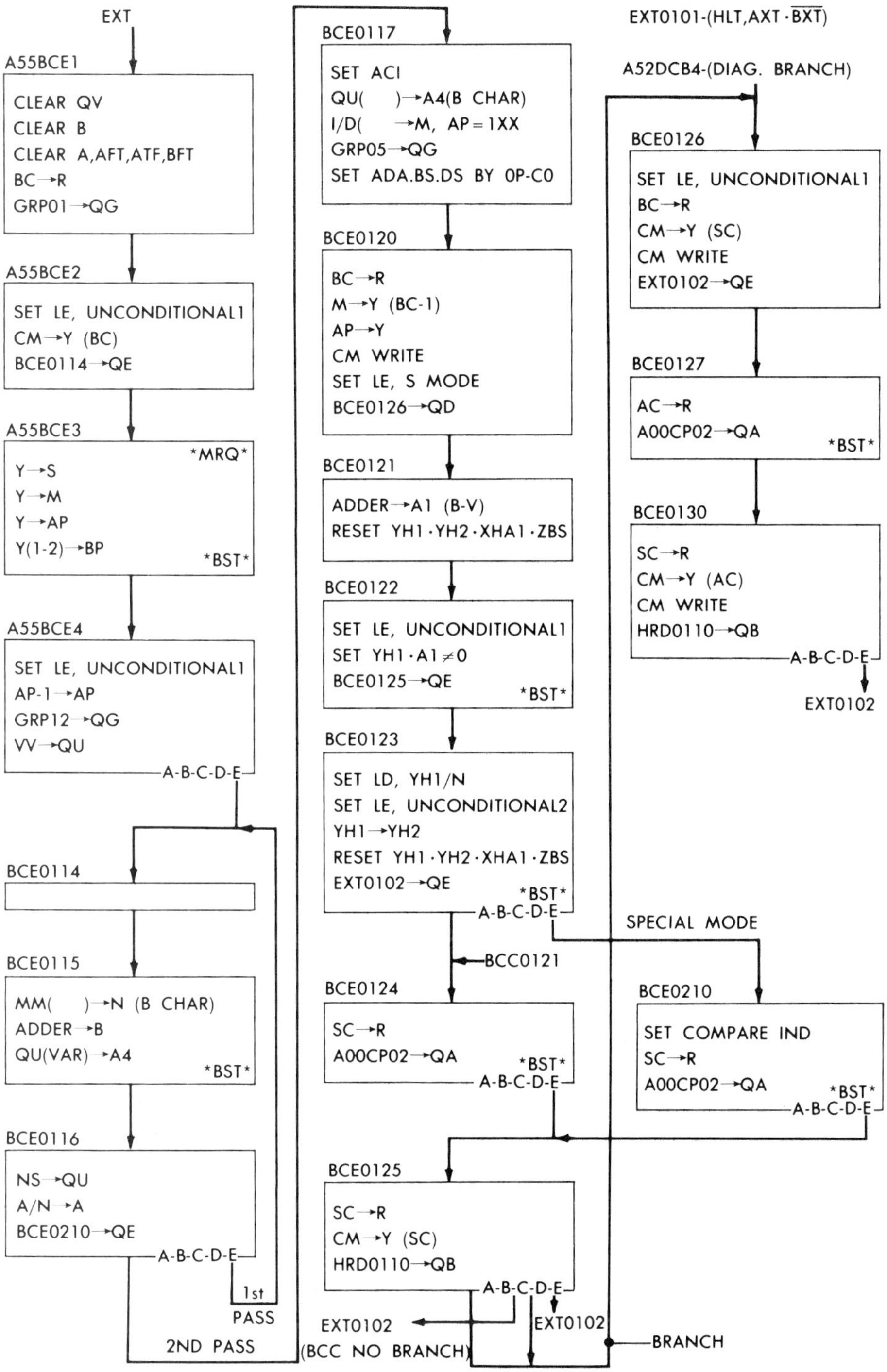

Figure 10.18 BCE execution flow.

EXT0102 to be entered from BCE0125 if the variant and B field characters are not equal (i.e., link flop D was not set).

The B address counter contents are decremented by one, and the result is returned to control memory in BCE0120. The decrement is accomplished by placing the upper 17 bits of the B address in the M register and the lower two bits of the B address in the A pointer register. The A pointer register contents are decremented in A55BCE4 and examined in BCE0117 to determine if the upper 17 bits of the address must be modified because of a borrow from bit three of the address. (A borrow is necessary if the lower two bits are both zero and is indicated by bit three of the A pointer register setting.) The M register contents are decremented if bit three of the A pointer register is set in BCE0117 to reflect the action of the borrow.

Link flop E is set in BCE0120 if the processor is operating in special mode. The setting of link flop E causes BCE0120 to be executed instead of BCE0124 and the comparison indicators (ALB and AEB) to be set to show how V compared to B (i.e., ALB sets if V is less than or equal to B defined by Overflow; AEB sets if they are equal).

Exiting BCE0125, the variant character has been compared to the B field character via binary subtraction and, if the processor is operating in special mode, functions ALB and AEB have been set to define how V compared to B.

The next cycle is as follows:

BCE0126 — Link D — if the variant and B field characters are equal, and
EXT0102 — Link E — if the variant and B field characters are not equal.

(c) BCE0126 through BCE0130. These cycles store the contents of the sequence counter in the B counter and cause a branch by transferring the contents of the A counter to the sequence counter.

The contents of the sequence counter are extracted from the control memory in BCE0126 and written into the B counter in this same cycle. The A counter contents are extracted from the control memory BCE0130 and written into the sequence counter.

Exiting BCE0130, the sequence counter contents have been modified to cause a Branch to the Op-code specified by the A address of the instruction just executed. The next cycle is EXT0102 (Link E, unconditionally).

Example 5*

Move character and edit

Introduction

The Edit instruction is used to insert identifying symbols and punctuation, and to suppress unwanted zeros in a data field. The A field contains the information to be edited. The B field contains an edit control word which provides a framework for the Edit operation. When the instruction is executed, the A field data is moved to the B field. The format and punctuation of the result (located in the **B** field) are determined by the control information contained in the original B field. The Edit instruction is very useful in setting up the format and punctuation of an information field which is to be printed.

Editing is performed according to the following rules:

1. Any character in the Series 200 character set can be used in the control word. The characters having special meaning are listed in Table 10.2. Any other character, if included in the edit control word, will remain in that same position in the result field.

2. A word mark in the high-order position of the B field terminates the first scan.

3. The number of replaceable characters in the edit control word must be at least as large as the number of characters in the A field.

4. Data is transferred from the A field, character-by-character, from right to left (decrementing). If a zero suppression symbol is not sensed in the edit control word, the operation terminates when the B field word mark is detected. A zero suppression symbol would cause a scan of the result field from left to right (incrementing). During this scan, high-order zeros and commas are replaced by blanks (unless asterisk protection is called for). Zero suppression is terminated by finding any of the following:
   (a) A decimal digit from one through nine.
   (b) A decimal point.
   (c) The location that originally contained the zero suppression symbol (this location is word marked on the first scan of the B field).

A call for zero suppression is registered if a zero is found in the control word.

5. An asterisk immediately to the left of the zero suppression symbol in the control word causes the high-order zeros and commas to be replaced by asterisks instead of blanks during the second scan. High-order blanks are also replaced by asterisks.

6. A dollar sign immediately to the left of the zero suppression symbol in the control word causes the edited result to be rescanned following the zero

*This example, its flow-chart and microinstruction trace are from Reference 4 to illustrate the concept of a complete microinstruction trace.

Table 10.2. MOVE CHARACTER AND EDIT CONTROL

Control character	*Description*
b (blank) 15_8	Blanks are replaced with A field characters such that the rightmost character in the A field replaces the rightmost blank in the edit control word and all high-order A field characters replace successively higher order blanks.
0 (zero)	This symbol specifies zero suppression. Its location in the control word is interpreted as the rightmost limit of zero suppression. It is replaced with an A field character.
(decimal point) 33_8	The decimal point remains in the edit field in the position where written.
(comma) 73_8	Commas remain in the edit field where written unless zero suppression is specified. Commas in control word positions to the left of the high-order character transferred from the A field are replaced by blanks.
C, R, C_R (credit) 23_8 -51_8 -75_8 0 (minus) -40 0 is printed as a minus symbol.	The credit or minus symbol is undisturbed if the sign in the units position of the A field is negative. If the sign is positive, the credit (or minus) symbol is blanked out. A credit (or minus) symbol transferred from the A field is not subject to sign control.
37_8	An octal 37 is replaced by a blank in the edit field.
(asterisk) 54	The asterisk remains in the edited field in the position where written unless it appears immediately to the left of the zero suppression symbol.
$ (dollar sign) 53_8	The dollar sign remains in the edited field in the position where written unless it appears immediately to the left of the zero suppression symbol.

suppression scan. During this scan, the dollar sign is floated to the left of the high-order significant digit in the result.

Note: Float the dollar sign or asterisk protection may be called for during an edit, but not both. Only the first character left of the zero suppression symbol is examined for a dollar sign or asterisk.

Both the A and B field must have defining word marks. The A field word mark terminates the transfer of data from the A field. The B field word mark terminates the edit operation if no zero suppression symbol is sensed in the edit control word. The B field word mark is cleared during execution.

Implementation

First Scan (Right to Left)

The A field sign is tested and stored. Successive characters in the B field are examined. Any location in the B field containing an octal 37 has its contents replaced by an octal 15 (blank).

If the sign of [A] is positive, then until a zero or an octal 15 is encountered, any location in the B field that contains C, R, C_R, or a minus sign (i.e., octal 23, 51, 75 or 40) has its contents replaced by octal 15 (blank). This replacement is discontinued as soon as a zero or octal 15 is found in the B field, but is resumed after transfer of the high-order character of the A field.

A B field location is available for transfer of a character, provided that it contains zero or octal 15, or that it is immediately to the left of the first (rightmost) zero encountered in the B field and contains an asterisk (octal 54) or dollar sign (octal 53). B field locations available for transfer will be referred to hereafter as containing replaceable characters. Each time a replaceable character is found in the B field, a character from the A field will be transferred into this location. The first character is to be transferred to the rightmost character of the A field. The sign positions of the first character will be replaced by zeros. Subsequent transfers of A characters into the B field will include the sign positions (bits five and six) of these characters.

Transfer of data from the A field terminates with the transfer of the word marked character. After the A field terminates, scanning of the B field continues (octal 37 being replaced by octal 15), and if the sign of A is positive, replacement of C, R, C_R by blanks is now resumed. After the A field has terminated, any commas (octal 73) found in the B field will be replaced by octal 15. The scan of the B field terminates after the word marked location is examined and the word mark is cleared.

If during the first pass a zero has been found in the B field, a call for zero suppression is registered and the location containing the zero is word marked. Only the rightmost zero found in the B field is treated in this manner. The character in the next location to the left of this zero is examined to determine if it contains an asterisk or dollar sign (octal 54 or 53). An asterisk in this location registers a call for asterisk protection. A dollar sign in this location registers a call to float the dollar sign. Either of these cases requires another scan of the B field. If no zero is found in the B field, the end of the first scan is the end of the operation.

Second Scan (Left to Right)

During the second scan, successive characters are examined starting with the high-order location of the B field. When zero suppression is on, all zeros, commas, and blanks are replaced either by a blank (no call for asterisk protection) or by an asterisk (if a call for asterisk protection has been registered). When a location in the B field is found to contain a decimal point (octal 33) or a numeric, one through nine, zero suppression is terminated. The scan of the B field continues until the location, which originally contained the zero suppression symbol, is read out. The word mark set during the first scan is cleared. If no call to float the dollar sign has been registered, the end of the second scan is the end of the operation.

Third Scan (Right to Left)

The third scan is necessary if the location left of the zero suppression symbol contained a dollar sign. After clearing the word mark in the location which contained the zero suppression symbol, this location and successive locations to the left are examined. When a location is found containing a blank (octal 15), a dollar sign is written into that location and the operation terminates.

The logical implementation of the edit rules is relatively simple, providing one thoroughly understands these rules. The following are three examples of edits showing the information (A) and control (B) fields before and after each scan.

1.

Locations	0100	0101	0102	0103	0104	0105	0106
A Field	(00)	00	00	05	01	03	45

Locations	200	201	202	203	204	205	206	207	210	211	212	213	214
B Field (before scan)	(15)	15	15	73	15	15	15	33	15	15	37	37	40
B Field (after scan)	00	00	00	73	00	05	01	33	03	05	15	15	40
Result =	15	0	0	·	0	5	1	·	3	5			—

In the first example, no call for zero suppression is registered. Only one scan is necessary.

2.

Locations	100	101	102	103	104	105	106
A Field	(00)	00	00	05	01	03	45

Locations	200	201	202	203	204	205	206	207	210	211	212	213	214
B Field (before scan)	(15)	15	73	15	54	00	15	33	15	15	37	37	40
B Field (end of first scan)	15	00	73	00	00	(05)	01	33	03	05	15	15	40
B Field (end of second scan)	54	54	54	54	54	05	01	33	03	05	15	15	40
Result =	*	*	*	*	*	5	1	·	3	5			—

In the second example, a call for asterisk protection is registered (zero in location 205, and an asterisk in location 204). Two scans are required for this edit.

3\.

Locations	100	101	102	103	104	105	106
A Field	(00)	00	00	05	01	03	45

Locations	200	201	202	203	204	205	206	207	210	211	212	213	214
B Field (before scan)	(15)	15	73	15	53	00	15	33	15	15	37	37	40
B Field (end of first scan)	15	00	73	00	00	(05)	01	33	03	05	15	15	40
B Field (end of second scan)	15	15	15	15	15	05	01	33	03	05	15	15	40
B Field (end of third scan)	15	15	15	15	53	05	01	33	03	05	15	15	40
Result =					$	5	1	.	3	5			—

In the third example, a call to float the dollar sign is registered (zero suppression symbol in location 205, and a dollar sign in location 204). This edit requires three scans.

Data Flow

First Scan. See Fig. 10.19(a) for the first scan data flow. The control field (B field) is read out of memory and right justified in the E register. The characters to be edited (A field) are read from and right justified in the A register. E4 is monitored by the edit control board (SEC). Under control of the rules previously mentioned the exit logic will generate the result character to be delivered to the B field. The result character is delivered back to E1. During the first scan this character is: (1) the character in E4 if that character is nonreplaceable, or (2) the character in A4 if the character in E4 is replaceable by an A field character, or (3) a blank (15_8) will be delivered back to E1 if the character in E4 is a nonreplaceable blank (37_8). Other conditions exist that also generate blanks. When the first zero is detected in E4, a word mark is sent back to E1 with the character that is in A4. When the E register has received enough characters for a B field write, the E register is transferred to the N register via the B register.

Second Scan. See Fig. 10.19(b). The high-order characters of the B field are left justified in the E register. E1 is monitored by the edit control logic. The

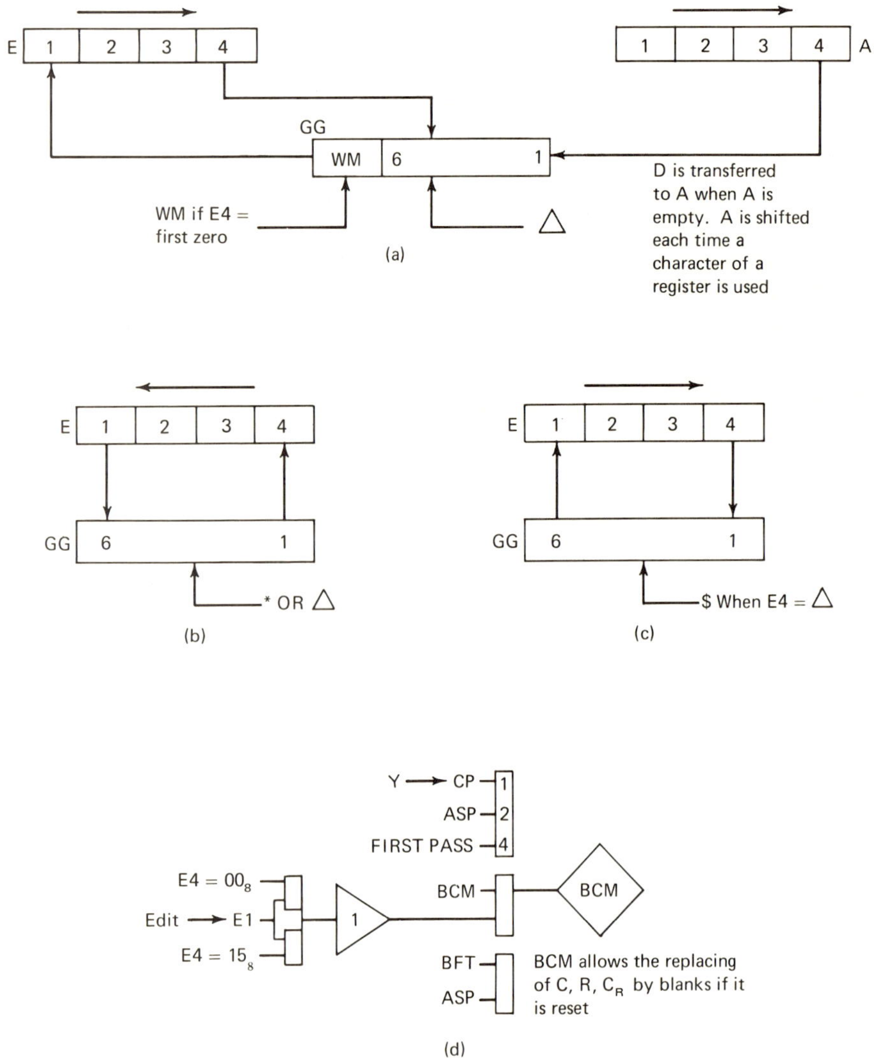

Figure 10.19 (a) First scan data flow; (b) second scan data flow; (c) third scan data flow; (d) BCM logic.

result character is delivered back to E4. The result character is: (1) the character in E1 if zero suppression is reset, (2) the character in E1 if zero suppression is set and the character in E1 is nonreplaceable, (3) a blank (15_8) if zero suppression is set, asterisk protect is reset, and the character in E1 is replaceable, or (4) an asterisk if zero suppression is set, asterisk protection is set, and the character in E1 is replaceable. Zero suppression was set in the first scan when a zero was found. It is reset in the second scan whenever a 1–9 or a decimal point or word mark shows up in the character in E1. Every time the E register has enough characters for a write into the **B** field, it is transferred to the N register via the B register. The next read of the B field is then made and the readout is again transferred to E. The scan terminates when the zero suppression word mark is found.

Third Scan. See Fig. 10.19(c). The data on the half-word location which contained the zero suppression word mark is right justified in the E register. E4 is transferred back to E1 if E4 is not a blank (15_8). In the case where E4 contains a blank, a dollar sign will be sent back to E1, and the third scan will terminate. The contents of the E register are written into the B field when it contains enough characters for the write. After the write, more **B** field characters are delivered to the E register.

Pointer Usage. AP is used to control the original right justification of the A register.

A History. AH contains the low two bits of the A address. It is decremented each time an A register character is used during the first scan. When AH = X00, the contents of the D register (next A field read) are transferred to the A register. When the first scan terminates, AH will be used to update the A address.

B Pointer. BP which contains the low two bits of the B address is used to determine the number of shifts required to left justify the contents of the E register during the execution of the first or third scan. It will be used to determine the number of shifts required to left justify the E register during the execution of the second scan.

B History. BH is always pointing at the next B field character to process. It is used to update BC at the end of the execution of the MCE instruction.

Work Location Usage. WL1 contains the address of the next read of the A field during the first scan. WL2 contains the address of the next read of the B field in all scans. BC contains the address of the next B field write during the execution of all scans.

Cycle Flow (Figs. 10.20 and 10.21). *A74MCE0–7, MCE0120–0126 (First Pass Setup).* These cycles issue two reads of the A field and right justify the first read in the A register (MCE0120–0122). The second read is transferred to the D register. A B field read is issued (A74MCE7) and the data of the B field is transferred to the E register (MCE0126). When MCE0124 is executed, the pass counter (C pointer) is cleared and A sign positive (ASP) inspects A4 for the sign of the A field. If the sign is positive, ASP is set. Since there are no sub-commands available which can be used to right justify the E register, groups of unconditional shifts are executed. The number of shifts required is indicated by the contents of the B pointer (low two bits of BC). Links A–D inspect the B pointer in MCE0124.

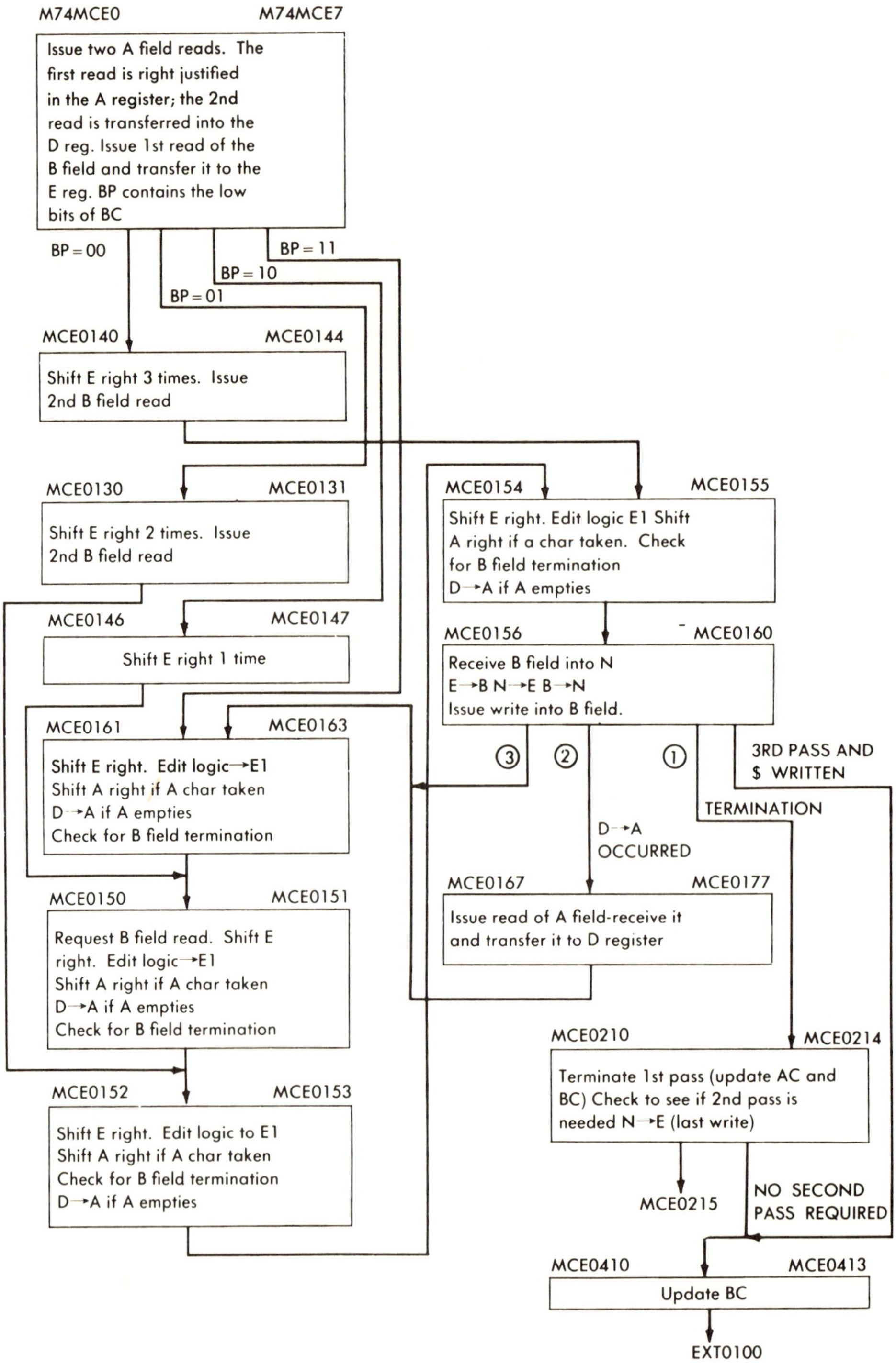

Figure 10.20 Move character and edit function description.

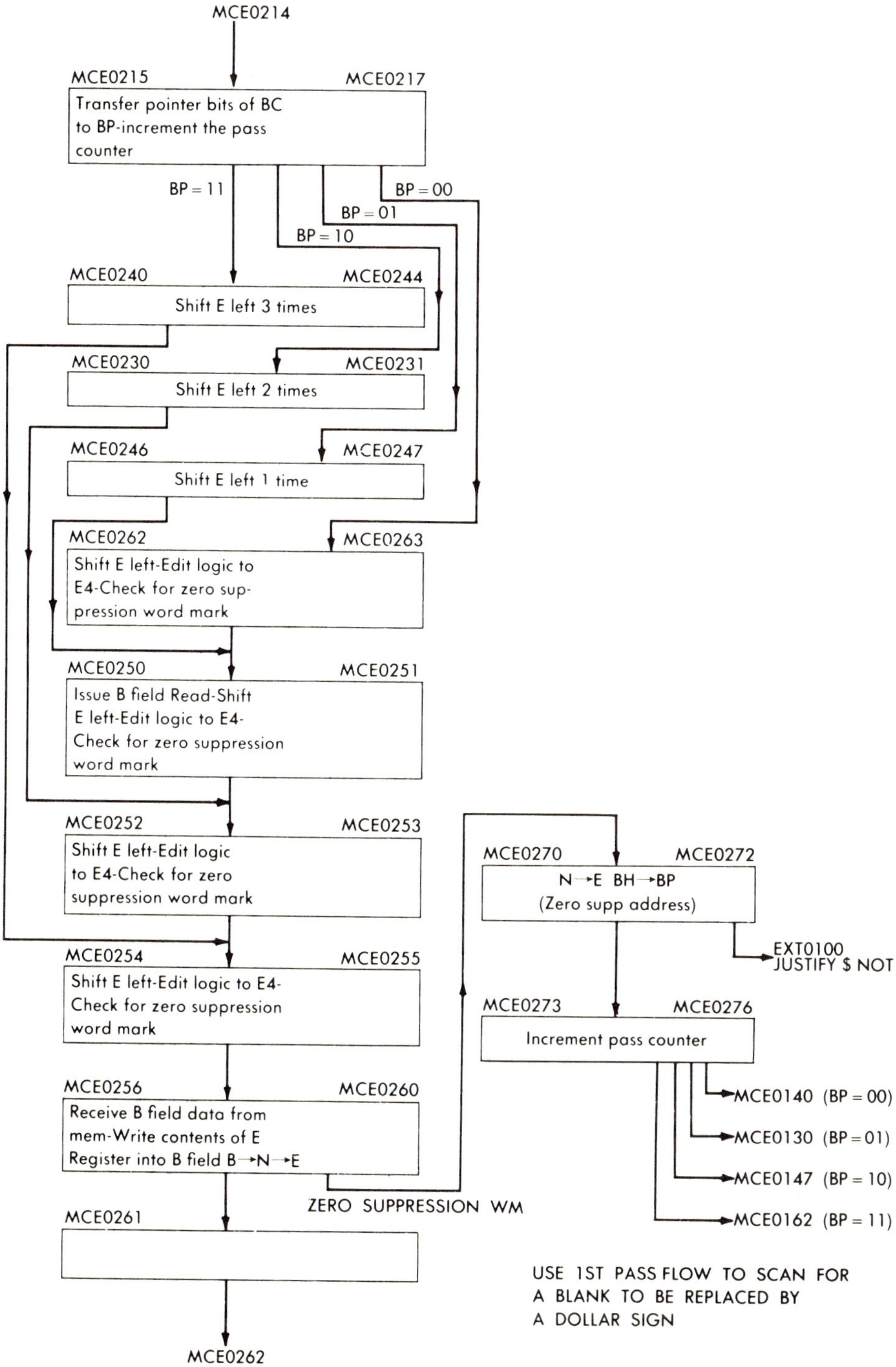

Figure 10.20 (*Cont.*)

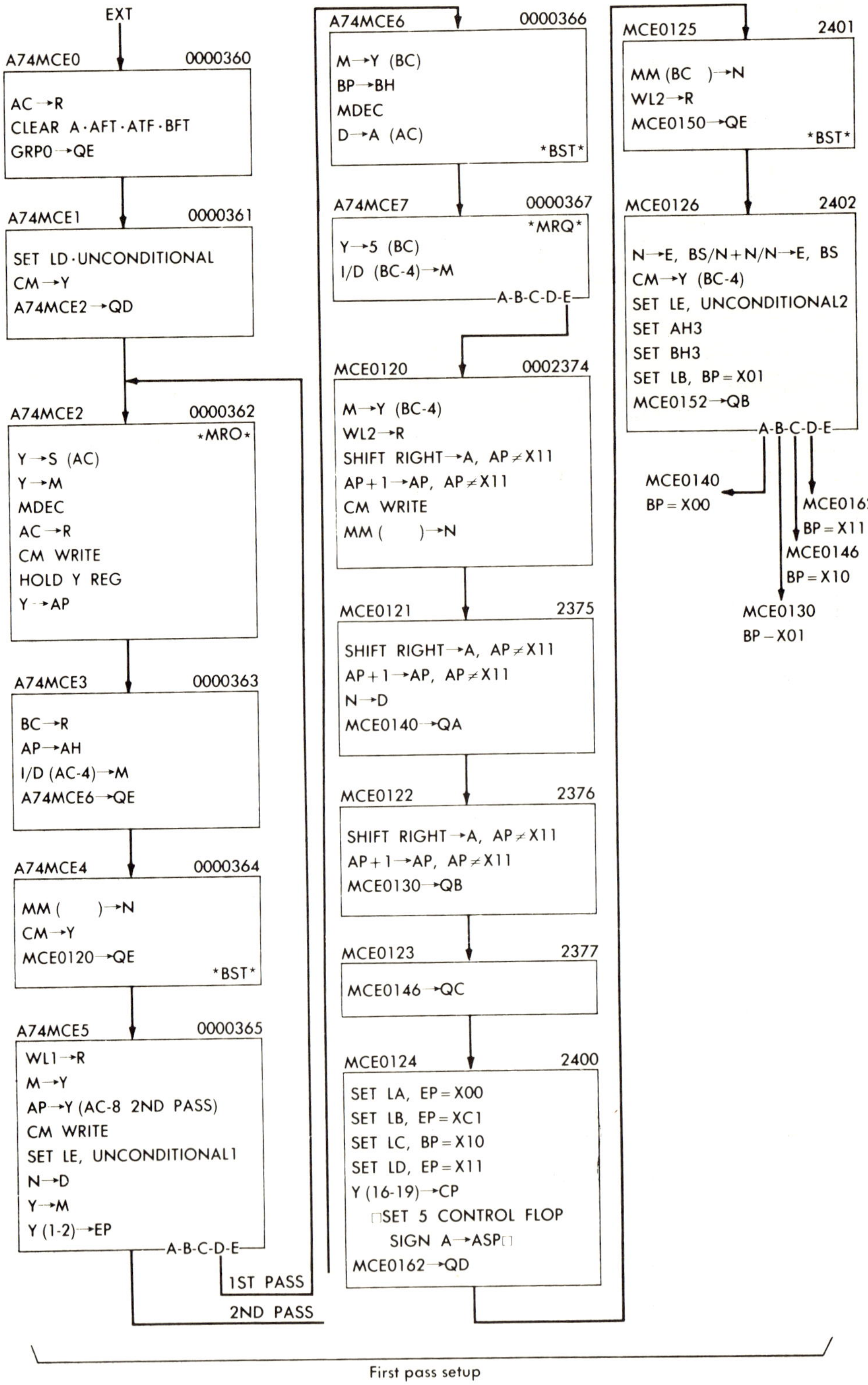

Figure 10.21 Move character and edit micro-order flow chart.

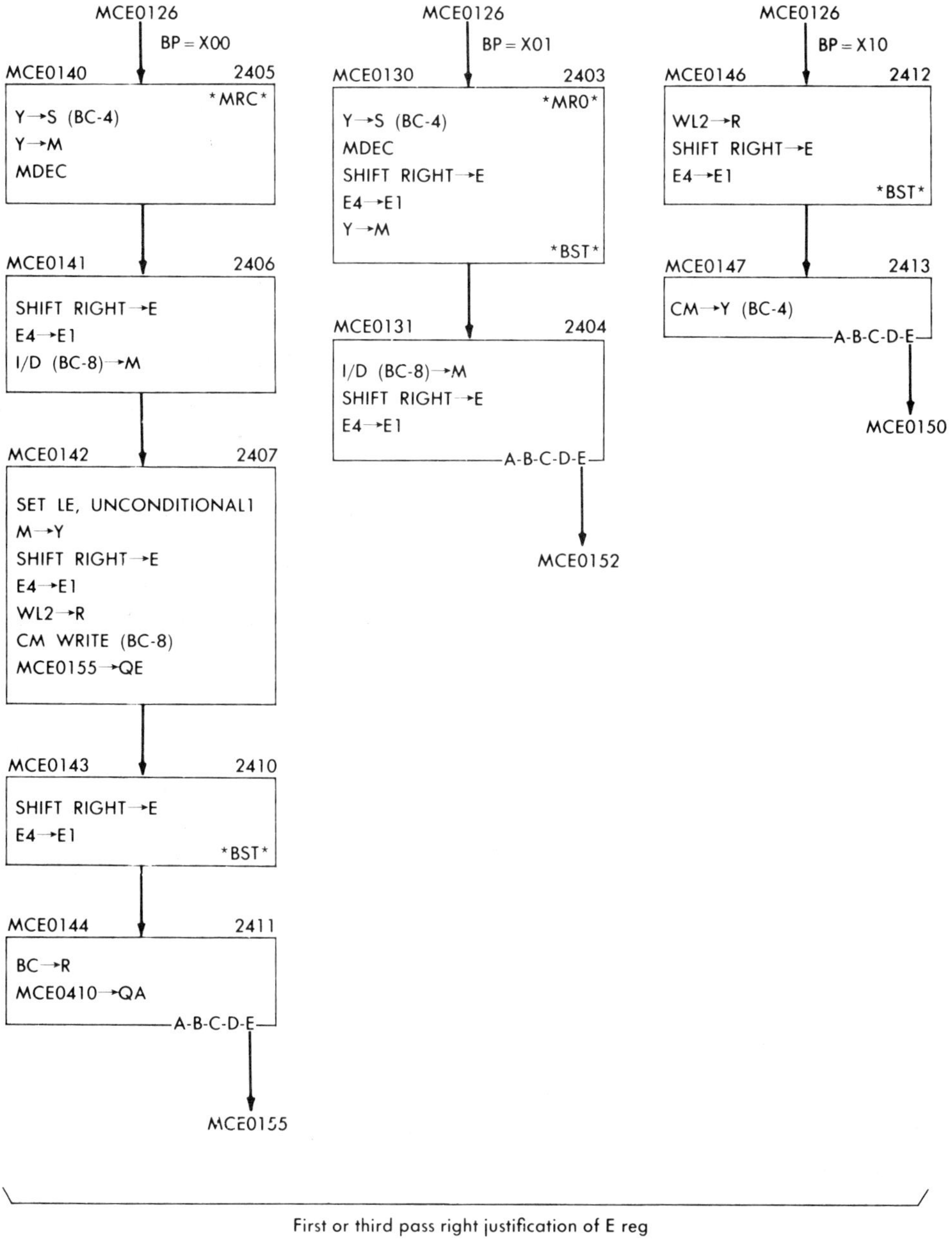

Figure 10.21 (*Cont.*)

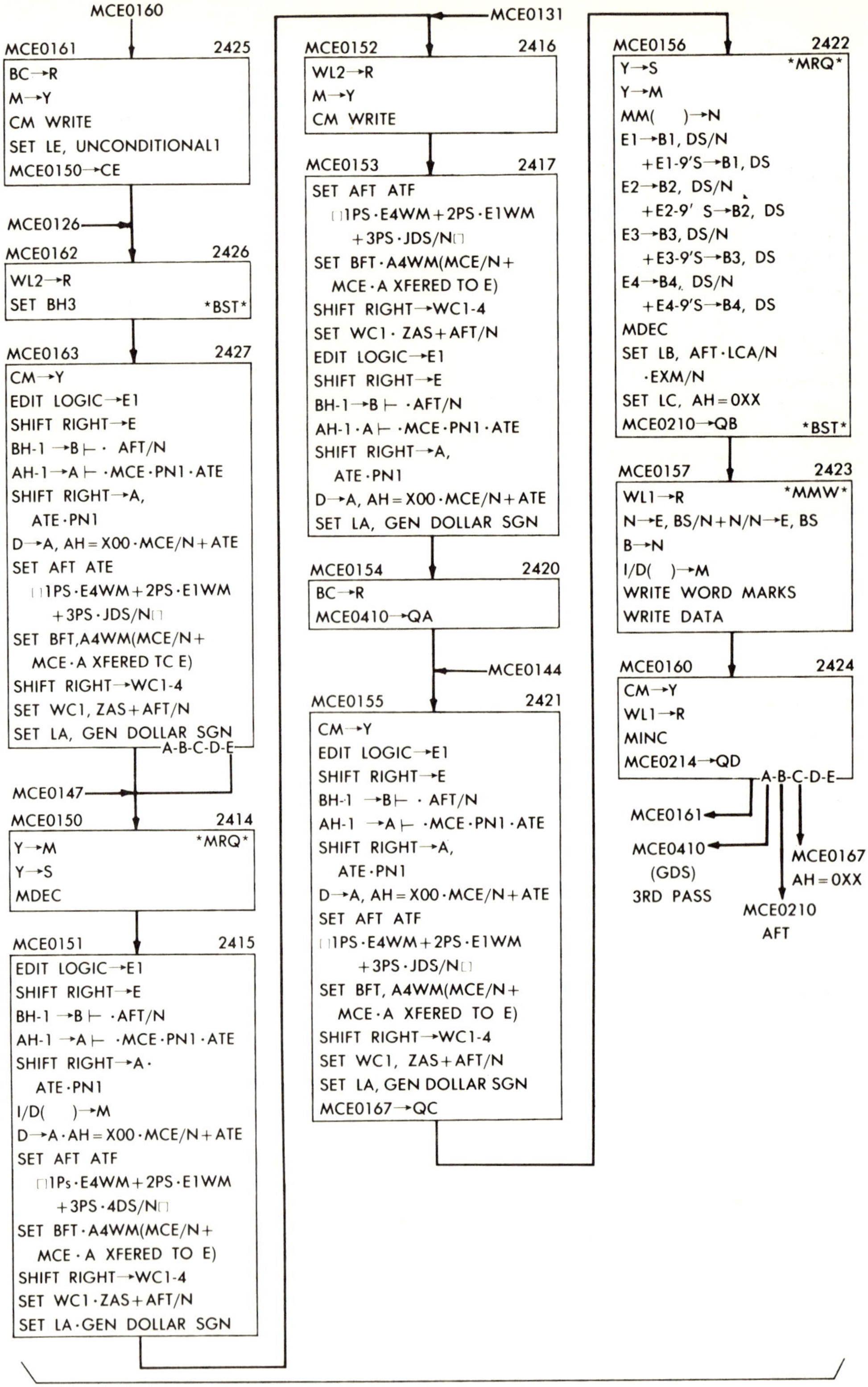

First or third pass processing loop

Figure 10.21 (*Cont.*)

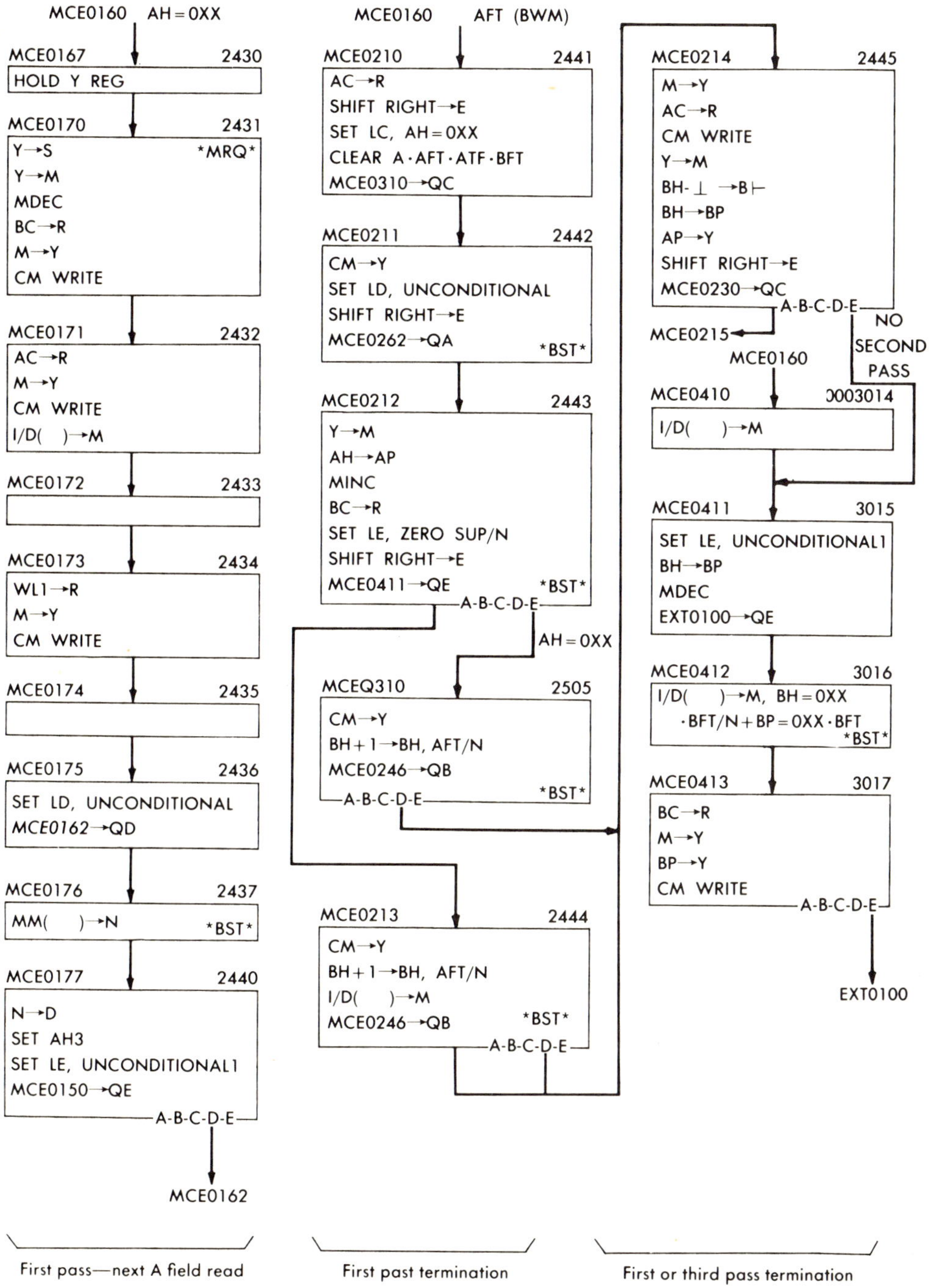

Figure 10.21 *(Cont.)*

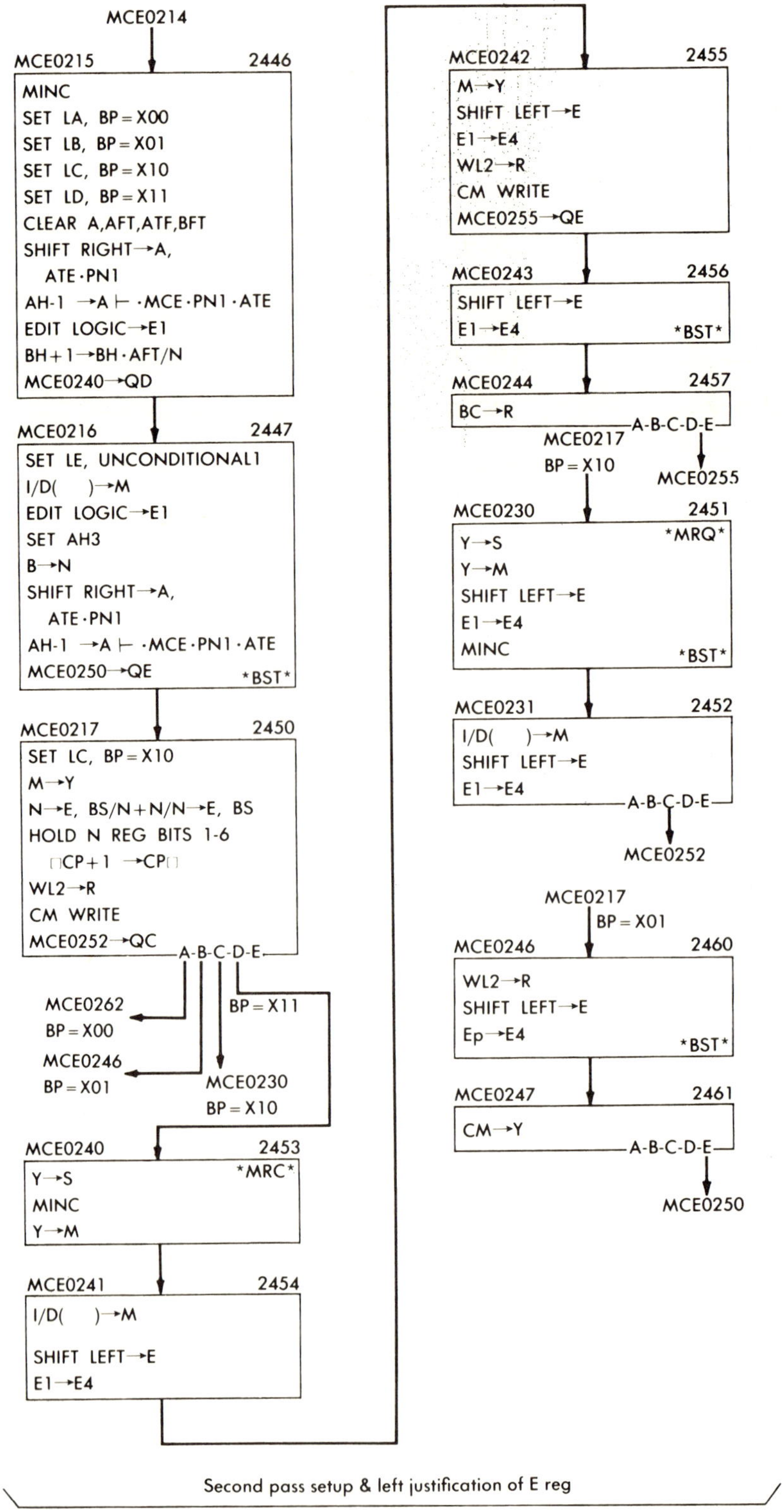

Figure 10.21 (*Cont.*)

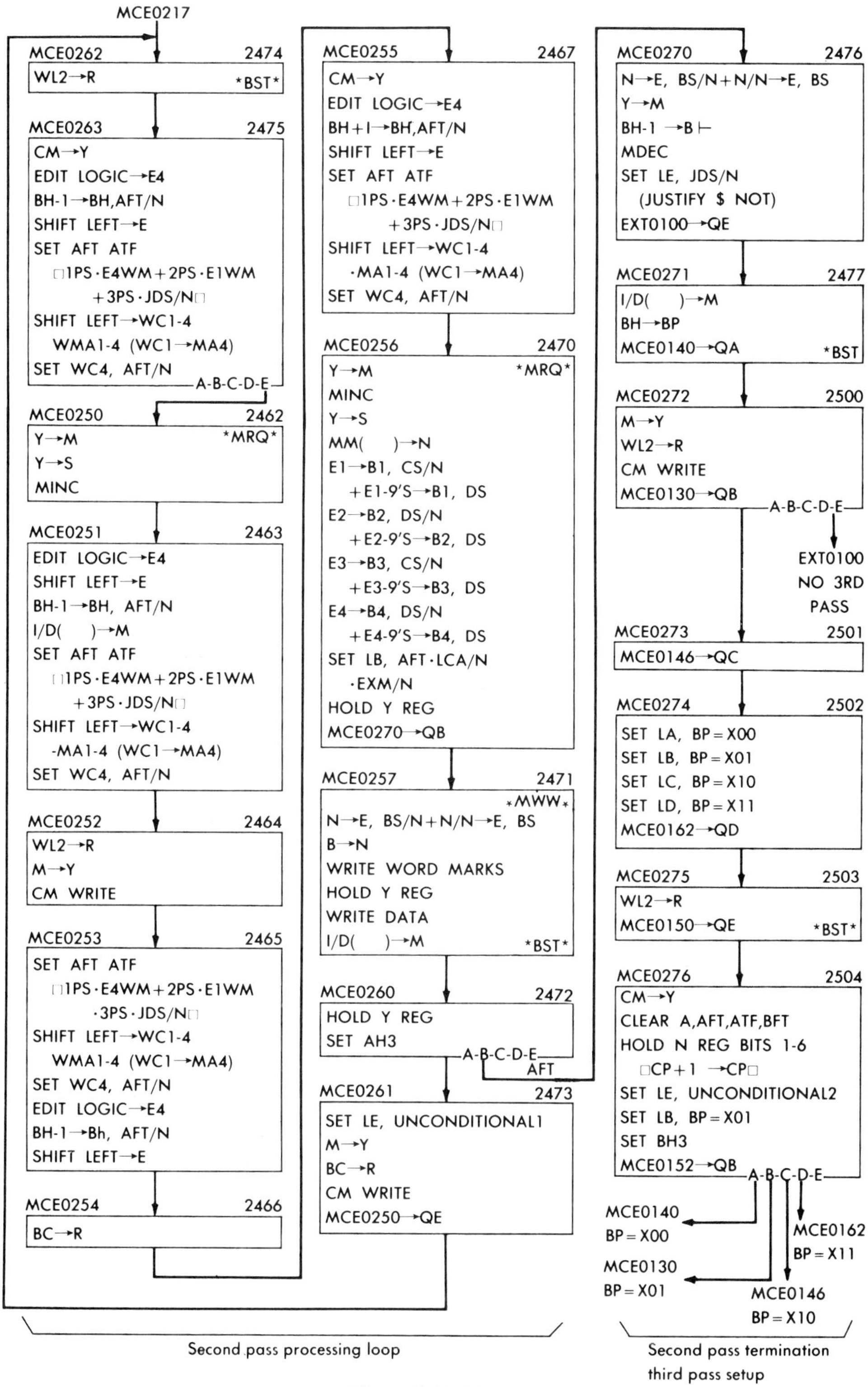

Figure 10.21 (*Cont.*)

The next cycle is:

MCE0140—if three shifts of E are required
MCE0130—if two shifts of E are required
MCE0146—if one shift of E is required
MCE0162—if E contains four characters

MCE0140-0144 (Right Justification of E Register for First and Third Scans). These cycles will shift the E register unconditionally to the right three times. The second read of the B field is issued in MCE0140.

The next cycle is MCE0155. This entry point into the first scan four-character processing loop allows one character in the E register to be processed prior to the first write into the B field.

MCE0130-0131 (Right Justification of E Register for First and Third Scans). Two shifts to the right are unconditionally issued by these cycles. The second read of the B field is issued.

The next cycle is MCE0152. This entry point into the first scan four-character processing loop allows two characters of the E register to be processed prior to the first B field write.

MCE0145-0147 (Right Justification of E Register for First and Third Scans). One shift right is issued to right justify the E register. No read of the B field is issued since it is issued in the cycle which this flow branches to.

The next cycle is MCE0150. This entry into the first scan main processing loop will allow three characters of the E register to be processed prior to the first write into the B field.

MCE0150-0163 (First or Third Scan Processing Loop). These cycles process four characters of the B field. E4 is checked for B field termination during the first scan. A4 is checked for A field termination. Each time the E register is shifted to the right, a delivery occurs to E1 from the GG register (collector on edit control board—SEC). If a character of the A field is allowed to pass through GG and transfer to E1, the A register is shifted to the right to allow the edit logic to inspect the next A field character. Each time A is shifted to the right, AH is decremented. When AH = X00, it indicates that the A register is ready for more characters and the D register is transferred to the A register. When setting up for the initial write, the entry into this flow is at four different points. This action allows MCE0156 to issue an unconditional E to B transfer. The B register is then transferred to the N register for the write. Since the N register contains the next read of the B field, an N to E transfer must take place at the same time the B to N transfer is issued. An exit of the loop can be made from MCE0160.

The next cycle is:

MCE0161—if no exit condition exists
MCE0210—if the B field word mark is found

MCE0167—if a D register to A register has taken place in the loop. This is indicated by AH-0XX. AH3 was forced set prior to entry into this loop. AH3 resets as the D register is transferred to to the A register. The next time the D register receives more A field data, AH3 will be forced set.

MCE0410—if the dollar sign has been generated when these cycles are executed in the third scan

MCE0167-0177 (Issue Next Read of A Field). These cycles are executed when the first scan processing loop has transferred the D register. The next read of the A field is issued by MCE0170. The data is received and transferred to D in MCE0177. AH3 is set in the same cycle.

The next cycle is MCE0162.

MCE0120-0214, 0310 (First Pass Termination). Basically these cycles update the A counter. In the previous flow AC was updated to the address of the half-word in the D register. This was done each time the new A field data was transferred to D. If these characters actually get to the A register, the A counter contains the proper address. However, if termination takes place prior to these characters being used, AC will have to be moved a half-word back to the right. If AH = 0XX, then it can be assumed that the characters of D have been transferred to A, and the A counter contains the proper A address. If AH = 1XX, MCE0310 is executed which increments the contents of M. The contents of AC are read from control memory in MCE0211 and transferred to M in MCE0212. MCE0214 transfers the contents of the M register and AP to AC. As a result of the four unconditional shifts of the E register which have taken place in these cycles, the E register is zeroed out. This action will be referred to in the next subsection.

The next cycle is:

MCE0411—if there is no second pass (zero Suppression Reset)

MCE0215—if there is a second pass

MCE0215-0217. BP, which contains the low two bits of the address of the high character of the B field, is inspected by links A-D in MCE0215. BCM (C_R control flop) on the SEC board may still be set if the A field is shorter than or equal to the length of the B field. BCM was originally forced set by MCE0124 and reset when the first zero or blank is detected. When the A field has terminated, BCM is set and held set. It fails to reset at the end of the first pass, and it must be reset before the second pass starts. Figure 10.19 (d) is a simplified diagram of the logic of BCM. The top gate is the initial set gate (MCE0124). The middle gate is the recirculation which is used to keep it set until the first zero or blank is detected. The bottom gate is the set gate which forces it set when the A field is terminated and the original sign of the A field is positive.

The bottom gate is inhibited by resetting BFT in MCE0215. BFT sets when the A field word mark is detected. However, this action will not break the recirculation of the middle gate. Recall that the E register has been zeroed out. By generating the subcommand edit logic → EI, the recirculation may be lost.

Other subcommands that come up in these cycles, such as the decrement of AH and the shift right of the A register, are similar to the edit logic → EI subcommand. Prior to exiting these cycles, the B register is transferred to the E register. The contents of the B register are the characters which were last written into the high position of the B field. This action saves the flow from issuing a read of these characters. These characters appear in memory in the same character positions they appear in in the B register. For this reason, the E register will have to be left justified prior to entering the second scan processing loop. The C pointer (pass counter) is incremented in MCE0217. The increment of this counter will inhibit first pass (scan) operations.

The next cycle is:

MCE0240—if E has to be shifted three times to the left
MCE0230—if E has to be shifted two times to the left
MCE0246—if E has to be shifted one time to the left
MCE0262—there are four valid characters already in E

MCE0240–0244 (Left Justification of E). The next request of the B field is issued and three unconditional left shifts of the E register are issued.

The next cycle is MCE0255 of the second scan processing loop.

MCE0230–0231 (Left Justification of E). The next request of the B field is issued and the register is shifted to the left twice.

The next cycle is MCE0252 of the second scan processing loop.

MCE0246–0247 (Left Justification of E). The E register is shifted to the left once. These cycles need not issue a read of the B field since the location which this flow branches to in the second scan processing loop issues the next read of the B field.

The next cycle is MCE0250.

MCE0250–0263 (Second Scan Processing Loop). These cycles scan the B field left to right checking for replaceable characters. If a replaceable character is found it is replaced by a zero or asterisk. Replaceable characters are: zeros, blanks, and commas if zero suppression is set. Zero suppression was turned on during the first scan when the first zero was found. It is reset when the first 1–9 or decimal point is detected or when the location which contained the zero suppression symbol is detected. AFT is set when the zero suppression word mark is detected in EI. When the flow has processed four characters, the E register is transferred to the B register (MCE0256). The B register is now transferred to the N register for the write request which was issued in MCE0256. The next read of the B field is issued in MCE0250. When B is transferred to N, N which contains the next read data is transferred to E. When termination is found, an exit takes place from MCE0260 to MCE0270.

MCE0270–0272 (Second Scan Termination). The B counter is updated. The M register which contains the address of the next write is decremented. This allows M to point to the address of the last write. BH is used as the low two bits. When termination is found, BH is inhibited from incrementing. When M and BH are written into BC, BC is pointing at the zero suppression location.

The next cycle is:

EXT0100—if no third scan is required (justify dollar sign reset)
MCE0273—if a third scan is required

MCE0273–0276 (Third Scan Setup). The B pointer, which contains the low two bits of the address of the location of the zero suppression word mark, is inspected by links A-D in MCE0274. The link which is set will determine the number of shifts required to right justify the E register. The E register contains the characters of the last write. They were placed there in MCE0270. The C pointer is incremented in MCE0276. This action inhibits first and second scan operations.

The next cycles are:

MCE0140—if three shifts are required to right justify the E register
MCE0130—if two shifts are required to right justify the E register
MCE0146—if one shift is required to right justify the E register
MCE0162—if there are four valid characters in the E register, entry is made directly into the first and third scan processing

The first and third scan processing loop functions are previously explained. Due to the fact that the C pointer is indicating the third scan, certain actions are inhibited. The A register is not used. Either E4 transfers back to E1 or if E4 equals a blank, a dollar sign is sent back to E1. When the dollar sign is generated, link A is set and the flow exits to MCE0410.

MCE0410–0413 (B Counter Update). Entry into these cycles can be made from first or third scan processing. Entry into MCE0410 from the third scan processing finds the M register sitting at the address of the next write. MCE0410 increments the M register so it points to the location of the last write issued in the third scan. When MCE0411 is entered from either MCE0410 or MCE0214 (first scan termination), the M register contains the address of the last write. If the last write of the third scan was a four-character write, BH3 is reset. This allows M to be decremented. BC should point to the left of the dollar sign when the third pass is complete. BH3 can never be reset when exiting from the first scan termination. When termination takes place from the first pass, the B address should be pointing at the high character of the B field. Thus, the address of the last write is always correct.

Example 6

Typical extraction loop—trace. This example is given in the form of an exercise for the reader. After a few introductory comments the reader is asked to trace the action for himself from the block diagram and the cycle flow chart for the four-character mode shown in Figs. 10.22 and 10.23, respectively.

In general, the central processor extracts the complete instruction from the memory before any address modification, if necessary, is accomplished. The address parameters which are stored in the QU and QV registers are checked during the extraction process to see if an address modification is called for. These address modifications are handled in the following order: Indirect A, Indirect B, Indexed A, and Indexed B.

In order to maintain the processing speed, the processor must always have characters available for the CPU to process without having to wait on the memory. To insure that, the following sequence is generally followed:

1. Extract a half-word (four characters) from memory and start processing.
2. Immediately retrieve the next sequential half-word and store it in a four-character register so it may be made available to the processor as soon as it has processed the first four characters.
3. Address memory to retrieve the third half-word.

By using this sequence, it is possible to be addressing locations in memory which may not be necessary for this operation. If, for example, the first half-word contained terminating punctuation, the processor may not have recognized it (depending upon which character position it was in) before it extracted the second half-word and addressed memory for the third half-word.

The processor keeps track of the number of half-words it actually processed in the two bit B history counter (BH). The sequence counter is stored sequentially each time it is used in WL3 through WL6 (i.e., WL3 = SC′, WL4 = SC + 4, WL5 = SC + 8, WL6 = SC12, and so on). When the extraction is complete, the correct sequence counter is retrieved from SPM by superimposing the BH and the WL3 address, and the correct sequence counter is transferred to the SC register located in SPM 77.

During the extraction, the A history (AH) register keeps track of the character of a half-word being processed. When AH = 11, the next word which is stored in the D register is advanced to the A register. At the same time the BH is incremented by one to indicate that a full half-word has been processed.

The first half-word is read into the N register and the Op-code is transferred to the JJ register. This data is also transferred to the A register and shifted left until the Op-code is lost. The data is then shifted left from A to B two, three, or four times, depending on the character mode. The B register is then stored in the appropriate address counter. As soon as the A register is empty, the next half-word is transferred from D to A. When the addresses have been extracted, the remaining characters are delivered to the VV register until a word mark is

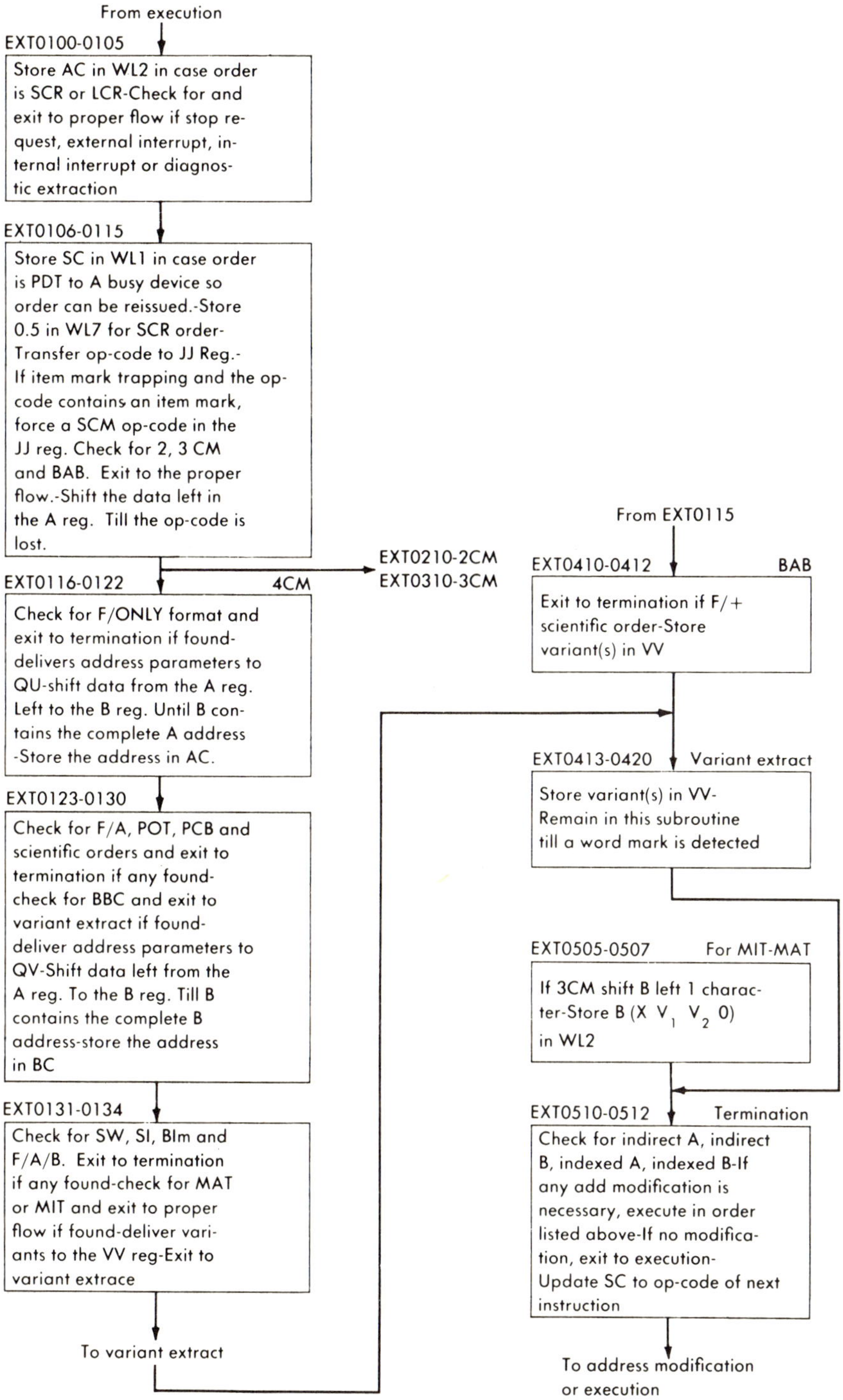

Figure 10.22 Block diagram of 4CM extraction.

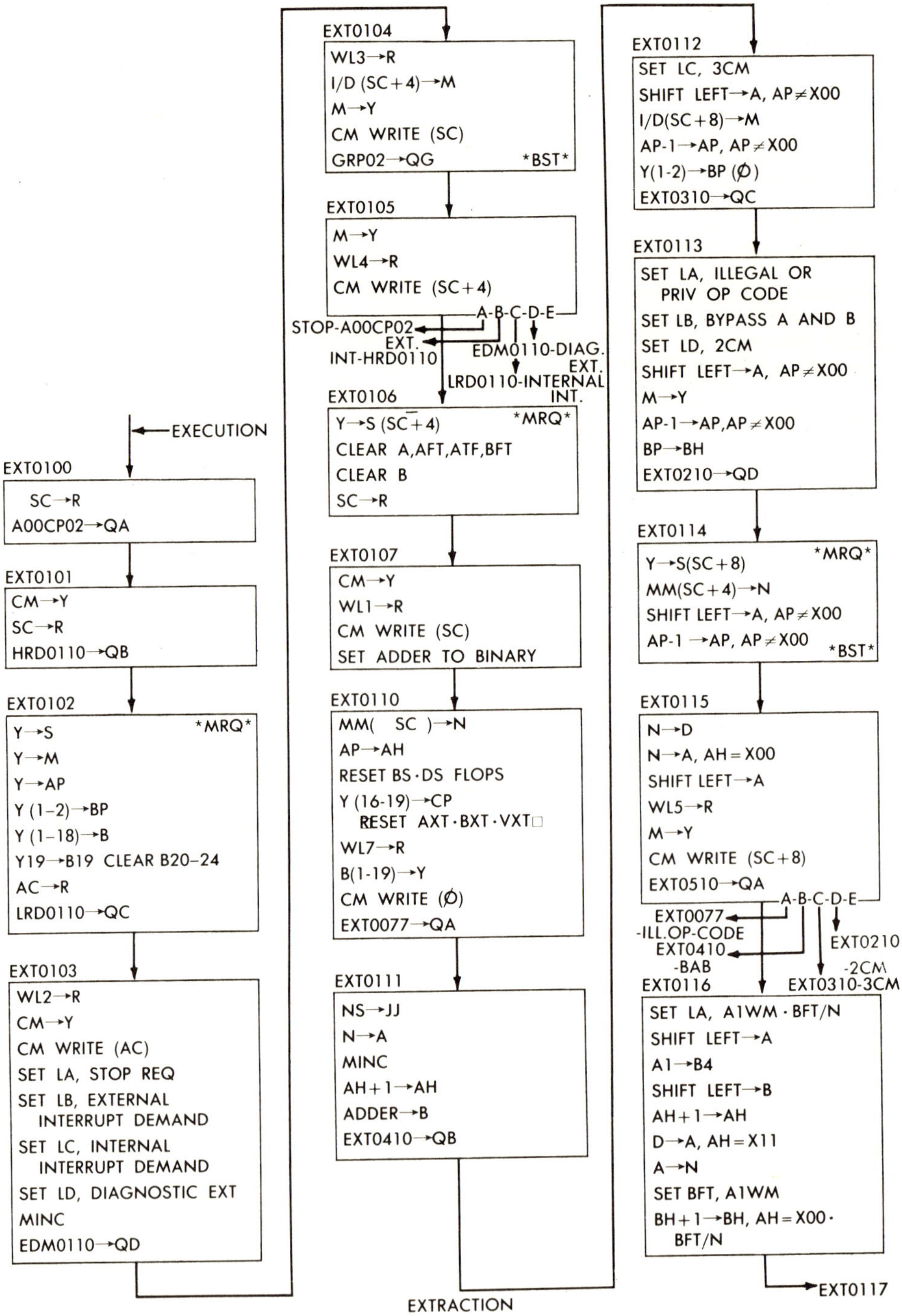

Figure 10.23 4CM extraction cycles.

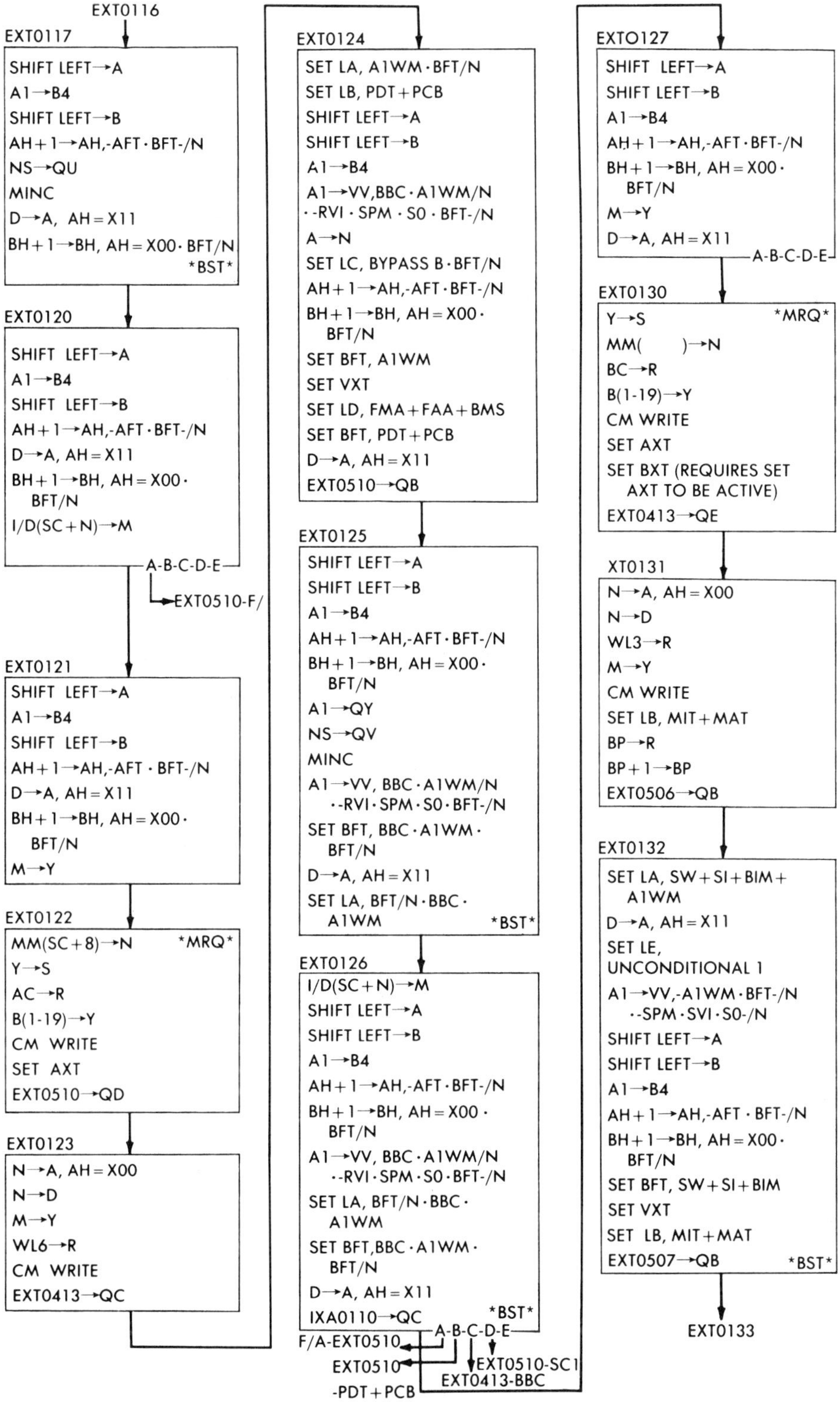

Figure 10.23 *(Cont.)*

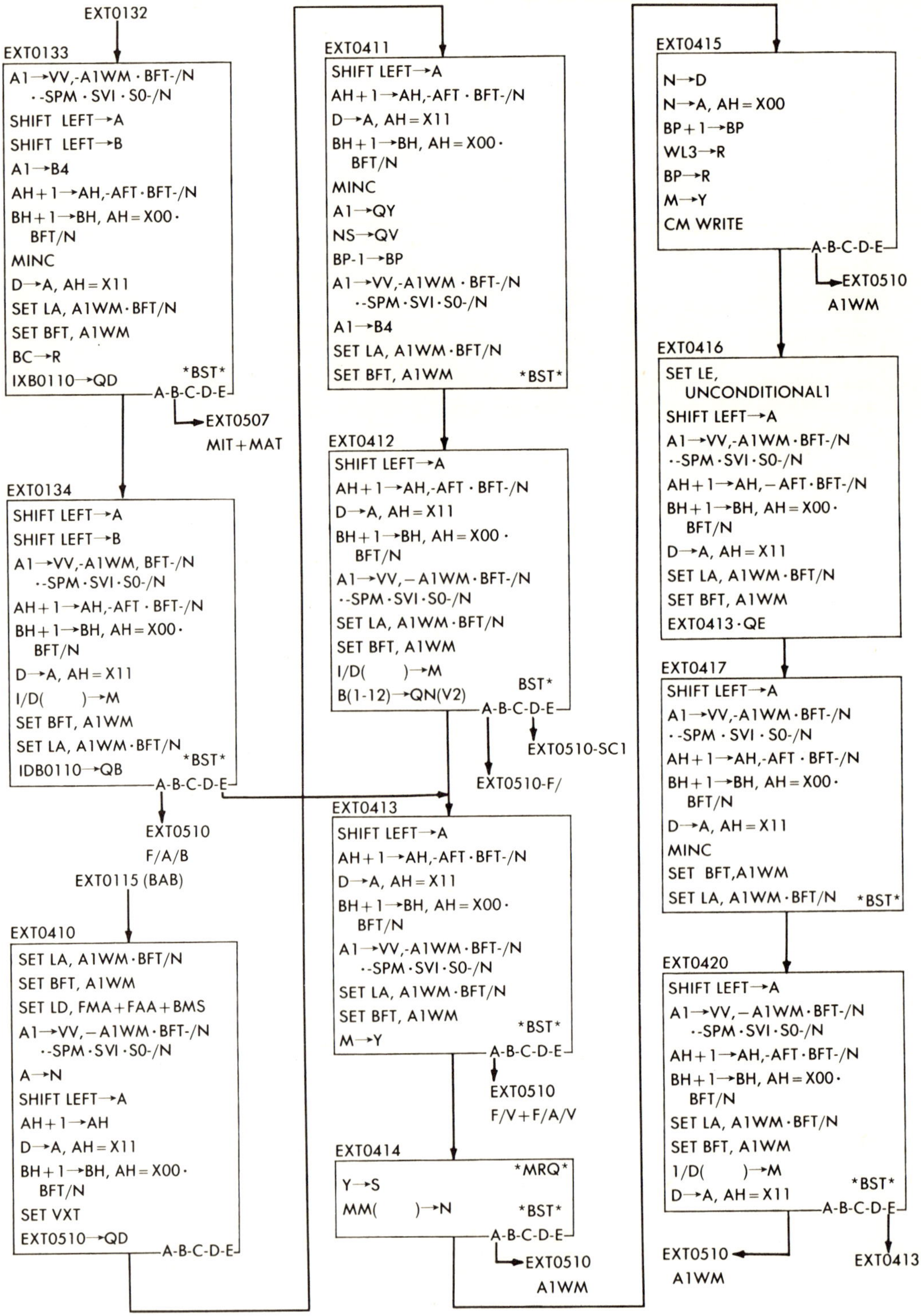

Figure 10.23 (*Cont.*)

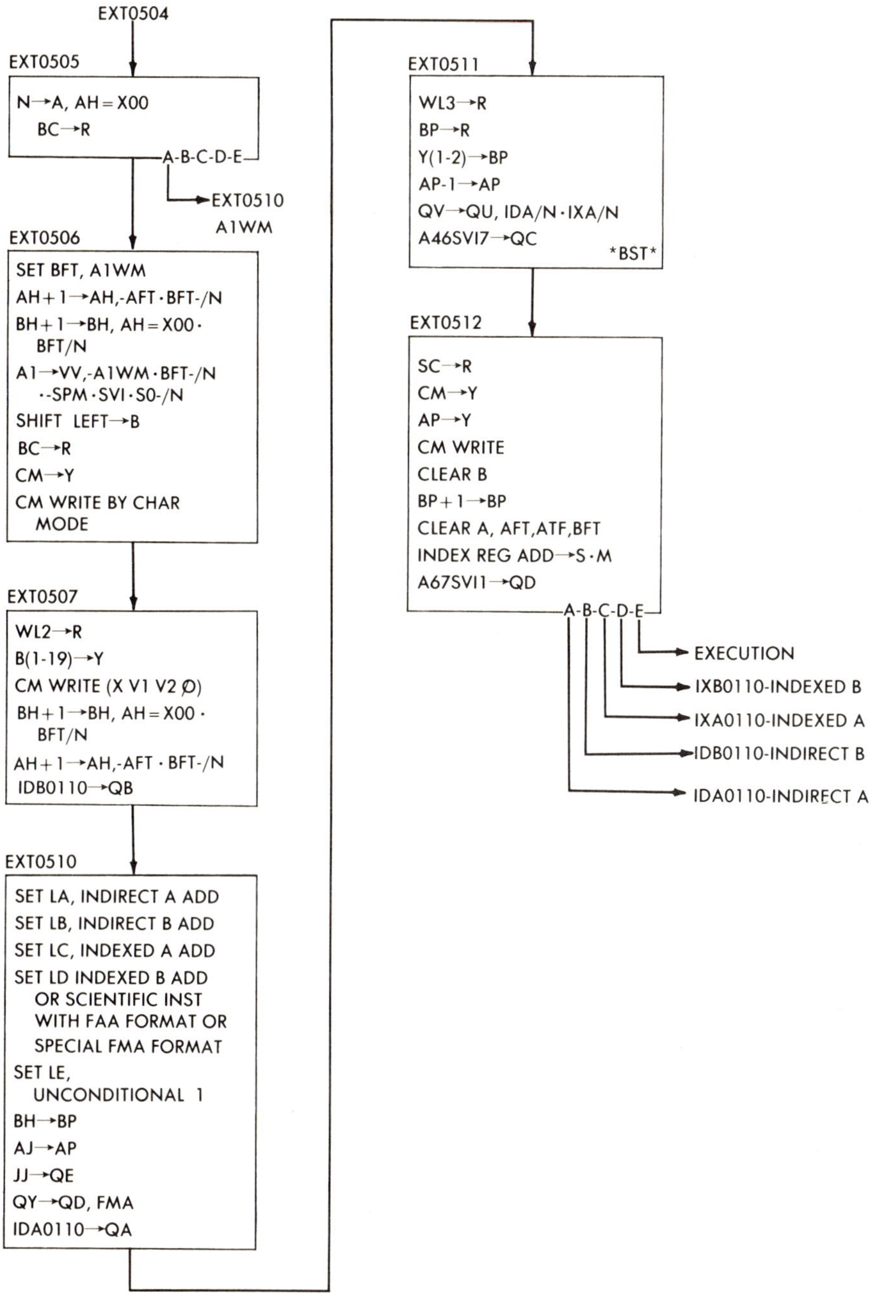

Figure 10.23 (*Cont.*)

found. If the order being extracted is MIT or MAT, the V1 and V2 variants are stored in the WL2 for use in execution.

If the address modification specified by the instruction being extracted is an indirect addressing, the appropriate address counter is used to fetch the first half-word of the indirect address from main storage. This first half-word is transferred to both the A and B registers. The pointer is examined to determine the number of characters of this half-word which are part of the indirect address. The next half-word is read from main storage to either the A or B registers depending on which of the two requires the least number of shifts to right justify the indirect address in the B register.

If the modifications specify an indexed addressing, the index register address always addresses the word containing IR1. This word is placed in the A register and the second word (index register address—4) is placed in the B register; one character left shift from A to B places the entire index register contents in the B register. The appropriate address counter is delivered to A, and the binary result of the adder is stored in the control memory under character mode control.

APPENDIX 10A: INSTRUCTION SUMMARY

Preliminary timing formulas for the Model 4200 are given in Table 10A.1. Since the internal operation of the Model 4200 processor differs from that of the other Series 200 processors in that data is moved in groups of four characters (a word) rather than singly, the 4200 timing formulas differ considerably from those of the other processors.

The formulas given in Tables 10A.1 and 10A.2 provide execution time in memory cycles. Equivalent expressions for symbols used in the tables are as follows:

Symbol	*Meaning*
A	Address of A-operand field.
B	Address of B-operand field.
h	The sum of the values of the multiplier digits which are less than or equal to five, plus the sum of the elevens complements of all digits whose values are greater than five.
N_a	Number of characters in the A-operand field.
N_{aw}	Number of words in the A-operand field.
N_b	Number of characters in the B-operand field.
N_{bw}	Number of words in the B-operand field.
N_{b1}	Number of words that the A field occupies in the B field, whether or not the operands have been modified by an arithmetic operation.
N_{b2}	Number of words in the B-operand field excluding N_{b1}.
N_c	Number of control characters in the instruction.

Symbol	*Meaning*
N_{cn}	Number of control characters following control character 3 (C3).
N_{dd}	Number of digits in the dividend.
N_i	Number of characters in the instruction.
N_{ia}	Number of words in the item to be translated.
N_{ib}	Number of words in the result item.
N_{ic}	Number of translation units (6-bit or 12-bit characters) to be translated.
N_j	Number of character locations bypassed to reach the next sequential op code.
N_{mr}	Number of digits in the multiplier.
N_q	Number of digits in the quotient ($= N_{dd} - Z_{1d} - N_a + Z_{1a} + 1$).
N_r	Number of characters referenced.
N_{st}	Number of characters stored.
N_w	Number of characters in the A- or B-operand field, whichever is shorter.
N_{wj}	Number of words bypassed to reach the next sequential op code.
N_{ws}	Number of words stored.
n	Number of items in the table or the number of times the A operand is compared against some portion of the B operand.
Q_i	The value of the ith digit of the quotient.
s	Sum of all multiplier digits.
SUM	Sum of the upwards-rounded values of all multiplier digits divided by 2.
V	Variant character.
W	Number of memory words used to store the data involved.
W_i	Number of four-character words used to store one more than the total number of characters in the instruction.
W_{mr}	Number of words in the multiplier.
X_0	Zero if no second scan (zero suppression); one if the scan is performed.
Y_0	Zero if no third scan (dollar-sign insertion); one if the scan is performed.
Z	Number of characters scanned during zero suppression.
Z_{1a}	Number of leading zeros in the A-operand field.
Z_{1aw}	Number of words containing leading zeros in the A-operand field.
Z_{1d}	Number of leading zeros in the dividend.
Z_{mr}	Number of zeros in the multiplier.
Z_{ta}	Number of trailing zeros (i.e., consecutive low-order zeros) in the A-operand field.
Z_{taw}	Number of words containing trailing zeros in the A-operand field.
Z_w	Number of words scanned during zero suppression.
Z_z	Zero if $Z_{1a} = 0$; one if $Z_{1a} \neq 0$.
\$	Number of characters scanned during dollar-sign insertion.
$\$_w$	Number of words scanned during dollar-sign insertion.

Note: The timing formulas presented in Tables 10A.1 and 10A.2 are based on the use of direct addressing. If address modification is used, the formulas in Table 10A.2 for the Models 200, 1200, and 2200 should be modified as follows:

1. *Indirect Addressing*—Add one memory cycle for each *character* extracted as a result of indirect addressing.
2. *Indexed Addressing*—Add three memory cycles for each indexed *address*.

Likewise, the use of address modification requires that the formulas in Tables 10A.1 and 10A.2 for the Model 4200 be modified as follows:

1. *Indirect Addressing*—Add 1.16 memory cycles for each indirect *address* formed *plus* one memory cycle for each *word* extracted as a result of indirect addressing.
2. *Indexed Addressing*—Add 3.167 memory cycles if one address is indexed, 5.16 memory cycles if both addresses are indexed.

Table 10A.1. INSTRUCTION TIMINGS FOR THE MODEL 4200

Mnemonic Op Code	*Instruction Name*	*Timing Formulas*	*Notes*
		Fixed-Point Arithmetic Instructions	
A	Decimal Add	No Recomplement $W_i + N_{aw} + 2N_{b1} + 2.5N_{b2} + K^1$ Recomplement $W_i + N_{aw} + 2.5N_{bw} + 2N_{b1} + 2.5N_{b2} + K^1$	BC K = 5 AV K = 6.5 WC K = 7
S	Decimal Subtract	No Recomplement $W_i + N_{aw} + 2N_{b1} + 2.5N_{b2} + K^1$ Recomplement $W_i + N_{aw} + 2.5N_{bw} + 2N_{b1} + 2.5N_{b2} + K^1$	
BA	Binary Add	$W_i + N_{aw} + 2N_{b1} + 2.5N_{b2} + K^1$	BC K = 5 AV K = 6.5 WC K = 7.5
BS	Binary Subtract	$W_i + N_{aw} + 2N_{b1} + 2.5N_{b2} + K^1$	
ZA	Zero and Add	$W_i + 2N_{b1} + N_{b2} + K^2$	N_{b1} value is even K = 6
ZS	Zero and Subtract	$W_i + 2N_{b1} + N_{b2} + K^2$	N_{b1} value is odd K = 7 S Mode: Add 2 cycles
M	Decimal Multiply	See Table 10A.2.	
D	Decimal Divide	See Table 10A.2.	

Table 10A.1. (Cont.)

Mnemonic Op Code	*Instruction Name*	*Timing Formulas*	*Notes*
		Logic Instructions	
EXT	Extract	$W_i + 3N_{b1} + 7$	
HA	Half Add	$W_i + 3N_{b1} + 7$	
SST	Substitute	$W_i + 7.34$	
C	Compare	$W_i + 2N_{b1} + N_{b2} + K^1$	BC K = 5.5 AV K = 7 WC K = 8
B	Branch (unconditional)	$W_i + 4.5$	
BCT	Branch on Condition Test	$W_i + 4.5$	
BCC	Branch on Character Condition	$W_i + 6$	
BCE	Branch on Character Equal	$W_i + 6$	
BBE	Branch on Bit Equal	$W_i + 6$	
		Control Instructions	
SW	Set Word Mark	$W_i + 5$	
SI	Set Item Mark	$W_i + 5$	
CW	Clear Word Mark	$W_i + 5$	
CI	Clear Item Mark	$W_i + 5$	
H	Halt	$W_i + 5$	
NOP	No Operation	$W_i + 4$	
MCW	Move Characters to Word Mark	$W_i + 2N_{b1} + K^2$	N_{b1} value is even K = 6 N_{b1} value is odd K = 7
LCA	Load Characters to A-field Word Mark	$W_i + 2N_{b1} + K^2$	N_{b1} value is even K = 5 N_{b1} value is odd K = 6
SCR	Store Control Registers	$W_i + W + 4.33$ $W_i + W + 13$	Non-I/O register I/O register
LCR	Load Control Registers	$W_i + W + 4.33$ $W_i + W + 24$	Non-I/O register I/O register

Table 10A.1. (Cont.)

Mnemonic Op Code	*Instruction Name*	*Timing Formulas*	*Notes*
		Control Instructions (Cont.)	
CAM	Change Addressing Mode	$W_i + 4$	
CSM	Change Sequencing Mode	$W_i + 4$	
EXM	Extended Move	$W_i + 2N_{bl} + K^2$	N_{bl} value is even $K = 5$ N_{bl} value is odd $K = 6$
MAT	Move and Translate	$W_i + 1.67N_{ia} + 1.67N_{ib} + N_{ic} + 6.5$	If $A < 9 + B$ then N_{ia} = Number of *characters* in the item to be translated. N_{ib} = Number of *characters* in the result item.
MIT	Move Item and Translate	$W_i + 1.67N_{ia} + 1.67N_{ib} + N_{ic} + 6.5$	If $B < 9 + A$ or if the translation is 6-bit to 12-bit (or 12-bit to 6-bit) then N_{ia} = Number of *characters* in the item to be translated. N_{ib} = Number of *characters* in the result item.
LIB	Load Index/Barricade Register	Basic storage protection $W_i + W + 4.5$ Storage protection with base relocation $W_i + 2W + 6$	
SIB	Store Index/Barricade Register	Basic storage protection $W_i + W + 4$ Storage protection with base relocation $W_i + 2W + 5$	
TLU	Table Lookup	$W_i + n(N_{aw}) + N_{bw} + 3n + 9$	

Table 10A.2. TIMINGS FOR DECIMAL MULTIPLY AND DIVIDE, MODELS 200, 1200, 2200, AND 4200

Function	*Model*	*Timing (Memory Cycles)*
Multiply	200 and 1200	$N_i + 5 + 2N_a + 2Z_{ta} + 5N_{mr} - Z_{mr} + s(N_a - Z_{ta}) + 2(N_a - Z_{ta})(N_{mr} - Z_{mr})$
	2200	$N_i + 8 + 2N_a + 2Z_{ta} + 5N_{mr} - Z_{mr} + \text{sum}(N_a - Z_{ta}) + 3(N_a - Z_{ta})(N_{mr} - Z_{mr})$
	4200	$w_i + 2N_{aw} + w_{mr} + h(3N_{aw} - Z_{taw} + 2) + 13$
Divide	200 and 1200	$N_i + 4 + 2N_a$ if divisor $= 0$ $N_i + 17.5 + 4.5N_a + 15.5Z_{1a} + 12.5N_{dd} + 15N_a(N_{dd} - N_a + Z_{1a})$ if $(N_a - Z_{1a}) \leq (N_{dd})$ and divisor $\neq 0$ $N_i + 7 + 4N_a$ if $(N_a - Z_{1a}) > (N_{dd})$
	2200	$N_i + 7 + 2N_a$ if divisor $= 0$ $N_i + 9 + 2Z_z + 5N_a + 3Z_{1d} + N_q + 15N_a - 2Z_{1a} + 18.25)$ if $(N_a - Z_{1a}) \leq (N_{dd} - Z_{1d})$ and divisor $\neq 0$ $N_i + 9 + 2N_a + 2N_{dd}$ if $N_a > N_{dd}$ and $(N_a - Z_{1a}) > (N_{dd} - Z_{1d})$ $N_i + 10 + N_a + 3N_{dd}$ if $N_a \geq N_{dd}$ and $(N_a - Z_{1a}) > (N_{dd} - Z_{1d})$
	4200	$W_i + 2N_{aw} + Z_{1aw} \sum_{i=1}^{N_q} (Q_i + 2)(3N_{aw} + 2) + 19$

APPENDIX 10B: SUBCOMMAND GLOSSARY FILE

```
FILE: 82140110     FOR CHECKING USE ONLY
 REV: 00.075.3              EDITED: 09/26/69

   RUN REQUEST:                                     LT: MACHLOG8
EDIT.FFP)PRINT1PLYORELSEBIGAL00001                        )'
                               '          8  RR82140110092569Z

FUNCTION

0000BAD:   /   /CMT(BRANCH ADDRESS FIELD              )
        = 0020011)

0000BDT:   /   /CMT(BRANCH ON DIRECT TEST FAST BRANCH       )
        = 0511001)

0000BPE:   /   /CMT( EVEN PARITY FOR BRANCH ADDRESS FLD     )
        = 0010001)

0000BST:   /   /CMT(BRANCH ON STORED TEST  )
        = 0131001)

000AORN:SU /   /CMT(IF SGNS OF OPRND:RSLT NEG.MAA.SAORN     )
        = 0163002)

000AOVL:   /   /CMT( SET LFA IF OVL                         )
        = 0163004)

000AA1W:   /   /CMT(SET LA IF WRD MK IN A1.BFT NOT.SAA1W10 )
        = 0163001)

000ABPO:   /   /CMT(SET LA IF B POINTER EQUALS 0.SFBP010   )
        = 0163001)

000ABPP:   /   /CMT(SET LA IF BP=X00.SFBP010         )
        = 0163007)

000ABWM:   /   /CMT(SET LA IF B REG HAS WORD MARK.SABWM10  )
        = 0163003)

000ACOB:   /   /CMT(SET LA IF B FLD NEEDS RECMPLTNG.BAOVF00)
        = 0163002)

000ACOM:   /   /CMT(SET LFA IF COMPARE.RJLB310      )
        = 0163001)

000ADCB:   /   /CMT(SET LA IF ERROR COND TSTD TRUE.SADCB10 )
        = 0163004)

000AFAS:SU /   /CMT(SET LA IF FLTNG ADD OR SUBTRACT.SAFAS10)
        = 0163002)
```

```
000AGDS:  /  /CMT(SET LA IF GENERATE DOLLAR SIGN CYCLE    )
      = 0163007)

000AIOC:  /  /CMT(SET LA IF ILLEGAL OP CODE.SAIOC10      )
      = 0163001)

000AIDA:  /  /CMT(SET LA IF A ADD IS INDIRECT.SAIDA10     )
      = 0163003)

000AIFC:SU /  /CMT(SET LA IF ILLEGAL FL.CODE.SAIFC10    )
      = 0163005)

000AINS:  /  /CMT(SET LA IF INSTRUCT.CINST30     )
      = 0163006)

000AMES:  /  /CMT(SET LA IF MAIN MEMORY ENTER.CENTM30     )
      = 0163006)

000AMPD:  /  /CMT(SET LA IF MULTPLY+DIVDE ODR.SAMPD10     )
      = 0163004)

000AN71:SU /  /CMT(SET LA IF QN GTR OR EQL TO 71.SAN7110  )
      = 0163003)

000ANSW:  /  /CMT(SET LFA IF NNS HAS WM                   )
      = 0163006)

000ANWM:  /  /CMT(SET LA IF NWM.KUMA1-4.SANWM10           )
      = 0163004)

000APAC:  /  /CMT(SET LA IF PERIPHERAL ACTIVITY.SIACT30   )
      = 0163006)

000AQNO:SU /  /CMT(SET LA IF QN=0.SAQN010                 )
      = 0163003)

000ASK3:  /  /CMT(SET LA IF SRVC AKN FRM IO IS 11.SASK310)
      = 0163007)

000ASPM:  /  /CMT(SET LA IF SW+SI+BIM ODER+WMA1.SASPM10   )
      = 0163005)

000ASTR:  /  /CMT(SET LA IF STOP REQUEST.CSTPR10 )
      = 0163005)

000AV1W:  /  /CMT(SET LA IF BBC.A1WM.BFT NOT.SAV1W10      )
      = 0163005)

000AXEO:  /  /CMT(SET LA IF X OPERAND IS ZERO.BXHA120     )
      = 0163002)

000B4CM:  /  /CMT(SET LB IF 4CM.CCCM430  )
      = 0193002)

000BA4W:  /  /CMT(SET LB IF WM IN A4 CHAR.BAAW430)
      = 0193001)

000BACZ:  /  /CMT(SET LB IF ACC.=0 AND RNZ NEG.SBACZ10    )
      = 0193001)

000BAEO:  /  /CMT(SET LB IF A1 DEC CHAR NOT ZERO.CSL1F10 )
      = 0193006)

000BAFT:  /  /CMT(SET LB IF AFT.LCA.EXM-NOT.SBAFT10       )
      = 0193002)

000BAP3:  /  /CMT(SET LFB IF AP=X11                       )
      = 0193007)

000BAVI:  /  /CMT(SET LB IF PROCEED.-HRM+LRM-.ADD VIOL    )
      = 0193001)

000BBAB:  /  /CMT(SET LB IF BYPASS A AND B ORDER.SBBAB10 )
      = 0193002)

000BBP1:  /  /CMT(SET LB IF B POINTER EQUALS 1.SFBP110    )
      = 0193003)

000BBRA:  /  /CMT(SET LFD IF BASE RELOCATION ACTIVE       )
      = 0193006)

000BBTS:  /  /CMT(SET LB IF BOOTSTRAP.CBTSP30    )
      = 0193007)

000BCMB:  /  /CMT(SET LB IF BP0+BP1.2.3CM.4CM.SBCMB10     )
      = 0193003)

000BCRD:  /  /CMT(SET LB IF CONTINUE INTERRUPT DEMAND     )
      = 0193007)

000BCRM:  /  /CMT(SET LB IF INTERRUPT MODF.CCCIM30        )
      = 0193004)

000BFBL:SU /  /CMT(SET LB IF FP IND TEST IS FALSF.SBFBL10 )
      = 0193003)

000BFMP:SU /  /CMT(SET LB IF FLOATING MULTIPLY.SBFMP10    )
      = 0193004)

000BIDB:  /  /CMT(SET LB IF B ADD IS INDIRECT             )
      = 0193004)

000BIVV:  /  /CMT(SET LB IF PROTECTED I/O SCR    )
      = 0193002)

000BLRD:  /  /CMT(SET LB IF LOW LVL INTRPT DMD-SBLRD10    )
      = 0193004)

000BMTT:  /  /CMT(SET LB IF MAT + MIT ORDER.SBMTT10       )
      = 0193005)

000BN23:  /  /CMT(SET LB IF QN=2 OR 3 OR UP.SBN2310       )
      = 0193005)
```

```
000BPDC:   /   /CMT(SET LFB IF PDT + PCB, SBPDC10            )
       = 0193003)

000BQV1:SU /   /CMT(SET LB IF QV1 IS ZERO.RQV0160            )
       = 0193005)

000BSK1:   /   /CMT(SET LB IF SVRCE AKN FRM IO IS 1.SBSK110)
       = 0193005)

000BVXT:   /   /CMT(SET LB IF VARIANT WAS EXTRACTD.SBVXT10 )
       = 0193007)

000BXUQ:SU /   /CMT(SET LB IF EXU,QN12=0.SBXUQ10             )
       = 0193006)

000BXYZ:SU /   /CMT(SET LB IF X OR Y OPERAND IS ZERO        )
       = 0193001)

000C0BS:SU /   /CMT(NO ONE BIT SENSED IN DTB                )
       = 0223007)

000C3CM:   /   /CMT(SET LC IF 3 CHAR MODE.CCCM330  )
       = 0223001)

000CA3R:   /   /CMT(SET LC IF AH3=0 AND MCE)
       = 0223007)

000CB0X:   /   /CMT(SET LC IF COND FOR TLU IS VALID.SCBCX30)
       = 0223006)

000CB1Z:SU /   /CMT(SET LC IF B1 IS XX1010 -ZERO CODE DTB  )
       = 0223001)

000CBBC:   /   /CMT(SET LC TEST BY BCCBBEBCE IS TRE.SCBBC10)
       = 0223002)

000CBFT:   /   /CMT(SET LFC IF BFT SET RCBF30       )
       = 0223005)

000CBNN:SU /   /CMT(SET LC IF B NOT NORMAL,BBBIN00          )
       = 0223003)

000CBP2:   /   /CMT(SET LC IF B POINTER EQUALS 2.SFBP210   )
       = 0223003)

000CBWM:   /   /CMT(SET LC IF B 4 HAS WORD MARK.SCB4W10     )
       = 0223001)

000CBYB:   /   /CMT(SET LB IF BYPS B:B FLD TRM NT.SCBYB10  )
       = 0223004)

000CCES:   /   /CMT(SET LC IF CONTROL MEM ENTER.CFNTC30     )
       = 0223005)

000CCH1:   /   /CMT(SET LC IF CH=0001.SCCH110       )
       = 0223002)

000CDSB:   /   /CMT(SET LD IF TRUE SUB IS SET               )
       = 0223006)

000CDUP:   /   /CMT(SET LC IF A EXT B EXT NOT DUP A.SCDUP10)
       = 0223002)

000CETS:   /   /CMT(SET LC IF DIAGNOSTIC DUMP               )
       = 0223007)

000CFBA:SU /   /CMT(SET LC IF ACC TEST IS TRUE.SCFBA10      )
       = 0223004)

000CFDV:SU /   /CMT(SET LC IF FLOATING DIVIDE.SCFDV10       )
       = 0223003)

000CFVL:   /   /CMT(SET LC IF FIELD OVRLP 1-4 CHAR.SCFVL10 )
       = 0223004)

000CHRD:   /   /CMT(SET LC IF HIGH LVL INTRPT DMD-SCHRD10  )
       = 0223004)

000CHRM:   /   /CMT(SET LC IF INTERRUPT MODF.CCEIM30        )
       = 0223005)

000CIXA:   /   /CMT(SET LC IF A ADD IS INDEXED.SCIXA10      )
       = 0223003)

000CJU5:   /   /CMT(SET LC IF TABLE LOOK UP        )
       = 0223007)

000CLRM:   /   /CMT(SET LC IF LOW LVL INTRPT MODE.CCIIM30  )
       = 0223005)

000CMDS:   /   /CMT(SET LD IF MEMORY DISPLAY                )
       = 0223006)

000CSK2:   /   /CMT(SET LC IF SERVICE ACKN FROM I/O IS 10  )
       = 0223005)

000CTRP:   /   /CMT(SET LC IF BCT VARIANT VIOLATION         )
       = 0223002)

000CZBI:   /   /CMT(SET LC IF ZBI IS SET AND DEC. OP CODE  )
       = 0223006)

000D2CM:   /   /CMT(SETT LD IF 2 CHAR MODE )
       = 0253001)

000DB3R:   /   /CMT(SET LD IF BH3 IS ZERO                   )
       = 0253002)

000DB4W:   /   /CMT(IF WM IN B4.= LCA.EXM-NOT               )
       = 0253007)

000DBCT:   /   /CMT(SET LD IF COND TST BY BCT ACTVE.SCBCT10)
       = 0253003)
```

```
000DBFZ:   /   /CMT(SET LD IF B FIELD ZONE                        )
       = 0253002)

000DBGA:   /   /CMT(SET LD IF BP IS GREATER THAN AP-BITS1-2)
       = 0253001)

000DBNT:   /   /CMT(SET LD IF AXT.BXT NOT-CABNT10                 )
       = 0253006)

000DBP3:   /   /CMT(SET LD IF B POINTER EQUALS 3.SRBP310    )
       = 0253001)

000DBPD:   /   /CMT(SET LD IF BPO + BP1:3CM.SDBZA10)
       = 0253001)

000DCDS:   /   /CMT(SET LD IF CONTROL DISPLAY                     )
       = 0253003)

000DDEC:   /   /CMT(SET LD IF DECIMAL OPERATION                   )
       = 0253006)

000DDLR:   /   /CMT(SET LD IF DATA LINK READ, I/O TO CP     )
       = 0253007)

000DEDM:   /   /CMT(SET LD IF IN DIAGNOSTIC MODE    )
       = 0253002)

000DFMA:   /   /CMT(SET LD IF FMA+FAA+BMS ODR=SCEN.SDFMA10 )
       = 0253002)

000DISR:   /   /CMT(SET LD IF IDX B+SEPARATE FMA+FAA EXEC   )
       = 0253004)

000DIXB:   /   /CMT(SET LD IF B ADD IS INDEXED       )
       = 0253003)

000DLDL:SU /   /CMT(SET LD IF LOAD LOW ORDER RESULT               )
       = 0253004)

000DNO1:SU /   /CMT(SET LD IF QN EQUALS 1.BQN0130                 )
       = 0253005)

000DNEM:   /   /CMT(SET LD IF NON-EXISTANT MEM DETECTED           )
       = 0253007)

000DQ60:SU /   /CMT(SET LD IF QN BIT 06 IS 0.BQN0600              )
       = 0253004)

000DQVO:   /   /CMT(SET LD IF QV IS ZERO.RQV1600    )
       = 0253003)

000DSBN:   /   /CMT(SET LFD IF SIGN B IS NEGATIVE                 )
       = 0253005)

000DSET:   /   /CMT(SET LINK FLOP D UNCOND )
       = 0253005)

000DV02:   /   /CMT(SET LD IF VV02=1         )
       = 0253006)

000DV06:   /   /CMT(SET LD IF VV06 IS SET.RVV0630                 )
       = 0253006)

000DYH0:   /   /CMT(LD IF YH1 IS RESET.BYHA120      )
       = 0253004)

000DZAS:   /   /CMT(SET LD IF ZA + ZS ORDER                       )
       = 0253005)

000EOV3:   /   /CMT(SET LE IF OVERLAP = 3- BBAA310                )
       = 0283005)

000EA3R:   /   /CMT(SET LE IF AH3 IS ZERO           )
       = 0283006)

000EAFT:   /   /CMT(SET LE IF AFT.-EXM + LCA        )
       = 0283001)

000EAH0:   /   /CMT(SET LE IF AH=X00                              )
       = 0283002)

000EAH3:   /   /CMT(SET LE IF AH EQUALS X11.RJBIM10)
       = 0283001)

000EBIM:   /   /CMT(SET LE IF BINARY INTEGER MULTIPLY             )
       = 0283001)

000EBIP:   /   /CMT(SET LE IF B IS POSITIVE         )
       = 0283001)

000ECMP:SU /   /CMT(SET LE IF COMPENSATE NOT RQD IN MAA+BIM)
       = 0283006)

000ECPC:   /   /CMT(SET LE IF CLEAR CM AND MM                     )
       = 0283005)

000EDIA:   /   /CMT(SET LE IF DIAG. SW. IS RESET RJDAJ20   )
       = 0283004)

000EDV2:   /   /CMT(SET LE IF DIAG VAR BIT 2 IS SET               )
       = 0283007)

000EDV4:   /   /CMT(SET LE IF DIAG VAR BIT 4 IS SET               )
       = 0283003)

000EEST:   /   /CMT(SET LFE                                       )
       = 0283006)

000EHAX:   /   /CMT(SET LE IF HALF ADD + EXTRACT                  )
       = 0283002)

000EJDR:   /   /CMT( SET LE IF JUSTIFY DOLLAR SIGN RESET    )
       = 0283004)
```

```
000ERUN:   /   /CMT(SET LE IF RUN IS SET                        )
       = 0283003)

000ESCR:   /   /CMT(SET LE IF SCR.RUN NOT                       )
       = 0283002)

000ESET:   /   /CMT(SET LE UNCONDITIONAL    )
       = 0283003)

000ESMD:   /   /CMT(SET LE IF S-MODE.-CCSMD30           )
       = 0283004)

000ESTT:   /   /CMT(SET LE UNCONDITIONAL                )
       = 0283005)

000EV01:   /   /CMT(SET LE IF VV01 = 1.                         )
       = 0283002)

000EV04:   /   /CMT(SET LE IF V04 IS RESET                      )
       = 0283003)

000EZSR:   /   /CMT(SET LE IF ZERO SUPPRESSION IS RESET         )
       = 0283004)

000QAD1:   /   /CMT(TEST LOGIC OF QUADRANT 1        )
       = 0142002)

000QAD2:   /   /CMT(TEST LOGIC OF QUADRANT 2        )
       = 0142001)

000QAD3:   /   /CMT(TEST LOGIC OF QUADRANT 3        )
       = 0142003)

000QAD4:   /   /CMT(TEST LOGIC OF QUADRANT 4        )
       = 0142000)

001CATR:   /   /CMT(CONT MEM ADD IS ACCOUNTING TIMER REG        )
       = 0345014)

001CMAC:   /   /CMT(CONTROL MEM ADD FLD OF K REG IS AC          )
       = 0345034)

001CMBC:   /   /CMT(CONT MEM ADD FIELD OF K REG IS BC           )
       = 0345030)

001CMCI:   /   /CMT(CONT MEM ADD FIELD IS CONT. INTRP REG       )
       = 0345032)

001CMCS:   /   /CMT(CONT MEM ADD FIELD OF K REG IS CSR          )
       = 0345024)

001CMDA:   /   /CMT(CM ADD IS DIAGNOSTIC DUMP ADD               )
       = 0345033)

001CMDW:   /   /CMT(CM ADD IS DIAG WORK LOCATION                )
       = 0345031)

001CMHR:   /   /CMT(CONT MEM ADD FIELD OF K REG IS EIR          )
       = 0345026)

001CMLR:   /   /CMT(CONT MEM ADD FIELD OF K REG IS IIR          )
       = 0345036)

001CMSC:   /   /CMT(CONT MEM ADD FIELD OF K REG IS SC           )
       = 0345037)

001CMW1:   /   /CMT(CONT MEM ADD FIELD OF K REG IS WL1          )
       = 0345035)

001CMW2:   /   /CMT(CONT MEM ADD FIELD OF K REG IS WL2          )
       = 0345027)

001CMW3:   /   /CMT(CONT MEM ADD FIELD OF K REG IS WL3          )
       = 0345020)

001CMW4:   /   /CMT(CONT MEM ADD FIELD OF K REG IS WL4          )
       = 0345021)

001CMW5:   /   /CMT(CONT MEM ADD FIELD OF K REG IS WL5          )
       = 0345022)

001CMW6:   /   /CMT(CONT MEM ADD FIELD OF K REG IS WL6          )
       = 0345023)

001CMW7:   /   /CMT(CONT MEM ADD FIELD OF K REG IS WL7          )
       = 0345025)

00AACL0:   /   /CMT(A REG CLR ALSO RESETS AFT.BFT.ATF           )
       = 1113004)

00AACL1:SU /   /CMT(IF EX IS LESS THAN EY.BXGEY21               )
       = 0701001)

00ACSR0:   /   /CMT(RESET CHECK SIZE INDICATOR-OVERFLOW         )
       = YXMUEE )

00AFA10:   /   /CMT(A4 FROM A1 CHAR XFR                         )
       = 0541001)

00AFA40:   /   /CMT(A1 FROM A4 CHARACTER XFR        )
       = 0731001)

00AFA41:   /   /CMT(A1 FRM A4 IF AP DOES NT = X11.SFAP303       )
       = 0781001)

00AFAD0:   /   /CMT(A1 FROM ADDER                               )
       = 0831001)

00AFAD1:   /   /CMT(IF ZA + ZS SDZAS10      )
       = 0401001)

00AFAN0:   /   /CMT(A REG FROM A REG NOT    )
       = 1113002)
```

```
00AFAN1:SU /   /CMT(IF FP SUBTRACT,RQY0130                    )
       = 0581001)

00AFAN2:SU /   /CMT(IF SNA=1 ,SGNAN30                         )
       = 0621001)

00AFAN3:SU /   /CMT(IF SNX=0 ,SGNBN40                         )
       = 0631001)

00AFAN4:SU /   /CMT(IF SIGNS OPPOSITE, BSGDF10                )
       = 1183004)

00AFB10:   /   /CMT(A 4 CHAR FROM B1 CHAR XFR                 )
       = 0391001)

00AFB11:   /   /CMT(IF AP DOES NOT EQUAL X00;AFT NOT          )
       = 0731001)

00AFB40:SU /   /CMT(A1 FROM B4 CHARACTER XFR                  )
       = 0491001)

00AFB41:   /   /CMT(IF AFT NOT,MPY NOT,RAFB410      )
       = 0621001)

00AFB42:   /   /CMT(IF AP DOES NT =X11,AFT NOT;MPY NOT:       )
       = 0661001)

00AFCL1:   /   /CMT(A4 FROM CONT IND AS SPECIFIED BY QV REG)
       = 0851001)

00AFDD0:   /   /CMT(A REG FROM D REG 36 BIT XFR               )
       = 0671001)

00AFDD1:   /   /CMT(A FROM D IF AH=X11,SEAH310      )
       = 0711001)

00AFDD2:   /   /CMT(IF AH=X00,EDIT NOT+EDIT,A IS TAKEN        )
       = 1013001)

00AFNN0:   /   /CMT(A REG FROM N REG=36 BITS        )
       = 0942002)

00AFNN1:   /   /CMT(IF AH EQUALS X00        )
       = 0913004)

00AFQU0:   /   /CMT(A4 FROM QU REG                            )
       = 1062002)

00AFSB0:   /   /CMT(A1 FROM RESULT SIGN-VIA ADDER             )
       = 0861001)

00AFTS1:   /   /CMT(SET AFT IF B CHAR SATISFIES V.OP CODE )
       = 0891001)

00AFTS2:   /   /CMT(AND IF AP DOES NOT =X11)
       = YXNUEE )

00AFTS3:   /   /CMT(AND IF AP DOES NOT =X00)
       = YXNUEE )

00AFTS4:   /   /CMT(FOR MOS A OPERAND TERM )
       = 0541001)

00AFTS5:   /   /CMT(FOR MOS B OPERAND TERM )
       = 0871001)

00AFW70:   /   /CMT(A01 FROM QW07                                        )
       = 0461001)

00AHIN0:   /   /CMT(INC A HISTORY CTR        )
       = 1113005)

00AHIN2:   /   /CMT(IF B FLD TERM NOT,AFT NOT       )
       = 0451001)

00AHIN3:   /   /CMT(IF AP DOES NOT EQUAL X00;AFT NOT              )
       = YXNUEE )

00AH3S0:   /   /CMT(SET A HISTORY COUNTER BIT 3                   )
       = 0411001)

00AHDE0:   /   /CMT(DEC A HISTORY CTR        )
       = 1162002)

00AHDE2:   /   /CMT(IF AP DOES NOT EQUAL X11;AFT NOT              )
       = YXNUEE )

00AHDE3:   /   /CMT(IF AFT NOT  RCAFT20     )
       = YXNUEE )

00AHDE4:   /   /CMT(IF A IS TAKEN FOR EDIT,PN1                    )
       = YXNUEE )

00AHFN1:   /   /CMT(SET AH 1 POS LEFT OF NWM IF WM IN N REG)
       = 0461001)

00AHFP0:   /   /CMT(A HISTORY FROM A POINTER                      )
       = 1013007)

00ALFY0:   /   /CMT(A REG LOWER 18 BITS FROM Y REG )
       = 1113003)

00AMFY0:   /   /CMT(A REG. BIT 19 FROM Y19,CLEAR 20-24            )
       = 1062003)

00APIN1:   /   /CMT(IF AP DOES NOT EQUAL X11                      )
       = 1113007)

00APIN2:   /   /CMT(IF AP DOES NOT = AH;BFT          )
       = 0791001)

00APIN3:   /   /CMT(IF AP DOES NOT EQUAL X00                      )
       = 0581001)
```

```
00APDE0:  /  /CMT(DEC A POINTER  )
      = 0771001)

00APDE1:  /  /CMT(IF AP DOES NOT EQUAL X00       )
      = 0751001)

00APDE2:  /  /CMT(IF 3CM                                    )
      = 0781001)

00APDE3:  /  /CMT(IF AP DOES NOT EQUAL X11       )
      = 0741001)

00APFH0:  /  /CMT(A POINTER FROM A HISTORY CTR   )
      = 1013004)

00APFY0:  /  /CMT(A POINTER FROM Y REG LOW ORDER 2 BITS )
      = 0942003)

00ARTA0:  /  /CMT(ADDER RESET ADVANCE CARRY             )
      = YXNUHDW)

00ASLB0:SU /  /CMT(A SHIFT LEFT ONE BIT                 )
      = 0691001)

00ASLC0:  /  /CMT(A SHIFT LEFT ONE CHARACTER      )
      = 1162003)

00ASLC1:  /  /CMT(IF AP DOES NOT EQUAL X00       )
      = 0481001)

00ASRB0:SU /  /CMT(A SHIFT RIGHT ONE BIT                )
      = 0771001)

00ASRB1:SU /  /CMT(IF EX IS LESS THAN FY BXGEY21        )
      = YXNUEE )

00ASRC0:  /  /CMT(A SHIFT RIGHT ONE CHARACTER           )
      = 1113006)

00ASRC1:  /  /CMT(IF BP DOES NOT EQUAL X11.SFBP300      )
      = 0611001)

00ASRC2:  /  /CMT(IF AP DOES NOT EQUAL X11. SFAP300     )
      = 0913002)

00ASRC3:  /  /CMT(IF A IS TAKEN FOR EDIT.PN1.SEATF10    )
      = 0661001)

00ASTA0:  /  /CMT(ADDER SET ADVANCE CARRY)
      = 0761001)

00ASTA1:SU /  /CMT(IF FP SUBTRACT RQY0130                )
      = YXNUEE )

00ASTA2:  /  /CMT(IF OVERFLOW                            )
      = 0441001)

00ASTA3:SU /  /CMT(IF SNA=1, OTHERWISE RESFT ACI.SGNAN30 )
      = 0741001)

00ASTA4:SU /  /CMT(IF SNX=0 .SGNBN40                      )
      = 0411001)

00ASTA6:SU /  /CMT(IF SIGNS OPPOSITE.BS8DF10              )
      = YXNUEE )

00ASTA7:SU /  /CMT(IF BASTC10 IS SET:35 DIVIDE DONE       )
      = 0831001)

00ASTA8:  /  /CMT(SET ACI IF DS + BS = FOR GROUP 5        )
      = 0561001)

00ATMOP:  /  /CMT(ACT TO B2+3 OR SET FI IF ACT OVL SET  )
      = 0491001)

00AUFY0:SU /  /CMT(A REG. BITS 19-36 FROM Y. BITS 1-18   )
      = 0671001)

00AVIS0:  /  /CMT(SET ADD VIOL IF PROCEED.-HRM+LRM-.VIOL )
      = 0731001)

00AXTS0:  /  /CMT(SET A EXTRACT FLOP                      )
      = 0541001)

00AZBR0:  /  /CMT(RESET ZERO BALANCE INDICATOR            )
      = YXNUEE )

00AZBS1:  /  /CMT(SET ZBI IF S MODE.ZAB.YH2 NOT           )
      = 0491001)

00AZBS2:  /  /CMT(SET ZBI IF YH1 IS RESET        )
      = 0501001)

00BAGC2:  /  /CMT(SHIFT ADDER OUTPUT CONTROLS.B-D-H-L   )
      = 0491001)

00BAGC3:  /  /CMT(SET ADDER CONTROLS TO BINARY            )
      = 0471001)

00BAGFT:  /  /CMT(ALLOW ZONE BITS THRU ADDER-DECIMAL     )
      = 0571001)

00BBCL0:  /  /CMT(B REGISTER CLEAR        )
      = 0913001)

00BBCL1:SU /  /CMT(IF EX IS GREATER THAN EY               )
      = YXNUEE )

00BBDI0:SU /  /CMT(BTD OR DTB I/O CONVERSION.BBDI010     )
      = 0561001)

00BBDRN:SU /  /CMT(BTD OR DTB RUN BY OPCODE.BBDRN10      )
      = 0891001)
```

```
00BDT1A:SU /    /CMT(QA TO QR IF L35=1 ANB SENT. BIT SET     )
        = 0211001)

00BDT1B:SU /    /CMT(QB TO QR IF OVF NOT AND B34+B33=0       )
        = YXNUEE )

00BDT1C:SU /    /CMT(QC TO QR IF B36=0.B35=0. AND B34=1      )
        = YXNUEE )

00BDT1D:SU /    /CMT(QD TO QR IF B33=1. AND B36.B35.B34=0    )
        = YXNUEE )

00BDT1E:SU /    /CMT(QE TO QR IF B36.B35.B34.B33=0           )
        = YXNUEE )

00BDT2B:SU /    /CMT(QB TO QR IF QN03+QN04+QN05+QN06=1       )
        = 0241001)

00BDT2C:SU /    /CMT(QC TO QR IF QN07=1                      )
        = YXNUEE )

00BDT2E:SU /    /CMT(QE TO QR IF QN02=1.QN01=0               )
        = YXNUEE )

00BDT3A:SU /    /CMT(QA TO QR IF L35=1 AND SFNT. BIT SET     )
        = 0271001)

00BDT3C:SU /    /CMT(QC TO QR IF B32=B33=1                   )
        = YXNUEE )

00BDT4A:SU /    /CMT(QA TO QR IF QN=0 +X=0 +Y=0              )
        = 0301001)

00BDT4B:SU /    /CMT(QB TO QR IF L03=L04=0                   )
        = YXNUEE )

00BDT4C:SU /    /CMT(QC TO QR IF L03=1.L04=0                 )
        = YXNUEE )

00BDT4D:SU /    /CMT(QD TO QR IF L03=0.L04=1                 )
        = YXNUEE )

00BDT4E:SU /    /CMT(QE TO QR IF L03=1.L04=1                 )
        = YXNUEE )

00BDT5A:SU /    /CMT(QA TO QR IF QN=0                        )
        = 0141001)

00BDT5B:SU /    /CMT(QB TO QR IF L02=1.L03=1                 )
        = YXNUEE )

00BDT5C:SU /    /CMT(QC TO QR IF L02=0.L03=0                 )
        = YXNUEE )

00BDT5D:SU /    /CMT(QD TO QR IF L02=1.L03=0                 )
        = YXNUEE )

00BDT5E:SU /    /CMT(QE TO QR IF L02=0.L03=1                 )
        = YXNUEE )

00BDT6A:   /    /CMT( QA TO QR IF QN =0                      )
        = 0151001)

00BDT6B:   /    /CMT( QB TO QR IF L03=1.L04=1                )
        = YXNUEE )

00BDT6C:   /    /CMT( QC TO QR IF L03=0.L04=0                )
        = YXNUEE )

00BDT6D:   /    /CMT( QD TO QR IF L03=1.L04=0                )
        = YXNUEE )

00BDT6E:   /    /CMT( QE TO QR IF L03=0.L04=1                )
        = YXNUEE )

00BFA10:   /    /CMT(B4 FROM A1 CHARACTER SHIFT              )
        = 0441001)

00BFA11:   /    /CMT(IF AP DOES NOT EQUAL X00                )
        = 0581001)

00BFA40:   /    /CMT(B1 FROM A4 CHARACTER SHIFT              )
        = 0551001)

00BFA41:   /    /CMT(IF BP DOES NOT EQUAL X11                )
        = 0411001)

00BFAD0:   /    /CMT(B4 FROM ADDER - 6 BITS                  )
        = 0451001)

00BFB10:   /    /CMT(B4 FROM B1 CHAR XFR     )
        = 0551001)

00BFB12:   /    /CMT(B4 FROM B1 IF OVERLAP =3                )
        = 0601001)

00BFB40:   /    /CMT(B1 FROM B4 CHAR XFR                     )
        = 0541001)

00BFB41:   /    /CMT(B1 FROM B4 IF AP DOES NOT EQUAL X11     )
        = 0521001)

00BFB42:   /    /CMT(B1 FROM B4 IF OVERLAP=3                 )
        = 0641001)

00BFBA0:   /    /CMT(B REG FROM BINARY ADDER UP TO 36 BITS   )
        = 1013006)

00BFE10:   /    /CMT(BFE1 IF DS FLOP NOT.BFE1-9'S IF DS,     )
        = 1013003)

00BFE20:   /    /CMT(BFE2 +E2-9'S.IDENTITY TO BFE1.BFE4      )
        = YXNUEE )
```

```
00BFE30:  /  /CMT(BFE3 +E3-9'S.IDENTITY TO BFE1.BFE4          )
      = YXNUEE )

00BFE40:  /  /CMT(BFE4 IF DS FLOP NOT+BFE4-9'S IF DS,         )
      = 0962003)

00BFEE1:  /  /CMT(IF BH=X00.BFE9'S IF DSF+BFE IF DSFN         )
      = 0451001)

00BFEE2:  /  /CMT(IF BH=X00.AP DOES NOT EQUAL X11)
      = YXNUEE )

00BFEE3:  /  /CMT(IF BH=X11.AP DOES NOT EQUAL X00             )
      = 0471001)

00BFEE4:  /  /CMT(IF BH = X11                                 )
      = 0461001)

00BFEE5:  /  /CMT(IF OVL=4 B FROM E IF DS NOT+E9'S IF DS )
      = 0551001)

00BFFL1:  /  /CMT(A1+A NEG+A9 TO B3 IF OVL=1,2+TO B2 IF=3)
      = YXNUEE )

00BFTS1:  /  /CMT(SET B FIELD TERM NWM.SANWM10    )
      = 0481001)

00BFTS2:  /  /CMT(SET BFT IF A1WM.RAAW130)
      = 0811001)

00BFTS3:  /  /CMT(SET BFT IF PCB+PDT.SCPDT10       )
      = 0621001)

00BFTS4:  /  /CMT(SET BFT IF BBC.A1WM:BFT NOT-SAV1W10      )
      = 0791001)

00BFTS5:  /  /CMT(SET BFT IF SW+SI+BIM.SASPM10    )
      = 0881001)

00BFTS6:  /  /CMT(IF A4WM.-MCE=NOT+MCE:ATE BAAW430           )
      = 0571001)

00BFTS7:  /  /CMT(IF B4TM.-MPY=NOT+MPY.QU07.RBFT710          )
      = 0611001)

00BHIN1:  /  /CMT(AH=X00.BFT NOT---BFLD TERM FLOP            )
      = 0681001)

00BHIN2:  /  /CMT(IF AFT NOT                                 )
      = 0781001)

00BHIN3:  /  /CMT(IF AP DOES NOT EQUAL X00:AFT NOT           )
      = YXNUEE )

00BHIN4:  /  /CMT(IF RUN.V04+DISPLAY+1    )
      = 0861001)

00BH3S0:  /  /CMT(SET B HISTORY COUNTER BIT 3     )
      = 0501001)

00BH3S1:SU / /CMT(IF OVF =0 + BAOVF00                                  )
      = YXNUEE )

00BHDE0:  /  /CMT(DEC B HISTORY CTR        )
      = 0881001)

00BHDE2:  /  /CMT(IF AP DOES NOT EQUAL X11:AFT NOT                     )
      = YXNUEE )

00BHDE3:  /  /CMT(IF AFT NOT      )
      = 0851001)

00BHDE4:  /  /CMT(IF RUN V04 NOT+DISPLAY-1        )
      = YXNUEE )

00BHFP0:  /  /CMT(B HISTORY CTR FROM B POINTER                         )
      = 1142002)

00BJVS1:  /  /CMT(SET BARRICADE VIOLATION IF BFT                       )
      = YXNUEE )

00BLFY0:  /  /CMT(B REG LOWER 18 BITS FROM Y REG )
      = 1013002)

00BLFY1:SU / /CMT(IF BH = X00.SEBH010                                  )
      = 0601001)

00BMFY0:  /  /CMT(B REG. BIT 19 FROM Y19.CLEAR 20-24       )
      = 1162001)

00BP1N0:  /  /CMT(INC B POINTER   )
      = 0942001)

00BP1N1:  /  /CMT(IF BP DOES NOT EQUAL X11                             )
      = 1042003)

00BP1N2:  /  /CMT(IF AP DOES NOT = AH:BFT         )
      = YXNUEE )

00BPDE0:  /  /CMT(DEC B POINTER   )
      = 0913005)

00BPDE1:  /  /CMT(DEC BP IF ATF IS SET+ADVANCE TERM FLOP )
      = 0521001)

00BPFH0:  /  /CMT(B POINTER FROM B HISTORY CTR    )
      = 1062001)

00BPFY0:  /  /CMT(B POINTER FROM Y REG LOW ORDER 2 BITS  )
      = 1191001)

00BSLB0:  /  /CMT(B SHIFT LEFT ONE BIT    )
      = 0391001)
```

```
00BSLB1:SU /   /CMT(IF B IS NOT NORM BBBINOO                         )
       = 0401001)

00BSLB2:SU /   /CMT(IF L35=0,B35=0,B36=0                             )
       = 1183001)

00BSLB4:SU /   /CMT(IF  B35=1,L35=0                                  )
       = 1183003)

00BSLB5:SU /   /CMT(IF L35=0,  BDVNC10                               )
       = 1183007)

00BSLC0:   /   /CMT(B SHIFT LEFT ONE CHARACTER        )
       = 0561001)

00BSLC1:   /   /CMT(IF AP DOES NOT EQUAL X00                         )
       = 0531001)

00BSRB0:   /   /CMT(B SHIFT RIGHT ONE BIT                            )
       = 0421001)

00BSRB1:SU /   /CMT(IF EX IS GREATER THAN EY BXGEY30                 )
       = 0431001)

00BSRB3:SU /   /CMT(IF MANTISSA OVERFLOW BMOVF10                     )
       = 0411001)

00BSRC0:   /   /CMT(B SHIFT RIGHT ONE CHARACTER       )
       = 1142003)

00BSRC1:   /   /CMT(IF BP DOES NOT EQUAL X11,SFBP301                 )
       = 0431001)

00BSRC2:   /   /CMT(IF AP DOES NOT EQUAL X11,SFAP303                 )
       = 0411001)

00BSRC3:   /   /CMT(IF FIELD OVERLAP=1,BBAA130        )
       = 0401001)

00BT011:   /   /CMT(FOR PARITY-IS SAME AS GRP02                      )
       = 0111001)

00BT014:   /   /CMT(FOR PARITY-IS SAME AS OQAD1                      )
       = 0141001)

00BT016:   /   /CMT(FOR PARITY-IS SAME AS OAMPD                      )
       = 0161001)

00BT022:   /   /CMT(FOR PARITY  SAME AS OCBYB                        )
       = 0221001)

00BT034:   /   /CMT(FOR PARITY-IS SAME AS 1CMW3                      )
       = 0341001)

00BT051:   /   /CMT(FOR PARITY BIT NOT USED IN GROUPS BDEJM)
       = 0511001)

00BT055:   /   /CMT(FOR PARITY BIT NOT USED IN GROUPS A      )
       = 0551001)

00BT060:   /   /CMT(FOR PARITY BIT NOT USED IN GRP CFGHKL  )
       = 0601001)

00BT061:   /   /CMT(FOR PARITY BIT USED ONLY IN DUMMY--    )
       = 0611001)

00BT063:   /   /CMT(FOR PARITY BIT NOT USED IN GROUPS DGH  )
       = 0631001)

00BT065:   /   /CMT(FOR PARITY BIT NOT USED IN GROUPS CKM  )
       = 0651001)

00BT076:   /   /CMT(FOR PARITY BIT NOT USED IN GROUPS J    )
       = 0761001)

00BT081:   /   /CMT(FOR PARITY BIT NOT USED IN GROUPS AF   )
       = 0811001)

00BT083:   /   /CMT(FOR PARITY BIT NOT USED IN GROUPS GHL  )
       = 0831001)

00BT087:   /   /CMT(FOR PARITY BIT NOT USED IN GROUPS BF   )
       = 0871001)

00BT099:   /   /CMT(FOR PARITY-IS SAME AS YFCMO            )
       = 0991001)

00BT104:   /   /CMT(FOR PARITY-IS SAME AS MINCO            )
       = 1041001)

00BT109:   /   /CMT(FOR PARITY-IS SAME AS MFYYO            )
       = 1091001)

00BT111:   /   /CMT(FOR PARITY-IS SAME AS AACLO            )
       = 1111001)

00BT114:   /   /CMT(FOR PARITY-IS SAME AS BHFPO            )
       = 1141001)

00BT119:   /   /CMT(FOR PARITY-IS SAME AS BPFYO            )
       = 1191001)

00BUFY0:SU /   /CMT(B REG. BITS 19-36 FROM Y. BITS 1-18    )
       = 0541001)

00BXTS0:   /   /CMT(USED WITH AXTSO TO SET B EXTRACT FLOP  )
       = 0501001)

00COVR0:   /   /CMT(RESET OVF INDICATOR IF TESTED          )
       = 0751001)

00CARB0:   /   /CMT(CHECK A REL TO B,SET PR FLOPS          )
       = 0771001)
```

```
00CBRS0:  /  /CMT( COMM. BUFFER SERVICE REQUEST              )
      = 0581001)

00CFYA0:  /  /CMT(CONTROL MEMORY WRITE ALL                    )
      = 0981001)

00CFYA1:  /  /CMT(CONTROL MEMORY WRITE CHARACTER MODE         )
      = 0801001)

00CFYA2:  /  /CMT(CONTROL MEMORY WRITE BITS 13-15             )
      = 0781001)

00CFYA3:  /  /CMT(WRITE ZCM FOR DIAGNOSTIC EXTRACTION         )
      = 0721001)

00CHCL0:  /  /CMT(CLEAR CH COUNTER     )
      = 0461001)

00CHDE0:  /  /CMT(DEC CH CTR     )
      = 0551001)

00CHFU0:  /  /CMT(CH FROM QU01=4                              )
      = 0881001)

00CIMOP:  /  /CMT(SET OR RESET CONTINUE INTERRUPT MODE        )
      = 0711001)

00CLFA1:  /  /CMT(IND FROM A1 AS SPECIFIED BY QV REG          )
      = 0511001)

00CLFV0:  /  /CMT(ITEM MARK TRAP,S-MODE,CHAR MODE FROM VV)
      = 0851001)

00CLFX0:  /  /CMT(CONTROL INDICATORS FROM AIR     )
      = 0811001)

00CM3S0:  /  /CMT(SET 3CM  RESET ITEM -MARK TRAPPING          )
      = YXNUEE )

00CPIN1:  /  /CMT(INC CP IF ADDER OVERFLOW       )
      = 0591001)

00CPFH0:  /  /CMT(CP FROM HISTORY        )
      = 0671001)

00CPFY0:  /  /CMT(C POINTER FROM Y REG BITS 16-19             )
      = 0431001)

00CXFL0:  /  /CMT(AIR FROM CONTROL INDICATORS                 )
      = 0611001)

00DFNN0:  /  /CMT(D REG FROM N REG 36 BITS       )
      = 0962002)

00DLRS0:  /  /CMT(SET SERVICE REQ FOR DATA LINK READ          )
      = 0631001)

00DMY01:  /  /CMT( BITS 13-21 FOR ALLONES LOCATION           )
      = 0139777)

00DMY02:  /  /CMT( BITS 22-30     )
      = 0229777)

00DMY03:  /  /CMT( BITS 31-39     )
      = 0319777)

00DMY04:  /  /CMT( BITS 40-48     )
      = 0409777)

00DMY05:  /  /CMT( BITS 49-57     )
      = 0499777)

00DMY06:  /  /CMT( BITS 58-66     )
      = 0589637)

00DMY07:  /  /CMT( BITS 67-75     )
      = 0679777)

00DMY08:  /  /CMT( BITS 76-84     )
      = 0789777)

00DMY09:  /  /CMT( BITS 85-93     )
      = 0859777)

00DMY10:  /  /CMT( BITS 94-102    )
      = 0949777)

00DMY11:  /  /CMT( BITS 103-111   )
      = 1039777)

00DMY12:  /  /CMT( BITS 112-120   )
      = 1129777)

00DVCS1:SU /  /CMT(SET DIVIDE CHECK FLOP IF X=0               )
      = 1083004)

00DVFV0:  /  /CMT(DIAG VARIANT REG FROM VV REG                )
      = 0751001)

00EFE10:  /  /CMT(E4 FROM E1 CHAR XFR     )
      = 0681001)

00EFE40:  /  /CMT(E1 FROM E4 CHAR XFR                         )
      = 0721001)

00EFED1:  /  /CMT(E1 FROM EDIT LOGIC                          )
      = YXNUEE )

00EFED4:  /  /CMT(E4 FROM EDIT LOGIC                          )
      = 0741001)

00EFNN1:  /  /CMT(EFN IF BS FLOP NOT+FFN NOT BS FP,BBSUB )
      = 1142001)
```

```
00EFNN3:   /   /CMT(SAME AS EFNN1 IF OVL=4-8;BBAA710        )
       = 0701001)

00EFNN4:   /   /CMT(E FROM N IF AH=X11,SEAH310      )
       = 0701001)

00ESLC0:   /   /CMT(E SHIFT LEFT ONE CHARACTER      )
       = 0491001)

00ESLC1:   /   /CMT(IF AP DOES NOT EQUAL X00 SFAP000        )
       = 0771001)

00ESRC0:   /   /CMT(E SHIFT RIGHT ONE CHARACTER     )
       = 0711001)

00ESRC1:   /   /CMT(IF BP DOES NOT EQUAL X11                )
       = 0871001)

00EXOS1:SU /   /CMT(SET EXO IF QN12 CHANGING FROM + TO -    )
       = 0571001)

00EXOTO:SU /   /CMT(EXO FROM TXO                            )
       = 0821001)

00EXCD1:SU /   /CMT(CHECK EXP CONDITION FOR DAA             )
       = 1083002)

00EXUR1:SU /   /CMT(RESET EXU OR TXO IF SIGN CORRECT        )
       = 0881001)

00EXUS1:SU /   /CMT(SET EXU IF QN12 CHANGING FROM - TO +    )
       = 0521001)

00EXUS2:SU /   /CMT(IF A12=1 , B12=1                        )
       = 0431001)

00EXUS3:SU /   /CMT(SET EXU IF QN12=1                       )
       = 1083006)

00EXUS4:SU /   /CMT(SET EXU IF QM12=0.QN12=1                )
       = 0841001)

00FPCRO:SU /   /CMT(DTB AND BTD CONTROL RESET               )
       = 0401001)

00FVLS1:   /   /CMT( SET OVERLAP FLP IF 6=1-8               )
       = 0811001)

00GRP01:   /   /CMT(PUT GROUP BIT A IN ADDRESS FIELD        )
       = 0121001)

00GRP02:   /   /CMT(PUT GROUP BIT B IN ADDRESS FIELD        )
       = 0111001)

00GRP03:   /   /CMT(PUT GROUP BIT C IN ADDRESS FIELD        )
       = 0101001)

00GRP04:   /   /CMT(PUT GROUP BIT D IN ADDRESS FIELD        )
       = 0091001)

00GRP05:   /   /CMT(PUT GROUP BIT E IN ADDRESS FIELD        )
       = 0081001)

00GRP06:   /   /CMT(PUT GROUP BIT F IN ADDRESS FIELD        )
       = 0071001)

00GRP07:   /   /CMT(PUT GROUP BIT G IN ADDRESS FIELD        )
       = 0061001)

00GRP08:   /   /CMT(PUT GROUP BIT H IN ADDRESS FIELD        )
       = 0051001)

00GRP09:   /   /CMT(PUT GROUP BIT J IN ADDRESS FIELD        )
       = 0041001)

00GRP0A:   /   /CMT(GROUP A=1 IS ACTIVE                     )
       = YXNUDMY)

00GRP0B:   /   /CMT(GROUP B=2 IS ACTIVE                     )
       = YXNUDMY)

00GRP0C:   /   /CMT(GROUP C=3 IS ACTIVE                     )
       = YXNUDMY)

00GRP0D:   /   /CMT(GROUP D=4 IS ACTIVE                     )
       = YXNUDMY)

00GRP0E:   /   /CMT(GROUP E=5 IS ACTIVE                     )
       = YXNUDMY)

00GRP0F:   /   /CMT(GROUP F=6 IS ACTIVE                     )
       = YXNUDMY)

00GRP0G:   /   /CMT(GROUP G=7 IS ACTIVE                     )
       = YXNUDMY)

00GRP0H:   /   /CMT(GROUP H=8 IS ACTIVE                     )
       = YXNUDMY)

00GRP0J:   /   /CMT(GROUP J=9 IS ACTIVE                     )
       = YXNUDMY)

00GRP0K:   /   /CMT(GROUP K=10 IS ACTIVE                    )
       = YXNUDMY)

00GRP0L:   /   /CMT(GROUP L=11 IS ACTIVE                    )
       = YXNUDMY)

00GRP0M:   /   /CMT(GROUP M=12 IS ACTIVE                    )
       = YXNUDMY)

00GRP10:   /   /CMT(PUT GROUP BIT K IN ADDRESS FIELD        )
       = 0031001)
```

```
00GRP11:   /   /CMT(PUT GROUP BIT L IN ADDRESS FIELD          )
       = 0021001)

00GRP12:   /   /CMT(PUT GROUP BIT M IN ADDRESS FIELD          )
       = 0011001)

00GRPCG:   /   /CMT( ONLY COMMON GROUP U=OPS AVAILABLE        )
       = YXNUDMY)

00HCTR0:   /   /CMT( HIT COUNTER RESET                        )
       = 0661001)

00HRCL0:   /   /CMT(HIGH LEVEL INTERRUPT MODE CLEAR)
       = YXNUEE )

00HRMS0:   /   /CMT(SET HIGH LEVEL INTERRUPT MODE             )
       = YXNUEE )

00JFNS0:   /   /CMT(J REG. FROM N SELECT BY B POINTER         )
       = 0651001)

00KBRFB:   /   /CMT(LOAD BASE RELOCATION REGISTER             )
       = 0481001)

00L01S2:SU /   /CMT(SHIFT 1 INTO L01                          )
       = 1083001)

00L36S0:SU /   /CMT(L36 FROM SIGN STORAGE                     )
       = 0811001)

00LF0F0:SU /   /CMT(L01 FROM BAOVF10,QUOTIENT BIT FOR DAA     )
       = YXNUEE )

00LFCS0:SU /   /CMT(E36 FROM A01,DTB CONVERSION               )
       = 0391001)

00LLCL0:SU /   /CMT(L REGISTER CLEAR                          )
       = 0461001)

00LLCL1:SU /   /CMT(L REG CLEAR IF UNDERFLOW                  )
       = 0581001)

00LLFB0:SU /   /CMT(L REG 36 BITS FROM B REG                  )
       = 0391001)

00LLFM0:SU /   /CMT(L REG LOW 12 BITS FROM QM                 )
       = 0591001)

00LLFY0:SU /   /CMT( L REG LOWER 18 BITS FROM Y REG           )
       = 0751001)

00LRCL0:   /   /CMT(LOW LEVEL INTERRUPT MODE CLEAR )
       = 0841001)

00LRMS0:   /   /CMT(SET LOW  LEVEL INTERRUPT MODE             )
       = 0841001)

00LSFBM:SU /   /CMT(SIGN OF L FOR BIM                          )
       = 0861001)

00LSLB0:SU /   /CMT( L SHIFT LEFT ONE BIT                      )
       = 0781001)

00LSLB1:SU /   /CMT(IF L35=0,  BDVNC10                         )
       = YXNUEE )

00LSLB2:SU /   /CMT(IF L35=0 AND B35=B36=0, BBH2B20            )
       = YXNUEE )

00LSLB4:SU /   /CMT(IF  B35=1,L35=0                            )
       = YXNUEE )

00LSLB5:SU /   /CMT(IF QV2=1,RQV0210                           )
       = 0781001)

00LSLB6:SU /   /CMT(IF QV3=1,RQV0311                           )
       = 0741001)

00LSLB7:SU /   /CMT(IF B NOT NORMAL,BBBIN00                    )
       = YXNUEE )

00LSRB0:SU /   /CMT(L SHIFT RIGHT ONE BIT                      )
       = 0731001)

00LSRB1:SU /   /CMT(IF QV2=1,RQV0210                           )
       = 0771001)

00LSRB2:SU /   /CMT(IF QV3=1 RQV0311                           )
       = 0561001)

00LSRB4:SU /   /CMT(IF MANTISSA OVERFLOW,BMOVF10               )
       = YXNUEE )

00LUFL0:SU /   /CMT(L REG HIGH 18 BITS FROM L REG LOW          )
       = 0481001)

00LUFY0:SU /   /CMT( L REG UPPER 18 BITS FROM Y REG            )
       = 0451001)

00MINC0:   /   /CMT(M REG INC                                  )
       = 1042002)

00MINC1:   /   /CMT(IF RUN.V04+=DISPLAY +1 )
       = 0831001)

00MCST0:   /   /CMT(SET MONITOR CALL DEMAND)
       = 0831001)

00MDEC0:   /   /CMT(M REG DEC                                  )
       = YXNUDMY)

00MDEC1:   /   /CMT(IF RUN.V04 NOT + DISPLAY-1                 )
       = YXNUDMY)
```

```
00MF1D0:   /   /CMT(M REG FROM INC-DEC LOGIC                    )
       = 1201001)
00MF1D1:   /   /CMT(IF AP EQUAL 1XX,SAP0330)
       = 0691001)
00MF1D2:   /   /CMT(IF BH =0XX,BFT NOT+BP=0XX,BFT   )
       = 0681001)
00MF1D3:   /   /CMT(IF AH=0XX,SAH0300       )
       = 0871001)
00MFYY0:   /   /CMT(M REG FROM Y REG=17 BITS                    )
       = 1091001)
00MPOS1:SU /   /CMT(SET MPO IF B01-B24 NOT ALL SAME,BMPOC10)
       = 0671001)
00NCFB0:   /   /CMT(NC FROM B3,B4-HARDWARE GENERATED         )
       = YXNUHDW)
00NCLA0:   /   /CMT(CLEAR A CHAR IN N AS SPECIFIED BY VV     )
       = 0611001)
00NCLB0:   /   /CMT(CLEAR B CHAR IN N AS SPECIFIED BY VV     )
       = 0801001)
00NCLD0:   /   /CMT(IF NONNUMERIC,DEC-CLR DEC        )
       = 0421001)
00NCLZ0:   /   /CMT(N REG CLR ZONE IF DEC OP CODE   )
       = 0901001)
00NFAA0:   /   /CMT(N REG FROM A REG 36 BITS                 )
       = 0913003)
00NFAA1:   /   /CMT(UNDER CONT OF WC1-4-CHAR XFRS            )
       = 0761001)
00NFBB0:   /   /CMT(N REG FROM B REG 36 BITS        )
       = 0913007)
00NFUU0:   /   /CMT(N REG FROM MM 36 BITS                    )
       = YXNUHDW)
00NHWT1:   /   /CMT(HOLD N WM,TM IF EXM.OVL=5-8-BASED ON VV)
       = 0691001)
00NNHD0:   /   /CMT(HOLD N NUMERIC,ZONE BASED ON V-MOS ONLY)
       = 0421001)
00QAFK0:   /   /CMT(QA REG FROM K REG- 11 BITS               )
       = 0313001)
00QBFK0:   /   /CMT(QB REG FROM K REG- 11 BITS               )
       = 0313002)
00QCFK0:   /   /CMT(QC REG FROM K REG- 11 BITS                 )
       = 0313003)
00QDFK0:   /   /CMT(QD REG FROM K REG- 11 BITS                 )
       = 0313004)
00QDFY1:SU /   /CMT(QD REG FROM QY IF FMA ORDER                )
       = 0571001)
00QEFJ0:   /   /CMT(QE REG FROM J REG                          )
       = 0741001)
00QEFK0:   /   /CMT(QE REG FROM K REG- 11 BITS                 )
       = 0313005)
00QFBR0:   /   /CMT( QU FROM KBR01-6                           )
       = 0671001)
00QFBR7:   /   /CMT( QU FROM KBR07                             )
       = 0421001)
00QFCP0:   /   /CMT(QU REG. FROM CP                            )
       = 0491001)
00QFVV0:   /   /CMT(QU FROM VV CHAR XFR               )
       = 0451001)
00QGFK0:   /   /CMT( GROUP FLOP REGISTER FROM K                )
       = 0313006)
00QMCL0:SU /   /CMT(QM REGISTER CLEAR                          )
       = 0531001)
00QMCL1:SU /   /CMT(IF EXU OR EXTENDED ADD RESULT = 0          )
       = 0441001)
00QMCL2:SU /   /CMT(IF X OR Y OPERAND=0 AND FMP                )
       = 0771001)
00QMF50:SU /   /CMT( QM REG FROM DEC 35 LOGIC                  )
       = 0661001)
00QMFN0:SU /   /CMT(QM REG FROM QN REG 12 BITS                 )
       = 0701001)
00QMFY0:SU /   /CMT(QM REG FROM Y REG LOW ORDER 12 BITS        )
       = 0491001)
00QMFY1:SU /   /CMT(IF EX-EY IS POSITIVE,BXGEY30               )
       = 0481001)
00QN1D0:SU /   /CMT(QN REG FROM INC-DEC LOGIC                  )
       = 0501001)
00QN1D1:SU /   /CMT(IF XGY IS RESET                            )
       = YXNUEE )
```

```
00QN1D2:SU /   /CMT(IF MANTISSA OVERFLOW                    )
     = 0471001)

00QN1D6:SU /   /CMT(IF B REG. NOT NORMAL                    )
     = YXNUEE )

00QN1N0:SU /   /CMT(INC QN REG                              )
     = 0871001)

00QNDE0:SU /   /CMT(DEC QN REG                              )
     = YXNUDMY)

00QNFB0:SU /   /CMT(Q REG FROM B REG LOW ORDER 12 BITS      )
     = 0891001)

00QNFM0:SU /   /CMT(QN REG FROM QM REG 12 BITS              )
     = 0851001)

00QNFN1:SU /   /CMT(QN FROM QN NEG. IF X&Y IS RESET         )
     = 0591001)

00QNZMC:   /   /CMT(CLEAR FOR QN=0 MONITOR-BQNIZ10          )
     = 0901001)

00QTFB0:   /   /CMT(QT FROM B1.B2-HARDWARE GENERATED        )
     = YXNUHDW)

00QTFF0:   /   /CMT( QT FROM K - FOR DIAGNOSTICS            )
     = 0313007)

00QUFD0:   /   /CMT(QU REG FROM QD-8 BIT XFR        )
     = 0391001)

00QUFN0:   /   /CMT(QU REG. FROM N SELECT BY B POINTER      )
     = 0401001)

00QUFV1:   /   /CMT(QU REG FROM QV IF IDA.IXA NOT           )
     = 0691001)

00QUFW0:   /   /CMT(QU REG FROM QW01-6                      )
     = 0621001)

00QVCL0:   /   /CMT(QV REG CLEAR                            )
     = 0681001)

00QVFN0:   /   /CMT(QV REG FROM N SELECT    )
     = 0641001)

00QWFU0:   /   /CMT(QW01-6 FROM QU QW07 FROM BNNS130        )
     = 0711001)

00QYFA0:   /   /CMT(QY REG FROM A1 )
     = 0761001)

00ROSR0:   /   /CMT(RESET RESETTABLE ONE-SHOT       )
     = 0461001)

00ROVF0:   /   /CMT(RESET OVERFLOW          )
     = YXNUEE )

00RCPF0:   /   /CMT(RESET CONTROL PANEL FUNCTIONS                  )
     = 0471001)

00RFAX0:SU /   /CMT(R REG FROM ACC X                               )
     = 0681001)

00RFAX1:SU /   /CMT(IF TAM + TAL ORDER                             )
     = 0691001)

00RFAY0:SU /   /CMT(R REG FROM ACC Y                               )
     = 0641001)

00RFAY1:SU /   /CMT(IF FAA + TMA ORDER                             )
     = 0791001)

00RFBP0:   /   /CMT(R REG BITS 1.2 FROM B POINTER   )
     = 1042001)

00RFKK0:   /   /CMT(R REG BITS 1-5 FROM K REG                      )
     = YXNUDMY)

00RFKK1:SU /   /CMT(IF MEM AND ACC OPERATION                       )
     = YXNUDMY)

00RFKK2:SU /   /CMT(IF NOT TMA OR TAM ORDER                        )
     = YXNUDMY)

00RFVV0:   /   /CMT(R REG BITS 1-5 FROM V REG BITS 1-5             )
     = 0701001)

00RTID0:SU /   /CMT(RESET TESTED FL.PT INDICATORS                  )
     = 0631001)

00SOCK0:   /   /CMT( SET OP CODE VIOLATION                         )
     = 0591001)

00SOVID:   /   /CMT(SET OVERFLOW INDICATOR IF OVF SET              )
     = 0601001)

00SAFN0:   /   /CMT(SIGN A STORAGE FROM N SELECT    )
     = 0621001)

00SAFY0:SU /   /CMT(SIGN A STORAGE FROM Y18                        )
     = 0801001)

00SBFN0:   /   /CMT(SIGN B STORAGE FROM N                          )
     = 0391001)

00SETOP:   /   /CMT(SET OPTIONS ACCORDING TO LIB VARIANT   )
     = 0531001)

00SFQC0:   /   /CMT(S.M REG FROM QU.QW-REQ. UOP FOR 2 PDA.S)
     = 0861001)
```

```
00SFYY0:   /   /CMT(S REG FROM Y REG-17 BITS                  )
       = 1101001)

00SGCL0:   /   /CMT(CLEAR THE SIGN FLOP                       )
       = 0521001)

00SLFL0:SU /   /CMT(SIGN L STORAGE FROM L36                   )
       = 0721001)

00SLNZ1:SU /   /CMT(SET LOR NOT 0 FLOP IF 1 IS SHIFTED        )
       = YXNUEE )

00SPMS1:   /   /CMT(SET WORD+ITEM BITS IN A IF SW+SI ORDER )
       = 0681001)

00STDR0:   /   /CMT( RESET BS AND DS FLOPS                    )
       = 0601001)

00STDS0:   /   /CMT(SET DEC SUB FLOP        )
       = 0651001)

00STDS1:   /   /CMT(SET ADDER AND BS.DS FLOPS BY OP.CO.       )
       = 0721001)

00STPS0:   /   /CMT(SET STOP REQUEST        )
       = 0531001)

00SXFY0:SU /   /CMT(SIGN X STORAGE FROM Y18                   )
       = 0651001)

00SXGY1:SU /   /CMT(SET XGY IF EX-EY IS POSITIVE              )
       = 0751001)

00TXOR3:SU /   /CMT(IF QM12=0                                 )
       = 1083005)

00TXOR4:SU /   /CMT(IF QN12 CHANGING FROM - TO +              )
       = YXNUEE )

00TXOS1:SU /   /CMT(SET TEMP. EXP OVF IF A12=B12=0            )
       = YXNUEE )

00TXOS4:SU /   /CMT(SET TXO IF QN12=1                         )
       = 0431001)

00UAFC0:   /   /CMT(MEM MASK AUX FROM WC1-4                   )
       = 0441001)

00UDAT0:   /   /CMT(MM WRITE DATA COMMAND  )
       = YXNUHDW)

00UDAT1:   /   /CMT(OP CODE AND VARIENT DEPENDENT             )
       = YXNUHDW)

00UFMA0:   /   /CMT(MEM MASK FROM AUX MASK )
       = 0761001)
```

```
00UMMW0:   /   /CMT(M M WRITE COMMAND       )
       = 0962001)

00UMRQ0:   /   /CMT(SET MM MODULE REQUEST                     )
       = YXNUHDW)

00UMSL0:   /   /CMT(MEMORY CHAR MASK SHIFT LEFT     )
       = 0901001)

00UMSL1:   /   /CMT(IF AP DOES NOT EQUAL X00                  )
       = 0641001)

00UMSR0:SU /   /CMT(WC1-4.MA1-4 SHIFT RIGHT                   )
       = 0831001)

00UMSR1:   /   /CMT(IF AP DOES NOT EQUAL X11        )
       = 0631001)

00UMSR2:   /   /CMT(WC1-4 SHIFT RIGHT                         )
       = 0821001)

00USMA0:SU /   /CMT(MEMORY SET WRITE CHAR MASK AUX            )
       = 0681001)

00USMST:   /   /CMT(STALL UNTIL MAIN MEM IS STARTED)
       = YXNUHDW)

00USW11:   /   /CMT(SET WC1 IF AFT NOT + ZAS                  )
       = 0841001)

00USW41:   /   /CMT(SET WC4 IF AFT NOT                        )
       = 0651001)

00USWC0:   /   /CMT(MEMORY SET WRITE CHAR MASK       )
       = 0913006)

00USWC1:   /   /CMT(SET WC4.WC3-WC2 IF 3+4CM-WC1 IF 4CM       )
       = 0821001)

00USWC3:   /   /CMT(SET WC1+WC2+WC3+WC4 WITH B POINTER        )
       = 0791001)

00UTWM0:   /   /CMT(TERMINATE MM WRITE ON WORD MARK IN MEM )
       = YXNUHDW)

00UWTM0:   /   /CMT(MEM WRITE ITEM MARKS COMMAND    )
       = YXNUHDW)

00UWTM1:   /   /CMT(OP CODE AND VARIANT DEPENDENT   )
       = YXNUHDW)

00UWWM0:   /   /CMT(MEM WRITE WORD MARKS COMMAND              )
       = YXNUHDW)

00UWWM1:   /   /CMT(OP CODE AND VARIANT DEPENDENT             )
       = YXNUHDW)
```

```
00VFA10:   /   /CMT(V REG. FROM A1                        )
        = 0521001)

00VFA11:   /   /CMT(V FROM A1 IF A1WM.BFT.SPM.SVI.S0-NOT      )
        = 0901001)

00VFA12:   /   /CMT(IF BBC-A1WM,RVI.SPM.S0.BFT-NOT            )
        = 0881001)

00VFRS0:   /   /CMT(V REG FROM R SWITCHES                 )
        = 0571001)

00VXTS1:   /   /CMT(SET VXT IF VV FROM A4 XFR             )
        = YXNUEE )

00XH1S0:   /   /CMT(SET XH1 FLOPIF Y REG IS NOT ZFRO          )
        = 0611001)

00YFAP0:   /   /CMT(Y01.2 FROM AP01.2                         )
        = 1013005)

00YFBB0:   /   /CMT(Y REG FROM B REG LOWER 19 BITS            )
        = 1001001)

00YFBB1:SU /   /CMT(IF EXU IS RESET AND YH1                   )
        = 1083003)

00YFBP0:   /   /CMT(Y01.2 FROM BP01.2                         )
        = 1181001)

00YFBU0:SU /   /CMT(Y REG FROM B REG UPPER 18BITS             )
        = 0711001)

00YFBU1:SU /   /CMT(IF EXU IS RESET AND YH1                   )
        = 1083007)

00YFCM0:   /   /CMT(Y REG FROM CONT MEM-19 BITS               )
        = 0991001)

00YFLU0:SU /   /CMT(Y REG FROM HIGH ORDER 18 BITS OF L REG )
        = 0581001)

00YFMM0:   /   /CMT(Y REG FROM M REG-17 BITS                  )
        = 1081001)

00YH1R0:   /   /CMT(RESET BYHA110.BYHA210.BXHA110.BFZBS10 )
        = 0851001)

00YH1S0:SU /   /CMT(SET YH1 FLOP IF Y REG IS NOT ZERO         )
        = 0551001)

00YH1S1:   /   /CMT(IF A1 CHAR IS NOT ZERO                    )
        = 0471001)

00YH2S0:   /   /CMT(SET YH2 FLOP IF YH1 IS SET      )
        = 0531001)

00YYHD0:   /   /CMT(HOLD Y REG      )
        = 0591001)
```

APPENDIX 10C: SYMBOLIC MICRO-ORDER DECODING CHART

Flow-chart equivalent	*Sub*	*ROM Bit*
BDT	00BDT	0511001
BDT	00BDT	0511001
BST	00BST	0131001
BST	00BST	0131001
MMW	UMMW0	0962001
MRQ	UMRG0	YXNUHDW
AQ1→L36(DTB)	LFCS0	0391001
A1→A4	AFA10	0541001
A1→B4	BFA10	0441001
A1→B4, AP≠X00	BFA11	0581001
A1→IND BY QV	CLFA1	0511001
A1→IND BY QV	CLFA1	0511001
A1→IND BY QV	CLFA1	0511001
A1→IND BY QV	CLFA1	0511001
A1→IND BY QV	CLFA1	0511001
A1→QY	QYFA0	0761001
A1→VV	VFA10	0521001
A1→VV,-A1WM.BFT-/N .-SPM.SVI.S0-/N	VFA11	0901001
A1→VV, BBC.A1WM/N .-RVI.SPM.S0.BFT-/N	VFA12	0881001
A4→A1	AFA40	0731001
A4→A1, AP≠X11	AFA41	0781001
A4→B1	BFA40	0551001
A4→B1, BP≠X11	BFA41	0411001
A→N	NFAA0	0913003
A→N	NFAA0	0913003

Flow-chart equivalent	*Sub*	*ROM Bit*
A→N	NFAA0	0913003
A→N	NFAA0	0913003
A→N BY WC1-4	NFAA1	0761001
A→N BY WC1-4	NFAA1	0761001
A→N BY WC1-4	NFAA1	0761001
A→N BY WC1-4	NFAA1	0761001
A/N→A	AFAN0	1113002
A/N→A, FLT PT SUB	AFAN1	0581001
A/N→A, SNX=0 (SIGN X POS)	AFAN3	0631001
A/N→A IF SIGN A NEG	AFAN2	0621001
A/N→A IF SIGNS DIFF	AFAN4	1183004
AC→R	ICMAC	0345034
ACC X(OV4,5)→R03,4 (INHIBIT WL→R)	RFAX0	0681001
ACC X(OV4,5)→R03,4 IF TAM+TAL	RFAX1	0691001
ACC Y(QV1,2)→R03,4 (INHIBIT WL→R)	RFAY0	0641001
ACC Y(QV1,2)→R03,04 IF FAA+TMA	RFAY1	0791001
AD+AD/N+AD9'S→B3, OVL=1+2''→B2, OVL=3	BFFL1	YXNUEE
AD+AD/N+AD9'S→B3, OVL=1+2''→B2, OVL=3	BFFL1	YXNUEE
ADDER→A1	AFAD0	0831001
ADDER→A1 IF ZA+ZS	AFAD1	0401001
ADDER→B4	BFAD0	0451001
ADDER→B	BFBA0	1013006
ADDRESS→QA	QAFK0	0313001
ADDRESS→QB	QBFK0	0313002
ADDRESS→QC	QCFK0	0313003
ADDRESS→QD	QDFK0	0313004
ADDRESS→QE	AEFK0	0313005
AH-1→AH	AHDE0	1162002
AH-1→AH, MCE.PN1.ATE	AHDE4	YXNUEE
AH-1→AH, AP≠X11.AFT/N	AHDE2	YXNUEE
AH+1→AH	AHIN0	1113005
AH+1→AH, -AFT.BFT-/N	AHIN2	0451001
AH-1→AH, AFT./N	AHDE3	YXNUEE
AH+1→AH, AP≠0.AFT/N	AHIN3	YXNUEE
AH→AP	APFH0	1013004
ALL ONES RCM LOCATION DMY01-DMY12	DMY01	0139777
AP-1→AP, 30M	APDE2	0781001
AP1→AP	APDE0	0771001
AP1→AP, AP≠X00	APDE1	0751001
AP-1→AP, AP≠X11	APDE3	0741001
AP+1→AP, AP≠AH.BFT	APIN2	0791001
AP+1→AP, AP≠X00	APIN3	0581001
AP+1→AP, AP≠X11	APIN1	1113007
AP→AH.	AHFP0	1013007
AP→Y	YFAP0	1013005
AUX IND→PRIM IND	CLFX0	0811001
AUX IND→PRIM IND	CLFX0	0811001
AUX IND→PRIM IND	CLFX0	0811001
AUX IND→PRIM IND	CLFX0	0811001
B01-07→BR1-7	KBRFB	0481001
B1→A4	AFB10	0391001
B1→A4, AFT/N.AP≠X00	AFB11	0731001
B1→B4	BFB10	0551001
B→B4, OVERLAP=3	BFB12	0601001
B1.B2→QT	QTFB0	YXNUHDW
B3.B4→NC	NCFB0	YXNUHDW
B4→A1	AFB40	0491001
B4→A1-AFT.MPY-/N.AP≠3	AFB42	0661001
B4→A1 IF AFT/N.MPY/N	AFB41	0621001
B4→B1	BFB40	0541001
B4→B1, AP≠X11	BFB41	0521001
B4→B1, OVERLAP=3	BFB42	0641001
B(19-36)→Y	YFBU0	0711001
B(19-36)→Y, YH1. EXU/N	YEBU1	1083007
B(1-12)→QN	QNFB0	0891001
B(1-19)→Y	YFBB0	1001001
B(1-19)→Y, YH1.EXU/N	YFBB1	1083003
B→L(1-36)	LLFB0	0391001
B→N	NFBB0	0913007

Flow-chart equivalent	*Sub*	*ROM Bit*
BC→R	1CMBC	0345030
BH-1→BH	BHDE0	0881001
BH-1→BH, AFT/N	BHDE3	0851001
BH-1→BH, RUN.V04/N +DISPLAY-1	BHDE4	YXNUEE
BH-1→BH, AP≠X11.AFT/N	BHDE2	YXNUEE
BH+1→BH, AFT/N	BH1N2	0781001
BH+1→BH, AH=X00.BFT/N	BH1N1	0681001
BH+1→BH, AP≠XOO.AFT/N	BH1N3	YXNUEE
BH+1→BH, RUN.V04 +DISPLAY+1	BH1N4	0861001
BH→BP	BPFH0	1062001
BP-1→BP	BPDE0	0913005
BP-1→BP, ATF	BPDE1	0521001
BP+1→BP	BP1N0	0942001
BP+1→BP, AP≠AH.BFT	BP1N2	YXNUEE
BP+1→BP, BP≠X11	BP1N1	1042003
BP→BH	BHFP0	1142002
BP→R	RFBP0	1042001
BP→Y	YFBP0	1181001
BR7→QU	QFBR7	0421001
BR(1-6)→QU	QFBR0	0671001
CH-1→CH	CHDE0	0551001
CH→CP	CPFH0	0671001
CLEAR A, XGY/N	AACL1	0701001
CLEAR A, AFT, ATF, BFT	AACL0	1113004
CLEAR B	BBCL0	0913001
CLEAR B, XGY	BBCL1	YXNUEE
CLEAR CH	CHCL0	0461001
CLEAR EXT INT MODE	HRCL0	YXNUEE
CLEAR INTERNAL INTERUP	LRCL0	0841001
CLEAR L	LLCL0	0461001
CLEAR L	LLCL0	0461001
CLEAR L, EXU	LLCL1	0561001
CLEAR L, EXU	LLCL1	0561001
CLEAR N, NUMERIC/N.DEC	NCLD0	0421001
CLEAR R BY VV (A FIELD DATA)	NCLA0	0611001
CLEAR N NUM, MOS/N	NCLB0	0801001
CLEAR N ZONE, DECIMAL	NCLZ0	0901001
CLEAR QM	QMCL0	0531001
CLEAR QM, -XH1+YH1/N	QMCL2	0771001
CLEAR QM, EXU+B.L REG=O	QMCL1	0441001
CLEAR QV	QVCL0	0661001
CLEAR SIGN FLOPS	SGCL0	0521001
CM→Y	YFCM0	0991001
CM WRITE(13-15)	CFYA2	0721001
CM WRITE, 2CM	CFYA3	0721001
CM WRITE	CFYA0	0981001
CM WRITE	CFYA0	0981001
CM WRITE BY CHAR MODE	CFYA1	0801001
CM WRITE BY CHAR MODE	CFYA1	0801001
CM WRITE BY CHAR MODE	CFYA1	0801001
COMM BUFF SVC REQST	CBRS0	0521001
CP+1→CP, OVERFLOW	CP1N1	0591001
CP→QU	QFCP0	0491001
CSC→R	1CMCS	0345024
D→A	AFDD0	0671001
D→A	AFDD2	0671001
D→A, AH=X00.MCE/N+ATE	AFDD2	1013001
D→A, AH=X00.MCE/N+ATE	AFDD1	1013001
D→A, AH=X11	AFDD1	0711001
D→A, AH=X11	AFDD1	0711001
DDA→R (DIAG DUMP ADD)	1CMDA	0345033
DWL→R (DIAG WORK LOC)	1CMDW	0345031
E1→B1, DS/N +E1-9'S→B1, DS	BFE10	1013003
E1→B1, DS/N +E1-9'S→B1, DS	BFE10	1013003
E1→E4	EFE10	0621001
E2→B2, DS/N +E2-9'S→B2, DS	BFE20	YXNUEE
E3→B3, DS/N +E3-9'S→B3, DS	BFE30	YXNUEE
E4→B4, DS/N +E4-9'S→B4, DS	BFE40	0962003
E4→B4, DS/N +E4-9'S→B4, DS	BFE40	0962003
E4→E1	EFE40	0721001
E4→E1	EFE40	0721001
E→B, BH=X00.AP≠X11	BFEE2	YXNUEE

Flow-chart equivalent	*Sub*	*ROM Bit*
E→B, BH = X00.AP≠X11	BFEE2	YXNUEE
E→B, BH = X00.AP≠X11	BFEE2	YXNUEE
E→B, BH = X00.DS/N .OVL≠1-4	BFEE1	0451001
E→B, BH = X00.DS/N .OVL>1-4	BFEE1	0451001
E→B, BH = X00.DS/N .OVL≠1-4	BFEE1	0451001
E→B, BH = X00.DS/N .OVL≠1-4	BFEE1	0451001
E→B, BH = X11	BFEE4	0461001
E→B, BH = X11	BFEE4	0461001
E→B, BH = X11	BFEE4	0461001
E→B, BH = X11.AP≠X00	BFEE3	0471001
E→B, BH = X11.AP≠X00	BFEE3	0471001
E→B, BH = X11.AP≠X00	BFEE3	0471001
E→B, OVL = 4.DS/N + E9'S→B, OVL = 4.DS	BFEE5	0551001
E→B, OVL = 4.DS/N + E9'S→B, OVL = 4.DS	BFEE5	0551001
E→B, OVL = 4.DS/N + E9'S→B, OVL = 4.DS	BFEE5	0551001
EDIT LOGIC→E1	EFED1	YXNUEE
EDIT LOGIC→E4	EFED4	0741001
EIR→R	ICMHR	0345026
ENABLE B & L REG CONV LOGIC	BBDRN	0891001
ENABLE XLATOR FLOPS	BBDI0	0561001
END WRITE, WM BY OP-CD	UTWM0	YXNUHDW
GRP01→QG	GRP01	0121001
GRP02→QG	GRP02	0111001
GRP03→QG	GRP03	0101001
GRP04→QG	GRP04	0091001
GRP05→QG	GRP05	0081001
GRP06→QG	GRP06	0071001
GRP07→QG	GRP07	0061001
GRP08→QG	GRP08	0051001
GRP09→QG	GRP09	0041001
GRP10→QG	GRP10	0031001
GRP11→QG	GRP11	0021001
GRP12→QG	GRP12	0011001
HOLD N PUNCT BY VARIANT, EXM.OVL = 5-8	NHWT1	0691001
HOLD N REG BITS 1-6	NNHD0	0421001
HOLD Y REG	YYHD0	0591001
I/D( )→M	MF1D0	1201001
I/D( )→M, AH = OXX	MF1D3	0871001
I/D( )→M, AP = 1XX	MF1D1	0691001
I/D( )→M, BH = OXX .BFT/N + BP = 0XX.BFT	MF1D2	0681001
I/D( )→QN	QN1D0	0501001
I/D( )→QN	QN1D0	0501001
I/D( )→QN, B36.B35 = 00 + 11	QN1D6	YXNUEE
I/D( )→QN, B36.B35 = 00 + 11	QN1D6	YXNUEE
I/D( )→QN, MANTIS OVERFLOW	QN1D2	0471001
I/D( )→QN, MANTIS OVERFLOW	QN1D2	0471001
I/D( )→QN, XGY/N	QN1D1	YXNUEE
I/D( )→QN, XGY/N	QN1D1	YXNUEE
IIR→R	ICMLR	0345036
IND→A4, BY QV CONTROL	AFCL1	0851001
IND→A4, BY QV CONTROL	AFCL1	0851001
INDEX REG ADD→S.M	SFQC0	0861001
INDEX REG ADD→S.M	SFQC0	0861001
INDEX REG ADD→S.M	SFQC0	0861001
JJ→QE	QEFJ0	0741001
K→QT (FOR DIAG)	QTFF0	0313007
L36→SIGN L STORAGE	SLFL0	0721001
L(19-36)→Y	YFLU0	0581001
L(1-18)→L(19-36)	LUFL0	0481001
M→Y	YFMM0	1081001
MA1-4→WC1-4	UFMA0	0761001
MA1-4→WC1-4	UFMA0	0761001
MA1-4→WC1-4	UFMA0	0761001
MDEC	MDEC0	YXNUDMY
MDEC.RUN.V04/N + DSPLY-1	MDEC1	YXNUDMY
MINC	MINC0	1042002
MINC. RUN.V04 + DSPLY + 1	MINC1	0831001
MM( )→N	NFUU0	YXNUHDW
N→A	AFNN0	0942002
N→A. AH = X00	AFNN1	0913004
N→D	DFNN0	0962002

Flow-chart equivalent	*Sub*	*ROM Bit*	*Name*
N→E, AH = X11	EFNN4	0701001	
N→E, BS/N + N/N→E, BS.	EFNN1	1142001	
N→E, BS/N + N/N→E, BS.	EFNN1	1142001	
N→E, BS/N.0VL = 4-8 + N/N- E, BS.OVL = 4-8	EFNN3	0701001	
N→E, BS/N.0VL = 4-8 + N/N- E, BS.OVL = 4-8	EFNN3	0701001	
NS→JJ	JFNS0	0651001	
NS→QU	QUFN0	0401001	
NS→QV	QVFN0	0641001	
NS→SIGN A STORAGE	SAFN0	0621001	
NS→SIGN B STORAGE	SBFN0	0391001	
OVF→L01	LF0F0	YXNUEE	
PRIM IND→AUX IND	CXFL0	0611001	
QA→QR, L34 + L33.B35/N.B34/N	BDT1A	0211001	
QA→QR, L35 + L34.B35 + L33.B35.B34	BDT3A	0271001	
QA→QR, QN = 0 FLOP + YH1/N + XH1/N	BDT4A	0301001	
QA→QR, QN = 0 FLP	BDT6A	0151001	
QA→QR, QN EQUAL 0 FLP	BDT5A	0141001	
QB→QR, L03.L02 = 11	BDT5B	YXNUEE	
QB→QR, L04.L03 = 00	BDT4B	YXNUEE	
QB→QR, L04.L03 = 11	BDT6B	YXNUEE	
QB→QR, 0VF/N .B36.B34/N	BDT1B	YXNUEE	
QB→QR, QN = 4-63	BDT2B	0241001	
QC→QR, B35-B33 = 111	BDT3C	YXNUEE	
QC→QR, B36-B34 = 001	BDT1C	YXNUEE	
QC→QR, L03.L02 = 00	BDT5C	YXNUEE	
QC→QR, L04.L03 = 00	BDT6C	YXNUEE	
QC→QR, L04.L03 = 01	BDT4C	YXNUEE	
QC→QR, QN QR = 64	BDT2C	YXNUEE	RPRC010
QD(6 BITS)→QU	QUFD0	0391001	QUFD010
QD→QR, B36-B33 = 0001	BDT1D	YXNUEE	RPRD010
QD→QR, L03.L02 = 01	BDT5D	YXNUEE	RPRD010
QD→QR, L04.L03 = 01	BDT6D	YXNUEE	RPRA010
QD→QR, L04.L03 = 10	BDT4D	YXNUEE	RPRD010
QE→QR, B36-B33 = 0000	BDT1E	YXNUEE	RPRE010
QE→QR, L03.L02 = 10	BDT5E	YXNUEE	RPRE010
QE→QR, L04.L03 = 10	BDT6E	YXNUEE	RPRA010
QE→QR, L04.L03 = 11	BDT4E	YXNUEE	RPRE010
QE→QR, QN = 2	BDT2E	YXNUEE	RPRE010
QM→L(1-12)	LLFM0	0591001	LLFM010
QM→QN	QNFM0	0851001	QNFM010
QM MINUS 35→QM	QMF50	0661001	QMF5010
QN→QM	QMFN0	0701001	QMFN010
QN/N→QN, XGY/N	QNFN1	0591001	QNFN110
QNDEC	QNDE0	YXNUDMY	QN NORMA
QNINC	QNIN0	0871001	QNIN010
QU(1-4)→CH	CHFU0	0881001	CHFU010
QU( )→A4	AFQU0	1062002	AFQU010
QU( )→A4	AFQU0	1062002	AFQU010
QU→QW1-6. NSI→QW7	QWFU0	0711001	QWFU010
QV→QU, IDA/N.IXA/N	QUFV1	0691001	QUFV110
QW07→A01	AFW70	0461001	AFW7010
WQ07→A01	AFW70	0461001	AFW7010
QW(1-6)→QU	QUFW0	0621001	QUFW010
QY→QD, FMA	QDFY1	0571001	QDFY110
R SWTICHES→VV	VFRS0	0571001	VFRS010
RESET ACI	ARTA0	YXNUHDW	GENERATE
RESET BS.DS FLOPS	STDR0	0601001	STDR010
RESET CONT PAN FUNCT	RCPF0	0471001	RCPF010
RESET CSI	ACSR0	YXNUEE	CXFL010
RESET CSI IF TESTED	C0VR0	0751001	C0VR010
RESET EXU. S12(GM12 + A12/N + FDV/N)	EXUR1	0881001	EXUR110
RESET NON-EXISTANT MEM HIT CNTR	HCTR0	0661001	HCTR010
RESET OVERFLOW	R0VF0	YXNUEE	BSLB510
RESET QN = 0 FLOP	QNZMC	0901001	QNZMC10
RESET TESTED FL PT IND	RTID0	0631001	RTID010
RESET TX0, QM12 = 0	TX0R3	1083005	TX0R310
RESET TX0 QN12 CNGING 1 TO 0	TX0R4	YXNUEE	TX0R410
RESET YH1.YH2.XHA1.ZBS	YH1R0	0851001	YH1R010
RESET ZBI	AZBR0	YXNUEE	CXFL010
SC- R	1CMSG	0345037	RFKK030
SET 3CM.RESET ITEM MARK TRAPPING	CM3S0	YXNUEE	CXFL010

Flow-chart equivalent	*Sub*	*ROM Bit*	*Name*
SET ACI	ASTA0	0761001	ASTA010
SET ACI, BASTCIO SET (ONLY 35 DIV DONE)	ASTA7	0831001	ASTA710
SET ACI, CVF	ASTA2	0441001	ASTA210
SET ACI, SNA=1 (SIGN A NEG)	ASTA3	0741001	ASTA310
SET ACI, SNX=0 (SIGN X POS)	ASTA4	0411001	ASTA410
SET ACI IF DS+BS	ASTA8	0561001	ASTA810
SET ACI IF FLT PT SUB	ASTA1	YXNUEE	AFAN110
SET ACI IF SIGNS DIFF	ASTA6	YXNUEE	AFAN410
SET ADA.BS.DS. BY OP-CO	STDS1	0721001	STDS110
SET ADDER TO BINARY SET BS FLOP	BAGC3	0471001	BAGC310
SET AFT, A FIELD	AFTS4	0541001	AFTS410
SET AFT, B FIELD	AFTS5	0871001	AFTS510
SET AFT ATF	AFTS1	0891001	AFTS110
SET AFT ATF, AP≠00	AFTS3	YXNUEE	AFB1110
SET AFT ATF, AP≠11	AFTS2	YXNUEE	AFB4210
SET AH3	AH3S0	0411001	AH3S010
SET AH LEFT OF N WH	AHFN1	0461001	AHFN110
SET AXT	AXTS0	0541001	AXTS010
SET BARRIC VIOL. BFT	BJVS1	YXNUEE	AHFN110
SET BFT, AIWM	BFTS2	0811001	BFTS210
SET BFT, A4WM (MCE/N+MCE.A XFERED TO E)	BFTS6	0571001	BFTS610
SET BFT, B4TM.MPY/N +QU07.MPY	BFTS7	0611001	BFTS710
SET BFT, NWM	BFTS1	0481001	BFTS110
SET BFT, PCT+PCB	BFTS3	0621001	BFTS310
SET BFT, SW+SI+BIM	BFTS5	0821001	BFTS510
SET BFT.BBC.A1WM.BFT/N	BFTS4	0791001	BFTS410
SET BH3	BH3S0	0501001	BH3S010
SET BH3, OVF=0	BH3S1	YXNUEE	BSLB510
SET BXT (REQUIRES SET AXT TO BE ACTIVE)	BXTS0	0501001	BXTS010
SET COMPARE IND.	CARB0	0771001	CARB010
SET DEC SUB FLOP	STDS0	0651001	STDS010
SET DIVIDE CHECK INDICATOR IF XH1/N	DVCS1	1083004	DVCS110
SET EXT INT MODE	HRMS0	YXNUEE	CXFL010
SET EXU, A12=B12=1	EXUS2	0431001	EXUS210
SET EXU, QN12=1	EXUS3	1083006	EXUS310
SET EXU, QN12 .QM12/N.TX0/N	EXUS4	0841001	EXUS410
SET EXU. TXO/N .QN12 CNGING 1 TO 0	EXUS1	0521001	EXUS110
SET I/0 SERV REQ TURN ON N XMITTERS	DLRS0	0631001	DLRS010
SET IAV, ADD VIOL .PROCEED.EIM/N	AVIS0	0731001	AVIS010
SET IND FOR CAM BY VV	CLFV0	0851001	CLFV010
SET INTERNAL INT IF PROTECT.BMA	S0CK0	0591001	S0CK010
SET INTERNAL INT MODE	LRMS0	0841001	CTIIM10
SET L01, SES=0	L01S2	1083001	L01S210
SET L36, SNA=SNB	LSFBM	0861001	LSFBM10
SET LA, A1WM.BFT/N	0AA1W	0163001 AND 0142002	
SET LA, B REG WM	0ABWM	0163003 AND 0142001	
SET LA, BFT/N.BBC.A1WM	0AV1W	0163005 AND 0142000	
SET LA, BP=X00	0ABP0	0163001 AND 0142001	
SET LA, BP=X00	0ABPP	0163007 AND 0142002	
SET LA, COMPARE	0AC0M	0163001 AND 0142003	
SET LA, DIAG ERROR	0ADCB	0163004 AND 0142003	
SET LA, FLTNG ADD+SUB	0AFAS	0163002 AND 0142003	
SET LA, GEN DOLLAR SGN	0AGDS	0163007 AND 0142000	
SET LA, I/C ACK=11 I/C ERROR DETECTED	0ASK3	0163007 AND 0142001	
SET LA, ILL FLPT OP-CD	0AIFC	0163005 AND 0142003	
SET LA, ILLEGAL CR PRIV OP CODE	0AI0C	0163001 AND 0142000	
SET LA, INDIRECT A ADD	0AIDA	0163003 AND 0142002	
SET LA, INSTRUCT BUTN	0AINS	0163006 AND 0142003	
SET LA, MM ENTER	0AMES	0163006 AND 0142002	
SET LA, MPY+DIV	0AMPD	0163004 AND 0142001	
SET LA, N REG WM	0ANWM	0163004 AND 0142002	
SET LA, N SELECT WM	0ANSW	0163006 AND 0142003	
SET LA, OPERNDS. RESLT SIGN NEGATIVE.QY02	0A0RN	0163002 AND 0142000	
SET LA, OVERLAP	0A0VL	0163004 AND 0142000	
SET LA, OVF/N	0AC0B	0163002 AND 0142001	
SET LA, PERIPH ACTIVE	0APAC	0163006 AND 0142001	
SET LA, QN >CR =71	0AN71	0163003 AND 0142003	
SET LA, QN=ZR0 FLOP	0AQN0	0163003 AND 0142000	
SET LA, STOP REQ	0ASTR	0163005 AND 0142001	

Flow-chart equivalent	*Sub*	*ROM Bit*
SET LA, SW+SI+BIM+A1WM	0ASPM	0163005 AND 0142002
SET LA, XH1/N	0AXE0	0163002 AND 0142002
SET LB, 4CM	0B4CM	0193002 AND 0142003
SET LB, A1 DEC BITS≠0	0BAE0	0193006 AND 0142002
SET LB, A4WM	0BA4W	0193001 AND 0142002
SET LB, ADD VIOL.PROCD	0BAVI	0193001 AND 0142001
SET LB, AFT.LCA/N .EXM/N	0BAFT	0193002 AND 0142002
SET LB, AP=X11	0BAP3	0193007 AND 0142000
SET LB, BOOTSTRAP	0BBTS	0193007 AND 0142003
SET LB, B&L REG=0	0BACZ	0193001 AND 0142003
SET LB, BP=0+3CM.BP=1 +4CM(BP=1+2)	0BCMB	0193003 AND 0142003
SET LB, BP=X01	0BBP1	0193003 AND 0142001
SET LB, BYPASS A AND B	0BBAB	0193002 AND 0142000
SET LB, EXTERNAL INTERRUPT DEMAND	0BHRD	0193004 AND 0142001
SET LB, EXTERNAL INTERRUPT MODE	0BHRM	0193004 AND 0142000
SET LB, EXU .QN12/N	0BXUQ	0193006 AND 0142000
SET LB, FL PT BCH/N	0BFBL	0193003 AND 0142000
SET LB, FLT MULT	0BFMP	0193004 AND 0142003
SET LB, I/C ACK=01 I/C NORMAL RESP	0BSK1	0193005 AND 0142001
SET LB, INDIRECT B ADD	0BIDB	0193004 AND 0142002
SET LB, MIT+MAT	0BMTT	0193005 AND 0142002
SET LB, PDT+PCB	0BPDC	0193003 AND 0142002
SET LB, PROTECT .BMA	0BIVV	0193002 AND 0142001
SET LB, QN> 1	0BN23	0193005 AND 0142003
SET LB, QV01/N	0BQV1	0193005 AND 0142000
SET LB, RELOCATION ACTIVE	0BBRA	0193006 AND 0142001
SET LB, VXT	0BVXT	0193007 AND 0142001
SET LB, XH1/N+YH1/N	0BXYZ	0193001 AND 0142000
SET LC, 3CM	0C3CM	0223001 AND 0142001
SET LC, ACC TEST TRUE	0CFBA	0223004 AND 0142000
SET LC, AH=0XX	0CA3R	0223007 AND 0142002
SET LC, AXT.BXT/N	0CDUP	0223002 AND 0142003
SET LC, B04-B01=1010	0CB1Z	0223001 AND 0142000
SET LC, B36.B35=11+00	0CBNN	0223003 AND 0142000
SET LC, B4WM	0CBWM	0223001 AND 0142003
SET LC, BFT	0CBFT	0223005 AND 0142002
SET LC, BP=X10	0CBP2	0223003 AND 0142001
SET LC, BRANCH TRUE	0CBCT	0223002 AND 0142000
SET LC, BYPASS B.BFT/N	0CBYB	0223004 AND 0142002
SET LC, CH=0001	0CCH1	0223002 AND 0142002
SET LC, CM ENTER	0CCES	0223005 AND 0142003
SET LC, END OF TRACE	0CETS	0223007 AND 0142000
SET LC, FLT DIVIDE	0CFDV	0223003 AND 0142003
SET LC, I/0 ACK=10 I/0 BRANCH RESP	0CSK2	0223005 AND 0142001
SET LC, INDEXED A ADD	0CIXA	0223003 AND 0142002
SET LC, INTERNAL INTERRUPT DEMAND	0CLRD	0223004 AND 0142001
SET LC, INTERNAL INTERUPT MODE	0CLRM	0223005 AND 0142000
SET LC, MM DISPLAY	0CMDS	0223006 AND 0142002
SET LC, NO BINARY 1'S	0C0BS	0223007 AND 0142001
SET LC, NO BRANCH	0CBBC	0223002 AND 0142001
SET LC, 0VL=1−4	0CFVL	0223004 AND 0142003
SET LC, TLU	0CJU5	0223007 AND 0142003
SET LC, TLU COMPLETE V1.A>B+V2.A=B+V3.A<B.	0CB0X	0223006 AND 0142000
SET LC, TRUE SUB	0CDSB	0223006 AND 0142003
SET LC, ZBI.DECIMAL	0CZB1	0223006 AND 0142001
SET LD, 2CM	0D2CM	0253001 AND 0142000
SET LD, AXT.BXT/N	0DBNT	0253006 AND 0142001
SET LD, B4WM.LCA/N .EXM/N	0DB4W	0253007 AND 0142002
SET LD, B FLD ZONE BIT	0DBFZ	0253002 AND 0142003
SET LD, BH=0XX	0DB3R	0253002 AND 0142000
SET LD, BP> AP BITS 1.2	0DBGA	0253001 AND 0142003
SET LD, BP=0+BP=1.3CM	0DBPD	0253001 AND 0142002
SET LD, BP=X11	0DBP3	0253001 AND 0142001
SET LD, CM DISPLAY	0DCDS	0253003 AND 0142003
SET LD, DATA LNK WRTE 42 MLR LOADED BY I/O	0DDLR	0253007 AND 0142001
SET LD, DEC ADD+SUB	0DDEC	0253006 AND 0142002
SET LD, DIAGNOSTIC EXT	0DEDM	0253002 AND 0142001
SET LD, FMA+FAA+BMS	0DFMA	0253002 AND 0142002
SET LD, INDEXED B ADD	0DIXB	0253003 AND 0142002
SET LD, INDEXED B ADD OR SCIENTIFIC INST	0D1SR	0253004 AND 0142002

Flow-chart equivalent	*Sub*	*ROM Bit*
SET LD, LOAD L REG	0DLDL	0253004 AND 0142003
SET LD, NON EXIST MEM	0DNEM	0253007 AND 0142000
SET LD, QN06/N	0DQ60	0253004 AND 0142000
SET LD, QN=1	0DN01	0253005 AND 0142003
SET LD, QV=0	0DQV0	0253003 AND 0142001
SET LD, SIGN B NEG	0DSBN	0253005 AND 0142002
SET LD, UNCONDITIONAL	0DSET	0253005 AND 0142000
SET LD, VV02	0DV02	0253006 AND 0142000
SET LD, VV06 CP CONTROL MEM	0DV06	0253006 AND 0142003
SET LD, YH1/N	0DYH0	0253004 AND 0142001
SET LD, ZA+ZS ORDER	0DZAS	0253005 AND 0142001
SET LE, AFT(EXM+LCA)	0EAFT	0283001 AND 0142002
SET LE, AH=OXX	0EA3R	0283006 AND 0142001
SET LE, AH=1XX+SNX	0ECMP	0283006 AND 0142003
SET LE, AH=X00	0EAH0	0283002 AND 0142002
SET LE, AH=X11	0EAH3	0283001 AND 0142001
SET LE, B36=0	0EBIP	0283001 AND 0142000
SET LE, BIM ORDER	0EBIM	0283001 AND 0142003
SET LE, CLEAR MM & CM	0ECPC	0283005 AND 0142003
SET LE, DIAG SW OFF	0EDIA	0283004 AND 0142000
SET LE, DV2 (TRACE EXECUTION)	0EDV2	0283007 AND 0142000
SET LE, DV4 (DIAGNOSTIC DUMP)	0EDV4	0283003 AND 0142001
SET LE, HA+EXT ORDER	0EHAX	0283002 AND 0142001
SET LE, JDS/N (JUSTIFY & NOT)	0EJDR	0283004 AND 0142002
SET LE, 0VL=3	0E0V3	0283005 AND 0142002
SET LE, RUN	0ERUN	0283003 AND 0142003
SET LE, RUN/N.SCR	0ESCR	0283002 AND 0142003
SET LE, S MODE	0ESMD	0283004 AND 0142003
SET LE, UNCONDITIONAL1	0ESET	0283003 AND 0142002
SET LE, UNCONDITIONAL2	OESTT	0283005 AND 0142001
SET LE, UNCONDITIONAL3	0EEST	0283006 AND 0142000
SET LE, VV01	0EV01	0283002 AND 0142000
SET LE, VV04/N	0EV04	0283003 AND 0142000
SET LE, ZERO SUP/N	0EZSR	0283004 AND 0142001
SET LNZ, 1 BIT→L36 (LOR≠0)	SLNZ1	YXNUEE
SET MA1-MA4	USMA0	0681001
SET MC DEMAND	MCST0	0831001
SET MP0, B1-24 NOT ALL SAME	MP0S1	0671001
SET OVF IND, ADA 0VFL	S0VID	0601001
SET 0VL FLOP, B=0-7	FVLS1	0811001
SET STOP REQUEST	STPS0	0531001
SET TX0, A12.B12=00	TX0S1	YXNUEE
SET TX0, QN12	TX0S4	0431001
SET TX0, EXU/N.QN12	EXCD1	1083002
SET TX0, QN12 CNGING 0 TO 1.EXU/N	EX0S1	0571001
SET VXT, XFER TO VV	VXTS1	YXNUEE
SET WC1, ZAS+AFT/N	USW11	0841001
SET WC1-WC4 BY BP	USWC3	0791001
SET WC1-WC4 BY BP	USWC3	0791001
SET WC1-WC4 BY BP	USWC3	0791001
SET WC1-WC4 BY BP	USWC3	0791001
SET WC4, AFT/N	USW41	0651001
SET WC4.WC3; WC1, 4CM;WC2, 3+4CM	USWC1	0821001
SET WC4.WC3; WC1, 4CM;WC2, 3+4CM	USWC1	0821001
SET WC4.WC3; WC1, 4CM;WC2, 3+4CM	USWC1	0821001
SET WC4.WC3; WC1, 4CM;WC2, 3+4CM	USWC1	0821001
SET WORD.ITEM BITS IN A, SW+SI	SPMS1	0681001
SET WRITE CHAR MASK	USWC0	0913006
SET WRITE CHAR MASK	USWC0	0913006
SET WRITE CHAR MASK	USWC0	0913006
SET WRITE CHAR MASK	USWC0	0913006
SET XGY, X EXP> OR= Y EXP	SXGY1	0751001
SET XH1, Y≠0	XH1S0	0611001
SET YH1, A1 = 0	YH1S1	0471001
SET YH1, Y≠0	YH1S0	0551001
SET ZBI, S MODE .ZAS.YH2/N	AZBS1	0491001
SET ZBI IF YH1/N	AZBS2	0501001
SHIFTADDER CONT (B→D→HA→LGP)	BAGC2	0601001
SHIFT LEFT→A	ASLC0	1162003
SHIFT LEFT→A, AP≠X00	ASLC1	0481001
SHIFT LEFT→B	BSLC0	0561001
SHIFT LEFT→B, AP≠X00	BSLC1	0531001

Flow-chart equivalent	*Sub*	*ROM Bit*
SHIFT LEFT→E	ESLC0	0491001
SHIFT LEFT→E, AP≠X00	ESLC1	0771001
SHIFT LEFT→WC1-4, AP≠X00	UMSL1	0641001
SHIFT LEFT→WC1-4, (WC1→MA4)	UMSLO	0901001
SHIFT LEFT BINARY→A	ASLB0	0691001
SHIFT LEFT BINARY→B	BSLB0	0391001
SHIFT LEFT BINARY→B,	BSLB2	1183001
−L35.L34.B36.B35−/N	BSLB2	1183001
SHIFT LEFT BINARY→B, B35.L35/N.L34/N	BSLB4	1183003
SHIFT LEFT BINARY→B, B36.B35=11+00	BSLB1	0401001
SHIFT LEFT BINARY→B, L35.L34=00	BSLB5	1183007
SHIFT LEFT BINARY→L	LSLB0	0781001
SHIFT LEFT BINARY→L, −L35.B36.B35−/N	LSLB2	YXNUEE
SHIFT LEFT BINARY→L, B35.L35/N	LSLB4	YXNUEE
SHIFT LEFT BINARY→L, B36.B35=11+00	LSLB7	YXNUEE
SHIFT LEFT BINARY→L, L35=0	LSLB1	YXNUEE
SHIFT LEFT BINARY→L, QV02=1	LSLB5	0781001
SHIFT LEFT BINARY→L, QV03=1	LSLB6	0741001
SHIFT RIGHT→A	ASRC0	1113006
SHIFT RIGHT→A, AP≠X11	AXRC2	0913002
SHIFT RIGHT→A, ATE.PN1	ASRC3	0661001
SHIFT RIGHT→A, BP≠X11	ASRC1	0611001
SHIFT RIGHT→B	BSRC0	1142003
SHIFT RIGHT→B, AP≠X11	BSRC2	0411001
SHIFT RIGHT→B, BP≠X11	BSRC1	0431001
SHIFT RIGHT→B, OVLP=1	BSRC3	0401001
SHIFT RIGHT→E	ESRC0	0711001
SHIFT RIGHT→E, BP≠X11	ESRC1	0871001
SHIFT RIGHT→MA1-MA4 .WC1-WC4 (MA4→WC1)	UMSR0	0831001
SHIFT RIGHT→WC1-4	UMSR2	0821001
SHIFT RIGHT→WC1-4 AP≠X11	UMSR1	0631001
SHIFT RIGHT BINARY→A	ASRB0	0771001
SHIFT RIGHT BINARY→B	BSRB0	0421001
SHIFT RIGHT BINARY→B MANTISSA OVERFLOW	BSRB3	0411001
SHIFT RIGHT BINARY→L, MANTISSA OVERFLOW	LSRB4	YXNUEE
SHIFT RIGHT BINARY→L IF QV03	LSRB2	0561001
SHIFT RIGHT BINARY→L QV02=1	LSRB1	0771001
SHIFT RIGHT BNRY→A, XGY/N	ASRB1	YXNUEE
SHIFT RIGHT BNRY→B, XGY	BSRB1	0431001
SHIFT RIGHT BNRY→L,	LSRB0	0731001
SIGN→A1 (VIA ADDER)	AFSB0	0861001
SIGN L STORAGE→L36	L3650	0811001
STOP SPECIAL CODE TRANSLATOR	FPCR0	0401001
TXO→EXO	EX0T0	0821001
VV(1-5)→R	RFVV0	0701001
VV→DV	DVFV0	0751001
VV→QU	QFVV0	0451001
WC1-4→MA1-4	UAFC0	0441001
WL1→R	1CMW1	0345035
WL2→R	1CMW2	0345027
WL3→R	1CMW3	0345020
WL4→R	1CMW4	0345021
WL5→R	1CMW5	0345022
WL6→R	1CMW6	0345023
WL7→R	1CMW7	0345025
WRITE DATA	UDAT0	YXNUHDW
WRITE DATA BY OP-CODE	UDAT1	YXNUHDW
WRITE ITEM BY OPCODE	UWTM1	YXNUHDW
WRITE ITEM MARKS	UWTM0	YXNUHDW
WRITE WORD BY CPCODE	UWWM1	YXNUHDW
WRITE WORD MARKS	UWWM0	YXNUHDW
Y18→SNA (SIGN A STORAGE)	SAFY0	0801001
Y18→SNX (Y18 TO SNB) RESET EXU.TXO	SXFY0	0651001
Y19→A19, CLEAR A20-24	AMFY0	1062003
Y19→B19 CLEAR B20-24	BMFY0	1162001
Y(16-19)→CP	CPFY0	0431001
Y(1-12)→QM	QMFY0	0491001
Y(1-18)→B(19-36)	BUFY0	0541001
Y(1-18)→B	BLFY0	1013002
Y(1-18)→B, BH=X00	BLFY1	0601001
Y(1-2)→BP	BPFY0	1191001

Flow-chart equivalent	*Sub*	*ROM Bit*
Y→A(19–36)	AUFY0	0671001
Y→A(1–18)	ALFY0	1113003
Y→AP	APFY0	0942003
Y→L(19–36)	LUFY0	0451001
Y→L(1–18)	LLFY0	0751001
Y→M	MFYY0	1091001
Y→QM, XGY SET	QMFY1	0481001
Y→S	SFYY0	1101001
Y→S	SFYY0	1101001
YH1→YH2	YH2S0	0531001
ZONE BITS THRU ADDER	BAGFT	0571001

REFERENCES

1. Honeywell EDP technical publication entitled "Series 200 Programmers' Reference Manual (Models 200/1200/2200/4200), Hardware Manual, File No. 113.0005.0000.1-139," November 1966.

2. Hoff, George, "Presentation on the development of a microprogramming system for the H4200." First Annual Workshop on Microprogramming, Bedford, Mass., Oct. 11, 1968.

3. Klein, Stuart, and Schwartz, Scott. "Model 4200/8200 Read Only Memory Control Logic." *Honeywell Computer Journal.* (Winter–Spring 1968): 25–33.

4. Paragraphs 3.1 through 3.7 from the "4201 Central Processor Operation and Maintenance Manual."

5. Private correspondence from George S. Hoff to the author.

BIBLIOGRAPHY

MICROPROGRAMMING—ALPHABETICAL LISTING BY AUTHOR

Akushsky, I. Y., Lensky, U.S., et al. "Methods of speeding-up the operation of digital computers." Information Processing (1960), p. 382–389.

Allen, M. W. "System design of Cirrus." Australian Computer Conference (1960), Sec. C5.2.

Allen, M. W., Pearcey, T., et al. "Cirrus, an economical multiprogram computer with microprogram control." IEEE Trans. on Electr. Comp. (December,1963), p. 663–671.

Amdahl, G. M., Amdahl, L. D., et al. "Fourth generation hardware." Datamation (1967), Vol. 13, No. 1, p. 25.

Amdahl, G. M., Blaauw, G. A., et al. "Architecture of the IBM System/360." IBM Journal of Research and Development (April, 1964), Vol. 8, No. 2, p. 87–101.

Amdahl, L. D. "Microprogramming and stored logic." Datamation (February, 1964), p. 24–26.

Auerbach, I. L. "Variable matrix for performing arithmetic and logic functions." Patent No. 2,844,812 (July, 1958).

Ault, C. F., et al. "No. 1 ESS program store." Bell System Tech. J. (September, 1964), Vol. 43, No. 5, p. 2097.

Bailey, S. J. "Faster computer control with a read-only memory." Control Engineering (September, 1967).

Baker, D. M. "Economic considerations of conversion." Datamation (June, 1966), Vol. 12, No. 6, p. 30–34.

Bankowski, J., Fiatkowski, K. "Instruction coding and checking problem for a horizontally microprogrammed computer." Bull. Acad. Polon. Sci. Ser. Tech. (1965), L 965, p. 119–122.

Bartee, T. C., Lebow, I. L., et al. "Theory and Design of Digital Machines." McGraw-Hill Book Company (1962).

Barton, R. S. "A new approach to the functional design of a digital computer." Proc. West. Joint Comp. Conf. (1961), Vol. 19, p. 393–396.

Bashkow, T. R., Sasson, A., et al. "System design of a Fortran machine." IEEE Transactions on Electronic Computers (September, 1967), EC Vol. 16, No. 4, p. 485–499.

Beard, A. D. "Spectra-70, basic design and philosophy of operation Wescon/65." (1965), Part 4, paper number 12.1.

Beaumont, J. O., et al. "The micro-code program." IBM Technical Paper, IBMR & D Lab., San Jose, Calif. (February, 1954).

Beck, L., & Keeler, F. "The C-8401 data processor." Datamation (February, 1964), p. 33–35.

Beckman, F. S., et al. "Developments in the logical organization of computer arithmetic and control units." Proc. IRE 49 (1961), p. 53.

Benjamin, R. I. "The Spectra 70/45 emulator for the RCA 301." Communications of the ACM (December, 1965), Vol. 8, No. 12, p. 748–752.

Benjamin, R. I. "The Spectra 70/45 emulator for the RCA 301." IEEE Trans. on Electronic Comp. (December, 1965). #4339, p. 283.

Bennett, R. W., & Juliusberger, H. Y. "A special microprogram for the pre-processing of input data." IBM Technical Report ASDD (November, 1964), p. 17–153.

Billing, H., Hopmann, W., et al. "Mikroprogramm-steuerwerk." Elektronische Rundschau (1955), Vol. 9, p. 349.

Bjurel, B. "Telecommunications in Sweden." IETC (June, 1967), p. 324–333.

Blaauw, G. A. "Indexing." McGraw-Hill Book Company (1962), p. 150–178.

Blankenbaker, J. V. "Logically microprogrammed computers—examples of small computers." Proc. International Conference on Information Processing, UNESCO (1959) p. 430.

Blankenbaker, J. V. "Logically microprogrammed computers." IRE Prof. Group on Elec. Com. (December, 1958), EC-7, 2, p. 103–109.

Boland, L. G. "Analysis of read-only memory for control." Master's dissertation, Graduate School of Syracuse University (June, 1963), 63-E6-4362.

Bosak, R. "An information algebra." Proc. ACM Conf. (September, 1961), Sec. 6B-2.

Boutwell, E. O., & Hoskinson, E. "The logical organization of the PB-440 microprogrammable computer." Proc. Fall Joint Comp. Conf. (November, 1963), p. 201–213.

Boutwell, E. O. "The PB 440 computer." Datamation (February, 1964), p. 30–32.

Braito, D., et al. "La calcolatrice CEP del centro studi calcolatrici elettroniche dell Universita di Pisa (CSCE), alta frequenza." (December, 1961), Vol. 30, p. 873–876.

Breuer, M. A. "Adaptive computers." Information and Control (October, 1967), Vol. 11, No. 4, p. 402–422.

Briley, R. E. "Picoprogramming, a new approach to internal computer control." FJCC (1965), Vol. 27, No. 2, p. 93–98.

Brooks, Jr., F. P., & Blaauw, G. A. "Architecture of the IBM System/360." IBM Journal (January, 1964), p. 87–101.

Brooks, Jr., F. P. "Recent developments in computer organizations." Advances in Electronics (1963), Vol. 18, p. 45–64.

Brown, J. L., et al. "IBM System/360 engineering." Proc. Fall Joint Comp. Conf. (1964), p. 231–250.

Buckingham, B. R. S., et al. "The controls automation system." Sixteenth Annual Symposium on Switching Circuit Theory and Logic (October, 1965), p. 279–288.

Buckingham, B. R. S., et al. "The controls automation system." TR-00. 1281 (IBM), (1967).

Campbell, C. R., Neilson, D. A., et al. "Micro-programming the Spectra 70/35." Datamation (September, 1966), Vol. 12, No. 9, p. 64–67.

Cantor, D., et al. "Logarithmic and exponential function evaluation in a variable structure digital computer." IRE Trans. on Elec. Comp. (April, 1962), EC-11, p. 155–164.

Carroll, J. W. "Computer programming and artificial intelligence." Univ. of Michigan (1958), p. 315.

Carroll, J. W., ed. "Microprogramming." Conf. Computer Programming and Artificial Intelligence, University of Michigan (1958), p. 319–325.

Cartledge, W. J. "Program and file conversion." Datamation (June, 1966), Vol. 12, No. 6, p. 22–27.

Cecchini, G., & Gerace, G. B. "L'organizzazione logica dell' unita di controllo della CEP." Internal Rept. CSCE, Series I, University of Pisa, Rome, Italy (June, 1963), No. 36.

Chu, Y. "An Algol-like computer design language." Commun. ACM (October, 1965), 8, p. 607–615.

Cliff, R. A. "Application of the stored program computer to small scientific spacecraft." NASA TN-D-3988, (June, 1967).

Cliff, R. A. "A stored program computer for small scientific spacecraft." Goddard Space Flight Center, NASA; TN-D-3640, (October, 1966).

Conroy, E. D. "Microprogramming." Proc. ACM National Conference (September, 1961), p. 603.

Conroy, E. D., & Meade, R. M. "A microinstruction system." Proc. ACM (May, 1961), p. 602.

Conti, C. J., Gibson, D. H., et al. "Structural aspects of the System/360, Model 85." IBM Systems Journal (1968), Vol. 7, No. 1, p. 2–21.

Creveling, C. "A unified information processing telemetry system." Adaptive Telemetry Conference, Goddard Space Flight Center (February, 1966).

Crowley, H. J. "Dicose—a new tool for communications studies." IETC (June, 1967), p. 360–364.

Curtis, H. A. "A functional canonical form." Journ. ACM (April, 1959), p. 245–258.

Datamation Staff. "The next generation." Datamation (January, 1967), Vol. 13, No. 2, p. 31–39.

Devonald, C. H., & Fotheringham, J. A. "The Atlas computer." Datamation (May, 1961), p. 23–27.

Diamond, T. L. "No. 5 crossbar AMA translator." Bell Labs Record (1961).

Dinneen, G. P., Lebow, I. L., et al. "The logical design of CG24." Proc. E.J.C.C. (December, 1958), p. 91–94.

Dreyer, L. "Principles of a two-level memory computer." Computers and Automation (May, 1968), Vol. 17, No. 5, p. 40–42.

Elmer, H. G., Hawley, H. M., et al. "Investigation of read only storage for RAMAC 305." (November, 1960), TM-02.140.

Emelyanov-Yaroslavsky, L. B., et al. "Microprogram control for System/360." Proc. IFIP Congress (1967), p. 567.

Emelyanov-Yaroslavsky, L. B., et al. "Microprogram control for digital computers." Proceedings IFIP Congress 1962 (1962), p. 567–569.

Fagg, P., et al. "IBM System/360 engineering." Proceedings FJCC (1964).

Fallkoff, A. D., Iverson, K. E., et al. "A format description of System/360." IBM Systems Journ. (1964), Vol. 3, Nos. 2 & 3, p. 198–263.

Farina, M. V. "Training for conversion." Datamation (June, 1966), Vol. 12, No. 6, p. 27–29.

Felton, G. "Multi-programming on the Ferranti Orion computer system." Proc. IFIP Cong., Munich, Germany (August, 1962), p. 570–571.

Flynn, M. J., MacLaren, M. D. "Microprogramming revisited." ACM Proc. 22 National Conference (1967), p. 457.

Forbes, R. E., Tung, L. H., et al. "Self diagnosable computers: Part II-Romic, a read-only memory input-output communicator." IBM Technical Report, S.G.C., Owego, N.Y. (April, 1963), 63-C4-2290, p. 825–829.

Frankel, S. P. "The logical design of a simple general purpose computer." IRE Trans. on Elec. Comp. (March, 1957), Vol. EC-6, p. 5–14.

Gerace, G. B. "Un sistema di controllo ad alta velocita per calcolatrici elettroniche digital I." Automazione e Strumentazione (November, 1961), No. 11, p. 533–538.

Gerace, G. B. "La temporizzazione, La minimizzazione dei tempi ed il controllo dinamico della CEP." University of Pisa, Rome, Italy, alta frequenza. (March, 1963), Vol. 32, p. 184–191.

Gerace, G. B. "Microprogrammed control for computing systems." IEEE Trans. on Elec. Comp. (December, 1963), p. 733–747.

Gilgoff, J. L. "Microprogramming the IBM System/360 Model 60/62." IBM-TR-00.1139 (May, 1964).

Gilmore, Jr., J. T., & Peterson, H. P. "A functional description of the TX-O computer." MIT, Cambridge, Mass., (October, 1958), Memorandum 6M-4789, Lincoln Laboratory.

Glantz, H. T. "A note on microprogramming." Journal ACM 3 (1956), p. 77.

Glushkov, V. M. "An application of the abstract theory of automatons for minimization of microprograms." Tech. Cybernetics (May, 1964), p. 1–19.

Glushkov, V. M. "O primenenii abstraktnoi teorli automatov, dlya mimimizatsii mikroprogramm." Akademiya Nauk, SSSR Izvestiya, Tekhnicheskaya Kibernetika (January, 1964), p. 3–8.

Glushkov, V. M. "Automata theory and microprogram transformations." Cybernetics (1965), Vol. 1, No. 5, p. 1–8.

Glushkov, V. M. "Automata theory and structural design problems of digital machines." Kibernetika (1965), Vol. 1, No. 1, p. 3–11.

Glushkov, V. M. "The application of the theory of automata to the minimization of microprograms." Engineering Cybernetics (1964), Vol. 1, No. 1, p. 1–5.

Gorn, S., Ingerman, P. Z., et al. "On the construction of microflowcharts," Comm. ACM (1959), p. 27–31.

Grasselli, A. "Automatic scanning and measuring of bubble-chamber photographs." Lawrence Rad. Lab., University of California, Berkeley. (June, 1960), Rept. UCRL-9231.

Grasselli, A. "The design of program-modifiable micro-programmed control units." IRE Trans. on Elec. Comp. (June, 1962), p. 336–339.

Grasselli, A. "Control units for sequencing complex asynchronous operations." PGEC (August, 1962).

Green, J. "Microprogramming emulators and programming languages." Comm. of ACM (March, 1966), Vol. 9, p. 230–232.

Greene, J., et al. "IBM System/360 engineering." FJCC (1964), Vol. 26, p. 205–231.

Greene, J. E., Dean, B. M., et al. "Microprogrammed implementation of the IBM System/360 model 30 machine organization." IBM-TR-01.813 (June, 1964).

Guntsch, F. R., & Handler, W. "Zur simultanarbeit bei digitalrechnern." Elektronische Rechenanlager (August, 1960), Vol. 2, p. 117–128.

Hack, F. J., & Shirk, R. W. "An integrated approach to automated computer maintenance." TR-CO.128 (April, 1965).

Hagiwara, H., Amo, K., et al. "The KT pilot computer, a microprogrammed computer with a phototransistor fixed memory." IFIP Cong., Munich, Germany (September, 1962), p. 684–689.

Harrand, Y. "Evolution of microprogramming concepts." Proc. 3rd AFCALTI Congress of Computing and Information Processing (1965), p. 187–190.

Hartmanis, J. "Loop-free structure of sequential machines." Information and Control (1962), Vol. 5, p. 25.

Hartmanis, J. "Further results on the structure of sequential machines." Journ. ACM (January, 1961), p. 78–88.

Hassett, R. P., & Miller, E. H. "Multithreading design of a reliable aerospace computer." Radio Corp. of America (July, 1966).

Hawryszkiewycz, I. T. "Microprogrammed control in problem-oriented languages." IEEE Transactions on Electronic Computers (October, 1967), EC-16;5, p. 652–658.

Hellerman, H. "Logical design of arbitrary sequence counters and microprogram sequence control circuits." IBM Res., Yorktown Heights (June, 1963), NC-289.

Hellerman, L., & Hoernes, G. E. "Control storage use in implementing an associative processor for time sharing." IBM TR-00. 1718 (April, 1968).

Henderson, G. L., Mehal, V. J., et al. "IBM System/360 Model 30-1620 Emulator program" (April, 1966).

Hill, R. H. "Stored logic programming and applications." Datamation (February, 1964), p. 36–39.

Hollobaugh, C. T., & Tunis, C. T. "Modular computer with read-only memory." IBM Technical Disclosure Bull. (March, 1965), p. 919–922.

Husson, S. S. "Microprogramming manual for IBM System 360 Model 50." TR-00. 479-1 (IBM Technical Report) (November, 1967).

IBM. "Data processing, microprogramming PI course." IBM #221-0083 (1965).

Ingalls, R. A. "Logical design of a microprogrammed special-purpose computer." U.S. Naval Postgraduate School (December, 1966).

Iverson, K. E. "The description of finite sequential processes." Proc. Fourth London Symposium on Info. Theory, RC-359 (August, 1960).

Iverson, K. E. "Microprogramming: a partial description of the 8106 computer." IBM Res. Report, Yorktown Heights, N.Y. RC-465 (May, 1961).

Iverson, K. E. "A transliteration scheme for keying and printing microprograms." IBM Technical Report NC-79; TP62-1193 (April, 1962).

Iverson, K. E. "Formalism in programming languages." Comm. ACM (February, 1964), p. 80–88.

Jeans, H. "Possible performance improvements by microprogramming for System/360 Models 40, 50, and 65." IBM SRDC Technical Report (July, 1965).

Jewett, D. L. "A fourth generation computer now?" Datamation (April, 1968), p. 37–38.

Kampe, T. W. "The design of a general-purpose microprogram-controlled computer with elementary structure." IRE Trans. (June, 1960), EC-9, p. 208–213.

Kane, J. V., & Spinrad, R. J. "A stored program computer as multiparameter radiation analyzer." Bell Telephone Laboratories (December, 1963), TP-64-4329; MONO4718.

Kauer, A. K., & Quinlan, M. J. "Programmed micro-instruction trace." (1967), TR-21.206.

Kilburn, T., Grimsdale, L., et al. "A digital computer store with very short read time." Proc. Inst. Elec. Eng. (November, 1960), Vol. 107, Part B, p. 567–572.

Kreynin, G. V., Lazarov, V. G., et al. "The synthesis of a type of microprogram automata." Engineering Cybernetics (1967), 3, p. 12–131.

Lawson, H. W. "Programming language-oriented instruction streams." IEEE Trans. (1968), C-17, p. 476.

Lazarev, V. G. "Matrichnyi metod minimization skhem mikroprogramm." Akademiya Nauk, Izvestiya, Tekhnicheskaya Kibernetika (March, 1965), p. 35–42.

Lazarev, V. G. "Matrix method of minimization of microprogram systems." Engineering Cybernetics (March, 1965), 2, p. 32–38.

Lazarev, V. G. "Matrix method of minimizing the number of internal states of asynchronems finite automata." Engineering Cybernetics (1964), 1, p. 26–28.

Levy, L. S. "State of the art of microprogramming." Aerospace Corp. (April 1965).

Liverance, H. H. "A System/360 profile—read-only storage." (June, 1965), IBM#ZNN-5170.

Loopstra, B. J. "Input and output in the X-1 system." Proc. IFIP, UNESCO, Paris, France (June, 1959), p. 342–344.

Maeller, R. I. "Use of flexible program computer." IBM Technical Paper, Owego, N.Y. (August, 1957).

Matsevity, L. V. "An algorithm for minimization of elementary operations in a microprogram." Engineering Cybernetics (1964), Vol. 1, p. 5–13.

McCormack, M. A., "1401 compatibility feature on the IBM system-360 Model 30." Comm. ACM, 8 (1965), p. 773.

McCormack, M. A., Schansman, T. T., et al. "1401 compatibility feature on the IBM System/360 Model 40." CACM (December, 1965), p. 773–776.

McCormack, M. A., Schansman, T. T., et al. "1402 compatibility feature on the IBM System 360/Model 30." CACM (1965), 8, p. 773.

McGee, W. C. "The TRW-133 computer." Datamation (February, 1964), p. 27–29.

McGee, W. C., & Petersen, H. E. "Microprogram control for the experimental sciences." AFIPS F.J.C.C. 27 (1965), p. 77.

McGee, W. C., & Petersen, H. E. "Microprogramming for data and control." IEEE Trans. Nuclear Sci. (USA) (August, 1965), 4, p. 310–320.

Meade, R. M. "A discussion of machine-interpreted macroinstructions." Proc. ACM National Conference (September, 1961), p. 601.

Mechelen, K. W. "The logical design of a microprogrammed computer." Memo. to R. G. Baron, IBM, Owego, N.Y. (December, 1959).

Meggitt, J. E. "Pseudo division and pseudo multiplication processes." IBM J. Res. and Develop. (April, 1962), p. 210–226.

Meggitt, J. E. "Microprogrammed machine to implement the Adam language." IBM Technical Report, Yorktown Heights, N.Y. (November, 1963).

Meggitt, J. E. "A character computer for high-level language interpretation." IBM Systems Journal (1964), Vol. 3, No. 1.

Melbourne, A. J., Pugmire, J. M., et al. "A small computer for the direct processing of Fortran statements." Computer Journ. (England) (April, 1965), p. 24–27.

Mercer, R. J. "Micro-programming." Journal ACM, 4 (1957), p. 157.

Mercer, Robert J. "Micro-programming." Numerical Analysis Research, Univ. of California, Los Angeles (1956).

Middleton, B. H. "A microprogram processing and simulation system." Joint IEE and IERE Colloquium, 10 (October, 1966).

Miller, E. H. "Reliability aspects of the variable instruction computer." IEEE Transactions on Electronic Computers (October, 1967), EC-16; 5, p. 596–602.

Morrison, G. E. "Investigation of microprogramming as a technique in the control of digital computing systems." Computing Reviews (January, 1968), 9; 1, p. 40–41.

Morrison, Gordon E. "Investigation of micro-programming as a technique in the control of digital computing systems." Univ. of California, Berkeley (thesis) (1958).

Mussell, H. A. "Revised stored program." IBM TR-00.05.015.058 (January, 1950).

Mussell, H. A., & Frizzell, C. E. "Stored Program." IBM-TR-00.05.015.037 (December, 1949).

Newman, E. G., & Winter, L. F. "Magnetically controlled variable logic." IEEE Trans. (June, 1965).

Newman, E. G., & Winter, L. F. "Magnetically controlled variable logic." IBM Journal (July, 1964), p. 308–317.

Opler, A. "New directions in software 1960–1966." Proceedings of the IEEE (December, 1966), Vol. 54, No. 12, p. 1757–1763.

Opler, A. "Fourth-generation software, the realignment." Datamation (January, 1967), Vol. 13, No. 1, p. 22–24.

Patrick, R. L. "A customizable computer." Datamation (May, 1964), p. 44–45.

Patzer, W. J., & Vandling, G. C. "Systems implications of microprogramming." Computer Design (December, 1967), p. 62–66.

Patzer, W. J., & Vandling, G. C. "Systems engineering implications of microprogramming." 66-825-1990, (June, 1966).

Phillips, E. L., & Melahn, G. C. "IBM System/360 Model 65 7070/7074 compatibility feature." PT-1393-F042866, (April, 1966).

Pitkowsky, S. H. "Control storage systems." TR-00. 1360 (November, 1965).

Radner, Z. M. "Read-only storage—an explanation." FS/DC-Dept-595 (January, 1966).

Ragland, T. "Spook design and microprogramming." SLT Architecture Memo., Advanced Systems, Poughkeepsie, N.Y. (August, 1962).

Ramamoorthy, C. V. "Discrete Markov analysis of computor programs." Assn. for Computing Machy-Nat Conf., 20th, Proc. paper; 17.2, 24–26, p. 386–392.

Renwick, W. "EDSAC II," Proc. Inst. Elec. Eng. (April, 1956), Vol. 103, Part B, Supp. 102, p. 277–278.

Renwick, W. "A magnetic core matrix store with direct selection using a magnetic core switch matrix." Proc. Inst. Elec. Eng. (October, 1956).

Rice, R. "Macrocoding—key to computer design." IBM Technical Report, Prod. Dev. Lab., Poughkeepsie, N. Y., 00.011.000.654, (February, 1958).

Richard, R. K. "New logical and systems concepts." Proc. EJCC (December, 1968), p. 51–55.

Rose, G. A. "Intergraphic, a microprogrammed graphical-interface computer." IEEE Transactions (December, 1967), EC-16, 6, p. 776–784.

Rosen, S. "Hardware design reflecting software requirements." Fall Joint Computer Conf. Proc. 1443–1449 (1968).

Roth, J. P. "Hardware implementation of microprograms." RD-1456, (August, 1965).

Schauer, R. F., & Meggitt, J. E. "Evaluating the logical design of the problem-oriented symbol processor." Thomas J. Watson Research Center (November, 1963), AD-600013, p. 147.

Schauer, R. F., Notz, W. A., et al. "Design mechanization of a problem-oriented symbol processor." RC-1227, (May, 1964).

Schlaeppi, H. P. "Notes on microprogramming of digital systems." IBM Res. Memo., Zurich, Switzerland RZ-33, (July, 1958).

Schlaeppi, H. P. "Microprogrammed computer control." IBM Tech. Disc. Bull. (1963), Vol. 5, No. 9, p. 45–46.

Schlaeppi, H. P. "Fast sequencing loop for computer control," IBM Tech. Disc. Bull. (February, 1963), Vol. 5, No. 9, 63B4-1319.

Schnabel, D. L. "Microprogrammed by the moves with word offset." IBM Tech. Disc. Bull. (August, 1965), p. 549–552.

Schnabel, D. L., & Thomas, J. "General microprogram for VFL byte handling." Tech. Disc. Bull. (August, 1965), p. 545–548.

Seckler, H. N. "Early NC 1 ESS field experiences—II four wire system for government and military applications." IETC (December, 1967), p. 751–754.

Semarne, H. M., & McGee, W. C. "Stored logic computing." Reprints ACM 16th National Conference (August, 1961), 6C-4.

Semarne, H. M., & Porter, R. E. "A stored logic computer." Datamation 2 (1961), No. 5, p. 33.

Senzig, D. N. "Suggested timing notation for the Iverson notation." IBM Technical Paper (July, 1962), TIC62-A9-028, p. NC-120.

Serrell, R., et al. "The evolution of computing machines and systems." Proc. of the IRE (May, 1962), Vol. 50, p. 1039–1058.

Shafritz, A. B., Miller, A. E., & Rose, K. "Multi-level programming for a real-time system." Proc. Eastern Joint Computer Conf. (1961), p. 1–16.

Slobodyanyuk, T. F. "A method of minimizing microprograms." Defense Documentation Center (February, 1968), 821, p. AD662.

Slobodyanyuk, T. F. "A method of minimizing microprograms." Foreign. Tech. Div. (July, 1967), AD662821; FTD-Ht-23-600-67.

Spence, A. L. "Hardware and software interactions, a machine organization solution—the variable instruction computer." Radio Corp. of America (September, 1966).

Sproul III, W. W. "Microprogram computers with minimum redundancy memories." IRAD Final Report, SGC, Owego, N. Y. (December, 1963), 63-533-014.

Steet, T. B. "The foundations of a theory of data processing." Proc. of ACM (September, 1961), Sec. 6B-2.

Stevens, W. Y. "The structure of System/360 Part II. System implementation." IBM Systems Journal, 3 (1964), p. 136.

Strachey, C. "Time-sharing in large, fast computers." Proc. IFIP, UNESCO, Paris, France (June, 1959), p. 336–341.

Sturman, J. N. "An iteratively structured general purpose digital computer." Bell Telephone Labs. (January, 1968), Vol. 2, No. 9.

Sumner, F. H., et al. "The central control unit of the Atlas computer." Proc. IFIPS Congress (1962), p. 657.

Teager, H. M. "A discussion of machine interpreted macroinstructions." Preprints 16th National ACM (September, 1961).

Tichner, E. J., & Tuffill, H. W. "CPU diagnostics using microprogram routine." IBM Tech Discl. (June, 1964), p. 49.

Tompkins, C. B. "Machine attacks on problems whose variables are permutations." Proc. for Sixth Symposium on Applied Mathematics. McGraw-Hill, N. Y. (1956).

Totaro, J. B. "Time sharing Spectra 70 style." Data Processing (July, 1967).

Tucker, S. G. "Microprogram control for digital computers." IBM Systems Journal 6 (1962), p. 222.

Tucker, S. G. "Emulation of large systems." Communications of the ACM (December, 1965), Vol. 8, No. 12, p. 753–761.

Tucker, S. G. "Microprogram control for system/360." IBM Systems Journal (1967), Vol. 6, No. 4, p. 222.

Valassis, J. G. "Modular computer design with picoprogrammed control." AFIPS 1967 Fall Joint Computer Conference (1967), p. 611–619.

Van Der Poel, W. L. "Microprogramming and trickology." Digitale Informationsevandler (1962), p. 269–311.

Van Der Poel, W. L. "The Logical Principles of Some Simple Computers." Published Doctoral Dissertation, University of Amsterdam (1956).

Van Der Poel, W. L. "Micro-programming and trickology." John Wiley and Sons, Inc. (1962), Digital Information Processors.

Van Der Poel, W. L. "Zebra, a simple binary computer." Proc. International Conference on Information Processing, UNESCO (1959), p. 361.

Wagner, E. G. "Sax-1: a simple universal computer." IBM Technical Report (August, 1959), SR-84.

Walter, G. "Description of proposed stored program controlled interconnected-tape processing system." IBM-TR-00.01.001.454 (March, 1953).

Watt, W. C. "A prerequisite to the utility of micro-grammers." Nat. Bureau of Standards (April, 1965), Vol. 19, p. AD-614258.

Weber, H. "A microprogrammed implementation of Euler on IBM System/360 Model 30." Communications of the ACM (September, 1967), p. 549–558.

Wheeler, D. J. "Read-only stores for the control of computers." Information Processing (1962), p. 561–569.

Wilkes, M. V. "The design of a control unit of an electronic digital computer." Proc. IEEE, 105 (1958), p. 121.

Wilkes, M. V. "Microprogramming." Proceedings of the EJCC (1959), p. 18–20.

Wilkes, M. V. "The best way to design an automatic calculation machine." Manchester University Computer Inaugural Conference Proceeding (1951), p. 16.

Wilkes, M. V. "Microprogramming." Proc. E.J.C.C. (1958), p. 18.

Wilkes, M. V., Renwick, W., et al. "The design of a control unit of an electronic digital computer." IEEE Proceedings (1958), Vol. 105, p. 121.

Wilkes, M. V., & Stringer, J. B. "Micro-programming and the design of the control circuits in an electronic digital computer." Proc. Camb. Phil. Soc. 49 (1953), p. 230.

Wilkes, M. V., et al. "Panel on priority problems in computer systems." Proc. IFIP Cong., Munich, Germany (August, 1962), p. 711–715.

Reference Sources Without Authors

"Bendix G-20 system." Comm. ACM (May, 1960), p. 325–328.

"TR4-Telefunken." Comm. ACM (October, 1960), p. 586–589.

"The Burroughs B2500 and B3500." Datamation (April, 1966), p. 76.

"System/360 Model 40,2040 processing unit." IBM Field Engineering Diagrams Manual (January, 1966), Document No. 0223-2842.

"System/360 Model 40,2040 processing unit." IBM Field Engineering Maintenance Manual (January, 1966), Document No. 0223-2841.
"System/360 Model 40, comprehensive introduction." IBM Field Engineering Manual of Instruction (December, 1965), Document No. 0223-2840.
"System/360 Model 40, functional units." IBM Field Engineering Manual of Instruction (December, 1965), Document No. 0223-2843.
"System/360 Model 40, theory of operation." IBM Field Engineering Manual of Instruction (December, 1965), Document No. 0223-2844.
"System/360 Model 50, comprehensive introduction." IBM Field Engineering Manual of Instruction (1966), Document No. 0223-2821.
"System/360 Model 50, functional units." IBM Field Engineering Manual of Instruction (1965), Document No. 0223-2822.
"Specification of the system Q1." Bull. GE division systems and ensembles (March, 1965), p. 65/22.131.
"Information outline, remote-access time-sharing system." Allen-Babcock Computing, Inc. (1967).
"Storage matrix for microprogramming." Electric and Musical Industries Ltd. (January, 1961), B.P. 859846,25.
"Logical circuit with micro-control." Phillips Electrical Industries Ltd. (April, 1963), B.P. 9224173.
"Microprogramming system." Texas Instruments Inc. (May, 1963), U.S. Pat. 3088667.
"2841 Storage Control unit." Available from Mechanicsburg only (IBM) (1965), 227-3614-3.
"IBM System/360 Model 40 functional units." Available from Mechanicsburg only (1965), 223-2843-0.
"IBM System/360 Model 40 machine status charts." Mechanicsburg only (April, 1966), 223-2858-0.
"Customizing CCROS document cards—machine level control." Available from Standards only (April, 1966), CMS 0-2540 003 0466.

MICROPROGRAMMING—CHRONOLOGICAL LISTING

1949, December. "Stored program." Mussell, H. A., Frizzell, C. E. IBM-TR-00.05.015.037.
1950, January. "Revised stored program." Mussell, H. A. IBM TR-00.05.015.058.
1951. "The best way to design an automatic calculation machine." Wilkes, M. V. Manchester University Computer Inaugural Conference Proceeding, p. 16.
1953. "Micro-programming and the design of the control circuits in an electronic digital computer." Wilkes, M. V., Stringer, J. B. Proc. Camb. Phil. Soc. 49, p. 230.
1953, March. "Description of proposed stored program controlled interconnected tape processing system." Walter, G. IBM-TR-00.01.001.454
1954, February. "The micro-code program." Beaumont, J. O., et al. IBM Technical Paper, IBM R & D Lab., San Jose, Calif.
1955. "Mikroprogramm-Steuerwerk," Billing, H., Hopmann, W. et al. Elektronische Rundschau, vol. 9, p. 349.
1956. "Machine attacks on problems whose variables are permutations." Tompkins, C. B. Proc. for sixth symposium on applied mathematics, McGraw-Hill.
1956. "The logical principles of some simple computers." Van Der Poel, W. L. Published Doctoral dissertation, University of Amsterdam.
1956. "A note on microprogramming." Glantz, H. T. Journal ACM, 3, p. 77.
1956. "Micro-programming." Mercer, Robert J. Numerical analysis research, U. of Calif., L.A.
1956, April. "EDSAC II." Renwick, W. Proc. Inst. Elec. Eng. Vol. 103, Part B, Supp. 102, p. 277–278.

1956, October. "A magnetic core matrix store with direct selection using a magnetic-core switch matrix," Renwick, W. Proc. Inst. Elec. Eng.

1957. "Micro-programming." Mercer, R. J. Journal ACM, 4, p. 157.

1957, March. "The logical design of a simple general purpose computer." Frankel, S. P. IRE Trans. On Elec. Comp. Vol. EC-6, p. 5–14.

1957, August. "Use of flexible program computer." Maeller, R. I. IBM Technical Paper, Owego, N.Y.

1958. "Microprogramming." Carroll, J. W., ed. Conf. computer programming and artificial intelligence, University of Michigan, p. 319–325.

1958. "The design of a control unit of an electronic digital computer." Wilkes, M. V. Proc. IEE, 105, p. 121.

1958. "Investigation of micro-programming as a technique in the control of digital computing systems." Morrison, Gordon E. Univ. of Calif., Berkeley (thesis).

1958. "Computer programming and artificial intelligence." Carroll, J. W. Univ. of Michigan, p. 315.

1958. "Microprogramming." Wilkes, M. V. Proc. EJCC, p. 18.

1958. "The design of a control unit of an electronic digital computer." Wilkes, M. V., Renwick, W., et al. IEE proceedings, 105, p. 121.

1958, February. "Macrocoding—key to computer design." Rice, R. IBM technical report, Prod. Dev. Lab., Poughkeepsie, N.Y. 00.011.000.654

1958, July. "Variable matrix for performing arithmetic and logic functions," Auerbach, I. L. Patent No. 2,844,812.

1958, July. "Notes on microprogramming of digital systems." Schlaeppi, H. P. IBM res. memo., Zurich, Switzerland. RZ-33.

1958, October. "A functional description of the TX-O computer," Gilmore, J. T., Jr. & Peterson, H. P. MIT, Cambridge, Mass. Memorandum 6M-4789, Lincoln laboratory.

1958, December. "The logical design of CG24." Dinneen, G. P., Lebow, I. L., et al. Proc. E.J.C.C., p. 91–94.

1958, December. "Logically microprogrammed computers." Blankenbaker, J. V. IRE Prof. Group on Elec. Com. EC-7; 2, p. 103–109.

1959. "On the construction of microflowcharts." Gorn, S., Ingerman, P. Z., et al. Comm. ACM, p. 27–31.

1959. "Logically microprogrammed computers—examples of small computers." Blankenbaker, J. V. Proc. International Conference on Information Processing, UNESCO, p. 430.

1959. "Microprogramming." Wilkes, M. V. Proceedings of the EJCC 1959, p. 18–20.

1959. "ZEBRA, a simple binary computer." Van Der Poel, W. L. Proc. international conference on information processing, UNESCO, p. 361.

1959, April. "A functional canonical form." Curtis, H. A. Journ. ACM, p. 245–258.

1959, June. "Input and output in the X-1 system." Loopstra, B. J. Proc. IFIP, UNESCO, Paris, France, p. 342–344.

1959, June. "Time-sharing in large, fast computers," Strachey, C. Proc. IFIP, UNESCO, Paris, France, p. 336–341.

1959, August. "Sax-1: a simple universal computer." Wagner, E. G. IBM Technical Report, SR-84.

1959, December. "The logical design of a microprogrammed computer." Mechelen, K. W. Memo. to R. G. Baron, IBM, Owego.

1960. "System design of Cirrus." Allen, M. W. Australian Computer Conference, Sec. CS.2.

1960. "Methods of speeding-up the operation of digital computers." Akushsky, I. Y., Lensky, U.S., et al. Information Processing, p. 382–389.

1960, May. "Bendix G-20 system." Comm. ACM, p. 325–328.

1960, June. "Automatic scanning and measuring of bubble-chamber photographs." Grasselli, A. Lawrence Rad. Lab., Univ. of California, Berkeley, Rept. UCRL-9231.

1960, June. "The design of a general-purpose microprogram-controlled computer with elementary structure." Kampe, T. W. IRE Trans., EC-9, p. 208–213.

1960, August. "The description of finite sequential processes." Iverson, K. E. Proc. Fourth London Symposium on Info. Theory, RC-359.

1960, August. "Zur simultanarbeit bei digitalrechnern." Guntsch, F. R., & Handler, W. Elektronische Rechenanlager 2, p. 117–128.

1960, October. "TR4-Telefunken." Comm. ACM, p. 586–589.

1960, November. "A digital computer store with very short read time," Kilburn, T., Grimsdale, L., et al. Proc. inst. elec. eng., Vol. 107, Part B, p. 567–572.

1960, November. "Investigation of read only storage for RAMAC 305." Elmer, H. G., Hawley, H. M., et al. TM-02.140.

1961. "A new approach to the functional design of a digital computer." Barton, R. S. Proc. West. Joint Comp. Conf., Vol. 19, p. 393–396.

1961. "A micro-instruction system." Conroy, E., Meade, R. M. Proc. ACM National Conference, p. 602.

1961. "Developments in the logical organization of computer arithmetic and control units." Beckman, F. S., et al. Proc. IRE 49, p. 53.

1961. "No. 5 crossbar AMA translator." Diamond, T. L. Bell Labs Record.

1961. "Multi-level programming for a real-time system." Shafritz, A. B., Miller, A. E. & Rose, K. Proc. Eastern Joint Computer Conf., p. 1–16.

1961, January. "Storage matrix for microprogramming." Electric and Musical Industries Ltd. B.P. 859846,25.

1961, May. "Microprogramming: a partial description of the 8106 computer." Iverson, K. E. IBM Research Report, Yorktown Heights, N.Y. RC-465.

1961, May. "A microinstruction system." Conroy, E. D., & Meade, R. M. Proc. ACM, p. 602.

1961, May. "The Atlas computer." Devonald, C. H., & Fotheringham, J. A. Datamation, p. 23–27.

1961, September. "Microprogramming." Conroy, E. D. Proc. ACM National Conference, p. 603.

1961, September. "A discussion of machine-interpreted macroinstructions." Meade, R. M. Proc. ACM National Conference, p. 601.

1961, September. "Stored logic computing." Semarne, H. M., McGee, W. C. Preprints ACM 16th National Conference, 6C-4.

1961, September. "An information algebra." Bosak, R. Proc. ACM Conf., Sec. 6B-2.

1961, September. "A discussion of machine interpreted macroinstructions." Teager, H. M. Preprints ACM 16th National Conference.

1961, September. "The foundations of a theory of data processing." Steet, T. B. Proc. of ACM conf., 1961 Sec. 6B-2.

1961, November. "Un sistema di controllo ad alta velocita per calcolatrici elettroni—che digital I." Gerace, G. B. Automazione e Strumentazione, No. 11, p. 533–538.

1961, December. "La calcolatrice CEP del centro studi calcolatrici elettroniche dell Universita di Pisa (CSCE), alta frequenza." Braito, D., et al. Vol. 30, p. 873–876.

1962. "Theory and design of digital machines." Bartee, T. C., Lebow, I. L., et al. McGraw-Hill Book Company.

1962. "Indexing." Blaauw, G. A., Planning a Computer System, McGraw-Hill Book Company, p. 150–178.

1962. "Micro-programming and trickology." Van Der Poel, W. L. John Wiley and Sons, Inc., Digital Information Processors.

1962. "Loop-free structure of sequential machines." Hartmanis, J. Information and Control, Vol. 5, p. 25.

1962. "Microprogram control for digital computers." Tucker, S. G. IBM Systems Journal 6, p. 222.

1962. "The central control unit of the Atlas computer." Sumner, F. H., et al. Proc. IFIPS Congress, p. 657.

1962. "Microprogram control for digital computers." Emelyanov-Yaroslavsky, L. B., et al. Proceedings IFIP Congress 1962, p. 567–569.

1962. "Microprogramming and trickology." Van Der Poel, W. L. Digitale Informationsevandler, p. 269–311.

1962. "Read-only stores for the control of computers." Wheeler, D. J. Information Processing, p. 561–569.

1962, April. "A transliteration scheme for keying and printing microprograms." Iverson, K. E. IBM Technical Report, NC-79; TP62-1193.

1962, April. "Logarithmic and exponential function evaluation in a variable structure digital computer." Cantor, D., et al. IRE Trans. on elec. comp EC-11, p. 155–164.

1962, April. "Pseudo division and pseudo multiplication processes." Meggitt, J. E. IBM J. Res. Develop., p. 210–226.

1962, May. "The evolution of computing machines and systems." Serrell, R., et al. Proc. of the IRE, 50, p. 1039–1058.

1962, June. "The design of program-modifiable micro-programmed control units." Gasselli, A. IRE Trans. on elec. comp., p. 336–339

1962, July. "Suggested timing notation for the Iverson notation." Senzig, D. N. IBM technical paper, TIC62-A9-028, p. NC–120.

1962, August. "Multi-programming on the Ferranti Orion computer system." Felton, G. Proc. IFIP Cong., Munich, Germany, p. 570–571.

1962, August. "Control units for sequencing complex asynchronous operations." Grasselli, A. PGEC.

1962, August. "Spook design and microprogramming." Ragland, T. SLT Architecture Memo., Advanced Systems, Poughkeepsie, N.Y.

1962, August. "Panel on priority problems in computer systems." Wilkes, M. V., et al. Proc. IFIP Cong., Munich, Germany, p. 711–715.

1962, September. "The KT pilot computer, a microprogrammed computer with a phototransistor fixed memory." Hagiwara, H., Amo, K., et al. IFIP Cong., Munich, Germany, p. 684–689.

1963. "Recent developments in computer organization." Brooks, Jr., F. P. Advances in Electronics, Vol. 18, p. 45–64.

1963. "Microprogrammed computer control." Schlaeppi, H. P. IBM Tech. Disc. Bull., Vol. 5, No. 9, p. 45–46.

1963, January. "Further results on the structure of sequential machines." Hartmanis, J. Journal ACM, p. 78–88.

1963, February. "Fast sequencing loop for computer control." Schlaeppi, H. P. IBM Tech. Disc. Bull., Vol. 5, No. 9, 63B4-1319.

1963, March. "La temporizzazione, la minimizzazione dei tempi ed il controllo dinamico della CEP." Gerace, G. B. University of Pisa, Rome, Italy, Alta Frequenza, Vol. 32, p. 184–191.

1963, April. "Self diagnosable computers: part II-ROMIO, a read-only memory input-output communicator." Forbes, R. E., Tung, L. H., et al. IBM Technical Report, S.G.C., Owego, N.Y. 63-C4-2290, p. 825–829.

1963, April. "Logical circuit with micro-control." Phillips Electrical Industries Ltd. B.P. 9224173.

1963, May. "Microprogramming system." Texas Instruments Inc. U.S. Pat. 3088667.

1963, June. "L'organizzazione logica dell' unita di controllo della CEP." Cecchini, G., & Gerace, G. B. Internal Rept. CSCE, Series I, University of Pisa, Rome, Italy, No. 36.

1963, June. "Analysis of read-only memory for control," Boland, L. G. Master's dissertation, Graduate School of Syracuse Univ., 63-E6-4362.

1963, June. "Logical design of arbitrary sequence counters and microprogram sequence control circuits." Hellerman, H. IBM res., Yorktown Heights, N.Y.

1963, August. "Microprogram control for digital computers." Emelyanov-Yaroslavsky, L. B., et al. North Holland Publishing Co. Proc. IFIP Congress 62, p. 567.

1963, November. "The logical organization of the PB-440 microprogrammable computer." Boutwell, E. O., Hoskinson, E. Proc. Fall Joint Comp. Conf., p. 201–213.

1963, November. "Microprogrammed machine to implement the Adam language," Meggitt, J. E. IBM Technical Report, Yorktown Heights, N.Y.

1963, November. "Evaluating the logical design of the problem oriented symbol processor." Schauer, R. F. & Meggitt, J. E. Thomas J. Watson Research Center AD-600013, p. 147.

1963, December. "Cirrus, an economical multiprogram computer with microprogram control." Allen, M. W., Pearcey, T., et al. IEEE trans. on electr. comp., p. 663–671.

1963, December. "Microprogrammed control for computing systems." Gerace, G. B. IEEE trans. on elec. comp., p. 733–747.

1963, December. "Microprogram computers with minimum redundancy memories." Sproul III, W. W. IRAD Final Report, SGC, Owego, N.Y., 63-533-014.

1963, December. "A stored program computer as multiparameter radiation analyzer." Kane, J. V., & Spinrad, R. J. Bell Telephone Laboratories, TP-64-4329; MONO4718.

1964. "IBM system/360 engineering." Brown, J. L., et al. Proc. Fall Joint Comp. Conf., p. 231–250.

1964. "A character computer for high-level language interpretation." Meggitt, J. E. IBM Systems Journal, Vol. 3, No. 1.

1964. "A format description of system/360." Fallkoff, A. D., Iverson, K. E., et al. IBM systems journ., Vol. 3, Nos. 2 & 3, p. 198–263.

1964. "IBM System/360 engineering." Fagg, P., et al. Proceedings FJCC.

1964. "The application of the theory of automata to the minimization of microprograms." Glushkov, V. M. Engineering Cybernetics 1, 1, p. 1–5.

1964. "Matrix method of minimizing the number of internal states of asynchronems finite automata." Lazarev, V. G. Engineering cybernetics, 1, p. 26–28.

1964. "An algorithm for minimization of elementary operations in a microprogram." Matsevity, L. V. Engineering Cybernetics, 1, p. 5–13.

1964. "The structure of System 360, Part II. System implementation." Stevens, W.Y. IBM systems, 3, p. 136.

1964. "IBM system/360 engineering." Greene, J., et al. FJCC, 26, p. 205–231.

1964, January. "O primenenii abstraktnoi teorli automatov, dlya minimizatsii mikroprogramm." Glushkov, V. M. Adademiya Nouk, USSR Izvestiya, Tekhnicheskaya Kibernetika, p. 3–8.

1964, January. "Architecture of the IBM system/360." Brooks, Jr., F. P., Blaauw, G. A. IBM Journal, p. 87–101.

1964, February. "The C-8401 data processor." Beck, L., Keeler, F. Datamation, p. 33–35.

1964, February. "The PB 440 computer." Boutwell, Jr., O. Datamation, p. 30–32.

1964, February. "Microprogramming and stored logic." Amdahl, L. D. Datamation, p. 24–26.

1964, February. "The TRW-133 computer." McGee, W. C. Datamation, p. 27–29.

1964, February. "Stored logic programming and applications." Hill, R. H. Datamation, p. 36–39.

1964, February. "Formalism in programming languages." Iverson, K. E. Comm. ACM, p. 80–88.

1964, April. "Architecture of the IBM system/360." Amdahl, G. M., Blaauw, G. A., et al. IBM journal of research and development 8; 2, p. 87–101.

1964, May. "A customizable computer." Patrick, R. L. Datamation, p. 44–45.

1964, May. "An application of the abstract theory of automations for minimization of microprograms." Glushkov, V. M. Tech. Cybernetics, p. 1–19.

1964, May. "Microprogramming the IBM System/360 Model 60/62." Gilgoff, J. L. IBM-TR-00.1139.

1964, May. "Design mechanization of a problem oriented symbol processor." Schauer, R. F., Notz, W. A., et al. RC-1227.

1964, June. "CPU diagnostics using microprogram routine." Tichner, E. J., & Tuffill, H. W. IBM Tech. Discl. Bull., p. 49.

1964, June. "Microprogrammed implementation of the IBM System/360 Model 30 machine organization." Greene, J. E., Dean, B. M., et al. IBM-TR-01.813.

1964, July. "Magnetically controlled variable logic." Newman, E. G., & Winter, L. F. IBM Journal, July 1964, p. 308–317.

1964, September. "No. 1 ESS program store," Ault, C. F., et al. Bell System Tech. J., Vol. 43, no. 5; p. 2097.

1964, November. "A special microprogram for the pre-processing of input data." Bennett, R. W., & Juliusberger, H. Y. IBM Technical Report ASDD, p. 17–153.

1965. "System/360 model 50, functional units." IBM Field Engineering Manual of Instruction, Document No. 0223-2822.

1965. "Spectra-70, Basic design and philosophy of operation." Beard, A. D. Wescon/65 Part 4, paper number 12.1.

1965. "1401 compatibility feature on the IBM system-360 model 30." McCormack, M. A., Comm. ACM 8, p. 773.

1965. "Microprogram control for the experimental sciences." McGee, W. C., & Petersen, H. E. AFIPS FJCC 27, p. 77.

1965. "Automata theory and microprogram transformations." Glushkov, V. M. Cybernetics, Vol. 1, No. 5, p. 1–8.

1965. "Automata theory and structural design problems of digital machines." Glushkov, V. M. Kibernetika, Vol. 1, No. 1, p. 3–11.

1965. "Matrix method of minimization of microprogram systems." Lazarev, V. G. Engineering Cybernetics, Vol. 2, p. 32–38.

1965. "1402 compatibility feature on the IBM system 360/model 30." McCormack, M. A., Schansman, T. T., et al. CACM, Vol. 8, p. 773.

1965. "Picoprogramming, a new approach to internal computer control." Briley, R. E. FJCC, Vol. 27, No. 2, p. 93–98.

1965. "Data processing, microprogramming PI course." IBM#221-0083.

1965. "Microprogram control for the experimental sciences." McGee, W. C., & Petersen, H. E. Proceeding AFIPS, 1965 Fall Joint Computer Conference, Vol. 27; Part 1, p. 77–91.

1965. "Instruction coding and checking problem for a horizontally microprogrammed computer." Bankowski, J., & Fiatkowski, K. Bull. Acad. Polon. Sci. Ser. Tech. L965, p. 119–122.

1965. "2841 storage control unit." Available from Mechanicsburg only (IBM). 227-3614-3.

1965. "IBM System/360 Model 40 functional units." Available from Mechanicsburg only, 223-2843-0.

1965, May. "Specification of the system Q1." Bull. GE Division Systems and Ensembles, p. 65/22.131

1965, March. "Modular computer with read-only memory." Hollobaugh, C. T., & Tunis, C. T. IBM Tech. Disc. Bull., p. 919–922.

1965, March. "Matrichnyi metod minimization skhem mikroprogramm." Larev, V. G. Akademiya Nauk, Izvestiya, Tekhnicheskaya Kibernetika, p. 35–42.

1965, April. "A small computer for the direct processing of Fortran statements." Melbourne, A. J., Pugmire, J. M. et al. Computer Journ. (England), p. 24–27.

1965, April. "State of the art of microprogramming." Levy, L. S. Aerospace Corp.

1965, April. "A prerequisite to the utility of micro-grammers." Watt, W. C. Nat. Bureau of Standards. 19, p. AD-614258

1965, April. "An integrated approach to automated computer maintenance." Hack, F. J., & Shirk, R. W. TR-Co. 128.

1965, June. "A System/360 profile—read-only storage." Liverance, H. H. IBM#ZNN-5170.

1965, June. "Magnetically controlled variables logic." Newman, E. G., & Winter, L. F. IEEE Trans. on Elec. Comp.

1965, July. "Possible performance improvements by microprogramming for System/360 models 40, 50, and 65." Jeans, H. IBM SRDC Technical Report.

1965, August. "Discrete Markov analysis of computer programs." Ramamoorthy, C. V. Assn. for Computing Machy-Nat Conf., 20th Proc. paper, 17.2, 24–26, p. 386–392.

1965, August. "Microprogramming for data and control." McGee, W. C., & Petersen, H. E. IEEE Trans. Nuclear Sci. (USA), 4, p. 310–320.

1965, August. "Hardware implementation of microprograms." Roth, J. P. RD-1456.

1965, September. "General microprogram for VFL byte handling." Schnabel, D. L., & Thomas, J. IBM Tech. Disc. Bull., p. 545–548.

1965, September. "Microprogrammed by the moves with word offset." Schnabel, D. L. IBM Tech. Disc. Bull., p. 549–552.

1965, October. "The controls automation system." Buckingham, B. R. S., et al. Sixteenth Annual Symposium on Switching Circuit Theory and Logic, p. 279–288.

1965, October. "An Algol-like computer design language." Chu, Y. Comm. ACM, 8, p. 607–615.

1965, November. "Control storage systems." Pitkowsky, S. H. TR-00.1360.

1965, December. "System/360 Model 40, comprehensive introduction." IBM Field Engineering Manual of Instruction, Document No. 223–2840.

1965, December. "System/360 Model 40, functional units." IBM Field Engineering Manual of Instruction, Document No. 223–2843.

1965, December. "System/360 Model 40, theory of operation." IBM Field Engineering Manual of Instruction, Document No. 223-2844.

1965, December. "1401 compatibility feature on the IBM system/360 Model 40." McCormack, M. A., Schansman, T. T., et al. Comm. ACM, p. 773–776.

1965, December. "The Spectra 70/45 emulator for the RCA 301." Benjamin, R. I. Comm. ACM. Vol. 8, 12, p. 748–752.

1965, December. "Emulation of large systems." Tucker, S. G. Comm. ACM. Vol. 8, No. 12, p. 753–761.

1965, December. "The Spectra 70/45 emulator for the RCA 301." Benjamin, R. I. IEEE Trans. on electronic comp. #4339, p. 283.

1966. "System-360 Model 50, comprehensive introduction." IBM Field Engineering Manual of Instruction, Document No. 223–2821.

1966, January. "System/360 Model 40, 2040 processing unit." IBM Field Engineering Diagrams Manual, Document No. 223–2842.

1966, January. "System-360 Model 40, 2040 processing unit." IBM Field Engineering Maintenance Manual, Document No. 223–2841.

1966, January. "Read only storage—an explanation." Radner, Z. M. FS/DC-Dept.-595.

1966, February. "A unified information processing telemetry system." Creveling, C. Adaptive Telemetry Conference, Goddard Space Flight Center.

1966, March. "Microprogramming emulators and programming languages." Green, J. Comm. ACM, 9, p. 230–232.

1966, April. "The Burroughs B2500 and B3500." Datamation, p. 76.

1966, April. "IBM System/360 Model 65 7070/7074 compatibility feature." Phillips, E. L., & Melahn, G. C. PT-1393-F042866.

1966, April. "IBM System/360 Model 40 machine status charts." Mechanicsburg only. 223-2858-C.

1966, April. "IBM System/360 Model 30—1620 emulator program." Henderson, G. L., Mehal, V. J., et al. PT-6637-F04116.

1966, April. "Customizing CCROS document cards—machine level control." Available from Standards only, CMS O-2540 CO3 0466.

1966, June. "Economic considerations of conversion." Baker, D. M. Datamation, Vol. 12, No. 6, p. 30–34.

1966, June. "Program and file conversion." Cartledge, W. J. Datamation, Vol. 12, No. 6, p. 22–27.

1966, June. "Training for conversion." Farina, M. V. Datamation, Vol. 12, No. 6, p. 27–29.

1966, June. "Systems engineering implications of microprogramming." Patzer, W. J., & Vandling, G. C. 66-825-1990.

1966, July. "Multithreading design of a reliable aerospace computer." Hassett, R. P., & Miller, E. H. Radio Corp. of America.

1966, September. "Micro-programming the Spectra 70/35." Campbell, C. R., Neilson, D. A., et al. Datamation, Vol. 12, No. 9, p. 64–67.

1966, September. "Hardware and software interactions, a machine organization solution—the variable instruction computer." Spence, A. L. Radio Corp. of America.

1966, October. "A microprogram processing and simulation system." Middleton, B. H. Joint IEEE and IERE Colloquium, 10.

1966, October. "A stored program computer for small scientific spacecraft." Cliff, R. A. Goddard Space Flight Center, NASA; TN-D-3640.

1966, December. "Logical design of a microprogrammed special-purpose computer." Ingalls, R. A. U.S. Naval Postgraduate School.

1966, December. "New directions in software 1960–1966." Opler, A. Proceedings of the IEEE 54, 12, p. 1757–1763.

1967. "Fourth generation hardware." Amdahl, G. M., Amdahl, L. D., et al. Datamation, Vol. 13, No. 1, p. 25.

1967. "Microprogramming revisited." Flynn, M. J., MacLaren, M.D. ACM Proc. 22 National Conference, p. 457.

1967. "Microprogram control for System-360." Emelyanov-Yaroslavsky, L. B., et al. Proc. IFIP Congress, p. 567.

1967. "The synthesis of a type of microprogram automata." Kreynin, G. V., Lazarov, V. G., et al. Engineering Cybernetics 3, p. 12–131.

1967. "Information outline, remote-access time-sharing system." Allen-Babcock Computing Inc.

1967. "Modular computer design with picoprogrammed control." Valassis, J. G. AFIPS 1967 Fall Joint Computer Conference, p. 611–619.

1967. "Programmed micro-instruction trace." Kauer, A. K., & Quinlan, M. J. TR-21.206.

1967. "Microprogram control for System/360." Tucker, S. G. IBM Systems Journal, Vol. 6, No. 4, p. 222.

1967. "The controls automation system." Buckingham, B. R. S., et al. TR-00.1281 (IBM).

1967, January. "The next generation." Datamation Staff. Datamation, Vol. 13, No. 2, p. 31–39.

1967, January. "Fourth-generation software, the realignment." Opler, A. Datamation, Vol. 13, No. 1, p. 22–24.

1967, June. "Application of the stored program computer to small scientific spacecraft." Cliff, R. A. NASA TN-D-3988.

1967, June. "Telecommunications in Sweden." Bjurel, B. IETC, p. 324–333.

1967, June. "Dicose—a new tool for communications studies." Crowley, H. J. IETC, p. 360–364.

1967, July. "Time sharing Spectra 70 style." Totaro, J. B. Data Processing.

1967, July. "A method of minimizing microprograms." Slobodyanyuk, T. K. Foreign Tech. Div. AD662821; FTD-HT-23-600-67.

1967, September. "Faster computer control with a read-only memory." Bailey, S. J. Control Engineering.

1967, September. "System design of a Fortran machine." Bashkow, T. R., Sasson, A., et al. IEEE transactions on electronic computers EC-16, 4, p. 485–499.

1967, September. "A microprogrammed implementation of Euler on IBM System/360 Model 30." Weber, H. Comm. ACM, p. 549–558.

1967, October. "Adaptive computers." Breuer, M. A. Information and Control, Vol. 11, No. 4, p. 402–422.

1967, October. "Microprogrammed control in problem-oriented languages." Hawryszkiewycz, I.T. IEEE transactions on electronic computers, EC Vol. 16, No. 5, p. 652–658.

1967, October. "Reliability aspects of the variable instruction computer." Miller, E. H. IEEE transactions on electronic computers, EC Vol. 16, No. 5, p. 596–602.

1967, November. "Microprogramming manual for IBM System 360 Model 50." Husson, S. S. TR-00.479-1 (IBM Technical Report).

1967, December. "Systems implications of microprogramming." Patzer, W. J., & Vandling, G. C. Computer Design, p. 62–66.

1967, December. "Intergraphic, a microprogrammed graphical-interface computer." Rose, G. A. IEEE transactions on electronic computers. EC Vol. 16, No. 6, p. 773–784.

1967, December. "Early NO 1 ESS field experiences—II four wire system for government and military applications." Seckler, H. N. IETC, p. 751–754.

1968. "Programming-language-oriented instruction streams." Lawson, H. W. IEEE Trans C-17, p. 476.

1968. "Structural aspects of the System/360, Model 85." Conti, C. J., Gibson, D. H., et al. IBM Systems Journal, Vol. 7, No. 1, p. 2–21.

1968. "Hardware design reflecting software requirements." Rosen, S. Fall Joint Computer Conf. Proc. 1443–1449.

1968, January. "An iteratively structured general purpose digital computer." Sturman, J. N. Bell Telephone Labs., IEEC Vol. 2, No. 9.

1968, January. "Investigation of microprogramming as a technique in the control of digital computing systems." Morrison, G. E. Computing Reviews, Vol. 9, No. 1, p. 40–41.

1968, February. "A method of minimizing microprograms." Slobodyanyuk, T. F. Defense Documentation Center 821, p. AD662.

1968, April. "Control storage use in implementing an associative processor for time sharing." Hellerman, L., & Hoernes, G. E. IBM#TR-00.1718.

1968, April. "A fourth generation computer now?" Jewett, D. L. Datamation, p. 37–38.

1968, May. "Principles of a two-level memory computer." Dreyer, L. Computers and Automation, Vol. 17, No. 5, p. 40–42.

1968, December. "New logical and systems concepts." Richard, R. K. Proc. EJCC, p. 51–55.

CONTROL STORAGE—ALPHABETICAL LISTING BY AUTHOR

Alexander, E. J., et al. "A permanent magnet twistor memory element of improved characteristics." J. Appl. Physics. (March, 1962), Vol. 33, No. 3, p. 1075.

Anderson, J. L., & Reynolds, J. L. "Light coupled diode, read only memory." (December, 1964), IBM TR-22.146.

Andrews, E. G. "The Bell computer, Model 6." Proc., Harvard Univ. (1951), p. 20.

Askin, H. O., Cirielli, C. A., et al. "C-90 CROS unit phase of System 360 Model 65." (September, 1965), IBM PT-4325-F.

Auerbach, J. L. "European electronic data processing—a report on the industry and the state-of-the-art." Proc. IRE (January, 1961), Vol. 49, No. 1, p. 330–331.

Ault, C. F., et al. "No. 1 ESS program store." Bell System Technical Journal 43 (September, 1964), No. 5, Part 1, p. 2097.

Aviles, J. F., Bourrez, J. M., et al. "A read only 70-NS waffle iron memory." IEMG (September, 1967), p. 321–324.

Barnett, W. A., et al. "A card changeable permanent-magnet-twistor memory of large capacity." IRE (September, 1961).

Beard, A. D. "RCA Spectra 70, basic design and philosophy of operation." Wescon, San Francisco, Paper (1965) Vol. 12, No. 1.

Biche, P. W., & Ruoff, C. E. "Modular design and analysis of a read only resistor storage." IBM TR-01.774 (March, 1964).

Biche, P. W., & Ruoff, C. E. "A survey of read only storage techniques." IBM-TD-01.303 (December, 1963).

Bobeck, A. H. "A new storage element suitable for large-sized memory arrays: the twistor." Bell System Technical Journal, Vol. 36, p. 1319–1340.

Boland, L. "Control memory." MS thesis (EE) Syracuse University (June, 1963).

Bowling, R. E., & Southard, C. D. "CCROS—a card capacitor read only storage." IBM-TR-01.864 (May, 1965).

Bramley, A. "Photoconductive switching devices." Proc. of Symposium on the Role of Solid State Phenomena in Electronics. (April, 1957), p. 289–301.

Brick, D. B., & Pick, G. G. "Microsecond word recognition system." IEEETEC (February, 1964), p. 27–35.

Briley, B. E. "MYRA-a new memory element and system." Proc. Intermag Conference (1964), Paper No. 14-8.

Brown, D. R., & Rochester, N. "Rectifier networks for multiposition switching." Proc. IRE (1949), p. 139–147.

Bruce, G. D. "Fiber optics large scale read-only memory." IBM Tech. Disclosure Bull. (October, 1961), Vol. 4, p. 76–77.

Burgess, G. S. "Read only memory systems." IBM-TR-12.036 (April, 1962), p. TIC62A4-0.

Burks, A. W. "Electronic computing circuits of the ENIAC." Proc. IRE (1947), Vol. 35, p. 756.

Butcher, I. R. "A pre-wired storage unit." IEEETEC (April, 1964), Vol. EC 13, No. 2, p. 106–111.

Carter, W. C., Montgomery, H. C., et al. "Design of serviceability features for the IBM System/360." IBM J. Res. Develop. (1964), Vol. 8, No. 115.

Clemson, D., & Kuttner, P. "Application of rope memory devices." Computer Design (October, 1964), p. 12–22.

Clover, Jr., R. B., Kollar, E. P. "Optical coincidence read only memory system." IBM-TR-SD-007 (September, 1962), p. TIC63C2-2.

Cole, J. N. "Read-only memory using non-linear resistive elements." IBM Tech. Disc. Bull. (September, 1963), Vol. 6, p. 31–32.

Computer Design Staff. "Core-Rope memory." Computer Design (June, 1963), p. 16–19.

Craft, J. L., et al. "A table look-up machine for processing of natural languages." IBM Journal (July, 1961), p. 192–203.

David, C. "Fixed memory with resistive coupling." Proc. ICMT (1965).

De Bot, L.A.P.M. "Linear read-only memories." Proc. ICMT.

Debuske, J. H., et al. "A card changeable non-destructive readout twistor store." Proc. WJCC (1959).

Diamond, T. L. "No. 5 crossbar AMA translator." Bell Lab. Record (1951), Vol. 29, p. 62.

Donnelly, J. M. "Card changeable memories." Computer Design (June, 1964), p. 20–30.

Duke, K. A. "Inductive read-only memory bit pattern." IBM Tech. Disclosure Bull. (March, 1963), Vol. 5, p. 1–75.

Endo, I., & Yamato, J. "The metal card memory—a new semi-permanent store." LCM-TCS (November, 1961), p. 213–230.

Fasano, G., et al. "Semi-permanent memories made of diodes and photosensitive elements." Large Capacity Memory Techniques for Computing Systems (1962), Macmillan.

Filippazzi, F. "A new approach to permanent memory." IEEC (June, 1967), p. 370–371.

Foglia, H. R., McDermid, W. L., et al. "Card-capacitor—a semipermanent read-only memory." IBM J. Res. Develop. (January, 1961), Vol. K, No. 1, p. 67–69.

Foglia, H. R., McDermid, W. L., et al. "Card capacitor memory." IBM Research Report (June, 1960), RC290.

Gianola, U. F. "Non-destructive memory employing a domain oriented steelwire." J. Appl. Phys. (May, 1958), Vol. 29, p. 849–853.

Gianola, U. F., et al. "Large capacity card changeable permanent magnet twistor memory." Large Capacity Memory Techniques for Computing Systems, p. 177.

Gianopulos, W. "Capacitor read only memory analysis." IBM TR 21.274 (February, 1967).

Gogos, B., & McClure, J. C., Jr. "Linear microferrites for read-only storage." IBM TR-00.1666 (October, 1967).

Goldberg, J., & Green, M. W. "Large files for information retrieval based on simultaneous interrogation of all items." Large Capacity Memory Techniques for Computing Systems, p. 63–77.

Grubb, H. R., Haskell, J. W., et al. "A drive with noise cancellation for the CROM." Invention Disclosure, IBM Endicott GPD. (October, 1962).

Gutenmakher, L. I. "Electronic information—logic machine." Interscience (1963), English Translation, p. 41–47.

Gutenmakher, L. I. "Prospects of using magnetic and capacitor blocks and elements in automatic systems." Conf. on Scientific Problems of Automatic Production (October, 1956), The Academy of Sciences, USSR.

Hagiwara, H., et al. "The KT pilot computer—a microprogrammed computer with a phototransistor fixed memory." Proc. IFIPS Congress, Muhaginich, North Holland Publishing Co. (1962), p. 684.

Halsted, R. E. "Luminescence effects having circuit applications." Proc. of Symposium on the Role of Solid State Phenomena in Electronics (April, 1957), p. 275–287.

Hampton, L. N., & Newsom, J. B. "The card translator for nation-wide dialing." Bell System Tech. J. (1953), Vol. 32, p. 1037.

Haskell, J. W. "Design of a printed card capacitor read-only store." IBM J. Res. Develop. (1966), Vol. 10, No. 2E, p. 142–157.

Haskell, J. W. "Capacitor read-only memory." IBM Tech. Disc. Bull. (March, 1963), Vol. 5 p. 96–97.

Haskell, J. W. "Printed cards for the cards memory." IBM J. Res. Develop. (October, 1962), Vol. 6, p. 462–463.

Haskell, J. W. "The printed card capacitor read only storage system." IBM TR 01.886 (October, 1965).

Hoffman, G. R., & Jeffreys, D. C. "A computer fixed store using light pulses for read-out." J. Brit. IRE (February, 1963), Vol. 25, No. 2, p. 99–106.

Hoffman, G. R., Jones, P. L. "An electroluminescent fixed store for a digital computer." Proc. IEE (1962), Vol. 109B, p. 177.

Hoffman, G. R., et al. "High-speed light output signals from electroluminescent storage systems." Proc. IEE (1960), Vol. 107B, Suppl. 20, p. 257.

Hoffman, Walter. "Russian capacitor-type read-only memory blocks." IBM Research Laboratory, Zurich, Switzerland (January, 1960), Report No. RZ-58.

Hoover, C. W., & Haugk, G. "The flying spot store." Large Capacity Memory Techniques for Computing Systems, p. 79–98.

Hoover, C. W. Jr., et al. "Fundamental concepts in the design of the flying spot store." Bell System Tech. Journal (September, 1958).

Hsia, Y. "Solenoid coupled read only memory." IEEC (June, 1967), p. 344–350.

Ishidate, T., Yoshizawa, S., et al. "Eddycard memory—a semi-permanent storage." Proc. EJCC (December, 1961), p. 194–208.

Johnson, A. R. "Resistor read only storage feasibility." IBM-TD-01.365 (May, 1965).

Johnson, C., et al. "Localized field permanent magnet array for thick-film read only store." IBM J. Res. Develop. (July, 1966), p. 343–395.

Johnson, C. Jr. "Electrodeposition of thick magnetic films." IBM Research, Yorktown Heights (April, 1963), Report No. RC-937.

Kilburn, T., & Grimsdale, R. L. "A digital computer store with very short read time." Proc. IEE (1960), Vol. 107B, p. 567–607.

Kilburn, T., Laithwaite, E. R. "Servo control of the position and size of an optical scanning system." Proc. IEE (1954), Vol. 101, Part IV, p. 129.

King, G. W., Brown, G. W., et al. "Photographic techniques for information storage." Proc. IRE (October, 1953), Vol. 41, No. 10, p. 1421–1428.

Kingston, B. W. "Cryogenic read-only memory." IBM Tech. Disc. Bull. (June, 1963) Vol. 6, p. 34–35.

Kingston, B. W., & Taub, D. M. "Analysis of worst case signals in read only register memory." IBM-TR-12.035-TIC62A9-0037 (April, 1962).

Kovalick, V. E., & Duffy, R. "Read only memories." Bibliography—51 items (March, 1964), 64-816-880.

Krylow, K. E., et al. "Semipermanent memory—latest use for twistors." Electronics (March, 1963), Vol. 36, No. 11, p. 80.

Kuttner, P. "The rope memory—a permanent storage device." Proc. 1963 Fall Joint Computer Conference (1963), p. 45–57.

Lewin, M. H. "A survey of read-only memories." Proc. FJCC (1965), Part 1, p. 775–787.

Lewin, M. H., et al. "Fixed resistor-card memory." IEETEC (June, 1965), p. 428–434.

Long, T. R. "Electro-deposited elements for a non-destructive memory." J. Appl. Phys (Supl.) (May, 1960), Vol. 31, No. 5, p. 123–124.

Looney, D. H. "A twistor matrix memory for semi-permanent information." Proc. WJCC (March, 1959) p. 36–41.

MacPherson, D. H., & York, R. K. "Semi-permanent storage by capacitive coupling." IRETEC (September, 1961), p. 446–451.

Maeda, H., et al. "A high speed woven read-only memory." Proc. FJCC (1965), Part I, p. 789–799.

Margenson, E. O. "Wired-core matrix memories." Instr. & Control Systems (August, 1962).

Marino, J. J., & Sirota, J. J. "Weaving a braided memory that's fast and inexpensive." Electronics (September, 1967), p. 121–126.

Matick, R. E. "A semi-permanent read-only memory using thick magnetic films." IBM Research, Yorktown Heights, N.Y. (February, 1962), Report No. NC-52.

Matick, R. E. "Operating characteristics of a thick film read-only memory model." IBM Research, Yorktown Heights, N.Y. (March, 1963), Report No. RC-903.

Matick, R. E. "Balanced drive-diode selection scheme for use with the thick film read only memory." IBM Research, Yorktown Heights, N.Y. (April, 1963), Report No. RC-926.

Matick, R. E. "A study of the switching properties of thick multidomain, magnetic films." IBM Research, Yorktown Heights, N.Y. (April, 1963). Report No. RC-911.

Matick, R. E. "Thick film read-only memory device." J. of Appl. Physics, (April, 1963), Vol. 34, No. 4, p. 1174.

Matick, R. E. "Read-only memory." IBM Tech. Disclosure Bull. (January, 1963), Vol. 5, p. 67.

Matick, R. E. "Thick film read only memory." IBM NC-13 (September, 1961).

Matick, R. E. "Semipermanent read only memory using thick magnetic films." IBM-NC-52-TIC62B2-0032 (February, 1962).

Matick, R. E. "A thick film read only memory without sneak paths." IBM-NC-246 (November, 1963).

Matick, R. E., et al. "High speed read only store using thick magnetic film." IBM J. Res. Develop. (July, 1966), Vol. 10, No. 4, p. 333–342.

Matick, R. "Fast nondestructive read, slow write memory device using thick magnetic films." IEEE Trans. on Elec. Comp. (September, 1966), EC-15, 4, p. 435–441.

McWhorter, A. L., & Rediker, R. H. "The Cryosar—a new low-temperature computer element." Proc. IRE (July, 1959), Vol. 47, p. 1207–1213.

Meulners, D. "CCROS (Card Capacitor Read Only Storage) for use in IBM 2841." IBM-PT-9108-F (March, 1964).

Morgenson, E. O., Jr. "Wired core matrix memories." Instruments and Control Systems (1962), Vol. 35, p. 75.

Moto, Jr., J. W. "Using the Hall generator, a new control and instrumentation component." Automatic Control (June, 1967), Vol. 14, No. 6, (Part I) & No. 7.

Near, C. W., & Watson, R. E. "Read-only memory adds to calculator's repertoire." Electronics (March, 1969), p. 70–77.

Neihoff, W. H. "Mechanical reliability analysis of CCROS." IBM-TR-01.850 (January, 1965).

No author listed. "Diode matrix in integrated circuit form." Computer Design (June, 1965), p. 12.

Norton, D. E. "A transistor differential amplifier." IBM Tech. Disc. Bull. (August, 1963), Vol. 6, No. 3.

Ordemann, F. A. "Read-only memory." IBM Tech. Disclosure Bull. (August, 1961), Vol. 4, p. 23–24.

Owen, C. E. "Read-only memory." IBM Technical Disclosure Bull. (June, 1962), Vol. 5, p. 38–39.

Owen, C. E., Proudman, A., et al. "Read-only memory." IBM Technical Disclosure Bull. (January, 1963), Vol. 5, No. 8.

Parker, D. J., et al. "Optical Memory." U.S. Patent 3,191,157 (January, 1960).

Petschauer, R. J., & Turnquist, R. D. "A nondestructive readout film memory." Proc. Western Joint Computer Conference (1961), p. 411–425.

Pick, G. G. "A semi-permanent memory using correlation addressing." AFIPS Conf. Proc. (April, 1963), Vol. 26, Part 1, p. 107.

Pricer, W. D. "Read-only memory." IBM Tech. Disclosure Bull. (June, 1963), Vol. 5, p. 26–27.

Procyk, F. J., & Young, L. H. "A highspeed card-changeable permanent magnet memory—the inverted twistor." Proc. Interman (1964), Paper No. 8-6.

Proudman, A., Draper, S. W., et al. "Solid logic technology: versatile, high-performance microelectronics." IBM British Patent Specification 1,037,633 (1964).

Raisanen, W. "Meet the read only memory." Motorola Monitor (1968), p. 4–9.

Rajchmann, J. A. "Computer memories—a survey of the state-of-the-art." Proc. IRE (January, 1961), Vol. 49, No. 1, p. 104–127.

Renard, A. M., & Newman, W. J. "Unifluxo: a permanent memory element." Proc. WJCC (May, 1960), p. 91.

Roberts, F., & Young, J. Z. "The flying spot microscope." Proc. IEE (1952), Vol. 99, Part IIIA, p. 747.

Schlaeppi, H. P. "Read only memory system for large capacity." IBM-NZ-31 (October, 1963).

Schuenzel, E. C. "Read-only memory fabrication." IBM Tech. Disclosure Bull. (August, 1962), Vol. 5, p. 43.

Shimp, E. M., & Palmer, R. L. "IBM 2030 repackaged CCROS." ICM PT-2228-F 050466 (May, 1966).

Shook, C. G. "A digital static magnetic wire storage with nondestructive read-out." (March, 1961), IRE Trans. on electronic computers, Vol. EC-10, No. 1, p. 56–62.

Shuba, J. P. "An Eddy card twistor memory for semi-permanent storage." Poc. Intermag Conference, Paper No. 14-2 (1965).

Sie, C., et al. "A high-speed thick film read-only store." Proc. IFIPS Congress, New York City (May, 1965).

Smith, D. H. "Electroluminescence—its characteristics and applications." Electronic Engineering (March, 1961), Vol. 33, No. 396 (Part II), p. 1397.

Smith, K. G. "Transformer read only store, technical description." IBM-TN-1000 (April, 1966).

Stammerjohn, L. W. "An evaluation of design and performance of the permanent magnet twistor memory." Proc. Intermag Conference, Paper No. 8-4 (1964).

Stapper, C. H. "Read-only memory." IBM Tech. Disc. Bull. (January, 1963), Vol. 5, p. 1–46.

Takahashi, S., & Watanabe, S. "Capacitance type fixed memory." LCMTCS (1962), p. 53–62.

Taren, W. J. "Read-only thin film memory." IBM Tech. Disc. Bull. (June, 1962), Vol. 5, p. 58–59.

Taub, D. M. "A short review of read-only memories." Proc. IEE (January, 1963), Vol. 110, No. 157.

Taub, D. M. "Analysis of sneak paths and sense line distortion in an improved capacitor read-only memory." Proc. IEEE (November, 1963), Vol. 51, No. 11, p. 1554.

Taub, D. M. "A survey of recent read only stores." IBM World Trade Labs (Great Britain), Ltd., p. TR-12.060.

Taub, D. M. "Short review of read only memories." IBM-TR-12.045 (April, 1962), p. TIC63B6-1.

Taub, D. M., & Kingston, B. W. "The design of transformer (Diamond ring) read-only stores." IBM Journal (September, 1964), p. 443–459.

Tsui, F., et al. "Neues konstruktionsprinzip fur einen wirtschaftlichen und betriebssicheren halbfestwertspeicher." Elektronische Rechenanlagen 5 (1963).

Van Goethem, J. "Operating principle, manufacture and application of a semi-permanent memory." Proc. ICMT.

Van Goethem, J. "The capacitive semi-permanent information store and its uses in telephone exchanges." Proc. IEE Brit. (1960), Vol. 107B, suppl. 20, p. 346.

Van Stuyvenberg, J. A. "A 1 US memory with semi-permanent facilities." Large Capacity Memory Techniques for Computing Systems (1962), Macmillan.

Wamsley, J. G. "Inductive coupled read only memory." IBM TR 00.826 (June, 1961).

Webb, E. J. "Diode card read-only memory." IBM Tech. Disc. Bull. (January, 1962), p. 16.

Weir, J. M. "A high speed permanent storage device." IRE Trans on elec. computers (March, 1955), Vol EC4, p. 16–20.

Wilkes, M. V., et al. "The design of the control unit of an electronic digital computer." Proc. IEEE (1958), Vol. 105B8, p. 121.

Wolff, M. F. "Conductive memory points to new displays." Electronics (August, 1963), 63-459-754, p. 10–11.

Yamato, J., & Suzuki, Y. "Forming semi-permanent memories with metal card storage." Electronics (November, 1961).

Young, J. E. R. "Read-only memory." IBM Tech. Disc. Bull. (June, 1962), p. 26.

Younker, E. L., et al. "Design of an experimental multiple instantaneous response file." Proc. 1964 Spring Joint Computer Conference (1964), p. 515–528.

Zeuthen-Heidam, N. "Read only memories." IBM-TN-18.056 (August, 1961).

CONTROL STORAGE—CHRONOLOGICAL LISTING

1947. "Electronic computing circuits of the ENIAC." Burks, A. W. Proc. IRE, Vol. 35, p. 756.

1949. "Rectifier networks for multiposition switching." Brown, D. R., & Rochester, N. Proc. IRE, p. 139–147.

1951. "No. 5 crossbar AMA translator." Diamond, T. L. Bell Lab. Record, Vol. 29, p. 62.

1951. "The Bell computer, Model 6." Andrews, E. G. Proc., Harvard Univ., p. 20.

1952. "The flying spot microscope." Roberts, F., & Young, J. Z. Proc. IEE, Vol. 99, Part IIIA, p. 747.

1953. "The card translator for nation-wide dialing." Hampton, L. N., Newsom, J. B. Bell System Tech. J., Vol. 32, p. 1037.

1953, October. "Photographic techniques for information storage." King, G. W., Brown, G. W., et al. Proc. IRE, Vol. 41, No. 10, p. 1421–1428.

1954. "Servo control of the position and size of an optical scanning system." Kilburn, T., & Laithwaite, E. R. Proc. IEE, Vol. 101, Part IV, p. 129.

1955, March. "A high speed permanent storage device." Weir, J. M. IRE Trans. on Elec. Computers, Vol. EC4, p. 16–20.

1956, October. "Prospects of using magnetic and capacitor blocks and elements in automatic systems." Gutenmakher, L. I. Conf. on Scientific Problems of Automatic Production, The Academy of Sciences, USSR.

1957, April. "Luminescence effects having circuit applications." Halsted, R. E. Proc. of Symposium on the Role of Solid State Phenomena in Electronics. p. 275–287.

1957, April. "Photoconductive switching devices." Bramley, A. Proc. of Symposium on the Role of Solid State Phenomena in Electronics. p. 289–301.

1958. "The design of the control unit of an electronic digital computer." Wilkes, M. V., et al. Proc. IEEE, Vol. 105B8, p. 121.

1958, May. "Non-destructive memory employing a domain oriented steelwire." Gianola, U. F. J. Appl. Phys., Vol. 29, p. 849–853.

1958, September. "Fundamental concepts in the design of the flying spot store." Hoover, C. W., Jr., et al. Bell System Tech. Journal.

1959. "A card changeable non-destructive readout twistor store." Debuske, J. H., et al. Proc. WJCC.

1959, March. "A twistor matrix memory for semi-permanent information." Looney, D. H. Proc. WJCC, p. 36–41.

1959, July. "The Cryosar—a new low-temperature computer element." McWhorter, A. L., & Rediker, R. H. Proc. IRE, Vol. 47, p. 1207–1213.

1960. "The capacitive semi-permanent information store and its uses in telephone exchanges." Van Goethem, J. Proc. IEE (British), Vol. 107B, Suppl. 20, p. 346.

1960. "A digital computer store with very short read time." Kilburn, T., Grimsdale, R. L. Proc. IEE, Vol. 107B, p. 567–607.

1960. "High-speed light output signals from electroluminescent storage systems." Hoffman, G. R., et al. Proc. IEE, Vol. 107B, Suppl. 20, p. 257.

1960, January. "Russian capacitor-type read-only memory blocks." Hoffman, Walter. IBM Research Laboratory, Zurich, Switzerland, Report No. RZ-58.

1960, January. "Optical Memory." Parker, D. J., et al. U.S. Patent 3,191,157.

1960, May. "Unifluxo: a permanent memory element." Renard, A. M., & Newman, W. J. Proc. WJCC, p. 91.

1960, May. "Electro-deposited elements for a non-destructive memory." Long, T. R. J. Appl. Phys (Suppl.). Vol. 31, No. 5, p. 123–124.

1960, June. "Card capacitor memory." Foglia, H. R., McDermid, W. L., et al. IBM Research Report, RC290.

1960, October. "Increasing the brightness-voltage nonlinearity of electroluminescent devices." O'Connell, J. A. & Narken, B. IBM Journal, Vol. 4, No. 4, p. 426–429.

1961. "A nondestructive readout film memory." Petschauer, R. J., & Turnquist, R. D. Proc. Western Joint Computer Conference, p. 411–425.

1961, January. "Card-capacitor—a semipermanent read-only memory." Foglia, H. R., McDermid, W. L., et al. IBM Journal of Research and Development, Vol. K, No. 1, p. 67–69.

1961, January. "European electronic data processing—a report on the industry and the state-of-the-art." Auerbach, J. L. Proc. IRE, Vol. 49, No. 1, p. 330–331.

1961, January. "Computer memories—a survey of the state-of-the-art." Rajchmann, J. A. Proc. IRE, Vol. 49, No. 1, p. 104–127.

1961, January. "Thin silicon wafers used in fixed digital memory." Electronic Design.

1961, March. "A digital static magnetic wire storage with nondestructive read-out." Shook, C. G. IRE Trans. on Electronic Computers, Vol. EC-10, No. 1, p. 56–62.

1961, March. "Electroluminescence—its characteristics and applications." Smith, D. H. Electronic Engineering, Vol. 33, No. 396 (Part II), p. 397, 164.

1961, June. "Inductive coupled read only memory." Wamsley, J. G. IBM TR 00.826.

1961, July. "A table lock-up machine for processing of natural languages." Craft, J. L., et al. IBM Journal, p. 192–203.

1961, August. "Read-only memory." Ordemann, F. A. IBM Tech. Disclosure Bull., Vol. 4, p. 23–24.

1961, August. "Read only memories." Zeuthen-Heidam, N. IBM-TN-18.056.

1961, September. "Semi-permanent storage by capacitive coupling." MacPherson, D. H., & York, R. K. IRETEC, p. 446–451.

1961, September. "A card changeable permanent-magnet-twistor memory of large capacity." Barnett, W. A., et al. IRE.

1961, September. "Thick film read only memory." Matick, R. E. IBM NC-13.

1961, October. "Fiber optics large scale read-only memory." Bruce, G. D. IBM Tech. Disclosure Bull., Vol. 4, p. 76–77.

1961, November. "Forming semi-permanent memories with metal card storage." Yamato, J., & Suzuki, Y. Electronics.

1961, November. "The metal card memory—a new semi-permanent store." Endo, I., & Yamato, J. LCM-TCS, p. 213–230.

1961, December. "Eddycard memory—a semi-permanent storage." Ishidate, T., Yoshizawa, S., et al. Proc. EJCC, p. 194–208.

1962. "Capacitance type fixed memory." Takahashi, S., Watanabe, S. LCMTCS, p. 53–62.

1962. "The KT pilot computer—a microprogrammed computer with a photo-transistor fixed memory." Hagiwara, H., et al. Proc. IFIPS Congress, Muhaginich, North Holland Publishing Co., p. 684.

1962. "Wired core matrix memories." Margenson, E. O., Jr. Instruments and Control Systems, Vol. 35, p. 75.

1962. "Study of a memory with inductive coupling." Basque, M., & Dandrel, P. Large Capacity Memory Techniques for Computing Systems, Macmillan.

1962. "Semi-permanent memories made of diodes and photosensitive elements." Fasano, G., et al. Large Capacity Memory Techniques for Computing Systems, Macmillan.

1962. "A 1 US memory with semi-permanent facilities." Van Stuyvenberg, J. A. Large Capacity Memory Techniques for Computing Systems, Macmillan.

1962. "Square-loop ferrite circuitry." Quartly, C. J. ILIFFE and Prentice-Hall, p. 109.

1962. "An electroluminescent fixed store for a digital computer." Hoffman, G. R., & Jones, P. L. Proc. IEE, Vol. 109B, p. 177.

1962, January. "Diode card read-only memory." Webb, E. J. IBM Tech. Disclosure Bull. 4, p. 16.

1962, February. "Semipermanent read only memory using thick magnetic films." Matick, R. E. IBM-NC-52-TIC62B2-0032.

1962, March. "A permanent magnet twistor memory element of improved characteristics." Alexander, E. J., et al. J. Appl. Physics, Vol. 33, No. 3, p. 1075.

1962, April. "Analysis of worst case signals in read only register memory." Kingston, B. W., & Taub, D. M. IBM-TR-12.035-TIC62A9-0037.

1962, April. "Read only memory systems." Burgess, G. S. IBM-TR-12.036-TIC62A4-0.

1962, April. "Short review of read only memories." Taub, D. M. IBM-TR-12.045, p. TIC63B6-1.

1962, June. "Read-only memory." Owen, C. E. IBM Tech. Disc. Bull., Vol. 5, p. 38–39.

1962, June. "Read-only thin film memory." Taren, W. J. IBM Tech. Disc. Bull., Vol. 5, p. 58–59.

1962, June. "Read-only memory." Young, J. E. R. IBM Tech. Disc. Bull., Vol. 5, p. 26.

1962, August. "Wired-core matrix memories." Margenson, E. O. Instr. & Control Systems.

1962, August. "Read-only memory fabrication," Schuenzel, E. C. IBM Tech. Disc. Bull., Vol. 5, p. 1–43.

1962, September. "Optical coincidence read only memory system." Clover, Jr., R. B., & Kollar, E. P. IBM-TR-SD-007, p. TIC63C2-2.

1962, October. "A drive with noise cancellation for the CROM." Grubb, H. R., Haskell, J. W., et al. Invention Disclosure, IBM Endicott GPD.

1962, October. "Printed cards for the cards memory." Haskell, J. W. IBM J. Res. & Develop., Vol. 6, p. 462–463.

1963. "Electronic information-logic machine." Gutenmakher, L. I. Interscience, English Translation, p. 41–47.

1963. "The rope memory—a permanent storage device." Kuttner, P. Proc. 1963 Fall Joint Computer Conference, p. 45–57.

1963. "Neues konstruktionsprinzip fur einen wirtschaftlichen und betriebssicheren halbfestwertspeicher." Tsui, F., et al. Elektronische Rechenanlagen 5.

1963, January. "A short review of read-only memories." Taub, D. M. Proc. IEE., Vol. 110, No. 157.

1963, January. "Read-only memory." Owen, C. E., Proudman, A., et al. IBM Technical Disc. Bull., Vol. 5, No. 8.

1963, January. "Read-only memory." Matick, R. E. IBM Tech. Disc. Bull., Vol. 5, p. 67–67.

1963, January. "Read-only memory." Stapper, C. H. IBM Tech. Disc. Bull., Vol. 5, p. 1–46.

1963, February. "A computer fixed store using light pulses for read-out." Hoffman, G. R., & Jeffreys, D. C. J. Brit. IRE, Vol. 25, No. 2, p. 99–106.

1963, March. "Operating characteristics of a thick film read-only memory model." Matick, R. E. IBM Research, Yorktown Heights, N.Y. Report No. RC-903.

1963, March. "Semipermanent memory—latest use for twistors." Krylow, K. E., et al. Electronics, Vol. 36, No. 11, p. 80.

1963, March. "Inductive read-only memory bit pattern." Duke, K. A. IBM Tech. Disc. Bull., Vol. 5, p. 1–75.

1963, March. "Capacitor read-only memory." Haskell, J. W. IBM Tech. Disc. Bull., Vol. 5, p. 96–97.

1963, April. "Balanced drive-diode selection scheme for use with the thick film read only memory." Matick, R. E. IBM Research, Yorktown Heights, N. Y., Report No. RC-926.

1963, April. "A study of the switching properties of thick multidomain, magnetic films." Matick, R. E. IBM Research, Yorktown Heights, N. Y., Report No. RC-911.

1963, April. "Electrodeposition of thick magnetic films." Johnson, C. Jr., IBM Research, Yorktown Heights, N. Y., Report No. RC-937.

1963, April. "Thick film read-only memory device." Matick, R. E. J. of Appl. Physics, Vol. 34, No. 4, p. 1174.

1963, April. "A semi-permanent menory using correlation addressing." Pick, G. G. AFIPS Conf. Proc., Vol. 26, Part 1, p. 107.

1963, May. "Read-only memory." Pricer, W. D. IBM Tech. Disc. Bull., Vol. 5, p. 26–27.

1963, June. "Cryogenic read-only memory." Kingston, B. W. IBM Tech. Disclosure Bull., Vol. 6, p. 34–35.

1963, June. "Control memory." Boland, L. MS thesis (EE) Syracuse University.

1963, June. "Core-rope memory." Computer Design Staff, Computer Design, p. 16–19.

1963, August. "A transistor differential amplifier." Norton, D. E. IBM Tech. Disc. Bull., Vol. 6, No. 3.

1963, August. "Conductive memory points to new displays." Wolff, M. F. Electronics, 63-456-754. p. 10–11.

1963, September. "Read-only memory using non-linear resistive elements." Cole, J. N. IBM Tech. Disc. Bull., Vol. 6, p. 31–32.

1963, October. "The rope memory: a permanent storage device." Kuttner, P. FJCC Abstract Computer Design, p. 9.

1963, October. "Read only memory system for large capacity." Schlaeppi, H. P. IBM-NZ-31.

1963, November. "Analysis of sneak paths and sense line distortion in an improved capacitor read-only memory." Taub, D. M. IBM World Trade Laboratories (Great Britain), Ltd., Technical Report TR 12.044.

1963, November, "Analysis of sneak paths and sense line distortion in an improved capacitor read-only memory." Taub, D. M. Proc. IEEE, Vol. 51, No. 11, p. 1554.

1963, November. "A thick film read only memory without sneak paths." Matick, R. E. IBM-NC-246.

1963, December. "A survey of read only storage techniques." Biche, P. W., & Ruoff, C. E. IBM-TD-01.303.

1964. "Solid logic technology: versatile, high-performance microelectronics." Proudman, A., Draper, S. W., et al. IBM British Patent Specification 1,037,633.

1964. "Design of serviceability features for the IBM System/360." Carter, W. C., Montgomery, H. C., et al. IBM J. Res. Develop., Vol. 8, No. 115.

1964. "Design of an experimental multiple instantaneous response file." Younker, E. L., et al. Proc. 1964 Spring Joint Computer Conference, p. 515–528.

1964. "MYRA—a new memory element and system." Briley, B. E. Proc. Intermag. Conference, Paper No. 14-8.

1964. "An evaluation of design and performance of the permanent magnet twistor memory." Stammerjohn, L. W. Proc. Intermag. Conference, Paper No. 8-4.

1964. "A highspeed card-changeable permanent magnet memory—the inverted twistor." Procyk, F. J., & Young, L. H. Proc. Intermag. Conference, Paper No. 8-6.

1964. "A semi-permanent memory using correlation addressing." 26 Part 1, Spartan, 1964.

1964, February. "Microsecond word recognition system." Brick, D. B., & Pick, G. G. IEEETEC, p. 27–35.

1964, March. "Modular design and analysis of a read only resistor storage." Biche, P. W., & Ruoff, C. E. IBM TR-01.774.

1964, March. "CCROS (Card Capacitor Read Only Storage) for use in IBM 2841." Meulners, D. IBM-PT-9108-F.

1964, March. "Read only memories." Kovalick, V. E., & Duffy, R. Bibliography—51 items, 64-816-880.

1964, April. "A pre-wired storage unit." Butcher, I. R. IEEETEC, Vol. EC 13, No. 2, p. 106–111.

1964, June. "Card changeable memories." Donnelly, J. M. Computer Design, p. 20–30.

1964, September. "The design of transformer (Diamond ring) read-only stores." Taub, D. M., & Kingston, B. W. IBM Journal, p. 443–459.

1964, September. "No. 1 ESS program store." Ault, C. F., et al. Bell System Technical Journal 43, No. 5, Part 1, p. 2097.

1964, October. "Application of rope memory devices." Clemson, D., & Kuttner, P. Computer Design, p. 12–22.

1964, December. "Light coupled diode, read only memory." Anderson, J. L., & Reynolds, J. L. IBM TR-22.146.

1965. "Fixed memory with resistive coupling." David, C. Proc. ICMT.

1965. "RCA Spectra 70, basic design and philosophy of operation." Beard, A. D. Wescon, San Francisco, Paper, Vol. 12, No. 1.

1965. "An Eddy card twistor memory for semi-permanent storage." Shuba, J. P. Proc. Intermag. Conference, Paper No. 14-2.

1965. "A survey of read-only memories." Lewin, M. H. Proc. FJCC, Part 1, p. 775–787.

1965. "A high speed woven read-only memory." Maeda, H., et al. Proc. FJCC, Part 1, p. 789–799.

1965, January. "Mechanical reliability analysis of CCROS." Neihoff, W. H. IBM-TR-01.850.

1965, May. "A high-speed thick film read-only store." Sie, C., et al. Proc. IFIPS Congress, New York City.

1965, May. "Resistor read only storage feasibility." Johnson, A. R. IBM-TD-01.365.

1965, May. "CCROS—a card capacitor read only storage." Bowling, R. E., & Southard, C. D. IBM-TR-01.864.

1965, June. "Fixed resistor-card memory." Lewin, M. H., et al. IEETEC, p. 428–434.

1965, June. "Diode matrix in integrated circuit form." No author listed, Computer Design, p.12.

1965, September. "C-90 cross unit phase of System 360 Model 65." Askin, H.O., Cirielli, C. A., et al. IBM PT-4325-F.

1965, September. "Design of a printed card capacitor read only store." Haskell, J. W. IBM Journal, March 1966, Vol. 10, No. 2E, p. 142–157.

1965, October. "The printed card capacitor read only storage system." Haskell, J. W. IBM TR 01.886.

1966. "Design of a printed card capacitor read-only store." Haskell, J. W. IBM J. Res. Develop., Vol. 10, No. 142.

1966, April. "Transformer read only store, technical description." Smith, K. G. IBM-TN-1000.

1966, May. "IBM 2030 repackaged CCROS." Shimp, E. M., & Palmer, R. L. IBM PT-2228-F 050466.

1966, July. "Localized field permanent magnet array for thick-film read only store." Johnson, C., et al. IBM J. Res. Develop., p. 343–395.

1966, July. "High speed read only store using thick magnetic film." Matick, R. E., et al. IBM J. Res. Develop., Vol. 10, No. 4, p. 333–342.

1966, September. "Fast nondestructive read, slow write memory device using thick magnetic films." Matick, R. IEEE Trans. on Elec. Comp. EC-15, 4, p. 435–441.

1967, February. "Capacitor read only memory analysis." Gianopulos, W. IBM TR 21.274.

1967, June. "Using the Hall generator, a new control and instrumentation component." Moto, J. W., Jr., Automatic Control, Vol. 14, No. 6, June 1961 (Part I) & No. 7.

1967, June. "Solenoid coupled read only memory." Hsia, Y. IEEC, p. 344–350.

1967, June. "A new approach to permanent memory." Filippazzi, F. IEEC, p. 370–371.

1967, September. "Weaving a braided memory that's fast and inexpensive." Marino, J. J., & Sirota, J. J. Electronics, p. 121–126.

1967, September. "A read only 70-NS waffle iron memory." Aviles, J. F., Bourrez, J. M., et al. IEMG, p. 321–324.

1967, October. "Linear microferrites for read-only storage." Gogos, B., & McClure, J. C., Jr. IBM TR-00. 1666.

1968. "Meet the read only memory." Raisanen, W. Motorola Monitor, p. 4–9.

1969, March. "Read-only memory adds to calculator's repertoire." Near, C. W., & Watson, R. E. Electronics, p. 70–77.

INDEX